GLOBAL ECOLOGY

A DERIVATIVE OF ENCYCLOPEDIA OF ECOLOGY

EDITORIAL BOARD

GLOBAL ECOLOGY

A DERIVATIVE OF ENCYCLOPEDIA OF ECOLOGY

Editor-in-Chief

SVEN ERIK JØRGENSEN

Copenhagen University,
Faculty of Pharmaceutical Sciences,
Institute A, Section of Environmental Chemistry,
Toxicology and Ecotoxicology,
University Park 2,
Copenhagen,
Denmark

Associate Editor-in-Chief

BRIAN D. FATH

Department of Biological Sciences,
Towson University,
Towson, Maryland,
USA

AMSTERDAM • BOSTON • HEIDELBERG • LONDON •
NEW YORK • OXFORD • PARIS • SAN DIEGO •
SAN FRANCISCO • SINGAPORE • SYDNEY • TOKYO
Academic Press is an imprint of Elsevier

Academic Press is an imprint of Elsevier
Radarweg 29, 1043 NX Amsterdam, The Netherlands
Linacre House, Jordan Hill, Oxford OX2 8DP, UK
525 B Street, Suite 1900, San Diego, CA 92101-4495, USA
30 Corporate Drive, Suite 400, Burlington, MA01803, USA

Material in the text originally appeared in the *Encyclopedia of Ecology*, edited by
Sven Erik Jørgensen (Elsevier B.V. 2008)

British Library Cataloguing in Publication Data
A catalogue record for this book is available from the British Library

Library of Congress Cataloging-in-Publication Data
Global ecology: a derivative of encyclopedia of ecology/edited by Sven Erik Jørgensen.
p. cm.
Includes index.
ISBN 978-0-444-63830-4
1. Ecology–Encyclopedias. 2. Biosphere–Encyclopedias. I. Jørgensen, Sven Erik,
1934-QH540.4.G54 2010
577.03–dc22

2010010188

ISBN: 978-0-444-63830-4

For information on all Academic Press publications
visit our website at elsevierdirect.com

Printed and bound in Italy

10 11 12 13 10 9 8 7 6 5 4 3 2 1

CONTENTS

LIST OF CONTRIBUTORS

G Alexandrov
National Institute for Environmental Studies, Tsukuba, Japan

R W Arnold
USDA Natural Resources Conservation Service, Washington, DC, USA

V N Bashkin
VNIIGAZ/Gazprom, Moscow, Russia

P J Boston
New Mexico Institute of Mining and Technology, Socorro, NM, USA

C Bounama
Potsdam Institute for Climate Impact Research, Potsdam, Germany

J G Bryce
University of New Hampshire, Durham, NH, USA

T P Burt
Durham University, Durham, UK

P Carl
Leibniz Institute of Freshwater Ecology and Inland Fisheries, Berlin, Germany

J Cebrian
Dauphin Island Sea Laboratory, Dauphin Island, AL, USA

J Chen
Tsinghua University, Beijing, People's Republic of China

S V Chernyshenko
Dnipropetrovsk National University, Dnipropetrovsk, Ukraine

W Cramer
Potsdam Institute for Climate Impact Research, Potsdam, Germany

C L De La Rocha
Alfred Wegener Institute for Polar and Marine Research, Bremerhaven, Germany

A V Eliseev
AM Obukhov Institute of Atmospheric Physics RAS, Moscow, Russia

J J Elser
Arizona State University, Tempe, AZ, USA

S Franck
Potsdam Institute for Climate Impact Research, Potsdam, Germany

G M Gadd
University of Dundee, Dundee, UK

A Ganopolski
Potsdam Institute for Climate Impact Research, Potsdam, Germany

P J Geogievich
AN Severtsov Institute of Ecology and Evolution, Moscow, Russia

P J Georgievich
Russian Academy of Sciences, Moscow, Russia

F W Gerstengarbe
Potsdam Institute for Climate Impact Research, Potsdam, Germany

J P Grover
University of Texas at Arlington, Arlington, TX, USA

C J Hoff
Potsdam Institute for Climate Impact Research, Potsdam, Germany

C J Hoff
University of New Hampshire, Durham, NH, USA

K A Hunter
University of Otago, Dunedin, New Zealand

C Jäger
Potsdam Institute for Climate Impact Research, Potsdam, Germany

S E Jørgensen
Copenhagen University, Copenhagen, Denmark

A D Kay
University of St. Thomas, St. Paul, MN, USA

A Kleidon
Max-Planck-Institut für Biogeochemie, Jena, Germany

R Klige
Moscow State University, Moscow, Russia

Z W Kundzewicz
RCAFE Polish Academy of Sciences, Poznań, Poland

H N Lee
US Department of Homeland Security, New York, NY, USA

Y Liu
Tsinghua University, Beijing, People's Republic of China

P A Loka Bharathi
National Institute of Oceanography, Panaji, India

D Lyuri
Russian Academy of Sciences, Moscow, Russia

H Matsuda
Yokohama National University, Yokohama, Japan

C P McKay
NASA Ames Research Center, Moffett Field, CA, USA

I I Mokhov
AM Obukhov Institute of Atmospheric Physics RAS, Moscow, Russia

L Olsen
NASA/GSFC, Greenbelt, MD, USA

S A Pegov
Russian Academy of Sciences, Moscow, Russia

S Pegov
Russian Academy of Sciences, Moscow, Russia

I V Priputina
Institute of Physico-Chemical and Biological Problems of Soil Science RAS, Moscow, Russia

J Puzachenko
Russian Academy of Sciences, Moscow, Russia

A Quigg
Texas A&M University at Galveston, Galveston, TX, USA

M A Reuter
Ausmelt Ltd, Melbourne, VIC, Australia

D W Schwartzman
Howard University, Washington, DC, USA

A Shvidenko
International Institute for Applied Systems Analysis, Laxenburg, Austria

I N Sokolik
Georgia Institute of Technology, Atlanta, GA, USA

G Stenchikov
Rutgers University, New Brunswick, NJ, USA

R W Sterner
University of Minnesota, St. Paul, MN, USA

R Strzepek
University of Otago, Dunedin, New Zealand

A Svirejeva-Hopkins
Potsdam Institute for Climate Impact Research, Potsdam, Germany

Y M Svirezhev
Potsdam Institute for Climate Impact Research, Potsdam, Germany

Y M Svirezhev
University of Lisbon, Lisbon, Portugal

Y Svirezhev
Potsdam Institute for Climate Impact Research, Potsdam, Germany

V O Targulian
Russian Academy of Sciences, Moscow, Russia

S A Thomas
University of Nebraska, Lincoln, NE, USA

S Unninayar
NASA/GSFC, Greenbelt, MD, USA

W von Bloh
Potsdam Institute for Climate Impact Research, Potsdam, Germany

A van Schaik
MARAS (Material Recycling and Sustainability), Den Haag, The Netherlands

T Vrede
Umeä University, Umeä, Sweden

T Vrede
Uppsala University, Uppsala, Sweden

P C Werner
Potsdam Institute for Climate Impact Research, Potsdam, Germany

P E Widdison
Durham University, Durham, UK

D J Wuebbles
University of Illinois at Urbana-Champaign, Urbana, IL, USA

G A Zavarzin
Russian Academy of Sciences, Moscow, Russia

PREFACE

The focus of global ecology is the biosphere or the ecosphere conceived as one unified cooperative system with numerous synergistic effects that explain the unique properties of this sphere.

Part A of the book presents these unique properties of the biosphere, which are able to explain its life-bearing role. The biosphere is open to all other spheres, which determine its composition. The compositions of all the spheres are also presented in this part.

The biosphere supports the global cycles of the elements that are crucial for life. A quantitative representation of the global balances of energy and matter is covered in Part B, in addition to the important flows of energy, matter, and information in the biosphere.

Part C presents the results of the global cycles and flows and the biosphere properties: formations of patterns of climatic factors and marine currents.

The climate is of utmost significance for the life on Earth, but due to the huge impact of human activities on the biosphere, changes in the global climate are foreseen. It is probably the hottest environmental issue of today. The biosphere–climate interactions and climatic changes and their consequences for the life on Earth are discussed in Part D.

Part E covers ecological stoichiometry, which focuses on the application of stoichiometry for the quantification of the various biogeochemical cycles in the biospheres and in ecosystems.

The book is a derivative of the recently published *Encyclopedia of Ecology*. Due to an excellent work by the section editor of Global Ecology, Yuri M. Svirezhev, and the section editor of Ecological Stoichiometry, James Elser, it has been possible to present a comprehensive overview of global ecology and ecological stoichiometry as a useful tool to couple the global and ecological processes. Yuri Svirezhev considered his editorial work with the Global Ecology section as a great challenge and did his utmost to achieve a profound and comprehensive coverage of this ecological field, which was very close to his heart. Yuri passed away in February 2007, when about 90% of the work was done. I would therefore like to dedicate this derivative book to his memory.

I would like to thank James Elser and all the authors of the Global Ecology and the Ecological Stoichiometry entries, who made it possible to produce this broad and up-to-date coverage of ecotoxicology.

Sven Erik Jørgensen
Copenhagen, November 2009

PART A

Global Ecology, The Biosphere and its Evolution

Introduction

S E Jørgensen, Copenhagen University, Copenhagen, Denmark

Further Reading

The focus of global ecology is the biosphere or the ecosphere conceived as one unified cooperative system with numerous synergistic effects that explain the unique properties of this sphere.

The biosphere has several unique properties that explain its function and role in supporting life on the Earth. The biosphere is open to the other spheres, and exchanges matter, energy, and information with the other spheres. The compositions of all the spheres are therefore important for life on the Earth.

We also use the term ecosphere for the part of the Earth that is bearing life and which includes both living and nonliving components. The composition of the ecosphere is important for its life-bearing ability and the composition of the ecosphere is dependent on the composition of all the other spheres, with which it exchanges matter, energy, and information.

The biosphere – like ecosystems – cycles the elements that are essential for life. The cycling of matter makes it possible to use again and again the matter to build up new biological components and is therefore a prerequisite for evolution. The global cycles and flows of elements are a result of a number of biological, physical, and chemical processes. It is important that we quantify the cycles and flows of the essential elements, because they determine whether the concentrations of biologically essential elements are in accordance with the functions and roles of the biosphere. They also determine the atmospheric and marine currents, which are decisive for the global pattern of the climate. The life conditions of all parts of the ecosphere are therefore rooted in a proper function and balance of the cycles and flows of the about 20 essential elements. A massive and steadily increasing impact of human activities on the biosphere has, however, reached a level where the global cycles and flows are influenced significantly by human activities. As one of the most important results we can foresee changes of the global climate, which will inevitably cause changes in the life conditions of all organisms on the Earth from microorganisms to humans. Moreover, the climatic changes will change the pattern of species and biodiversity on the Earth, which will influence the life conditions further.

The book Global Ecology presents the latest results of these dramatic global changes. Part A of the book presents the unique properties of the biosphere, which help to explain its life-bearing function and role. The compositions of all the spheres are presented in this part and all the spheres are open and determine the composition of the biosphere. Part A also discusses the crucial question in astrobiology, 'can life be found outside the Earth?', and presents the controversial Gaia hypothesis, which presumes that the ecosphere is working as one cooperative unit with numerous synergistic effects.

The biosphere supports global cycles of the elements that are crucial for life. A quantitative representation of the global balance of energy and matter is covered in Part B, in addition to the important flows of energy, matter, and information in the biosphere. This part reveals the imbalances in the global balance of the biologically essential elements.

The formations of different patterns of climatic factors and marine currents, which are the results of global cycles and flows and the biosphere properties, are presented in Part C. The agricultural pattern, which is a result of the pattern of climatic factors, is included in this part. Furthermore, Part C also covers the processes of global significance.

The climate is of utmost significance for the life on Earth, but due to the massive impact of human activities on the biosphere and as a matter of fact on all the spheres, changes in the global climate are foreseen. The interactions between the biosphere and the climate and the climate change and the consequences for the biosphere are covered in Part D.

Part E covers ecological stoichiometry, which focuses on the application of stoichiometry for the quantification of the various biogeochemical cycles in the biospheres and in ecosystems. The ecological stoichiometry gives the elementary interactions and interdependence of the various global cycles, balances, flows, and processes.

Further Reading

Jørgensen SE (2008) *Evolutionary Essays. A Thermodynamic Interpretation of the Evolution*, 210pp. Amsterdam: Elsevier
Jørgensen SE, Fath BD, Bastianoni S, *et al.* (2007) *A New Ecology: Systems Perspective*. 288pp. Amsterdam: Elsevier.
Jørgensen SE and Svirezhev YM (2004) *Towards a Thermodynamic Theory for Ecological Systems*, 366pp. London: Elsevier.

Abiotic and Biotic Diversity in the Biosphere

P J Geogievich, AN Severtsov Institute of Ecology and Evolution, Moscow, Russia

Introduction

The phenomenon 'diversity' is related to the reflection of any natural phenomena through a set of elements (particles, material points) with different classes of property states observed in space. The elements are confirmed to interact potentially with each other. This is a thermostatistical model of the world acceptable for a wide set of phenomena, from the atomic level to the social–economic one. As in physics, within the framework of a model of a particular phenomenon, an element is considered, as that is invariable in the process of all the imaginable transformations. The invariability is nothing more than an assumption simplifying the model. In general, if physical essence is given to an element, the very element is implied as an integral system supported by internal negative and positive relations between the parts forming it. The proven universality of fractality of nature, that is, its correspondence to the model of continuous–discontinuous set enables to determine an element as a cell of certain size in the accepted scale of space–time.

Model

Gene, allele, chromosome, cell, individual, chemical element, compound of elements, mineral, rock, community of organisms described on a sample plot selected, pixel of a cosmic image, car, plant, settlement, town, country, and so on may be elements of models. In all the cases, we have n elements, and each of them may be referred to one of the k classes according to its properties. In the process of interactions, the elements belonging to different classes may be assumed to form structures locally stable in time. It is unknown *a priori* what structures are stable or unstable, but their whole diversity is described by the formula $I = n_1!n_2!n_3!\ldots n_m!$, where n_i is the number of elements in class i.

This is a large value. Using simple rearrangements, we obtain that this value is in general agreement with

$$\ln(I) = N\ln(N) - \sum_{i-1}^{m} n_i \ln(n_i) = -N\sum_{i=1}^{i=m}\frac{n_i}{N}\ln\frac{n_i}{N}$$

where n_i is the number of elements of class i, $S = -Kp_i\sum_{i=1}^{i=m} p_i$ is the Gibbs–Shannon's entropy ($p_i = n_i/N$ is the probability of elements of class i in sample N, K is the analog of Planck's constant).

Under equilibrium (derivatives are close to zero), in a linear case, the Gibbs's distribution has resulted. A. Levich in 1980 supposed the nonlinearity of relations to the property space and obtained the rank distributions:

- $p_i = \mu\exp(-\lambda_i)$ – the Gibbs' rank distribution, that is, condition of linear dependence of a system on a resource;
- $p_i = \mu\exp(\lambda\log(i)) = \mu i^{-\lambda}$ – the Zipf's rank distribution, that is, logarithmic dependence on a resource;
- $p_i = \mu\exp(\lambda\log(a+i)) = \mu(a+i)^{-\lambda}$ – the Zipf–Mandelbrot's rank distribution, that is, logarithmic dependence on an resource, where a is the number of unoccupied (vacant) state with an unused resource;

- $p_i = \mu \exp(\lambda \log(\log i)) = \mu \log i^{-\lambda}$ – the MacArthur's rank distribution (the broken stick), at twice logarithmic dependence on a resource.

Simple transformations on making the assumption that there is some class with only one element allow finding widespread relationships of the number of species with the volume of sampling N or with the area, where the sampling was made. Such relations obtained in island biogeography are true for any phenomenon.

If these relations are nonequilibrium, members with order >1 are included into rank distributions. These forms of distributions are typical in nature. If a system is nonstationary, Kulback's entropy is a measure of nonstationarity. Under the same conditions, entropy of the nonstationary system is less than entropy nonequilibrium one, and the entropy of the nonequilibrium system is less than that of equilibrium one.

According to the model, diversity (entropy) of a system is the function of power or diversity of the environment and evolutionary parameters. The first parameter is identical to free energy of Gibbs (exergy in a nonstationary case), the second one to temperature. Thus, in the closed space, evolution of diversity corresponds to the thermodynamic model, and entropy increases in time.

Living Matter

If a system is open and dissipative, its diversity and nonstationarity is supported by the flow of information and energy from the environment. The system selects an order from the environment and increases its entropy (disturbs its own environment).

Living matter differs from abiotic one. As V. I. Vernadsky in 1926 wrote, "living organisms change the course of the biosphere equilibrium (unlike abiotic substance) and represent specific autonomic formations, as if special secondary systems of dynamic equilibria in the primary thermodynamic field of the biosphere." According to Jorgensen's ideas, they also increase their own exergy (useful work) supporting their local stability in aggressive medium. Probably, the maximization of stability via increasing exergy is not the single way of survival. Many organisms make the stability maximum at very low energy expenditures via the complexity of their own structure that decrease the destructive action of the environment.

Evolution of living systems appears to be founded on mechanisms that do not fit the framework of three principles of thermodynamics. Nowadays, a satisfactory physical model of this evolution is absent. An empirical fact is the growth of biological diversity in time according to hypergeometric progression. The mode of the statistical model shows that in the course of evolution, the dimension of the space as well as the volume of resources increase (**Figure 1**).

- *Model 1.* log(number of families) $= (-0.078\,61 + 0.031\,733 \log T)T \log T$, where T is the time (unit of measurement is 1 million years).
- *Model 2.* Number of families $= \exp(0.030\,053\,(1.038\,47^{T})T)$.

The younger the taxon, the faster the growth of its diversity. The rate (ΔT) of evolution increases in time

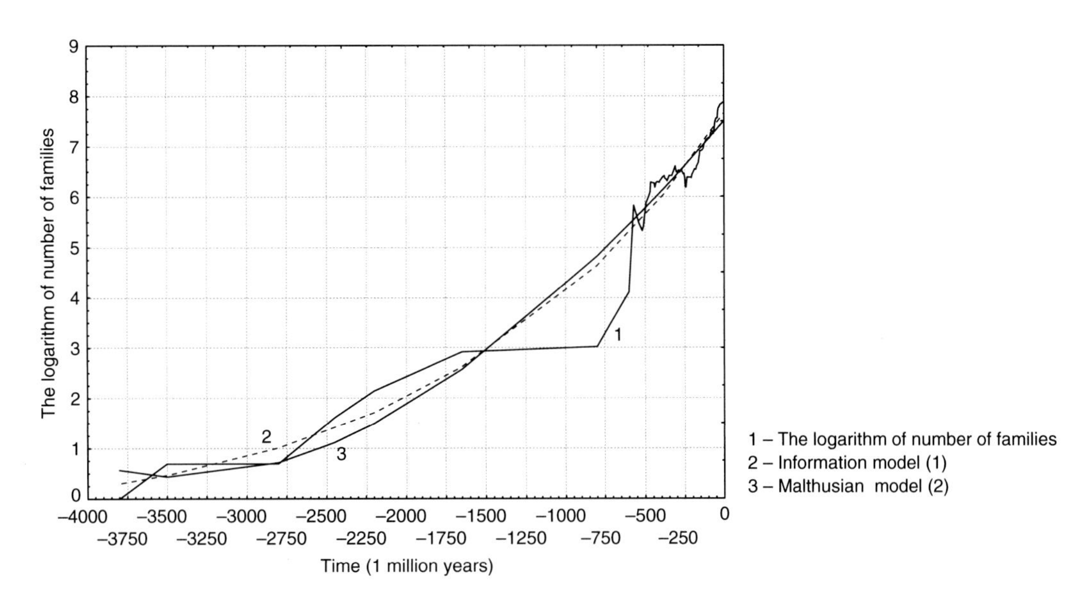

Figure 1 Changes of a global biodiversity biological variety at a level of families on a database (Fossil Record 2). Based on Puzachenko Yu G (2006) A global biological variety and his (its) spatially times changes. In: Kasimov NS (ed.) *Recent Global Changes of the Natural Environment*, vol. 1, pp. 306–737. Moscow: Scientific World (in Russian).

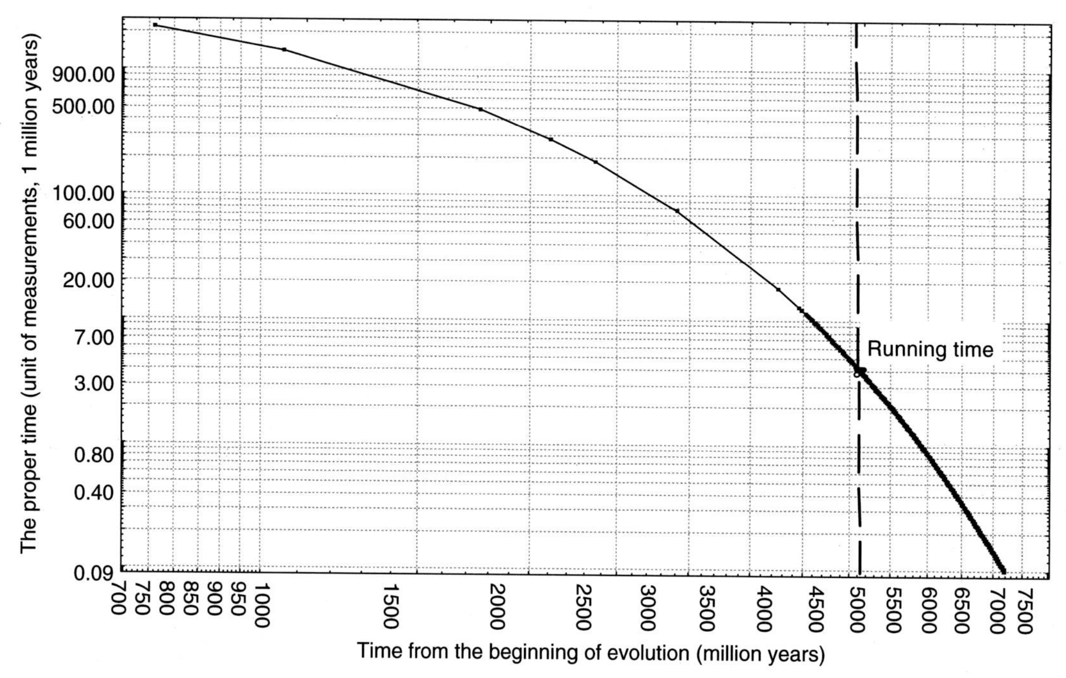

Figure 2 The proper time (ΔT) change.

(**Figure 2**) as $\Delta T = \text{constant} \times T^{-3.7}$ ($R^2 = 0.53$). In order to explain this phenomenon, the memory about the past successes and failures in the synthesis of new structures and variability that allow opening new possibilities of the environment should be added. The thermodynamic law of evolution for living matter appears to reduce to a decrease of expenditures per unit of complexity (1 bit). Such structures extracting energy and substance from the environment can keep the area far from equilibrium for a long time.

Phenomenology of changes in the number of species as a function of environmental quality with regard to the time of continuous development is within the framework of this model.

Landscape Diversity

Unlike biological diversity, landscape diversity combines biotic and abiotic constituents. As the landscape diversity is assessed using cosmic images, it is maximum for territories without vegetation and minimum for rainy tropical forests of Amazonia. This effect is determined by a more complete absorption of solar radiation by plants that transform it into energy spent for evaporation, production, internal energy, and heat flow. Upon the transformation of solar radiation, vegetation (due to the species diversity) lowers the diversity of reflection (in each particular variant of the environment, there is found a plant species with the most efficient absorption). In this case, Ashby's 'law of the necessary diversity' manifests itself. The same effect is also true for the diversity of the soil cover and other abiotic factors. Autofluctuations described by the Holling's model of panarchy are imposed upon the general trend of evolution of living matter and *socium.*

Conclusion

The phenomenon of diversity is a basic property of any forms of matter, being observable via the locally stable state of particles (elements). The behavior of a set of particles in space of their material properties follows the principles of nonequilibrium dynamics. Living matter, unlike abiotic substance, expands its thermodynamic possibilities via a search for structures that use spaces with increasing volume and dimension and, accordingly, with a high flow of energy. Evolution of abiotic substance obeys the second principle of thermodynamics – the growth of entropy as a measure of disorder. Evolution of living matter obeys the opposite growth of order, also upon increase in the total entropy, that is, upon self-organization in Foerster's opinion.

Further Reading

Benton MJ (ed.) (1993) *The Fossil Record 2*, 845pp. London: Chapman & Hall. http://www.fossilrecord.net/fossilrecord/index.html(accessed December 2007).

Holling CS and Gunderson LH (2002) Resilience and adaptive cycles. In: Gunderson LH and Holling CS (eds.) *Panarchy: Understanding*

Transformations in Human and Ecological Systems, pp. 25–62. Washington, DC: Island Press.
Jorgensen SE (2000) 25 years of ecological modelling by ecological modelling. *Ecological Modelling* 126(2–3): 95–99.
Jorgensen SE and Svirezhev Iu M (2004) *Towards a Thermodynamic Theory for Ecological Systems*, 366pp. Amsterdam: Elsevier Science.
Levich AP and Solov'yov AV (1999) Category-function modeling of natural systems. *Cybernetics and Systems* 30(6): 571–585.
Puzachenko Yu G (2006) A global biological variety and his (its) spatially times changes. In: Kasimov NS (ed.) *Recent Global Changes of the Natural Environment*, vol. 1, pp. 306–737. Moscow: Scientific World (in Russian).
Tribus M (1961) *Thermostatics and Thermodynamics*. New York: Van Nostrand/Reinhold.
Vernadsky VI (1998) *The Biosphere*, 192pp. New York: Copernicus (first published in Russian in 1926).
von Foerster H (1960) On self-organizing systems and their environments. In: Yovits MC and Cameron S (eds.) *Self-Organizing Systems*, pp. 31–50. London: Pergamon.

Anthropospheric and Antropogenic Impact on the Biosphere

S Pegov, Russian Academy of Sciences, Moscow, Russia

Introduction

Industrial growth proceeded at such a fast pace that in the second half of the eighteenth century it became globally important and resulted in what was called the industrial, or second technological, revolution. Approximately 100 years later, the use of new sources of raw materials and energy brought to life high-efficiency technologies of mass production to produce machine tools and consumption goods. In the later part of the twentieth century, scientific and technical progress stimulated development of high technologies and the advent of space, petrochemical, electronic, pharmaceutical, and other industries. Further progress has brought enormous achievements in the field of information technologies. The rates of dissemination of new technological achievements and economic growth were amazing. Unparalleled high rates of technological development led to a multifold increase in industrial production and consumption of energy resources. The gross world product increased from about US$ 60 million up to US$ 39.3 billion (more than 650 times) between 1900 and the end of the twentieth century. If it took several millennia for agriculture to win the world, then the industrial revolution became a global phenomenon within 1.5–2 centuries.

There were unprecedented rates achieved of burning fossil fuels that had been created by ancient biospheres during a long geological history. For the period from 1950 to 1998, the consumption of various kinds of fossil fuels, expressed in the oil equivalent, increased by 2.1 times for coal, 7.8 times for oil, and 11.8 times for natural gas. While per capita energy consumption was 4000 kcal d^{-1} in the Stone Age, it rose to 12 000 kcal d^{-1} during the era of agricultural technologies, and reached 23 000–250 000 kcal d^{-1} at present. Technogenic interventions in the environment began to compete with many natural processes. Extraction of solid minerals and, hence, the massive impact on the lithosphere sharply increased. About 100 billion tons of raw material is excavated from the Earth's crust annually, or 15 t per inhabitant of our planet.

Studies of ice cores taken from depths of glaciers in Antarctica and Greenland show that such rates of change in biogenic concentrations in the atmosphere did not happen for more than 150 000 years during the overall modern Holocene period.

Studies of carbon isotopes, C^{13} and C^{14}, show that the growth in CO_2 concentrations in the atmosphere for the recent decades is connected with combustion of mineral fuels (**Figure 1**). Thus, a huge amount of carbon – up to 180 Gt – had been emitted in the atmosphere as a result of various forms of human land use since its establishment as a planetary phenomenon before 1980, while industrial emissions from the period of industrial revolution to 1980

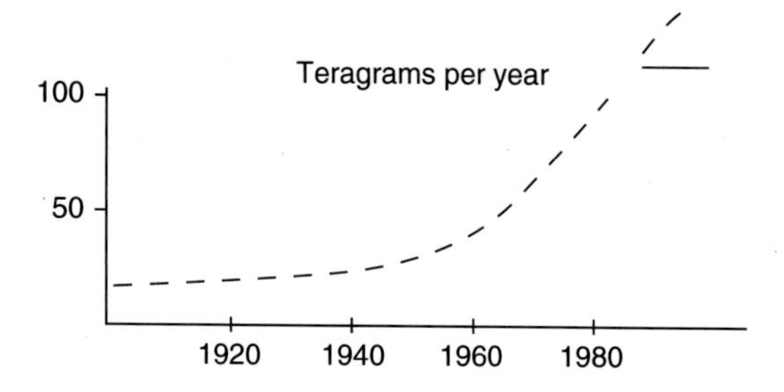

Figure 1 Natural (solid line) and anthropogenic (dashed line) nitrogen fluxes in the twentieth century. From Vitousek PM (1994) Beyond global warming: Ecology and global change. *Ecology* 75(7): 1861–1876.

contributed only 160 Gt of carbon. Thus, a share of land use in CO_2 concentration changes in the atmosphere exceeds 50%.

However, if one compares anthropogenic contribution to the basic biogeochemical cycles, which constitute 'biosphera machina' (see more about it below), they do not appear to be too great. At the same time, we feel that there is something odd in our human environment, which leads; us to be concerned about a potential ecological crisis. What is the impact of a dominant anthroposphere on the ecosphere? Is harmonious coexistence of the anthroposphere and the ecosphere possible?

Let us note that unlike such biosphere components as the atmosphere, biota, soils, hydrosphere, and stratosphere, each of which has had more or less clear spatial localization, the anthroposphere has lacked it and has always permeated the above media, even penetrating in the Earth's crust.

World Human Population, Energy Food Demand, and Energy Consumption

It is natural that the intensity of anthropogenic impact on the ecosphere depends (not usually in a linear way) on the size of human population, which grows as shown in **Figure 2**.

Two thousand years ago, there were a quarter of a billion people living on the planet. This had doubled to about half a billion by the sixteenth to seventeenth centuries. The next doubling required two centuries (from the middle of the seventeenth century to 1800), the following doubling occurred over only 100 years, while the last one took only 39 years.

Homo sapiens belongs to both the biosphere and anthroposphere. If we consider humans as animals, then all human energy requirements are satisfied through food, and the annual energy food demand per individual is 4×10^9 J. Thus, in the year 2000, the annual energy food demand that determines the annual trophic flow to species *H. sapiens* in the world ecosystem must be 2.4×10^9 J.

The Earth receives 3.5×10^{24} J of solar energy annually, providing the work of the 'green cover' with net primary production (NPP) equal to 5.5×10^{21} J yr^{-1} of new biomass. This energy flow also provides a steady state for 1.84×10^{18} g of living biomass (or 3.5×10^{22} J), and animal biomass constitutes only 0.8% of it, that is, 1.46×10^{16} g. Animals consume only 3% of the NPP (7.35×10^{19} J yr^{-1}). *Homo sapiens* is one of the animal species with biomass 4.2×10^{14} g (in the year 2000), constituting 2.8% of the total biomass of animals. Therefore, humans can use only $2.8 \times 3 = 0.084\%$ of the NPP, that is, 2×10^{18} J. Thus, the food demand of mankind is more by almost 1 order of magnitude than the trophic flow, that is, the trophic chains including

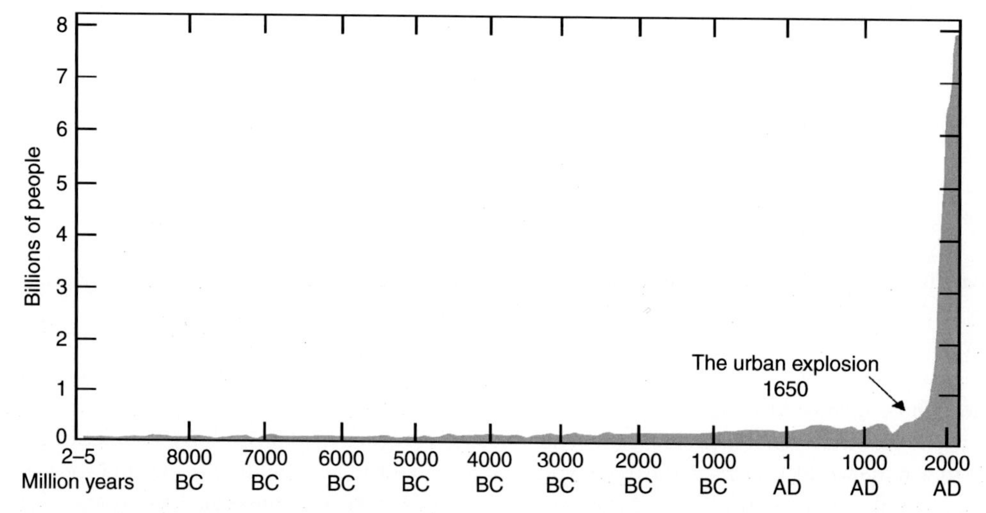

Figure 2 Dynamics of the world population. From Heinke GW (1997) The challenge of urban growth and sustainable development for Asian cities in the 21st century. *AMBIO* 8: 130–143.

H. sapiens are very strained. It may bring in turn either global starvation or destruction of this chain, elimination of many species from the chain (or its elimination in the whole from the global ecosystem).

In 1650, human population was approximately 600 million, that is, an order of magnitude less than today (**Figure 2**). From this, it follows that that the trophic flow was equal to food demand, and the corresponding trophic chain was not strained. In other words, humans were still one of many species, coexisting within the biosphere.

On the other hand, if we consider the fate of *H. sapiens* from the point of view of physical theory of fluctuations, the probability of fluctuation, which could cause the elimination of *H. sapiens*, is equal to

$$\text{Pr} = \exp\left[-\frac{\text{energy demand for human population}}{\text{energy supply for all animals}}\right].$$

At the time of the Neolithic revolution, the human population consisted of around 4×10^6 individuals, and required an energy supply of $1.6 \times 10^{16}\,\text{J}\,\text{yr}^{-1}$, then $\text{Pr} = \exp[-1.6 \times 10^{16}/7.35 \times 10^{19}] \approx 99.98\%$. If we estimate this probability for the year 2000, we get $\text{Pr}' = \exp[-2.4 \times 10^{19}/7.35 \times 10^{19}] \approx 72.2\%$. Looking at these numbers one can say that *H. sapiens* as a biological species was very fortunate that it has not been eliminated before the anthroposphere arose. Namely, the industrial and accompanying agricultural revolution could mask the consequences of growing strain in the trophic chain.

One of the main characteristics of the anthroposhere is the use of fossil fuels (traces of the past biospheres), and (at present) such 'nonbiosphere' energy as nuclear, with an accelerating rate (see **Figure 3**).

At the present time, the anthroposphere spends about $3 \times 10^{20}\,\text{J}\,\text{yr}^{-1}$ to provide for its functioning. This is mainly energy of fossil fuels and nuclear energy (fraction of the 'pure' biosphere energy – hydropower station and firewood – in this balance is $\sim 5\%$), and it constitutes about 13% of the global NPP, $2.3 \times 10^{21}\,\text{J}\,\text{yr}^{-1}$. Nevertheless, this percentage is enough for the biosphere and anthroposphere to strongly compete for common resources, such as land area and freshwater. Contamination of the environment and reduction of biotic diversity are typical consequences of the competition.

Since the biosphere (considered as an open thermodynamic system) is at a dynamic equilibrium, all entropy flows have to be balanced as well. Therefore, the entropy excess, which is created by the anthroposphere, has to be compensated by means of two processes: (1) degradation of the biosphere, and (2) changes in the work of the Earth's climate machine (in particular, through increases in the Earth's average temperature).

The energy of dissipation, corresponding to the full destruction of biota (equivalent to its complete

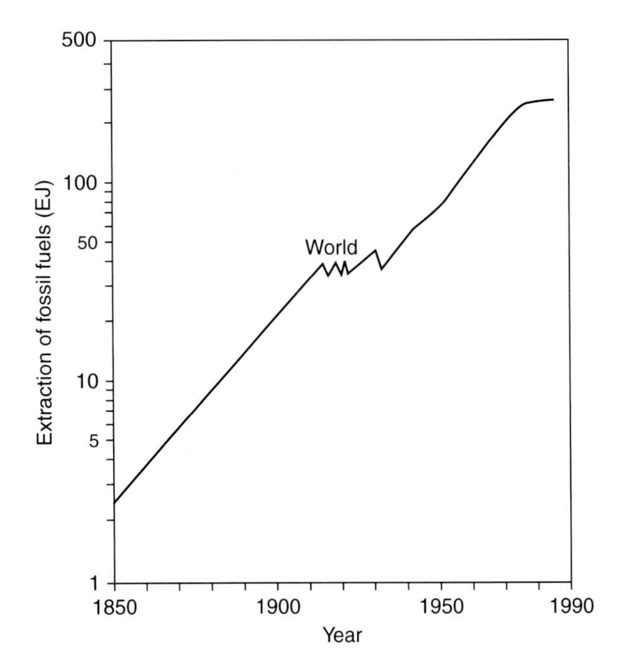

Figure 3 Accelerating rate of use of fossil fuels and nuclear energy.

combustion), is equal to $3.5 \times 10^{22}\,\text{J}$, while the energy dissipated by the anthroposphere is $3 \times 10^{20}\,\text{J}$. Even if the rate of the energy consumption in the anthroposphere does not increase, then this 'anti-entropy storage' of biota can make up for the entropy, produced by the anthroposphere, in the next 120 years. If this 'technogeneric' entropy could be compensated by soil destruction, then the agony would continue in the course of 300–400 years, since the storage of organic matter in soil is three- to fourfold larger than in biota.

Anthropogenic Impact on the Global Biogeochemical Cycles

It is known that all biogeochemical work of the biosphere is performed by the global biogeochemical cycles. The principal ones, which are, in particular, responsible for the contemporary global climate change, are the global carbon, nitrogen, and sulfur cycles.

Carbon. Gaseous carbon compounds of the global cycle include carbon oxides (CO_2, CO), methane (CH_4), and a great amount of different volatile hydrocarbons that are released as a result of vegetation metabolism and fuel combustion. The main problem here is to estimate flows of the main 'greenhouse gases', such as carbon dioxide and methane, into the atmosphere, and their anthropogenic components.

The CO_2 flow into the atmosphere from anthropogenic sources results mainly (75%) from organic fuel combustion (coal, oil, gas) and also from other kinds of economic activities (cement production, flue gas burning), making

20 billion tons yr^{-1}. One should add about 7 billion tons of CO_2 due to annual destruction of forests and loss of vegetative cover. The overall CO_2 anthropogenic flow into the atmosphere reaches about 27 billion tons yr^{-1}, that is, less than 0.01% from the CO_2 total amount in the atmosphere. According to earlier data, the CO_2 anthropogenic emission into the air amounted to 21.3 billion tons yr^{-1} in 1990. Thus, estimating the proportion of anthropogenic and natural components in the CO_2 flux into the atmosphere, one should note that the natural component is approximately 25–30 times more than the human-made one.

Methane inflows to the atmosphere are subdivided into two groups:

- natural biogenic and abiogenic;
- anthropogenic that consists of two subgroups: sources relating to human activity as a biological species and technogenic sources.

An analysis of different data by Adushkin *et al.* in 1998 allows us to conclude that:

1. natural biogenic sources are responsible for an annual average flow of methane equal to about 540 million tons yr^{-1};
2. abiogenic natural sources from lithosphere and hydrosphere make up *c.* 1360 million tons of methane annually (therefore, a ratio between biogenic and abiogenic methane is 1:2.5 in natural sources);
3. anthropogenic sources, including methane resulting from human agricultural activity, losses of methane during extraction of fossil fuels, and its industrial emissions produce an average annual flow of methane equal to about 1100 million tons yr^{-1}.

Therefore, the natural component of methane in the atmosphere estimated at 1900 million tons yr^{-1} is 1.7 times larger than its anthropogenic component.

Nitrogen. There are three kinds of nitrogen oxides – nitrous oxide (N_2O), nitrogen oxide (NO), nitrogen dioxide (NO_2) – and some ammonia. Nitrous oxide has the greatest concentration in the atmosphere (=270–280 ppbv).

Nitrogen oxides reach the atmosphere from different natural sources, such as decomposition of nitrogen-based compounds in the ground by anaerobic bacteria, forest and peat fires, hydrolysis, and sedimentation of nitrates. Nitrogen oxides give rise to aerosols of nitric acid, which is one of the basic components of acid deposits. Total emissions of nitrogen oxides from natural sources are estimated to be 310 million tons yr^{-1}, 540 million tons yr^{-1}, or 1090 million tons yr^{-1} depending on the source.

Sources of the anthropogenic flux of nitrogen oxides are industrial emissions of thermal power stations, chemical and iron and steel industry enterprises, waste dumps of coal and sulfur mines, motor transport, burning of biomass, etc. Total emissions of nitrogen oxides from anthropogenic sources are estimated to be from 30–55 million to 100–110 million tons yr^{-1}.

Therefore, a ratio of anthropogenic and natural components in a flux of nitrogen oxides is 1:10, that is, the anthropogenic flux is 10 times less than the natural one.

Sulfur. In nature, sulfurous gas, hydrogen sulfide, and other gaseous compounds containing sulfur are formed in large quantities as a result of processes of biological decomposition, decomposition of sulfur-containing ores, volcanic activity, and geothermal sources. Hydrogen sulfide getting in the atmosphere is quickly oxidized to make sulfurous gas; therefore, it can be considered one of the significant sources of SO_2.

A wide spectrum of gaseous sulfur compounds is released in the atmosphere after eruptions of volcanoes. Over a 25-year period, annual SO_2 emissions by subareal volcanoes changed from 10 to 30 million tons yr^{-1}. Volcanoes are responsible for approximately 7% of sulfur compounds getting to the atmosphere.

Thus, a total flux of gaseous sulfur compounds from natural sources (mainly gaseous sulfur dioxide) is estimated at 200–300 million tons yr^{-1}.

Anthropogenic sources of gaseous sulfur compounds are metallurgical enterprises, thermal power stations, cheminasescal and coke plants, oxidated landfills of collieries and sulfidic ores, transport, and explosive works. In addition, anthropogenic hydrogen sulfide is formed at factories manufacturing kraft pulp, mineral oil and natural gas treatment facilities, and enterprises making artificial silk and nylon. Global emissions of anthropogenic sulfur dioxide increased during 1950–90 from 20 to 160 million tons yr^{-1}.

The total emissions of anthropogenic sulfur oxides in the world are estimated at 130–200 million tons yr^{-1}. As a result, we observe that the anthropogenic flux of sulfur oxides is practically same, as its natural counterpart. Hence, an impact of anthropogenic sulfur oxide emissions on the environment, in particular, as regards atmospheric pollution, is comparable to the one from natural sources (**Table 1**).

Table 1 Global gas fluxes in the atmosphere from biosphere and anthroposphere

Source	*CO (bln. t yr^{-1})*	*CH_4 (10^6 t yr^{-1})*	*SO_2 (10^6 t yr^{-1})*	*NO_2 (10^6 t yr^{-1})*	*Total fluxes (bln. t yr^{-1})*
Natural	700	1900	200–300	310–1090	707.41–708.29
Anthropogenic	21.3–27	1100	130–210	30–110	22.92–29.12
Common	721.3–727	3000	330–510	340–1200	730.33–737.41

Anthropogenic Impact on Chemical Composition of the Biosphere

The biosphere represents an immense equilibrium system of chemical reactions. Perturbation of the equilibrium at one site may provoke uncontrolled change in the whole system, in spite of the fact that there are different compensating mechanisms (Le Chatelier's principle). We can say that chemical activity of mankind is almost compared now with the chemical work of all living matter. For instance, about 10^{17} g of minerals are excavated annually from the Earth; this value already constitutes 5.5% in relation to 1.84×10^{18} g of all living biomass.

This is in regard to the so-called 'gross' characteristics; if we look at 'information' ones, in particular atomic composition of excavated matter, then one can see that its composition significantly differs from the compositions of living matter, soil, and oceanic waters. Note that all these minerals are dispersed finally over the Earth surface. The impact on the metal cycles is most significant (**Table 2**).

Our technocivilization is a civilization of iron. About 10% of iron used is destroyed as a result of corrosion, friction, etc. If the amount of lost iron increases by a factor of 2, then, in accordance with our table, soil concentrations of lead increase more than tenfold, and mercury concentrations by 100 times, with toxic contamination of these substances.

Global Land Use: Agriculture and Urbanization

One of the main spatial factors of anthropogenic impact on the biosphere is the rapid growth of agricultural lands, with accompanying change in their land use. Human activity to produce food leads to the reduction of areas of habitat for natural organisms and to a sharp increase in the area of marginal ecosystems. Improvement of agricultural technologies and wide application of fertilizers led to a fourfold rise in land productivity and sixfold rise of agricultural yield in the twentieth century. However, this was accomplished by reducing populations of organisms and biodiversity of natural ecosystems (**Figure 4**). The biomass of agrocenoses never reaches the biomass of forests, while agrocenosis productivity is lower than that of natural ecosystems. Replacement of natural ecosystems by agrocenoses results in an 11.7% loss of the net primary product, while about 27% of NPP is lost in all human-degraded ecosystems.

About 23% of all usable lands in the world are subject to degradation, which leads to a reduction in its productivity. Agricultural technologies also lead to the destruction of a mid-term reservoir of biogenes, that is, soils. Significant amounts of soil are washed away. As a result of desertification, about 3% of NPP is lost, but soil organisms essentially suffer since they perish due to soil erosion and compression by agricultural implements, plowing, and application of fertilizers. For example, administration of nitrogen in the ground amounting to $3\,g\,m^{-2}$ a year, with an unchanging amount of other fertilizers, would reduce the population of species by 20–50% (**Figure 5**).

Cities exert a spatially concentrated impact on the environment. While the world population has grown, since 1976, by 1.7% a year on average, population of cities increased by 4% annually. Accelerated urban growth leads to pollution of water, soil, and the air, making their inhabitants live in an unfavorable ecological and social environment. In addition, urbanization is accompanied by a sharp decrease in resistance of urban area territories to technogenic and

Table 2 Relation of metals in soil, ocean, living matter, and world economy with respect to iron concentration

Element	*Soil*	*Ocean*	*Living matter*	*World economy*
Fe	1	1	1	1
Al	1.8	1	0.5	1.5×10^{-2}
Be	1.5×10^{-4}	6×10^{-5}	Traces	2×10^{-5}
Cr	5×10^{-3}	2×10^{-3}	1×10^{-2}	2×10^{-2}
Mn	2.1×10^{-2}	2×10^{-1}	1×10^{-1}	4×10^{-3}
Co	2.5×10^{-4}	5×10^{-2}	2×10^{-3}	3×10^{-4}
Ni	1×10^{-3}	2×10^{-1}	5×10^{-3}	4.5×10^{-4}
Cu	5×10^{-4}	3×10^{-1}	2×10^{-2}	1×10^{-2}
Zn	1×10^{-3}	1	5×10^{-2}	5×10^{-3}
Mo	5×10^{-5}	1	1×10^{-3}	3×10^{-5}
Ag	2.5×10^{-6}	3×10^{-2}	Traces	1.7×10^{-4}
Sn	2.5×10^{-4}	3×10^{-2}	5×10^{-3}	1.3×10^{-4}
Sb		5×10^{-2}	Traces	3×10^{-5}
W		10	Traces	2.5×10^{-5}
Hg	2.5×10^{-5}	3×10^{-3}	1×10^{-5}	1×10^{-5}
Au		4×10^{-4}	Traces	3×10^{-6}
Pb	2.5×10^{-4}	3×10^{-3}	5×10^{-3}	4.5×10^{-3}

Vinogradov AP (1959) *Chemical Evolution of the Earth*. Moscow: USSR Academy Scientific Publisher.

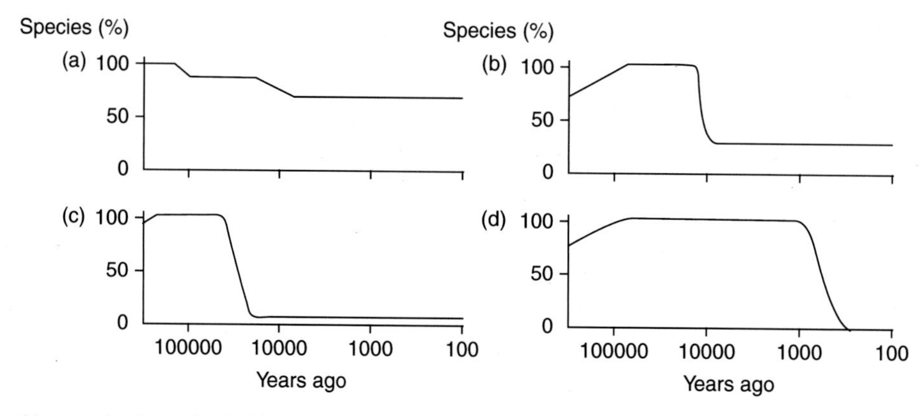

Figure 4 Loss of large animal species in Africa (a), North America (b), Australia (c), Madagascar and New Zeeland (d) (The World Environment, 1992).

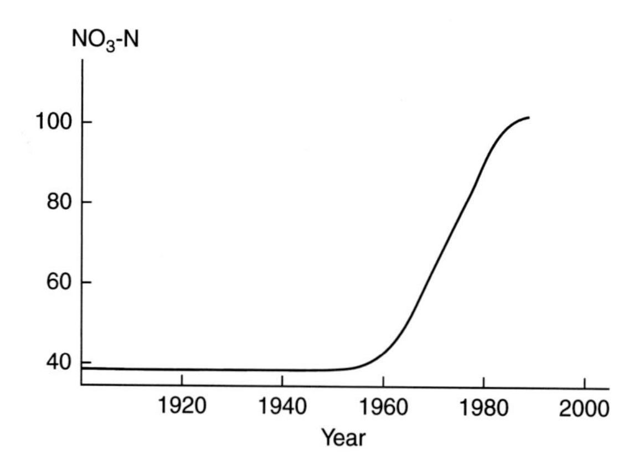

Figure 5 Change in concentration of nitrogen compounds in estuary of the Mississippi River since the beginning of the twentieth century. From Vitousek PM (1994) Beyond global warming: Ecology and global change. *Ecology* 75(7): 1861–1876.

technonatural hazards. This raises risks of urban dwellers and requires huge efforts of municipal authorities to maintain viability of urban infrastructure.

Industrial Revolution, Anthropocentrism, and the Biosphere Degradation

Industrial revolution unequivocally established an anthropocentric ideology in the human–nature relations. Humans placed themselves at the center of the biosphere, giving it a role of a huge pantry from which it is possible to extract resources beyond all bounds and, in return, store resulting waste. From the point of view of preservation of the global ecosystem, such relations are unpromising. Calculations show that the twenty-first century will see the exhaustion of many kinds of natural resources of our planet with perhaps unrealistic expectations that further technological advances and economic growth will open up new vistas for solving environmental problems.

Table 3 Human-disturbed terrestrial ecosystems (not including glaciers and bare lands)

Land area	*Undisturbed area*	*Partly undisturbed areas*	*Totally disturbed area*
134 904 471 km^2	27%	36.7%	36.3%

Environmental degradation in the latter part of the twentieth century reached global scales. Notwithstanding that about US$ 1.2 billion was spent over the 20 years between the UN conferences in Stockholm (1972) and Rio de Janeiro (1992) on environmental protection, the state of the Earth's environment was worsening. Industrial development that should have strengthened economic advances went into contradiction with the environment since it failed to take into account real limits to biosphere sustainability. Two opposite trends prevail in the global economy: gross world income is growing while the global wealth (first of all, life-supporting resources) is shrinking.

Industrial revolution has led to further pressure of technically and technologically equipped humans on the environment and has created conditions for a new ecological crisis. The consequences of such processes are hard to predict. It is clear that the coming crisis will essentially differ from the previous crises.

Data on disturbed ecosystems is also given in **Table 3**.

Conclusion: Philosophy of the Biosphere

A concern over an imminent catastrophe is growing in the enlightened sectors of society. One of the first among the outstanding thinkers who have realized all the gravity of consequences of industrial revolution was Vernadsky, who developed a scientific concept about the biosphere as a synthesis of knowledge about humans, biology, and sciences about nature, closely connected historically.

Dominant in this doctrine is belief in an indestructible power of scientific ideas as a planetary phenomenon capable to reconstruct the biosphere in a noosphere – the sphere of reason.

Many scientists and public and political leaders have understood this idea as a philosophical doctrine of the future development of the world. At the same time, the doctrine about a noosphere remains hardly worked out even at the conceptual level. At the world summit in Rio de Janeiro (1992), an attempt was made to suggest a global program of development of civilization. The document accepted at the conference was named as a concept of sustainable development.

The biosphere as a self-developing system for all its history has gone through a large number of local and global crises, every time reviving and continuing its development at a new evolutionary level. Humans as any biological species are temporary inhabitants on the Earth. Studies of biologists show that mechanisms of constant change of species incorporated in evolution of fauna provide existence in the biosphere of one species during about 3.5 million years on average. Therefore the modern human – Cro-Magnon man – that appeared 60 000–30 000 years ago as a biological species is at its initial stage of development. However, his activity for rather a short term placed him against the biosphere and he created conditions for an anthropogenic crisis.

Considering prospects of the postindustrial development of society, it is necessary to return to ecological understanding of sustainable development. Development can be considered sustainable if it remains within the limits of economic capacity of the biosphere, and maintains its functions as a self-organized and self-adjusted system.

Further Reading

Barnola JM, Pimienta P, and Korotkevich YS (1991) CO_2 climate relationship as deduced Vostok ice core: A re-examination based on new measurements and re-evolution of the air dating. *Tellus* 43B(2): 83–90.

Coldy ME (1990) *Environmental management in development: The evolution of paradigm. World Bank Discussion Paper No. 80*. Washington, DC: The World Bank.

Dobrecov NL and Kovalenko VI (1995) Global environmental changes. *Geology and Geophysics* 36(8): 7–29 (in Russian).

Golubev GN (2002) Global Ecological Perspective-3: Past, Present, Future. UNEP Moscow Interdialekt (in Russian).

Hannah L, Lohse D, Hutchinson Ch, Carr JL, and Lankerani A (1994) A preliminary inventory of human disturbance of world ecosystems. *AMBIO* 4–5: 246–250.

Heinke GW (1997) The challenge of urban growth and sustainable development for Asian cities in the 21st century. *AMBIO* 8: 130–143.

Jorgensen SE and Svirezhev Yu M (2004) *Towards a Thermodynamics Theory for Ecological Systems*, 370pp. Amsterdam: Elsevier.

Laverov NP, *et al.* (1997) *Global Environment and Climate Change*, 430pp. Moscow: Minnauki of Russia, RAN (in Russian).

Pegov SA and Homiakov PM (2005) *Influence of the Global Climatic Change on the Economy and Human Health in Russia*, 424pp. Moscow: URSS (in Russian).

Tolba MK, El-Kholy OA, El-Hinnawi E, Holdgate MW, and McMichael DF (ed.) (1992) *The World Environment 1972–1992*, pp. 884. London: Chapman and Hall.

Vernadsky VI (1998) *The Biosphere*, 192pp. New York: Copernicus.

Vinogradov AP (1959) *Chemical Evolution of the Earth*. Moscow: USSR Academy Scientific Publisher.

Vitousek PM (1994) Beyond global warming: Ecology and global change. *Ecology* 75(7): 1861–1876.

Vitousek PM, Erlich PR, Erlich AHE, and Matson PA (1986) Human appropriation of the products of photosynthesis. *Bioscience* 36: 368–373.

Zavarzin GA (1995) Circulation of methane in the ecosystems. *Nature* 6: 3–14 (in Russian).

Zimmerman PR, Greenbery JP, Wandiga SO, and Crutzen PJ (1982) Termites: A potentially large source of atmospheric methane, carbon dioxide and molecular hydrogen. *Science* 218(4572): 563–565.

See also: Biosphere. Vernadsky's Concept.

Astrobiology

C P McKay, NASA Ames Research Center, Moffett Field, CA, USA

Introduction

Earth is characterized by its global ecology and the widespread effects that life has on the environment. The fossil record as well as the tree of life, both indicate that life was present on Earth from very early in its history. The assumption that life has been continuously present on Earth is supported by the carbon-isotope record and the deep branches of the tree of life. The ecological correlate of life on Earth is liquid water. Liquid water is widespread on Earth and has been over its history and thus life has been global and persistent on this planet.

The fundamental goal of 'astrobiology' is to understand if Earth is rare, possibly even unique or if life is widespread throughout the universe. Astrobiology also considers the future of life and the possibility that global ecosystems can be created, or recreated as the case might be, on other worlds.

Our Solar System

Spacecraft and telescopic investigation of the other worlds of our solar system has not given any indication of a flourishing global biosphere like Earth's. Indeed, there is no another world with liquid water present on its surface. If there is other life in our solar system, it is not a global ecology but cryptic, subsurface life. Mars, the Jovian satellite Europa, and the Saturnian satellite Enceladus provide the most likely sites for present or past water and hence life. Titan, the largest Saturnian satellite, has a liquid on its surface but it is methane, not water.

Mars

Mars today is a cold, dry desert world with a thin atmosphere. There is no firm evidence for liquid water on its surface at any place or any season. There has been recent evidence of activity on Mars, notably gullies on the side of crater walls; however, while these features might be explained by liquid water, they can also be explained by the movement of dry materials.

Although there is no evidence for liquid water presently there is extensive evidence for liquid water on the surface of Mars in the past. **Figure 1** shows a sinuous canyon on Mars. This canyon, Nanedi Vallis, is the best evidence we have that some of the fluvial features on Mars were carved by liquid water in stable flow on the surface for an extended interval. Note in particular the presence of a channel on the bottom of the canyon which presumably reflects the flow path of a liquid. Explanations other than liquid water have been suggested for the fluvial features on Mars, including ice flow, lava flow, wind, and carbon dioxide flow. None of these can explain the morphology of Nanedi Vallis.

The low pressure on Mars today is inconsistent with the stable flow of liquid water on the surface. Thus, the water-carved features on Mars attest to an early climate with a thicker atmosphere and at least slightly warmer conditions. We do not know how long this cold thick atmosphere persisted, but climate models suggest that liquid water habitats would have been present on Mars for longer than the time associated with the earliest evidence for microscopic life on Earth.

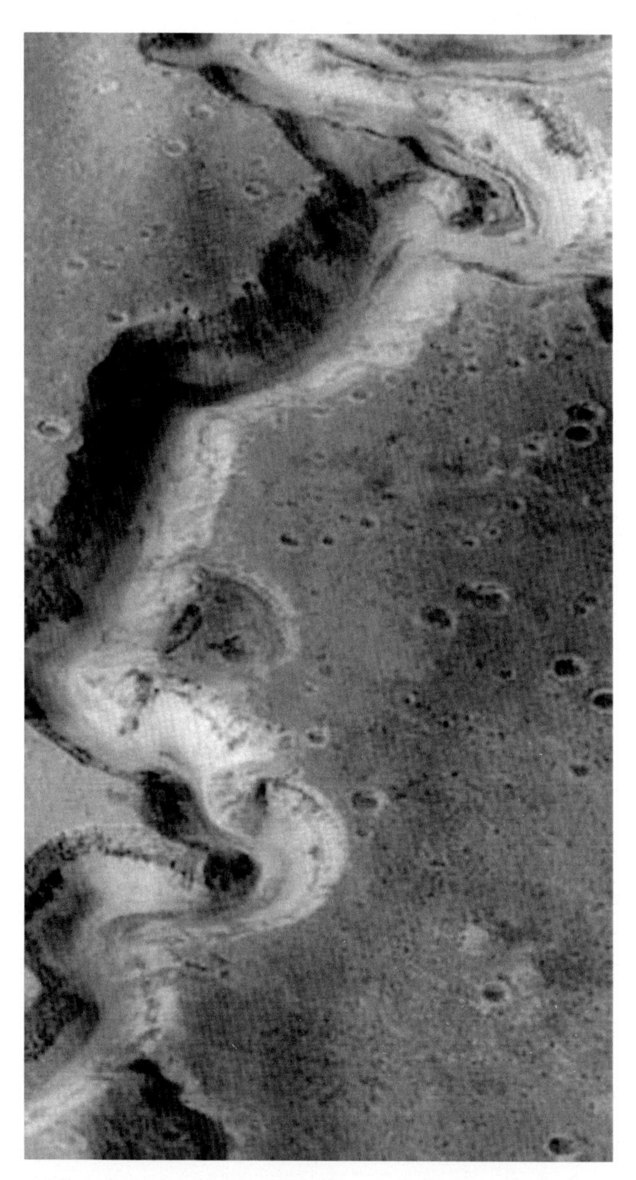

Figure 1 Liquid water in the past on Mars. Mars Global Surveyor image showing Nanedi Vallis in the Xanthe Terra region of Mars. Image covers an area $9.8 \times 18.5\,km^2$; the canyon is about 2.5 km wide. Photo from NASA/Malin Space Sciences.

If there had been a time on Mars when liquid water was widespread, we can expect that life, if present, would have also been widespread. Even if life had been global, we would only expect evidence of this early biosphere to be preserved until today in the polar permafrost. Organisms preserved for billions of years in the martian permafrost would probably be dead due to accumulated radiation from crustal levels of radioactive elements. However, these dead microbes would retain the biomolecules of martian life. Thus, unlike fossils, these frozen dead remains could be used to determine if martian life shared a common origin with life on Earth or represents a second genesis. It is possible that life on Earth and Mars shared a common origin through the exchange of meteorites.

Europa

Europa is one of the Galilean moons of Jupiter and is interesting for astrobiology because of the presence of an ocean under its icy surface. There are two lines of evidence that indicate an ocean: the frozen surface of iceberg-like features and the magnetic disturbance as Europa moves through the Jovian field. The former indicates the depth to the ocean is about 10 km and the latter indicates that the ocean is still present today.

Life on Earth may have originated in hot deep sea vents and Europa may have had similar deep sea vents, thus it is plausible that life may have also originated in Europa's seas. The same hot circulation could provide a continued energy source for life. Europa is more likely than Mars to have been free of any meteorites from Earth, so if there is life it is less likely to have been transported from Earth, hence more likely to be a second genesis.

The ocean of Europa is difficult to access but if the linear features seen on the surface are cracks then these may be locations where water from the ocean has been deposited on the surface. Any life in the water would remain, frozen and dead, on the surface. Samples of this material might allow us to investigate the biochemistry and genetics of a second example of life.

Enceladus

Enceladus is perhaps the most interesting astrobiology target in the outer solar system. This small satellite of Saturn has jets of water ice particles emanating from its South Pole. In addition to water ice, the jets include methane, propane, acetylene, and nitrogen gas or carbon monoxide. The likely source of Enceladus' jets is a pressurized subsurface liquid reservoir. If nitrogen is present it may reflect thermal decomposition of ammonia associated with the subsurface liquid reservoir and may imply that the water is in contact with hot rocks – providing a source of heat as well as mineral surfaces for catalyzing reactions. If this scenario proves correct, then all the ingredients are present on Enceladus for the origin of life by chemoautotrophic pathways – a generally held model for the origin of life on Earth in deep-sea vents. In this case, the Enceladus South Polar jets would include the waste products of such an ecosystem – that is, its biomarkers.

Titan

Titan is the largest moon of Saturn and is the only moon in the solar system with a substantial atmosphere. The main constituent of the atmosphere is nitrogen with methane forming several percent. Sunlight and electrons from Saturn's magnetosphere dissociate the nitrogen and methane and thereby start a cascade of reactions that produce organic compounds including the solid organic haze that fills the atmosphere and shrouds the surface. The organic chemistry on Titan may be a model for abiotic organic synthesis. However, the surface temperature is −180 °C, so no liquid water is present. Thus biological systems based on liquid water are not possible on the surface of Titan.

There is a liquid present on Titan; liquid methane and ethane are present in lakes in the polar region and as a moist film at the equatorial landing site of the Huygens Probe. Life in that liquid methane on Titan would be able to derive energy from atmospheric gases, in particular combining acetylene and hydrogen to form methane.

Planets around Other Stars

There may be planets orbiting other stars that have life and also have globally extensive ecosystems. It is generally thought that for a planet to be habitable it must have an average surface temperature between 0 and 30 °C and to maintain its habitability over geologically long periods of time, and it must have a mass between 0.5 and 10 times the mass of Earth. Until recently, extrasolar planets of this size were not detectable. However, Earth-based telescopes have now detected at least one such planet and telescopes soon to be placed in space should be able to detect many more.

The most direct evidence for a global ecology is the presence of an oxygen-rich atmosphere. This could be detected by spectroscopic identification of either oxygen or ozone. Pigments such as chlorophyll may also be detectable and indicate life on the surface.

Mars Future

Although our solar system currently has only one world with a global ecology, this may not always be the case. There has been serious discussion of planetary ecosynthesis on Mars. The fact that Mars once supported widespread liquid water, and possibly life, motivates the question of restoring such conditions on Mars by artificial means.

The fundamental challenge of restoring habitable conditions on Mars is to warm up the planet from its current −60 °C to over 0 °C, and perhaps as warm as Earth, +15 °C. Humans have demonstrated, and implemented, the technology to warm planets with Earth as our first target. The level of human-induced warming on Earth is debated but is probably of order a few degrees. On Mars the warming needed would be tens of degrees – many times larger than on Earth – but the extrapolation from Earth to Mars is conceptually straightforward. Energy balance calculations suggest that warming Mars might be achieved in 100 years or less. However, producing an oxygen-rich atmosphere would take more than 100 000 years. Thus,

Table 1 Habitability

Parameter	*Limits*	*Note*
Global temperature	0–30 °C	Earth = 15 °C
Composition for plants, algae, microorganisms		
Total pressure	>1 kPa	Water vapor pressure plus O_2, N_2, CO_2
CO_2	>0.015 kPa	Lower limit set by photosynthesis
		No clear upper limit
N_2	>0.1–1 kPa	Nitrogen fixation
O_2	>0.1 kPa	Plant respiration
Composition for breathable air		
Total pressure		
Pure O_2	>25 kPa	Lung water vapor plus CO_2, O_2
Air mixture	>50 kPa	Based upon high elevation
	<500 kPa	Buffer gas narcosis
CO_2	<1 kPa	Set by toxicity
N_2	>30 kPa	Buffer gas
O_2	>13 kPa	Lower limit set by hypoxia
	<30 kPa	Upper limit set by flammability

Adapted from McKay CP, Toon OB, and Kasting JF (1991) Making Mars habitable. *Nature* 352: 489–496.

warming Mars is within current technology and this fact frames the discussion about Mars in a fundamentally different way than planetary-scale environmental alteration on any other world of the solar system.

We tend to think of the present oxygen-rich Earth as the only model for a global ecology. However, there are two alternative possibilities for life supporting states for Mars: one with oxygen and one without. These two alternative states are listed in **Table 1**.

If there was life on Mars and it is now extinct beyond recovery, then planetary ecosynthesis can be viewed as a type of 'restoration ecology'. If there is life on Mars, or recoverable life, but it shares a common ancestor with life on Earth then it seems plausible that planetary ecosynthesis can proceed using Earth life forms as needed.

Perhaps the most interesting and challenging case is that in which Mars has, or had, life and this life represents a distinct and second genesis. I would argue that if there is a second genesis of life on Mars, its enormous potential for practical benefit to humans in terms of knowledge should motivate us to preserve it and to enhance conditions for its growth – a second genesis in a second global ecology.

See also: Biosphere. Vernadsky's Concept; Climate Change 2: Long-Term Dynamics; Coevolution of the Biosphere and Climate; Gaia Hypothesis; Phenomenon of Life: General Aspects.

Further Reading

de Duve C (1995) *Vital Dust*. New York: Basic Books.
McKay CP (2004) What is life – and how do we search for it on other worlds? *PLoS Biology* 2: 1260–1263.
McKay CP, Toon OB, and Kasting JF (1991) Making Mars habitable. *Nature* 352: 489–496.
Ward PD and Brownlee D (2000) *Rare Earth*. New York: Springer.

Biogeocoenosis as an Elementary Unit of Biogeochemical Work in the Biosphere

J Puzachenko, Russian Academy of Sciences, Moscow, Russia

Introduction

Biogeocoenosis belongs to a class of ecological concepts such as phytocoenosis, landscape, units, and sites based on the ideas of spatial uniform units that are distinguished in a given area and separated by visible boundaries. The concept originated by realizing the necessity to study and display the interactions among soil-forming rocks, soil (edaphotope), atmosphere (climatope) with vegetation (phytocoenosis), animal population (biocoenosis), and microorganisms (microbocoenosis). The author of this concept is a Russian geobotanist and paleogeographer V. N. Sukachev.

Biogeocoenosis

The ideas of Sukachev as a geobotanist were close to those of Clements, although Sukachev never recognized phytocoenosis as an organism motivated by the fact that, unlike an organism, elements and parts of phytocoenosis and biogeocoenosis can exist out of the whole. However, he also did not accept the individualistic concept of Glizon-Ramenskii concerning the plant cover organization. According to the author's definition, on a specific area of the Earth's surface, biogeocoenosis is a combination of homogenous natural phenomena (atmosphere, rocks, vegetation, animal and microorganisms, and soil and water conditions). These components possess specific types of interactions and a definite type of interchange of matter and energy occurs between them and with other natural phenomena, thus representing an internally contradictory dialectical unity, being in constant movement and development. N.V. Timofeev-Resovskii determined biogeocoenosis as a biochorological unit, within which there exist no biocoenotical, geomorphological, hydrological, climatic, or pedological–geochemical boundaries. Biogeocoenosis is implied as an integral discrete elementary natural cell of the biosphere that realizes the function of matter and energy transformation. Although the boundaries of each biogeocoenosis may be distinguished according to any of its components, practically it is better to accomplish it using boundaries of the best-observed component, namely vegetation, that is, according to the boundaries of phytocoenosis. Different biogeocoenoses interact with one another in space forming the biogeocoenotic cover. Sukachev did not consider specially the spatial dimension of the biogeocoenotic cover, but as it follows from the context, it corresponds to a rather vast territory commensurable with a floristic district or area. Sukachev discussed in detail the correlation of his concept with Tansley's concept of an ecosystem, different variants of its definition, and the concept of landscape and its morphological units, mainly in the interpretation of the adherents of the Russian school. He paid attention rightly to the fact that an ecosystem is considered (according to Tansley) as an abstract physical system uniting organisms with their environment. It is worthwhile to recall that Tansley actively objected to Clements's holistic concept of organism and considered ecosystem as a set of relations within different spatial–temporal intervals and at different hierarchical levels rather than a reality. Later on, this methodological content of the ecosystem concept disappeared almost completely, and ecosystem has been considered as a natural unit representing a totality of biotic and abiotic elements and as a functional system. Nevertheless, the concept of ecosystem maintains its general meaning along with its traditional interpretation as a chorological unit. Sukachev insisted that the concept of biogeocoenosis as a strictly territorial unit was more definite than the uncertain concept of ecosystem. One can accept this to be true to some extent, but the history of development of science showed that precisely some uncertainty inherent to the concept of ecosystem ensured its viability and incorporation into the general scientific basis. In the light of general system concepts, biogeocoenosis may be considered as a kind of ecosystem which possesses relatively spatially homogenous or stable (random or specific quasi-regular variation) properties in terms of its components within the framework of their observed boundaries. At the same time, the reality and commonness of the distinguished boundaries are not proved specially, but accepted *a priori*, assuming that these boundaries are relatively gradual.

Comparing the concept of biogeocoenosis with the modern concepts of landscape, it is worthwhile to note that the latter are interpreted differently. The concept of biogeocoenosis is most likely to be close to that of units accepted in the Canadian and Australian schools. However, a unit in landscape science is a functional unit rather than an operational one. In American forest science, the notion of biogeocoenosis is comparable territorially with that of stand.

Sukachev, who fully accepted the concept of the biosphere proposed by Vernadsky, regarded biogeocoenosis as an elementary cell of the biosphere.

Researchers who accepted the concept of biogeocoenosis differentiated between the spatial structural elements of biogeocoenosis: vertical layers and horizontal occasionally or quasi-regularly alternating parcels (parts), which are usually distinguished by the shrub, grass, and moss layers commensurable with microassociations. The genesis of parcels was mainly related to the heterogeneity of the tree layer, and they may be associated with gaps. Different parcels are often related to different pedons of soil. Sometimes, parcels are determined by the initial pattern of the nanorelief and soil-forming rocks.

Biogeocoenotic Process

Sukachev considered studies of biogeocoenoses as an independent science – biogeocoenology, which studies the biogeocoenotic process. The idea of the biogeocoenotic process proper, being formed after the definition of the notion 'biogeocoenosis', is the content of the concept and supplements the Vernadskii ideas for the local level of the biosphere organization. In addition, the elaboration of the concept of biogeocoenotic process was based on the ideas of materialistic dialectics (in this aspect, Sukachev was close to Tansley), kibernetics, and systems theory affecting greatly the development of science in the 1960s. The abundant experience of Sukachev himself as a paleobotanist, geobotanist, geographer, and naturalist was also of great importance.

The dialectic law of the unity and conflict of opposites is postulated as the basis of self-development of biogeocoenosis, with its existing discontinuities or disruptions, destruction of the old, and initiation of the new. Although a biogeocoenosis is an open system, all of its components together still form a certain integral dialectical unity characterized by internal contradictory interactions, which never produce a state of equilibrium within that unity (system). Climatope, edaphotope, phytocoenosis, zoocoenosis, and microbocoenosis are considered as components of biogeocoenoses.

The action of these interior forces leads to self-development, whereas the effect of the external ones leads to some variation and disturbance of the developmental process proper. It is practically useful to consider the mechanisms of nonequilibrium thermodynamics discussed by I. Prigogene, and the dynamics of nonlinear dissipative systems with positive and negative feedbacks capable of innovations discussed by G. Hacken.

The biogeocoenotic process is understood as a change in the matter and energy exchange due to the interaction of organisms with each other and with the environment, as well as between components of biogeocoenosis. The biogeocoenotic process includes not only interactions and exchange of matter and energy between biogeocoenose components, but also interactions and exchange of matter and energy between biogeocoenoses and their surroundings – the environment, in which they exist, and other biogeocoenoses (both adjacent and more remote ones). Since the process of interaction of a biogeocoenosis with its environment is partly expressed in terms of the incessant outflow of energy into space, it has, as it were, an entropic character. But, at the same time, new matter and energy are constantly entering the biogeocoenosis. A biogeocoenosis is considered as an elementary cell, and the biogeocoenotic process in each biogeocoenosis is typical due to specific relations between the biogeocoenosis components and the interaction with their environment. Under similar environmental conditions, biogeocoenoses with similar composition and structure also realize similar biogeocoenotic processes. Evidently, this model is the basis for the development of spatial hierarchical organization of biogeocoenosis, and the author of the concept suggests that the biogeocoenotic cover is a set of interacting biogeocoenoses over a rather vast territory.

The biogeocoenotic process unites four relatively independent processes:

1. The interactions of biogeocoenosis components and elements among themselves, which do not remain constant, but change in time and alter the course of the biogeocoenotic process. This process is a purely internal one and may be called 'endocoaction'.
2. The introduction of microorganism germs, plants or new species of organisms by wind and water, and of some organisms from outside, which can change somewhat the biogeocoenotic process. This process was proposed to be called 'inspermation'.
3. The introduction of mineral and partly organic matter with dust, surface, and intrasoil runoff. This process is called 'inpulverization'.
4. The removal of mineral and organic matter by water and other organisms. This process is called 'expulverization'.

The process of internal interactions never ceases; it slows down or accelerates to some extent. The slowing down is determined by a gradual increase in the resilience of biogeocoenosis, but the acceleration is determined by the disturbance of this stability via both settling of new species and changes in the structure of the interactions in the course of self-development. The second and third processes change at the level of the biogeocoenotic cover resulting from climatic and geodynamic fluctuations and asynchronous self-development of neighboring biogeocoenoses as well. The fourth process may be considered as an irreversible one to a considerable degree, and if it is not compensated for the third process, the changes in biogeocoenosis are determined by slow but permanent removal of mineral and organic substances from it. Finally, within the biogeocoenotic cover, the process of formation related to the origin of new phenotypes, genotypes, and morphofunctional forms of organisms is also realized.

A rather strict definition of the biogeocoenotic process as a change of states determined by different mechanisms allowed Sukachev to construct a harmonious classification of the dynamics of biogeocoenoses and biogeocoenotic cover on the following basis: equilibrium process with natural reversibility; nonequilibrium irreversible process; self-development (autogenous or endogenous processes), processes under the influence of external forces (exogenous); according to variation in time and space.

The classification of types of dynamics of the forest biogeocoenoses elaborated by Sukachev is given below:

A. Cyclic (periodic) dynamics of forest biogeocoenoses (reversible changes in forest biogeocoenoses).
   (1) Daily changes in biocoenoses.
   (2) Seasonal changes in biocoenoses.
   (3) Annual (weather) changes in biocoenoses.
   (4) Changes in biocoenoses due to the process of regeneration and growth of woody and other vegetation:
      (a) regular regeneration of woody plants;
      (b) irregular (wave) regeneration of tree stands;
      (c) synusial dynamics, especially parcel dynamics (these variants of the dynamics were likely to be associated with a gap dynamics model; models of these types of relationships reproduce usually restricted quasi-cyclic fluctuations of productivity, biomass, and species composition).

B. Dynamics of the forest biogeocoenotic cover of the earth, or successions of forest biogeocoenoses.
   I. Autogenous (irreversible) successions of biogeocoenoses (developments of the forest phytogeosphere, of forest biogeocoenogenesis).
      (1) Syngenetic succession of biogeocoenoses.
      (2) Endogenous (endodynamic) successions of biogeocoenoses.
      (3) Phylocoenogenetic successions of biogeocoenoses:
         (a) phytophylocoenogenetic successions of biogeocoenoses;
         (b) zoophylocoenogenetic successions of biogeocoenoses.
         (*Note.* Syngenetic processes are irreversible ones that proceed only due to alterations in the species structure without irreversible environmental changes (typical processes are the development of high bogs, progressive development of eluvial and illuvial horizons of soils). Phylogenetic successions imply processes determined by the origin of new forms. Probably, such processes are useful to be included into the dynamics determined by phylocoenogenesis of viruses and bacteria, including also the saprophytic microorganisms).
   II. Exogenous (reversible and irreversible) successions of biogeocoenoses.
      1. Hologenetic (irreversible) successions of biogeocoenoses:
         (1) climatogenic successions of biogeocoenoses;
         (2) geomorphogenic successions of biogeocoenoses;
         (3) selectocoenogenetic or areogenic successions of biogeocoenoses;
            (a) phytoareogenic successions of biogeocoenoses;
            (b) zooareogenic successions of biogeocoenoses.
            (*Note.* Hologenetic processes are realizable at the regional level of the biogeocoenotic cover organization. It is worth noting that Sukachev did not extend the principle of actualism and reversibility to climatogenic, that is, paleoclimatic successions. Selectocoenogenetic successions may appear due to the invasions of alien species. Changes determined by invasions of agents of feral herd diseases of plants, animals, and saprophytic microorganisms are expedient to be included into this type of dynamics. A typical example is the mass and, most likely, irreversible death of American chestnut (*Castanea dentata* (Marsh.) Borkh.) in the Appalachians. If species change their properties in the process of settling, selectocoenogenetic successions are indistinguishable from phylocoenogenetic ones).
      2. Local (reversible and irreversible) catastrophic successions of biogeocoenoses.
         (a) anthropogenic successions of biogeocoenoses;
         (b) zoogenic successions of biogeocoenoses;
         (c) pyrogenic successions of biogeocoenoses;
         (d) windfall successions of biogeocoenoses;
         (e) successions of biogeocoenoses produced by mud streams, landslides, sudden inundations, and other causes.

Probably, this classification of the dynamics may be recognized as the most complete. For the modern ecology, it contains all the bases for particular and integrating models of dynamics and research programs (e.g., programs directed to the accumulation of data on the irreversibility of self-development processes). However, in order that the concept of biogeocoenosis might create the necessary bases for studies and simulation of biogeochemical cycles, it should contain some concrete system definition and refinement of ideas of the spatial–temporal hierarchy and elimination or weakening of contradictions between individualistic and organism concepts of spatial organization of the biosphere and its components.

Biogeocoenosis and the Biosphere

The system that specifies the biogeocoenosis concept is rigorously introduced in works by Vernadsky, who was not only a naturalist, but also a physicist and chemist;

he possessed knowledge in thermodynamics and thermostatics. In complete accordance with concepts of thermostatics, he determined an object and its elements in the following way: "I will call a set of organisms participating in geochemical processes living matter. Organisms composing this set will be elements of living matter. With all this going on, we will pay attention not to all the properties of the living matter, but only to those which are related to its mass (weight), chemical composition, and energy. In such a comprehension, living matter is a new scientific notion". Later, Vernadsky directly associates individuals with molecules of gases and suggests to consider living matter as a statistical ensemble of elements. Thus determining the concept of the biosphere, he states that laws of equilibrium (equilibrium process) in general mathematical form as revealed by J. Gibbs (1884–87) (who reduced them to relationships between independent variables, such as temperature, pressure, physical state, and chemical composition, which characterize the chemical and physical processes and participate in system processes) could be applied to a living system of bodies. According to this statement, one can distinguish "thermodynamic spheres as areas of equilibrium of thermodynamic variables that are determined by values of temperature and pressure; phase spheres that are characterized by the physical state (solid, liquid, etc.) of bodies in their composition, chemical spheres different in the chemical composition. Only one sphere distinguished by E. Suess – the biosphere – remained aside. Undoubtedly, all the reactions of the biosphere follow the laws of equilibrium, but they include a new characteristic, new independent variable which was not taken into account by J. Gibbs and is very important in other equilibrium forms (in the context of thermodynamics). A special reaction is the phenomenon of photosynthesis, with radiant light energy as an independent variable. Therefore, "living organisms, introducing the radiant light energy to physicochemical processes of the earth crust, drastically differ from other independent variables of the biosphere. Like these variables, living organisms change the course of equilibrium, but unlike them, they represent specific autonomous formations as specific secondary systems of dynamic equilibrium in the primary thermodynamic field of the biosphere. The autonomy of living organisms reflects the fact that the thermodynamic field, which inherently has quite other parameters than those observed in the biosphere. Therefore, organisms retain their own temperature (many organisms do so strongly) within the medium at another temperature and have their interior pressure. They are isolated in the biosphere, and its thermodynamic field is important for these organisms only due to the fact that it determines the area of existence of these autonomous systems, but not their interior field. From the chemical standpoint, their autonomy is expressed in the fact that chemical compounds produced in these systems cannot be synthesized beyond them under usual inanimate conditions of the biosphere. Being fallen into the conditions of this medium, they turned out to be unstable, are decomposed, transformed to other bodies, and in that way, they become disturbers of the equilibrium and represent a source of free energy in the biosphere". Vernadsky discusses in detail all the properties of living matter known by that time, including basic mechanisms of its evolution. Generalizing his writings and using the modern terminology, one can define living matter as a stationary dissipative system of organism elements, which is far from thermodynamic equilibrium with free energy and exergy. The stationary state of this system is supported by the absorption of solar energy, which is responsible for the permanent conversion of the chemical element flux into a new organic form, realizing the cycle with a release of free energy to the environment, and transforming the latter as a result of useful work (exergy). The simplest example of this work is the intensification of the water cycle in the biosphere with appropriate contribution to climate control, that is, changes of equilibrium correspond to thermodynamic variables that change climate. So, when combining the concept of biogeocoenosis with the concept of 'living matter', we obtain rather strict thermodynamic bases for the characterization of the biogeocoenotic process, as well as all the necessary fundamentals for consideration of their autochthonous (endogenous) dynamics and self-development of biogeocoenoses and biogeocoenotic cover as a nonequilibrium, stationary thermodynamic dissipative system. However, all this is insufficient to consider biogeocoenosis as an elementary cell of the biosphere, within which nonliving matter is converted to living one and, conversely, incomplete transformation of the former to mobile chemical compounds occurs with a release of free energy and changes in the environment (thermodynamic variables of the atmosphere, hydrosphere, and lithosphere).

Reductionism and Holism

According to the Gleason–Ramenskii continuum individualistic concept (reductionism), there are no necessary bases for the initiation of spatial cells as relatively discrete formations without any additional conditions. The existence of such cells is the basis of Clements's organism concept (holism). It is worthwhile to note that, strictly speaking, Clements may not be its original author. Even at the dawn of the development of geography, in 1811, Butte stated that none of the scientists had any doubts regarding the existence of earth organisms. Within any specific field, a combination of all the phenomena is not a simple set; they represent a holon. Butte assumed individual countries and districts (including humans), as 'organisms', which, as any organism, may be considered both in terms of their physical and psychical aspects. He wrote that "areas as a holon assimilate the human population", and "population assimilates these areas not less constantly". At the same time, opponents of the hyperholon paid attention to the fact that it was difficult to find districts the boundaries of which could be determined as the basis of all the phenomena. The most complete criticism of this integral concept was given by A. L. Bucher in 1827. As a result, he concluded that there was no necessity to study boundaries, and regions might be distinguished in any arbitrary manner. He proved that geography should study relations between particular phenomena in any area of the earth's surface. Even now the same contradictions exist: on the one hand, the individualistic concept has been fully recognized; on the other, Gaia's superorganism concept is very popular. The criticism of this concept rests on traditional bases and factually repeats the discussion that has been continued for almost 200 years. If to leave aside these disputes, we can state that the two models of living matter – individualistic (reductionism) and organism (holism) – may be considered as those reflected in real natural phenomena. Developing these models up to possible logical limits, in both cases we obtain incompatible constructions. In the first variant, it is a construction similar to Dawkins's selfish gene; in the second one, it is a superorganism with its own purposeful development and superstability similar to Gaia's model. The individualistic model has been well substantiated theoretically and realized in microcosm and perfusion cultures. For the simplest linear variant, a theorem has been proved asserting that, in the homogenous environment, the number of stable coexisting populations is equal to the number of resources or, in general, to the number of any operating factors. The relations following this model were obtained by the methods of ordination for a wide diversity of plants and animals in direct terrain investigations. Particularly, such relations between different layers of a forest community and main tree species were shown for the Eurasian forest zone.

To prove the integrity of biogeocoenosis, ecosystem or plant community and their emergent properties should be understood more completely. Raised bogs may be referred to the formations of this type, the progressive growth of which is supported by the positive feedback between the groundwater table at the territory adjacent to the bog and development of sphagnum mosses. The accumulation of dead parts of mosses raises the groundwater table, and this process promotes the further moss growth and peat accumulation. Raised bogs form their own dynamic spatial structure, and minimization of moisture evaporation in hot summer months may be accepted as its emergent feature. Such a raised bog in fact resembles a superorganism, which occupies slowly (up to 10 cm per year) the neighboring territories displacing forest communities. True, this superorganism exists primarily due to the almost complete cessation of the cycle of matter, representing an essential deceleration of the water cycle at the exergy lowest for the forest zone. Such organism features are difficult to find for many typical cases. If not to ignore the traditional experience to distinguish phytocoenoses as relatively homogenous spatial formations that indicate biogeocoenosis, their integrity may be accepted as an empirical fact. At the same time, it is admitted that the corresponding mechanisms are poorly known.

From the standpoint of postmodern science, there is no necessity to create a single eternal theory. The most topical concept is one that initiates research and provides foundations for verification of competitive hypotheses, as well as stimulates their diversification and does not eliminate their joint acceptability. From these positions, *a priori* denial of these two models is identical to a nonacceptance of liberal or social views in the organization of human society. It is evident that the individualistic concept is mainly close to the thermodynamic model of the world in its movement to equilibrium and higher entropy. The basis of the individualistic model is maximization of independency of each component and its resilience within the holon that is a rather satisfactory strategy for its survival. But at states far from equilibrium, positive correlation and effects of self-organization and relatively discrete spatial structures arise in the thermodynamic system in accordance to the theory of nonequilibrium thermodynamics. This is a good hypothesis, which is useful to be verified for the biosphere. If to lean upon the theory of dynamic systems, the entire biosphere and its patches may be considered with certainty as nonlinear oscillators of high dimension. From these positions, the efficiency of the fractal model for characterizing the diverse natural processes is well explained. Formally, a fractal set is continuous but undifferentiated, and displays a cascade of bifurcations in the spatial–temporal dynamics of nonlinear oscillators. In nature, it manifests itself in the possibility to distinguish between different-scaled and hierarchically subordinate relatively homogenous formations, boundaries of which may also be divided into such structures. The formal fractal model assumes a self-similar division into indefinitely small units. Real natural objects do not possess this specific feature – their fractality has a finite range of dimensions. Taking into account this property, the model is sufficient for the theoretical definition of a biogeocoenosis as a spatial–temporal cell commensurable with linear dimensions of dominant plant species in it, that is, including some minimal population stable at least in one generation. Direct measurements of the fractal landscape cover the structure using data of remotely sensed investigations and three-dimensional models of relief show that, almost everywhere, a fractal spectrum connecting the amplitude of spatial variation of the variables measured with the spatial wave number reveals quasi-harmonic fluctuations, but describes only some percentage of the spatial variation. However, the relative peaks of the spectrum, corresponding to definite linear sizes, allow correcting a choice of scale for different hierarchical levels. The nature of local spatial homogenous structures may be determined by the organization of relief, soil-forming rocks, soil, dynamics of vegetation, effects of animals, tree windfalls, fires, and so on. The spatial–temporal dynamics of each of these components are stipulated by both their own fluctuations and those originated due to their interactions. As a result, the spatial structure is fractal, and distinguishing the relatively even territories is possible and strictly realizable on the basis of classification of multispectral images, in particular. In the framework of the model of a nonlinear oscillator, the individualistic concept does not contradict the different-scale discontinuity, and although nonlinear oscillators produce effects of self-organization under definite conditions, these models do not contain mechanisms of structural stability. On the basis of these models, a holistic model is impossible to construct. On the other hand, using the multifractal model, the proportion between total energy, free energy, and entropy is deduced resulting in the natural generalization of two models of reality. Following this method, the concept of biogeocoenosis as an elementary cell of the biosphere may be of constructive importance in both the organization of terrain investigations for assessment of biogeocoenotic and biosphere processes and the elaboration of corresponding models. The fractal scheme of organization of the biogeocoenotic cover gives prerequisites for the recalculation of parameters obtained in large-scale studies to those corresponding to the high level of organization. In order to obtain the behavior similar to that of an organism, it is necessary to add contours of positive feedback providing relationships between components of the system and supporting system resilience under conditions far from the thermodynamic equilibrium in the environment to the model of nonlinear dynamics. The fact that such relations are realizable in organizing the components of the biosphere was shown from the example of the bog. A similar type of relation holds for a tropical forest that evaporates moisture intensely. The same is true for boreal spruce forest that evaporates more moisture than a deciduous forest and supports low temperatures favorable for spruce due to expenditures of heat for evaporation. The positive feedback is characteristic of mycorrhizae fungi and their hosts. There are many examples of positive feedbacks in a plant community (mutualism). However, the conditions under which they determine holistic features of biogeocoenosis and those of higher levels of its organizations are not evident and need special investigation. At the same time, their potentially significant role in the maintenance of homeostasis in an aggressive medium is evident, as is the nature of spasmodic and catastrophic transformations at small disturbances, primarily in the margin areas of the system tolerance.

Summary

A discussion of the general concepts of ecology and attempting to specify their physical sense are *a priori* ungrateful tasks. In ecology, as in any natural science, notions or definitions had a quite uncertain content at the time of their introduction and they determine an approach to studies rather than their object. Later, these notions and definitions were redetermined many times and differently by different researchers. The multidimensional subject of ecology stipulates such an uncertainty. The uncertainty causes periodically renewed discussions of the theoretical bases of science that are inevitable in formulating the concept of biogeocoenosis. The latter may be considered as a specific one in relation to a more general concept of an ecosystem. In the framework of the modern theory of thermodynamics and nonlinear dynamical systems, accepting the existence of self-similar quasi-discrete territorial units, the concept of biogeocoenosis is interpreted via living matter as a thermodynamic variable. On the other hand, the ideas of the dynamics of biogeocoenosis elaborated by Sukachev represent good bases to formulate verifiable hypotheses. They allow combining the concepts of reductionism and holism as interconnected (but not contradictory) models, the contribution of which to the spatial–temporal dynamics depends on geographic conditions, the current status, and the time of self-development.

See also: Iron Cycle.

Further Reading

Abrosov NS, Kovrov BG, and Cherepanov OA (1982) *Ecological Mechanisms of Co-Existence and Species Regulation*, 287pp. Novosibirsk: Nauka.

Alcock J (2003) Positive feedback and system resilience from graphical and finite-difference models: The Amazon ecosystem – an example. *Earth interactions* 7: 23pp. Paper No.5.

Hargrove WW, Hoffman FM, and Schwartz PM (2002) A fractal landscape realizer for generating synthetic maps. *Conservation Ecology* 6(1): 2.

Hartshorne R (1939) *The Nature of Geography*. Lancaster, PA: Association of American Geographers.

http://www.consecol.org/vol6/iss1/art2 (accessed December 2007).

Ilya P (1997) *The End of Certainty: Time, Chaos, and the New Laws of Nature*. New York: Free Press.

Jorgensen SE (2000) 25 Years of ecological modelling by ecological modelling. *Ecological Modelling* 126(2–3): 95–99.

Lovelock JE (1979) *Gaia: A New Look at Life on Earth*, 252pp. Oxford: Oxford University Press.

Prigogine I and Stengers I (1990) *Order Out of Chaos: Man's New Dialogue with Nature* (First publ. 1984). London: Flamingo.

Puzachenko, Yu G, D'yakonov KN, and Aleshenko GM (2002) Diversity of landscape and methods of its measurement. Geography and biodiversity monitoring. Series of manuals. *Conservation of Biodiversity*,143–302. Moscow: NUMTs.

Puzachenko Yu G and Skulkin VS (1982) *The Structure of Forest Vegetation*, 320pp. Moscow: Nauka.

Sagoff M (2003) The plaza and the pendulum: Two concepts of ecological science. *Biology and Philosophy* 18: 529–552.

Schroeder M (1991) *Fractals, Chaos, Power Laws: Minutes from an Infinite Paradise*, 429pp. New York: W.H. Freeman and Company.

Shugart HH (1984) *A Theory of Forest Dynamics. The Ecological Implications of Forest Succession Models*, 278pp. New York: Springer.
Sukachev VN and Dylis NV (eds.) (1964) *Fundamentals of Forest Biogeocoenology*, 574pp. Moscow: Nauka (Science).
Tansley AG (1935) The use and abuse of vegetational concepts and terms. *Ecology* 16: 284–307.
Turcotte DL (1997) *Fractals and Chaos in Geology and Geophysics*, 2nd edn., 367pp. Cambridge: Cambridge University Press.
Vernadsky W (1929) *La Biosphere*. Paris: Librairie/Feliz Alcan.
Vernadsky VI (1926) *The Biosphere*. Leningrad: Nauchtekhizdat (in Russian). English version: Vernadsky VI (1998) *The Biosphere* (complete annotated edn.), 192pp. New York: Copernicus

Biosphere. Vernadsky's Concept

[†]**Y M Svirezhev**, University of Lisbon, Lisbon, Portugal

A Svirejva-Hopkins, Potsdam Institute for Climate Impact Research, Potsdam, Germany

Introduction

Vladimir Ivanovich Vernadsky (1863–1945) (**Figure 1**), the great Russian scientist and thinker, the founder of modern concept of the biosphere has shown that during all geological epochs on Earth, the life was developing as interconnected group of organisms (as he calls it 'living matter') that provided and provides the continuous flow of elements in biogenic turnover of matter and energy on the surface of our planet. This could be easily called scientific revolution at that time. Indeed, Thomas Kuhl points out that scientific revolutions occur when someone creates a new perspective, a model used for understanding reality. Only after introduction of such an idea, great progress, that was previously impossible, could be made, opening new ways of thinking. In recent years it becomes even clearer that all the works of Vernadsky were devoted to the further development of scientific thought as a planetary phenomenon. The significance of his ideas could be compared to the teachings of great philosophers of Ancient Greece, Roman Empire and Eastern world, of the Renaissance; the developers of the basics of mathematics and physics, such as I. Newton; the creators of system thinking about the origin and functioning of life on Earth, C. Linnaeus, G. Buffon, J.-B. Lamarck, C. Darwin, A. Humbolt, G. Mendel, etc.; as well as famous Russian scientists, M. V. Lomonosov, D. I. Mendeleev, I. M. Sechenov, I. N. Mechnikov, V. V. Dokuchaev, and others. As A. E. Fersman, Vernadsky's devoted pupil and successor in the area of development of geochemistry wrote about his teacher: "His general ideas will be studied and elaborated during centuries and one will discover new pages in his works which will serve as the source for new searches. Many scientists will learn his creative thought, which is acute, stubborn and articulated, always genial, but sometimes poorly understood. As for young generations, he always will be a teacher in science and a striking example of a fruitfully lived life."

Figure 1 Vladimir Ivanovich Vernadsky.

[†] Deceased.

Basic Idea

In Vernadsky's book *Biosfera*, first published in 1926, in accordance with dialectic principle, the process of cosmogonic evolution of Earth is considered, in the light of dynamics of the environment, which includes the system of many different forms of matter turnover, while its highest form, the life, is determining other planetary processes. The latter being the very central idea in Vernadsky's teachings. Namely this concept served as a necessary and desired base for the development of modern ecology.

Living and Nonliving Matter, Their Interaction and Cosmogonic View

The first step toward changing the world's picture, as natural scientists see it now, was the introduction of the concept of living matter by Vernadsky, and the second step was considering it as a cosmoplanetary phenomenon. Vernadsky has defined the living matter as "the existing at present time unity of organisms with the mass, chemical composition and energy" connected with its environment by constant processes such as breathing, feeding, and procreation, but we shall further address it sometimes as organic matter. Vernadsky has classified the dead organic matter as 'bio-generic matter', belonging to sediments, or, how one can call it 'the remnants of past biospheres'. The processes of interaction between the living (organic) and nonliving (inorganic) matter, considered as the most important initial stage of cosmoplanetary evolution, could be observed since the very first stages of planet's existence. As an example, proving Vernadsky's generalizations, the recent determination of time of the beginning of formation of primary sedimentary rocks (the *Stratispehere*, as Vernadsky named it) is 3.7 billions years ago, while slightly younger age (3.4–3.5 billions years ago) is determined for the formation of first organic compounds, that is, 'islands of living matter'. The origin of life on our planet is directly connected to the origin of the biosphere; and evolution, as we perceive it, always takes place inside biosphere, involving exclusively living matter. Vernadsky has expressed the fundamental importance of eternal interconnection between the living and nonliving matter, in the following very significant paragraph: "The Earth cover, Biosphere, while fully embracing the globe, has limits that are strictly determined by the existence of living matter in it – it is populated by it. Between its inorganic 'lifeless' and living parts, inhabiting it, exists continuous exchange of matter and energy, expressed by atomic movement caused by living matter. With the time course, this exchange is expressed by constantly changing and tending to steady-state equilibrium. This equilibrium threads through the entire biosphere and this biogenic atomic flux to a large extent creates and maintains it. Hence, in this manner and during all geological epochs, Biosphere is connected with the living matter that populates it. And namely by this biogenic flux of atoms and energy, the strong planetary cosmic significance of living matter is determined." This view through the 'cosmic prism', so to speak, radically changes our understanding of dialectical interconnection of living and nonliving matter, which originally differ in their composition of elements. As one can add, nowadays there is certainly more data and hypotheses about the influence of solar and other cosmic radiation on the living organisms at different levels of their structural development. For example, quite a lot of data have been accumulated on the influence of weak electromagnetic fields on the information exchange between living cells. Cosmic rays could in certain way influence the information exchange that is conducted by means of weak electromagnetic fields between cells of a living organism, and therefore alter functioning of the multicell structures.

It is necessary also to say a few words here about the two main principles of interactions between living and nonliving matter that Vernadsky has formulated, namely two biogeochemical principles, that describe the nature of energy fluxes in the biosphere:

1. Geochemical biogenic energy of the biosphere tends to maximum.
2. During the evolution of species, only the organisms that increase this biogenic geochemical energy in a process of their life will survive.

Vernadsky also writes about the irreversibility of life's processes and the increase in life's free energy, expressed in dissymmetry of composition of living matter.

In connection to the first biogeochemical principle, it becomes important to mention the similar work of Russian theoretical biologist, E. Bauer, who has formulated the fundamental principle of the permanent inequilibrium of living matter and the principle of maximum effect of external work. These principles, describing the thermodynamics of evolution and organization of living matter, are called 'the law of Bauer-Vernadsky'.

Concept of the Biosphere, Definition of Term, and Method of Analysis

At the beginning of the nineteenth century J.-B. Lamarck had introduced the term 'biosphere'. He considered it as the 'scope of life' and some sort of external cover for the Earth. In 1875 the same term was introduced in geology by E. Süss, who distinguished the biosphere as one of the Earth's covers. But V. Vernadsky was the person who first created the modern concept of the biosphere. This concept was first stated in his two lectures in Paris, published in 1926. In the sequel it was developed by Vernadsky himself, and by V. Kostitzin, V. Sukhachev, N. Timofeev – Resovsky, and other Russian scientists.

Biosphere includes all the hydrosphere, troposphere to the height of 30 km, and the upper part of the Earth's crust down to a depth of 2–3 km, for living bacteria still may be found at this depth in the underground waters and in the oil. It is an open thermodynamic system that exists with a permanent flow of solar energy (1.2×10^{22} kcal yr^{-1}) since the very beginning of the Earth's history.

According to Vernadsky, the biosphere is an external Earth cover, the 'scope of life' (as Lamarck named it). He also notes that this definition (as just the 'scope of life') is not complete. The Vernadsky's biosphere includes

- living matter;
- 'biogeneric matter', that is, organic and mineral substances, created by living matter (for instance, coal, peat, litter, humus, etc.); and
- 'bioinert matter', created by living organisms together with inorganic nature (water, atmosphere, sediment rocks).

Empirical Generalization Method

There are two components in Vernadsky's concept of the biosphere. The first is the proper biosphere concept, which can be called a verbal model of the biosphere. The second component is the method of study of such a complex system as the biosphere, which he called the 'empirical generalization method' (EGM). Vernadsky opposes the reliance on mere hypotheses, repeatedly insisting that the better suited method for a scientist is the Baconian system of accumulation of facts, as the generalizations become apparent from the data. That is essentially an inductive method and it indeed lies at the heart of the modern science. A perfect example of this method is the Periodic Table of the elements by Mendeleev. Certainly, the EGM is essentially wider than a method for the study of biosphere processes; it is a general scientific method. Let us remember Descartes' principle that "Science is a method." Speaking in modern terms, the EGM is a typical method of systems analysis.

The empirical generalization is based on real facts collected in an inductive way, keeping in mind not to leave the domain of these facts. At this first stage all possible scientifically established facts about studied phenomenon must be collected. The next stage is the aggregation of collected facts into some more general categories, called proper empirical generalizations. Basically, it gives us the possibility to pass from a huge number of accumulated facts to a considerably lesser number of statements that, in turn, allows us to truly speak about the possibility to describe the studied large (complex) system quantitatively. This stage does not allow formulating any kind of hypothesis, on which there is inevitably a mapping not only of scientific ideas, but also of nonscientific ones.

Really an empirical generalization is a system of axioms, reflecting our level of empirical knowledge, which could be used as a basis for any formal theory developed in the future.

Hence, having the system of empirical generalizations, we can follow either of two ways, when we are constructing models. Either we remain within the framework of this system, constructing so-called 'phenomenological' models, or complementing some hypotheses relating to the existing empirical generalizations, we shall get some new models. In accordance with Vernadsky's opinion, the choice on the set of these models, hypotheses must be produced by the coincidence of predicted and observed again facts. If this coincidence takes place, then the hypothesis becomes an empirical generalization of a higher level. From this point of view, for example, the practical astronomy of Ancient World was a typical empirical generalization, and ancient astronomers were successfully using the phenomenological model created on its basis. The same underlying empirical generalization is the basis of two principally different cosmogonic hypotheses by Ptolemee and Copernicus. If and only if new facts had appeared, the Copernicus cosmogony would have become a new empirical generalization. Therefore, the same empirical generalization can be a basis of different models.

However, the reciprocal picture can be possible, when an empirical generalization exists separately, without some kind of hypotheses and explanations from the viewpoint of contemporary science. For example, the radioactivity phenomenon could not be explained in frameworks of the physics of nineteenth century.

The System of Axioms

What kind of empirical generalizations lies at the base of Vernadsky's biosphere? In this case we will call this system of axioms 'Vernadsky's biosphere'; however, these axioms will be presented in slightly more formal way than in Vernadsky's original work.

1. *During all geological periods on Earth, living organisms have never been created directly from inorganic matter.* This is the homogeneity axiom. Note that in mathematics the operators that transform a zero into zero are called homogeneous, too. There is the analog of this axiom in biology, called the Redi's law ('life comes only from life').

2. *The existing facts cannot answer the question about the origin of life on Earth.* To get an answer, we must go beyond the framework of the EGM and use different speculations. There is only one way to resolve this contradiction, namely, to postulate the following: whatever the prebiosphere history of the Earth was, the evolution of the biosphere during all geological periods must give the contemporary biosphere as a result. This is the ergodicity axiom. It postulates that to a large degree the process of the biosphere's evolution is deterministic and stable in respect to the initial periods of its history.

3. *There were no lifeless geological epochs.* This means that the contemporary living matter is genetically connected with living matter of all the previous epochs. It is natural to call this axiom the continuity axiom. The following empirical generalizations are, actually, a form of conservation law. On the other hand, since they generalize some equilibrium properties of the biosphere, it is natural to call them: the axioms of stationary state.

4. The chemical composition of living matter was, on average, the same as it is now.

5. *The amount of living matter, on average, was the same for all geological time.* These generalizations of Vernadsky cause a lot of objections at the present time. However, there are not enough new facts to formulate new empirical generalizations. Therefore it is quite possible to consider the changes of the total amount of living matter, observed in different geological epochs, as fluctuations around some constant average level. (The same can be said also about the chemical composition of living matter and the terrestrial crust.) And, lastly, here are generalizations that determine the principles of functioning for biosphere mechanisms.

6. *The energy which is being stored and emitted by living organisms is the solar energy.* With the help of organisms this energy is controlling the chemical processes on the Earth (in particular, the global biogeochemical cycles).

7. *Vegetation plays the main role in the assimilation and allocation of the solar energy.* If we agree with the axiom about the constancy of the total amount of living matter during the whole lifetime of the biosphere, then we have to assume that its evolution only followed the path of the structural complication of living matter, either by increasing the number of species (there are 3106 species on Earth), or by complicating the structure of biological communities.

The Future of the Concept: Noösphere and the Modern Perspective

While Vernadsky's concept can be considered as maximally aggregated (it is like a view on the biosphere from the outside), the concept of the biogeocoenosis (BGC) first suggested by V. Sukhachev in 1945 and later developed by N. Timofeev-Resovsky, relates to the elementary units of the biosphere, and is a concept that is basically atomistic in nature. In accordance with the Timofeev–Resovsky's definition, the BGC is the part (area) of the biosphere, which has no essential ecological, geomorphologic, hydrological, microclimatic, or any other boundary inside itself. By this the whole biosphere of the Earth is divided into elementary systems, naturally separated from one another. According to Vernadsky, the biogeocenosis should have appeared immediately upon the start of existence of the biosphere.

In his last years of life, the works of Vernadsky were directed toward the future development of scientific thought as a planetary phenomenon. He perceived civilization as a form of a new geological force – scientific force performance. Vernadsky wrote:

> In the 20th Century man for the first time in the history of the Earth knew and embraced the whole Biosphere...That mineralogical rarity, native (pure) iron, is now being produced by billions of tons. Native aluminum, which never before existed on our planet, is now produced in any quantity. The same is true with regard to the countless number of artificial chemical combinations newly created on our planet. Chemically, the face of our planet, the biosphere, is being sharply changed by man... New species and races of animals and plants are being created by man.

The biosphere is a powerful geological force that has transformed this planet and its geochemistry in a most spectacular way. Toward the end of his life, Vernadsky planned to consider the Noösphere – the supremacy of scientific thought that always existed (term introduced in 1922 by a French philosopher and mathematician Edouard Le Roy) in more detail but he had no time.

Vernadsky's ideas act on the contrary to doomsday scenarios since he views our civilization as a form of a new geological force – scientific thought and therefore it cannot destroy itself. However at present, despite its growing industrial power, our civilization is not yet able to reconstruct the biosphere in the desirable way, if the biosphere is near the critical conditions. It is practically impossible to predict the new quasi-stationary state, as any other possible states of the biosphere's equilibrium are not known. The model of coevolution of 'man' and the 'biosphere' as the main principle of global coexistence was suggested by Yu. Svirezhev. The basic idea is to study the dynamics of the biosphere as the entity and its reactions to human impact. In the framework of this research the study of possible ways of the development of human society as a natural component of the biosphere is of special importance. The problem of coexistence of human society and the biosphere, which is actually the problem of mutual coevolution and harmonic coupling of humans and their environment, is now becoming one of the most important scientific and social problems of the modern society. This problem is multiscale and very complex, integrating natural dynamic processes operating within the biosphere with human dynamics.

In conclusion, we again would like to stress the significance of Vermadsky's ideas for the future and say that the concept of the biosphere and the following new state of it, Noösphere (from the Greek noó(s) (mind) and sphere), has shaped, and is continuing to do so, the global understanding of the origin and evolution of mankind, since as the teacher himself points out: It "is a new geological phenomenon on our planet. In it for the first time man becomes a large-scale geological force... Wider and wider creative possibilities open before him. It may be that the generation of our grandchildren will approach their blossoming... Fairy-tale dreams appear possible in the future; man is striving to emerge beyond the boundaries of his planet into cosmic space. And he probably will do so."

See also: Biogeocoenosis as an Elementary Unit of Biogeochemical Work in the Biosphere; Energy Flows in the Biosphere; Information and Information Flows in the Biosphere; Noosphere.

.

Further Reading

Bauer E (1935) *Theoretical Biology*, 206pp. Moskva-Leningrad: M.-L. Izd. Vsesoiuz nogo Instituta Eksperimentalnoi Mediciny (VIEM).
Kuhn TS (1996) *The Structure of Scientific Revolutions*, 222pp. Chicago: University of Chicago Press.
Svirezhev YuM (1998) Globalistics: A new synthesis philosophy of global modelling. *Ecological Modelling* 108(1–3): 53–65.
Timofeev-Resovsky NV (1968) Biosphere and Mankind. *UNESCO Bulletin.* 1: 3–10.
Vernadsky VI (1926) *Biosphere*. Leningrad: Gostekhizdat.
Vernadsky VI (1945) The biosphere and the noösphere. *American Scientist* 33: 1–12.
Vernadsky VI (1991) *Scientific Thought as Planetary Phenomenon*, 271pp. (in Russian). Moscow: Nauka.
Vernadsky VI (1997) *Biosfera*. Langmuir D (trans) and McMenamin MAS (revised by). New York: Springer.
Vernadsky VI (2007) *Geochemistry and the Biosphere (First English Translationfrom the Russian edition of selected works edited by Frank Salisbury F(ed.))*. Santa Fe, New Mexico: Sciences Synergetic Press.

Deforestration

A Shvidenko, International Institute for Applied Systems Analysis, Laxenburg, Austria

Introduction

Conversion of forest to other land-use is an inherent feature in the history of human civilization. During the last 8000 years, the planet has lost ~40% of its original forest cover. By estimates, mature primary tropical forests once occupied 1600×10^6 ha; today their area is about 900×10^6 ha. Latin America and Asia have already lost 40% of their original forest cover, and Africa almost 50%. Most of this loss has occurred during the last two to three centuries. By the beginning of the third millennium, forests have completely disappeared in 25 countries, and forest of 57 countries covers less than 10% of their total land area. Historically, converting forest land to agricultural, infrastructural, industrial, and urban uses was ubiquitous process for development and progress. However, human interactions with forests have often resulted in conversion to unsustainable land uses that led to substantial environmental, social, and economic losses. During the last decades, deforestation and degradation of forests, mostly in the Tropics, have been continuing at an alarming rate. Increasing forest plantations do not change this trend – they account for less than 5% of the world's forest area. Between 1980 and 2005, the Tropics experienced much more intensive forest-cover loss than other regions, with the largest concentration of deforestation occurring in the Amazon Basin, Southeast Asia, and the Congo Basin.

Deforestation and other processes of impoverishment of the world's forests are one of the major drivers of undesirable transformation of the planet. It causes substantial losses of different nature (ecological, economic, social, etc.), dramatically accelerates global biogeochemical cycling, and negatively impacts the Earth system.

Major Definitions

Forest Cover

The global Forest Resource Assessment (FRA), which is provided regularly by the United Nations Food and Agricultural Organization (UN FAO), uses three major terms related to tree cover. Forest is defined as a land class with tree cover more than 10%, and the trees should be able to reach a minimum height of 5 m. Other wooded land (OWL) has either a tree canopy cover of 5–10%, or is presented by a combined cover of shrubs, bushes, and trees above 10%. Other land with tree cover (OLTC) should meet the above criteria for forest but is related to land classified as 'other land' (e.g., groups of trees on agricultural land, parks, gardens, etc.). All three definitions require a minimum area of 0.5 ha. The last FRA (2005) estimated the world's forest area as 3952×10^6 ha of which the Tropics (~46%) and the boreal domain (~29%) are major forest biomes. About 30% of land surface is covered by forest. In addition, 1376×10^6 ha was classified as OWL. Global data on OLTC are incomplete – the area of this land-cover category was estimated at 76×10^6 ha in 61 countries out of 229. These estimates are mostly based on national forest statistics and individual country's reports. Another information source on the world's forests is observation of the Earth from satellites. Four different global remote sensing (RS) land-cover products, which have been reported during the two last decades, indicated the area of the world's forests on average as ~20% smaller than the FAO estimate. Reasons for this inconsistency stem from coarse resolution of the imagery used (1 km), not completely compatible definitions of forest, differences in classification, fragmentation of forests in many regions, long-period cloudiness in some (particularly tropical) regions, and lack of satisfactory ground truth data for proper validation of RS imagery. On the other side, national forest inventories are neither complete nor reliable for many developing countries.

Deforestation

Trends in dynamics of global forest cover are defined by two groups of processes. A major driver of decreasing the world forest area is deforestation. Two definitions of deforestation are widely used in their respective international frameworks. By the definition accepted by the UN Framework Convention for Climate Change (UNFCCC) this is an anthropogenic process: deforestation is defined as the direct human-induced conversion of forest to nonforest land. The FRA did not distinguish natural loss of forest from that caused by human impacts: deforestation was defined as the conversion of forest to another land-use or the long-term reduction of the tree canopy cover below the minimum 10% threshold. Both definitions refer to long-term or permanent change from forest to nonforest. Major alternative processes to deforestation are afforestation (establishment of forest on previously nonforest land), reforestation (natural or artificial development of forest on recently forest land), and natural expansion of forests into previously nonforest land. A superimposition of these four major processes results in net change of forest areas.

Here we apply the FAO definition which is used in most assessments and inventories: deforestation includes areas of forest converted to agriculture, natural nonforests such as shrubs and savannas due to anthropogenic impacts or natural disasters, nonvegetative land (e.g., water reservoirs and urban territories), etc. The term does not include areas where the trees have been removed as a result of forest management activities (e.g., logging) or due to natural disturbances (fire, insects), and where forest is expected to regenerate in a natural way or by silviculture activities. Deforestation also includes areas where some permanently impacted drivers (e.g., disturbance, overutilization, pollution, or other changing environmental conditions) do not allow for maintaining the tree cover above the 10% threshold during a long period of time.

Some other processes like degradation and fragmentation also contribute to the impoverishment of the world's forests and impact global biogeochemical cycling. Three mostly used definitions of degradation (of FRA, International Timber Trade Organization (ITTO), and UN Convention to Combat Desertification (UNCCD)) are comparable with respect to the main clusters. Forest degradation means a process leading to a temporary or permanent decline in the density or structure of forest cover or its species composition, and thus leading to a lower capacity of forest to supply products and/or services, and finally to reduction or loss of the biological productivity of the land. Many reasons can contribute to forest degradation, including diverse human-induced disturbances; unsustainable, excessive forest exploitation; insufficient logging; short rotation periods; etc. In many regions of the world, particularly where forest ecosystems are impacted by accelerated regimes of natural and human-induced disturbances, deforestation and degradation are usually closely interconnected. Fragmentation of forests often leads to decreasing vitality of remaining pieces of forest cover, acceleration of disturbance regimes, and, under the lack of integrated land management, can be a component of structural degradation of forest cover, impoverishment of ecosystem structure, and decline of productivity. Only 22% of the world's forests are classified now as 'intact' forests; 82 of 148 countries lying within the forest zone have lost all their intact forest landscapes.

Desertification

The process of (forest) desertification is a specific type of combining deforestation and degradation. The UNCCD defines desertification as land degradation in the arid, semiarid, and dry subhumid areas resulting from various factors, including climatic variations and human activities. Current drylands occupy 5.4×10^9 ha (~40% of the world land's area) of which $3.5–4.0 \times 10^9$ ha (57–65%) is either desertified or prone to desertification. Recently, a term of 'green desertification' was introduced for the boreal biome, meaning a long-term (more than the life span of major forest-forming tree species) replacement of forests by grass-, shrub-, and wetlands, mostly due to disturbances (e.g., fire), the extent, frequency, and severity of which exceed the restoration capacity of forest ecosystems.

Major Drivers of Deforestation and Degradation

Tropical deforestation is driven by a sophisticated combination of direct and indirect drivers of different nature (social, ecological, economic, environmental, biophysical), which interact with each other, often synergistically; the specific combinations of drivers vary within a region of the globe, by countries, and across localities within countries.

Direct drivers are basically human activities at the local level and can be broadly categorized into those related to agricultural expansion, wood extraction, and infrastructure extension. Agriculture expansion is the most important direct driver of deforestation in practically all tropical regions and includes shifting cultivation, permanent agriculture, pasture creation, and resettlement programs, following converting the forest to other land uses. Wood extraction includes commercial logging, fuelwood harvesting, and charcoal production. A substantial negative effect is provided by illegal harvest: over 70 countries have problems with illegal logging that leads to dramatic ecological and economic losses. Commercial logging is an important direct driver in Asia and Latin America while fuelwood gathering is one of the most important drivers in Africa. Infrastructure extension includes construction of transport ways; development of new industrial enterprises; settlement expansion; and a variety of other activities (oil exploration and extraction, mining, construction of hydropower stations, pipeline and electric grids). The construction or paving of roads in forested areas is among the principal causes of deforestation. For example, in the Brazilian Amazon, above 80% of deforestation occurs in a 100 km band along major roads. During recent decades, wildfires have been recognized as a new actor of deforestation and degradation in the Tropics, as a rule following land-use change and fragmentation of forest cover. Exceptional fires took place in east South Asia and the Amazon in 1997–98, provoked by the severe droughts due to the El Niño event. In Indonesia alone, these fires enveloped 2.4 million ha of forest and peatland. Other drivers can be important in different regions of the globe, such as insect damage, drainage or other forms of alteration of wetlands, permafrost destruction in high latitudes, etc.

Indirect drivers of deforestation are caused by fundamental social processes which are usually revealed as a sophisticated interplay of factors of different nature. Economic factors (e.g., rapid market growth and incorporation into the global economy, commercialization, urbanization and industrialization, growth of demand for forest-related consumer goods, poverty, etc.) are crucial across many tropical regions. Institutional factors (taxation, subsidies, corruption, property rights, etc.) are frequently tied to economic drivers. Cultural and sociopolitical factors like lack of public support for forest protection and sustainable use, low educational level, and low perception of public responsibilities also play a substantial role. Population growth, density, and spatial distribution are usually not a primary driver of deforestation: these are always combined with other factors. Nevertheless, in a number of studies, population density has been shown to be highly correlated with the determination of certain land-use patterns often connected to deforestation. Impacts of some of the above factors are often difficult to separate.

Ecological Consequences of Deforestation

The primary ecological consequences of deforestation are decline in biodiversity, invasion of exotic species, destruction of hydrological cycle, increase in water runoff and decrease in water quality, and acceleration of soil erosion. Tropical forests contain between 70% and 90% of all of the world species, and as a result of deforestation the planet is losing between 50 and 130 animal and plant species each day. Deforestation dramatically impacts runoff and hydrological regime, which threatens, for example, about 2000 known species in the waters of the Amazon Basin – 10 times the number found in Europe. Clearing of tropical forests substantially impacts fertility of soil. About 80% of soil in the humid Tropics is acid and infertile. Once the soil temperature exceeds 25 °C, volatile nutrient ingredients like nitrogen can be lost. Due to intensive rainfalls and soil specifics in the Tropics, a single storm can remove up to 100–150 t topsoil per hectare after deforestation. Deforestation can also decrease the social, esthetic, and spiritual values of forested landscapes. The extent and magnitude of these impacts are influenced by the size, connectivity, shape, context, and heterogeneity of the forest patch remnants. A critical point of negative change in landscape functionality – when fragmentation increased rapidly – on average occurs when mature forest declined to 30–35% of the landscape area.

Forest-cover decline alters regional and potentially global climate system by affecting surface energy, water, and greenhouse gas (GHG) fluxes. Deforestation of temperate and boreal forests has a cooling effect on near-surface climate by increasing surface albedo because cultivated fields generally have a higher surface albedo than natural forests. In tropical regions, deforestation generally leads to the opposite response where the prevailing effect is a decrease of evapotranspiration due to lower surface roughness and a shallower rooting zone. The associated decrease in the latent heat flux suggests a warming trend. Change of evapotranspiration and sensible heat flux impacts the low-level atmosphere, regional, and, potentially, global-scale atmospheric circulation. For instance, higher rainfall and warmer temperature are already observed due to recent large-scale deforestation in the Amazon Basin. However, the length of the dry season increases due to deforestation-induced rainfall inhibition, which can be accelerated by rainfall reduction in future due to global warming. The changes in forest cover have consequences far beyond the Amazon Basin. Regional-scale deforestation in the Tropics has been observed in a number of modeling results to lead to remote temperature and precipitation changes. Simulations for the twenty-first century give regional anomalies (due to human-induced land-cover change) as ± 2 K in magnitude.

There are already recognized impacts of current climate change on tropical forests. Biomass and production of pristine tropical forests is increasing but it expected to be reversed. Climate change and fragmentation substantially increased vulnerability of tropical forests to fire. The species composition is changing even in remote areas. Strong negative relationship is recognized between changes of precipitation regime and net primary productivity (NPP) in the humid Tropics.

Global simulations show a clear decline of vegetation productivity with increasing values of its fraction that is appropriated to human use. This decline is the consequence of two effects: reduction of biomass and climatic differences associated with a reduced vegetation cover. An important biospheric feedback of decreasing biomass is changes in the strengths of dissipative processes in terrestrial ecosystems.

Territories prone to deforestation, degradation, and desertification contain a huge amount of organic carbon: world tropical forests contain 220–270 Pg C in vegetation and 220 Pg C in soil (down to a depth of 1 m); drylands and boreal forests contain, respectively, *c.* 240 Pg C and 470 Pg C in soil. Human-induced land-use–land-cover change (LULCC) destroys the equilibrium state of these carbon pools and eventually impacts stability of the Earth system due to large emissions of major GHGs to the atmosphere because the biomass stock per hectare in standing forest is much higher than in any replacement use, including tree crops and silviculture, and as a result of substantial decrease of soil carbon after the conversion.

Understanding the Extent of Deforestation and Degradation of Forests

Lengths of retrospective periods of documented LULCC, as well as reliability of data, vary by continents and countries. Relatively reliable 'reconstructions' of global land-use dynamics have been done since the 1850s. According to these estimates, between 1850 and 1990, the area of cultivated lands, worldwide, has been estimated to have increased by more than a factor of 4, from 320×10^6 ha in 1850 to 1360×10^6 in 1990. The most rapid increase occurred in tropical regions after the 1940s – about half of the increase of $\sim 1000 \times 10^6$ ha occurred during this period. About 730×10^6 ha of agricultural lands was cleared from forests and woodlands that reduced the area of the world forests by 17% since 1850. The total net flux of carbon to the atmosphere from changes in land use was estimated to be 124 Pg C over the period 1850–1990. Changes in the forest area accounted for almost 90% of the net long-term carbon flux.

Currently, there are two major sources for monitoring LULCC including deforestation and degradation: RS and data of national forest inventories. RS was provided either at global or biome scale using imagery of coarse spatial resolution, or by statistical sampling of fine resolution. The basic tradeoff between these two groups of RS instruments is between spatial and temporal resolution. For example, Landsat's revisit time is 16 days – for satellites of coarse resolution, near-daily. Due to the high probability of cloud cover in many regions and the presence of smoke from vegetation fires, instruments of fine resolution cannot provide a satisfactory complete coverage. However, fine-resolution sensors cannot be avoided due to patchy structure of tropical deforestation with many small plots. Currently, a number of different satellites (optical bands) are used: SPOT (20 m resolution), IRS-2 (6–56 m), Landsat 5 and 7 (30 m), Terra (250–1000 m), ENVISAT (300 m), some others. Use of radars, which can penetrate the cloud cover, is one of the alternatives for estimating deforestation by satellites. Applications of radars from new satellites (e.g., SAR from ENVISAT, 75 m resolution, or ALOS, 50 m) show promising results. However, identification of small patches of deforested land, distinguishing degraded forests, and indicating regrowth can still not reliably be done from space for large areas. Thus, available estimates of global deforestation are not consistent and reliable enough yet.

Table 1 Annual average rates of tropical deforestation (10^6 ha yr^{-1})

Region	Average annual rates of deforestation			Net loss of forest area	
	1980s[a]	1990s[a]	2000–05[b]	1990–2000[c]	2000–05[c]
America	4.4–7.4	4.0–5.2	4.6	−4.5	−4.6
Asia	2.2–3.9	2.7–5.9	3.8	−0.8	+1.0
Africa	1.5–4.0	1.3–5.6	4.1	−4.4	−4.0
Total	8.1–15.3	8.0–16.7	12.5	−10.3	−7.6

[a]Range due to available publications and results of surveys.
[b]FAO (2005) data estimated as the total area for countries with net loss of forest area; data for Asia additionally include 0.4×10^6 ha for Oceania.
[c]FAO (2005) Global Resources Assessment 2005. Progress towards sustainable forest management. *FAO Forestry 147*, 350pp. Rome: Food and Agriculture Organization of the United Nations; the area of net change of forest area for USA and Canada are deducted from the total area for the American continent.

Reported areas of deforestation in the Tropics vary substantially (**Table 1**). Several subsequent estimates of dynamics of the global forest cover were provided by FAO FRA, mostly using country surveys which are based on compilation and standardization of data of national statistics. Recently, FRA-2005 presented revised estimates for 1990–2000 and new results for 2000–05. As a total conclusion for 1990–05, the global deforestation rate was estimated at 13×10^6 ha per year, almost completely in the Tropics. Taking into account increased areas of forest plantations and natural expansion of forests, particularly in temperate and boreal zones, the global net change of forest area was estimated at -8.9×10^6 ha in 1990–2000 (equivalent to loss of 0.22% to remaining area annually) and -7.3×10^6 ha in 2000–05 (−0.18%). The largest net loss of forests in 2000–05 was estimated at 4.6×10^6 ha yr^{-1} for Central and Southern America followed by Africa, which lost 4.0×10^6 ha annually. During the last 5 years, the previously negative trend in Asia has reversed due to large-scale plantations established mostly in China and India. Other estimates of tropical deforestation for the last two decades of the twenty-first century vary from 8×10^6 to $>16 \times 10^6$ ha yr^{-1}. RS estimates report that over 20 years (from the 1970s to the 1990s) the area of global forest decreased by 6%. On average, the RS estimates report lesser areas of tropical deforestation than FAO estimates. Likely, the above estimates of deforestation rate are slightly overestimated due to the fact that national inventories and RS data do not adequately record the regrowth. However, from another side, small deforested patches and selective logging, as a rule, are not included in the reported area. Probably, an aggregated conclusion on the current level of deforestation in the Tropics of *c.* 10×10^6 ha yr^{-1} can be considered as 'the best' conservative estimate of this process.

Considering the regional aspect, Brazil reported 21% of the net global loss for 1990–2000 and 24% for 2000–05, but this country has probably the best national RS system of deforestation monitoring: since 1997, the Brazilian National Institute of Space Research (INPE) has been monitoring deforestation down to 6.25 ha. Estimated areas for the three years 2002–05 (August to August) were on average 2.37×10^6 ha yr^{-1} with reported error ±4%. Overall, during the last 25 years, the Brazilian Amazon lost an area of forest greater than the size of Germany. For ten countries of Southeast Asia, about 2.3×10^6 ha of forests was cleared every year between 1990 and 2000 and transferred to other forms of land use. Annual deforestation in Indonesia was estimated some 1.7×10^6 ha in 1987–97 with the increase to 2.1×10^6 ha in 2003.

Estimation of Carbon Emissions

Major results for assessing emissions due to land-use change were received using inventory-based approaches or models of different type. Inventory-based models consider all or some of the basic processes: (1) the immediate release of carbon to the atmosphere from organic matter burned at the time of clearing, (2) postdisturbance flux of carbon from decay of slash, (3) accumulation of carbon during regrowth, and (4) changes in soil carbon.

Table 2 contains data on carbon emissions caused by deforestation in the Tropics. The estimates differ substantially: the average annual carbon emissions for 1990–2005 are estimated in the range 0.8–2.2 Pg C yr^{-1} (15–35% of the annual global emissions from fossil fuels approximately during this period) with the overall average at about 1.5 Pg C yr^{-1}. This estimate corresponds well to the estimate of the third IPCC assessment of 1.6 ± 0.8 Pg C yr^{-1} for the period 1987–98 and to recent estimates for 2000–06. Simulations done with the model IMAGE 2.1 estimated C emissions from deforestation from 0.83 Pg C yr^{-1} in 1995, 1.04 in 2000, 1.58 in 2005, to 2.16 Pg C yr^{-1} in 2015. Several estimates of aggregated carbon fluxes from tropical land given by inverse modeling vary from 1.2 to1.5 Pg C yr^{-1}, if both fluxes to the atmosphere and hydrosphere are accounted for.

Table 2 Annual carbon emissions from tropical deforestation

Region	Carbon emissions due to deforestation ($Pg\,C\,yr^{-1}$)		Total emissions in 1990–2005 (Pg C)
	1990s[a]	2000–05[b]	
America	0.55 (0.35–0.75)	0.55	8.3
Asia	0.72 (0.35–1.09)	0.64	10.4
Africa	0.24 (0.12–0.35)	0.29	3.8
Total	1.5 (0.8–2.2)	1.5	22.5

[a]Range due to available estimates.
[b]Emissions are calculated based on the average estimate for 1990–2000.

These estimates do not include carbon emissions from wildfire which could be very high, particularly during years of severe droughts. For instance, recent estimates put global carbon emissions from fires during 1997–98 El Niño event at 2.1 ± 0.8 Pg C, particularly in Indonesia.

The carbon stocks in forests may change without a change in forest area (e.g., selective harvest, forest fragmentation, non-stand-replacing disturbances, shifting cultivation, browsing, and grazing) and accumulation of biomass in growing and recovering forests. During the last two decades, the area of primary natural forests decreased or modified through human intervention by 6×10^6 ha yr^{-1}. Due to FAO estimates, degraded and secondary forests in Africa, America, and Asia covered about 850×10^6 ha in 2002. While deforestation can be measured from space with relatively high accuracy, this is not the case for degradation and secondary regrowth; usually regrowth is spectrally indistinguishable from mature forests as early as after 15–20 years. Forest inventories, as a rule, do not contain any specific data on forest degradation. FAO (2000) estimated the area of disturbances that can be labeled as forest degradation at 24×10^6 ha yr^{-1} in the period 1990–2000; another recent estimate is at 10×10^6 ha yr^{-1}. Estimates of carbon emissions from the degradation of forests (expressed as a percentage of the emission from deforestation) vary greatly – from 5% for the world's humid Tropics to 25–42% for tropical Asia and above 100% for tropical Africa. Another study reports the global net emissions from land-use change in the Tropics including emissions from conversion of forest to other land use (71%) and loss of soil carbon after deforestation (20%), emissions from forest degradation (4.4%), emissions from the 1997–98 fires (8.3%), and sinks from regrowth (−3.7%).

Uncertainties of the above data are high. A number of reasons impact reliability of carbon emissions from deforestation and forest degradation: (1) accuracy of recognizing the areas of tropical deforestation and degradation; (2) weak knowledge of the amount of biomass and soil carbon on areas impacted by the land-use change; (3) fate of deforested land, that is, how much is reverting to secondary forests; (4) how much forests are burnt; and (5) how forest disturbance is affecting soil and forest floor carbon stores. In a number of studies, uncertainties on the amount of CO_2 released are estimated to be 25–50%. For the Brazilian Amazon, for example, a range of 150–280 Mt C yr^{-1} was reported.

The greenhouse impact of deforestation is greater than the difference in carbon stock between the forested and replacement landscapes due to releases of other GHGs, basically methane (CH_4) and nitrous oxide (N_2O) (ozone, carbon monoxide, and some other gases which are produced by deforestation are not direct GHGs; nevertheless, they impact concentrations of CO_2 and CH_4 in the atmosphere). The emissions of these gases do not occur directly with deforestation, but basically with the following land use such as rice cultivation, cattle breeding, application of fertilizers, etc. IPCC-2001 assesses the following contribution of the major GHGs to the enhanced greenhouse effect in 1750–2000: CO_2 – 60%, CH_4 – 20%, and N_2O – 6% (the other 14% are caused by halocarbons which are not produced by the biosphere). The contribution of deforestation to the global greenhouse effect is estimated in the range of 25–35%. Of this total, the contribution of CO_2 is about 15% (or about one-fourth of the global CO_2 emissions), CH_4 9–11% (40–50% of the global methane emissions), and N_2O 2% (from one-fifth to one-third of the global nitrous oxide emissions). Available regional estimates are of a similar magnitude. In the case of Brazilian Amazonia, for example, gases other than CO_2 increase the greenhouse effect by about 35%.

For decades, deforestation and degradation were considered as an almost exceptional phenomenon of the Tropics and arid lands. However, recent years have brought much evidence of possible damage to forests due to ongoing and expected global change in the boreal biome. Forest degradation and deforestation here mostly relate to the increase in frequency and severity of large-scale disturbances, change of hydrological regimes mostly related to permafrost destruction, industrial pressure on landscapes, pollution, and unsustainable logging. For instance, wild vegetation fires enveloped 23×10^6 ha (of which 17×10^6 ha on forest land) in Russia in 2003; during the first years of this century, outbreaks of dangerous insects in boreal forests exceeded 20×10^6 ha in the circumpolar boreal zone, of about the same area in American and Asian continents. The increase in the area of 'green desertification' in the Russian taiga zone is estimated to be about 5×10^6 ha during the last two decades. The direct carbon emissions due to a fire in 2003 are estimated to be about 200 Tg C yr^{-1}. Very likely, the expected dramatic warming in high latitudes (up to 6–10 °C) will substantially accelerate processes of northern deforestation and degradation.

Conclusion: Managing Deforestation

While many governments try to provide a legislative basis and to realize measures to slow deforestation, the most recent and thorough deforestation studies offer no suggestion that deforestation is decreasing, either of its own accord or in consequence of policy interventions. On the contrary, increasing global integration of markets and growing demand for agricultural commodities and fuelwood in many regions of the developing world appear to be driving substantial increases in deforestation rates that will result in unsustainable forest management and further declining diverse forest services.

Some models and scenarios predict a substantial 'baseline' deforestation, for example, for 2005–2015 ($\times 10^6$ ha yr^{-1}): South America 3.9, Central America 1.2, Southeast Asia 2.6, Africa 5.2, and the total 12.9, with the average annual carbon efflux at the level of 1.2–2.0 Pg C yr^{-1} during the next two decades. The Special Report of IPCC on LULCC (2001) predicts the average annual accounted carbon stock change due to deforestation at −1.8 Pg C yr^{-1} of which −1.6 Pg C yr^{-1} is expected in the Tropics, and the global result of ARD activities between −1.2 and −1.6 Pg C yr^{-1}. The ongoing climatic change will accelerate negative consequences of the human-induced deforestation: the expected significant warming by the end of the century suggests dangerous implications for forests and human welfare. Studies report that the warming turns more and more tropical rainforest into steppe, and will transform up to 60% of this forest into dry land, dramatically impacting the region's richest biodiversity. Very likely, a similar process will be accelerated in the forest–steppe ecotone of the Northern Hemisphere, with substantial (up to 30%) increase in the area of the desertified steppe.

Tropical countries can reduce deforestation through adequate funding or programs designed to enforce environmental legislation; support for economic alternatives to extensive forest clearing, including carbon crediting; building institutional capacity in remote forest regions; and increase in areas of protective forests. Planted forests provide an opportunity to sequester carbon in vegetation and soils: afforestation and reforestation potentially could achieve annual carbon sequestration rates in live biomass in tropical regions 4–8 versus 0.4–1.2 t C ha^{-1} yr^{-1} in boreal regions, and 1.5–4.5 t C ha^{-1} yr^{-1} in temperate regions. An IPCC scenario (2000) predicts that the maximal amount of carbon that can be sequestered by global afforestation and reforestation activities is 60–87 Pg C on 344×10^6 ha during the first half of the twenty-first century with 70% in tropical, 25% in temperate, and 5% in boreal forests, provided the average annual carbon uptake is at 1.1–1.6 Pg C yr^{-1}. Of course, vast areas of forests converted to agriculture use, particularly to pastures, cannot be expected to recover forests of the original type on a timescale relevant to human planning: secondary forests differ in structure, composition, and productivity from their predecessors.

Reducing the rate of deforestation is another major way to decrease GHG emissions. However, neither the UNFCCC nor the Kyoto Protocol has introduced a satisfactory mechanism reducing GHG emissions from deforestation. Avoided deforestation was excluded from the Clean Development Mechanism, and the current international climate policy regime does not provide incentives for developing countries to reduce carbon emissions from tropical deforestation. This problem is under intensive international debates. One of the relevant ways how to curb emissions from deforestation is a so-called compensated reduction of tropical deforestation – the idea that tropical countries might reduce national deforestation under a historical baseline and be allowed internationally tradable carbon offsets having demonstrated reductions. Recent estimates assume that net deforestation would continue until the price of 1 t of sequestered carbon will be less than \$100 t^{-1} C. Such a price could give a possible decrease of carbon fluxes due to avoided deforestation at 300–650 Tg C yr^{-1}.

Tropical deforestation may be decisive in global efforts to stabilize GHG concentrations at levels that avoid dangerous interference in the Earth system. However, it will require substantial international and national efforts in many aspects, for many nations, at all times.

Further Reading

Achard F, Eva HD, Mayaux P, Stibig H-J, and Belward A (2004) Improved estimates of net carbon emissions from land cover change in the Tropics for the 1990s. *Global Biogeochemical Cycles* 18: GB2008 (doi:10.1029/2003GB002142).

DeFries RS, Houghton RA, Hansen MC, *et al.* (2002) Carbon emission from tropical deforestation and regrowth based on satellite observations for the 1980s and 90s. *Proceedings of the National Academy of Sciences of the United States of America* 99: 14256–14261.

FAO (2005) Global Resources Assessment 2005. Progress towards sustainable forest management. *FAO Forestry 147*, 350pp. Rome: Food and Agriculture Organization of the United Nations.

Fearnside PM (1997) Greenhouse gases from deforestation in Brazilian Amazonia: Net committed emissions. *Climatic Change* 35(3): 321–360.
Fearnside PM (2000) Global warming and tropical land-use change: Greenhouse gas emissions from biomass burning, decomposition and soils in forest conversion, shifting cultivation and secondary vegetation. *Climatic Change* 46: 115–158.
Geist HJ and Lambin EF (2002) Proximate causes and underlying driving forces of tropical deforestation. *BioScience* 52: 143–150.
Hirsch AI, Little WS, Houghton RA, Scott NA, and White JD (2004) The net carbon flux due to deforestation and forest re-growth in the Brazilian Amazon: Analysis using a process-based model. *Global Change Biology* 10: 908–924.
Houghton RA (2003) Revised estimates of the annual flux of carbon to the atmosphere from changes of land use and land management 1850–2000. *Tellus* 53B: 378–390.
Houghton RA, Joos F, and Asner GP (2004) The effect of land use and management on the global carbon cycle. In: Gutman G, Janetos AC, Justice CO, *et al.* (eds.) *Remote Sensing and Digital Processing Series, Vol. 6: Land Change Science*, pp. 237–256. Amsterdam: Kluwer Academic.
Lambin EF, Geist H, and Lepers E (2003) Dynamics of land use and cover change in tropical regions. *Annual Review of Environment and Resources* 28: 205–241.
Lepers E, Lambin EF, Janetos AC, *et al.* (2005) A synthesis of information on rapid land-cover change for the period 1981–2000. *BioScience* 55(2): 115–124.
Moutinho P and Schwartzman S (eds.) (2005) *Tropical Deforestation and Climate Change*, 132pp. Washington, DC: Amazon Institute for Environmental Research.
Phillips OL, Malhy J, Vinceti B, *et al.* (2002) Changes in growth of tropical forests: Evaluating potential biases. *Ecological Applications* 12: 576–587.
Santili M, Moutinho P, Schwartzman S, *et al.* (2005) Tropical deforestation and the Kyoto Protocol: An editorial essay. *Climatic Change* 71: 267–276.
Shvidenko A, Barber CV, Persson R, *et al.* (2005) Forest and woodland systems. In: Hassan R, Scholes R, and Ash N (eds.) *The Millennium Ecosystem Assessment Series, Vol. 1: Ecosystems and Human Well-Being: Current State and Trends*, pp. 585–621. Washington, DC: Island Press.
Watson RT, Nobble IR, Bolin B, *et al.* (eds.) (2000) *Special Report of the Intergovernmental Panel on Climate Change: Land Use, Land-Use Change, and Forestry*. Cambridge: Cambridge University Press.

Environmental and Biospheric Impacts of Nuclear War

P Carl, Leibniz Institute of Freshwater Ecology and Inland Fisheries, Berlin, Germany
†Y Svirezhev, Potsdam Institute for Climate Impact Research, Potsdam, Germany
G Stenchikov, Rutgers University, New Brunswick, NJ, USA

Introduction

Apprehensions about uncontrolled thermonuclear fusion arose early in the Manhattan Project: Could explosion of a hydrogen bomb trigger a global physical catastrophe in starting chain reactions that seize the light elements and thus wipe out all life on Earth? It was not without grave anxiety that Emil Konopinski, Cloyd Marvin, Jr., and Gregory Breit ruled out this possibility. Since the end of World War II, public knowledge about 'the unthinkable', the consequences of nuclear war, was largely shaped by the horrible direct health effects of the 1945 atomic bombing of Hiroshima and Nagasaki, by the devastations and disruptions of life and infrastructure due to heat, blast and electromagnetic waves, prompt ionizing radiation, and radioactive fallout.

Public awareness of worldwide fallout risks was triggered by the fatal outcome of the atmospheric test Bravo, the largest weapon exploded by the United States (Bikini atoll, 28 February 1954; 15 Mt TNT equivalent). Temporary geophysical effects of atmospheric tests, such as planetary pressure waves, magnetic field distortions, or ionospheric disruptions causing blackout in radio communication, were frequently recorded. A signature in worldwide weather was not found, however. The largest weapon ever tested, the 'Tsar of Bombs', a more than 50 Mt 'clean' bomb (with a nonfissionable mantle), was exploded on 30 October 1961, above the northern Soviet

† We are mournful about the loss of our friend and colleague Yuri Mikhailovich Svirezhev who passed away during the time of working on this paper. We dedicate our own contribution to his memory. (GS & PC)

test site at Novaja Zemlya. Its pressure wave circled the Earth several times. The most obvious long-term, large-scale direct geophysical effect of nuclear explosions appears to have been caused by the Starfish Prime test on 9 July 1962, a 1.4 Mt detonation some 400 km above the Johnston Island area in the tropical central North Pacific. An artificial (mini-van Allen radiation) belt of charged particles was trapped by the Earth's magnetic field and traced for a couple of years. High-altitude nuclear explosions may 'blind' reconnaissance satellites, impair electronics over vast areas, and even inflict on a missile attack, by their electromagnetic pulse (EMP).

Radioactive tracers from the 539 atmospheric test explosions until 1980, with an aggregate yield of about 440 Mt, will remain identifiable worldwide for millennia. Hot spots at test sites and the unresolved issue of low-dose radiation effects notwithstanding, though, their health effects are far from endangering the species of man. In a nuclear war, the number of warheads and their total explosive yield might exceed these figures by an order of magnitude, and the period of 35 years would reduce to a couple of days, if not hours. Such a 10^5-fold 'compaction' of interacting effects rules out extrapolation from the test series, as do consequences of a decisive distinction in targeting: The deadly logic of 'mutual assured destruction' (MAD) bears attacks on large population centers. Cities would also not escape 'countervalue' and 'counterforce' strikes against the economic and military potential, notably the command, control, communication, and intelligence (C^3I) structures. Not only does this turn 'warfare' into 'exchange', it also gives birth to a new quality of risks – the long-term, worldwide indirect aftereffects that add to and interfere with the disastrous direct effects of nuclear explosions.

Studies on Indirect Effects of Nuclear War

Multiple upper-atmosphere nuclear bursts might be scheduled for a ballistic missile defense (BMD) system's 'terminal phase defense' under the threat of a 'decapitation' strike. This would endanger the Earth's 'ozone screen' in generating high amounts of nitrogen oxides (NO_x). Rising fireballs of tropospheric bursts would have the same effect. The biologically active part of the solar ultraviolet radiation (UV-B), which causes structural change in amino acids and is normally absorbed by stratospheric ozone molecules, would then reach the surface. According to the US National Academy of Sciences (NAS; 1975), a 10 000 Mt exchange could cause a hemispheric ozone loss by 30–70% that would extend globally, with potentially grave impacts on terrestrial and aquatic ecosystems, and recovery over years. The amount of stratospheric dust, injected by megaton-yield near-surface explosions, may resemble the aerosol load due to the Krakatau eruption in 1883. NAS thus took a surface cooling of only a few tenths of a degree centigrade for plausible, but expressed another concern: "It is not known whether climatic variables have stable equilibrium values and a tendency to relax after an impulse disturbance such as that generated by a nuclear exchange." This led the authors to conclude that irreversible climatic shifts cannot be ruled out.

Criticism by the Federation of American Scientists (FAS) for too 'optimistic' NAS conclusions about potential impacts on remote, noncombatant countries became justified 7 years later. Invited to contribute on ozone impacts to a first international, comprehensive nuclear war risk study commissioned by the Royal Swedish Academy of Sciences and first published in its environmental journal *AMBIO* (1982), Paul Crutzen and John Birks were surprised to find unknown, severe effects due to the smoke from 'postnuclear' fires. A key mechanism like this was missing since the early speculations on changes of weather and climate due to nuclear war.

To grip consequences of the changing military policy away from 'assured destruction', the US Office of Technology Assessment (OTA; 1979) had just analyzed a range of scenarios in an influential war risk study, from attacks on cities and oil refineries to one-sided counterforce and countervalue strikes. High-altitude bursts were mentioned as critical but not addressed, though doubts were cast on massive stratospheric ozone depletion. The chemical system was better known then, and high-yield weapons had given way to missiles with multiple warheads of lower yield each. Focusing on direct effects of nuclear attacks, civil defense, economic breakdown, recovery and societal impacts, OTA suggests that extreme uncertainties, and certainty about disastrous 'minimum' consequences, both "play a role in the deterrent effect of nuclear weapons".

Other than OTA, the *AMBIO* study used a global reference exchange that comprises ground and tropospheric bursts with a total yield of about 5750 Mt. Crutzen and Birks confirmed both the NAS estimates of ozone depletion and the OTA doubts, that is, their advanced ozone model did not qualitatively alter the results, but the evolving strategic arsenals apparently did. In addition, they identified large-scale forest fires and intense urban and industrial conflagrations as sources of a long-lasting photochemical smog over large areas of the Northern Hemisphere. The sunlight needed for its formation, however, could be blocked by high amounts of smoke, notably when oil and gas fields or refineries were targeted. Smoke absorbs the short-wave solar flux and heats up, but interferes much less with the outgoing long-wave thermal radiation. The surface cools therefore, which would also suppress convection and reduce precipitation. Cold and darkness after a nuclear war would be severe for terrestrial ecosystems, but especially grave for oceanic food chains given the quick consumption of phytoplankton at the very base of the trophic web.

The anticipated climatic disruption triggered inquiry by Richard Turco, Owen Toon, Thomas Ackerman, James Pollack, and Carl Sagan (TTAPS), who were able to marshal the data for urban mass fires and fire storms not available in time to Crutzen and Birks, and to quantify the effect they coined 'nuclear winter' for a broad range of scenarios (100 Mt 'countercity', 5000 Mt 'baseline', 10 000 Mt full-scale exchange, for example): a massive thermal inversion of the planetary atmosphere and its climatic consequences. A line of research which turned out to be essential addresses mass extinctions in Earth history, including hypotheses of extraterrestrial impacts and those that blame geological periods of enhanced volcanism or worldwide forest fires, maybe even all in combination. An authoritative circle of biologists and ecologists, led by a group including Paul Ehrlich, John Harte, Mark Harwell, Peter Raven, and George Woodwell, concluded that the extinction of man after a large nuclear war could no longer be ruled out. A public conference "The World after Nuclear War" (Washington, 31 October–1 November 1983) attracted unprecedented attention by communicating these findings, by the participation of scientists of both superpowers, and by a technique used on this occasion: a satellite TV bridge (Moscow–Washington) between both academies of sciences. Two research groups from either side of the globe exchanged their first 'nuclear winter' results, based on global climate models, via this public 'Moscow link:' Vladimir Aleksandrov and Georgiy Stenchikov, and Curt Covey, Steven Schneider and Starley Thompson.

Rethinking the Unthinkable

An unparalleled, open activity toward a worldwide process of research and education, which goes beyond the traditional understanding of scientific responsibility in terms of specialist's denial, the public action of the scientific elite, and the work in closed circles, was launched after delivery of the *AMBIO* study by the International Council of Scientific Unions (ICSU). Steered by the project "Environmental Consequences of Nuclear War" (ENUWAR) of the Scientific Committee on Problems of the Environment (SCOPE), resulted it by the end of 1985 in the two-volume report SCOPE-28. This study cannot offer an overall view on the complex entity 'ecosphere' after nuclear war. None of the major physical control parameters (light, temperature, water) are expected to be disturbed that hard or weak at planetary scale so as to justify 'simple' conclusions. A substantial impact on the stratospheric ozone budget, however, maintained by the US National Research Council (NRC; 1985), holds the more, since smoke-induced heating would change all chemical reaction rates.

Long-term problems due to radioactive exposure are borne in the selective vulnerability of forest communities. Nuclear war might change the biogeography of vast areas, notably where coniferous temperate forests dominate and the 'radiation shock' combines with large-scale fire, climatic change, UV-B effects, etc. SCOPE-28 notes specific sensitivities: "Temperature effects would be dominant for terrestrial ecosystems in the Northern Hemisphere and in the Tropics and subtropics; light reductions would be most important for oceanic ecosystems; precipitation effects would be more important to grasslands and to many Southern Hemisphere ecosystems." Such a 'distributed vulnerability' structure and its more subtle patterns, the other side of the 'biogeography' medal, reflects a range of stabilizing feedbacks against gradual or abrupt transitions in atmospheric or oceanic conditions and related compositional, thermal, meteorological, and hydrological regimes. The 'acute' phase of nuclear winter would bear structural changes in the physical environment of a dimension that might transgress those stability limits, maybe at hemispheric scale. The crucial question is the one posed by the NAS 10 years before: Will new feedbacks take over to stabilize the system in a regime different from present day, or will it return? Latest, the transition to 'chronic' response would be influenced by climate–biosphere feedbacks. The result of acute-phase environmental devastations, like the patchiness of surviving communities, may thus attain a structural role in shaping a postwar environment.

The vulnerability of the 'noosphere' – taken as the complex entity of man's society and managed environment – against the direct effects of nuclear war is much higher than that of the (natural) ecosphere. Agricultural systems may only exist due to human maintenance. Worldwide disruption of functioning agriculture and food supply after nuclear war would expose the majority of survivors to the risk of starvation. The OTA study suggests that this would even hold without severe environmental aftereffects. Climatic impacts that hit agriculture at vulnerable spots (length of the growing season, hydrological change, etc.) may bear just that sort of feedback, however, which keeps surviving humans stuck to marginal subsistence for any period relevant to societal restoration. The impact on the Southern Hemisphere of nuclear war in the north is a key issue in view of a postwar noosphere. Beyond the 'import' of climatic effects due to interhemispheric smoke transport, with their ecological and agricultural consequences, a major disturbance would be caused by interruption of the lifeline of international trade, even if agricultural productivity could be maintained at a level of sufficiency. The risk consists in a large societal setback to which a modern society may not adapt without existential disturbance.

A convincing approach to environmental impacts addresses productivity limits and 'convolves' this knowledge with a realistic range of stresses derived from climate model output. The resulting 'response surfaces' to stress factors as expected after nuclear war (changes in temperature, light level, precipitation) are not of a simple shape for grassland ecosystems, notably when secondary productivity of herbivores is considered. Regulatory feedbacks act together, that is, stresses are not generally additive or even mutually enhancing ('synergistic'). The 'nonsevere factor space' of functioning ecosystem response may have rather sharp boundaries, however, beyond which the 'message' becomes simple: ecosystem productivity reduces almost abruptly to a level that would not support a human population. An important conclusion is on uncertainty again: a group of survivors who found an ecological niche in a postwar environment may be "plunged into destruction by seemingly minor drift away from those conditions." Ecosystems become unpredictable when driven to marginal existence, in the vicinity of critical transition, by a changing climate.

In two authoritative assessments, the World Meteorological Organization (WMO) confirmed the risk of a severe smoke-induced climatic impact. Among the pertaining uncertainties, a potential modification of the hydrological cycle was emphasized. This concerns the 'Hadley circulation' above all, a double-cell of upward and poleward circulation (in the zonal mean) flanking the meteorological equator, which is driven by the strongest heating there. As the season advances, the southern 'Hadley cell' shifts northward and blows up to form the 'monsoon cell' in boreal summer, with upward legs as far north as the Tibetan and Mexican plateaus. A smoke veil above would attenuate this structure, that is, weaken or disrupt the monsoons, but depending on season, injection heights, and location of the smoke source, the Hadley circulation may also become enhanced. The WMO assessments posed into doubt that climate models may reach the required realism and reliability soon, but confirmed both a potential monsoon disruption in boreal summer and smoke lofting as well as transport into the Southern Hemisphere.

A minimum demand, the realistic simulation of seasonally varying rainfall, is not easily met by global climate models. Even more challenging is the agriculturally important intraseasonal activity, notably the active–break cycles of the major monsoon branches and their dynamic interplay. Key knowledge about monsoon dynamics, and thus about the atmospheric hydrological cycle and its interactions, did just settle when the scientific consensus formed about major climatic effects of nuclear war. This bears potential for surprise, and the consensus may deserve further development just where it directly concerns half the world population. Abrupt onset and retreat of the boreal summer monsoon are known for decades, its oscillatory (interhemispheric) nature since the mid-1970s. These are features typical of a dynamic system that passes a critical transition. Long-term consequences for man and the biosphere of the climatic response to nuclear war may thus be borne in the potential for structural recovery of the present-day 'monsoon climate' on Earth. This includes monsoon interactions with the El Niño–Southern Oscillation (ENSO) system. Both dynamic subsystems mediate climatic and environmental impacts on the Southern Hemisphere, beyond the more direct effects of interhemispheric smoke transport.

SCOPE successor studies to ENUWAR (1982–88) took another turn: the projects RADPATH (1988–93) and RADTEST (1993–99) addressed the pathways of radionuclides across the environment, exemplified by field studies into the consequences of the 1986 Chernobyl nuclear reactor accident and at selected nuclear test sites worldwide. These projects mounted a substantial database, improved the knowledge of processes, and identified gaps in understanding the biogeochemical dispersion of radionuclides. All three projects were led by Sir Frederick Warner. Their documented results, SCOPE report nos. 28, 50, and 59, are reference sources for the state of scientific knowledge toward the turn of the twentieth century about the gravest risk and challenge of its second half – the 'doomsday' of man in an all-out nuclear conflict.

Behind and Beyond the Scenarios

The idea of a 'doomsday machine' as an *ultima ratio* of deterrence is due to Herman Kahn. As a 'terminal' retaliation should deterrence fail, such a hypothetical device was thought to automatically kill the majority of mankind, if not the species of man or all life on Earth. MAD was a sort of 'homicide pact' indeed, settled by the Antiballistic Missile Treaty of 1972 (which allowed one BMD system at either side). Negotiations could give MAD a frame as long as it was accepted as a matter of fact and a relatively stable island was sought within the sea of inherent risks. In a severe crisis, however, a strategic exchange could have been initiated just by technical failure, misinterpretation, false information, or madness. Aimed to balance Soviet conventional forces, the US nuclear guarantee for Western Europe established the principal context of the doctrine of extended deterrence. The ability to control escalation, a prerequisite of this posture, was its dilemma as well. The myth, the adversaries in a nuclear war could climb a fictitious 'escalation ladder' up and down at will, is not backed by any realistic view on the dynamics of escalation, be it only due to the vulnerability of the very means of control, the C^3I systems, which are primary targets in the earliest phase of war. Moreover, tactical nuclear weapons are an escalation-prone arming *per se*. Their massive deployment along the European front made an early, uncontrolled use in any armed conflict nearly certain. Postures other than MAD were also delusionary due to the 'third power problem': the nuclear forces of Britain and France, maintained in part in recognition of the US dilemma with extended deterrence, were 'MAD forces' by intention, with a substantial destruction potential. 'Escalation control' and 'limited nuclear war' were sold by a 'nuclear utilization theory' (NUT) as alternatives to MAD. Soviet strategic forces in 'launch on warning' alert, however, and a doctrine of earliest possible, massive infliction of (not just response to) any nuclear attack would have left no space for bargaining after crossing the threshold to war. NUT did not replace MAD, but increased the risk of strategic instability.

When Herman Kahn died on 3 July 1983, a revision of his classic *Thinking about the Unthinkable* had been caught up with 'nuclear winter'. In a comment, the editors admit strategic consequences, excepting the 'war fighting' postures. A similar view was held in a brief report delivered by the US Secretary of Defense, Caspar Weinberger. It focuses on the early TTAPS study and uncertainties discussed there, cites with the same bias the reasoning of the NRC study, 'massacres' Soviet contributions as 'propaganda', and praises escalation control as one of the means to avoid nuclear winter. That resistance of the military bureaucracy to new knowledge drives the 'overkill' arsenals beyond any justification shows also the example of the Pacific-Sierra Research Corporation, where smoke emissions and fire effects have been studied with a primary view on target planning for nuclear war: the 'blast model' of casualty estimation survived any 'fiery' challenge. For a 10 000 Mt war, the World Health Organization (WHO) estimated a short-term toll of 2.2–2.5 billion casualties, with a ratio of deaths to injured from 1.1 to 1.6. Lacking appropriate medical care, many of the injured would be doomed to die. Immediate casualty estimates of the Greater London Area War Risk Study (GLAWARS; 1986), the most comprehensive public assessment of the impact of nuclear war on a region, range from 1 to 6.2 millions (97%) of the London population. At a symposium at the NAS Institute of Medicine (IOM; September 1985) such estimates were challenged by a new model that takes 'postnuclear' fires into account. Immediate fatalities had been underestimated by a factor of 2 or more when 'only' prompt radiation, heat and blast waves, as well as local radioactive fallout were considered (blast model). The lower-edge figures for London increase substantially when using the 'conflagration model'.

Difficulties in 'translating' climatic into health effects are partly due to missing local information, neither provided by climate models nor easily derived: fog or haze, storminess, chemical and radioactive load of precipitation, etc. For the longer term, GLAWARS' gravest concern is food supply for survivors. Genuine medical aspects include enteric diseases and those spread by insects or due to poor sanitation and nutrition, all favored in victims who became 'immunocompromised'. The key point here, also identified at the IOM symposium, is just the combined action of stresses in the nuclear aftermath to impair the immune system. Factors causing immune suppression include radioactive and UV-B radiation, malnutrition, burns and trauma, as well as psychosocial stress. Clinical evidence indicates that these factors all converge in their action on a single element of the immune system, the T-lymphocyte, of which also the 'helper-to-suppressor ratio' is crucial. The Acquired Immune Deficiency Syndrome (AIDS) is characterized by deficiencies of the T-lymphocyte variety similar to those expected due to the combined stresses after nuclear war. The list of factors is certainly not exhaustive. The 'clinical record' of today's monsoon and ENSO variability, from lasting hot-dry to torrential flooding, should bear medical implications of structural impacts on the tropic–subtropical climate. Coming to grips with these dynamic systems challenges climate modeling today, as did a smoky atmosphere in the 1980s.

'Nuclear Winter' Modeling – A Sketch

To follow solar and thermal radiations through a smoke- and dust-laden atmosphere, TTAPS had used a height-resolved (one-dimensional; 1-D) radiative–convective model (RCM) with annual mean insolation. An RCM describes these processes in greater detail than general circulation models (GCMs) do but misses their horizontal motions. Further 1-D and 2-D studies, at the Lawrence Livermore National Laboratory (LLNL), the US National Aeronautics and Space Administration (NASA), and the University of Maryland, helped clarifying basic effects and feedbacks including smoke uplift and changes in the snow-ice albedo, both of which may protract climatic effects. Like Covey, Schneider, and Thompson of the US National Center for Atmospheric Research (NCAR), who used a GCM of Australian origin with 7 atmospheric layers, Aleksandrov and Stenchikov of the Moscow Computing Center of the USSR Academy of Sciences (CCAS) confirmed TTAPS' major results. They used a coarse-resolution two-layer tropospheric GCM that had been adapted for different purposes in cooperation with Lawrence Gates of the Oregon State University (OSU), and equipped it with a simple ocean model. In addition to severe surface air temperature drops in continental interiors (mitigated near oceanic coasts) and large-scale thermal inversions of the atmosphere, clear signs of inter-hemispheric smoke transport due to a structural response of the Hadley circulation were noted by both groups.

These early 3-D 'nuclear winter' studies were admittedly quick shots: their immobile smoke stayed uninfluenced by atmospheric motions, did not interact with the hydrological cycle to become washed out, and could not buoyantly rise by solar heating. Though state-of-the-art in the early 1980s, artificial model climates had also to be left behind for more realistic assessments. Michael MacCracken and John Walton of the LLNL and the CCAS team introduced more realistic feedbacks into their two-layer GCMs, whereas the NCAR group focused on the model 'physics' first to keep firm footing. A visit in Moscow, coincidentally just before the 1983 Washington conference, triggered a study series by Stenchikov and Carl that addressed a 'minimum' disturbance (without minimizing the problem), traced conditions for Southern Hemisphere impacts, and explored the transient response for hints to answer the 1975 NAS question on 'postnuclear' climate relaxation. This induced a closer view on the complexity of the acute phase of perturbation and its implications. Just during startup of this common work in Berlin, on 31 March 1985, Vladimir Aleksandrov vanished without a trace in Madrid. The shock and irritation about his disappearance and fate (which made him even an 'unperson' for a couple of months) drove the first of those studies into an unexpected tension field. Nevertheless, it helped to overcome the Soviet 'hard scenario' attitude and to tear down a barrier to public information at the German east side of the Iron Curtain.

The most detailed results were due to Thomas Malone and co-workers of the Los Alamos National Laboratory (LANL), who extended the NCAR GCM in the vertical to address the smoke transport more precisely. They confirmed the expected lofting into the lower stratosphere and thus a much prolonged residence and forcing. Though the NCAR authors fixed the important issue of 'quick freeze' beneath smoke clouds, notably in the subtropics and Tropics at startup of interhemispheric transport, a controversy arose from their inquiries suggesting change of the popular metaphor into 'nuclear fall'. Until 1987, persistent efforts to deblur longer-term effects due to the oceanic response have only been undertaken at the CCAS. In a more realistic atmosphere–ocean GCM study, virtually the last 'nuclear winter' publication for 15 years, Steve Ghan confirms Alan Robock's (1-D) finding that the acute-phase ocean and sea-ice response may bear climatic impacts for years.

Regional Conflicts and Their Global Effects

The 'doomsday scenario', executed by retreating Iraqi troops in February 1991 in setting the Kuwaiti oil fields alight, was meant as a modern version of Kahn's ultimate deterrent – an idea that failed. Two climate modeling responses, from the United Kingdom Meteorological Office and the Max Planck Institute for Meteorology, denied an attenuating impact on the Indian monsoon (a concern that had been expressed before). Successors of their GCMs did correctly represent the major Asian rainbelts as part of a planetary system, and their seasonal migration, but not the seasonal mean distribution, to say nothing about intraseasonal activity. Just those 30–60-day active–break monsoon cycles, including realistic motions of the major Asian systems, were now found in the Berlin version ('CCAS-B') of the CCAS GCM in an own Kuwait oil well fire study. This GCM version is a completely regenerated, flexible tool of dynamic systems analysis. Its boreal summer monsoons turned out indeed to be part of an interhemispheric, oscillatory seasonal climate regime between critical transitions in June and September. The Kuwait oil fire smoke caused a regional lower-troposphere heating anomaly, and thus an 'exciting' disturbance that fanned the GCM's dynamics in a way not dissimilar to the observed 1991 season. Such a type of monsoon climate may thus be the 'playing ground' for martial adventures seizing the source regions of the atmospheric water cycle. Its structural robustness is unknown.

The theme is again put on the agenda by recent studies into the climatic effects of a potential regional nuclear conflict of 1.5 Mt 'size' in Southeast Asia, using the full atmosphere–ocean GCM with high vertical resolution of NASA's Goddard Institute of Space Studies (GISS), which has been successfully applied to study the climatic impact of volcanic eruptions. The GISS model shows extremely long smoke residence times, up to a decade, due to efficient lofting into the upper stratosphere, all year round in these latitudes. The surface air temperature drop is much less than in the 'nuclear winter' case, of course, but still considerable if compared with the climate record: a global cooling from 1.25 to 0.5 K over a decade, with minima of several kelvin (degrees centigrade) over large areas of North America and Eurasia. A 10% weakening of the global precipitation is concentrated in the Tropics, but substantial (seasonal mean) reductions of the Asian subtropical summer monsoons are also found, with potentially serious human impacts.

Final Remarks

The scientific consensus as settled on the pages of SCOPE-28 was a snapshot taken from a dynamic research process. Shortly after the second edition, the Cold War ended, and modeling the climatic response to massive smoke injections was terminated just when it had reached more firm grounds. The theme was picked up not before another recent revision using the GISS model. Questions like that of the 1975 NAS study on climate relaxation remain unanswered as yet. Summarizing the status of the smoke source term discussion in their last common paper, though, TTAPS had shown that figures which were finally used in the climate model studies of the 1980s remained in the vicinity of earlier assumptions – a consequence of mutually balancing changes in detail. It has been learnt, for example, that smoke consists of fractal aggregates which have little in common with the earlier picture of largely spherical objects. This reduces the rate at which their short-wave absorptivity decreases and prolongs the direct radiative forcing of climatic effects. The debate about nuclear 'winter' or 'fall' occupied the community but did not fundamentally change the perspective as well. A detailed study of atmospheric coastal flow fields did not even confirm a mitigating oceanic impact on the surface air temperature drop over land.

We do not mirror and discuss the points here that have been made with due justification concerning the political response to nuclear winter. Science itself is the addressee of a disturbing question: Was there a potential to substantially influence public and strategic thinking by timely, deliberate inquiry? A 'doomsday potential' was inherent to nuclear deterrence since the 1960s, at the latest, and it may be questioned that the 'policy war' between MAD and NUT was predicated to end at the terms of the warfighting strategists. Game theory was abused to justify NUT, risky nuclear weapons tests were conducted, and the Cuban missile crisis made humankind totter at the brink of its ultimate catastrophe. Though lately a result of the arms race, MAD was a vulnerable and immoral posture. Remarkable activities of the 1960s notwithstanding, though, did nuclear deterrence and nuclear war become great themes for the general scientific community only during the 1980s. A largely unmonitored evolution toward 'wars of the twenty-first century' is likewise a risky habit. It has 'tradition' in military politics to occupy gray zones of knowledge, and in scientific 'surveillance' to lag behind the arsenals and strategies of war.

See also: Monitoring, Observations, and Remote Sensing – Global Dimensions; Noosphere.

Further Reading

Ball D (1981) Can nuclear war be controlled? *Adelphi Paper No. 169*, 51pp. London: International Institute for Strategic Studies.

Bergström S, Bochkov NP, Leaf A, *et al.* (1987) Effects of nuclear war on health and health services, 179pp. *Report A40/11,* 2nd edn., Geneva: World Health Organization.

Carl P, Worbs KD, and Tschentscher I (1995) On a dynamic systems approach to atmospheric model intercomparison. *Report of the World Climate Research Programme (WCRP–92), WMO/TD-No. 732*, pp. 445–450. Geneva: World Climate Research Programme.

Carrier GF, Moran WJ, Birks JW, *et al.* (1985) *The Effects on the Atmosphere of a Major Nuclear Exchange*, 193pp. Washington, DC: National Research Council.

Ehrlich A, Gunn SW, Horner JS, *et al.* (1986) *London under Attack*, 397pp. Oxford: Basil Blackwell.

Ehrlich PR, Sagan C, Kennedy D, and Roberts WO (eds.) (1984) *The Cold and the Dark: The World after Nuclear War*, 229pp. New York: Norton & Co.

Gadgil S and Sajani S (1998) Monsoon precipitation in AMIP runs. *Report of the World Climate Research Programme (WCRP-100), WMO/TD-No. 837*, 86pp. Geneva: World Climate Research Programme.

Golitsyn GS and MacCracken MC (1987) Atmospheric and climatic consequences of a major nuclear war: Results of recent research. *Report of the World Climate Research Programme (WCP-142), WMO/TD-No. 201*, Geneva: World Climate Research Programme.

Harwell MA, Hutchinson TC, Cropper WP, Jr., Harwell CC, and Grover HD (1985) *SCOPE 28 – Environmental Consequences of Nuclear War, Vol. 2: Ecological and Agricultural Effects*, 523pp. Chichester, UK: Wiley (2nd edn. with a 31pp. updating preface, 1989).

Johns LS, Sharfman P, Medalia J, *et al.* (2005) *The Effects of Nuclear War*, 151pp. Washington, DC: Congress of the U.S., Office of Technology Assessment.

Kahn H (1984) *Thinking about the Unthinkable in the 1980s*, 250pp. New York: Simon & Schuster.

McNaughton SJ, Ruess RW, and Coughenour MB (1986) Ecological consequences of nuclear war. *Nature* 321: 483–487.

Nier AOC, Friend JP, Hempelmann LH, *et al.* (1975) *Long-Term Worldwide Effects of Multiple Nuclear-Weapons Detonations*, 213pp. Washington, DC: National Academy of Sciences.

Peterson J and Hinrichsen D (eds.) (1982) *Nuclear War: The Aftermath, 196pp.; AMBIO*, 11(2/3); Oxford: Pergamon.

Pittock AB, Ackerman TP, Crutzen PJ, *et al.* (1986) *SCOPE 28 – Environmental Consequences of Nuclear War, Vol. 1: Physical and Atmospheric Effects*, 359pp. Chichester, UK: Wiley (2nd edn. with a 36pp. updating preface, 1989).

Robock A, Oman L, and Stenchikov GL (2007) Nuclear winter revisited with a modern climate model and current nuclear arsenals: Still catastrophic consequences. *Journal of Geophysical Research* 112: D13107 (doi:10.1029/2006JD008235).

Sagan C and Turco R (1990) *A Path Where no Man Thought: Nuclear Winter and the End of the Arms Race*, 499pp. New York: Random House.

Solomon F and Marston RQ (eds.) (1986) *The Medical Implications of Nuclear War*, 619pp. Washington, DC: Institute of Medicine, National Academy of Sciences.

Stenchikov GL (1985) Climatic consequences of nuclear war. In: Velikhov Ye P (ed.) *The Night After ... Climatic and Biological Consequences of a Nuclear War*, pp. 53–82 (Russian edn.: Nauka, Moscow 1986). Moscow: Mir Publishers.

Svirezhev Ju M, Carl P, *et al.* (1990) *Götterdämmerung. Globale Folgen eines atomaren Konflikts* (substantially revised and extended German edn. of Svirezhev YM, Alexandrov GA, Arkhipov PI, *et al.* (1985) *Ecological and Demographic Consequences of Nuclear War*, 267pp. Moscow: USSR Academy of Sciences, Computer Center), 261pp. Berlin: Akademie-Verlag.

Thompson SL, Aleksandrov VV, Stenchikov GL, *et al.* (1984) Global climatic consequences of nuclear war: Simulations with three dimensional models. *AMBIO* 13: 236–243.

Toon OB, Robock A, Turco RP, *et al.* (2007) Consequences of regional-scale nuclear conflicts. *Science* 315: 1224–1225.

Turco RP, Toon OB, Ackerman TP, Pollack JB, and Sagan C (1990) Climate and smoke: An appraisal of nuclear winter. *Science* 247: 166–176.

Evolution of Oceans

R Klige, Moscow State University, Moscow, Russia

Early Stage of Water Formation on Earth

On the Earth's surface, the World Ocean forms the major part of the hydrosphere which is among the most ancient external shells of our planet.

The volume of the hydrosphere was mainly formed through smelting and degassing of mantle matter and was governed by in-depth geophysical processes. The degassing is a result of the active gravitational differentiation of mantle matter near the Earth nucleus, which caused the convective circulation in the mantle with the period corresponding to the global tectonic cycles.

Recent geological studies indicate that the oceans existed on the Earth practically in all geological epochs. This is supported by the presence of the most ancient sedimentary rocks aged *c.* $3.76 \pm 0.07 \times 10^9$ yr BP, which were found in southwestern Greenland.

Formation of oceans as huge reservoirs of surface waters began simultaneously with that of the oceanic crust, the surface of which is now at the average depth of 4.42 km. The oceanic crust was formed in close relation to volcanism, which is particularly active at present within the global system of the mid-ocean ridges. In these zones, there is a permanent input of basalt matter and juvenile waters and at the same time the oceanic crust is formed.

At present, the total length of the mid-ocean ridges is about 60 000 km, and the average rate of floor spreading is 5 cm yr^{-1}. The oceanic crust (without sedimentary layer) is 6.5 km thick and its average density is 2.88 g cm^{-3}. Annually about 56×10^{15} g of basalt is generated as oceanic crust and the same amount of matter (plus sediments) sinks in subduction zones, thus providing for the balance of matter at the bottom of the ocean. Probable variations in spreading rates could result in different juvenile water inputs to the ocean through time.

The study of global evolution of the Earth with due consideration for the processes of gravitational differentiation of matter and mantle degassing suggests that there was a gradual acceleration of the hydrosphere formation and accumulation of the oceanic water with the maximum probably dating back to the Late Riphean (Mezoproterozoic), about 1.5×10^9 yr BP. Simultaneously the Earth's crust developed through the growth of geosynclinals and mountain-building and weathering processes. The cores of future continents were formed and gradually expanded, and relief of the Earth's surface became more and more contrasting. The Earth's crust was differentiated into oceanic and continental.

By considering the Earth crust evolution and the increase in the total volume of the hydrosphere, models have been developed for the description of water volume formation and changes on the Earth's surface:

$$m_w = c_0(H_2O)\, m_g\left(1 - e^{-\alpha n(t)/n(\infty)}\right)$$

where c_0 is concentration of water in primary matter and the mantle, m_g is mass of the Earth, a is parameter of mobility of H_2O component in the active layer of upper mantle, and $n(\infty)$ is the normalization coefficient ($n(\infty) = n(t)$ if $t = \infty$).

Fluctuations of the sea level during the course of geological time could be partially caused by the changes in the size of oceanic depressions accompanied by the increase of the total amount of water and steady deepening of the ocean. Various correlations of these factors governing the tectonic and sedimentation processes resulted in global transgressions and regressions.

It is necessary to consider the process of ocean volume formation in view of the changing relief of the Earth's surface. This could be illustrated by a specially developed dynamic model of hypsographic curve (**Figure 1**). This curve can be represented as an integral equation accounting for varying frequency distribution (a), mean square deviation (5), and altitude of the surface (h_m).

The hypsographic curve can be described as the sum of three integral functions:

$$N(h, a, \sigma) = \frac{1}{\sqrt{2\pi}\sigma}\int_{\infty}^{h} \exp\left(-\frac{(u-a)^2}{2\sigma^2}\right) du$$

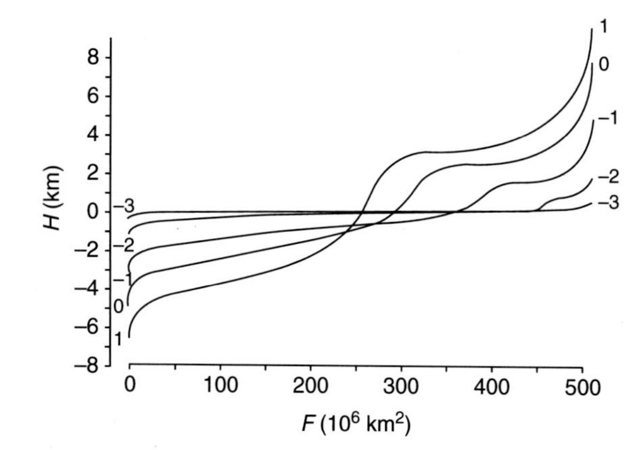

Figure 1 Hypsographic curve showing change in ocean volume with respect to change in relief of the earth's surface.

In short, it can be represented as

$$S(h) = 0.5N(h-4.6;\ 0.8) + 0.4N(h-2.9\ 2.5)$$

The analysis of hydrosphere evolution at the early stages of the Earth's geological history on the basis of the developed model allows calculating the changes of the total area of the oceans and the trends of the gradual rise of sea level in the geological past (**Table 1**; **Figure 2**).

Seawater Formation and Origin of Life

Changes in the ocean area, depth, and volume of the oceans were accompanied by significant qualitative transformation of seawater composition.

During the Early Archaean (*c.* 4–3 × 10^9 yr BP), the dissolved volcanic products with accompanying gases (HCl, HF) predominated in seawater, besides boric acid, CO_2, CH_4, and other hydrocarbons, and SiO_2. Seawater was acidic, with pH *c.* 1–2.

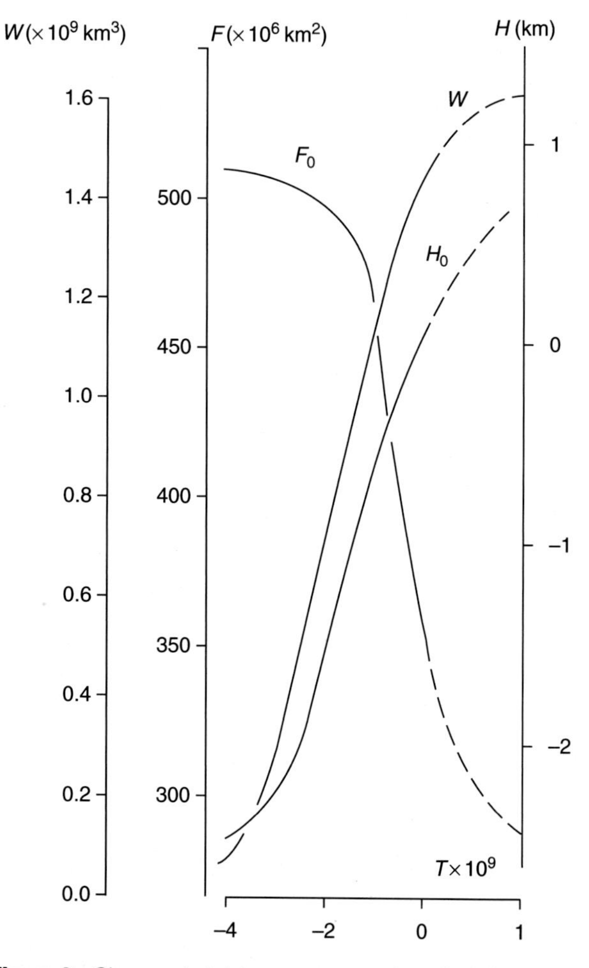

Figure 2 Changes in total ocean area and gradual rise of sea level.

Gradually the composition of seawater underwent changes as a result of neutralization of acids by carbonates (K, Na, Ca, Mg), which formed on land during weathering of volcanic rocks and were washed down to the oceans. Little by little the seawater became a chloride solution with

Table 1 Changes of the basic parameters of surface waters and the oceans

Time (10^9 years)	*Total mass of water in hydrosphere (10^{24} g)*	*Total volume of hydrosphere (10^6 km^3)*	*Total volume of oceans (km^2)*	*Total area of oceans (10^6 km^2)*	*Average depth of the ocean (km)*	*Sea level in relation to the average elevation of the Earth crust H_1 (km)*	*Sea level in relation to its actual level H_2 (km)*
−4.0	0.02	0.02	0.02	509	0.04	0.01	−2.49
−3.5	0.09	0.09	0.09	508	0.18	0.10	−2.40
−3.0	0.22	0.22	0.22	506	0.44	0.25	−2.25
−2.5	0.43	0.42	0.42	504	0.83	0.53	−1.97
−2.0	0.66	0.64	0.63	499	1.26	1.00	−1.50
−1.5	0.90	0.88	0.86	488	1.76	1.50	−1.00
−1.0	1.10	1.07	1.04	462	2.25	1.88	−0.62
−0.5	1.27	1.24	1.20	418	2.87	2.18	−0.32
0.0	1.42	1.39	1.34	361	3.71	2.50	0.00

Al, Fe, Mn, and a small amount of sulfates. As a result of further chemical weathering on the land surface seawater acquired chloride–carbonate composition and sedimentation of $CaCO_3$, $MgCO_3$, $FeCO_3$, and $MnCO_3$ began.

At the end of the Archaean (*c.* 1 3 × 10^9 yr BP), the first green plants appeared, whose life-supporting processes were based on photosynthesis, with resulting production of oxygen. It was a turning point in the evolution of the atmosphere leading to its transformation into the oxygen-type one.

In the Early and Middle Proterozoic (*c.* 2 × 10^9 yr BP), rather numerous traces of plant organic life became evident. The case in point is calcareous algae, or stromatolites, known from the carbonate formations of Huronian series of the Lower Proterozoic in Canada. Free oxygen being available, the oxidation of sulfur and hydrogen sulfide began, seawater received a sulfate ion; therefore, its composition gradually became chloride–carbon–sulfate.

The graphic evidence of transition from reduction to oxidation in the atmosphere and the ocean in the Early Proterozoic was mass accumulation of thick strata of banded ferruginous quartzite (jaspilites), many of which aged from 1.9 to 2.2 × 10^9 yr BP.

It is thought that the amount of free oxygen in the atmosphere equaling one per mille of its present-day concentration (the Jury point) was reached *c.* 1.2 × 10^9 yr BP. Since then the formation of thick acid-leached weathering crusts enriched with iron hydroxides and thus red- and brown-colored soil began on land.

Due to the presence of even a small amount of free oxygen in the atmosphere and hydrosphere, first oxygen-consuming living organisms appeared in the oceans. The most ancient fossils (worm tubes) were found in the Middle Proterozoic sedimentary rocks. The Vendian sedimentary formations embed the fossils of at least 20 genera of sea animals, mainly Coelenterata (jellyfish) and Arthropoda.

At the turn of the Phanerozoic, an important milestone had been achieved. Around 600 × 10^6 yr BP, practically between the Vendian and the Paleozoic, the amount of free oxygen in the atmosphere exceeded 1% of the present-day concentration (the so-called Paster point). It was this circumstance that caused the evolutionary explosion in the beginning of the Phanerozoic, when practically all types of marine animals except chordates were widely distributed. About 400 × 10^6 yr BP, the concentration of free oxygen was already 10% of the present-day values, providing for the formation of the atmospheric ozone layer and the penetration of life on the land. Further on, during just several dozens of million years, the terrestrial vegetation evolved rapidly and the present-day concentration of oxygen in the atmosphere was achieved due to the process of photosynthesis.

The composition of hydrochloric sediments and buried brines of sea origin gives evidence that already in the Cambrian the composition of seawater was the same as nowadays. In other words, final stabilization of the present-day salt composition of the World Ocean took place between 1.5 and 0.5 × 10^5 yr BP.

Factors Governing Water Regime Changes

Global changes of the ocean regime during the Earth's history can be reconstructed in detail by analyzing the fragments of preserved sedimentary rocks that were formed as a result of water action. Such an important indicator as the areas of marine sediments makes it possible to restore spatial-temporal dynamics of water regime of the oceans and reconstruct the total area of seas, the relative altitude of the mean sea level, and the average depth of the oceans. It is well known that during billions and millions of years major changes of the oceans were caused mainly by tectonic processes and evolution of the continents.

Different interrelations of such factors as the increase of the total volume of water and steady deepening of the oceans combined with the development of tectonic processes and sedimentation resulted in global marine transgressions and regressions.

The distribution of sedimentary rocks over the continents testifies that numerous transgressions of the sea took place (**Figure 3**) due to the development of geosynclinal processes, thus leading to the reduction of ocean area in the geological past. The reduction of ocean area against the background of continuous degassing could be

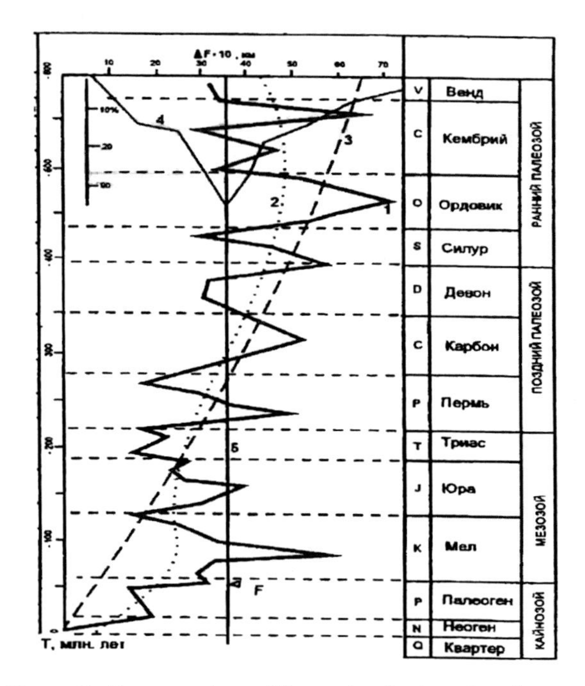

Figure 3 Transgressions of the sea leading to reduced ocean area.

explained only by changes in the relief of the Earth's surface, that is, gradual rising of the continents and deepening of ocean depressions.

During the Phanerozoic, the gradual reduction of the ocean area and the increase of its level were interrupted by large marine transgressions and regressions. The oceans could then cover more than 50% of the present-day land area and the rate of change of the sea level could reach more than 10 m 10^{-6} yr. On an average, transgress and regress phases altered every 60–70 × 10^6 yr. A clear concurrence was revealed between the largest regress phases and the periods of more intensive tectonic activity and orogenic processes. Large glaciations could also influence the ocean phases change.

During the geological history of the Earth, horizontal movements of the lithosphere plates could probably result several times in the consolidation of the Earth's crust and formation of 'supercontinents'. It is supposed that such 'supercontinents' were formed at least 3 times during the Phanerozoic; they were called Gondvana (570–440 × 10^6 yr), Pangea (280–200 × 10^6 yr), and Laurasia (160–100 × 10^6 yr). Lesser Earth heat flux through the consolidated blocks of the continental crust could probably cause the increase of temperature and expansion of the underlying mantle. The consequent rise of continental blocks above the ocean floor and increase of the ocean size could lead to considerable lowering of the mean sea level. The amplitude of sea-level fluctuations from the consolidation of 'a supercontinent' to its disintegration is estimated at about 500 m.

The average rate of bottom sediment accumulation calculated according to the age parameters of different marine ground layers makes about 1 × 10^{-3} mm per year. The rate of sedimentation can change from 0 in the areas of high bottom erosion up to 1 mm yr^{-1} in delta areas. The deep-sea (4500–5000 m) carbonate clay accumulates with an average rate of 1–10 mm per 100 years. At large depths, the sedimentation rate changes from 0.01 mm per 1000 years up to 0.5 mm per year. At present the larger part (~36%) of sediments – nearly 21.3 × 10^9 t of suspended matter per year – comes to the ocean with the runoff. The coastal erosion is the second important source of sediments. Detailed evaluation of this process suggests that the volume of solid matter can amount to 16.7 × 10^9 t yr^{-1}, or about 28% of the total.

An important role in filling the oceanic depressions belongs to the eolian processes. Based on the assessment of continental shelf deposits and Quaternary sediments of abyssal plains, the rate of eolian deposition is estimated at 1 up to 80 m 10^{-6} yr. At present the input of eolian matter in the oceans is about 11 10^9 t yr^{-1} (18% of the total).

Other processes closely related to the continental crust evolution (input of dissolved matter, volcanic sedimentation, etc.) are also important for filling of the ocean. The increasing amplitude of relief augmented the role of sedimentation. Thus the average rate of sedimentation in the Holocene could not be taken as standard for the whole geological history of the Earth. The available data suggest considerably higher rates of sedimentation in the Quaternary period with particular variations during glacial epochs. During the decay of the last Upper Pleistocene ice sheet, the inflow of suspended matter into the oceans could be 15 times greater than at present.

Total volume of accumulated matter can be estimated using the volumetric weight of solid marine deposits. Density of sediments is estimated at 1.5–2.7 g cm^{-3}. Bearing in mind the compaction of sediments, the average annual filling of the oceanic depressions should be about 30 km^3, thus leading to the sea-level rise by 83 mm 10^{-3} yr. Without spreading and subduction of the Earth crust and changes of relief, the complete filling of the oceans would take *c.* 45–50 × 10^6 yr. Data on the recent rate of sedimentation in the oceans show that it is an order higher than that of sea transgressions of the geological past. Therefore it is possible to suggest that during inter-orogenic periods with no essential deformations of the Earth crust, the general sea-level rise was mainly the result of sedimentation processes. At the same time, changing water exchange conditions were also of importance.

During the recent 150 × 10^6 yr, *c.* 225 × 10^6 km^3 of sediments was removed from the ocean floor by subduction processes. Thus the average rate of this process is probably about 1.5 × 10^6 km^3 every 10^6 yr, or 1.5 km^3 yr^{-1}. It is worth noting that during the geological history the intensity of this process could vary depending on tectonic and volcanic activity; the balance of matter could also vary considerably through time. Under the present-day rate of sedimentation in the oceans (30 km^3 yr^{-1}), the equilibrium of the oceanic depressions size could be preserved if the rate of subsidence is at least 12 cm yr^{-1} (the total area of subsidence being 60 × 10^3 km × 4 km).

General Organic Features in the Phanerozoic

Sea level changes can be reliably described for the last 600 × 10^6 yr. Using data on the total area of inundated territories of the present-day continents and considering possible transformation of the hypsographic curve of the Earth's surface (**Figure 4**), it is possible to reconstruct sea-level fluctuations during the Phanerozoic. The results reveal the rising trend of the sea level for almost a billion years with the average rate about 0.5 m 10^6 yr. The general tendency was, however, complicated by significant transgressions and regressions:

$$H_{oc} = 60 + 0.334T - 0.27 \times 10^{-3} T^2$$

Thus, the Vendian period (*c.* 600 × 10^6 yr BP) is usually regarded to be a geocratic era, during which the sea level

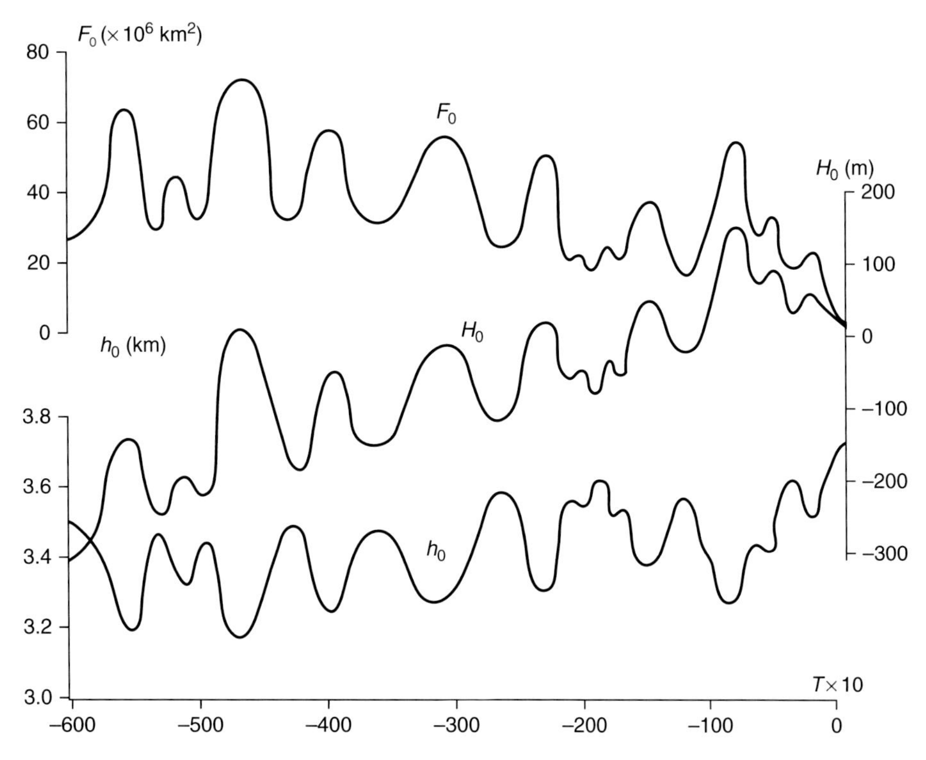

Figure 4 Hypsographic curve showing sea-level fluctuations during the Phanerozoic.

was rather low. In this period, the total area of the oceans could exceed the present-day one by *c.* 32×10^6 km. With the corresponding volume of all oceans taken as $1352 \times 10^6\,\mathrm{km}^3$ and due account of its gradual increase as a result of degassing, the average depth should be about 3.44 km (**Figure 4**). It is worth noting that possible changes of water-exchange processes at the Earth's surface could cause sea-level fluctuations of rather high amplitude (about 100 m) and shorter period, as evidenced by the traces of the Early Cambrian glaciation.

The highest sea-level rise took place in the Ordovician (*c.* 500×10^6 yr BP). The most extensive sea transgression in the Earth's history marked by the marine sediments reached its maximum in the middle of the period. More than $72 \times 10^6\,\mathrm{km}^2$ of present-day continents, or 50% of land, has been flooded. During the Ordovician transgression, sea-level rise was probably more than 250 m and its rate amounted to 8 m 10^{-6} yr. More than 83% of the total surface of our planet was under water. Average depth of the oceans declined to 3.12 km, probably due to reduced size of the oceanic basins.

The end of the Ordovician was marked by sea regression during the Taconian phase of the Caledonian orogeny. The sea level rapidly became more than 200 m lower and the area of the oceans declined to about $391 \times 10^6\,\mathrm{km}^2$. According to geological data, during the Caledonian orogeny, the transition zones of the oceans could be transformed into young platforms, which adjoined the ancient shields, thus increasing the land area. A tendency toward the consolidation of separate continental shields in Central and Southern Asia was also obvious.

Global paleogeographic and stratigraphic data suggest that after the Caledonian (Taconian) orogeny within Europe and Northern America, the Devonian should be characterized by the general transition from epicontinental marine conditions to continental ones. Old red sandstone was accumulated within intermontane depressions of Scotland, Asia, and Northern America. On a boundary of Siluric and Devon periods, maximum of a significant transgression fell on the turn of the Silurian and the Devonian (*c.* 395×10^6 yr BP).

During the epoch of Hercynian orogeny, further transformation of continental margins took place, resulting probably in the consolidation of Europe and Asia and the expansion of other continents. Increase of the land area and reduction of the area of oceans in combination with the ongoing input of juvenile water from the mantle could cause the deepening of the World Ocean.

In the Early Carboniferous, a significant part of the east European platform was drained to become a wetland with numerous lakes and rivers where coal beds were formed. The accumulation of coal-bearing strata under the particularly humid climate of the Carboniferous was interrupted by the Hercynian orogeny. During the next Perm-Triassic epoch, the new red continental facies became widespread from New

Jersey to Tasmania. The traces of continental glaciation of that time are found in Antarctica, India, Australia, and Southern America.

During the Saalian phase of the Hercynian orogeny (the Perm), a new cycle of transgressional sea-level rise occurred with mean rate up to 8.5 m 10^6 yr have been revealed. The transgression had ended about 240×10^6 yr BP when the area of oceans totaled about 411×10^6 km^2.

The end of the Perm and the beginning of the Triassic was a geocratic period accompanied by regression of the sea from about 32×10^6 km^2 of land area. During the Triassic, there were only minor fluctuations of sea level, which was generally rather low.

The onset of the Jurassic was marked by another sea transgression, which coincided with the initial phase of the Alpine tectonic and volcanic epoch. The sea level was on the rise for *c.* 40×10^6 yr with a different intensity (2–6 m 10^{-6} yr). The second half of the Jurassic was the time of the ocean regression, which continued till the beginning of the Cretaceous. The sea level lowered by 100–120 m, which could be partly explained by possible accumulation of compatible amount of water in the inland drainage water bodies and epicontinental seas. In the Late Devonian (370–350×10^6 yr BP), the area of such seas, including geosynclinal ones, could amount to $50–60 \times 10^6$ km^2. The larger parts of the Russian Plain, western Siberia, and the Far East were marine basins at those times, and the geosynclinal seas occupied Southern Europe, Kazakhstan, the Urals, and northeastern Asia. Continental regime was characteristic to the extreme northwest of the Russian Plain, several uplands in its central part, as well as middle Siberia.

During the Cretaceous, the second largest transgression of the Phanerozoic took place. The Cretaceous transgression, the probable maximum of which took place at $90–97 \times 10^6$ yr BP, was accompanied by sea-level rise up to 150 m. About 38×10^6 km^2 of land area was inundated, that is, 36% of the present-day land. The total area of the oceans was approximately 415×10^6 km^2, or 81% of the Earth's surface (**Figure 5**).

In the last quarter of the Cretaceous period, another sea regression began (*c.* 100 m), which was interrupted by a short and rather small transgression in the Early Paleogene and then proceeded approximately to the Middle Paleogene. This regression distinctly coincided with several phases of the Alpine orogeny.

Distribution of sedimentary rocks of marine origin on the continents illustrates the fact that the sea level fell more than 150 m since the Cretaceous. The rate was about 1.5 mm 10^{-3} yr. This suggests the progressive

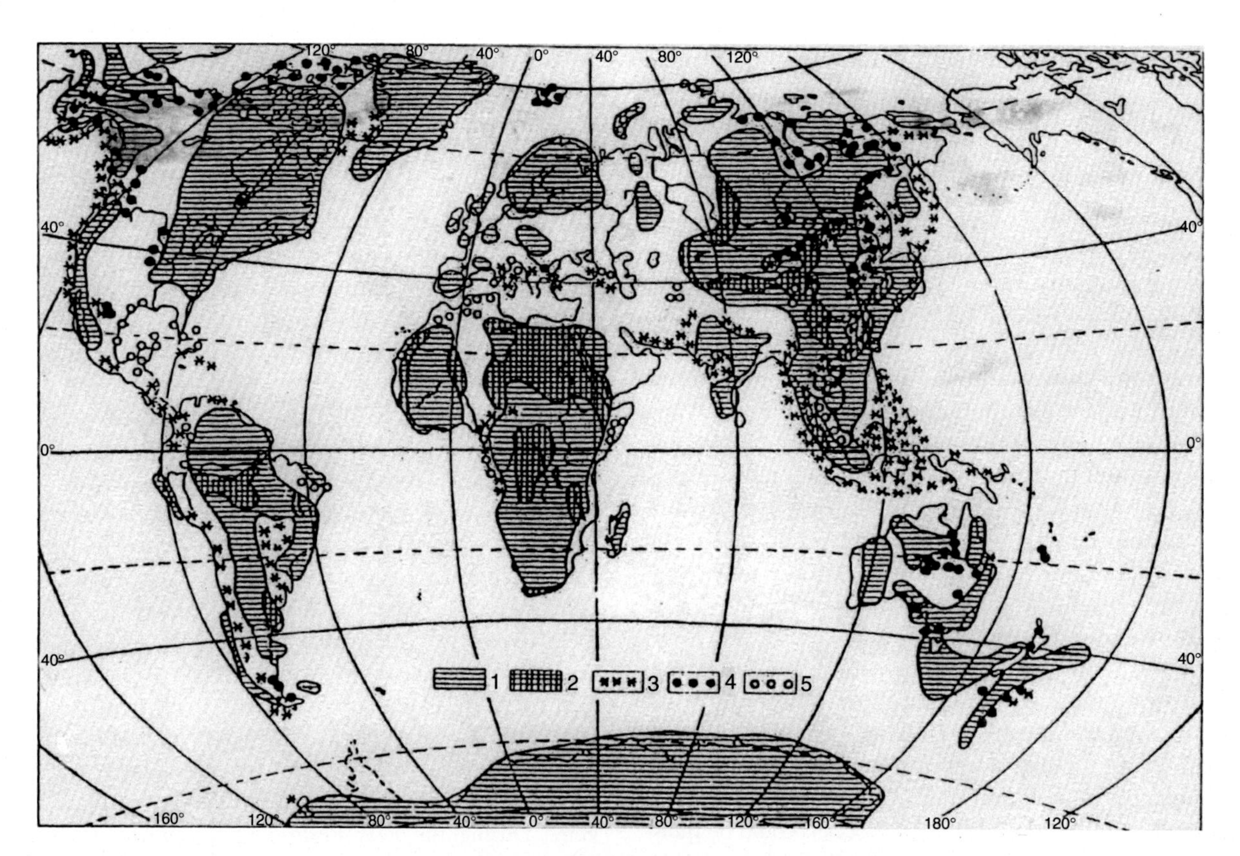

Figure 5 Transgression during the cretaceous period with 81% of the Earth's surface occupied by oceans.

deepening of the oceanic depression, which increased its volume by nearly 0.5 $km^3\ yr^{-1}$ despite rather high rates of marine sedimentation.

Investigations of the sea level dynamics in the geological time show a rather close correlation between its pronounced fluctuations and tectonic and igneous processes. As a rule, the regressions of the oceans occurred during the orogenic periods, which led to the significant reorganization of the Earth's surface, increased relief contrasts, and thus augmented the volume of oceanic depressions.

At the same time, the general tendency of land area growth for the last 600×10^6 yr developed. Investigations of the probable sea-level fluctuations during the Phanerozoic indicate that according to the generalized geological data the maximum rate of regression could exceed 10×10^{-6} m yr^{-1}. Thus the volume of oceanic depressions could increase by up to 4 $km^3\ yr^{-1}$, on average by 1.1 $km^3\ yr^{-1}$. At the same time the accumulation of sediments could reduce the intensity of this process.

According to calculations, the mean rate of transgressions typical for inter-orogenic periods of the Phanerozoic could be as high as 14 m 10^{-6} yr, averaging about 4 m, or 4×10^3 mm yr^{-1}. High rates of the sea-level rise could be to a large extent the result of sedimentation.

Sea Level in the Mesozoic and Cenozoic

The ongoing process of land expansion and increase of its average elevation accompanied by the reduction of the ocean area and increase of its average depth was typical for the Cenozoic. For example, the Tethys Sea, which separated Europe and Asia from Africa, became dry during the Alpine orogeny.

Long-continued subsidence of the ocean floor could be of particular importance for the development of sea regressions. Thus, the present-day occurrence of shallow sea facies and evaporates at the depth of 1.5–2 km in different parts of the World Ocean, particularly in the Atlantic Ocean, and the presence of flat-top underwater mountains (gayots), sometimes with old coral structures, at the depth of about 1300 m suggest the considerable subsidence of the ocean floor during the Mesozoic and Cenozoic.

By analyzing the depth of coral sediments on Pacific atolls, as well as the indirect data on the heat flow for the islands of different age, one can suppose that the subsidence of the oceanic depressions floor could accelerate during the Cenozoic. The rate of this process during the last $1–3 \times 10^6$ yr is estimated at about 0.15–0.23 m 10^{-3} yr while the average rate for $15–25 \times 10^6$ yr is 0.03–0.04 m 10^{-3} yr.

During particular periods of the Cenozoic and probably at the beginning of the Late Riphean, *c.* 1.1×10^9 yr BP, there could have been rhythmic fluctuations of sea level with the characteristic time from 40 to hundreds of thousands of years and the amplitude of dozens of meters. These fluctuations could be largely related to the glacioeustatic processes.

The increasing volume of oceanic depressions, general rise of the land, and other factors caused the overall decrease of the relative sea level during the Cenozoic, thus leading to the considerable deepening of river valleys. A large Sarmatian Sea (lake) was formed in the south of Eastern Europe and its level was several hundreds meters above the ocean level. The lake was from time to time connected to the ocean through the Mediterranean Sea and drained.

During the sharp fall of the sea level, the Mediterranean Sea intercepted the flow of the Danube River and several other large rivers of Europe. In combination with tectonic processes, this led to the disintegration of the vast lacustrine–marine water body and formation of three individual basins (Caspian, Black Sea, and Pannonian) at *c.* $7–8 \times 10^6$ yr BP. The levels of these seas were several hundred meters below the present-day one. In the course of such profound transformation of water balance, the sea level could fall by 0.27 m yr^{-1}. When the Mediterranean Sea was linked with the ocean again and the ocean level became higher, the rivers' flow was redirected to the Black Sea. The rapid rise of its level by several meters during 100 years resulted in the short-term reintegration of the Black Sea and the Caspian Sea (the so-called Pontic Basin). At *c.* 5×10^6 yr BP, these seas were definitely disintegrated.

At the same time, a deep regression (probably down to 300 m and more) took place at the Arctic coast of Eurasia. The Arctic Basin was completely isolated and the area of ice cover increased considerably. After the Early Pliocene transgression of the oceans, a pronounced sea down-drop occurred in the middle of the Pliocene ($3.7–3.3 \times 10^6$ yr BP). Overdeepening of river valleys in the north of the east European platform, west Siberian lowland, and Far East regions was 200–300 m.

Since the Early Pliocene, periodic fluctuations of sea level in isolated and semienclosed seas occurred along with the above-discussed irregular changes. For the Black Sea, the periods of such fluctuations were 40–50 and *c.* 200×10^3 years and their amplitude amounts to 20–25 m. Periodic oscillations could be correlated with the global water-exchange processes and, probably, with changing intensity of the continental glaciation and amount of water resources. The decrease in the total area of oceans by 15% during the Cenozoic contributed to the differentiation of global climatic conditions and deceleration of water-exchange processes.

In the Early Cenozoic, total area of the oceans was much larger than at present, mainly due to the inundation of vast areas of the continental platforms and wide occurrence of geosynclinal seas. Data of historical geology, geological maps, and other sources of information allow estimating the area of the Late Mesozoic sea at about $416 \times 10^6\,km^2$, that is, $55 \times 10^6\,km^2$ more than at present. The average depth of the oceans was about 3 km.

One should take into account the formation of the Antarctic ice shield about 40×10^6 yr BP (Eocene-Oligocene). As a result, *c.* $24 \times 10^6\,km^2$ of water was withdrawn from the global water cycle for a long period causing the decrease of sea level by more than 60 m and a certain reduction of the oceans' area.

During the Cenozoic, there was a relatively high synchronism between the large sea regressions and periods of high tectonic activity, such as Austrian (95×10^6 yr BP), Danubian (25×10^6 yr BP), and Attic (9×10^6 yr BP).

It is worth noting that during the Cenozoic 43% of the ocean floor area subsided to the depth of more than 1 km, and 13% to more than 2 km. This is well confirmed by the data of deep-sea drilling within the atolls, which were submerged by approximately 1300 m.

Besides the deepening of the seafloor, sea-level fluctuations could be caused by a wide range of factors. In the Mesozoic, the oceanic depressions were $230\,km^3$ smaller than at present. Their growth could lead to the sea down-drop by more than 600 m. In addition, a certain amount of water was accumulated in the Cenozoic ice shields of Greenland and Antarctic. However, the sea-level rising factors were also active at the same time. The oceans were supplied with the juvenile water, bottom sediments and volcanic material were accumulated, geosynclinal and shelf seas were drained, and the amount of water on the land decreased, particularly within the enclosed water bodies and the underground aquifers.

As a result, there was a general trend toward sea-level down-drop by about 200 m (**Figure 6**).

Oceanic Water Regime in the Pleistocene

During the Pleistocene, the ongoing altitudinal differentiation of the Earth's surface, expansion of the total land area, and gradual cooling of land surface were accompanied with increasing climatic contrasts and alteration of global cooling and warming epochs. As a result, a certain cyclicity of natural processes on the Earth's surface was typical for the period. Specific features of the evolution of natural conditions were reflected in the global water balance, particularly in the sea-level fluctuations.

The periodic occurrence of global cooling epochs was mainly induced by changes in solar radiation coming to the Earth's surface. They were a result of changes in the astronomic factors governing the position and movement of the Earth in space (the eccentricity of the Earth's orbit, the longitude of the perihelion, the angle of the Earth's rotational axis relative to the plane of its orbit, etc.).

The influence of the astronomical factors on the Earth's climate is well confirmed by changes of temperature conditions, which were reconstructed based on the analysis of bottom sediments from several areas of the World Ocean. It was found that the concentration of ^{18}O isotope in carbonates deposited from water solutions

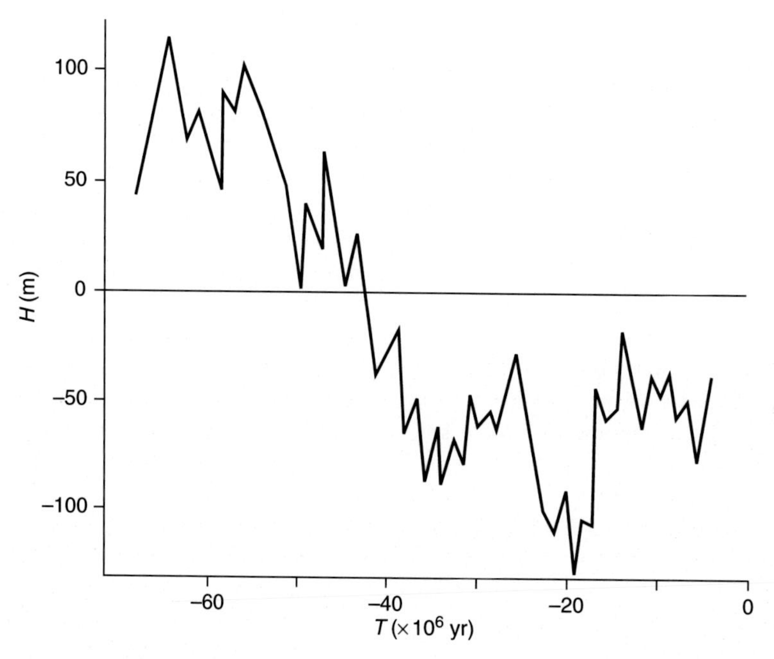

Figure 6 Sea-level down-drop in the Mesozoic.

depends on the temperature. Thus the changes of the oxygen-isotope composition along the bottom sediment columns made it possible to reconstruct the temperature variations over recent >100 × 10^3 years.

The general pattern of temperature variations suggests the alteration of prolonged cooling epochs accompanied with large continental glaciations of about 100 × 10^3 yr cycle and the relatively warm periods which took place every 20 × 10^3 yr. The results of this study point to the fact that rather high temperatures like those we observe on the Earth now occurred for just 5% of the time.

The analysis of elevations and age of sea terraces along the coasts in different parts of the world revealed that their age increases from the lowest up to the highest one, which is generally 100 m high. This could seemingly prove the regressive trend of sea-level changes during the Pleistocene (**Figure 7**).

The areas with unstable tectonic conditions could demonstrate even more considerable changes of sea level in relation to the shorelines. The study of marine sediments in Japan suggested the amplitude of sea-level fluctuations to be more than 200 m. This is indicative of very active recent vertical movements of the Earth crust in the area of Japanese islands. Oceanic islands, particularly atolls, the majority of which have rather stable tectonic conditions, are good indicators of sea-level fluctuations in the Pleistocene. During the glacial–eustatic regressions, they came from under the seawater, while during the interglacial epoches they became atolls again. Because of their altitudes, atolls could probably escape the effects of geocratic regressions during the submergence of the ocean floor. It is thus possible to reconstruct the pattern of sea-level fluctuations in the Pleistocene, which puts away the influence of global tectonic evolution of the Earth's surface. General pattern of sea-level fluctuations for the oceanic islands and atolls suggests that there was probably no regressive submergence of the ocean floor during the Pleistocene and the uplifting of the continents was most likely.

Despite significant variations, all curves of the sea-level fluctuations during the Pleistocene show the gradual decrease averaging to *c.* 0.1–0.4 mm yr^{-1}. A principal factor governing the sea-level trend is probably the globe-wide recent tectonic movements resulting from the Cenozoic uplifting of the main continental structures.

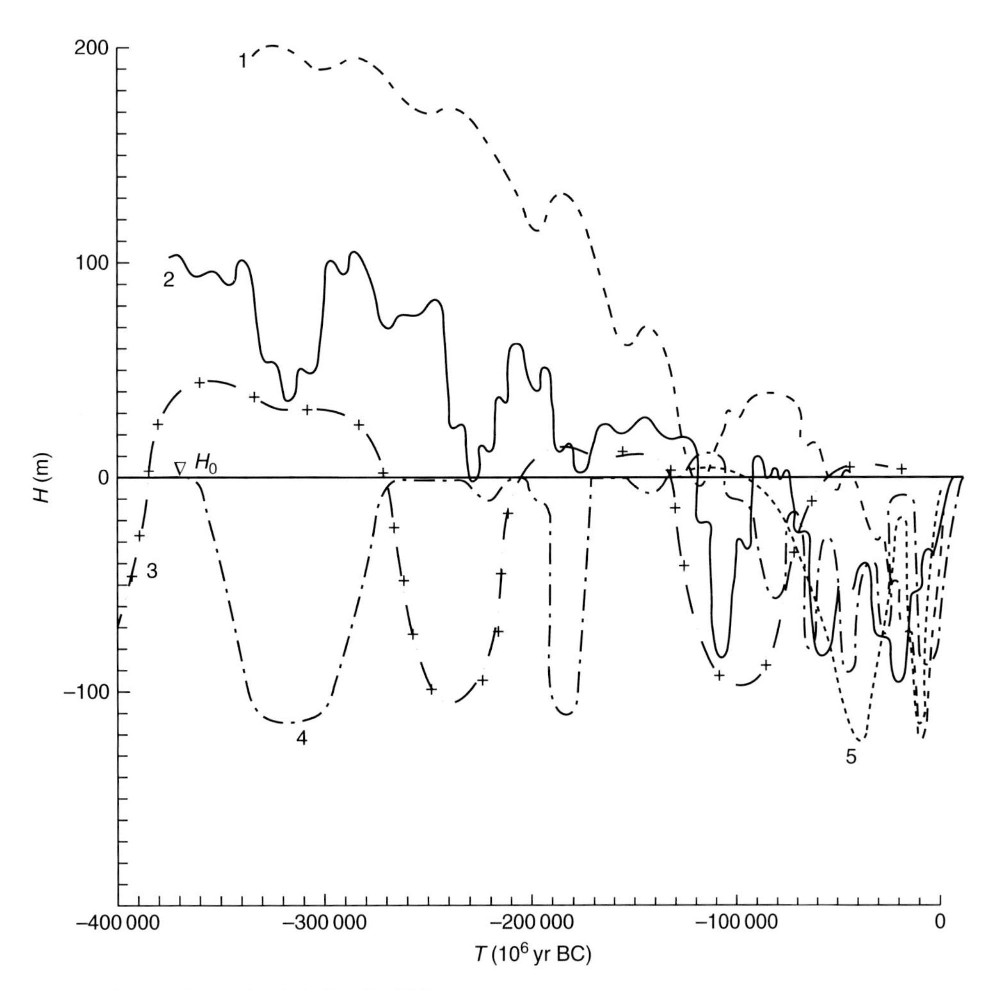

Figure 7 Regressive changes in sea-level during the Pleistocene.

Changes of the global water balance also influenced the large-scale fluctuations of the sea level during the Pleistocene. During the epochs of global cooling, large continental ice sheets were formed that accumulated enormous masses of water. It is supposed that at 230–300 × 10^3 yr BP the Middle Pleistocene glaciation could reach its maximum in Eastern Europe, western Siberia, and Northern America. Total volume of water accumulated in the ice sheet could exceed 60 × 10^6 km^3, thus resulting in the sea-level down-drop by more than 100 m.

Correlation of R. Fairbridge's data on the sea-level dynamics (if the general tectonic trend is excluded) with the mean annual temperature scale reconstructed on the basis of the oxygen-isotope composition of foraminifers from the bottom sediments and the data on solar radiation income to the Earth's surface reveals the surprisingly close synchronism of radiation changes, variations of temperature conditions, and water-balance fluctuations in the oceans (**Figure 8**).

Fluctuations of the sea level could also be induced by isostatic compensation. Movement and concentration of significant amounts of water in certain areas transformed the isostatic load on the surface of the Earth's crust both on land and in the ocean. As a result, compensatory vertical changes should occur lasting for several hundreds to dozens of thousand years.

Water-balance calculations show that during rather long periods of climate warming and almost total reduction of ice sheets, the sea level could stay >60 m higher than at present. During such periods, the sea level became radically less variable too, since the powerful glacioeustatic factor was inactive. Apart from tectonic movements, sea level was considerably influenced by the changes of land water resources (in rivers, lakes, bogs, and under ground), particularly under warm and moist climatic conditions when the total amount of water on the continents increased.

At the beginning of the Late Pleistocene glaciation, the average level of the oceans decreased rapidly, though the rate of this process was a little bit lower, than that of the subsequent rise. It is likely that in the time interval from 80 to 70–65 × 10^3 yr, the level of the oceans lowered by 50–80 m. The ocean water volume decrease rate could account for 3–4 × 10^3 km^3 yr^{-1}. The mean rates of the sea-level down-drop were 7–8 mm yr^{-1}. Particularly high rates of the sea-level down-drop during the Late Eemian interglacial are illustrated by the data on ancient shorelines of the Ryukyu, the Barbados, and the Bermuda Islands. According to the majority of models, the largest Late Pleistocene ice sheets of the Northern Hemisphere could form during 10–15 × 10^3 yr.

During the maximal stage of the last glaciation (18 × 10^3 yr BP), the temperature of superficial oceanic waters was on an average 2.3 °C below the present-day one. It is supposed that the maximum fall of seawater temperature (by more than 10 °C) occurred in the North Atlantic. Significant changes took place on the Earth's surface as a whole (**Figure 9**).

Evolution of natural conditions within the shelf area and the coast of the Black Sea during the latest large regression (about 18 × 10^3 yr BP) when the sea level was 90 m below the present-day one is illustrated by a paleogeographical scheme (**Figure 10**). It represents the extensive areas of terraced slopes, watersheds, and river valleys formed during the Middle Pleistocene. It is also possible to trace the fragments of marine terraces of the earlier transgressions. Wide occurrence of turbidity flows and landslides was probably typical for the period under discussion and the paleochannels are clearly detectable on the sea bottom.

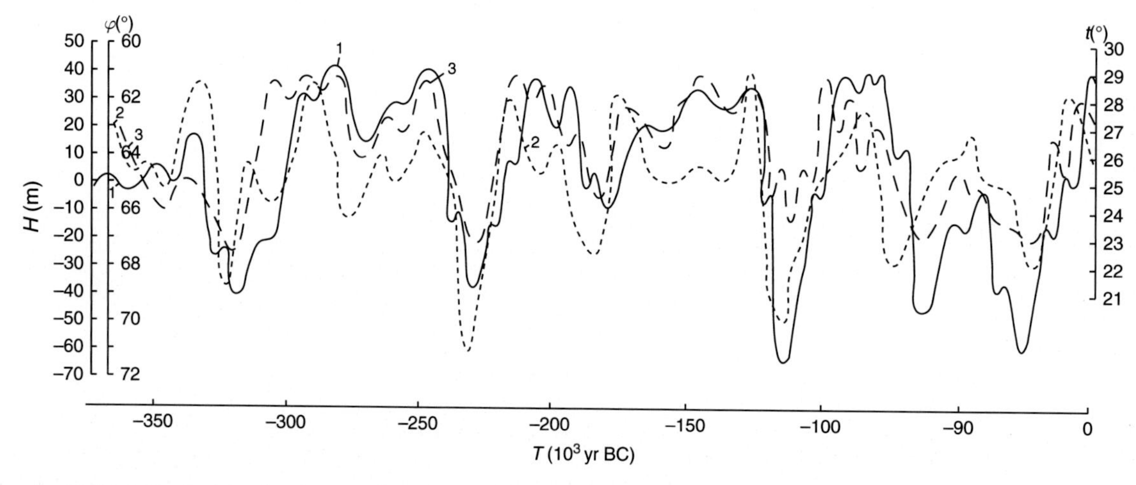

Figure 8 R. Fairbridge's data showing synchronism of radiation changes, variations of temperature conditions and water - balance fluctuations in the ocean.

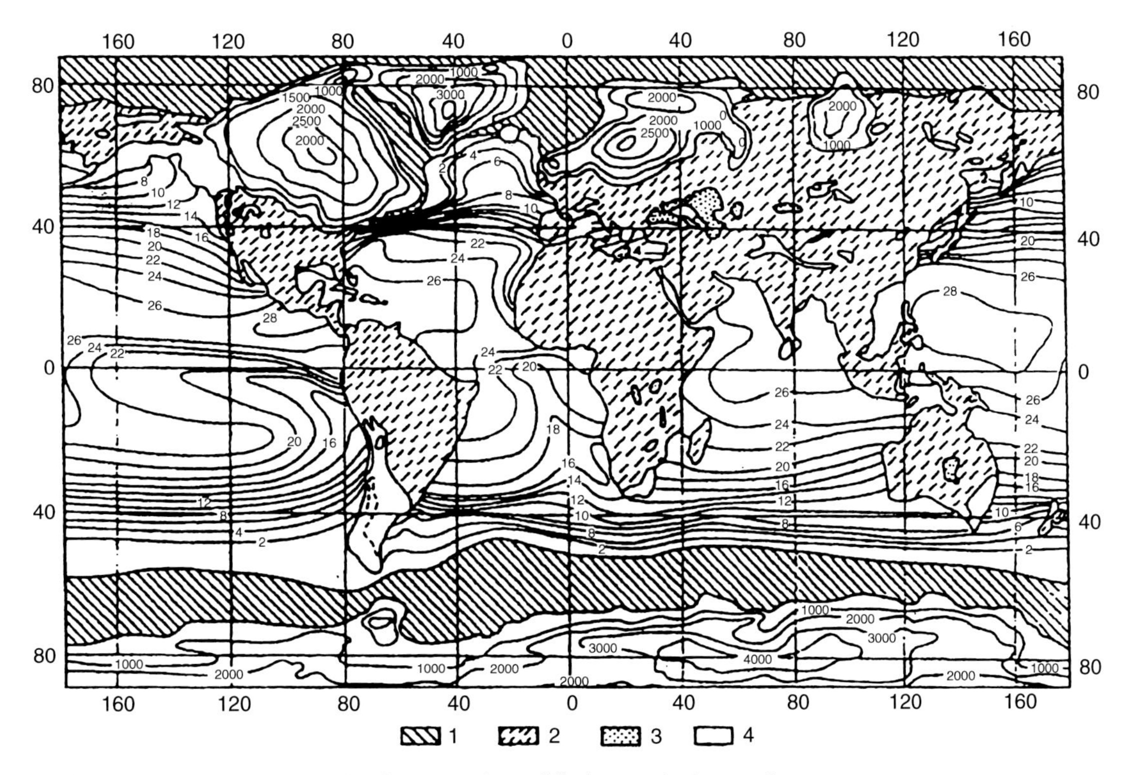

Figure 9 Changes during last glaciation leading to maximum fall of sea water temperature.

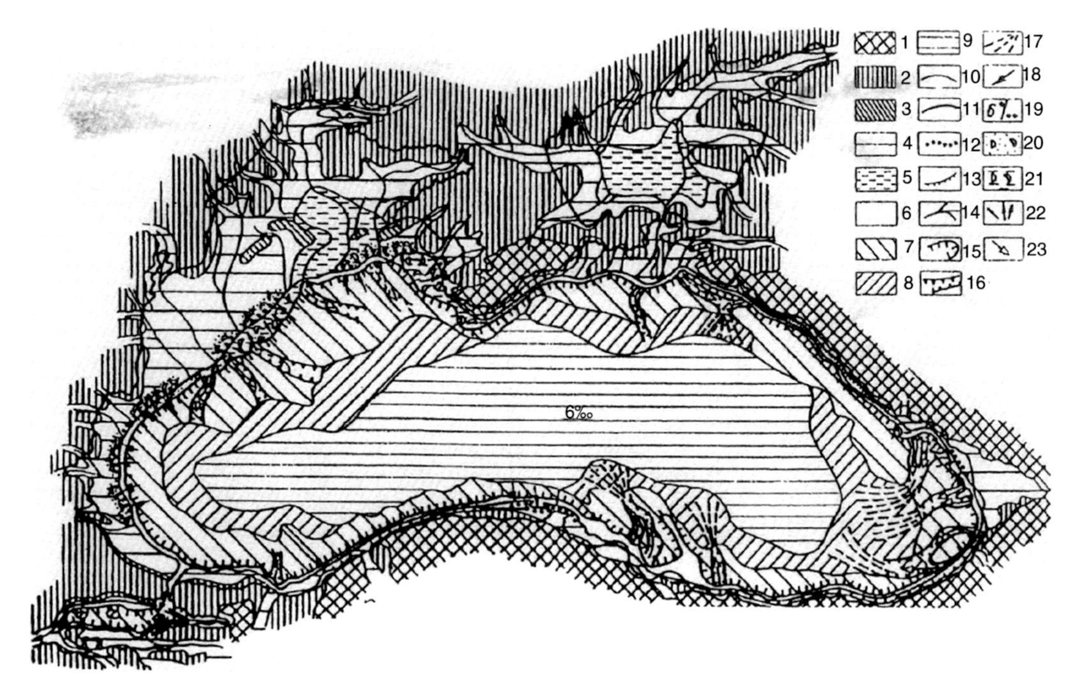

Figure 10 Paleogeographical scheme showing large regression within the shelf area and coast of Black sea.

The statistical analysis of more than 600 most reliable datings of the Pleistocene coastal deposits suggests that they were mainly formed 70–90, 110–140, 200–230, as well as 300–370 and 500–600 10^3 yr BP.

In the Early Pleistocene, the sea level (if the recent tectonic factor is excluded) could hardly exceed +20–55 m, in the Middle and Late Pleistocene +10–12 m (**Figure 11**). At the same time, the presence of sea basins

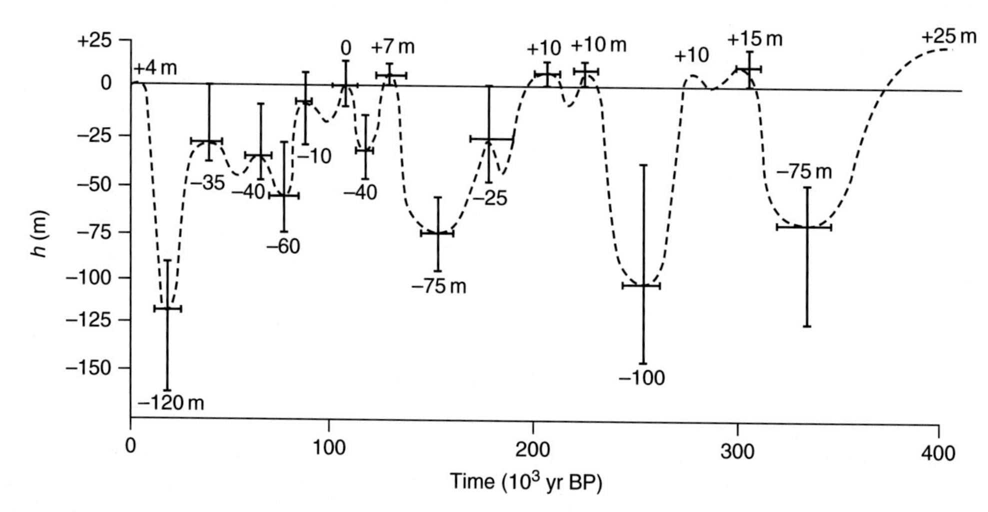

Figure 11 Sea levels during barly, Middle, and Late Pleistocene.

in the areas which were exposed to tectonic, glacioeustatic, and other deformations could contribute to much more significant rise of the sea level. For example, during the Mikulinian Interglacial, the paleosea occupied a significant part of the northern east European plain (at present the elevations of the area are +80–100 m, and up to +120 m in some places). Some authors even suggest the possibility of interconnection between the Arctic Basin and the Baltic Sea.

Sea-Level Fluctuations in the Holocene

Use of the isotope methods for the analysis of the geological history of the Earth gives an opportunity to construct more objectively the curves of sea-level change in the latest geological past, that is, for the recent $30–35 \times 10^3$ yr BP. The latest regression of the oceans by 100–130 m took place at $15–20 \times 10^3$ yr BP (see **Figure 9**). It was the result of global water cycle disturbances caused by the changes in thermal conditions on the Earth's surface that the transformation of the global water balance occurred.

At *c.* 16×10^3 yr BP, the rapid rise of the sea level began due to input of large amounts of water from the thawing ice sheets to the oceans. The rate of sea-level rise in this period was approximately 10 mm yr^{-1}; during particular periods, it could exceed 20–25 mm, and sometimes even 50 mm yr^{-1} (**Figure 12**).

According to the results of the isotope study of bottom sediments, the process of ice water input in the oceans, and thus the decrease of ice sheets, became sharply accelerated from 16 to 13×10^3 yr BP and from 10 to 7×10^3 yr BP.

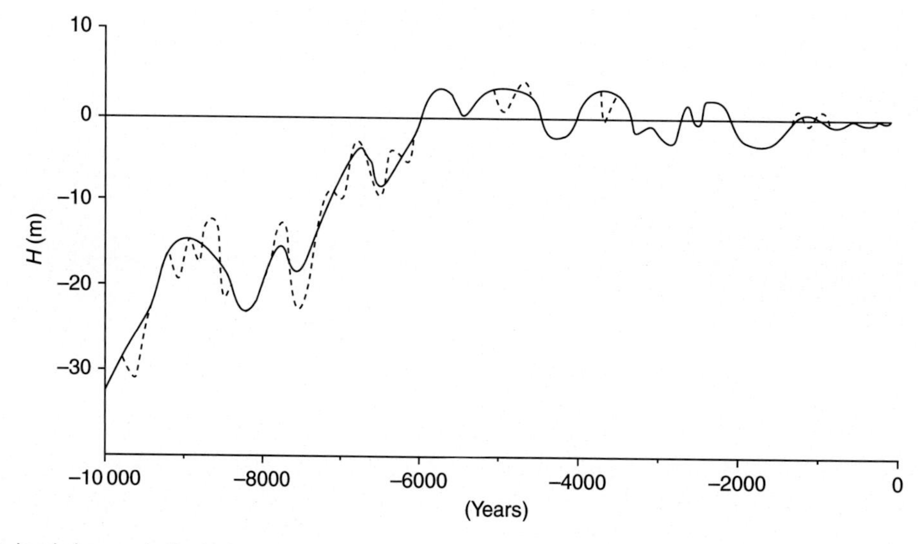

Figure 12 Sea-level changes in the Holocene.

Glaciers were the main factors that influenced these changes; their volume could change depending on temperature variations and the ratio between ablation and accumulation. Mountain glaciers that experienced both augmentation and decline during the Holocene are the most indicative in these terms. Changes of glacier masses resulted in their advance and retreat with approximately 2×10^3 yr period. The mass of mountain glaciers is relatively small, so they are just one of the indicators of ice volume changes. It is the fluctuations of the water balance of the Antarctic and Greenland ice sheets that contributed significantly to sea-level fluctuations during the Holocene.

The rise of the global sea level caused the rapid inundation of coastal lowlands, which in combination with the large input of sediments resulted in the formation of large accumulative features, such as bars, spits, sand barriers, etc. The most intensive formation of coastal barriers took place under the relatively lower rate of sea-level rise. These forms are partially preserved and provide many details of the sea-level rise in particular seas. Generalized data on such complexes all over the World Ocean suggest their obviously uneven distribution in terms of depth. Most of them are concentrated at depths of 20–25, 55–60, 80, 100, and 120 m below the present-day sea level.

The review of extensive data made it possible to identify particular zones of the World Ocean with principally different trends of changes of sea level during the Holocene. The models of uneven spreading of ice water and the development of glacial–isostatic and hydro-isostatic processes were elaborated, as well as of the uneven spreading of water from the thawing glaciers caused by the deformations of the Earth's body. If the difference between these zones is generalized, it becomes clear that in the tropical belts of Northern and Southern Hemispheres sea level was several meters above the present-day one during the Middle Holocene. In the equatorial zone, the sea level was progressively on the rise. The areas of the Late Pleistocene glaciation have been experiencing the general down-drop of the sea level till now. In the adjacent areas, a relative uplifting of land gave way to its submergence in the Middle and Late Holocene.

During the recent 10^4 years, the highest sea level was 5 m above its present-day one. This occurred during the climatic optimum of the Holocene. This period was marked by global climate warming by 2.5 °C, which resulted in the decline of ice volume and the relevant recharging of the oceans.

Conclusion

Summarizing and analyzing the data of the different earth sciences and modeling the processes governing the evolution of the hydrosphere and the Earth's relief made it possible for us to reconstruct global features of the oceans' formation and evolution, as well as to analyze the causes of particular global paleohydrological and environmental events during different stages of the Earth's history from the origin of the oceans till nowadays.

See also: Climate Change 2: Long-Term Dynamics; Ocean Currents and Their Role in the Biosphere; Water Cycle.

Further Reading

Fairbridge RW (1963) Mean sea level related to solar radiation during the last 20 000 years. In: *Changes of Climate*, Proceedings of the Rome Symposium, pp. 229–242. Brussels: UNESCO.
Fairbridge RW (1980) Holocene sea-level oscillations. *Strae* 14(1–4): 23–60.
Kaplin PA (1973) *Recent History of the World Ocean Coasts*. Moscow: MGU (in Russian).
Klige RK (1980) *Sea Level in the Geological Time*. Moscow: Nauka (in Russian).
Klige RK (1985) *Changes in the Global Water Exchange*. Moscow: Nauka (in Russian).
Klige RK (2006) Formation of the Earth's water. In: *Modern Global Changes of the Natural Environment*, pp. 210–222. Moscow: Nauchny Mir (in Russian).
Klige RK, Danilov IA, and Konishchev VN (1998) *History of the Hydrosphere*. Moscow: Mir (in Russian).
Ku T-L, Ivanovich M, and Shangde L (1990) U-series dating of last interglacial high sea stands: Barbados revisited. *Quaternary Research* 3(2): 129–147.
Lisitsyn AP (1980) History of the oceanic volcanism. In: *Oceanology: Geology of the Oceans*, pp. 278–319. Moscow: Nauka (in Russian).
Matthews RK (1990) Quaternary sea-level. In: Revelle RR (ed.) *Sea-Level Change*, pp. 88–103. Washington, DC: National Head Press.
Milliman JP and Syvitski JPM (1992) Geomorphology and tectonic control of sediment discharge of the ocean. *Journal of Geology* 100: 525–544.
Monin AS (1977) *History of the Earth*. Leningrad: Nauka (in Russian).
Rubey WW (1964) *Geologic History of Sea Origin and Evolution of Atmospheres and Oceans*. New York: Wiley.
Selivanov AO (1996) *Fluctuations of the World Ocean Level in the Pleistocene and Holocene and the Evolution of Marine Coasts*. Moscow: RAN, Institute of Water Problems (in Russian).
Sorokhtin OG and Ushakov SA (2002) *Evolution of the Earth*. Moscow: MGU (in Russian).
Termier H and Termier G (1952) *Histoire Geologique de la Biosphere*. Paris: Masson.
Vinogradov AP (1959) *Chemical Evolution of the Earth*. Moscow: Akademiya Nauk SSSR (in Russian).

Evolution of 'Prey–Predator' Systems

H Matsuda, Yokohama National University, Yokohama, Japan

Published by Elsevier B.V., 2010.

'Prey–Predator' System

Predation is often defined as an interspecific interaction in which an individual of one animal species kills an individual of another species for dietary use. As a broader definition, predation can include an interaction between an animal and seeds or between a parasitoid and a host. However, predation rarely includes disease-causing organisms or herbivores that do not kill their food.

Predation is one of the most important interactions between species, ranking with parasitism, competition, and mutualism. Predation can affect changes in population sizes, traits, or phenotypes, and consequently promote the evolution of underlying genetic traits. These interactions are termed 'prey–predator' or 'predator–prey' systems. The existence of such interactions creates a link between the prey and predator species, termed a 'trophic link'. The assembly of trophic links within a community forms a 'food web'. Predation probably plays a major role in determining the life-history pattern of every species, and organismal complexity may increase due to predation. Here we consider the co-dynamics of prey and predator populations and coevolution of prey and predator species. We also consider the relationship between population dynamics and evolutionary change in trait values.

Hereafter we focus on a system involving one predator and one prey species. The following dynamic model describes temporal changes in the predator and prey populations:

$$\begin{aligned} dN/dt &= -r(N)N - f(N,P)P \\ dP/dt &= [-d(P) + g(N,P)]P \end{aligned} \qquad [1]$$

where P and N are the population sizes (or densities) of the predator and prey; t is time; $d(P)$ and $r(N)$ are the intrinsic death rate of the predator and the intrinsic growth rate of the prey; and $f(N, P)$ and $g(N, P)$ are the per capita rate of predation and the contribution of predation to the predator's per capita growth rate, respectively. These factors are often considered independent of predator abundance P. Conditions where f and g depend on the ratio N/P are analyzed (this type of predation is called 'ratio-dependent predation').

Functions f and g are characterized by the predatory interaction; they are termed the functional response and the numerical response, respectively. Three variations of $f(N)$ exist: (1) a linear relationship ($f(N) = aN$); (2) a convex curve ($f(N) = aN/(1 + abN)$); and (3) a sigmoid curve (e.g., $f(N) = aN^2/(1 + abN^2)$), where b is handling time and a is the predation coefficient. For the mathematically simplest model with a type 2 functional response:

$$\begin{aligned} dN/dt &= [r - kN - aP/(1 + abN)]N \\ dP/dt &= [-d + baN/(1 + abN)]P \end{aligned} \qquad [2]$$

where k is the magnitude of an intraspecific density effect from growth of the prey population, and b is the conversion efficiency of ingested prey into the predator. This system can produce either a stable or unstable equilibrium (**Figures 1a** and **1b**). Note that increases in the predator population lag one-quarter of a period behind increases in the prey population (**Figure 1c**). This is fairly intuitive, since the predator population decreases to the left of the predator's 'null cline' (the vertical line in **Figures 1a** and **1b**) and increases on the right side of the line, whereas the prey population increases below the prey null cline (the parabolic curve in **Figures 1a** and **1b**) and decreases above the curve. These null clines are obtained by the solution of the equations $dP/dt = 0$ and $dN/dt = 0$, respectively. The equilibrium of coexistence is obtained by the intersection of these null clines.

Especially in host–parasitoid dynamics, time-discrete models, such as the famous Nicholson–Bailey model, are more reasonable because the hosts reproduce seasonally, and the life cycle of a parasitoid often synchronizes with that of its host. Because of time discreteness, these models are less likely to produce a stable equilibrium.

Using the Nicholson–Bailey model, many studies have incorporated factors representing the spatial distributions and behavioral characteristics of both the host and parasitoid. If parasitoids focus on the center of the host distribution, the risk of parasitism is low for hosts that are far from the population center. This aggregative response has a stabilizing effect on the host–parasitoid

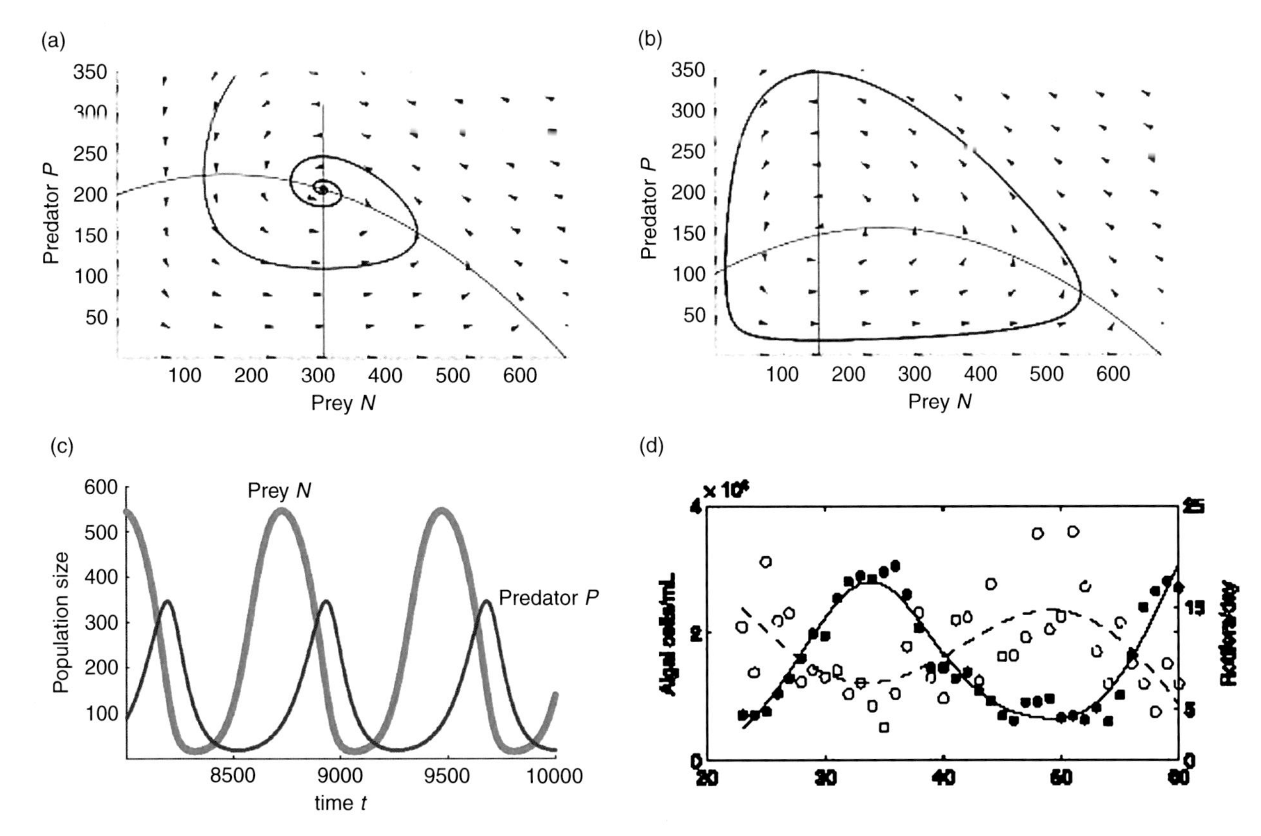

Figure 1 Prey–predator dynamics of model [2] and related empirical data. Parameter values are: $r = 2$, $b = 1$, $k = 0.003$, $d = 1.6$, $h = 0.3$; $a = 0.02$ in panel (a) and $a = 0.01$ in panels (b) and (c). (d) From Shertzer KW, Ellner SP, Fussmann GF, and Hairston NG, Jr. (2002) Predator–prey cycles in an aquatic microcosm: Testing hypotheses of mechanism. *Journal of Animal Ecology* 71: 802–815.

system. If parasitoids that share an individual host interfere mutually, a type 3 functional response is again possible because the potential for interference increases as the host density decreases. Another important factor in prey–predator systems is stochasticity. Many stochastic factors have been studied. For example, demographic stochasticity and genetic drift usually destabilize the equilibrium of prey–predator systems, despite a few examples to the contrary.

Hereafter we focus on time-continuous models. The Lotka–Volterra predator–prey model is an extremely simplified model of a prey–predator system, involving the case where $k = 0$ and $h = 0$ in model [2]. Even in this simple model, a time-dependent analytical solution has not been obtained. The Lotka–Volterra model has an interior equilibrium, $(N, P) = (d/ba, r/a)$. Because this equilibrium is neutrally stable, an interior equilibrium of variants of the Lotka–Volterra model can produce either stability or instability, as shown in **Figures 1a and 1b**. Type 3 functional responses and the density effects from both prey and predator growth rates produce stabilizing effects on the prey–predator dynamics. Type 2 functional responses and time lags between predation and population growth have destabilizing effects.

Increasing productivity of the prey usually has a destabilizing effect on the equilibrium. This is called the 'paradox of enrichment'. This is an intuitive result because the null cline of the prey (parabolas in **Figures 1a** and **1b**) shifts to the right; therefore, the equilibrium (the intersection of the parabola and the vertical line) point occurs to the left of the parabola's peak, as in **Figure 1b**.

Changes in the predator population lag behind changes in the prey population by one-quarter of a period (**Figure 1c**); however, few examples of such prey–predator cycles have been observed in the field. The prey population often regulates the predator population, whereas the latter less frequently regulates the former. One of the best examples is presented in **Figure 1d**, showing a one-half period lag between prey and predator cycles, as discussed later.

Strategy and Tactics of Prey–Predator Systems

Predators and parasites usually have traits that improve their predatory efforts, such as reduced handling time, increased search ability, or increased capture rate. However, a tradeoff can exist between predation ability and the per capita death rate of the predator population.

If the predator can process more than one species or type of prey, then prey choice becomes another important factor in the prey–predator dynamics. If the predator focuses its foraging effort on one prey species, the other prey species may experience a reduced risk of predation.

Frequency-dependent prey choice is called switching predation. Positive and negative switching refer to positive and negative relationships, respectively, between relative prey density and the per capita risk of predation for that prey species. Prey switching is a typical mechanism that enhances a type 3 functional response. Many theoretical works have shown that switching has a stabilizing effect on the relative frequency of the two prey populations, but does not always stabilize total prey density.

There are many empirical examples of the existence of switching predation; there are equally many examples of nonswitching and negative switching. Switching predation has been reported in a variety of predators and even herbivores. There are two types of indices for prey choice. A classic index is Ivlev's index (denoted here by I), which is defined as $I_i = (I_i' - 1)/(I_i' + 1)$, where I_i' is Shorygin's index, $I_i' = \left(N_i/\sum_i N_i\right)/\left(X_i/\sum_i X_i\right)$, and N_i and X_i, respectively, represent the number of preys of type i in the environment and the number of preys of type i in the predator's stomach contents. If I is positive for a particular prey species, the relative number of preys eaten by the predator is larger than the relative number of preys in the environment. However, I varies depending on which species are included as food items. In cases 1 and 2 in **Table 1**, a predator does not consume species 3 ($X_3 = 0$). I for prey 1 is 0.1 if species 3 is excluded from prey items, whereas a value of 0.3 results if species 3 is included. If a predator only rarely consumes prey 3, I will change considerably. I also depends on the abundances of rarely consumed species. In cases 3 and 4 in **Table 1**, I for prey 1 is highly dependent upon the abundance of prey 2. Jean Chesson's index, which is defined as $(X_i/N_i)/\left(\sum X_j/N_j\right)$, is more robust for the abundance of species that are rarely or never consumed, as shown in **Table 1**.

Prey and hosts typically have traits that help them avoid predation and parasitism. Some examples of such antipredator traits are: being vigilant, seeking refuges, having hard or thorny skin, and producing toxic chemicals. Tradeoffs usually exist between escape from predation and other factors that affect fitness, for example, growth rate, fecundity, or survival rate. If such tradeoffs do not exist, antipredator efforts could evolve infinitely. If more than one species or type of predator shares a common prey species, the prey may have multiple antipredator traits. If some antipredator efforts are effective against one predator but not against other predators, a tradeoff is likely to exist between antipredator traits that are effective against different predators.

Evolution of Prey–Predator Systems

To describe the strategies of prey and predators, we consider a in eqn [2] as a function of C and E, where C is the vulnerability of the prey to the predator (opposite of the antipredator effort) and E is the average capture effort of the predator. Because of the assumptions of tradeoffs between antipredator effort and growth rate for the prey, and between predation rate and death rate for the predator, we also assume r as a function of C, and d as a function of E in eqn [2]. In addition, C and E may be evolutionary variables that increase their own fitness. Two types of dynamic models can be used to describe temporal changes in trait values by adaptive evolution: (1) a quantitative genetic model; and (2) adaptive dynamics. Quantitative genetic models describe intergenerational changes in the population mean trait value in proportion to the selection differential of its fitness, with respect to an individual trait value. In adaptive dynamics, populations consist of asexual clones with mutations. Here we use a quantitative genetic model as explained below.

Several models have investigated the effects of evolutionary change in either prey or predator species

Table 1 Preference indices for a predator's choice of prey

	Abundance of prey in the habitat			*Abundance of prey consumed by a predator*				
Case	N_1	N_2	N_3	X_1	X_2	X_3	*Ivlev's index* I_1	*Chesson's index* a_1
1	10	10	10	5	3	0	0.3	0.6
2	10	10		5	3		0.1	0.6
3	9	91	0	10	0	0	0.8	1
4	9	1	0	10	0	0	0.1	1

on prey–predator systems, although only some authors have considered coevolution of prey and predators. First, we introduce the effects of predator evolution on prey–predator dynamics.

The evolution of the predator's capture rate is modeled by:

$$\frac{dP}{dt} = w_P(E)P = \left(-d_1 - d_2E + \sqrt{\frac{bEN}{1+bN}}\right)P$$
$$\frac{dN}{dt} = \left(r - kN - \frac{EP}{1+bN}\right)N \qquad [3]$$
$$\frac{dE}{dt} = u\frac{\partial w_P}{\partial E} = u\left(-d_2 + \frac{1}{2E}\sqrt{\frac{bEN}{1+bN}}\right)$$

where E is the foraging effort, $wP(E)$ represents predator fitness, $d_1 + d_2E$ is the predator death rate, and u is the additive genetic variance of the foraging effort. dE/dt is proportional to the derivative of the predator's fitness with respect to the foraging effort E. We assumed a saturated numerical response $\sqrt{[bEN/(1+bN)]}$. If we assume $w_P(E) = [-d_1 - d_2E + bEN/(1+bN)]$, there is no equilibrium, because $dP/dt = -d_1P$ when $dE/dt = 0$. Therefore, a nonlinear tradeoff between foraging effort and predator death rate is needed.

Why do predators not evolve a highly efficient capture ability that would result in prey extinction? In the late 1960s, Lawrence Slobodkin explained that predators avoid overexploitation of prey because prey extinction is not beneficial to predators themselves. This is called 'prudent predation'. There is no evolutionary mechanism by which evolution of a species avoids causing the extinction or reducing the population of another species. Avoidance of overexploitation evolves because excessive foraging by a predator does not benefit that predator.

The predator death rate increases as energetic costs increase or the risk of predation by a species at a higher trophic level increases. The optimal foraging effort is obtained by $\partial wP(E)/\partial E = 0$, where $wP(E)$ is given by the first equation in model [3]. Therefore, the optimal foraging effort is $bN/4d_2^2(1+bN)$. The optimal foraging effort decreases as d_2 increases. For a general form of the numerical response, the optimal foraging effort decreases as the number of enemies increases. In contrast, the optimal foraging effort can either increase or decrease with an increasing number of prey, depending on the form of the numerical response. In the case of model [3], foraging effort increases as the number of prey increases. The sensitivity of the prey density to the foraging effort is controversial.

Predator evolution has a stabilizing effect on the prey–predator dynamics (**Figure 2**). Without evolution ($u = 0$), the interior equilibrium is unstable. However, no empirical evidence matches this theoretical prediction. Even with predator evolution, a one-quarter period lag is generally obtained (**Figure 2**).

In conclusion, the evolution of antipredator traits by a prey species is more likely to cause instability than the evolution of foraging traits by the predator. When the evolution of prey traits destabilizes a prey–predator system, it produces cycles in both population size and trait values.

Prey can change antipredator traits. For the mathematically simplest case of prey–predator dynamics with prey evolution,

$$dP/dt = P[-d + bCN/(1+bCN)]$$
$$dN/dt = Nw(\hat{C}, C) = N[R + qC - kN - CP/(1+bCN)] \qquad [4]$$
$$dC/dt = v\left[\partial w(\hat{C}, C)/\partial \hat{C}\right]_{\hat{C}=C} = v[q - P/(1+bCN)]$$

where $w(\hat{C}, C)$ is the logarithmic fitness of an individual prey with an antipredator effort of $\hat{C}$ and a population average antipredator effort of C, r from model [2] is

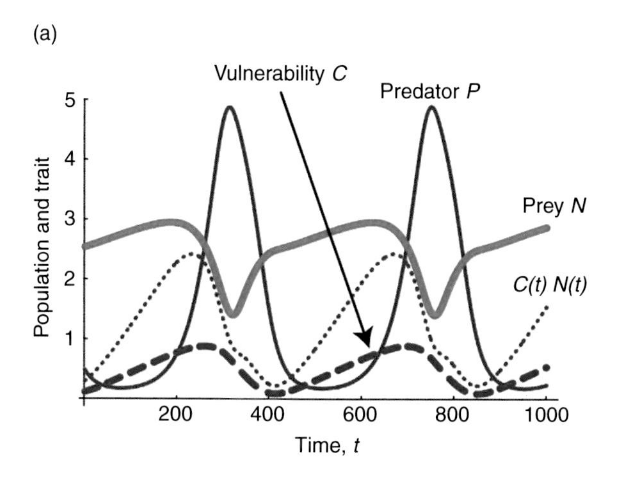

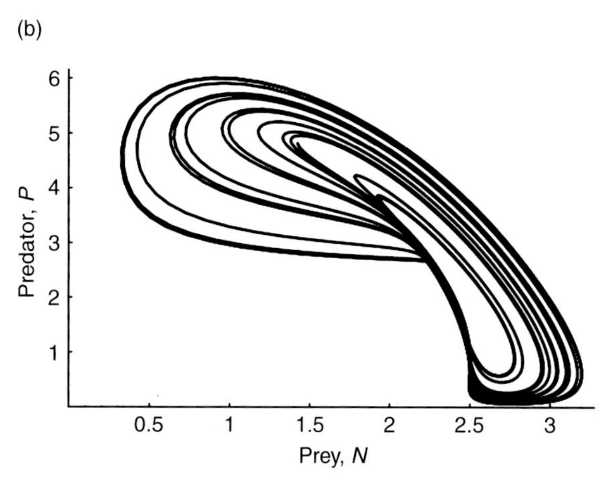

Figure 2 Prey–predator dynamics with the predator evolution of model [3]. Parameter values are: $h = 0.5$, $b = 0.654$, $d_1 = 0.18$, $d_2 = 0.18$, $r = 1$, $k = 0.3$; $u = 0.4$ in panel (a) and $u = 1.5$ in panel (b).

replaced by $R+qC$ as an increasing function of prey vulnerability (the tradeoff between the antipredator effort and the growth rate of the prey population), v represents the rate of evolutionary change in trait value C.

In this quantitative genetic model, the rate of evolutionary change in trait value is proportional to the additive genetic variance and the selection differential of an individual with a slightly higher trait value than the average. Here we simply assume that v is a positive constant, although it is possible that the additive genetic variance may change with trait values.

We have assumed that $w(\hat{C}, C)=[R+q\hat{C}-kN-\hat{C}P/(1+hCN)]$. Note that the numerator of the functional response, $\hat{C}P/(1+hCN)$, is proportional to the individual trait value $\hat{C}$, although the denominator is a function of the population average C. This is because a predator's handling time is affected by the average prey vulnerability, but each prey experiences a risk of predation dictated by its own vulnerability. This fitness therefore depends on both the population average and individual trait values. This is called frequency-dependent selection. Frequency-independent selection usually increases population abundance because no gap exists between individual optimization and group optimization. In contrast, frequency-dependent selection often decreases population abundance and can drive a population to extinction.

Model [4] exhibits effects of prey evolution on prey–predator cycles. Prey evolution often destabilizes prey–predator systems (**Figure 3a**). In addition, there is a one-half period lag between prey and predator cycles. Some empirical evidence suggests that multiple prey population clones play an important role in prey–predator cycles and in the occurrence of the one-half period lag (**Figure 1d**). This occurs because prey availability represents the number of prey individuals that are not vigilant, CN. Therefore, prey availability oscillates with a one-quarter period lead on the predator's oscillation (**Figure 3a**). Prey availability decreases with either decreasing prey density or vulnerability. In **Figure 3a**, prey density begins to decrease while population vulnerability is still increasing and when predator density is still small. Prey density decreases when the product of predator density and prey vulnerability, PC, is large. The rate of prey evolution determines the stability of the prey–predator system. A small rate of prey evolution stabilizes the equilibrium. With higher rates of prey evolution, the prey–predator system destabilizes and shows a stable limit cycle or chaos (**Figure 3b**).

To evaluate contributions of evolutionary change and population dynamics to prey–predator systems using empirical data, we examined the relationship between mean population fitness, trait values, and density; namely w in model [4]. To avoid mathematical complexity, fitness is assumed to be a function of trait value and density, or frequency-independent selection. Mean fitness of the population $w(C, N)$ changes with $\mathrm{d}w/\mathrm{d}t=(\partial w/\partial C)(\mathrm{d}C/\mathrm{d}t)+(\partial w/\partial N)(\mathrm{d}N/\mathrm{d}t)$. During empirical observations by Shertzer *et al.*, temporal changes of density and trait values were obtained. If the functional form $w(C, N)$ is given, the contribution of evolutionary change and that of population dynamics can be compared between $(\partial w/\partial C)(\mathrm{d}C/\mathrm{d}t)$ and $(\partial w/\partial N)(\mathrm{d}N/\mathrm{d}t)$. A few empirical studies have incorporated time series for both density and trait values, but very few empirical efforts have estimated the fitness function. In some chemostat systems, the magnitudes of these factors are comparable. During Darwin's fieldwork with finches in the Galapagos Islands, evolutionary change in prey size had a larger effect on fitness than did population dynamics.

The rate of prey evolution, v, is likely very slow when it is driven by genetic point mutations. The mutation rate of a single gene is usually 10^{-4} or smaller per generation per genome, and most mutants are lethal or selectively disadvantageous. However, the rate of

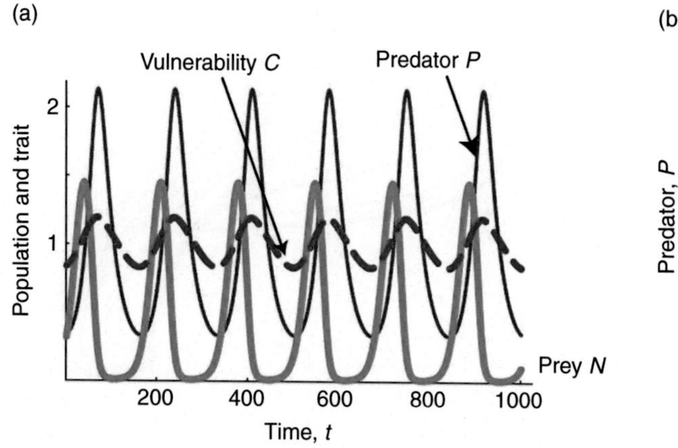

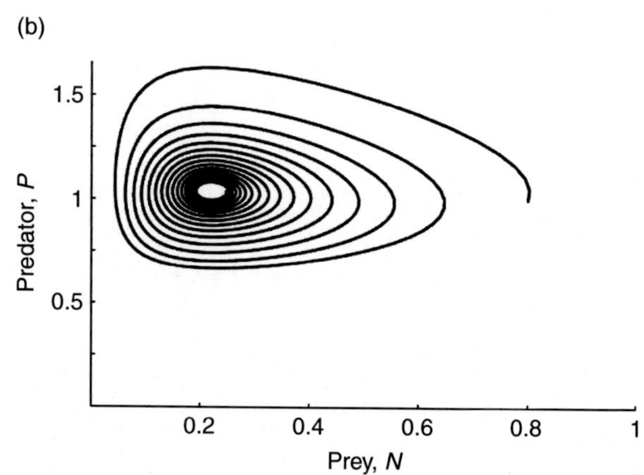

Figure 3 Prey–predator dynamics with the predator evolution of model [4]. Parameter values are: $h=1$, $b=1$, $q=0.8$, $d=0.5$, $R=2.5$, $k=1$; $v=0.05$ in panel (a) and $v=0.07$ in panel (b).

evolution is comparable to the rate of population dynamics if the population mean trait value depends on the gene frequency of allele polymorphism or a polygenic quantitative trait. In **Figure 1d**, the prey trait value varies with the gene frequency of clones. In addition, the rate of evolution is much higher than the population dynamics if the prey's behavior changes due to phenotypic plasticity of a single strategy.

Evolutionary change due to either point mutation, genetic polymorphism, or behavioral plasticity is described by model [4] for a range of values for the rate of evolution, v. Evolutionary traits change to increase individual fitness, and the velocity of change is likely proportional to the gradient of the individual fitness with respect to its trait value, $\partial w(\hat{C}, C)/\partial C$ in model [4]. **Figure 1d** suggests that evolutionary change may play an important role in the characteristics of prey–predator cycles.

Coevolution in Prey–Predator Systems

We introduced the effect of evolutionary change in the value of either predator or prey traits on prey–predator dynamics. It is definitely possible that trait values for both prey and predators will change. With a combination of prey and predator evolution as explained above, the prey–predator dynamics may cycle, whereas all four variables (population densities and mean trait values for prey and predator) may not change temporally. In some cases, either trait values or predator density do not change. As mentioned above, predator evolution often stabilizes the prey–predator system.

Coevolutionary cycles in prey and predator trait values may occur in models with fixed population sizes. Another theoretical model suggests that either one or two alternative stable equilibria are possible. In a theoretical study that included speciation in quantitative genetic traits, prey speciation likely occurred because of a reduction in the risk of predation. This occurs because disruptive selection by the predators on the prey results in prey speciation. Another theoretical study suggested that pairwise speciation of prey and predators may evolve because of prey–predator coevolution, which concurs with several empirical examples of pairwise coevolution.

Prey–predator coevolution sometimes results in the escalation of both prey and predator phenotypes. As predator capture ability increases, prey–antipredator traits all improve, producing an 'arms race'. If these traits are costly, the arms race often decreases the fitness of both prey and predators. Unlike the predictions of some theoretical works, arms races rarely result in infinitely increasing traits. Regardless, these escalations often increase the risk of extinction for both the prey and predator populations.

It is widely considered that 'adaptive' evolution of a particular trait may result in a reduction in the population size of that species. Some theoretical studies have suggested that increased capture abilities by predators can lead to a decreased predator population because of a reduction in prey density. Consider a three-trophic-level system that includes a top predator, a consumer, and a prey species. If the abundance of the top predator increases, the optimal foraging time of the consumer decreases because the benefit of food intake reaches a point of saturation and the risk of being killed by the top predator increases proportionally with foraging time. Decreased foraging time results in increased prey abundance, which consequently increases the consumer population. The top predator eventually prevents the overexploitation of the consumer.

Some evidence exists that the responses by prey to predators are often larger than the responses by predators to prey. This is referred to as the 'life–dinner principle', although some biologists have criticized this speculative expression. The principle is simply a consideration that the prey loses its life while the predator simply obtains one meal during a single act of predation.

In the case of prey–predator coevolution, prey vulnerability does not always monotonically increase with antipredator traits or foraging effort. The capture rate of prey by a predator may increase by matching the prey's phenotype. As a simple example, prey may have a bidirectional axis of vulnerability to a predator with a particular foraging behavior. The risk of predation is reduced for prey whose phenotype values are either larger or smaller than those targeted by the predator's phenotype. The 'worst' prey phenotype depends on the predator's phenotype. In the bidirectional axis of vulnerability, prey–predator coevolution may show cyclical changes in values. Prey–antipredator efforts do not always impose a cost.

Prey–Predator Systems as Food Web Components

Although we have focused on a two-species system, almost all communities have more than two species. Prey–predator interactions play a major role in food web structure.

There is some evidence to support the hypothesis that a predator's choice of prey is affected by relative prey abundance. Because of spatial heterogeneity, a predator may not encounter all possible prey species in its feeding area. For example, if species 1 and 2 exist in different areas, the predator will consume each prey in its respective area. In each area, there is a simple relationship between one prey and one predator, whereas a two-prey, one-predator system exists in the total habitat. In a similar way, if a predator consumes prey 1 in the summer and prey 2 in the winter, a simple relationship exists in each season. Therefore, it is important to discriminate temporal and spatial scales for food web structures. Population dynamics depend on the long-term structure of food webs in the overall habitat. Evolutionary changes due to genetic traits work on the same scale as population dynamics. In contrast, optimal behavior likely depends on the local and temporal conditions in the feeding area. Food web structure obtained by a long-term field survey is usually much more complicated than food webs obtained by short-term or local experiments.

One of the biggest problems in the ecology of prey–predator systems is a lack of empirical evidence. Population densities are rarely assessed from fossil records. Many empirical studies of prey–predator systems avoid using long-lived, large-bodied predators. There is a gap between the species that are used in studies of optimal behavior and studies of prey–predator systems, despite the fact that adaptive strategies depend on changes in population size, and that prey–predator dynamics depend on behavioral changes in either prey or predators. However, geographic, chemical, and genetic techniques will help to bridge the gap between evolution and ecology in empirical studies of prey–predator systems.

Further Reading

Abrams PA (1986) Adaptive responses of predators to prey and prey to predators: The failure of the arms race analogy. *Evolution* 40: 1229–1247.

Abrams PA (2000) The evolution of predator–prey interactions: Theory and evidence. *Annual Review of Ecology and Systematics* 31: 79–105.

Chesson J (1978) Measuring preference in selective predation. *Ecology* 59: 211–215.

Higashi M, Takimoto G, and Yamamura N (1999) Sympatric speciation by sexual selection. *Nature* 402: 523–526.

Matsuda H and Abrams PA (1994) Timid consumers: Self-extinction due to adaptive change in foraging and antipredator effort. *Theoretical Population Biology* 45: 76–91.

Matsuda H, Kawasaki K, Shigesada N, Teramoto E, and Ricciardi LM (1986) Switching effect on predation of the prey–predator system with three trophic levels. *Journal of Theoretical Biology* 122: 251–262.

Murdoch WW and Oaten A (1975) Predation and population stability. *Advances in Ecological Research* 9: 1–131.

Shertzer KW, Ellner SP, Fussmann GF, and Hairston NG, Jr. (2002) Predator–prey cycles in an aquatic microcosm: Testing hypotheses of mechanism. *Journal of Animal Ecology* 71: 802–815.

Yoshida T, Jones LE, Ellner SP, Fussmann GF, and Hairston NG, Jr. (2003) Rapid evolution drives ecological dynamics in a predator–prey system. *Nature* 424: 303–306.

Fungi and Their Role in the Biosphere

G M Gadd, University of Dundee, Dundee, UK

Introduction

The most important perceived environmental roles of fungi are as decomposer organisms, plant pathogens, symbionts (mycorrhizas, lichens), and in the maintenance of soil structure due to their filamentous branching growth habit and exopolymer production. However, a broader appreciation of fungi as agents of biogeochemical change is lacking, and apart from obvious connections with the carbon cycle because of their degradative abilities, they are frequently neglected in contrast to bacteria. A much wider array of metabolic capabilities are found within prokaryotes and while geochemical activities of bacteria and archaea receive considerable attention, especially in relation to carbon-limited and/or anaerobic environments, fungi are of great importance in aerobic environments. Although fungi can inhabit deep subsurface and anaerobic environments, rather less information is so far available about the biogeochemical transformations they mediate in such locations. While fungi are found in all manner of freshwater and marine ecosystems, the bulk of research has been concerned with decomposition, pathogenicity, and taxonomy. However, the significance of anaerobic and aquatic fungal communities as agents of biogeochemical change is probably limited in comparison to other microbiota. It is within the terrestrial aerobic ecosystem that fungi exert their profound influence on biogeochemical processes on the biosphere, especially when considering soil, rock, and mineral surfaces, and the plant root–soil interface (**Figure 1** and **Table 1**). For example, symbiotic mycorrhizal fungi are associated with ~80% of plant species and are responsible for major mineral transformations and redistributions of inorganic nutrients, for example, essential metals and phosphate, as well as carbon flow, while free-living fungi have major roles in decomposition of plant and other organic materials, including xenobiotics, as well as mineral transformations (**Figure 1**). Fungi are often dominant members of the soil microflora, especially in acidic environments, and may operate over a wider pH range than many heterotrophic bacteria. Fungi are also major biodeterioration agents of stone, wood, plaster, cement, and other building materials, and it is now realized that they are important components of rock-inhabiting microbial communities with significant roles in mineral dissolution and secondary mineral formation. The ubiquity and significance of lichens, a fungal growth form, as pioneer organisms in the early stages of mineral soil formation is well appreciated. The purpose of this article is to outline the important roles of fungi as biogeochemical agents in the biosphere and the significance of these processes for environmental cycling of elements on global and local scales.

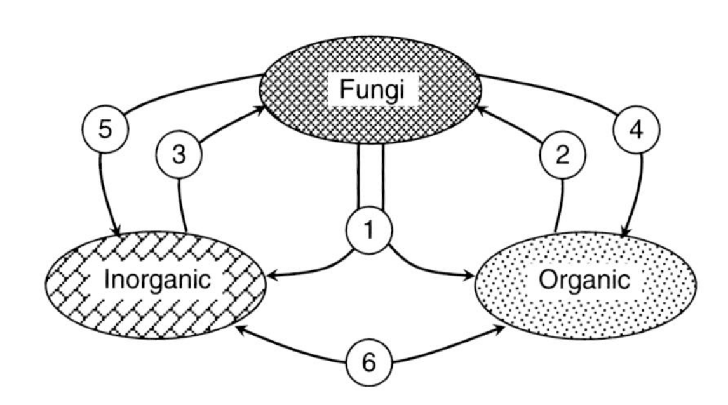

Figure 1 Simple model of fungal action on naturally occurring and/or anthropogenically derived organic and inorganic substrates. (1) Organic and inorganic transformations mediated by enzymes and metabolites, for example, H^+, CO_2, and organic acids, and physicochemical changes occurring as a result of metabolism; (2) uptake, metabolism or degradation of organic substrates; (3) uptake, accumulation, sorption, metabolism of inorganic substrates; (4) production of organic metabolites, exopolymers, and biomass; (5) production of inorganic metabolites, secondary minerals and transformed metal(loid)s; (6) chemical interactions between organic and inorganic substances, for example, complexation and chelation, which can modify bioavailability, toxicity, and mobility. Translocation phenomena may also be associated with the fungal component of this model.

Organic Matter Degradation and Biogeochemical Cycling

Most attention has been given to carbon and nitrogen cycles, and the ability of fungi to utilize a wide spectrum of organic compounds is well known. These range from simple compounds such as sugars, organic acids, and amino acids which can easily be transported into the cell to more complex molecules which are first broken down to smaller molecules by extracellular enzymes before cellular entry. Such compounds include natural substances such as cellulose, pectin, lignin, lignocellulose, chitin and starch to anthropogenic products like hydrocarbons, pesticides, and other xenobiotics.

Some fungi have remarkable degradative properties, and lignin-degrading white rot fungi, such as *Phanerochaete chrysosporium*, can degrade several xenobiotics including aromatic hydrocarbons, chlorinated organics, polychlorinated biphenyls, nitrogen-containing aromatics and many other pesticides, dyes, and xenobiotics. Such activities are of potential in bioremediation where appropriate ligninolytic fungi have been used to treat soil contaminated with substances like pentachlorophenol (PCP) and polynuclear aromatic hydrocarbons (PAHs), the latter being constituents of creosote. In many cases, xenobiotic-transforming fungi need additional utilizable carbon sources because although capable of degradation, they cannot utilize these substrates as an energy source for growth. Therefore, inexpensive utilizable lignicellulosic wastes such as corn cobs, straw, and sawdust can be used as nutrients to obtain enhanced pollutant degradation. Wood-rotting and other fungi are also receiving attention for the bleaching of dyes and industrial effluents, and the biotreatment of various agricultural wastes such as forestry, pulp and paper by-products, sugar cane bagasse, coffee pulp, sugar beet pulp, apple and tomato pulp, and cyanide.

Fungi are also important in the degradation of naturally occurring complex molecules in the soil, an environment where the hyphal mode of growth provides several advantages, and also in aquatic habitats. Since 95% of plant tissue is composed of carbon, hydrogen, oxygen, nitrogen, phosphorus, and sulfur, the decomposition activities of fungi clearly are important in relation to redistribution of these elements between organisms and environmental compartments. In addition to C, H, O, N, P, and S, another 15 elements are typically found in living plant tissues – K, Ca, Mg, B, Cl, Fe, Mn, Zn, Cu, Mo, Ni, Co, Se, Na, and Si. However, all 90 or so naturally occurring elements may be found in plants, most at low concentrations although this may be highly dependent on environmental conditions. These include Au, As, Hg, Pb, and U, and there are even

Table 1 Summary of some of the important roles and activities of fungi in biogeochemical processes

Fungal role and/or activity	*Biogeochemical consequences*
Growth	
Growth and mycelium development	Stabilization of soil structure; soil particulate aggregation; penetration of pores, fissures, and grain boundaries in rocks and minerals; mineral tunneling; biomechanical disruption of solid substrates; plant colonization and/or infection (mycorrhizas, pathogens, parasites); animal colonization and/or infection (symbiotic, pathogens, parasites); translocation of inorganic and organic nutrients; assisted redistribution of bacteria; production of exopolymeric substances (serve as nutrient resource for other organisms); water retention and translocation; surfaces for bacterial growth, transport, and migration; cord formation (enhanced nutrient translocation); mycelium acting as a N reservoir of N and/or other elements (e.g., wood decay fungi)
Metabolism	
Carbon and energy metabolism	Organic matter decomposition; cycling and/or transformations of component elements of organic compounds and biomass: C, H, O, N, P, S, metals, metalloids, radionuclides (natural and accumulated from anthropogenic sources); breakdown of polymers; altered geochemistry of local environment, e.g., changes in redox, O_2, pH; production of inorganic and organic metabolites, e.g., H^+, CO_2, organic acids, with resultant effects on the substrate; extracellular enzyme production; fossil fuel degradation; oxalate formation; metalloid methylation (e.g., As, Se); xenobiotic degradation (e.g., PAHs); organometal formation and/or degradation (note: lack of fungal decomposition in anaerobic conditions caused by waterlogging can lead to organic soil formation, e.g., peat)
Inorganic nutrition	Altered distribution and cycling of inorganic nutrient species, e.g., N, S, P, essential and inessential metals, by transport and accumulation; transformation and incorporation of inorganic elements into macromolecules; alterations in oxidation state; metal(loid) oxido-reductions; heterotrophic nitrification; siderophore production for Fe(III) capture; translocation of N, P, Ca, Mg, Na, K through mycelium and/or to plant hosts; water transport to and from plant hosts; metalloid oxyanion transport and accumulation; degradation of organic and inorganic sulfur compounds
Mineral dissolution	Rock and mineral deterioration and bioweathering including carbonates, silicates, phosphates, and sulfides; bioleaching of metals and other components; MnO_2 reduction; element redistributions including transfer from terrestrial to aquatic systems; altered bioavailability of, e.g., metals, P, S, Si, Al; altered plant and microbial nutrition or toxicity; early stages of mineral soil formation; deterioration of building stone, cement, plaster, concrete, etc.
Mineral formation	Element immobilization, including metals, radionuclides, C, P, and S; mycogenic carbonate formation; limestone calcrete cementation; mycogenic metal oxalate formation; metal detoxification; contribution to patinas on rocks (e.g., 'desert varnish'); soil storage of C and other elements
Physicochemical properties	
Sorption of soluble and particulate metal species	Altered metal distribution and bioavailability; metal detoxification; metal-loaded food source for invertebrates; prelude to secondary mineral formation
Exopolysaccharide production	Complexation of cations; provision of hydrated matrix for mineral formation; enhanced adherence to substrate; clay mineral binding; stabilization of soil aggregates; matrix for bacterial growth; chemical interactions of exopolysaccharide with mineral substrates
Symbiotic associations	
Mycorrhizas	Altered mobility and bioavailability of nutrient and inessential metals, N, P, S, etc.; altered C flow and transfer between plant, fungus, and rhizosphere organisms; altered plant productivity; mineral dissolution and metal and nutrient release from bound and mineral sources; altered biogeochemistry in soil–plant root region; altered microbial activity in plant root region; altered metal distributions between plant and fungus; water transport to and from the plant
Lichens	Pioneer colonization of rocks and minerals; bioweathering; mineral dissolution and/or formation; metal accumulation and redistribution; metal accumulation by dry or wet deposition, particulate entrapment; metal sorption; enrichment of C, N, etc.; early stages of mineral soil formation; development of geochemically active microbial populations; mineral dissolution by metabolites including 'lichen acids'; biophysical disruption of substrate
Insects and invertebrates	Fungal populations in gut aid degradation of plant material; invertebrates mechanically render plant residues more amenable for decomposition; cultivation of fungal gardens by certain insects (organic matter decomposition and recycling); transfer of fungi between plant hosts by insects (aiding infection and disease)

(*Continued*)

Table 1 (Continued)

Fungal role and/or activity	*Biogeochemical consequences*
Pathogenic effects	
Plant and animal pathogenicity	Plant infection and colonization; animal predation (e.g., nematodes) and infection (e.g., insects); redistribution of elements and nutrients; increased supply of organic material for decomposition; stimulation of other geochemically active microbial populations

Such activities take place in aquatic and terrestrial ecosystems, as well as in artificial and man-made systems, their relative importance depending on the populations present and physicochemical factors that affect activity. Clearly, the terrestrial environment is the main locale of fungal-mediated biogeochemical change, especially in mineral soils and the plant root zone, and on exposed rocks and mineral surfaces. There is rather a limited amount of knowledge on fungal biogeochemistry in freshwater and marine systems, sediments, and the deep subsurface. Fungal roles have been arbitrarily split into categories based on growth, organic and inorganic metabolism, physicochemical attributes, and symbiotic relationships. However, it should be noted that many, if not all, of these are inter-linked, and almost all directly or indirectly depend on the mode of fungal growth (including symbiotic relationships) and accompanying heterotrophic metabolism, in turn dependent on a utilizable carbon source for biosynthesis and energy, and other essential elements, such as N, O, P, S, and many metals, for structural and cellular components. Mineral dissolution and formation are outlined separately although these processes clearly depend on metabolic activity and growth form.

plants that accumulate relatively high concentrations of metals like Ni and Cd. In fact, plant metal concentrations may reflect environmental conditions and provide an indication of toxic metal pollution or metalliferous ores. Such plants are also receiving attention in bioremediation contexts (=phytoremediation). Animals likewise contain a plethora of elements in varying amounts. For example, the human body is mostly water and so 99% of the mass comprises oxygen, carbon, hydrogen, nitrogen, calcium, and phosphorus. However, many other elements are present in lower amounts including substances taken up as contaminants in food and water. A similar situation occurs throughout the plant, animal, and microbial world and therefore, any decomposition, degradative, and pathogenic activities of fungi must be linked to the redistribution and cycling of all these constituent elements, both on local and global scales (**Figure 2**).

Organometals (compounds with at least one metal–carbon bond) can also be attacked by fungi with the organic moieties being degraded and the metal compound undergoing changes in speciation. Degradation of organometallic compounds can be carried out by fungi, either by direct biotic action (enzymes) or by facilitating abiotic degradation, for instance, by alteration of pH and excretion of metabolites. Organotin compounds, such as tributyltin oxide and tributyltin naphthenate, may be degraded to mono- and dibutyltins by fungal action, inorganic Sn(II) being the ultimate degradation product. Organomercury compounds may be detoxified by conversion to Hg(II) by fungal organomercury lyase, the

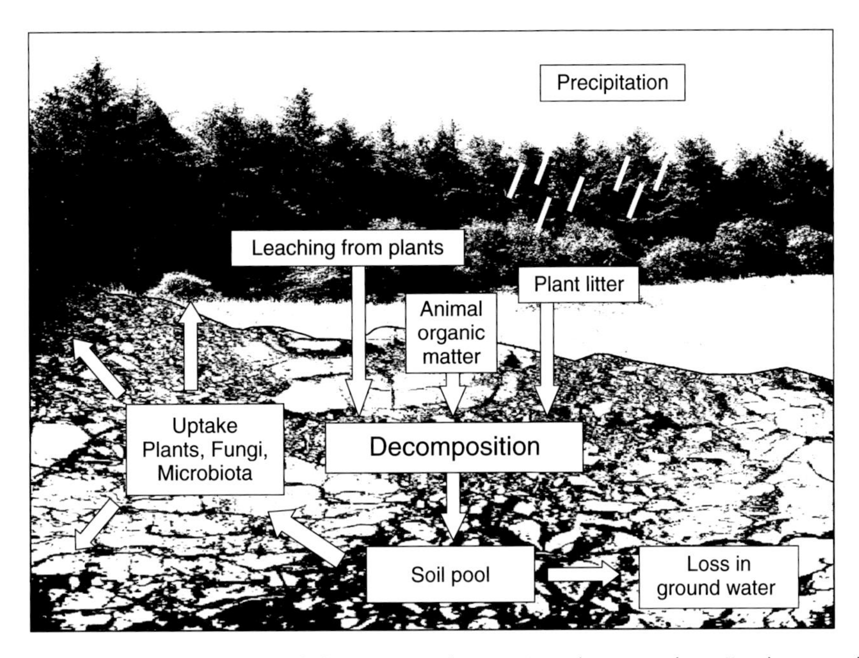

Figure 2 Simplified elemental biogeochemical cycle in a vegetated ecosystem where organic matter decomposition processes, and therefore a prime fungal role, leads to cycling of many other elements besides C. The cycle depicted could be of Ca or K for example. Organic matter could also arise from anthropogenic sources.

Hg(II) being subsequently reduced to Hg(0) by mercuric reductase, a system broadly analogous to that found in mercury-resistant bacteria.

Transformations of Rocks and Minerals

Minerals are naturally occurring inorganic solids of definite chemical composition with an ordered internal structure; rocks can be considered to be any solid mass of mineral or mineral-like material and may therefore often contain several kinds of minerals. The most common minerals are the silicates, with nonsilicates constituting less than 10% of the Earth's crust, the most common being carbonates, oxides, sulfides, and phosphates. Rocks and minerals represent a vast reservoir of elements, many of which are essential to life, and which must be released into forms that may be assimilated by the biota. These include essential metals as well as anionic nutrient species like sulfate and phosphate.

Bioweathering by Fungi

Bioweathering can be defined as the erosion, decay, and decomposition of rocks and minerals mediated by living organisms. One of the most important processes of bioweathering is weathering mediated by microorganisms, including fungi. Fungi are well suited as weathering agents since they can be highly resistant to extreme environmental conditions such as metal toxicity, UV radiation, and desiccation; they can adopt a variety of growth, metabolic and morphological strategies; they can exude protons and metal-complexing metabolites, and form mutualistic symbiotic associations with plants, algae, and cyanobacteria. Most fungi exhibit a filamentous growth habit which gives an ability to increase or decrease their surface area, to adopt either exploration or exploitation strategies. Some fungi are polymorphic occurring as filamentous mycelium and unicellular yeasts or yeast-like cells, for example, the black meristematic or microcolonial rock-dwelling fungi. The ability of fungi to translocate nutrients within the mycelial network is another important feature for exploiting heterogeneous environments.

Subaerial rock surfaces may be thought an inhospitable habitat for fungi due to moisture deficit and nutrient limitation although many species are able to deal with varying extremes in such factors as light, salinity, pH, and water potential, over considerable periods of time. Many oligotrophic fungi can scavenge nutrients from the air and rainwater which enables them to grow on rock surfaces. In the subaerial rock environment, they can also use organic and inorganic residues on mineral surfaces or within cracks and fissures, waste products of other microorganisms, decaying plants and insects, dust particles, aerosols and animal faeces as nutrient sources. Fungi may achieve protection by the presence of melanin pigments and mycosporines in their in cell wall, and by embedding colonies into mucilaginous polysaccharide slime that may entrap clay particles providing extra protection. It is likely that fungi are ubiquitous components of the microflora of all rocks and building stone and have been reported from a wide range of rock types including limestone, marble, granite, sandstone, basalt, gneiss, dolerite, and quartz, even from the most harsh environments.

The elements found in soil reflect the composition of the Earth's crust, though some modification occurs by weathering, biogenic and anthropogenic activities which on a local scale may be pronounced: chemical changes include dissolution of rock minerals while biological activity causes enrichment of C, N, and S. Elements and minerals that remain can reorganize into secondary minerals. In the soil, fungus–mineral interactions are an integral component of environmental cycling processes (**Figure 3**). Mycorrhizal fungi in particular are one of the most important ecological groups of soil fungi in terms of mineral weathering and dissolution of insoluble metal compounds. Fungi are also important components of lithobiotic communities (associations of microorganisms forming a biofilm at the mineral–microbe interface), where they interact with the substrate both geophysically and geochemically and this can result in the formation of patinas, films, varnishes, crusts, and stromatolites.

Biomechanical deterioration of rocks can occur through hyphal penetration and burrowing into decaying material and along crystal planes in, for example, calcitic and dolomitic rocks. Cleavage penetration can also occur with lichens. Spatial exploration of the environment to locate and exploit new substrates is facilitated by a range of sensory responses that determine the direction of hyphal growth. Thigmotropism (or contact guidance) is a well-known property of fungi that grow on and within solid substrates with the direction of fungal growth being influenced by grooves, ridges, and pores. However, biochemical actions are believed to be more important processes than mechanical degradation. Microbes and plants can induce chemical weathering of rocks and minerals through the excretion of, for example, H^+, organic acids and other metabolites. Such biochemical weathering of rocks can result in changes in the mineral micro-topography through pitting and etching of surfaces, and even complete dissolution of mineral grains. Fungi generally acidify their micro-environment via a number of mechanisms which include the excretion of protons via the plasma membrane proton translocating ATPase or in exchange for nutrients. They can also excrete organic acids, while respiratory activity may result in carbonic acid formation. In addition, fungi excrete a variety of other primary and secondary metabolites with metal-chelating properties (e.g., siderophores, carboxylic acids, amino acids, and phenolic compounds). The weathering of sandstone

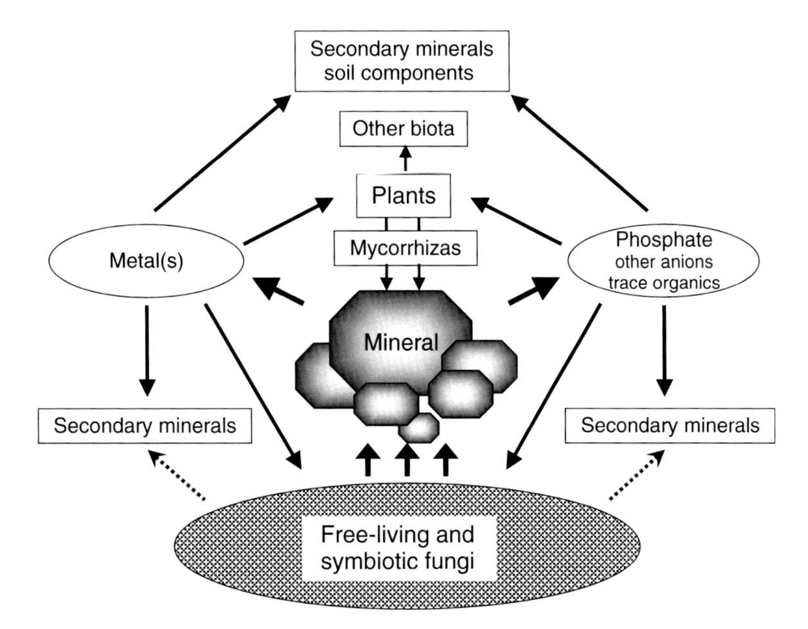

Figure 3 Action of fungi on insoluble metal minerals in the terrestrial environment resulting in release of mineral components – metal(s), anionic substances, trace organics, and other impurities – which can be taken up by the biota as well as forming secondary minerals with soil components or fungal metabolites/biomass, and also be sorbed or otherwise removed by organic and inorganic soil components. The dashed arrows imply secondary mineral formation as a result of excreted metabolites as well as fungal action on nonbiogenic minerals. Possible losses to groundwater are not shown.

monuments by fungi has been attributed to the production of, for example, acetic, oxalic, citric, formic, fumaric, glyoxylic, gluconic, succinic, and tartaric acids.

Formation of Secondary Mycogenic Minerals

The formation of secondary organic and inorganic minerals by fungi can occur through metabolism-independent and -dependent processes. Precipitation, nucleation, and deposition of crystalline material on and within cell walls are influenced by such factors like pH and wall composition. This process may be important in soil as precipitation of carbonates, phosphates, and hydroxides increases soil aggregation. Cations like Si^{4+}, Fe^{3+}, Al^{3+}, and Ca^{2+} (that may be released through dissolution mechanisms) stimulate precipitation of compounds that may act as bonding agents for soil particles. Hyphae can enmesh soil particles, alter alignment, and also release organic metabolites that enhance aggregate stability.

Carbonates

Microbial carbonate precipitation coupled with silicate weathering could provide an important sink for CO_2 in terrestrial environments. In limestone, fungi and lichens are considered to be important agents of mineral deterioration. Many near-surface limestones (calcretes), calcic and petrocalcic horizons in soils are secondarily cemented with calcite ($CaCO_3$) and whewellite (calcium oxalate monohydrate, $CaC_2O_4{\cdot}H_2O$). The presence of fungal filaments mineralized with calcite ($CaCO_3$), together with whewellite (calcium oxalate monohydrate, $CaC_2O_4{\cdot}H_2O$), has been reported in limestone and calcareous soils from a range of localities. Calcium oxalate can also be degraded to calcium carbonate, for example, in semi-arid environments, where such a process may again act to cement preexisting limestones. During decomposition of fungal hyphae, calcite crystals can act as sites of further secondary calcite precipitation. Chitin, the major component of fungal cell walls, is a substrate on which calcite will readily nucleate. Other experimental work has demonstrated fungal precipitation of secondary calcite, whewellite, and glushkinskite ($MgC_2O_4{\cdot}2H_2O$) (**Figure 4**).

Oxalates

Fungi can produce metal oxalates with a variety of different metals and metal-bearing minerals (Ca, Cd, Co, Cu, Mn, Sr, Zn, Ni, and Pb) (**Figure 4**). Calcium oxalate dihydrate (weddelite) and the more stable calcium oxalate monohydrate (whewellite) are the most common forms of oxalate associated with various ecophysiological groups of fungi. Depending on physicochemical conditions, biotic fungal calcium oxalate can exhibit a variety of crystalline forms (tetragonal, bipyramidal, plate-like, rhombohedral or needles) (**Figure 4**). Precipitation of calcium oxalate can act as a reservoir for calcium in the ecosystem and also influences phosphate availability. The formation of toxic metal oxalates may provide a mechanism whereby fungi can tolerate high concentrations of toxic metals. It has been reported that

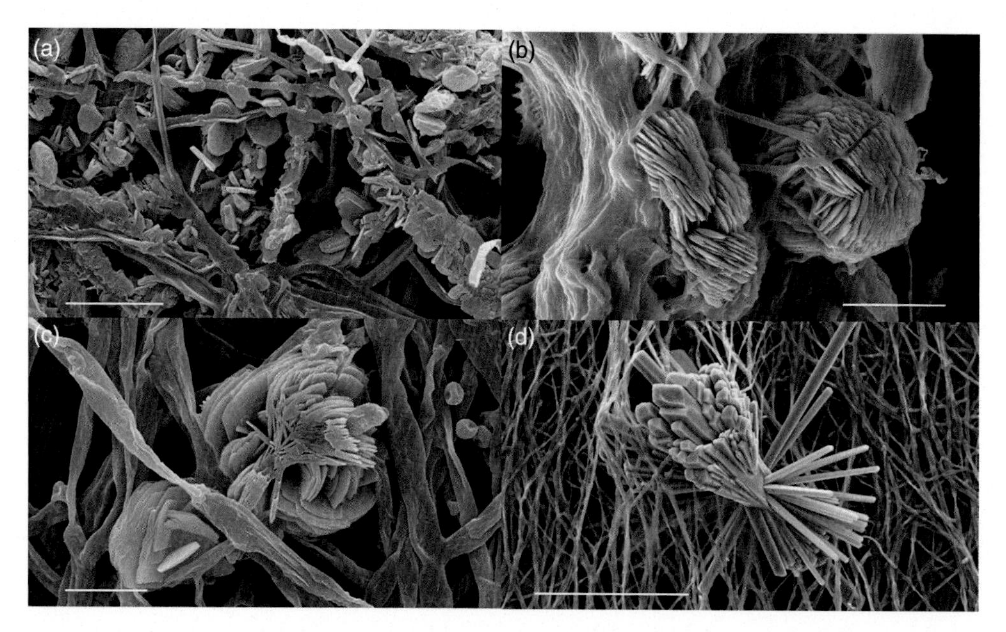

Figure 4 Mycogenic minerals associated with fungal biomass after growth in laboratory microcosms on various mineral substrates. (a) Glushinskite ($Mg(C_2O_4) \cdot 2H_2O$) and hydromagnesite ($Mg_5(CO_3)_4(OH)_2 \cdot 4H_2O$) on *Penicillium simplicissimum*. (b) Moolooite ($Cu^{2+}(C_2O_4) \cdot nH_2O$) ($n < 1$) on biomass of *Beauveria caledonica*. (c) Whewellite ($Ca(C_2O_4) \cdot H_2O$) on biomass of *Penicillium corylophilum*. (d) Strontium oxalate dihydrate ($Sr(C_2O_4) \cdot 2H_2O$) on biomass of *Serpula himantioides*. Scale bar = 20 μm (a, b); 10 μm (c); 100 μm (d).

oxalate excretion by fungi is enhanced with NO_3^- as a nitrogen source in contrast to NH_4^+, and also by the presence of HCO_3^-, Ca^{2+} and some toxic metals (e.g., Cu, Al) or minerals (e.g., pyromorphite and zinc phosphate).

Reductive and oxidative precipitation

Reduced forms of metals and metalloids (e.g., elemental silver, selenium, and tellurium) within and around fungal cells can be precipitated by many fungi. The reductive ability of fungi is manifest by black coloration of fungal colonies precipitating elemental Ag or Te, or red coloration for those precipitating elemental Se. An oxidized metal layer (patina) a few millimeters thick found on rocks and in soils of arid and semi-arid regions, called desert varnish, is also believed to be of microbial origin with some proposed fungal involvement. Fungi can oxidize manganese and iron in metal-bearing minerals such as siderite ($FeCO_3$) and rhodochrosite ($MnCO_3$) and precipitate them as oxides and also form dark Fe(II)- and Mn(II)- patinas on glass surfaces.

Other mycogenic minerals

A specific combination of biotic and abiotic factors can lead to the deposition of a variety of other secondary minerals associated with fungi, for example, birnessite, MnO and FeO, ferrihydrite, iron gluconate, calcium formate, forsterite, goethite, halloysite, hydrocerussite, todorokite, moolooite, and montmorillonite. Precipitation immobilizes metals in the soil environment and therefore limits bioavailability.

Metal and Metalloid Transformations

Fungi can transform metals, metalloids (elements with properties intermediate between those of metals and nonmetals, e.g., arsenic, selenium, and tellurium), and organometallic compounds by reduction, methylation, and dealkylation, again processes of environmental importance since transformation of a metal(loid) may modify its mobility and toxicity. For example, methylated selenium derivatives are volatile and less toxic than inorganic forms while reduction of metalloid oxyanions, such as selenite or tellurite to amorphous elemental selenium or tellurium respectively, results in immobilization and detoxification. The mechanisms by which fungi (and other microorganisms) effect changes in metal speciation and mobility are survival determinants but also components of biogeochemical cycles for metals, and many other associated elements including carbon, nitrogen, sulfur, and phosphorus.

Metals and their compounds interact with fungi in various ways depending on the metal species, organism and environment, while fungal metabolism also influences metal speciation and mobility. Many metals are essential for life, for example, Na, K, Cu, Zn, Co, Ca, Mg, Mn, and Fe, but all can exert toxicity when present above certain threshold concentrations. Other metals, for example, Cs, Al, Cd, Hg, and Pb, have no known biological function but all can be accumulated by fungi. Metal toxicity is affected by environmental conditions and the chemical behavior of the particular metal species in question. Despite apparent toxicity, many fungi survive, grow, and flourish in apparently metal-polluted locations and a variety of mechanisms, both active and incidental, contribute to tolerance. Fungi have many properties which influence metal toxicity including the production of metal-binding proteins, organic and inorganic precipitation, active transport and intracellular compartmentalization, while major constituents of fungal cell walls, for example, chitin and melanin, have significant metal-binding abilities.

Metal Mobilization

Metal mobilization from rocks, minerals, soil, and other substrates can be achieved by protonolysis, respiratory carbon dioxide resulting in carbonic acid formation, chelation by excreted metabolites and Fe(III)-binding siderophores, and methylation which can result in volatilization. In addition, other excreted metabolites with metal-complexing properties, for example, amino acids, phenolic compounds, and organic acids, may also be involved. Fungal-derived carboxylic acids can play an integral role in chemical attack of mineral surfaces and these provide a source of protons as well as a metal-chelating anion. Oxalic acid can act as a leaching agent for those metals that form soluble oxalate complexes, including Al and Fe. Solubilization phenomena can also have consequences for mobilization of metals from toxic metal containing minerals, for example, pyromorphite ($Pb_5(PO_4)_3Cl$), contaminated soil, and other solid wastes. Fungi can also mobilize metals and attack mineral surfaces by redox processes: Fe(III) and Mn(IV) solubility is increased by reduction to Fe(II) and Mn(II), respectively. Reduction of Hg(II) to volatile elemental Hg(0) can also be mediated by fungi.

The removal of metals from industrial wastes and by-products, low-grade ores and metal-bearing minerals by fungal 'heterotrophic leaching' is relevant to metal recovery and recycling and/or bioremediation of contaminated solid wastes. Although fungi need a source of carbon and aeration, they can solubilize metals at higher pH values than thiobacilli and so could perhaps become important where leaching with such bacteria is not possible and in bioreactors. Leaching of metals with fungi can be effective although a high level of organic acid production may be necessary. Other possible applications of fungal metal solubilization are the removal of unwanted phosphates, and metal recovery from scrap electronic and computer materials.

The ability of fungi, along with bacteria, to transform metalloids has been utilized successfully in the bioremediation of contaminated land and water. Selenium methylation results in volatilization, a process which has been used to remove selenium from the San Joaquin Valley and Kesterson Reservoir, California, using evaporation pond management and primary pond operation.

Metal Immobilization

Fungal biomass provides a metal sink, either by metal biosorption to biomass (cell walls, pigments, and extracellular polysaccharides), intracellular accumulation, and sequestration, or by precipitation of metal compounds onto and/or around hyphae. Fungi are effective biosorbents for a variety of metals, including Ni, Zn, Ag, Cu, Cd, and Pb and this can be an important passive process in both living and dead biomass. The presence of chitin, and pigments like melanin, strongly influences the ability of fungi to act as sorbents. In a biotechnological context, fungi and their by-products have received considerable attention as biosorbent materials for metals and radionuclides.

Fungi can precipitate several inorganic and organic compounds, for example, oxalates, oxides, and carbonates and this can lead to formation of biogenic minerals (mycogenic precipitates) as discussed previously. Precipitation, including crystallization, immobilizes metals but also leads to release of nutrients like sulfate and phosphate.

Fungal Symbioses in Mineral Transformations

One of the most remarkable adaptations of fungi for exploitation of the terrestrial environment is their ability to form mutualistic partnerships with plants (mycorrhizas) and algae or cyanobacteria (lichens). Symbiotic fungi are provided with carbon by the photosynthetic partners (photobionts), while the fungi may protect the symbiosis from harsh environmental conditions (e.g., desiccation and metal toxicity), increase the absorptive area, and provide increased access to mineral nutrients.

Lichens

Lichens are really fungi that exist in facultative or obligate symbioses with one or more photosynthesizing partners, and play an important role in many biogeochemical processes. The symbiotic lichen association with algae and/or cyanobacteria, where photosynthetic symbionts provide a source of carbon and surface protection from light and irradiation, is one of the most successful means for fungi to survive in extreme subaerial environments. Lichens are pioneer colonizers of fresh rock outcrops, and were possibly one of the earliest life forms. The lichen symbiosis formed between the fungal partner (mycobiont) and the photosynthesizing partner (algal or cyanobacterial photobiont) enables lichens to grow in practically all surface terrestrial environments: an estimated 6% of the Earth's land surface is covered by lichen-dominated vegetation. Globally, lichens play an important role in the retention and distribution of nutrient (e.g., C and N) and trace elements, in soil formation, and in rock weathering. Lichens can readily accumulate metals such as lead (Pb), copper (Cu), and others of environmental concern, including radionuclides, and also form a variety of metal-organic biominerals, especially during growth on metal-rich substrates. On copper sulfide bearing rocks, precipitation of copper oxalate (moolooite) can occur within the lichen thallus.

Mycorrhizas

Nearly all land plants depend on symbiotic mycorrhizal fungi. Two main types of mycorrhizas include endomycorrhizas where the fungus colonizes the interior of host plant root cells (e.g., ericoid and arbuscular mycorrhizas) and ectomycorrhizas where the fungus is located outside the plant root cells. Mycorrhizal fungi are involved in proton- and ligand-promoted metal mobilization from mineral sources, metal immobilization within biomass, and extracellular precipitation of mycogenic metal oxalates.

Biogeochemical activities of mycorrhizal fungi lead to changes in the physicochemical characteristics of the root environment and enhanced weathering of soil minerals resulting in metal cation release. It has been shown that ectomycorrhizal mycelia may respond to the presence of different soil silicate and phosphate minerals (apatite, quartz, potassium feldspar) by regulating their growth and activity, for example, colonization, carbon allocation, and substrate acidification.

During their growth, mycorrhizal fungi often excrete low molecular weight carboxylic acids (e.g., malic, succinic, gluconic, and oxalic) contributing to the process of 'heterotrophic leaching'. In podzol E horizons under European coniferous forests, the weathering of hornblendes, feldspars, and granitic bedrock has been attributed to oxalic, citric, succinic, formic, and malic acid excretion by ectomycorrhizal hyphae. Ectomycorrrhizal hyphal tips could produce micro- to millimolar concentrations of these organic acids. Ectomycorrhizal fungi (*Suillus granulatus* and *Paxillus involutus*) can release elements from apatite and wood ash (K, Ca, Ti, Mn, and Pb) and accumulate them in the mycelia.

Ericoid mycorrhizal and ectomycorrhizal fungi can dissolve a variety of cadmium-, copper-, zinc-, and lead-bearing minerals including metal phosphates. Mobilization of phosphorus is generally regarded as one of the most important functions of mycorrhizal fungi. An experimental study of zinc phosphate dissolution by the ectomycorrhizal association of *Paxillus involutus* with Scots pine (*Pinus sylvestris*) demonstrated that phosphate mineral dissolution, phosphorus acquisition, and zinc accumulation by the plant depended on the mycorrhizal status of the pines, the zinc tolerance of the fungal strain and the phosphorus status of the environment.

Concluding Remarks

Fungal populations are intimately involved in biogeochemical transformations at local and global scales, such transformations occurring in aquatic and terrestrial habitats. Within terrestrial aerobic ecosystems, fungi may exert an especially profound influence on biogeochemical processes, especially when considering soil, rock, and mineral surfaces, and the plant root–soil interface. Of special significance in this regard are lichens and mycorrhizas. Key processes include organic matter decomposition and element cycling, rock and mineral transformations, bioweathering, metal and metalloid transformations, and formation of mycogenic minerals. Some fungal transformations have beneficial applications in environmental biotechnology, for example, in metal leaching, recovery and detoxification, and xenobiotic and organic pollutant degradation. They may also result in adverse effects when these processes are associated with the degradation of foodstuffs, natural products, and building materials, including wood, stone, and concrete.

Acknowledgments

The author gratefully acknowledges financial support from the Biotechnology and Biological Sciences Research Council, the Natural Environment Research Council, and British Nuclear Fuels plc. Thanks also to Euan Burford and Marina Fomina for the fungal biomineral images.

See also: Calcium Cycle; Carbon Cycle; Climate Change 1: Short-Term Dynamics; Climate Change 2: Long-Term Dynamics; Material and Metal Ecology; Matter and Matter Flows in the Biosphere; Microbial Cycles; Nitrogen Cycle; Phosphorus Cycle; Radionuclides: Their Biochemical Cycles and the Impacts on the Biosphere; Sulphur Cycle; Trace Elements; Weathering; Xenobiotics Cycles.

Further Reading

Burford EP, Fomina M, and Gadd GM (2003) Fungal involvement in bioweathering and biotransformation of rocks and minerals. *Mineralogical Magazine* 67: 1127–1155.
Burford EP, Kierans M, and Gadd GM (2003) Geomycology: Fungal growth in mineral substrata. *Mycologist* 17: 98–107.
Fomina M, Burford EP, and Gadd GM (2005) Toxic metals and fungal communities. In: Dighton J, White JF, and Oudemans P (eds.) *The Fungal Community. Its Organization and Role in the Ecosystem*, pp. 733–758. Boca Raton, FL: CRC Press.
Frankland JC, Magan N, and Gadd GM (eds.) (1996) *Fungi and Environmental Change*. Cambridge: Cambridge University Press.
Gadd GM (1993) Interactions of fungi with toxic metals. *New Phytologist* 124: 25–60.
Gadd GM (1993) Microbial formation and transformation of organometallic and organometalloid compounds. *FEMS Microbiology Reviews* 11: 297–316.

Gadd GM (1999) Fungal production of citric and oxalic acid: Importance in metal speciation, physiology and biogeochemical processes. *Advances in Microbial Physiology* 41: 47–92.
Gadd GM (ed.) (2001) *Fungi in Bioremediation*. Cambridge: Cambridge University Press.
Gadd GM (2004) Mycotransformation of organic and inorganic substrates. *Mycologist* 18: 60–70.
Gadd GM (2005) Microorganisms in toxic metal polluted soils. In: Buscot F and Varma A (eds.) *Microorganisms in Soils: Roles in Genesis and Functions*, pp. 325–356. Berlin: Springer.
Gadd GM (ed.) (2006) *Fungi in Biogeochemical Cycles*. Cambridge: Cambridge University Press.
Gadd GM (2007) Geomycology: Biogeochemical transformations of rocks, minerals, metals and radionuclides by fungi, bioweathering and bioremediation. *Mycological Research* 111: 3–49.
Gadd GM, Dyer P, and Watkinson S (eds.) (2007) *Fungi in the Environment*. Cambridge: Cambridge University Press.
Gadd GM, Semple K, and Lappin-Scott H (eds.) (2005) *Microorganisms in Earth Systems – Advances in Geomicrobiology*. Cambridge: Cambridge University Press.

Gaia Hypothesis

P J Boston, New Mexico Institute of Mining and Technology, Socorro, NM, USA

Introduction

The Gaia hypothesis, named after the ancient Greek goddess of Earth, posits that Earth and its biological systems behave as a huge single entity. This entity has closely controlled self-regulatory negative feedback loops that keep the conditions on the planet within boundaries that are favorable to life. Introduced in the early 1970s, the idea was conceived by chemist and inventor James E. Lovelock and biologist Lynn Margulis. This new way of looking at global ecology and evolution differs from the classical picture of ecology as a biological response to a menu of physical conditions. The idea of co-evolution of biology and the physical environment where each influences the other was suggested as early as the mid-1700s, but never as strongly as Gaia, which claims the power of biology to control the nonliving environment. More recently, the terms Gaian science or Gaian theory have become more common than the original Gaia hypothesis because of modifications in response to criticisms and expansion of our scientific understanding.

Gaia – Original Versions

In the late 1960s, James Lovelock was working for NASA on life detection methods for Mars. With his chemistry training, this experience caused him to think deeply about what makes Earth different from Mars or her other neighbors in the solar system and the role that life might be playing in those differences. The imprint that life leaves on the chemistry of our own atmosphere stood out as a significant fingerprint of Earth's ecosystems. These musings led to the formulation of the first incarnation of the Gaia hypothesis. The early notion advanced by Lovelock is summarized in his 1972 paper, "Life regulates the climate and the chemical composition of the atmosphere at an optimum for itself." Novelist William Golding, who lived near Lovelock, suggested naming the idea after the Greek goddess. This was lovely and poetic, but probably contributed to early perceptions that the concept was cultic or New Age, not scientific.

After significant initial criticism, Lovelock and Margulis realized the flaws in the initial version that laid them open to criticism. Biologist Ford Doolittle was particularly helpful in pointing out that the hypothesis as stated required foresight and planning on the part of collections of organisms toward a common goal. This appeared to be a teleological (purposeful or designed) notion that is not in keeping with the scientific view of causality.

Later, the revised formulation appeared in a number of written and oral presentations that can be paraphrased as: "The whole system of life and its material environment is self-regulating at a state comfortable for the organisms." This was eventually restated by Lovelock in 1988 in his book *The Ages of Gaia* as "Living organisms and their material environment are tightly coupled. The coupled system is a superorganism, and as it evolves there emerges a new property, the ability to self-regulate climate and chemistry." Lynn Margulis, the innovator of the endosymbiotic theory of eukaryotic cell origins, emphasizes the role of symbiosis in biology. Her statements about Gaia usually include the phrase superorganismic system.

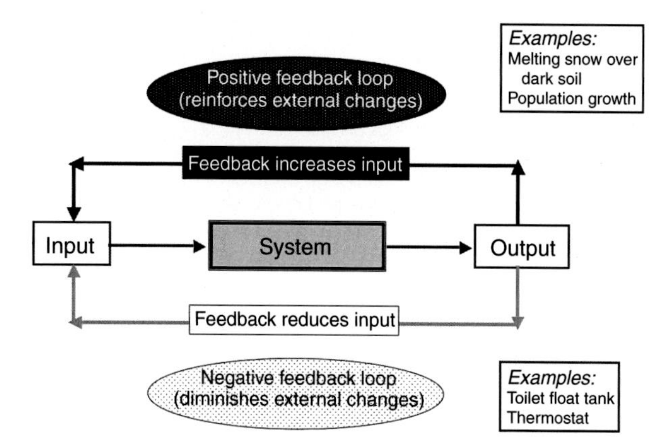

Figure 1 Logic of negative and positive feedback loops.

Table 1 Proposed Gaian-controlled parameters

- Temperature, gas balance, greenhouse feedback
- Plant-albedo feedback (e.g., Daisyworld-like mechanisms)
- Evapotranspiration, latent heat, climate feedback
- Photosynthetic manipulation of air composition
- Dimethyl sulfide (DMS), marine cloud, algae association
- Microbial respiration rates and the carbon cycle
- Methanogenesis and greenhouse warming
- Carbon dioxide levels and carbonate cycle
- Carbonate-shelled organisms as long-term carbon sink
- Continental weathering rates via lichen, other microorganisms, etc.
- Oxygen levels, biomass burning feedback
- Ocean salinity levels

In her view, evolution is the result of cooperation, not competition, and this is in keeping with the Gaian interpretation of global ecology.

The initial conception involved the idea of homeostasis, that is, regulation around a narrow range of physical variables and resistance to perturbation via cybernetic feedback loops. However, Margulis particularly argued that Gaian systems are rather homeorhetic, meaning that the Earth's atmosphere, hydrosphere, and lithosphere are regulated around set points that can change in time as the whole system evolves essentially through a life cycle. The basic logic of negative and positive feedback loops is illustrated in **Figure 1**.

Possible Evidence of Gaian Mechanisms

Biogeochemistry and Gaia

The chemical disequilibrium of the Earth's atmosphere is the feature that first captured Lovelock's attention. He noticed that on Venus and Mars (planets apparently with no life at least on the surface) the atmospheres are primarily CO_2. On Earth, the dominant constituents are reactive species of nitrogen, oxygen, and minor constituents(methane, ammonia, nitrous oxide, etc.). In the absence of other factors, over time, one would predict that Earth would resemble her neighbor planets but she does not. According to Gaia, life is the factor that maintains this disequilibrium over time. **Table 1** shows various parameters that have been suggested as possibly Gaia-controlled. The chemical compounds involved and their various reactions and fluxes control the large-scale biogeochemical cycles that enable Earth to constantly recycle materials and make them available for succeeding generations of life.

Bioweathering and Gaia

In 1989, Tyler Volk and Dave Schwartzman showed convincingly that the rock weathering rate increased by three orders of magnitude in the presence of life compared to the lifeless case. Gaian proponents viewed this as a major piece of supporting evidence in their contention that CO_2 effects on climate, known to be very powerful, could be significantly affected, even controlled, by the biologically enhanced rate of weathering. Lovelock has said of their work "This is much more than is needed to enable a powerful physiological regulation of climate and carbon dioxide. We think it could account for the 300-fold decline in carbon dioxide since life began on Earth."

A Controversial Idea from the Beginning

Early Criticisms

Since its inception, the Gaia hypothesis has been controversial. For a few years, it was simply ignored. Further papers and presentations caught attention and the notion was widely castigated. It was criticized as being merely a restatement of ideas that already had a long history, as early as the work of James Hutton (1727–97), the founder of modern geoscience, who suggested that the study of Earth should be considered geophysiology. Further, Earth viewed as a single entity is in conflict with the fundamental ecological ideas of organisms engaged in Darwinian competition and narrowly defined survival and reproductive success. It has also been pointed out that maybe we do not need to invoke Gaia because proposed geochemical mechanisms can adequately explain many aspects of the Earth system without biological processes. Besides genuine weaknesses in the arguments, initial negative reactions to Gaia may in part be blamed on the lack of common language between Earth sciences and biology in the early 1970s.

Kirchner's Formulations

The best critical analysis of Gaian ideas was done by Jim Kirchner, at a Chapman Conference (American Geophysical Union) in San Diego in 1988 that was

Table 2 The many types of Gaia according to Kirchner[a]

Hypothesis type	*Properties*	*Likely consensus*
Influential	Biology exerts significant influence over some aspects of the planetary system	Testable and supported by evidence
Co-evolutionary	Darwinian process in which biota affects nonliving systems, in turn they affect biota	Testable and under active debate
Homeostatic	System is stabilized by negative feedback loops involving biota and physical/chemical systems	Testable and under active debate
Teleological	Conditions maintained by the biosphere for its own benefit	Testable, refuted by the Daisyworld demonstration
Optimizing	The biosphere directly manipulates its environment to provide optimum conditions for itself	Skeptically received, possibly not testable, not self-consistent

[a]Strength of statement in order from highest (influential) to lowest (optimizing).

devoted to the scientific consideration of Gaia. He separated the jumble of ideas into four clear levels in order of increasing strength of claims from weak to strong (**Table 2**). These ranged from (1) 'co-evolutionary Gaia' that merely claimed life and Earth had evolved together over time affecting each other; (2) 'homeostatic Gaia' involving self-regulation around set points; (3) 'geophysical Gaia', which overlapped significantly with the physical Earth sciences; and (4) the most extreme claim of 'optimizing Gaia' that life was molding the planet's behavior into a state most favorable toward all life. The latter notion came in for the most criticism as being scientifically untestable and the most radical Gaian idea.

Further General Criticisms

Our planet has a long and dynamic history. How narrowly can Gaia be said to have constrained conditions? As we learn more about Earth's history, it is clear that huge changes have occurred in the climate, position of land masses, ocean currents, and other global-scale properties. For example, several times in the planet's history, we believe that it has been largely covered with ice. During the Mesozoic period, it appears that the planet was much warmer than it has been since. The atmosphere has evolved from anaerobic to a high level of free oxygen and many other major chemical changes have occurred. Against the dramatic backdrop of these changes, it is hard to claim that Gaia has held conditions constant and the window of variability seems very large even to qualify as homeorhesis.

Gaia as an organism has foundered on another point. Organisms reproduce. How can an entity the size of a whole planet reproduce? Gaia has not yet done so, but it has been suggested by some that space colonization may be the biosphere's first attempt to reproduce itself on other planetary bodies. The notion of Earth as superorganism may be specious and not central to the idea of global homeorhesis; thus, this may be a fairly trivial semantic criticism.

Because the notion has evoked visions of Gaia as the 'mother goddess', as a benign entity and protector of life, it is appealing to people outside the scientific community. This has also been another point of attack for its critics, who view it as overly romanticized, more philosophical in nature, and scientifically untestable, thus, not of value in the strict scientific sense. If Gaia is the mother goddess, then her first-born were probably bacteria-like organisms who would be poisoned by our current oxygen-containing atmosphere. Her enormous family includes countless species whose individual needs and welfare conflict with each other. This has resulted in the natural extinctions of the bulk of all species that have ever arisen. The interpretation of such a system as benign or nurturing is stretching it too far.

Specific Criticisms

There are numerous specific criticisms that have been leveled against examples that Lovelock, Margulis, and other proponents have put forth as evidence for Gaia. Only one is given here for brevity. For example, Lovelock invoked Gaia to explain life's survival during the rise of oxygen in the early Earth atmosphere. Photosynthetically produced high levels of oxygen (the so-called 'oxygen crisis') were lethal for the largely anaerobic lifeforms of the day. Ultimately, the oxygen atmosphere probably enabled a net increase in biodiversity and total biomass over time, but it spelled doom for many of the organisms then alive. If Gaia favors life, how can changes that favor some life but destroy large numbers of other organisms be reconciled? How can a mechanism be proposed that would amount to altruistic suicide on the part of many organisms on behalf of unrelated organisms? Kirchner summed up the dilemma in his 1989 paper, "If the most destabilizing biotic event in Earth's history can be construed as evidence for Gaia, and the relative stability since then can also be cited as evidence for Gaia, one wonders what conceivable events could not be interpreted as supporting the Gaia hypothesis. If there are none, Gaia cannot be tested against the geologic record.... If Gaia stabilizes and Gaia destabilizes... is there any possible behavior which is not Gaian?"

If Gaia cannot be disproved in any case, then it does not meet the criterion for falsifiability developed by philosopher of science, Karl Popper, in the 1930s. An idea must, in principle, be able to be proved false for it to be considered testable. Popperian falsifiability has itself been attacked, notably by Alan Sokal and Jean Bricmont in their 1998 book, *Fashionable Nonsense.* Some argue that Popperian falsifiability is already biased toward only the methodology of reductionism, and that it may be inherently unable to fully define the essence of extremely complex and closely coupled systems, especially those that change over time in some sort of ontological process. Nevertheless, it is a useful indicator of whether a concept can be considered scientifically testable at least in the narrow sense. The major result of Kirchner's criticism has resulted in attempts at crisper formulations of the ideas and has pushed Gaia to the weaker versions.

The highly reductionist geneticist Richard Dawkins believes that individual genes are in control of evolution and has entirely dismissed Gaia on that basis. Genes are grouped together into replicator packets. These are the functional units that Dawkins contends were both the first form of life and still remain the functional unit of selection. He dubs cells and organisms survival machines, and claims that they serve only to help the replicators propagate. Certainly any superorganismic concept violates Dawkins' notion of a single exclusive level of selection. Such an intensely reductionistic view does not take into account higher-order properties that may emerge from complex system interactions, and some scholars working on the mathematics of complex systems have in turn dismissed Dawkins' views as overly simplified.

Countering the reductionist view, J. Z. Young pointed out in the book *Doubt and Certainty in Science* that "Biology, like physics, has ceased to be materialist. Its basic unit is a non-material entity, namely an organization." Here the emphasis is on pattern, because matter is frequently replaced and thus, transient, in biological processes. An organism is not a particular chunk of matter, but a persistent pattern through which material flows. With such a definition, the notion of the Earth system as a superorganism becomes less strained and does not require a slavish point-by-point comparison to the properties of individual organisms.

Daisyworld

Lovelock and his collaborators' greatest effort to counter the teleological criticism and show that undirected negative feedback can result in homeostatic regulation came in the form of a computer model, Daisyworld (**Figure 2**). In the first and simplest version of the model, the imaginary planet Daisyworld has only one species of plant, white daisies. It has soil that is darker than the daisies. The star of this planet grows more and more luminous as we believe our sun to have done during the early history of the Earth. The relative abundance of daisies versus soil controls the temperature environment of the planet. The daisies have a physiological temperature window within which they are viable and reproductive. Using very simple rules, the differential albedos of soil and daisies combine to enable Daisyworld to remain habitable even as its sun is growing brighter. Of course, later versions of the model have added more biological variables and more complex physics of the environment, but the essence of the demonstration remains the same.

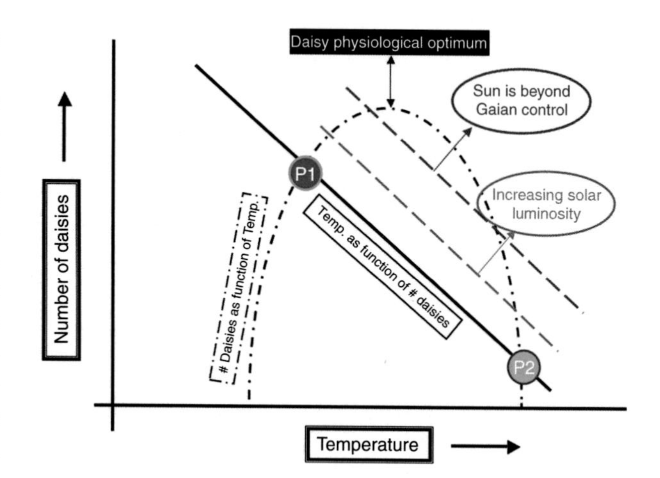

Figure 2 The Daisyworld computer model.

Where Is Gaian Science Headed?

Because of the global scale of Gaian processes, field observations of potential evidence supporting or refuting Gaia is difficult to obtain. Efforts continue sporadically to advance on this front, usually as a by-product of investigators' more mainstream activities. Possibly the most promising arena for testing Gaian ideas currently available lies in modeling and the understanding of complex systems. An early paper by Tregonning and Roberts in 1979 looked at how simple models of complex systems could develop homeostasis, and was seminal in early thinking about Gaia. Recently, the study of complex adaptive systems (CAS) has begun to advance our general understanding of the behavior of massively coupled complex systems. As this science progresses, insights applicable to testing Gaian predictions may well emerge.

See also: Carbon Cycle.

Further Reading

Charlson R, Lovelock J, Andreas M, and Warren S (1987) Oceanic phytoplankton, atmospheric sulfur, cloud albedo, and climate. *Nature* 326: 655–661.

Kirchner JW (1989) The Gaia hypothesis: Can it be tested? *Reviews of Geophysics* 27(2): 223–235.

Kirchner JW (2003) The Gaia hypothesis: Conjectures and refutations. *Climatic Change* 58(1–2): 21–45.

Lovelock J (1972) Gaia as seen through the atmosphere. *Atmospheric Environment* 6: 579–580.

Lovelock JE (1979) *Gaia: A New Look at Life on Earth*. Oxford: Oxford University Press.

Lovelock J (1983) Daisy world: A cybernetic proof of the Gaia hypothesis. *Coevolution Quarterly* 38: 66–72.

Lovelock JE (1995) *The Ages of Gaia. A Biography of Our Living Earth*, 2nd edn. Oxford: Oxford University Press.

Lovelock JE and Margulis L (1974) Atmospheric homeostasis by and for the biosphere: The Gaia hypothesis. *Tellus* 26: 2–9.

Schneider SH and Boston PJ (eds.) (1991) *Scientists on Gaia*. Cambridge, MA: MIT Press.

Schneider SH and Londer R (1984) *Coevolution of Climate and Life*. Berkeley, CA: Sierra Club Books.

Schneider SH, Miller JE, Crist E, and Boston PJ (eds.) (2004) *Scientists Debate Gaia: The Next Century*. Cambridge, MA: MIT Press.

Schwartzman DW and Volk T (1989) Biotic enhancement of weathering and the habitability of Earth. *Nature* 340: 457–460.

Tregonning K and Roberts A (1979) Complex systems which evolve towards homeostasis. *Nature* 281: 563–564.

Volk T (1998) *Gaia's Body: Toward a Physiology of Earth*. New York: Springer.

Watson AJ and Lovelock JE (1983) Biological homeostasis of the global environment: The parable of Daisyworld. *Tellus* 35B: 284–289.

Hydrosphere

Z W Kundzewicz, RCAFE Polish Academy of Sciences, Poznań, Poland

Introduction

The hydrosphere (Greek word *hydor* (υδωρ) means water), also called the water sphere, includes all water on the planet Earth. The Earth is indeed a blue planet, since the oceans cover nearly 71% of its surface, that is, over 361 million km^2, while the continents and islands – the solid surface of the Earth – make up only 29% of the total Earth area.

Water is the most widespread substance in the natural environment of our planet. It is available everywhere on Earth, albeit its abundance largely differs in space and time. Water exists on Earth in three states: liquid, solid, and gaseous (vapor). Liquid phase, being intermediary between solid and gaseous states, is dominant on Earth, in the form of oceans, seas, lakes, rivers on the ground, soil moisture, and aquifers under ground. In a solid state, water exists as ice and snow cover during winter in higher latitudes and during all the seasons in polar and alpine regions. Some amount of water is contained in the air as water vapor, water droplets, and ice crystals. Water is a constituent of the biosphere, the habitat of life, extending up to 10 km height into the high troposphere (migratory bird routes over the Himalaya) and down in the oceans to the depth of 10 km and deeper. Plant and animal tissues contain large proportion of water. Huge amounts of bonded water are present in the composition of different minerals of the Earth.

The abundance of liquid water on Earth distinctly distinguishes our unique planet from other planets in the solar system, where no liquid water can be found. The Earth is the only planet in the solar system with the right distance from the Sun, the right composition of the atmosphere, the right mass (gravity) and chemical composition, permitting water to exist in all three phases, but predominantly in the liquid form. The Venus is too warm for presence of liquid water. It is too near to the Sun and it has too strong greenhouse effect due to dense atmosphere. The Mars, being further to the Sun, is too cold. However, dendritic channels and deep canyons on Mars's surface were probably sculpted in the past by free-flowing water, when the climate was warmer.

We cannot satisfactorily explain the origin of the Earth's hydrosphere. One of the important processes was the outgassing of water vapor from the interior of the Earth, which took place as extrusion of material in volcanoes and ocean upwellings. Furthermore, the early Earth was bombarded by 'snowballs' of comets and asteroids, which were rich in water. Much of the Earth's water is likely to have originated from the outer parts of the solar system.

Water is the basic element of the life-support system of the planet, being essential for self-reproducing life. Water cannot be substituted by any other substance. By its capacity to dissolve and carry substances, water plays an essential role in the chemistry of life. Most life on the planet takes place in the saltwater of the oceans. It is estimated that the oldest life on Earth started in oceanic waters already 3.5 billion years ago. Most evolution has taken place in water. However, it is freshwater that is indispensable for much life on Earth, including the life of humans. Humans need freshwater and salt, but separately rather than together, as contained in salty water. The humans depend on regular availability of freshwater for drinking. Water is indispensable, in large quantities, virtually in every human activity, in particular in agricultural production. Water and solar radiation is the driving source behind the plants' primary productivity. Water is indispensable for plant growth. Some water is incorporated in plant tissues and much is transpired.

When looking for possibilities of extraterrestrial life, the focus is on the search for liquid water. Existence of liquid water on a celestial body now, or in the past, is interpreted as a necessary condition of life. Moreover, existence of water on other planets and moons is important in human's search for habitable places, where spacemen could live without having to bring large volumes of water with them. It is hypothesized that a hydrosphere may exist on Europa and Ganymede, two of the four large moons of the Jupiter, where the water is frozen on the surface, but may remain liquid under the surface.

Physical Properties of Water

Water, hydrogen oxide, is the simplest durable chemical compound of hydrogen and oxygen. Its molecule consists of two hydrogen atoms bonded to one oxygen atom, H_2O. Pure water is transparent and colorless, odorless, and tasteless. The range of surface temperatures and pressures on the Earth permit water (as the only substance) to exist naturally in all three states on our planet. Water in liquid state is predominant, but occurrence of water in the solid and gaseous states is also common. Water molecules undergo state changes: from liquid to gaseous phase – by evaporation (evapotranspiration); from gaseous to liquid phase – by condensation; from liquid to solid state – by freezing; and from solid to liquid state – by melting. Direct phase change between the solid and gaseous phase is also possible, in the process of sublimation.

Due to its molecular structure, water is a unique substance in that no other substance has similar physical and electrochemical properties. Understanding the physical properties of water is indispensable to interpret the functions of the hydrosphere. A summary of basic physical characteristics of water is compiled in **Table 1**.

The molecular polarity and the dipole structure of a water particle are responsible for the high surface tension and the solvent properties. Liquid water has a tetrahedral structure, which breaks down in the process of evaporation. There is a change in the arrangement of molecules corresponding to phase changes. When freezing, water molecules arrange themselves in such a way that water expands its volume (by about 9% for rapid freezing), becoming lighter than liquid water. Water is the only known substance where the maximum density does not occur in the solid state. The water density attains its maximum in the liquid phase, at 4 °C. Water becomes lighter when warming up above 4 °C or cooling down below this temperature.

There are further unique properties of water, making it an astonishing substance, which behaves in an anomalous way. These features play a crucial role in many processes in the geosphere and biosphere. The liquid water is an excellent, and universal, solvent, able to

Table 1 Physical characteristics of water for three temperatures (0, 20, and 100 °C)

Temperature (°C)	*Surface tension (erg cm^{-2})*	*Dynamic viscosity (kg m^{-1} s^{-1})*	*Saturation vapor pressure (kPa)*	*Density (g cm^{-3})*
0	75.6	0.001 792	0.611	Liquid 0.9999 Solid 0.9150
20	72.8	0.001 003	2.339	Liquid 0.9982
100	61.5	0.000 282	101.3	Liquid 0.9584 Gaseous 0.0006

At the temperature of 4 °C, the maximum water density of 1.0 g cm^{-3} is observed.
Temperature of melting: 0 °C (at pressure of 1013 hPa).
Temperature of boiling: 100 °C (at pressure of 1013 hPa).
Specific heat: 1.000 cal g^{-1} °C^{-1} (at 14.5 °C).
Latent heat of melting 79.7 cal g^{-1}(at 0 °C).
Latent heat of evaporation (vaporization) 597.3 cal g^{-1} (at 100 °C).
Latent heat of sublimation 677.0 cal g^{-1} (at 0 °C).
Water is a poor conductor of electricity and its compressibility is very low.

dissolve many chemical compounds, for example, mineral salts. Having a neutral pH (i.e., being neither acidic nor basic) in a pure state, water changes its pH when dissolving substances, being slightly acidic in rain (due to dissolution of carbon dioxide and sulfur dioxide, present in the air). Some 97.5% of all water on Earth is salty oceanic waters, containing dissolved natrium chloride (NaCl), with concentration of 33–37 g kg^{-1}. Water on the move carries dissolved and particulate substances (e.g., in hydrological processes of precipitation, runoff and river flow, infiltration and groundwater flow). The liquid water is adhesive and elastic due to high surface tension, which counteracts the downward pull of the gravity force. It aggregates in drops rather than spreading out as a thin film over a surface. Water conducts heat easier than any liquid, with the exception of mercury. Water has a high specific heat (higher than other liquids, except liquid ammonia), and a high latent heat of freezing/melting and evaporation/condensation. All these features play a significant role in the heat exchange processes in the Earth's system.

Hydrosphere in the Earth System

The hydrosphere is interconnected with all the other 'spheres' in the Earth system, that is, the geosphere (lithosphere and atmosphere), biosphere, and human-related anthroposphere (which includes technosphere). A temporarily immobilized part of the hydrosphere – ice and snow – is sometimes called the cryosphere, while the domain of salty water is sometimes called the oceanosphere.

Water is abundant in all the 'spheres' of the Earth system, in liquid, solid, and vapor states. There is water stored over the Earth's surface and in the atmosphere. There is abundance of water on the Earth's surface (hydrosphere: oceans and seas, polar ice, lakes, rivers and streams, wetlands and marshes, snow pack and glaciers; containing liquid and solid water) and in the lithosphere (solid Earth), under the Earth's surface (in the rocks and soil, including permafrost, and deeper in the ground, down to the Earth crust – in liquid, solid, and gaseous phases), and in the biosphere (in plants and animals). The water is on a perpetual move; it partakes in processes of exchange of mass and energy between the various spheres of the Earth system. The main, in volumetric terms, water transfer takes place between the hydrosphere and the atmosphere in processes of evaporation and precipitation. The evaporation process purifies (distills) salty oceanic water into freshwater. Water moves not only in the processes of evaporation, precipitation, and infiltration, or flow in rivers and streams, plants and animals, but also in oceans, seas, and lakes, in snow pack, and in even seemingly immobile glaciers.

The total volume of water in the hydrosphere is nearly constant over a longer timescale, with negligible changes due to gain of juvenile water through emergence of vapor via volcanic eruptions or seepage in sea floor, and loss of water bonded and buried in crustal sediments and by chemical breakdown (hydrolysis).

Presence of water, and its movement, is responsible for the chemical and mechanical breakdown of rocks in the lithosphere. Successive freezings of water (with volumetric expansion), and thawings, crack even the most durable rocks. In this process, called weathering, the rocks are partitioned into smaller pieces, and finally into stone, gravel, sand, and soil. Erosion induced by precipitation falling on the Earth's surface is responsible for sculpting the surface of the Earth. Geomorphological processes induced by running water form stream channels. Water transports the solid material to surface water bodies (rivers, lakes), seas, and oceans. Global sediment fluxes are very high, and may amount to 9.3–64.0 Gt yr^{-1}, depending on the source of estimates.

Water plays an essential role in the functioning of the biosphere. The solvent properties of water are indispensable in the life processes of transport of nutrients in organisms. In result of the water transfer between the hydrosphere and the biosphere, plants take water (with dissolved nutrients) from the soil. High surface tension of water explains the capillary movement of water (carrying nutrients) from the ground, in plants from their roots through their vascular system to stems and leaves. Further, there is water transfer from plants to the atmosphere via the leaf surfaces in a process called transpiration, which is of critical importance for the thermoregulation.

It is indeed a paradoxical property that warm water (4 °C) in lakes, ponds, and rivers is located near the bottom, under ice cover (lighter than water), which separates a warmer water body from a much colder environment. The existence of the ice cover prevents many lakes and rivers from freezing to the bottom during cold winters (with disastrous consequences to aquatic life).

The term 'anthroposphere' relates to the existence of 6.5 billion active human beings populating the Earth, and the human-created technosphere. Man has significantly impacted the water cycle on Earth, in both quantity and quality aspects, in particular through the water withdrawal for agricultural irrigation and industrial processes, including energy production. Man is responsible for widespread contamination of surface and ground waters.

Water and Climate

Climate and water on the planet Earth are closely linked. Water takes part in a large-scale exchange of mass and heat between the atmosphere, the ocean, and the land surface, thus influencing the climate, and also being influenced by the climate.

In the history of Earth's climate, there were time periods when much of the hydrosphere on the surface of the planet was in the solid form of glacial ice. Possibly, during the Cryogenian period, the range of sea ice extended nearly to the equator. There have been several ice ages in the history of the Earth, and the most recent retreat of glaciation is dated at some 10 000 years ago. Range and extent of ice sheets, glacier, and permanent snow areas remain a sensitive indicator of changes in the Earth's climate. After expansion during the Little Ice Age, they have been shrinking recently in response to the ongoing global warming.

Under normal pressure, water exists as a liquid over a large range of temperature from 0 to 100 °C; hence, water remains as a liquid in most places on the Earth. Because water has a high specific heat (heat capacity) defined as the amount of energy required to increase the temperature of 1 g of a substance by 1 °C, a water body can absorb (or release) large amounts of heat when warming (or cooling). This large hidden energy is released in the atmosphere when water vapor condenses. Latent heat (water vapor) transport is a major component of the Earth's heat balance. Some 23% of the solar radiation that reaches the Earth is used for evaporating water. Solar engine lifts about 500 000 km^3 of water a year, evaporating from the Earth's surface, therein 86% (430 000 km^3) from the ocean and 14% (70 000 km^3) from land.

Water plays a pivotal role in the redistribution of heat in the Earth's atmosphere, and in the Earth's thermal system. Due to high specific and latent heat, water moderates the Earth's climate, acting as air-conditioner in the Earth system. Most (1.338 billion km^3, i.e., 96.5% of all the Earth's waters) is contained in the oceans and the very high heat capacity of this large volume of water buffers the Earth surface from strong temperature changes such as those occurring on the waterless Moon. Ocean acts as the principal heat storage component in the Earth system, a regulating flywheel in the Earth's heat engine. The principal characteristics that affect density and motion (currents) of ocean's water are its temperature and salinity. Since warm water is less dense (lighter) than cold water and salty water is heavier (more dense) than freshwater, the combination of temperature and salinity of the oceanic water determines whether a water particle sinks to the bottom, rises to the surface, or stays at some intermediate depth. Thermohaline circulation can be interpreted as a conveyor belt of heat, responsible for the relatively mild climate of Europe. It is driven by the density of oceanic water, which, in turn, is impacted by freshwater influx to the ocean. Besides oceans and seas, surface water bodies, such as lakes, wetlands, and large rivers, also affect the local, or regional, climate and partake in temperature regulation processes. Enhanced evaporation in large water storage reservoirs is an important component of a water balance, especially in arid and semiarid areas, being a very essential part of the total water consumption in individual regions.

The hydrological cycle affects the energy budget of the Earth. Clouds alter Earth's radiation balance. Atmospheric water vapor (along with carbon dioxide and methane) is a powerful greenhouse gas, playing a significant role in the greenhouse effect. This effect, which can be described as absorbing the long-wavelength infrared radiation emitted by the Earth's surface, is responsible for maintaining the mean surface temperature about 33 °C higher than would be the case in the absence of the atmosphere. Condensation of water in clouds provides thermal energy, which drives the Earth's circulation. The atmospheric transport of water from equatorial to subtropical regions (where latent heat is released from water vapor) serves as an important mechanism for the transport of thermal energy. During 8–10 days that a water molecule resides, on average, in the atmosphere, it may travel about 1000 km.

Earth's climate has always been changing, reflecting regular shifts in its orbit and solar activity and radiation, and volcanic eruptions. However, a large part of the climate change being observed recently is due to human activity. The humankind has been carrying out a planetary-scale experiment, disturbing the natural composition of the atmosphere by increasing the contents of greenhouse gases. This takes place because of the increasing burning of fossil carbon (coal) and hydrocarbons (oil and natural gas), and large-scale deforestation (reduction of carbon sink). In consequence, carbon dioxide concentration in the Earth's atmosphere increases and the greenhouse effect becomes more intense, leading to global warming. The global mean temperature of the Earth has already visibly increased by over 0.74 °C since 1860 and further increase is projected, by up to 1.1–6.4 °C by 2100, depending on the socioeconomic (and – in consequence – carbon dioxide emission) scenarios. Apart from the warming, there are several further manifestations of climate change and its impacts, of direct importance to the hydrosphere.

Many climate-change impacts on freshwater resources have already been observed, and further (and more pronounced) impacts have been projected. There is a poleward shift of the belt of higher precipitation. Increased midsummer dryness in continental interiors has been observed. The effect of climate change on streamflow, lake levels, and groundwater recharge, which varies regionally, largely follows changes in the most important driver, precipitation. Effects of future climate change on average annual river runoff across the world in contemporary projections indicate increases in high latitudes and the wet tropics, and decreases in mid-latitudes and some parts of the dry tropics. The latter translates into lower water availability (lower river flows and stages, lake and groundwater levels, and soil moisture contents).

The weight of observational evidence indicates an ongoing intensification of the water cycle – very dry or very wet areas have increased, globally, from 20% to 38% in the last three decades. There is more water vapor in the atmosphere, and hence there is potential for more extreme precipitation. Based on the results of the climate models, it is projected that the water cycle will further intensify, with possible consequences to rendering extremes more extreme.

Warmer temperatures generate increased glacier melt; hence, widespread glacier retreat has been already observed, and many small glaciers disappear. High reductions in the mass of Northern Hemisphere glaciers are expected in the warming climate. As these glaciers retreat, rivers, which are sustained by glacier melt during the summer season, feature flow increase, but the contribution of glacier melt will gradually fall over the next few decades.

Water quality is likely generally to be degraded by higher water temperature, but this may be offset regionally by the dilution effect of increased flows. Warming-enhanced sea-level rise can lead to saltwater intrusion into fresh groundwater bodies. Thus, freshwater availability in coastal areas is likely to decrease in the warmer climate.

Water Resources of the World

Water is the most abundant substance at the Earth's surface, with most (almost all) of it contained in the oceans, which cover nearly 71% of the surface area of the Earth. Oceans are by far the Earth's largest reservoir, but their water is salty (with salinity from 33 to 37 g kg^{-1}). If evenly distributed on the Earth-sized uniform sphere, water would form a layer of a depth of ~2.7 km.

The global water resources constitute approximately 1.385 billion km^3 (**Figure 1**). This makes up 0.17% of Earth's volume. About 97.5% of global water resources are saline and only 2.5% are fresh. Saltwater stored in oceans is the prevailing portion (96.5%) of Earth's water resources (1.338 billion km^3). The average ocean's depth is 3794 m and the mass of the oceans is approximately 1.35×10^{18} t (about less than a quarter of a permille of the total Earth's mass). The second largest water store on Earth – glaciers and permanent snow cover – is very much smaller than the oceans, containing 24.4 million km^3 of water (*c.* 1.72% of global water resources), that is over 50 times less than the ocean water. However, this solid water store (whose prevailing part is ice and permanent snow cover in the Antarctic, the Arctic, and mountainous regions) contains freshwater, making up most (about 69%) of the total freshwater resources. The third largest global water store is groundwater, containing 23.4 million km^3 of water (1.7% of global water resources), but more than half of groundwater is not fresh. Fresh groundwater resources amount to approximately 10 530 000 km^3 (0.76% of total global resources but 30.1% of total freshwater resources). Since the frozen hydrosphere (cryosphere), being the largest reservoir of freshwater, is not easy to reach by the humans, groundwater is the largest source of freshwater, which is readily available. All the lakes on Earth contain 176.4 thousand km^3 of water (0.013% of total water) with freshwater constituting more than half of the total volume and 0.26% of total freshwater. Some 16.5 thousand km^3 of water is stored in

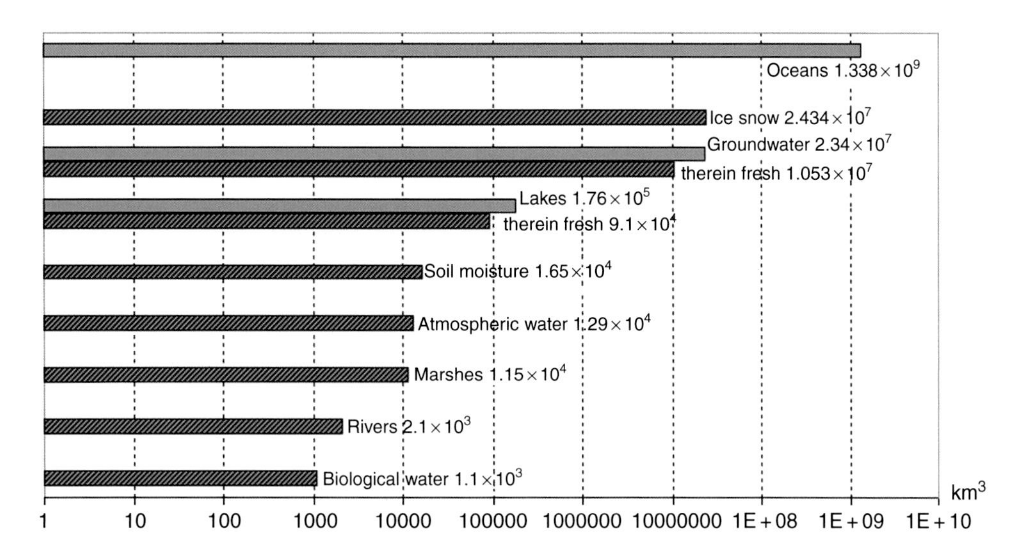

Figure 1 Global water resources: distribution into water stores, in cubic kilometers. Gray bars represent saltwater, and stripped bars represent freshwater.

the soil (0.001 2% of total water, 0.05% of global freshwater), while, on average, all the rivers of the world carry in any time instant approximately 2120 km^3 of water, that is only 0.006% of freshwater. The atmosphere itself is a large water store, with the total volume of stored water being about 13 000 km^3 (0.001% of total water, 0.04% of freshwater). Wetlands contain approximately 11 500 km^3 of water, that is, 0.000 8% of total water and 0.03% of global freshwater resources. Biological water has the global volume of 1120 km^3 that is 0.0001% of total water and 0.003% of freshwater. Total freshwater resources are estimated to be in excess of 35 million km^3.

Water is in a permanent motion converting from liquid to solid or gaseous phase, or back, with the principal processes being evaporation and precipitation, which are distributed very unevenly in space and time. Annual precipitation total largely depends on the latitude. Globally averaged latitudinal precipitation is highest near the equator and relatively high at the latitude around 60° (where upward lift of air masses is dominating). It is lower at the latitude around 30° and near the poles (where downward movement of air masses dominates). Also the altitude above the sea level is an important control of the amount of precipitation. Among further factors of importance are distance from source of water, exposition to prevailing wind, and large-scale landscape structure.

Surface Waters

Surface waters, rivers and lakes, are the most easily accessible water source for economic needs of humans, being of paramount importance for water ecosystems. The term 'renewable' means not only quantitative recharge, but also the possibility of in-river restoration of water quality (self-purification).

The distribution of river water in space is quite uneven in space and time; some locations have plenty of it while others have very little. River discharges in continents vary in time, following a periodic rule. For instance, a major part of river runoff in Europe occurs from April to June, in Asia from June to October, while in Australia and Oceania from January to April.

The global river runoff is estimated, on average, as about 40 000 $km^3\,yr^{-1}$, but it varies in time. The largest river discharges are in Asia and South America (respectively, 13 500 and 12 000 $km^3\,yr^{-1}$), while the smallest are in Europe and Oceania (respectively, 2900 and 2400 $km^3\,yr^{-1}$).

The year-to-year variability of water resources can be quite significant and considerably departs from the average values. This especially pertains to the arid and semiarid regions, where the water resources themselves are generally low. Here, in individual years, values of river discharges can be only half of the long-term averages, whereas for humid regions this difference is lower. Apart from between-year variability, important is the within-year variability, and seasonal and monthly patterns of discharge. Often, river runoff distribution is not uniform in time: a flooding season, which may last 3–4 months, is responsible for even 80% of annual discharge total, while during the low flow period, lasting 3–4 months, the river runoff may amount to a small portion (below 10%) of annual total.

The highest river runoff potential is concentrated in six countries: Brazil, Russia, Canada, the USA, China, and India, where nearly half of the total annual river runoff is formed. The greatest river of the world, Amazon, carries about 7000 km^3 of water, that is, 16% of annual global river runoff, while 11% of the total runoff is due to the four other large river systems: Ganges with Brahmaputra, Congo, Yangtze, and Orinoco.

Many river basins belong to the so-called endorheic (drainless) runoff regions that are not connected to oceans. The total area of endorheic runoff regions is about 30 million km^2 (20% of the total land area). However, only 2.3% ($\sim$1000 $km^3\,yr^{-1}$) of annual global river runoff is formed in these regions, much of whose area is covered by deserts and semi-deserts with a very low precipitation. The largest endorheic regions include the Caspian Sea basin, much of Central Asia, northeastern China, Australia, Arabian Peninsula, and North Africa. In endorheic regions, much of water is lost for evaporation and does not reach river mouths.

Approximately half of the total river water inflow to the world ocean (19 800 km^3) feeds into the Atlantic, where four of six largest rivers of the world flow into (Amazon, Congo, Orinoco, and Parana). The smallest amount of river water (5000 $km^3\,yr^{-1}$) flows into the Arctic Ocean; however, river waters are of most importance for the regime of this ocean. While containing only 1.2% of total oceanic water storage, the Arctic Ocean receives 12.5% of global river runoff. On average, much of the total river runoff (about 42%) enters the ocean in the equatorial region between 10° N and 10° S.

The values given above describe the average situation for a long-term period. For shorter time intervals (e.g., for an individual year), the values of water amounts in different stores in the hydrosphere may considerably depart from the long-term average.

All the lakes on Earth store approximately 91 000 km^3 of freshwater – much more than the rivers. Most lakes are young in geological terms (being 10–20 ky old), except for much older lakes of tectonic or volcanic origin (such as the Lake Baikal or lakes of the East African Rift). The Lake Baikal is the largest (by volume) and the oldest freshwater lake in the world, containing 91 000 km^3 of freshwater. Similar volume of freshwater is stored in the North American Great Lakes.

Uncertainty

Assessments of global water resources are uncertain. This refers to all data, but in particular to groundwater and water in permafrost areas. Vast volumes of water contained deeply in the Earth crust are not included in typical assessments. However, even the continental runoff cannot be reliably estimated. The differences between water resources assessments for continents done by different experts reach several tens of percent, being particularly strong for South America (highest estimates are of the order of 150% of lowest estimates). Older assessments are still quoted in recent works, because these stem from the time when hydrological observation networks were better developed and contained many more stations than today. There is an acute lack of newer data from several countries, due to the growing inadequacy of declining hydrological networks. Hydrological data collection and analysis worldwide are not keeping pace with the actual water development and management needs, despite the increasing demands for water and the growing water stress, calling for improvement in efficiency of water management. Hence, according to some experts, we are guessing rather than assessing the water resources.

More than half of the river gauges of the world are located in Europe and North America, where observation series are longest. In contrast, water resources estimates are most erroneous in a number of regions of Africa and Asia, where hydrological networks are weakly developed, and the situation is not improving.

Anthropopressure

Humans have always interacted with the hydrosphere, drinking freshwater, and using it for various purposes. However, until a century ago, the number of people on Earth was not high, and human impact on water resources was generally insignificant and local rather than global. Thanks to the renewal process of the water cycle and its self-purification properties, on average, the quantity and quality of fresh waters had not changed much (except for climate-driven natural variability at different timescales). The process of evaporation and surface water systems (rivers, lakes, and, in particular, wetlands) remove a large portion of pollutants from the water, in liquid or gaseous state. There had been an illusion that water resources are infinite, inexhaustible, and perfectly renewable, free goods. The situation has dramatically changed over the last century, when water withdrawals strongly increased due to the dynamic population growth and socioeconomic development driving the increase of human living standards. There has been a dramatic expansion of irrigated agricultural areas, growth of industrial water use (including the power sector), and intensive construction of storage reservoirs worldwide.

The characteristics of water resources, in both quantity and quality aspect, which used to be driven by natural conditions (climate, geology, soils, and resultant natural land cover) are now dependent, to an ever-increasing extent, on human economic activities. In many areas of the world, water resources have been adversely affected in quantitative and qualitative terms, by increasing water withdrawal and water pollution, respectively. Problems are particularly acute in arid regions.

Irrigated agriculture consumes, globally, 70% of the world water withdrawals. More and more water is needed to produce food for the ever-increasing population of the globe. Since projections for the future foresee further growth of population, the consequences to food and fiber production are clear and the global demand for water will grow further. Faster growth is expected in less developed countries: in the whole of Africa and much of Asia.

Poor water quality is another severe, and global, water problem. Traditionally, the water quality was mostly related to natural composition of water (salinity). Now, human has changed the quality of the world's water to a large degree. The structure of human-caused water pollution problems has changed in time, with fecal coliform bacteria and organic pollution being the oldest. Later, water pollution included salinization of freshwater (groundwater, rivers, lakes), for example, caused by irrigation or groundwater overexploitation and saltwater intrusion, pollution by metals, radioactive material, organic micropollutants, and acidification. It is estimated that only 5% of the world's wastewater is treated. Important water quality problems are caused by nutrients (nitrogen, phosphorus), whose abundance leads to eutrophication and toxic algae blooms. Remains of agricultural chemistry products, artificial fertilizers, pesticides, and herbicides, are particularly difficult to eliminate, due to the distributed nature of the source. Some synthetic chemicals, for example, organochlorines (organohalides) have a long half-life time: 8 years in the case of DDT.

In order to improve the quality of water in the countries of the European Community, the Water Framework Directive entered into force in December 2000, setting out a framework for actions in the field of water policy in the European Union (EU). The key objective of the directive, which imposes legal obligations on the authorities in EU member states, is to achieve a 'good water status' for all waters of the EU by 2015.

Even when perennial surface water source is available in a given location, water consumption in untreated state may present a risk to human health because of contamination by pathogens or waste. The number of people dying each year of water-related diseases is of the order of millions. Particularly burning water supply problems occur in informal human settlements, for example, slums around mega-cities, where the poor have no access to public, safe, tap water. They have to buy lower-quality water from vendors and pay much more than the price charged to more wealthy citizens who have access to the public supply of safe water.

Water is not a free goods any more. A future-oriented water resources management should emphasize shaping demands rather than supply extension. It is a must to improve the efficiency of water use, trying to "do more with less" ("more crop per drop"). Financial instruments, such as the water pricing not only granting full cost recovery but also accounting the cost of the resource, in the sense of foregone opportunities, can generally improve the efficiency of water use.

Global water consumption has increased nearly sixfold since the beginning of the twentieth century, that is twice stronger than the population growth. Facing the increasing pressures, the business-as-usual approach to water development and management cannot be globally sustainable. The problems of water shortage are likely to be aggravated in the twenty-first century, which was baptized 'the age of water scarcity'. Population growth, economic development, and increasingly consumptive lifestyle impact on the hydrological cycle, boosting water withdrawals and increasing the hazard of water stress and water scarcity.

The need for protection of the aquatic ecosystems is being increasingly recognized. Despite the rising human demand for water, it is necessary to allocate a share of water to maintain the functioning of freshwater-dependent ecosystems, thus meeting conditions of environmental water requirements. This would allow (if flows are regulated) to maintain the water regime within a river or a wetland, that suits aquatic and riparian ecosystems. However, earmarking water for environmental requirements is very difficult in some areas – even large rivers in China and Central Asia run dry, at times. River flow does not reach the sea due to excessive human water withdrawal.

See also: Water Cycle.

Further Reading

Arnell N and Liu Chunzhen (coordinating lead authors) (2001) Hydrology and water resources. In: IPCC (Intergovernmental Panel on Climate Change) (2001) *Climate Change 2001: Impacts, Adaptation and Vulnerability* McCarthy JJ, Canziani OF, Leary NA, Dokken DJ, and White KS (eds.) Contribution of the Working Group II to the Third Assessment Report of the Intergovernmental Panel on Climate Change. Cambridge: Cambridge University Press.

Chahine MT (1992) The hydrological cycle and its influence on climate. *Nature* 359: 373–380.

Eagleson PS (1970) *Dynamic Hydrology*, 462+xvi p. New York: McGraw-Hill.

German Advisory Council on Global Change (1999) *World in Transition: Ways towards Sustainable Management of Freshwater Resources*, 392 + xxv p. Berlin: Springer.

Herschy RW and Fairbridge RW (eds.) (1998) *Encyclopedia of Hydrology and Water Resources*, 803 + xxvii p. Dordrecht, The Netherlands: Kluwer.

Jones JAA (1997) *Global Hydrology: Processes, Resources and Environmental Management*, 399 + xiv p. Harlow, UK: Longman.

Kabat P, Claussen M, Dirmeyer PA, *et al.* (eds.) (2004) *Vegetation, Water, Humans and the Climate. A New Perspective on an Interactive System*, 566+xxiii p. Berlin: Springer.

Shiklomanov IA (1999) *World Water Resources and Their Use* (a joint SHI–UNESCO project; open database). http://webworld.unesco.org/water/ihp/db/shiklomanov/.

Shiklomanov IA and Rodda JC (eds.) (2004) *World Water Resources at the Beginning of the Twenty-First Century*. Cambridge: Cambridge University Press.

World Water Assessment Programme (2003) *Water for People. Water for Life. The United Nations World Water Development Report*. Paris: UNESCO Publishing/Berghahn Books.

Noosphere

C Jäger, Potsdam Institute for Climate Impact Research, Potsdam, Germany

The Noosphere Concept

The noosphere concept is best developed before the background of the related concept of ecosphere. The ecosphere is usually understood to be the space inhabited by living beings. It comprises the living organisms (biosphere), the lower atmosphere, the hydrosphere (oceans, lakes, glaciers, etc.), and the highest layer of the lithosphere (topsoil as well as various kinds of rocky ground). The word biosphere was invented by the Austrian geologist Eduard Süß, who used it more or less in passing, in an influential textbook on the formation of the Alps. In 1911, Süß met the Russian-Ukrainian mineralogist and geochemist Vladimir Vernadsky, who gave the word its current meaning. This meaning includes the fact that the biosphere is connected in space and time, that all living beings are related to each other by evolution, and that not only the biological, but also the chemical and physical, processes in the biosphere are shaped to a considerable extent by the functioning of living beings. A major example is the oxygen content of the atmosphere resulting from photosynthesis.

In the 1920s, Vernadsky was staying in Paris where he met the philosopher and mathematician Edouard LeRoy, whose lectures on biogeochemistry he attended. Through LeRoy, Vernadsky got exposed to a concept that Teilhard de Chardin, who also attended LeRoy's lectures, was developing in those days: the concept of noosphere. (The term noosphere, is derived from the Greek root *nous* meaning mind.)

Teilhard, a French geologist and Catholic priest, saw the emergence of the human species out of biological evolution as the beginning of a far-reaching transformation of the world we live in. The human mind would gradually learn to shape the world to a larger and larger extent, transforming the biosphere into the noosphere. Vernadsky related the concept to the historical dimension he had experienced in World War II. In his mind, this war showed that humankind was beginning to act on a global scale, but was not yet able to do so in a responsible way. The development of nuclear physics – that Vernadsky had been following already before World War I – presented the same challenge in an even more dramatic form. The transition from the biosphere to the noosphere, then, was to be the process in which humankind would learn to consciously and responsibly shape the ecosphere. This idea has been taken up in various forms by current authors interested in global environmental change.

Related Concepts

According to Venadsky, "The Noosphere is the last of many stages in the evolution of the biosphere in geological history" (Vernadsky, 1945, p. 10). The word "evolution" here does not refer to the interplay of variation and selection that Darwin saw at work in the evolution of biological species. Rather, it hints at a process in which new realities emerge in the course of time without any need for inheritance of traits between biological generations. This line of thinking is related to the idea of 'emergent evolution' proposed by the psychologist Lloyd Morgan and further developed by LeRoy. Today, the emergence of new realities in the course of time is often described as a process of self-organization in complex systems. Evolutionary history then becomes an overarching narrative telling the story of the world as a whole. It tells how physical matter rearranged itself up to the point where portions of it became the first living organisms, how these then evolved into species of increasing organic complexity, how the complexity of some organisms enabled them to develop the mental faculties that characterize humankind, and how humankind is now beginning to understand its own global environmental impacts.

The concept of the noosphere is also related to the concept of Gaia proposed by Lovelock and Margulis. The Gaia concept pictures the Earth as a complex, self-regulating system, a kind of organism that maintains conditions favorable to life despite a variety of disturbances. The emergence of the noosphere then means that some living beings – humans – became aware of this larger organism they are part of, of their capability to modify it by technological means, and of their responsibility to develop these means in ways that do not disrupt Gaia.

Closely related is a new concept of Earth system. Traditionally, Earth scientists considered as the Earth system those physical and chemical processes taking place on planet Earth that shaped oceans and continents, forming rocks, causing earthquakes, etc. Living beings were seen as playing a rather peripheral role (although for obvious reasons fossil fuels always were a big topic for the Earth sciences), and the influence of human beings on the Earth system was considered negligible. The debate about global environmental change and sustainability has changed this situation. As a result, a broader concept of Earth system has been proposed by Schellnhuber and others. In this perspective, the Earth system is seen as a complex system including physical, chemical, biological, as well as social and mental processes. Some sort of emergent evolution is seen as leading from a purely physicochemical system first to a biogeochemical system and then to one including human beings and their interactions. The first transition can be described as the emergence of the ecosphere, the latter as the emergence of the noosphere.

Finally, the role of humankind in shaping the face of the Earth has been used to propose a new geological epoch, the Anthropocene, supposed to start more or less with increased control over natural resources due to application of fossil energies during the industrial revolution in the nineteenth century. So far, geological epochs were defined to be periods of millions of years, and the last such epoch, the Holocene, has been defined to start just about 10 000 years ago. The concept of the Anthropocene marks a clear break with the previous practice of structuring a geological timeline. However, others have suggested that humankind significantly altered the climate system already some 8000 years ago by clearing forests. On a timescale of millions of years, this would make the beginning of the Holocene and the Anthropocene indistinguishable. On a conceptual level, of course, there still is a major difference between defining the current geological epoch in terms of an ice age that came to an end independent from any human action or in terms of the emergence of humankind as a new geological force. It is the latter approach that clearly relates to the concept of the noosphere.

Mechanisms and Institutions

As Vernadsky realized, the concept of the noosphere implies a causal chain from human thoughts to large-scale physical effects. This poses two challenges for research. First, there is the question of how the movements of human hands, legs, and bodies can be amplified so as to have effects that are observable at a planetary scale. And second, there is the question of how human thoughts can cause movements of hands, legs, and bodies.

As for the first question, fire has been a key amplification mechanism of human action since prehistorical times. Clearly, the burning of fossil fuels with the resulting emission of greenhouse gases is a related mechanism today. Vernadsky was particularly impressed by an amplification mechanism that was developed during his lifetime. The human capability to think had led to an understanding of subatomic processes that enabled human beings to build atomic bombs as well as to generate electricity from nuclear power plants. Vernadsky had studied radioactive materials already before World War I; during World War II he played a key role in triggering the nuclear weapons program under Stalin, and he forcefully supported the Soviet nuclear energy program. It is noteworthy that Lovelock, champion of the Gaia concept, strongly advocates nuclear power as the way to meet the challenge of anthropogenic climate change.

Nuclear physics is a prime example of human thoughts whose material impacts – while clearly being huge – depend mainly on political decisions. However, it is clear that the market institution is one of the most effective mechanisms to enlarge the range of human actions. The market economy has enabled human beings to develop global patterns of division of labor, of cooperation and competition. So far, research drawing directly or indirectly on the noosphere concept has not paid much attention to the economic links in the causal chain from thoughts to material impacts. This clearly is a major research challenge for the future. It includes the task of distinguishing those impacts of the market economy that change our global environment without impairing it from those that jeopardize properties of our environment that we value and need. Will the noosphere concept be helpful in new discoveries about how markets work and how key instances of market failure can be addressed?

Body and Soul

Vernadsky was fully aware of the fact that the second question – how human thoughts can cause changes in the material environment – was a key research challenge posed by the noosphere concept. Nowadays, brain research holds promise of important elements to address that question. However, when imagining that these elements will be sufficient to answer the question, a simple fact is ignored: what can be found in the human skull are neurons, synapses, electrochemical reactions, but no thoughts. One may expect that some day we will be able to establish a one-to-one relation between certain brain processes and certain thoughts, a bit as playing music from notes is based on a correspondence between certain marks on paper and certain sounds. But this does not mean that marks on paper and sounds are the same things. The noosphere concept challenges environmental research to reflect on one of the weak points of contemporary scientific culture: the difficulty in developing coherent arguments about the relations between movements of the human body and what was once called the human soul.

Research in logic has helped to clarify the role of domains of discourse for the development of arguments. For logical inference to be possible, participants in a debate must share the ability to refer to individuals – stones, dreams, numbers, rainbows, people, whatever – in some reasonably well-defined domain. This ability has a price, however: the domain itself must be presupposed; attempts to refer to it within the logical discourse it supports lead to paradoxes and eventually contradictions. Discourse A can refer to the domain of discourse B, but not to its own domain. The domains of discourse used in biogeochemistry, however, are quite different from the ones needed to talk about human thoughts. Perhaps a new domain of discourse needs to be established before the intuition conveyed by the noosphere concept can be used in reliable professional research. Using a word like 'noosphere' as if one had a great unified domain of discourse at hand, however, can be not only inspiring, but also seriously confusing.

The world as a whole is not a possible subject of logical inferences. This led Wittgenstein to suggest that accepting silence, mysticism if one wishes, was the appropriate stance toward the world in its entirety. Later, however, he realized that this silence was interwoven with a different kind of speech. In a letter to his friend Drury, a psychiatrist who at one stage wondered whether it would not have been better to become an academic, he wrote: "Look at your patients more closely as human beings in trouble and enjoy more the opportunity you have to say 'good night' to so many people" (Rhees, 1984, p.109f). We may call this way of using words – as in honestly wishing 'good night' to somebody in trouble – poetic. Developing a domain of discourse is a poetic craft, a way of world-making, perhaps. Of course, the argumentative and the poetic use of words are not mutually exclusive; but sometimes the former is more appropriate, sometimes the latter. And this can lead one to wonder whether the noosphere concept does not fit a poetic use of language more than an argumentative one.

See also: Anthropospheric and Antropogenic Impact on the Biosphere; Biosphere. Vernadsky's Concept; Global Change Impacts on the Biosphere; Urbanization as a Global Process.

Further Reading

Crutzen PJ (2002) The Anthropocene: Geology of mankind. *Nature* 415: 23.
Jaeger C (2003) A note on domains of discourse. Logical know-how for integrated environmental modelling. *PIK-Report No 86*. Potsdam: Potsdam Institute of Climate Impact Research.
LeRoy E (1928) *Les Origines Humaines et l'évolution de l'intelligence*. Paris: Bolvin.
Lloyd Morgan C (1923) *Emergent Evolution*. London: William & Norgate.
Lovelock JE and Margulis L (1974) Atmospheric homeostasis by and for the biosphere: The gaia hypothesis. *Tellus* 26: 2–10.
Rhees R (1984) *Ludwig Wittgenstein, Personal Recollections*. Oxford: University Press.
Ruddiman WF (2003) The anthropogenic greenhouse era began thousands of years ago. *Climatic Change* 61: 261–293.
Samson PR and Pitt D (eds.) (1999) *The Biosphere and Noosphere Reader: Global Environment, Society and Change*. London: Routledge.
Schellnhuber HJ and Wenzel V (1999) *Earth System Analysis. Integrating Science for Sustainability*. Berlin: Springer.
Schneider SH, Miller JR, Crist E, and Boston PJ (eds.) (2004) *Scientists Debate Gaia: The Next Century*. Cambridge, MA: MIT Press.
Süß E (1875) *Die Entstehung der Alpen (The Origin of the Alps)*. Vienna: W. Braunmuller.
Teilhard De Chardin P (2004) *The Future of Man* (first published during 1920–1952). Garden City, NY: Doubleday.
Vernadsky VI (1945) The biosphere and the noosphere. *Scientific American* 33(1): 1–12.
Vernadsky (1997) *The Biosphere* (first published in 1926). New York: Springer.
Wittgenstein L (2001) *Tractatus Logico-Philosophicus*. London: Routledge, (first published in 1921).
Wittgenstein L (2001) *Philosophical Investigations*. London: Routledge, (first published in 1953).

Pedosphere

V O Targulian, Russian Academy of Sciences, Moscow, Russia
R W Arnold, USDA Natural Resources Conservation Service, Washington, DC, USA

Concepts

The pedosphere is the soil mantle of the Earth. This concept evolved from the basic scientific concept of soils as specific bodies in nature that developed in time and space *in situ* at the land surface due to processes resulting from long-term interactions of soil-forming factors. These factors are the lithosphere, atmosphere, hydrosphere, biosphere, and the landforms or relief of local terrain.

This basic concept of soils was described by V. V. Dokuchaev in the nineteenth century and has generally been accepted worldwide. Humans as components of the biosphere have increasingly become a significant factor interacting with the other spheres; consequently, the anthroposphere (realm of human society) is now considered to be a major influence. A comprehensive definition of a soil using a system approach indicates that 'a soil is a complex, open, bio-abiotic, nonlinear, multifunctional, multiphased, vertically and horizontally anisotropic structural system formed *in situ* within the surficial part of the land lithosphere'.

Soils cover much of the Earth's land surface and the bottom of shallow waters as part of a continuum or mantle. This continuum called the pedosphere (from Greek *pedon* meaning ground) serves as the Earth's biogeomembrane, which is somewhat analogous to biomembranes of living organisms. As a biogeomembrane, the pedosphere facilitates and regulates the exchange of substances and fluxes of energy among the land biota, atmosphere, hydrosphere, and lithosphere. Additions, translocations, transformations, and removals occur in the soils of the pedosphere depending on the interplay of local environmental conditions and the inherent

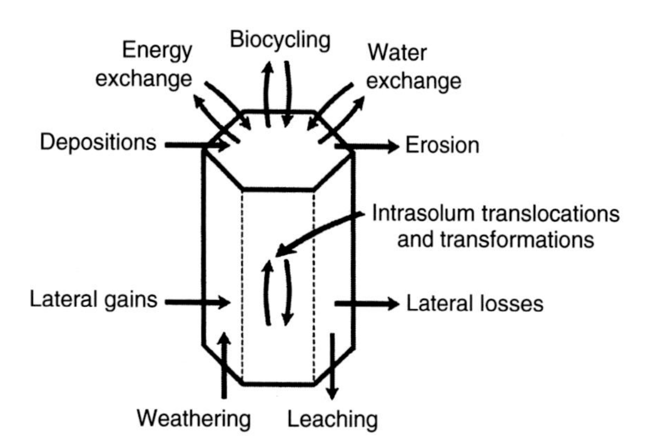

Figure 1 Generalized processes active in developing soil features and horizons. Adapted from Certina G and Scalenghe R (eds.) (2006) *Soils: Basic Concepts and Future Challenges*, fig. 2.1. Cambridge: Cambridge University Press.

properties within the soil bodies (**Figure 1**). Paleosols are found in early geologic periods, and it is expected that in Mesozoic and Paleozoic eras some extinct types of soils and pedogenic processes could be found. Emphasis has more commonly been given to the major climatic and geomorphic effects on the pedosphere that existed during the Pleistocene and Holocene epochs. Currently, pedogenic properties and functions influenced by extensive exploitation of soils by humans during the last two centuries (the Anthropocene) are receiving more attention.

Processes

Most processes of the pedosphere functioning operate in an open system, and although some appear to be cyclic and reversible, for example, biogeochemical cycling of C and N, many of them are unidirectional and irreversible, such as weathering of silicates in a soil and leaching of substances out of a soil. Due to the open and irreversible nature of the processes, there are many residual products, especially solid-phase materials, both organic and mineral, that are produced and retained in the parent materials. The annual formation of such components is very small and hardly detectable; however, when the soil-forming processes occur for a long time (10^2–10^6 years), the gradual long-term accumulation of pedogenic solid compounds alters parent materials in soil horizons and profiles. Such processes of solid-phase macrofeature formation during long-term multiphase functioning of a soil system can be perceived as a synergetic self-organization of the system – pedogenesis. Pedogenic features making up the solid-phase structure and composition of portions of the pedosphere are more pronounced where the upper unconsolidated layer of the lithosphere has been neither renewed by erosion or sedimentation nor mixed with deeper layers. Where landscapes have been stable and have had long-term functioning of soil-forming processes, gradual accumulation of pedogenic products occurs and well-differentiated soils form. The general development of the pedosphere is conceptually a sequence. There is an accumulation of earthy materials that over time are altered by processes of interaction with the atmosphere, hydrosphere, and biosphere. A general rule of pedogenesis is: interacting factors → open system processes → formation of pedogenic properties and features (**Figure 2**). Eventually a three-dimensional anisotropic structure, the pedosphere, covers the terrestrial and shallow aqueous land areas.

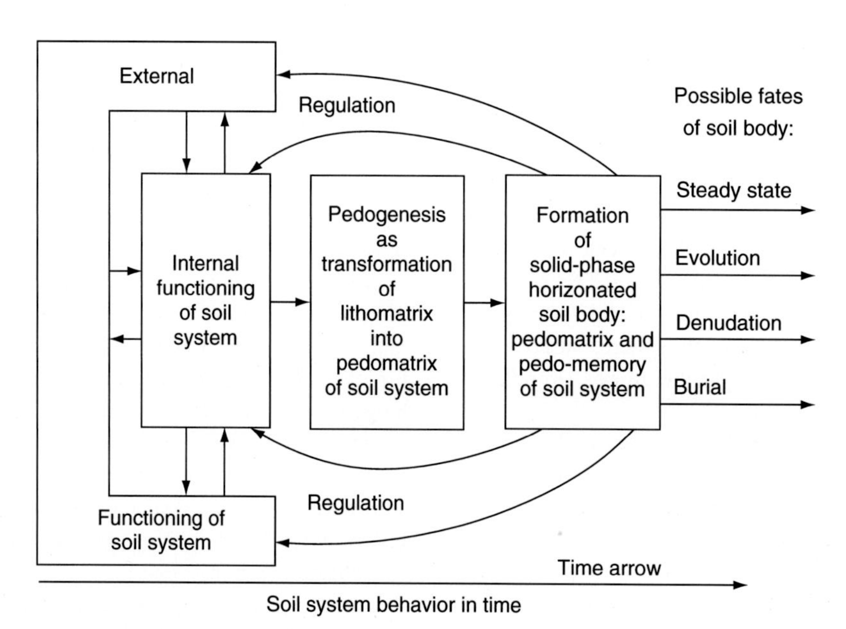

Figure 2 Functioning of a soil system and possible future condition of a soil body.

Structure

The pedosphere has its own specific structure. Vertical variability is the result of internal processes altering parent materials *in situ* into pedogenic features and properties that make up horizons and soil profiles; see **Figure 3**. These processes are usually called soil-forming, or specific pedogenic, processes. Many variations are possible due to the wide range of environmental conditions and scope of the factors themselves interacting to form and develop soils. The natural sequence of *in situ*-formed layers in a natural soil is a system of genetic soil horizons – a genetic profile or soil body. During the Anthropocene, human activities have already modified most of the land, so that few truly natural soils exist (in tundra and boreal taiga zones, high mountains, tropical rainforests, and extreme deserts). New kinds of anthropic features and soil horizons are being identified, described, and recognized as significant features of the pedosphere. The World Reference Base for Soil Resources now reflects such changes. Refinements of the concept of the pedosphere will be, and are, being made as improved techniques for their examination and measurement become available.

Soil, as a multiphase body in the pedosphere, has several kinds of depth distributions at any moment. There are temperature profiles, moisture profiles, gaseous ones, soil solution and nutrient profiles, macro- and microbiota ones, and solid-phase profiles. The first three or four are mainly functional, that is, they are very labile and change quickly (10^{-1}–10^{1} years). The solid-phase profile is more stable, changes slowly (10^{1-2}–10^{5-6} years), and is characterized by interrelated horizons with variable texture, structure, and mineralogical and chemical composition (**Figure 4**). Many kinds of diagnostic features and horizons are recognized, and their combinations give rise to a large number of unique soils throughout the pedosphere. Classification systems such as the World Reference Base for Soil Resources and Soil Taxonomy are based on combinations of defined pedogenic properties, mainly solid-phase ones. The organization of these systems facilitates small-scale representations of the pedosphere, as noted by the color patterns in **Figure 5**.

The lateral combinations of individual soil bodies comprise the continuous soil cover of land, the pedosphere. Spatial patterns or structures of soil cover exist at all scales of observation; however, there are differences of opinion about what and how to define the combinations at different scales. Soil surveys of portions of the pedosphere are made at different scales, usually depending on the nature and

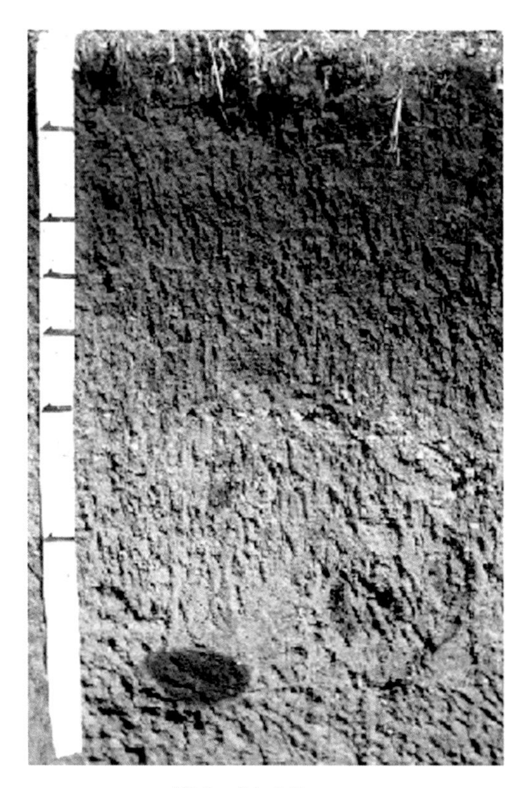

Wet soil in tall grass prairie (steppe)

Figure 3 Vertical variability revealed as genetic horizons (layers) in a drained and cultivated Chernozem soil derived from calcareous glacial till in Iowa, USA. Photo credit: R. W. Arnold.

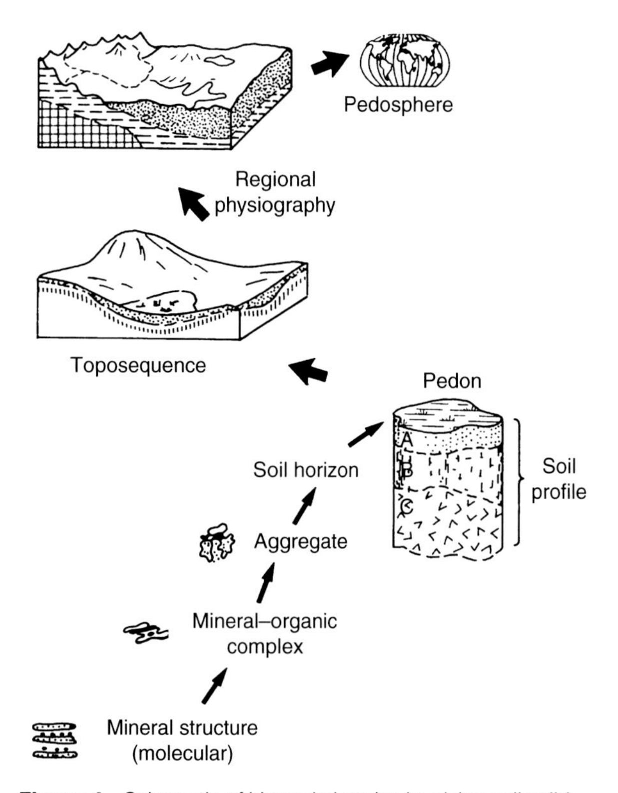

Figure 4 Schematic of hierarchal scales involving soil solid-phase components that combine to form horizons, profiles, local and regional landscapes, and the global pedosphere. Adapted from Sposito G and Reginato RJ (eds.) (1992) *Opportunities in Basic Soil Science Research*, p. 11. Madison, WI: Soil Science Society of America.

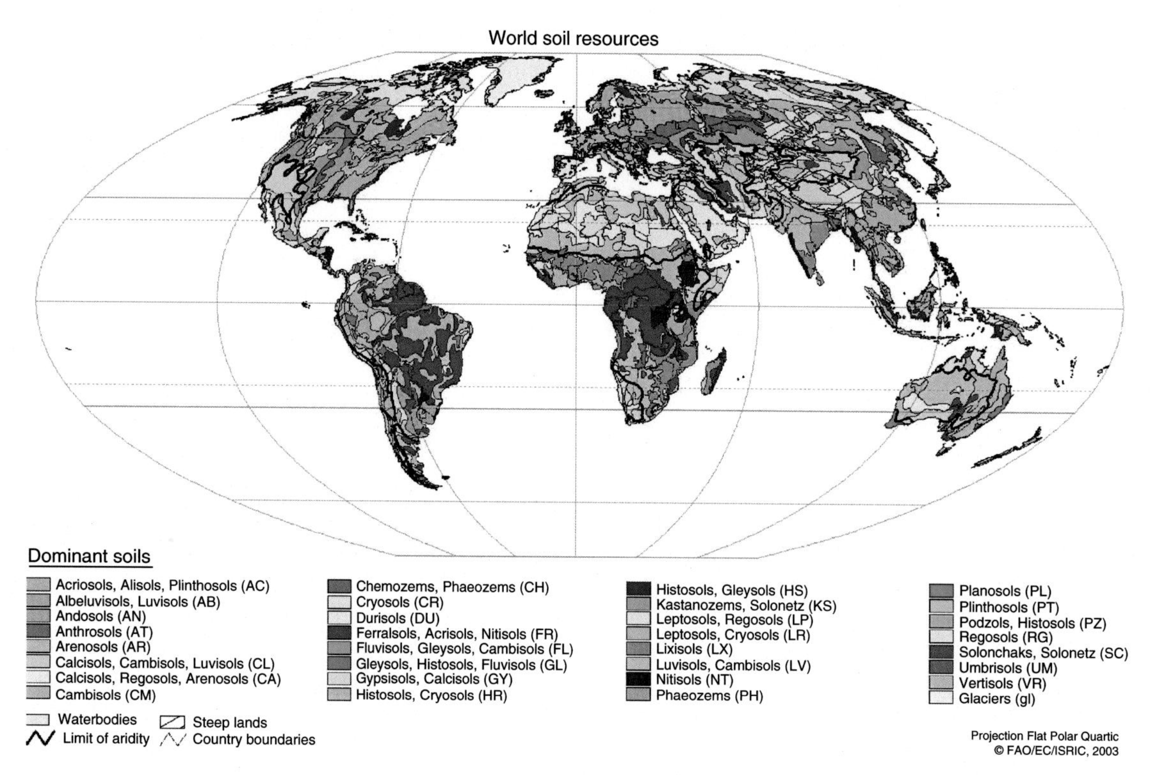

Figure 5 Map of world soil resources using the World Reference Base for names of major soil groups. Map produced by FAO, EC, and ISRIC.

genesis of soil patterns and need for detailed information about land use and management.

Pedo-Memory

Most soils are organized, structured, natural entities whose pedogenic properties have recorded the main features of environments and ecosystems that existed during their formation and subsequent changes. Soil, therefore, is commonly a product and a memory of long-term interactions and functioning in nature. During the past two or three centuries, much of the pedosphere has also recorded many anthropotechnogenic impacts and those portions now have memories of complex biosphere–geosphere–anthroposphere interactions.

Soils have different capacities for recording past and present environments depending on the time required for processes to come into quasi-equilibrium with environmental conditions (characteristic times, CTs). General CTs are: for gaseous phase, $CT \sim 10^{-1}–10^{1}$ years; liquid phase, $CT \sim 10^{-1}–10^{2}$ years; micro- and macrobiota, $CT \sim 10^{-1}–10^{3}$ years; and solid phase, $CT \sim 10^{1}–10^{6}$ years. These orders of magnitude are only indicative of the wide ranges involved. Although solid-phase features reflect environmental changes slower than the other phases, they retain the changes much longer and are the major recorders of prior environmental conditions.

Pedogenic solid-phase properties also have different characteristic memory retention times; the more quickly formed properties may record changes for years, decades, and even centuries. Litter leaching and decomposition, soil structure formation and degradation, salinization and desalinization, and reduction and oxidation are examples. The slower-formed properties may record changes for millennia to millions of years, for example, deep and strong weathering, transformation and translocation of clays, and alteration and accumulation of iron-rich compounds. The age of soil memory depends on the duration and interactions of soil-forming and weathering processes that occurred at a specific place.

Soils of the existing pedosphere generally consist of complex combinations of inherited properties of pre-Pleistocene and Pleistocene weathering, landscape evolution, and pedogenesis, as well as more recent Holocene and Anthropocene impacts. Some features of pedogenic properties are partially erased by erosion and other degrading processes such as excessive leaching or acidity, and later processes of landscape and soil evolution overprint properties and memories (a palimpsest phenomena). Usually local knowledge of geomorphology, sedimentation, and hydrology provide a foundation on which to base pedogenic interpretations. The complicated records of the pedosphere are slowly being read by pedologists to provide more information about past environmental conditions. Understanding soil components as carriers of pedo-memory and the rates of change of solid-phase properties remains a challenge to understanding and predicting future changes of the pedosphere.

Functions

The pedosphere is an extremely active terrestrial and subaqueous layer surrounding the Earth whose functions are closely linked with other spheres. The biospheric function is the major production function as it provides soil fertility and a suitable habitat for most species of organisms, thereby supporting land biodiversity. By this function, biomass transformations occur, nutrients are supplied and cycled, and the myriad microorganisms in soil enable sustainable biological productivity, diversity, and activity. Their metabolism is the primary basis for regulation and production functions in soils. Most biogenic substance fluxes are known as biogeochemical turnovers. The Millennium Assessment indicates that more land was converted to cropland since 1945 than in the eighteenth and nineteenth centuries combined, and that agricultural land uses now cover a quarter of the terrestrial surface.

Because the pedosphere is the zone of interaction between the biosphere and the atmo-hydro-lithospheres, it is commonly thought of as a reactor and regulator that functions to mediate and control fluxes of energy and substances. For example, temperatures are modified by the pedosphere and make most life, as we know it, a possibility.

The atmospheric function includes energy and moisture exchanges, respiration, and transfer of gases, including oxygen and the greenhouse gases, and is the force that transports and deposits dust derived from soils. Because of porosity and permeability, soils have a hydrospheric function to partition water in, through, and out of the pedosphere. The geochemistry of the Earth's waters are mainly determined by the influences of the pedosphere. Where resistance thresholds are exceeded, water erodes surface particles from soils and deposits sediments downstream. Soil erosion degrades soil quality and often jeopardizes sustainable uses of soils.

The lithospheric function of soils is that of a dynamic geoderma protecting landscapes and the deeper lithosphere and mitigating destructive actions of exogenous forces such as wind and water erosion, landslides, and tectonic and volcanic disturbances.

The pedosphere has an important utilization or carrier function manifested as building sites for communities and transportation networks. Soils supply materials for many types of construction, and also are critical areas for waste disposal.

Last but not least is the cultural and historical function of the pedosphere. Society's interactions with soils were initially for agricultural purposes and the lore is rich with stories and myths of the power of unseen forces to help sustain soil fertility. Soils also serve as a respository of archeological artifacts, stratigraphic markers, and memory of ancient settlement environments. In general, human attitudes that define 'self' in a context and in relation to nature result in religious beliefs as ways of bringing order into the seeming chaos of nature. The biogeochemical cycling of life, from dust to dust, is such a concept. Sanctity and stewardship of resources have their roots in the pedosphere.

Some Limiting Conditions

The Atlas of the World Reference Base for Soil Resources illustrates the striking variability of soils in the pedosphere, reminding us that there is a lot of uncertainty in the details of spatial patterns and explanations of soil evolution.

Because soil conditions such as fertility, drainage, and topography can be artificially modified and changed by external activities, it is often assumed that the pedosphere is a renewable resource. However, experience has demonstrated that maintaining soil functions desired by society is not ecologically sustainable; rather, they must be reinforced with external energy and substances. Many ecologically and agriculturally important soil features have their characteristic times of formation and/or resilience much longer than human lives and even longer than some civilizations. The interactions of environmental conditions in natural ecosystems produce modifications much more slowly in soils than needed by modern society to provide expected products and services. During the next 50 years, demand for food crops is projected to grow by 70–85% under the Millennium Assessment scenarios, and demand for water by 30–85%.

The pedosphere with its functional and structural features has its own space and time limitations. Thickness and area are spatial limitations, whereas temporal functions and soil processes vary so widely that incongruencies and inconsistencies often make successful management or control very difficult.

Soil thickness is not the thickness of the rooting zone, rather it is the unspecified thickness of the upper layer of the lithosphere involved in regular bio-litho-atmo-hydrosphere interactions. All of the interactions and resulting processes are relevant to defining the functional thickness of soils. This pedosphere thickness strongly controls and regulates the interactions – it is a real biogeomembrane of the Earth. The shallowness of fertile topsoil limits agricultural use and is susceptible to contamination by pollutants, in addition to degradation and destruction due to human-induced erosion.

Assuming the ice-free land area is about 131 Mkm^2, it has been estimated that about 93 Mkm^2 is biologically productive land, of which forests are about 33%, pastures 32%, and cropland 11%. Only about a third of the land surface has pedosphere components that can reasonably be expected to provide sufficient food to support our current human civilization. Major limitations for agriculture include drought, nutrient deficiency, pollution, shallow depth, excess water, and permafrost. Other use limitations involve expansion of urban areas and transportation networks, small isolated tracts of suitable land, traditional parceling of land ownership, and high costs of preparing land for cultivation.

Why are temporal functions a limitation? As mentioned, natural changes of the pedosphere occur at rates too slow to satisfy the desires of modern society. Rates and characteristic times of soil functions, formation, and evolution processes cover at least 9 orders of magnitude (from 10^{-3} to 10^{6} years). During the Anthropocene, humans have exploited the pedosphere's 'treasure trove' that accumulated over millennia and hundreds of thousands of years of natural soil formation and evolution, creating a modern-day dilemma.

Many of these ideas were originally presented by the same authors in the book, *Global Soil Change*, published by IIASA in 1990 which has recently been made available online. We thank them for permission to revise and update the section on the pedosphere.

Further Reading

Arnold RW, Szabolcs I, and Targulian VO (eds.) (1990) Global soil change. *Report of an IIASA-ISSS-UNEP Task Force on the Role of Soil in Global Change. CP-90-2*. Laxenburg, Austria: IIASA. http://www.iiasa.ac.at/Admin/PUB/Documents/CP-90-002.pdf (accessed December 2007).

Buol SW, Southard RJ, Graham RC, and McDaniel PA (2003) *Soil Genesis and Classification*, 5th edn. Ames, IA: Iowa State University Press.

Certina G and Scalenghe R (eds.) (2006) *Soils: Basic Concepts and Future Challenges*. Cambridge: Cambridge University Press.Eswaran H, Rice T, Ahrens R, and Stewart BA (eds.) (2003) *Soil Classification: A Global Desk Reference*. Boca Raton, FL: CRC Press.

ISSS Working Group RB (1998) In: Bridges EM, Batjes NH, and Nachtergaele FO *(eds.). World Reference Base for Soil Resources: Atlas*. Leuven, Belgium: ISRIC-FAO-ISSS.

IUSS Working Group WRB (2006) *World Reference Base for Soil Resources: A Framework for International Classification, Correlation and Communication*. Rome: FAO.

Sposito G and Reginato RJ (eds.) (1992) *Opportunities in Basic Soil Science Research*. Madison, WI: Soil Science Society of America.

Targulian VO and Krasilnikov PV (2007) Soil system and pedogenic processes: Self-organization, time scales and environmental significance. *Catena* 71(3): 373–382.

Ugolini FC and Spaltenstein H (1992) The pedosphere. In: Charlson R, Orions G, Butcher S, and Wolf G (eds.) *Global Biogeochemical Cycles*, pp. 85–153. San Diego, CA: Academic Press.

Phenomenon of Life: General Aspects

S V Chernyshenko, Dnipropetrovsk National University, Dnipropetrovsk, Ukraine

Introduction

The life phenomenon is one of the basic problems of understanding the universe. It is extremely important for both natural sciences (physics, chemistry, biology, etc.) and humanities (philosophy, psychology, etc.). The process of perception is a loop leading through inorganic nature, life, and consciousness back to reflection of the foundations of nature; so its understanding cannot be complete without answering the question: "What is life?"

The life phenomenon problem can include two important aspects concerning life, as a general concept, a logical scheme, on the one hand; and as the real object, special natural realization, on the other. This article is devoted to the first approach (the second one is considered in Structure and History of Life). The universal definition of life (including, e.g., its potential electronic forms) cannot be complete at the moment in absence of the practical experience of dealing with extraterrestrial or artificial life. However, it can be obtained by way of extrapolation of stored knowledge and is useful for the study of real life forms as a theoretical background, helping interpretation of real observations and giving general perspectives of life science development.

General Principles of Life

Life is a form of matter organization. It is an extremely complex phenomenon, which is still poorly comprehended by both science and common sense. The main features of life as a general phenomenon are the following:

- It is a dynamic process. It is impossible to stop it (even mentally) for investigation. Stopped life is death.
- It is superposition of many different scales. One cannot understand life without understanding of the different level processes: from the microlevel (down to quantum processes) to the macrolevel (up to planetary and space processes). The levels are in permanent interaction. During the evolution of life, both corpusclarization and globalization took place; they have been consistent in both directions.
- It is a hierarchical system of numerous elements. Biological systems can be described by laws of the systems theory and cybernetics. They have abilities for homeostasis, adaptation, use of information, self-organization, and evolution.

Figure 1 The Wheel of Life.

Dynamic Nature of Life

Life is not a structure, it is a process. Life units are similar to waves, they permanently renew their composition. The normal state of a biological system is a state of 'dynamic equilibrium', when inflow and outflow of matter compensate each other. Metabolism is one of indispensable conditions of a living organism. There is entry of the matter as a source of energy and constructional material, its use (assimilation), and excretion of decay product.

Balance of synthesis and destruction is one of the explanations of the cyclic nature of life. A more general explanation is that, the necessity for a stationary dynamic process to be cyclic, it must coil up in bounded space. At the level of a cell or organism, the cycle is shown as metabolism; at the level of ecosystems it is biogeochemical cycling. The concept of the cyclic character of natural processes is a part of many philosophical and religious doctrines. A good illustration of this fact is the well-known Buddhist Wheel of Life (see **Figure 1**). The most interesting details are the central circle, where one can see a naive image of closed nutrient cycling and the figure of the demon, personifying time, which gobbles all that is existing.

The next step of development of cyclic movement is iterative dynamics. Recursion, unlimited repetition of itself, can be considered as an important form of nonlinearity. It begets fractals in structure and iterations in dynamics. The main form of iterations in biological systems is replication or reproduction. It is extremely important at least in two aspects. First, it is a way to transmit information from micro- to macrolevel. Second, it is a prerequisite for evolution on the basis of Darwinian natural selection.

Multilayer Character of Life

One of the peculiarities of biology is the fact that it embraces many levels of matter organization, from molecules to biosphere. It results in a large complexity of life, and sometimes complexity and diversity are considered as important characteristics of biological systems. But complexity as such is not a solution; uncontrolled growth of complexity either leads to the reduction of stability, or does not influence it. The stability of real biological systems is a result of very specific interactions between its elements; complex systems must be very well organized. In accordance with the pronouncement of W. Weaver (1948), the subject of biology is 'organized complexity', contrary to classical physics ('organized simplicity') and statistical physics ('chaotic complexity'). Dynamic laws should be appropriate for

Figure 2 Pyramid of differentiation of living matter.

ensuring self-organization. Subjects of biological processes (cells, specimens, etc.) behave not chaotically, but coordinately.

Life development is, particularly, a process of matter differentiation. Step-by-step life makes the world more complex, changes 'the space of abilities', creates potential wells in this space (new niches for itself), and fills the wells with new species. In **Figure 2**, the niches are shown as a set of more and more narrow trapeziums, one originating from the other.

While 'inventing' new levels of matter organization, life keeps previous ones. Usually new forms of matter differentiation cannot exist without older forms, which are parts of their usual environment from their origin. Each step to deeper differentiation needs huge amount of less differentiate matter. Essential progress in producing new abilities usually accelerates the development of living matter, but this acceleration concerns a decreasingly small part of the matter.

Although sometimes a new form can essentially transform or even annihilate a previous one, the latter usually continues to exist as a basis and environment for the former. Life forms itself as a multilayer object. Each new layer emerges by using energy of the predecessors and establishing new forms of connections between the previous layers' elements.

Such an elementary type of formation does not give an optimal result. New objects' functions can duplicate functionality of lower layers or even be at variance with it. The design of the objects would be more rational in the case of starting from the very beginning, without context of previous stages. But nature prefers to build on the old basements from available bricks, which were not initially planned for forming new buildings. Such a choice has some advantages. Losing optimality, nature saves time and gets reliability. Systems with duplicated (and coordinated) functions are more stable; keeping of low-level reactions is useful in case of temporary degradation of environment and so on.

Interaction of different layers is not trivial; their structure and functions are in the permanent process of mutual coordination. The formation of the life multilayer structure was not a unidirectional movement from the lowest level to the highest one. In the course of history of the Earth, after the first chemical layer, the planetary layer of biosphere was formed. All the other levels (cellular, organism, etc.) were wedged between these extreme layers in the course of the process of 'discretization'.

Life, Death, and Immortality

The borders between living and inanimate objects are intuitively clear, but not very strict. Such micro-objects as viruses or plasmids are evidently a part of life, but at the same time they are chemical molecules (or a static group of several molecules) only. These obligatory parasites cannot exist as independent organisms, but they should be considered as living because they are part of biological macrosystems and even play an important role in their evolution. It is a good illustration of the fact that living matter sometimes cannot be divided for separate organisms.

Viruses form crystals, which can be disassembled and assembled again. Their individuality is interrupted; life and death lose their usual meaning. Death is not underside of life, and the syllogism "If there is no life on Mars, it means that there is no death there" is only a joke. The idea of death corresponds to high form of life only; it is inapplicable to unicellular organisms, which reproduce by division.

For life as a global phenomenon, death does not exist (at least, we know nothing about its imminence for biosphere). Death of individuals is a peculiarity of the life dynamics; it is explained by inexpediency to continue life of organisms, which have functioned their reproductive period.

Immortality, naturally, is impossible; there is nothing eternal in this world. However, individual life of multicellular organisms can be prolonged. Physiological limits of a lifetime are connected with a restriction for the number of divisions of somatic cells, which is connected, in its turn, with genome spoiling. There have now appeared the first ideas of how to struggle against this spoiling; present-day people have a chance for essential prolongation of their lives.

Extraterrestrial and Artificial Life

Our understanding of life is limited by its earth forms. Unfortunately, we do not know about extraterrestrial life (see Astrobiology), although Epicure spoke about it more than 2000 years ago, and J. Bruno was fagoted in 1600 because of his propagation of ideas about its possibility.

Now it is clear that the solar system planets are not really appropriate places for life, at least in its known forms. Jupiter's atmosphere is, probably, similar to that of the ancient Earth, and life can take its first steps there. Venus is too hot because of the greenhouse effect, and life is possible at some height in the atmosphere only. Mars is too cold, but in rocks found in the Antarctica and, probably, originating from Mars, scientists found microstructures resembling structures of leftovers from bacteria on the Earth. This indication of the existence of ancient life on Mars is very controversial, and the fact is only one 'collateral evidence' of extraterrestrial life.

Searching for sentient life in space was started in 1960 by the project OZMA, which was followed by the Cyclops program in 1971 and many others later. The search for artificial radio radiation and other indirect indications of life is still unsuccessful, provoking pessimistic opinion that mankind is alone in the universe.

We are also quite far from the origination of artificial life produced by man. In principle, it is possible to design self-assembling robots, but they cannot be reliable and self-sufficient. Modern electronic devices have some properties of living beings, but they are a part of the global noosphere system (combined biological and technical elements) and cannot exist for a long time without the environment of human civilization. Even if the perspective of electronic life exists, it is a very remote one.

Life as a System

Modern view on biological objects as complex systems is proposed by the prominent Austrian biologist L. von Bertalanfy (1901–84), who established a new scientific discipline – 'systems theory'. According to his definition, "system can be determined as a complex of interacting elements." Bertalanfy proposed to consider the role of the systems theory regarding living matter as similar to the role of physics regarding abiotic world.

The systems analysis plays the role of methodological background of biology. Fundamental laws of life (such as the law of natural selection) can be interpreted as universal laws for complex dynamical systems. And, vice versa, systems laws are organic for biology and allow for solving many of its theoretical and practical problems.

The effectiveness of the systems approach in biology is closely connected with high level of emergency, which is typical for biological systems. A biological object, as well as all stable complex systems, cannot be understood as a set of separate elements only. Each new layer of hierarchy is a new special object with its own properties, which are based on properties of its elements, but are not their direct consequence. There are two aspects which can partially explain the phenomenon of emergency:

- A higher level is a very special result of self-organization processes in the lower one. It is a summary of huge current and past processes at the lower level. At the same time, in biology, the higher level plays the role of regulatory mechanism and can radically influence low-level processes. Thus, both levels determine each other.
- Nature prefers 'economy' in principles of system organization. Systems can have similar structure, irrespective of elements' nature, and vice versa. It gives a possibility to study systems, abstracting internal elements' organization.

Nonlinearity of Biological Systems

Biological systems can have both linear and nonlinear properties; during their evolution they used all possible types of dynamics to increase their effectiveness and stability. Nevertheless, most biological processes are nonlinear. One can mention the following nonlinear effects: system's state jumping (bifurcation or 'transformation of quantity to quality'); system's transition between deterministic and chaotic behaviors; hysteretic effect, that is, the system 'remembers' its history; self-organization (purposeful decrease of the system entropy). Examples of evidently nonlinear biological processes are autocatalysis, reproduction, evolution of species, etc.

The analysis of critical regimes and singularities of the parametric space can be used for revealing 'acupuncture points', where small local perturbations provoke great large-scale metamorphoses of the system. A spectrum of quasistationary solutions is realized as a set of possible forms of morphogenesis. The discarded forms are still within system's reach but remain dormant, unknown to observers in the course of evolution.

Nonlinear dynamics of living beings is often intuitively incomprehensible; admiration of nonlinear algorithms of life produces paradoxical ideas about intelligence of cytoplasm or bacteria.

Structure and Hierarchy

One of the important characteristics of a system is its structure – a set of links between elements and their space distribution. There are two main structural forms of matter: centralized (hierarchical) and distributed (skeleton). Physical fields have a distributed organization, whereas atoms are centralized. A cell has the center (nucleus); a colony of cells is homogenous; and the organism is centralized again. Centralized organization is rational in the case of high-level differentiation of elements; the distributed one is more typical for systems with homogenous elements.

Often the structure of biological systems is tree-like; they are hierarchical systems. It is the result of two important processes: differentiation of living matter and bunching (oligomerization) of its elements. An important feature of hierarchical systems is the fact that each level is characterized by new emerging properties not presented at lower levels. A scheme of the general hierarchy of biological systems is represented in **Table 1**.

The structure of real systems can be changed, but its dynamics is very slow in comparison with other processes, called functionality of the system. A set of characteristic times can also form a hierarchy. For each process, other ones can be considered as part of the environment: slower ones because of their relative stability, and faster ones because of rapid running to equilibrium.

Cybernetic Principles in Biosystems

Adaptation and self-organization are impossible without information and controlling processes, that is, without realization of cybernetic principles. Cybernetic mechanisms can be found at all the levels of life, from biochemical processes to biosphere.

Self-regulation is a process of changing functionality of the system directed at its conservation. It is development of property of inanimate systems expressed by the Le Chatelier principle (1884) – external influence on the system's state is compensated by internal processes, influenced in the opposite direction. The law's version for open systems can be formulated as the following: an increase of the system's input leads to corresponding increase of its output. This reaction is passive and does not need energy.

For biological systems, it is very typical to use active methods to keep the system's steady state (maintain 'homeostasis'). One of the ways is to follow the cybernetic principle of 'negative feedback'; the output of some part of the system must influence its input – if the output is too large, the input is decreased, and vice versa.

Nature of Life: Mathematical, Physical, and Chemical Approaches

The multilevel and multimedium nature of life necessitates considering it in different aspects, in the framework of different sciences. Really, many definitions of life have no structure as "Life is . . .," but only as "Life can be considered as. . .." Probably, one can expect gradual synthesis of various approaches to the life problem, but, for the time being, the integrated picture has not been formed.

Table 1 Hierarchy of biological systems

Science	*System*	*Elements*	*Interactions*	*Elements' state*
Biochemistry	Chemical reaction	Organic macromolecules	Chemical	Form, position, energy, etc.
Cellular biology and genetics	Cell	Organelles and genome	Endoplasmic and nuclear	Kind, size, position
Morphology and anatomy	Organ	Types of cells and tissues	Intercellular, chemical, and electrical	Kind, vitality, phase of development, etc.
Physiology	Specimen	Organs	Interorgans, by hormones and neural impulses	Kind, vitality, state of health, etc.
Population ecology	Population	Specimens	Cooperative and competitive	Age, sex, physiological state, etc.
Global ecology, biogeocoenology	Ecosystems, biogeocoenose	Populations	Trophic, competitive, and cooperative	Size, age, sexual, genetic structure
Biosphere ecology	Biosphere	Regional ecosystems	Through climate, atmosphere, etc.	Productivity, sustainability, disturbance

Mathematical View on Life

Mathematics is a tool for abstraction, a way to the core of scientific knowledge. Mathematics is not interested in details; for it, life is a kind of complex systems with special relations between elements. For the description of different properties of life, there are various mathematical models. Probably, it is impossible to design a universal model of life; each model has its own field of application and level of approximation. Attractiveness of mathematical models does not consist in their complexity, but in their lucidity and explanatory power. According to Einstein, "Models should be as simple as possible, but not more so."

One of the first biological models was the Malthus model of exponential growth (1800). It was developed for the field of population dynamics by the Lotka–Volterra models (1925–31). Models of life were proposed by J. von Neuman, R. Tom, H. Meinhard, and others; mostly they were differential models. Their use was very productive; in particular, they are a basis for the nonlinear analysis.

Another interesting mathematical tool for life description is the theory of cellular automata. This kind of discrete model has a property to be chaotic at the micro-level and ordered at the global level. In principle, one can imagine the world as a cellular automaton with elements – physical particles. The well-known Game of Life of Conway (1970), which really reflects some features of real life, is also a cellular automaton.

Usually mathematical methods are numerical, but it seems to be a very perspective way to use topological and algebraic approaches also. The above-mentioned topological theory of ecological niches can be considered as an example of this way.

Physical Principles of Life

According to J. S. Mill (1806–73), laws of life cannot be something other than laws of behavior of molecules, interacting as parts of a living organism. But, because of emergence of biological systems, it is not easy to reduce biological laws to physical ones. Such a way, called 'reductionism', does not always give practical results, but it is important as the theoretical basis for searching borders of the possible for living objects. It is not easy to predict fundamental consequences from fundamental laws; each forecast of possible effects is a discovery.

Biology is a continuation of physics and chemistry and chemistry is a continuation of physics. One can understand biology as 'new physics' and pose a problem to find its form, which corresponds to physical traditions. Particularly, physics of life is possible only under very special values of the world constants; traditional physics 'does not know' what life is, and cannot explain these values. It is necessary to use the 'anthropic principle': our existence as intellectual beings, studying the world, presupposes its features ensuring origin of man.

In biology, as well as in the other sciences, the problem of energetic balance is very essential. It gives a general estimation of the process of life functioning. As open systems, living objects need permanent energy income; they use it step by step and finally transform it to thermal energy of the environment. The main source of energy for life as a whole is the radiation of the Sun (and, insufficiently, energy of the Earth's interior: chemical, thermal, and, probably, radioactive). Plants ('phototrophs') use the solar energy for chemical synthesis of organic substances (the process of photosynthesis), supporting their own existence and providing chemical energy for all other forms of life: 'heterotrophs' (herbivores and carnivores) and 'saprotrophs'. Physically one can say that the solar energy in the course of photosynthesis raises energetic levels of electrons in some atoms of living matter; then the electrons gradually and purposefully descend, executing chemical and mechanical work.

Life directs energy flux to itself and uses it. According to the I. Prigogine theorem, an open system, in the case of linearity of the energy flux through it, produces minimum likely entropy. Life is an inconvertible process, going in a linear area of forces–flow rates; it endeavors to keep this linearity. But irreducible small nonlinearity produces stochastic noise, finally destroying each living organism.

Contrary to general physical tendency, postulated in the second thermodynamic law, life as a global process is characterized by gradual decrease of entropy. (Separate organisms also decrease its entropy during most periods of their life, but after their death the entropy 'gains revenge'.) The paradox was already pointed out by the father of statistical physics L. Boltzmann (1844–1906), and later was deeply analyzed by A. J. Lotka (1880–1949). In 1944, the Nobel Prize winner physicist E. Schrödinger (1887–1961) published his famous book *What Is Life?* devoted to this problem.

The general explanation why entropy can be decreased in living systems is evident; these systems are open; they use external energy to decrease their own entropy and, at the same time, increase entropy of the environment. In general, both first and second thermodynamic laws hold true. But the ways of converting the energy income to entropy reduction (or maintaining order) is not so clear. According to E. Schrödinger, organisms 'drink orderliness' from a suitable environment. He explains about flux of 'negative entropy' (negentropy) to organism, which compensates natural increasing entropy. He does not explain the process in detail, but stresses that life's tools for this aim are 'aperiodic solids' – the chromosome molecules. Schrödinger's book had an essential influence on molecular biology; particularly, it stimulated J. D. Watson and F.

Crick to discover the DNA structure (1953) and explore in that way, the physical explanation of life.

It is not very clear yet what Schrödinger's negentropy is – free or stored energy, information, organization, or something else? Probably, a perspective conception is the idea about necessity for life of two coupled processes. The first (energetic) one accepts energy from environment and provides it to the second (information) process, which is responsible for the living system's development. A disproportion of entropy takes place; the second process presupposes decrease of entropy; the first one, correspondingly, increases it. Such processes are observed in inanimate nature; for example, explosion of an ultranew star transforms it into a primitive clot of neutrons, but, at the same time, heavy elements of the periodic system (prerequisites of life) are synthesized and spread in the universe. High-ordered entropy disproportion in living organisms presupposes very exact coordination of biological processes; in accordance with Schrödinger's opinion, information DNA molecules play the role of the coordination center.

Life is not contradictory to the second thermodynamic law, but uses it in a special way. Excluding from reproduction all the descendants of a couple except two of them, death of prey killed by predator, extinction of species in the course of evolution – all these events on the one hand increase entropy, but on the other hand they lead to general progress, to ordering matter in some local areas (from which, because of reproduction, the new forms spread as widely as possible).

As for inanimate nature, many scientists see in unidirectionality of the entropy change the basis of the time phenomenon; the tendency of entropy reduction in living systems can give a key to understanding of the general laws of living matter evolution. A. J. Lotka in the article 'Contribution to the energetics of evolution biology', published in 1922, proposed to consider energetic power of organisms as the main criterion maximized in the course of evolution. Later, he called this maximum power principle, the 'fourth thermodynamic law'. The approach is still under discussion; it was supported and developed by such prominent scientists as V. I. Vernadsky and H. T. Odum.

The law is based on the consideration of species' evolution, when in conditions of "the struggle for existence, the advantage must go to those organisms whose energy-capturing devices are most efficient in directing available energy into channels favorable to the preservation of the species" (A. J. Lotka). A capability of better assimilation of solar energy or energy collected by other organisms is a prior evolutional advantage.

It is quite right at the level of ecosystems, when stochastic fluctuations and individual peculiarities at the level of species are integrated and averaged out. More and more effective populations are involved into biological cycling, increasing its intensity. As a result, the ecosystem power (consumed energy per unit time) permanently grows. It is mainly the result of competition from plants (producers), which are forced to maximize production for keeping their place in the ecosystem. Another extremely important factor is the activity of animals (consumers). They withdraw producers' biomass and additionally intensify cycling. Probably, the global role of consumers in biosphere consists exactly in the spinning up of ecological cycles.

At the level of concrete species, classical power is not the only parameter determining its evolutionary perspectives. One should take into account, for example, the efficiency of the species in limitation of entropy growth. As a result, it is more reasonable to speak not about all the available energy, but about 'exergy' (entropy-free energy). The latter shows an ability of the organism to make the work relative to the surrounding; it is the 'co-property' of a system and a reservoir.

Another important aspect, influencing vitality of the species, is the integrated character of energetic abilities of living organisms. H. T. Odum proposed a concept of emergy (embodied energy) as "a measure of energy used in the past" and stored in the system's structure. The concept is being developed by S. E. Jørgensen and others. The maximum 'empower principle' is proposed by H. T. Odum as "a unifying concept that explains why there are material cycles, autocatalytic feedbacks, succession stages, spatial concentrations in centers, and pulsing over time." Generalization of the approach is possible by way of taking into consideration 'population strategies' of species. For example, one can base on the r/K concept or its modification the r/C model (in the context of which population preferences in division of its energetic recourses between the processes of growth and competition are considered).

Information Basis of Life

The notion about the information nature of life is generally accepted. At the same time, even the term 'information' is interpreted in biological literature in various ways. Starting from the classical works of the founders of the information theory, C. Shannon (1948) and J. von Neumann (1951), different directions of the generalization of the term were proposed.

Concerning the information character of internal biological processes, a reasonable approach is based on I. I. Shmalhausen's views (1968) about the resonance nature of biological information. Most of the relations between elements in biological systems are based on special resonance organization of living objects. Very often, an energetically weak influence of one element on another one produces its powerful reaction. This interaction cannot be interpreted as pure energetic; we call it 'information interaction'. In this context, information cannot be transmitted; it is a relation between two elements. There is connection between exergy as the object's energy with relation to its surrounding and the possibility of the object to realize information actions.

For realization of an information action, the initial influence (signal) must exceed some critical value: 'excitability threshold' or 'reobase'. For example, the maximal tension of cell electric discharge is 0.1 V; its reobase is 10% of this value. Sporadic resonance effects take place in inanimate nature too, but for living matter it is its basis; there are special mechanisms of energy charging for the creation of prerequisites for resonance (information) action. From the level of cells, there are extremely complex structures of signal relations, separated from processes of energy and matter transmission. In multicellular organisms and ecosystems, special information subsystems were originated. They influence each other by means of special media (substances and fields), bearing precisely signal character. Special systems of coding and decoding become more and more sophisticated. Information is the main instrument in all homeostatic processes; it is the main way to organize negative feedbacks in living systems.

Physically, the resonance is realization of potential energy; mathematically the same is bifurcation, nonlinear effect of steady-state (attractor) change. One can interpret information processes in a living system as a sequence of 'internal bifurcation'. As life is an information process, it exists near separatrixes, divided areas of steady states' attraction in the space of system parameters. It is a mathematical illustration of life's fragility: directed small changes of system critical parameters can easily upset its dynamic equilibrium.

Quite often, the concept of information is used in biological literature for designation of a measure of living matter ordering. In this case, information is the opposition to entropy, something like Schrödinger's negentropy. In the course of life development, its information content gradually grows. The philosophic question about the origin of information in the universe is still open. There are two main opinions: either the Big Bang in the result of a symmetry breach created all existing information and it is gradually sowing up, or its quantity equaled zero at the beginning and it is in the process of permanent growth. In the second case, the information creation is bound with Darwinian natural selection. According to G. Kastler "the information creation is storage of random selection." It happened also in the inanimate nature, but became much more intensive in the living one. The history of the universe is characterized by exponential growth of information.

General Chemical Principles of Life

Living matter is a direct sequel of the chemical level of organization. It is not invariant to its chemical composition; the material determines key properties of life.

Chemical reactions in living systems are cyclic and autocatalytic. According to the hypothesis of S. Kauffman (1993), big chemical systems of interacting polymers, which reach a critical level of complexity, necessarily become autocatalytic and self-replicating. Their elaboration can be ordered by natural selection. As examples of autocatalytic reactions, one can consider Calvin's cycle ('propagation of sugar phosphates') or replication of ATP, also connected with photosynthesis. The cyclic chemical reactions discovered by A. Szent-Györgyi and represented by the famous Belousov–Zhabotinsky reaction can play the role of 'soft clocks' in living organisms.

Life is a positive connection between information molecules and proteins. The most important chemical cycle is the following: DNA produces enzymes, which, in their turn, ensure its replication.

Organic macromolecules are not thermodynamic objects; they have no aggregate state and are naturally far from a steady state. It is a prebiological stage of matter development. According to Schrödinger's simile, a living organism as well as a pendulum clock are not thermodynamic objects; because of solidity of the clock and stability of the hereditary substance of the organism, room temperature for them is practically equivalent to zero.

Polymerization as well as enzymes decrease entropy, and thereby decrease molecules' freedom to move and select their co-reactants. A principal difference of life chemistry from ordinary one is the matrix synthesis. For living matter, contrary to inanimate one, a strict order of extremely long chain of reactions is possible.

Life Is a Way of Matter Self-Organization

It is common knowledge that life is the manifestation and result of a general tendency of matter to self-organization. The difference between opinions consists in understanding of the predetermination of self-organization steps. Really, this difference is not so essential: on the one hand, random mutations in pure Darwinism lead to realization of more or less predetermined process of adaptation to existing conditions, whereas on the other hand, a 'vital force' is needed in some mechanisms (why not stochastic?) to conduct their programs. The self-organization algorithms must have some physical basis (self-organization can be considered as a physical principle according to M. Eigen); it is not an opposition to Darwinian selection.

Self-organization is a process of forming order from disorder. It is possible in the following conditions: (1) there is a big quantity of simple components; (2) the components can constitute mutual relations; (3) there is a source of energy, supporting the relations formed; (4) external conditions are suitable for stability of the new system; and (5) the systems can play the role of elements in the process of forming systems of a next level. Evolution of living matter is a combination of rising tendency, connected with growth of complexity, and descending one, connected with tendency to stability.

There are no special biological fundamental forces or elementary particles. Life is the continuation of the inanimate world; it uses all opportunities given by the laws of nature, particularly nonlinear scenarios such as phase transformation, hysteresis effect, the formation of dissipative structures, etc. All the possibilities should be groped for, probably by accident.

Biological Self-Organization as a Process of Niche Creation and Filling

Life can be considered as existing in its own space of possibilities (a 'space of life'). Mathematically, this space can be called as a 'phase space'. Potential wells in this nonlinear space are occupied by different forms of life; they are the so-called ecological 'niches'. Usually the niche of species is defined as a set of environmental factors corresponding to the needs of the species. A linear world (without sources of energy) would have no niches. The inflow of energy creates or allows to create wells in the life space, where life can consolidate. The situation is similar to the physical picture of the world; elementary particles, probably, exist in potential wells and the world space is furrowed by some fundamental processes (e.g., the energy conservation law can be approximate, and leaking energy furrows space).

Evolution is a process of optimization. Biological systems evolve to a steady state or, in other words, optimize some criterion (Lyapunov's function), looking for a potential well (a niche) in a phase space. One of the optimization criteria is entropy production, which must be minimal in a state of dynamic equilibrium in accordance with the Glansdorff–Prigogine theorem.

In a niche, the system possess properties of homeostasis; after small distortions, it will return to the initial state. The evident discrete character of the steady states' set produces separateness of niches. Heterogeneity of the life space explains existence of strictly separated species and other taxonomic units. Living organisms are substantially casts of their niches. There are many examples of convergence, when similarity of niches leads to similarity of organization of living beings.

Tendency of nonlinear systems to keep in equilibrium is well known as a general systems property of 'equifinality', postulated by L. Bertalanfy. It is a spontaneous reaction of the system; it is an important factor of evolution, but it does not direct the life progress. Optimality does not mean progressive character; the level of amoeba fitness is not lower than that of man. Evolution uses the gradient method, which allows finding a local extremum only; for improving living conditions, species must jump over unacceptable ones. According to I. Prigogine, self-organization is a process of step-by-step loss of stability. The change of a steady state can be a result of either essential external influence upsetting the system from a previous equilibrium and forcing it to transit to a new one, or a gradual change of system parameters leading to a change of the phase space topology, to disappearance of the current steady state. Both of these ways were combined during life history; the model corresponds to the saltation conception about evolution as a sequence of catastrophes. External influences (geological, cosmic, anthropogenic) cause disappearance and reorganization of niches. New niches stimulate evolution of their potential hosts, which fill them because of reproductive abilities producing 'pressure of life' (see **Figure 3**). Species, which have lost their niches, should progress or become extinct.

In accordance with another topological model, catastrophes (and accompanying biological innovations) create new dimensions of the life space. In the niche (a point of local minimum), a new dimension appears; correspondingly, the host of the niche loses its stability and gets a possibility to find a better position in a new direction. At the result of such step-by-step changes, the development trajectory of the species is formed as a chain of orthogonal line segments. The path of progress can be compared with a 'bobsleigh track'.

Can evolution proceed under the constant conditions of environment? Darwinism does not exclude this possibility; the development of life organization can go on in the course of successive small improvements (decreasing entropy), but this movement is very slow. Usually it does not presuppose essential change of phase space topology;

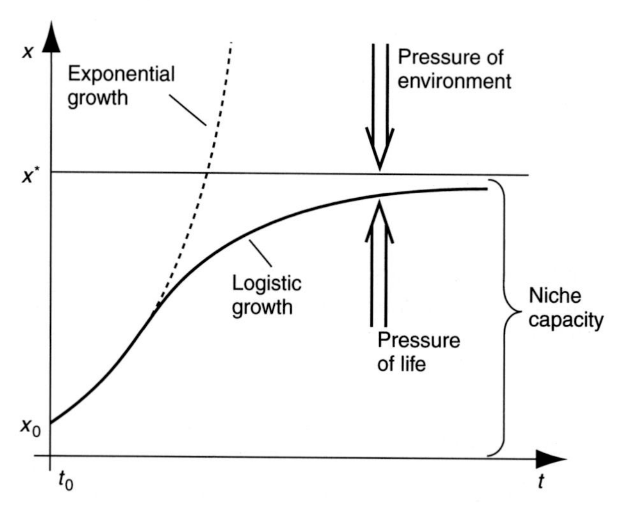

Figure 3 Niche capacity as a result of dynamic equilibrium between pressure of life and pressure of environment.

although it can sometimes stimulate radical changes: for example, the development of photosynthesis brought to the oxygen revolution and total reorganization of the system of niches. As a rule, a host of a stable niche slowly evolves and can meet competition only from the side of kindred species or, quite rarely, introduced ones. The Black Queen hypothesis about the necessity of permanent improvement of all species ("to stop means to die") is not true for stable niches. This fact is illustrated by the existence of a great number of primitive species formed millions and billions of years ago. For continuation of the race, it is necessary to change topology of the phase space, to disturb the system.

Relatively fast evolution of life has become possible because of general instability of niches. The history of life is a sequence of actions of forming and filling niches. As a result of divergence, niches bifurcate; the process of 'niche proliferation' takes place. Quite fast processes of niches interaction are ecological successions, when niches regularly change and supplement each other. Often infill of a niche creates a number of new ones, and, importantly, the complexity of the derived niches is usually higher than that of the initial one. It creates prerequisites for progressive evolution of life, an increase of its complexity. Newer forms of life produce newer local worlds, up to virtual worlds in the human mind, which are entirely real as both processes in brains and a plan for the real world change. Man with his imaginary worlds creates principally new powerful niches, particularly, for the development of artificial 'electronic beings'.

Principally, the evolution process concerns not separate organisms, but the biosphere as a whole. The optimality of organisms is not an absolute value; it has a sense in the context of environment including other organisms. In other worlds, self-organization of life cannot be understood at the level of organisms; it is necessary to consider the general life space, separated species in corresponding niches, and their interference.

General Evolutionary Rules

1. Evolution is irreversible (the principle of L. Dollo, 1893). Why does the return to former conditions not lead to recurrence of old forms? There are at least two explanations. First, energetic innovations found during the previous stages of the evolution will not be lost if they are effective. Second, change of a living form is reflected in surrounding ones; the form can regress only together with its community. It is not so probable, although sometimes an ecosystem can degrade – not because of degradation of developed forms, but as a result of their elimination and return to the forefront of primitive forms.

2. Evolution is a movement from simple to complex, but the simple is not destroyed – it is included in the new structure of life. According to A. Szent-Gyorgyi, life never revises what has been made, but it builds above the existing. The cell is similar to a lot of archeological diggings, where one can see a number of strata – the older, the deeper. The structure is not optimal, but is reliable; the damage of upper layers is not fatal – ancient mechanisms can smooth the blow.

3. Complexity growth stimulates self-organization process. As is shown by J. von Neumann's model of self-organizing automata, the ability to self-organize depends on complexity of the object. There is a critical level, beginning from which automata can reproduce more complex forms than themselves. During its development, life also passed through critical levels, which forced its tendency to progressive evolution.

4. Evolution has often a binary character. Evolving systems can consist of two closely connected parts such as DNA–proteins, cell–nucleus, male–female, etc. During active stages of evolution one part outstrips the other; the first one becomes an object for experiments, and the second one guarantees preservation of attained level.

5. Life uses for its self-organization, natural frequencies (modes) of component systems. Some modes can intercept incoming energy from other ones and intensify themselves. This effect has resonance nature and is close to information phenomenon. Octave principles can take place – structures are formed by series on multiple frequencies. It is one of the ways of forming hierarchy of living systems.

6. Spatial self-organization of life can be considered as a process of dissipative structures' formation – one of the nonlinear mechanisms of originating order from disorder.

7. Life evolution is directed from profound symmetry to absolute asymmetry. The main stages are ball, radial, axial, and bilateral symmetries, and triaxial asymmetry. The asymmetry is a reaction on anisotropy of the living space, for example, gravity anisotropy.

8. Evolutionary process is not so much an invention of new forms as a search for effective combinations of existing ones. Various living beings are built from the same standard bricks, which were formed during early stages of evolution.

9. Evolution evolves itself. In the history of life, there were several 'evolutionary formations', characterized by special evolutionary factors and features of self-organization forms.

Basics of Darwinism

Although people have used selection of domestic animals for a long time, and first guesses about the development of life were stated in the antiquity (e.g., according to Aristotle it is driven by a special living force – entelechy), the theory of 'transformism' (about changeability of living forms), which opposed the theory of creationism (about constancy of organisms, created by God), was formed only in the eighteenth century. The first fundamental theory of biological evolution was proposed by J. B. Lamarck (1744–1829) in 1809. This theory, progressive for its time, did not include the idea of natural selection and assumed inheritance of acquired characters. In 1858, Charles Robert Darwin (1809–82) proposed a new evolutionary theory based on the mechanism of natural selection.

Darwinism includes two key ideas: undirected variation ('mutations') of discrete hereditary codes ('genes') passed from parents to children, and elimination of less-adapted individuals in the course of 'struggle for existence'. The codes, better reflecting the external conditions, gradually become dominant; average characteristics of individuals are changed. Selection transfers information about environment in the hereditary code. Useful negative fluctuations of entropy are spread over the species.

The Darwinian idea of natural selection can be considered as a very general explanation of matter self-organization – the tendency of entropy decrease in particular objects. This thought corresponds to the opinions of the founders of mathematical genetics, such as R. E. Fisher (1890–1962): "Natural selection is a mechanism for generation improbability," and A. Lotka: "The principle of natural selection reveals itself as capable of yielding information which the first and second laws of thermodynamics are not competent to furnish."

Inheritance, Variability, and Natural Selection

For Darwinian self-organization, life must be structured for discrete generations and lifetime of each generation must be limited. To be effective, transfer of hereditary information from a previous generation to the next one cannot be absolutely free. Each species is divided into more or less isolated populations, where panmixia takes place. New specimens' characters are examined at the level of populations; only in the case of success, they spread for the whole species. Sometimes there are additional levels of such hierarchy: subspecies, races, subpopulations, etc. Structural units of species, uniting closely related individuals, are partly reproductively isolated. It is still not clear whether it is a natural result of hereditary remoteness or it is a manifestation of special isolation mechanisms forming an optimal structure of the hereditary field.

An important fact is discreteness of the hereditary code; genes are indivisible. This consideration eliminates Jenkin's nightmare: a useful character cannot resolve in descendants of reiterative coenobium.

Inherited characters cannot be absolutely independent; some of them are more or less correlated. On the one hand, it is a destabilizing factor; characters are selected in the context of other ones. On the other hand, it leads to tendentiousness of genes' variability, joint manifestation of correlated characters in accordance with the homologous series law of N. I. Vavilov (1920).

Darwinian variability is a principally random phenomenon. It is impossible to predict dynamics of external conditions; evolutionary perspective living forms must have multidirectional hereditary deviations. Initially directed evolution (as Berg's nomogenesis) is not sufficiently flexible to be effective.

Genes' variability (mutations) must be within reasonable limits. If it is too small, the progress will be too slow and can stop far from a local extremum. If the variability is too big, it will lead to system's chaotic behavior. These parametric effects are well illustrated by mathematical models based on the well-known 'genetic algorithm'.

Darwinian natural selection examines the character of different specimens: stability, amativeness, reproductive potential, etc. It is not always a struggle for existence; often it is a struggle for leaving sufficient number of descendants. The main criterion is birth rate. If it is less than one, the species is doomed.

There are three levels of natural selection; only such forms of life can exist, which: (1) are stable and can physiologically give a breed; (2) allow origin of intellectual man (the anthropic principle); and (3) survive in the course of competition with other species (Darwinian selection). Evolution eliminates evidently defective individuals; other ones are not really exterminated, but rather 'squeezed' from the ecosystem because of low birth rate.

Natural selection can be classified into three forms: stabilizing (supporting existing adaptations in stable environment); motivating (producing new adaptations); and disruptive (leading to separation of the population in condition of heterogenic environment).

Selection leads to harmony of the organism and its environment. Similar conditions produce similar organism's forms of adaptation to them. This effect is called 'convergence'. Adaptation can follow a limited number of ways; in particular cases, it is not evident whether this form is a result of selection, or it is a direct consequence of physical laws.

Evolution can be divided into micro- and macroevolution. Microevolution consists in accumulation of hereditary changes in the population. At this stage, the most essential effects are the change of statistical distribution of different genes within the population and a search for their optimal combinations. The main factors of population evolution are maintenance of genetic heterogeneity, population size fluctuations, reproductive isolation, and natural selection.

Macroevolution is the evolution at the level of ecosystems; its result is the origin of new species. Its nature is not absolutely clear yet. It can be explained by either multistep microevolution or a random essential change, appearance of 'perspective freaks'.

New Tendencies in Darwinism

The modern evolutionary theory is an elaboration of the classical Darwinian theory. Apart from such extremely important fields as 'genetics', Darwinism accepted a number of new ideas such as genetic drift and recombination, cooperative evolution, and global biospheric context of evolution.

Mutations do not so often revolutionize populations; mostly, they support its genetic heterogeneity. In accordance with modern views, genetic mutations are not obviously new forms of genes. Usually they are recombinations of existing hereditary material, at the molecular, cellular, organism, and ecosystem levels. Such recombinations can produce a fast evolutionary leap forward. There is a common genetic pool of life; similarity of genes does not necessarily mean cognation.

According to Kimura's theory of 'genetic drift', genes can exist and even breed in a latent form and then rapidly declare themselves. The effect does not contradict Darwinism; it only proposes a broadened understanding of mutations and their formation.

Variability as a result of genetic change is quite typical at the lower level of life. In cells, there is dissipated genetic material, which does not influence its characters, in silent parts in DNA (intrones), in cellular parasites, viruses and plasmids (DNA molecules in the cellular protoplasm, which can be transmitted not only to descendants, but also to neighboring cells).

Although multicellular organisms are protected from genetic material damage by special mechanisms, cases of genetic material transfer can take place for them as well: by viruses, as a result of distant hybridization, etc. Besides, genes, reflecting environment, reflect genes of neighboring organisms. Thus, hereditary units (genes) of all living beings form a closely connected system, a general 'gene pool' of biosphere.

One of the factors of horizontal genetic transfer, great importance of which has been understood lately, is the formation of symbiotic beings. A well-known example of symbiotic organisms is lichen, consisting of two species, fungi and alga, which, in principle, can exist separately. Because of long-standing coevolution, these two species have adjusted their biochemistry and synchronized reproduction. There is a small green sea worm *Convoluta roscoffensis*, feeding on symbiotic algae; algae germs transfer to the next worm generation through gametes. Most animals need symbiotic microorganisms for effective digestion, luminescent organs of animals are a result of symbiosis with bacteria, etc. One of the most important revolutions in the history of life was the origin of eukaryotic cells as a result of step-by-step symbiotic integration of several prokaryotic ones that finally shared their genes.

A subject of competitive selection can be a symbiotic system, unifying initial forms, which before had struggled for existence separately. The symbiotic evolutionary theory complements Darwinism with a deeper appreciation of the fundamental cooperative processes, which accompanied the origin and evolution of life.

Each organism has its own place in the biosphere. It cannot evolve separately; it should coordinate its change with connected species. Particularly, evolution of separate organisms should not damage biogeochemical cycles. In the course of evolution, the cycles, as well as the biosphere, reproduce themselves as comprehensive wholes.

Although the natural selection operates at the level of individuals, increase in information takes place only at the levels of species and ecosystems. Evolution of the biosphere is a grand process of information collection. The main source of the information storage is the biospheric gene pool.

Summary

Although life is an extremely complex phenomenon, including processes of different levels, nature, and duration, modern science has elaborated general approaches to its understanding. First of all, it is the systems approach which allows describing multilevel structure of life and modeling some general principles of its dynamics. The physical view on life is interesting due to understanding of biological laws as continuation of physical ones and the energy approach to living systems, particularly, investigation of role of entropy, exergy, emergy, and so on in biological evolution. Another interesting subject is the information nature of life, which is, finally, the main manifestation of universal information processes.

Dynamics of life is a permanent process of self-organization. A few very general principles of this process can be formulated: equifinality of biological processes; development of life as a process of niches proliferation; and Darwinian natural selection. There are a lot of questions science has no answers for, but general scope of the life phenomenon problem becomes more or less clear now.

See also: Structure and History of Life.

Further Reading

Bonner JT (1988) *The Evolution of Complexity by Means of Natural Selection*. Chicago: University of Chicago Press.

Brooks DR and Wiley EO (1988) *Evolution as Entropy: Toward a Unified Theory of Biology*. Chicago: The University of Chicago Press.

Camazine S, Deneubourg JL, and Franks NR (eds.) (2001) *Self-Organization in Biological Systems*. Princeton, NJ: Princeton University Press.
Eigen M and Schuster P (1979) *The Hypercycle. A Principle of Natural Self-Organization*. Berlin: Springer.
Jørgensen SE, Brown MT, and Odum HT (2004) Energy hierarchy and transformity in the universe. *Ecological Modelling* 178: 17–28.
Kauffman SA (1993) *The Origins of Order: Self-Organization and Selection in Evolution*. Oxford: Oxford University Press.
Lotka AJ (1925) *Elements of Physical Biology*. Baltimore, MD: Williams and Wilkins.
Margulis L and Sagan D (2000) *What Is Life?: The Eternal Enigma*. Princeton, NJ: University of California Press.
Odum HT (1994) *Ecological and General Systems: An Introduction to Systems Ecology*. Niwot, CO: Colorado University Press.
Pahl-Wostl C (1995) *The Dynamic Nature of Ecosystems: Chaos and Order Intertwined*. New York: Wiley.
Pimm SL (1991) *The Balance of Nature?* Chicago: University of Chicago Press.
Rossi E (1992) What is life: From quantum flux to the self. *Psychological Perspectives* 26: 6–22.
Rosen R (1967) *Optimality Principles in Biology*. New York: Plenum.
Rowe G (1994) *Theoretical Models in Biology*. New York: Springer.
Schrödinger E (1944) *What Is Life*? Cambridge: Cambridge University Press.
Seifert J (1997) *What Is Life? The Originality, Irreducibility*, and Value of Life. Amsterdam: Rodopi.
Sheldrake AR (1981) *A New Science of Life: The Hypothesis of Formative Causation*. London: Blond and Briggs.
Svirezhev Yu M (2000) Thermodynamics and ecology. *Ecological Modelling* 132: 11–22.
Ulanowich R (1986) *Growth and Development: Ecosystems Phenomenology*. New York: Springer.
von Bertalanfy L (1952) *Problems of Life*. New York: Wiley.

Structure and History of Life

S V Chernyshenko, Dnipropetrovsk National University, Dnipropetrovsk, Ukraine

Introduction

Earth life is a particular realization of general self-organizing abilities of matter (see Phenomenon of Life: General Aspects). There are no facts concerning other forms of life for comparison, but, probably, some features of Earth life have random character. In the same time, only on the basis of biological data, one can draw a generalization about nature of life.

Life has multilevel hierarchical structure. All the layers are closely connected and form a single whole; for understanding of global ecosystems, it is necessary to understand lower ones and vice versa. Recently, it has become more and more clear that life can be studied as a global phenomenon only. Separate organisms and biological systems are parts of the whole; they cannot be considered in isolation from their biological environment. In this article, which is devoted to ecological understanding of life, exactly such a global approach will be applied.

Global approach gives also the possibility to study general tendencies of life development. In the situation of the recent ecological crisis, it is extremely important to estimate the current situation and generate ideas concerning its improvement.

Microlevels of the Life Organization

The Earth's life has its own peculiarity, realizing general principles of life organization. At each level, beginning from biochemical to biospheric, special forms and mechanisms were elaborated. Modern biology is on the path to reconnect itself as a whole with all its parts, forming a science called systems biology. The microlevels of life belong to traditional biological disciplines and have been studied in depth, but their mutual integration and research on their connections with the macrolevels are very real problems, which are at the very beginning of their investigation.

Chemistry of Life

Chemical elements forming life were synthesized in the interior of stars. Their relative content in different forms of inanimate and living matter is represented in **Table 1**. The main life elements (hydrogen, carbon, nitrogen, and oxygen – from the first and second periods of Mendeleev's table) are also most common in space. The concentration of heavy elements is much higher in living matter. Sulfur and phosphorus are from the third period; as the first four elements, they can form multiple bonds. Apart from the common elements, life uses rare ones for special aims.

Table 1 Content of different elements in inanimate and living matter

Chemical elements	Content (% per weight)						
	Solar matter	*Atmosphere*	*Ocean*	*Earth's crust*	*Soil*	*Plants*	*Animals*
Hydrogen (H)	72		10.7	1.6	3.1	10.0	11.0
Helium (He)	27						
Oxygen (O)	0.28	20.97	85.8	56.2	66.8	70.0	65.0
Carbon (C)	0.12	0.01			1.2	18.0	19.0
Nitrogen (N)	0.05	78.08		0.26	0.06	0.9	3.0
Magnesium (Mg)	0.01		0.13	1.46	0.37	0.08	0.05
Silicon (Si)	0.01			23.05	19.41	0.35	0.24
Sulfur (S)	0.01		0.09	0.24	0.05	0.14	0.18
Iron (Fe)	0.01			3.63	2.24	0.02	0.02
Aluminum (Al)				6.3	4.18	0.01	
Natrium (Na)			1.03	1.95	0.37	0.03	0.05
Potassium (K)			0.04	1.95	0.8	0.03	0.02
Calcium (Ca)			0.04	2.58	0.81	0.03	0.03
Chlorine (Cl)			1.93	0.02	0.01	0.01	0.02

An extremely important substance for life is water. It is a universal medium for almost all chemical processes of life. According to the expression of R. Dubois, "life is animated water."

It is improbable that carbon can be substituted by silicon, and water by ammonia in some extraterrestrial forms of life, as it is described in science fiction. Siliceous polymers are unstable in solutions, and oxide of silicon is solid and inert substance.

The basis of life is formed by organic polymers. There are proteins, nucleic acids, carbohydrates, and lipids. All living beings use the same kinds of macromolecules; it is an illustration of commonalties of life. Polymers convey such life functions as metabolism, genetic inheritance, growth and reproduction, energy storage, and conversion.

'Metabolism' is a circular process of extracting, converting, and storing energy from nutrients. It is a complex network of chemical reactions, such as group transfer and oxidation–reduction reactions, dehydration, carbon–carbon bonding, etc. Groups of reactions form 'catabolism' (the oxidative degradation of molecules) and 'anabolism' (the reductive synthesis of molecules), maintaining metabolic 'homeostasis' (a steady state of organism).

Two key groups of macromolecules are nucleic acids ('legislative body' – storage and development of information) and proteins ('executive body' – metabolism and maintenance of nucleic acids).

In ordinary chemistry, reactions proceed as a result of heat motion. In biochemistry, proteins, called enzymes, evolve catalytic chemical reactions, proceeding in a very specific and efficient way. They have special active centers, geometrically stimulating proximity of algoristic molecules and their interaction. Under the influence of enzymes, reactions can run in conditions of low temperature, although usually they proceed under high temperature only: outside of organism, lipids and carbohydrates oxidize under the temperature 400–500 °C; synthesis of ammonia from molecular nitrogen proceeds under the temperature 500 °C and pressure 300–350 atm.

Enzymes catalyze every biological process of life, but proteins also play other roles in living organism. They form physical basis of tissues (collagen), transporting (hemoglobin), protecting (immunoglobulin), regulatory (hormones) agents. In human organism, there are more than 5 million different proteins.

Proteins are the result of polymerization of amino acids: from several dozens to many hundreds. Amino acids are connected by covalent peptide bonds and form the primary structure. The primary thread is packed in the spatial secondary and tertiary structures by hydrogen bonds. Several proteins (protomers) can form quaternary structure (oligomer). There are 20 different amino acids in proteins of living organisms.

Nucleic acids have two forms in living organisms: ribonucleic acid (RNA) and deoxyribonucleic acid (DNA). They are polymers of nucleotides, made up of the nitrogen bases: two purines (adenine and guanine) and two pyrimidines (thymine in DNA or uracil in RNA and cytosine). RNA has a spiral primary structure and more complex secondary ones. DNA forms a double helix from two complementary macromolecules.

The hereditary role of nucleic acids became clear in 1944, when the transfer of hereditary characters was discovered. A triplet of nucleotides ('codon') in DNA or RNA codes amino acid. The hereditary code was deciphered by J. D. Watson and F. Crick in 1953. Sixty-four different triplets code 20 amino acids in accordance with some rules. Human genome includes about 3.3×10^9 pairs of nucleotides.

One of the most important processes in living organisms is a cyclic process of joint replication of DNA and proteins:

nucleic acids store information about structure of enzymes, which, in their turn, catalyze replication of DNA.

RNA has some abilities to self-catalyze, and there is idea about 'RNA world' as one of the first steps of the life formation. The self-replication is not very reliable (1–10% of mistakes), and it is an explanation of forming modern catalytic replication (much less than 10^{-9}% of mistakes), but natural selection of RNA could take place.

There are many other organic substances, extremely important for life: carbohydrates, lipids, adenosinephosphates, etc. Theories about 'sugar model' by A. Weber in 1997 or 'lipid world' by D. Segre in 2001 reflect the base role of these substances for life functioning. It is necessary to emphasize the fundamental role of the energetic molecules well-known as adenosinetriphosphate (ATP). The beginning of cyclic reproduction of RNA or the system proteins DNA needed energy. A process of ATP synthesis, probably, preceded these cycles; from this point of view, ATP can be considered as the first molecule of life.

Organisms can produce most of the necessary chemical substances, although there are several dozens of them which should come from outside. All substances, especially proteins, are constantly destroying and synthesizing, renewing their composition.

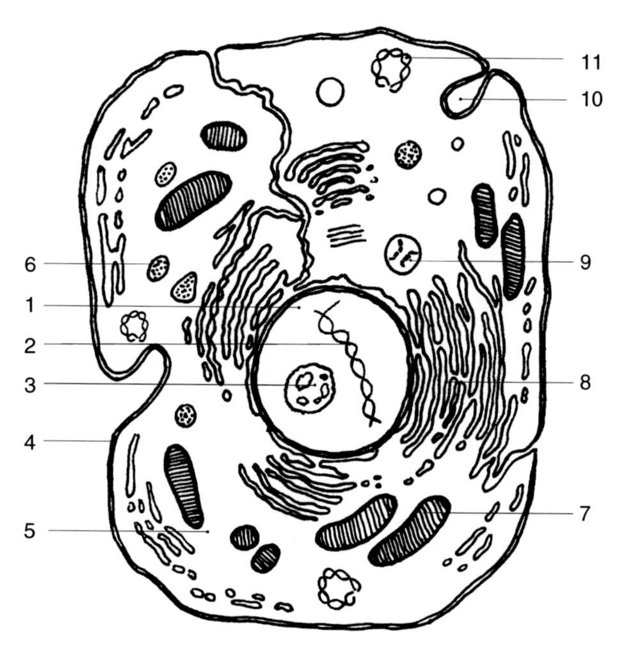

Figure 1 The cell structure (1, the nucleus; 2, DNA; 3, nucleolus; 4, membrane; 5, cytoplasm; 6, lysosome; 7, mitochondrion; 8, cytoplasmic reticulum; 9, centriole; 10, pinocytic vesicle; 11, plasmid).

Level of Cell

The next level of the life organization is a cell. Excluding some primitive forms, such as viruses, living matter is divided into separate cells. As an object of concentration of all coordinated chemical reactions of life, the cell must be big enough to be independent of fluctuations of the heat motion.

The simplest structure corresponds to prokaryotic cells. It contains only absolutely necessary parts – membrane, DNA, and cytoplasm – where chemical processes of protein synthesis (in special particles – 'ribosomes'), energy production, respiration, reproduction, and others are arranged. This type of cells is used by bacteria, archaea, cyanobacteria, actinomycete, and others. Some parasitic forms of prokaryotes such as *Mycoplasma*, Chlamydiae, and *Rickettsia* lost certain cellular mechanisms.

The eukaryotic cells are much larger (10–100 times). They contain multiple internal 'organelles', including the 'nucleus' (storage of hereditary information), the 'mitochondria' and, in plant cells, 'chloroplasts' (energy transformation), 'lysosomes' (digestion), 'kinetosomes' (movement), 'cytoplasmic reticulum' (redistribution of chemical substances). The cell structure is represented in **Figure 1**.

Eukaryotes are the result of symbiosis: chloroplasts are progeny of cyanobacteria, mitochondria descend from bacteria, etc. Mitochondria, chloroplasts, and kinetosomes have their own DNA and can reproduce partly independently (using proteins, synthesizing by the cell). Coding of proteins in mitochondria is different from nuclear one. But all processes in cell are coordinated – it is a united complex system.

The cell is separated from the environment by membrane; its internal space is also divided by membranes into compartments. The fluid mosaic model of membrane defines a phospholipid bilayer with hydrophilic part from the outside and hydrophobic layer inside. Membrane proteins float in the phospholipids and control exchange of matter with the environment.

Chemical synthesis in the cell is a complex process with multiple feedbacks. Proteins are synthesized permanently, but only from active sections of DNA. Some sections can be repressed by special regulatory proteins, histones; to be active, they must be 'derepressed'. For starting synthesis, a special 'inductor' has to interact with corresponding DNA section. Inductors and derepressors play information role (metabolites or hormones).

Information about protein structure is copied ('transcripted') to iRNA molecules and transported to cytoplasm, where it is processed by ribosomes. The ribosomes move along iRNA and read the information, synthesizing simultaneously the protein from amino acids transported by tRNA molecules.

A post-transcription and post-ribosome regulation of protein activity can take place. Under the influence of enzymes, proteins can be chemically transformed: for example, proline is oxidized to oxyproline. Other enzymes, 'allosteric effectors' (hormones, adenosine monophosphate (AMP), etc.), can change tertiary structure of proteins, influencing on their active centers. Finally, such agents as 'inhibitors' can temporarily deactivate enzymes.

The presence of the number of ways of regulation gives a possibility to establish a complex and reliable 'cybernetic structure' of the chemical synthesis process in the cell. In a typical cell, there are more than 1000 systems of control for production of different enzymes. They are in permanent interaction. The simplest way is to use the reaction product as a repressor of own synthesis, or its substratum can play the role of its inductor; but often the scheme is much more complex.

For all single-celled organisms (and somatic cells of multicellular organisms), 'asexual reproduction' is typical. Simple cell division involves duplicating the genetic material and separating into the two ensuing daughter cells. The process is called 'mitosis' for eukaryotic cells, and 'binary fission' for prokaryotic cells.

Level of Organism

Multicellular organisms have a number of eukaryotic cells. Contrary to protozoa colonies, their cells are varied; it is a single whole, the next level of living matter organization. Although there are intermediate forms, in adverse conditions a population of amoeba *Dictyostelium discoideum* can form from separate specimens of *Plasmodium*, having embryos with organs of movement and reproduction.

Tendency to form multicellular organisms caused problem of concentration of all hereditary information in a single cell nucleus. This information must include all the program of the individual's development. Modern views are closer to the concept of 'preformation' of A. Leeuwenhoek (1632–1723), than to 'epigenesis' of R. Descartes (1596–1650). Inheritance began to combine with 'ontogenesis' – the process of organism's formation from a gametal cell.

Differentiation of cells is guided by autocatalytic molecular reactions, similar to mechanisms caused by the evolutionary progress. It is an explanation of the biogenetic Haeckel–Muller law (1866) about the repetition of phylogenesis by ontogenesis.

The common genetic code of all cells of an organism creates prerequisites for their cooperative behavior. Intercellular interaction oversteps the cellular egoism, which is considered as pathology and can produce such diseases as cancer. The being of a cell is directed for prosperity of its organism, for protection, and spreading of the common genetic code. All cells have a specialization (neurons in the neural tissue, myocytes in muscles, etc.). There are programs of cellular self-destruction; their lifetime (for human erythrocyte – 100–120 days) is defined by interactions of the organism and species. Special 'stem cells' (0.2–1% of the total number of cells) have been discovered lately; they are not really specialized and have no restriction for divisions. Furthermore, they can form any kind of tissues; so they are used for treatment of many diseases.

Sexual reproduction is very typical for multicellular organisms. Its advantage is an increase of genetic variation of offspring in comparison with asexual reproduction. It stimulates evolution and is especially important in times of intensive species formation. For stable conditions, it is not so important, and now for such successful group as flowering plants there is a tendency to turn to asexual forms of reproduction.

A multicellular organism's cell usually contains two complete genetic sets ('genomes'); it is 'diploid'. Since sexual reproduction involves participation of hereditary material from two gametes, the number of genomes has to be halved in the gametes. As a part of the reproductive process, organisms of males and females produce 'haploid' cells (containing single genome) from diploid ones in a process of 'meiosis'.

At the new level, multicellular organisms have repeated the way of unicellular ones in coordination of internal processes. It was necessary to realize at least four main functions of an organism: homeostasis (first of all, maintenance of stability of internal environment, in accordance with the law of K. Bernard, 1872); growth (in a certain sense conflicting with homeostasis); adaptation; and reproduction. All these processes require united control systems for their regulation. In this way, the 'hormonal system' (of chemical nature, extension of the cellular one) was formed. In organisms, there is a system of interacting endocrine glands, producing special effectors ('hormones') and reacting on incoming chemical agents. Hormones can influence cells and organs as inductors, derepressor, etc. The cell membrane has receptors, reacting to chemical agents. In accordance with these signals, the cell coordinates its activity with the whole organism. Later the 'neural system' of animals was originated. It can influence cells both directly (electrically) and through the hormonal system. A special organ 'hypothalamus' plays the role of the coordinating center of the neural and hormonal systems.

The neural system is the most developed control system of life. It is organized on cybernetic principles, and it is small wonder, because the development of the nervous system gave ultimately the birth to science and, particularly, cybernetics. Control of all the biological processes has an information basis. Each reaction of the organism is based on two kinds of information: external from organs of sense, and internal from the organism's memory. An elementary unit of behavior is test-operate-test-exit (TOTE); it consists in comparison of real (from external information) and necessary (from internal information) states before and after the operation.

Memory should be big (there are a great number of possible influences and reactions) and fast (sometimes speed of response is critical). Usually organisms have both these types of memory separately. An organism deals with four kinds of information: (1) structural (fixed in its structure); (2) genetic; (3) mnemonic (from its memory); (4) external (from organs of sense).

Information flow to human organism is estimated as 10^7 bps; its lifetime is 10^9 s; the brain cortex contains about 10^{10} cells. Very probably, that information is stored at molecular level, for example, in RNA molecules. Some experiments confirm this theory. Theoretically, it can be explained by different probabilities of different nucleotide types' joining to forming RNA under the influence of external neural impulse.

Life as Ecological Phenomena

The next levels of life organization can be called the ecological levels. There are population, species, consortium, ecocycle, biogeocoenose (ecosystem), and biosphere levels. Historically, biosphere (protobiosphere) was the oldest form of ecological macrosystems; it was a direct extension of the geological cycle. Probably, from the very beginning, biosphere consisted of separate cycles (protobiogoenoses). As a result of corpusclezation, small pieces of living matter took shape of cells, which developed to populations of living organisms.

Ecological Macrosystems

The term 'biosphere' was proposed (in the present sense) by E. Suess (1831–1914) in 1875. The author said that the term expresses ideas of C. Darwin and J. B. Lamarck about the unity of life. A great contribution to the development of the biosphere theory was made by V. I. Vernadsky (1863–1945). A. Humboldt (1769–1859) in his book *Cosmos* (1848) first viewed the Earth as a whole. The modern version of the idea is known as the science of Gaia.

Biosphere is separated from environment much better than other biosystems. It is under the influence of space (the solar radiation – the main source of energy, gravity of the Moon – the flows and deceleration of the Earth's revolution) and geological processes in the Earth (volcanism, continental drift, sea transgression and regression, orogeny). The influence of life on geological processes exists, but it is very small and broadened in time. Its connection with environment can be considered as one-way.

Biosphere is a global open system with properties of homeostasis. Its input is the solar radiation and some substances from the Earth's interior; output is biogenic matters, leaving the cycle for a long time. Biosphere is a centralized cybernetic system with life ('biota') as a central controlling subsystem.

The basis of biosphere is the global biogeochemical cycle (see Matter and Matter Flows in the Biosphere). The big abiotic cycle is, first of all, the water cycle; from 5×10^{20} kcal, arriving from the Sun to the Earth, approximately one-half is spent for maintenance of this cycle. The small biological cycle is based on the abiotic one. It uses only 0.1–0.2% of the solar energy, but it is very effective because it has very special, information structure. The biological cycle is a multilayer object, consisting really of a lot of different cycles. Particularly, its oldest part and basis is matter cycling between unicellular synthesizers and destructors.

The previous level of the biological systems hierarchy is 'ecosystem' or 'biogeocoenose'. It is a stable self-reproducing system of interrelated populations of different species from some territory and their abiotic environment. The internal environment of ecosystems is relatively stable, all the species are co-adapted.

Ecosystems form biosphere; they are connected (e.g., in winter we breathe oxygen from the tropics and the other hemisphere), but not very closely. Potentially each type of biogeocoenoses can occupy all the Earth, but the competition of other ecosystems (more effective energetically in certain conditions) does not allow this. Existence of different biogeocoenoses exerting 'biological pressure' upon neighbors makes biosphere stable and adaptive. Data on main kinds of ecosystems are represented in **Table 2**.

A biogeocoenose is characterized by its own type of biological cycling, but really it is a combination of cycles of different nature, which are called 'ecocycles' or 'coenomes'. They supplement and duplicate each other; their interweaving creates a strong substance of the biogeocoenose. Similarly to ecosystem, each coenome can become dominant, but it is restrained by the competition with other coenomes for solar radiation and mineral resources (contrary to ecosystems, they are not separated geographically and share a territory). A kernel of the coenome is a species of plant producer; it introduces energy, revolving the cycle. Other important participants of the process are species of reducers (mainly bacteria and fungi), which decompose dead organic matter and return it to the cycling. On average, plants produce annually 10% of their biomass; biomass of reducers is a hundred times less, that is, they decompose annually 10 times more than their weight.

Species of consumers (mainly animals), feeding on other organisms (in other words, connected with them by 'trophic relations'), form 'food webs' of the ecosystem and stimulate the cycling. Usually, there are not more than four to six trophic levels. Energy transfer between the levels is about 10%. Trophic relations are not so much the struggle for existence; basically, it is the process of co-adaptation.

Table 2 Biomass (dry weight) of different ecosystems

Type of ecosystem	*Average biomass ($kg\,m^{-2}$)*	*Total biomass (10^9 t)*	*Annual production ($kg\,m^{-2}$)*
Tropical forests	52	[illegible]	[illegible]
Tropical and subtropical seasonal forests	40	300	2.2
Savannah	4.5	66	1.9
Deserts and semideserts	0.6	11	0.15
Steppe and forest-steppe	2.6	44	0.85
Forests of the temperate zone	37	408	1.2
Bogs	6.5	13	0.45
Taiga	21	253	0.8
Tundra	0.9	7	0.2
Agricultural lands	2	21	0.7
Continental ecosystems, total	13.3	1990	1.25
Open ocean	0.003	1.0	0.15
Zones of upwelling	0.025	0.01	0.5
Shelf	0.02	0.53	0.4
Sargasso and reefs	2	1.2	2.3
Estuaries	1.5	2.1	1.3
Sea ecosystems, total	0.013	4.8	0.18
Total	3.9	1995	0.44

The competition of coenomes is mainly a competition between their dominant producers. There are different kinds of the producers: powerful 'competitors', low-powerful 'pioneers', and medium-power suppliers 'opportunists'. The following three competitive situations can take place: evident preference for some species; 'hard competition', when powerfulness of populations is similar, but one population should win; and 'soft competition', when populations can coexist.

The competition of producers for ecological resources is a fundamental motivating force of ecosystems global dynamics. Disturbed biogeocoenose returns to its steady state not smoothly, but in discrete steps (stages), repeating partially the process of the biogeocoenose formation. This process, manifesting the ecosystems property of homeostasis, is called 'succession'. It is a nonlinear process of step-by-step changing dominant plant association, accounted for by insufficient solar energy utilization by first 'undemanding' succession stages, leaving 'energetic space' for the development of next, more effective stages. The evolution of biogeocoenose proceeds in the direction of maximal utilization of the solar energy, reaching a final state (state of 'climax'). As a rule, the final stage cannot develop during first stages of the succession. During the evolution of producers, more and more effective associations evolved, but they adapted to conditions of some previous biogeocoenoses. Their development is possible after creation at some succession stage of the corresponding ecological conditions.

The succession process is a process of producing information, if the information is understood as complexity of system response to variations of environmental conditions; the information measure achieves an extreme in the climax state. One of the manifestations of the progress in the course of successions (as well as evolution) is increase of 'biodiversity', multiplicity of different forms of life (an evolutionary point of view: God is 'generator of diversity').

According to W. R. Ashby (1955), diversity is necessary for system adaptability; development of life illustrates this statement. As a result of living forms' divergence, creation and infill of niches, a huge number of species have arisen. At present, there exist about 2 million species, and their number during all the history of life is about 1 billion. The data for different taxonomic groups are represented in **Table 3**. The numbers are approximate: many species are not discovered, and boarders between species are quite relative.

The level of species is more integrated than the biogeocoenotic one; it is based on common ancestry and allied bonds. The ability of populations to exponential growth, limited by competitive and trophic relations, creates internal energy of the ecosystem, its resilience. The interpopulation relations are not so much struggle for existence as permanent frontier wars, supporting high level of energy of all their participants.

Life as Geological Factor

The highest level of biological systems, biosphere, can be considered as a part of more global geological system. Geochemical processes gave rise to life; in its turn, life became an active participant of these processes.

Table 3 Number of species in taxonomic groups of eukaryotes

Organisms	*Number of species, thousands*
Protozoa	26
Sponges	5
Coelenterates and ctenophores	9.1
Worms	36
Flat worms	9
Round worms	15
Annelids	11
Mollusks	107
Arthropoda	1200
Insects	1100
Tentaculata	4.7
Pogonophora	0.1
Echinodermata	6
Semichordates	0.1
Chordates	39
Tunicates	1.1
Cyclostomes	0.1
Fishes	19
Amphibians	2.9
Reptiles	2.7
Birds	8.6
Mammals	3.7
Animals, total	1440
Algae	20
Diatoms	10
Red algae	2.5
Brown algae	1
Green algae	6.5
Mosses	24
Club-mosses	0.4
Pteridophytes	9
Gymnosperms	0.6
Metasperms	250
Plants, total	305
Fungi	105
Lichens	30
Total	1880

Speaking about life as a geological factor, it is possible to put aside biological peculiarities of living systems and concentrate on geochemical characteristics of life and its influence on geological objects. 'Living matter', in accordance with the views of V. I. Vernadsky, is "a totality of organisms, reduced to its weight, chemical composition and energy." At the geological level, life operates through these global parameters. From the biogeochemical point of view, the swarm of locusts is a moving dispersed rock, which is very active chemically.

Biosphere is a global form of living matter, a manifestation of its continuity. It consists of different, but closely connected forms of life. Biosphere is an ocean of genes; informative genes of higher organisms are a very small part of it. The reproductive part of living matter evolves its iterative self-reproduction; the somatic one supports its metabolism and growth. According to V. I. Vernadsky, biosphere tends to maximal intensity of biogenic migration of elements. It is close to Lotka's maximum power principle: the intensification of global biogeochemical cycling can be based only on increase of energetic power of biosphere.

The properties of living matter as a geochemical substance are the following:

- its chemical composition is stable only dynamically, while organisms are living;
- the chemical composition is very diverse (2 million organic substances in living objects as against 2000 minerals in animated nature);
- it has a huge free energy – exergy (similar to fresh lava, but much more durable);
- it is quite mobile: actively (because of active movement of such organisms as animals) and passively (as a result of reproduction and growth, the pressure of life: direct offspring of the gigantic puffball, producing 7.5 billion spores, can exceed the mass of the Earth by 800 times);
- it is dispersed matter, which consists of separate particles (living organisms) with the size from 20 nm to 100 m (the difference is in billion times).

The processes in living organisms, important for geological processes, are (1) formation of bodies; (2) excretion of products of metabolism; (3) transportation of indigested substances through the digestive tract; (4) movement; and (5) mechanical influence on environment.

The geochemical functions of living matter are energetic (photo- and chemosynthesis, energy transmission in food webs); transport (matter transportation against gravity and horizontally); extracting (involving matter from environment to the biogeochemical cycling); concentrating (selected accumulation of chemical substances in bodies or excrements); and environmental (influence on physical or chemical parameters of environment).

J.-B. Lamarck asserted already that practically all natural landscapes and the Earth's crust rocks are results of living matter functioning. V. I. Vernadsky added to this thesis the idea about life as an integral part of geological processes. Primary rocks, originated without participation of biosphere, are massive and homogenous; the existing diversity of rocks was created by life.

The Earth's crust consists of sediment rocks. The main stages of their formation are hypergenesis (destruction of parent materials, mainly as a result of living organisms' activity); sedimentogenesis (accumulation, under essential influence of biofiltrating organisms, of sediments in friable forms: clay, silt, peat); diagenesis (sediment compaction with participation of bacteria, mud-eaters, fossorial animals); and katagenesis (further compaction, sediment rocks transform to metamorphic ones).

The main biogenic sediment rocks are carbonate rocks (formed mainly by phytoplankton); siliceous rocks (sponges, radiolarians, diatoms); combustible organic substances (coal from peat, shale oil from sapropel, oil from plankton organic substances); and phosphate rocks (excrements, mass suffocations of marine organisms). Such minerals as bauxites, ferruginous and manganous rocks are formed as a result of activity of bacteria.

Under the sediment stratum, there is also biogenic matter, but melted in the course of the lithogenesis. From the ocean bottom, sediments deepen under continents, following the global Earth's crust movement.

Biosphere controls composition of the atmosphere; the content of oxygen and carbonic gas is a direct result of living matter activity. One can say the same about composition of ocean water. Life created such new components of environment as soil with hybrid (not pure biological, not pure abiotic) nature. It influences even such an energywise important global parameter as the reflectivity factor of different parts of the Earth's surface.

Geography of Life

Distribution of life within the Earth is limited by some physical and chemical parameters. In some conditions, living organisms can realize all their functions; in some others, their functionality is not full (usually the most sensitive process is reproduction). The first part of biosphere can be called a 'field of life'; the second one is a 'field of survival', or 'parabiosphere' in terminology of J. Hutchinson (1884–1972).

A very important parameter is temperature: too high one leads to protein degradation, too low temperature prevents activity of enzymes. Water is necessary in liquid form.

Some 'psychrophilic' (cold-loving) forms of bacteria, algae, and fungi can live and reproduce under *c.* –7 °C; the rootage of such higher plants as the Kamchatka rhododendron operates under 0 °C. At the same time, there are 'thermophilic' (heat-loving) species of eels, worms, and insects which live in thermal springs (50–80 °C). Spores of bacteria can survive in a temperature of 160–180 °C. Life has filled territories with all thermal conditions: from the Antarctic (the summer air temperature is 0 °C, the winter one is –40 °C) to the most inclement deserts (45 °C in summer, 0 °C in winter).

One of the factors determining the upper board of terrestrial life is the atmospheric pressure and availability of oxygen and carbonic gas. At a height of 6200 m, a partial pressure of carbonic gas is 2 times less than normal one, and plants cannot grow. The upper board of ecosystems of Alpine belt is at the height of 2000–3500 m; in the Pamir mountains, it is 4800 m. In eternal ice at the height of 6000 m, one can find snow fleas and flightless grasshoppers.

The near-surface Earth stratum can be inhabited at a depth of 1–3 km. Although if groundwater is too salted (more than 270 g l^{-1}, that is, 10 times more than in the seawater) or its reservoir is isolated, life can be lacking even at a depth of 0.5 km.

The main limiting factors in the oceans are the deficiency of mineral nutrition for plants and the impossibility of photosynthesis in ocean depths. Sometimes, a chemical contamination can take place (e.g., the Black Sea is polluted by hydrogen sulfide). High pressure is not so essential; under 1000 atm at the depth of 10 km, there are several hundred species. Sediments on the sea bottom can be inhabited at the depth of 120 m.

Atmosphere is, generally speaking, only a field of survival. Flying insects and birds can spend a lot of time in the air, but some part of their life is obviously connected with the land. Condor can be observed at the height of 7 km; viable bacteria were found at the height of 77 km.

Thus, the field of life in the ocean is water stratum and bottom sediments of quite various thicknesses. In the continents, it is thin ground and thick underground layers.

The vertical films of life are the following: surface (plankton) and bottom (benthos) in sea, and surface and soil in land. The horizontal concentrations are littoral, reef, sargasso, upwelling, abyssal rift (marine); coastal, riverside, tropic forest, lake (terrestrial). Some of them are represented in **Figure 2**.

Abyssal rift concentrations of life are based on chemotrophic producers and do not depend on the solar radiation. They can be considered as a reserve of biosphere in case of some global cataclysm, and, according to one hypothesis, are an initial form of living matter. It is also an illustration of the possibility of life existing without entry of solar radiation (e.g., under ice in the big planets of the solar system).

History of Life

Life is a process, and it is not only self-reproduction, but also permanent development, increase of complexity and organizational levels. This process is not smooth and linear; it is a chain of alternate rapid changes and periods of stability. The history of life is characterized by several great revolutions, and the first of them is the transition from chemical reactions to biochemical ones. Understanding of life origin is, besides other aspects, the best way to obtain an appropriate definition of life.

Origin of Life

From ancient time it was recognized that self-generation of living organisms (usually from rotting organic substances) can take place. Such thinkers and scientists as Aristotle, F. Bacon, R. Descartes, and G. Galilee shared this opinion. Correspondingly, the problem of initial origin of life was irrelevant. Only in 1688 F. Redi (1626–98)

Figure 2 The vertical films of life (I, surface (plankton); II, bottom (benthos)) and horizontal concentrations (1, littoral; 2, sargasso; 3, reef; 4, upwelling, 5, abyssal rift; 6, coastal) in marine and neighboring ecosystems.

experimentally demonstrated that maggots appear in carrion meat only in the condition of access of flies. The statement "all living beings originate from living beings" is called the principle of Redi.

Another view, incompatible with the idea of initial self-origin of life, is the opinion about the impossibility of synthesis of organic substances, which contain a special living force – *vis vitalis*. However, in 1828, F. Wöhler (1800–82) synthesized carbamide from ammonium cyanide. During the nineteenth century, *c.* 75 000 different organic substances from inorganic ones were synthesized (and a billion of them are received now).

The first theories concerning origin of life were proposed by E. F. W. Pflüger (1829–1910), H. F. Osborn (1857–1935), and others. The theories about genesis of organic matter proposed by A. I. Oparin (1924) and J. B. S. Haldane (1929) are still of great interest.

Self-organization took place from the very beginning of the formation of the world. Other prerequisites of the life origin were formation of planetary systems at the macrolevel and heavy elements at the microlevel. A perspective planet must have circular orbit and mass, which is similar to the Earth's (its atmosphere can lose hydrogen, but not carbonic gas and oxygen). A central star must have stable radiation.

One of the important prerequisites of the origin of life was synthesis of simple organic molecules in space. Such substances as ammonia, cyanhydric acid, methyl acetylene, and others are widespread in interstellar space; formaldehyde forms even clouds with concentration of 1000 molecules per cubic centimeter. The Earth gained organics during its formation and is still obtaining it through meteorites.

The initial atmosphere consisted mainly of hydrogen, helium, and methane. The anoxic conditions allowed synthesis of simple organic substances, similar to the ones originated in space. A very important point was the transition of water to liquid state. In water, much more complex organic molecules can be generated. As was shown by experiments of S. L. Miller, electric discharge in mixture of methane, ammonia, hydrogen, and water can produce a number of complex organic molecules, for example, amino acids. During 2 weeks, 15% of methane transformed to organics. All 20 amino acids, forming proteins, can be obtained inorganically, as well as nucleotides, fats, sugars, etc.

Organic substances are so multifarious by their functional properties that the selection of proper elements for the formation of biochemical systems was not so difficult (but could be random). Probability of life origin was quite high, although there were, probably, bottlenecks in the formation of some important reactions.

The second prerequisite of life was the availability of external energy sources. During first stages of the life formation, it was, probably, ultraviolet solar radiation. In accordance with the data of K. Sagan, 2400–2900 Å wavelength radiation can evolve 1% solution of organic substances in ocean water, failing the ozone shield. Corresponding energy is sufficient for existence of stable populations of bacteria. Molecules of ATP can operate in this case as primitive chemical accumulators of the solar energy. This substance may have played a similar role in the course of the origin of life.

The third important prerequisite of the origin of life is the abiotic water cycling. Dynamic equilibrium of life in the planetary scale can be reached only with the help of such global intermixing machine. Cosmic factors (territorial and temporal irregularity of heating by a central star, tidal influence of neighboring celestial bodies, etc.) determining the cycling are widespread in space.

Life originated as a biological cycling, united biochemical synthesis, and destruction in one process. According to K. Bernard, life is always a combination of these two tendencies. From the very beginning, reactions of primordial photosynthesis (initial producing) were accompanied by reactions using the energy of synthesized substances for new synthesis and, finally, for their destruction and return of elementary substances to environment (initial consuming). In the course of the corpusclezation process, these two groups of reactions begot first producers and reducers. Life as a unity of synthesis and destruction was not based at first on separate organisms. According to the statement of J. D. Bernal (1969), life appeared earlier than living organisms.

However, very often life is defined at the cellular level. Really, compartmentalization of groups of connected chemical reactions can evolve much more effective and ordered realization of the living-like processes, and origin of cell produced a revolution of life functioning. Theories of the origin of life attempt to explain a mechanism of formation of a primordial single cell from which all modern life originates.

Experiments have shown that formation of cell-like objects is quite usual for organic blends. First candidates for the role of initial cells are 'coacervates', discovered by G. H. Bunoenberg de Yong (1949). They are separated from the blend by a surface lipid film and can selectively absorb different substances from environment. The absorption can lead to dividing a drop into two similar parts.

S. Fox in 1970 found another potential pre-cell, 'microsphere'. Its properties are similar to coacervate ones, but it has a fixed size (about 10^{-6} m), is quite stable, and does not tend to merge with other microspheres. Its lipid membrane is bilayer. G. Tibor (1980) has considered microspheres (containing reagents of special autocatalytic reactions) as the first form of life and called them 'chemotones'.

The 'autocatalytic' origin of life is now universally recognized. It is not evident, what type of autocatalytic reaction was initial; now the general cellular autocatalytic process is very complex and includes many intermediate steps. The main players of this process are nucleic acids and proteins; its very simplified scheme is represented in **Figure 3**. Nucleic acids can store information and replicate, but need catalysts for this; proteins are good catalysts, but cannot replicate. Two groups of proteins catalyze self-reproduction of DNA (polymerases) and synthesis on the DNA matrix of both protein groups (synthetases).

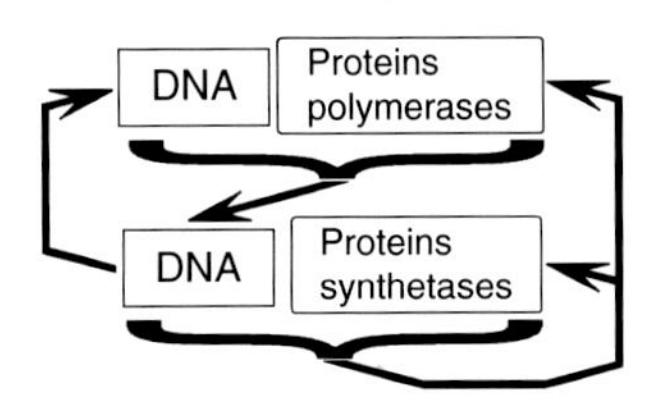

Figure 3 Simplest scheme of autocatalytic reproduction of cells.

There are two main hypotheses about the formation of the autocatalytic loop. The first one was deeply elaborated by M. Eigen and his co-authors in 1979 in the model of 'hypercycle'. It is based on the assumption about initial origin of many-component autocatalysis, including at least two components: RNA and a protein. The second theory, developed by G. F. Joyce in 1989, presupposes a simpler self-catalysis of RNA. In both approaches, RNA is considered as natural predecessor of DNA.

Thus, the origin of pre-cell consists in compartmentalization in separate volumes, bounded by lipid membranes, reagents of reactions, aimed at realization of at least two key functions: assimilation of energy and self-replication, leading to dividing the cell into two. Origin of the hereditary code and cell reproduction launch the process of Darwinian selection.

Modern science has explained many peculiarities of life development, but it is still impossible to check corresponding hypotheses practically. It is supposed, that some events should happen accidentally; even in laboratory, occurrence of a propitious condition can take thousand years, whereas in nature a million years. Anyway, it now does not look so plausible that life has an extraterrestrial origin, as it was supposed by W. Thomson (1871), G. L. F. Helmholtz (1872), S. A. Arrhenius (1915), and many others. Space still sends to the Earth organic substance – material for life – but nobody finds organisms in meteorites. All earthly organisms are very closely connected; it is not so evident that a single organism can give rise to life in a planet (not only one-time explosion) even in case of proper conditions.

Main Stages of the Life Evolution

Several decades ago, it was recognized that the age of life is *c.* 1 billion years. But last discoveries in Africa have shown that as long as 3.2 billion years ago there were bacteria-like forms of life. It is possible, reasoning from isotopic composition of oldest carbonates from Greenland, that life reached the cell level 3.9 billion years ago. The Earth's age is *c.* 4.6 billion years; if one excludes the first stage of intensive meteorite bombing (about 0.5 billion years), the way from simple organic molecules to life lasted only a few million years.

Probably, the first type of chemosynthesis was synthesis of methane from carbonic acid and hydrogen by organisms, similar to present-day methane bacteria. Under the influence of ultraviolet radiation, methane transformed to organic substances, consumed by other organisms, which returned carbonic acid to water. It was the first biological cycling; it came to naught in the course of decrease of hydrogen in atmosphere.

Concurrently, first forms of photosynthesis were originated. Oldest discovered organisms (*c.* 3.5 billion years old) are similar to cyanobacteria, that is, they were phototrophs. First phototrophic organisms did not produce oxygen, but 2.5 billion years ago the photosynthesis on the basis of chlorophyll arose. It was the start of the oxygen revolution, which led 0.5 billion years later to the total reorganization of biosphere. After some regression, explained by toxicity of oxygen for oldest organisms, life formed a new mechanism of oxygen use – respiration.

The oxygen revolution, probably, stimulated the further great step – origin of eukaryotic unicellular organisms 1.9–2 billion years ago. It was huge progress in structural and functional organization of cells; special organs (organelles) for different functions (inheritance, photosynthesis, respiration, etc.) arose, mainly as a result of symbiotic incorporation of other prokaryotic cells.

A little later, the next revolution took place: the first multicellular organisms originated about 1.8 billion years ago. The path to multicellular forms led through colonies of unicellular organisms, which are typical for many groups: flagellates, infusorians, algae, etc. Cells gradually differentiated for reproducing, alimentary, impellent, etc., purposes. Transition to multicellularity took place independently in different groups of unicellular organisms. By the beginning of Cambrian period, all kingdoms and subkingdoms, excluding higher plants, had existed, although practically only the zone of continental shelf was inhabited.

A very important point of the life history is the border between Cryptozoic and Phanerozoic eons. Many different groups of animals (sponges, arthropoda, echinodermata, mollusks, etc.) acquired exoskeleton. It is explained by the growth of oxygen concentration in the atmosphere that facilitated synthesis of collagen. Life explosion in Cambrian period is an explosion of extant fossils.

Approximately at the same time the inhabiting of lands started. As usual, new niche development provoked intensive process of new form building. The last big group – higher plans – originated and began its intensive evolution. Landscapes became more similar to modern ones: after origin of first forests in later Devonian period, they started to regulate surface-water flows, and rivers and lakes obtained the modern form.

Geochronology of the evolution of life is represented in **Table 4** and **Figure 4**. Last thousand years is nothing in the geological scale, but they are accompanied by the extremely quick development of principally a new global factor – the human mind.

Human Mind as a Stage of the Life Evolution

There are two main positions concerning the problem of the human mind origin; it is considered as a result of either gradual development of animal abilities or drastic change of nervous system's functioning. For example, C. Darwin considered the difference between thinking of human and that of animals as not qualitative, but only quantitative. At the same time, followers of the behavioristic theory proclaim that there is a gulf between people and animals. As usual, the truth is somewhere in between. On the one hand, we cannot consider intelligence as an immanent property of life. Its development is characterized by evident leaps, the most essential of which was the origin of the human mind. On the other hand, most of the intellectual abilities of people can be observed in behavior of higher animals. Experiments on fosterage of chimpanzee babies in human families, which were started in 1913, have shown that their intellect can reach the level of a 2.5-year-old child.

Intellect is an ability to make decision in the situations where reflexive decisions are underspecified. It is a process of logic operation with some abstract symbols, expressing knowledge of the individual (or, in other words, use of a model of the world, formed in his mind). It is interesting that intellectual abilities of animals and birds formed similarly, but independently and on a different morphological basis: for example, birds have no cerebral cortex.

Besides all the evolutionary innovations, mind gives to a species some advantages in competition with other ones. The human mind proved to be so effective that it spared mankind from competition with other species. The human population is out from the ecosystem regulation, and there can be danger for its future.

Possibilities of intellectual development appeared after the origin of nervous system (as long ago as Cryptozoic eon). A network of specialized commutation cells ('neurons') is a reflection in the living organism morphology of the information constituent of the world. It gives possibility to collect, transmit, process, and store information. The human nervous system consists of about 10^{10} neurons, and each of them has thousands of 'synapses', connecting it with other neurons. All of them together form a 'neural network', which is well-known in informatics as a universal basis of self-organizing information-processing systems. At the same time, abstract mathematical neural networks are insufficient for understanding of a real nervous system and its center, the brain, because the latter has a very special, evolutionarily determined structure.

Increasingly complex algorithms of information processing led to both morphological progress and improvement of using abstract languages for information coding. The human language, developed in the last millennia, has given a principally new possibility to transmit a huge amount of information between generations passing over genome. It has incredibly accelerated the evolution, and created new ways of its realization. Having an image of the world in his mind, man can influence his own evolution.

Table 4 Geochronology of life evolution

Eon	*Era*	*Period*	*Beginning and end, million years*	*Forms of life*
Cryptozoic	Archean		4500–2600	Origin of life (more than 3.5 billions years ago). Primordial anoxic organisms. Appearance of prokaryotes
	Proterozoic	Lower Proterozoe	2600–1600	Primitive unicellular photosynthetic and nitrofixing organisms
		Riphey	1600–570	Appearance of eukaryotes. Origin of multicellular organisms. Wide expansion of bacteria, fungi, and algae
Phanerozoic	Paleozoic	Cambrian	570–500	Appearance and expansion of marine invertebrates
		Ordovician	500–440	Origin and expansion of lower terrestrial plants. Appearance of terrestrial invertebrates
		Silurian	440–410	Maximal development of marine invertebrates
		Devonian	410–350	Origin of terrestrial vascular plants. Appearance of insects, first vertebral animals
		Carbonic	350–285	Maximal development of gigantic mosses and horsetails. Development of amphibians. Formation of coal as a result of disposal of plant residues
		Permian	285–230	Origin of gymnospermous plants, extinction of pteridophytes. Appearance of big reptiles
	Mesozoic	Triassic	230–195	Development of gymnospermous plants. Expansion of big reptiles
		Jurassic	195–137	Further development of gymnospermous plants. Origin of immediate ancestors of birds
		Cretaceous	137–67	Origin of angiosperm plants. Extinction of big reptiles. Disposal of carbon in the form of chalk (calcium carbonate)
	Cainozoic	Paleogene	67–25	Wide spreading of angiosperm plants. Intensive development of birds and mammals
		Neogene	25–1.5	Formation of modern flora and fauna. Origin and development of primordial anthropoids
		Quaternary	1.5–0	Origin of *Homo sapiens*. Development of the human society

As an information system, the nervous system superseded the older chemical one – the hormonal system. It does not replace the latter system completely, but it became a superstructure on its basis. Usually the two systems function consistently, but sometimes they can give different recommendations; in this case, one can say that there is a contradiction between one's mind and feelings.

The multilayer character of the human information system is one of the aspects of a vital issue about scientific explanation of the phenomenon of consciousness. The mind–body problem is still unsettled; it is one of the last secrets of the universe. There are no serious ideas, in both natural science and philosophy, about a general approach to the problem investigation.

Computer science has given a brilliant possibility to model information processes; it is quite imaginable to make a robot, realizing all the functions of man, including his intellectual abilities. But can we secure a feeling of pain for this robot, or give him self-consciousness? Can artificial intellectual devices develop their mind, at least after long-term evolution? This matter is beyond the fields of both electronics and cybernetics. Probably, the nature of the

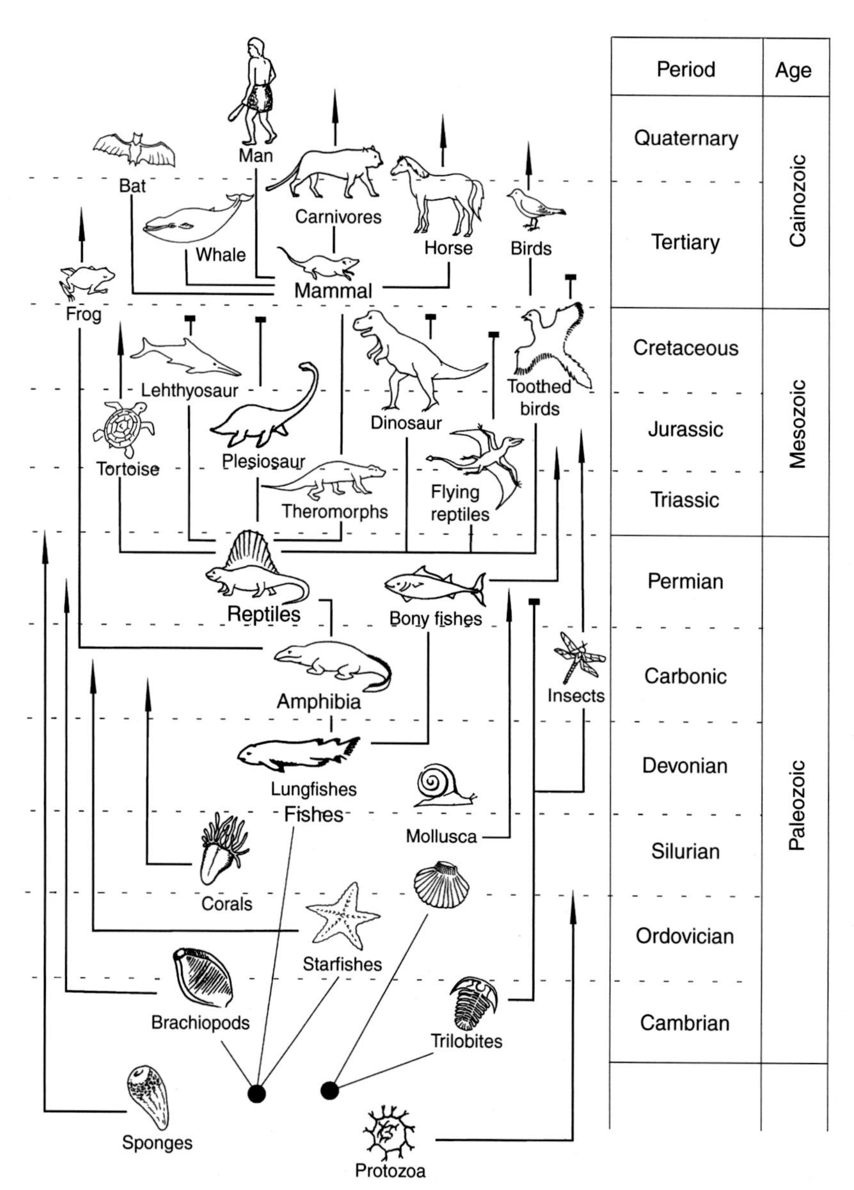

Figure 4 Diagram of animal evolution.

human mind is bipartite and has equally important cybernetic and chemical components. But the way of their synergy in the human body is still inscrutable.

Perspectives of Life

The philosophical problem of life is a problem of origin and being of man; it is another understanding of life, not as a biological phenomenon. However, these two approaches supplement each other: on the one hand, man is a biological being; on the other hand, science sees the world through human eyes, and it is necessary to take into considerations the peculiarities of the measuring instrument. The origin of consciousness is a principally new stage of life evolution, and the future of life depends now on tendencies of the human evolution.

According to F. Engels' (1820–95), in man the nature comes to self-understanding. N. F. Fedorov (1828–1903) developed this idea and said that it also found in man a new mechanism to self-control. As life integrated formerly geological processes, the human society is integrating life into its functions. Biosphere becomes a part of a larger system – 'noosphere'.

The term noosphere (from Greek word *noos* – mind) was proposed by E. Le Roy (1870–1954) and then essentially developed by other scientists. According to V. I. Vernadsy, noosphere is the next stage of biosphere, includes mankind and products of its activity, and is controlled by it to an essential degree. At present, mankind is not a good manager; there are a lot of ecological problems produced by human activity. Attention of people is focused mainly on technosphere, artificial product of human mind parasitized on biosphere. Perspectives of life depend on the success of people to find optimal ways of cooperation with nature and, particularly, include technosphere to the natural global cycling. Human influence is now much more essential for the destiny of life, than potentially dangerous geological (S. A. Arrhenius) or cosmic (S. Hocking) factors.

Summary

Natural evolution of Earth life has formed biological systems at different levels: biochemical, cellular, organism, ecosystem. Science has collected a lot of data about biological processes, but the process of their synthesis is only at its beginning.

Life is a global planetary process and involves in its functions all its forms of life. It intensively transforms its environment on all the levels: from biochemical processes in a single cell up to the global biogeochemical cycling on the scale of the Earth. Biosphere with its potential can help mankind solve many of its problems, but at the moment they are still in contradiction.

Understanding of life presupposes a knowledge concerning its history and laws of its evolution. The latest discoveries have made the process of the life origin more understandable, but further evolution of life is still unclear. Anthropogenic global change in combination with natural tendencies can give unexpected results. Further existence of life now depends on the ability of people to analyze the current situation and make right decisions.

See also: Biosphere. Vernadsky's Concept; Gaia Hypothesis; Noosphere; Phenomenon of Life: General Aspects.

Further Reading

Calder WA (1984) *Size, Function and Life History*. Cambridge, MA: Harvard University Press.
Forman RTT and Gordon M (1986) *Landscape Ecology*. New York: Wiley.
Hutchinson GE (1965) *The Ecological Theater and the Evolutionary Play*. New Haven, CT: Yale University Press.
Jablonka E and Lamb MJ (2005) *Evolution in Four Dimensions. Genetic, Epigenetic, Behavioral and Symbolic Variation in the History of Life*. Cambridge, MA: MIT Press.
Jørgensen SE (1992) *Integration of Ecosystem Theories: A Pattern*. Dordrecht: Kluwer.
Kareiva PM, Kingsolver JG, and Huey RB (eds.) (1993) *Biotic Interactions and Global Change*. Sunderland, MA: Sinauer.
Lehninger A (1982) *Principles of Biogeochemistry*. London: Worth Publishers.
Lovelock JE (1989) Geophysiology, the science of Gaia. *Reviews of Geophysics* 27: 215–222.
Margulis L and Sagan D (1986) *Microcosmos: Four Billion Years of Microbial Evolution*. New York: Simon & Schuster.
Marler PR and Hamilton WJ (1966) *Mechanisms of Animal Behavior*. New York: Wiley.
Maynard Smith J (1982) *Evolution and the Theory of Games*. Cambridge: Cambridge University Press.
Odum HT (1971) *Environment, Power, and Society*. New York: Wiley.
Oparin AI (1957) *The Origin of Life on Earth*. New York: Academic Press.
Partidge L and Harvey PH (1988) The ecological context of life history evolution. *Science* 241: 1449–1455.
Pert CB (1999) *Molecules of Emotion*. New York: Touchstone.
Pianka ER (1994) *Evolutionary Ecology*. New York: Harper Collins.
Pickett STA, Kolasa J, and Jones CG (1994) *Ecological Understanding: The Nature of Theory and the Theory of Nature*. San Diego, CA: Academic Press.
Schneider SH and Boston PJ (eds.) (1991) *Scientists on Gaia*. Cambridge, MA: MIT Press.
Watson JD and Berry A (2003) *DNA: The Secret of Life*. New York: Random House.
Wilson EO (1992) *The Diversity of Life*. New York: Norton.

PART B

Global Cycles, Balances and Flows

Calcium Cycle

C L De La Rocha, Alfred Wegener Institute for Polar and Marine Research, Bremerhaven, Germany

C J Hoff and J G Bryce, University of New Hampshire, Durham, NH, USA

Introduction

Although all of Earth's major biogeochemical cycles have been impacted by human activities, the calcium cycle has been one of the first to display significant changes. Harvesting of crops and the presence of acids and increased levels of carbon dioxide in rainwater from the burning of fossil fuels have altered the weathering rates of minerals and stripped calcium ions from soils, altering the primary sources of microbe- and plant-available calcium and upsetting community structures. The carbon dioxide is also depressing the saturation state of seawater with respect to calcium biominerals like aragonite and calcite, affecting the growth of organisms such as corals, coccolithophores, and pteropods. The ensuing shifts in limiting nutrients and competitive advantages within terrestrial communities, alteration of food webs, shifts in the balance between calcareous and noncalcareous plankton in the ocean, and diminishment of the reef-building ability of corals will in turn alter the delivery and cycling of nutrients and other elements in the terrestrial and marine biospheres and provide further perturbations to atmospheric concentrations of CO_2. As a result, anthropogenic impacts on the calcium cycle have ecological consequences that reach far beyond those ecosystems which are proximally impacted.

Importance of Calcium to Ecosystems

Terrestrial Ecosystems

Calcium is an element whose careful regulation within every living organism is critical to its survival. Eukaryotic cells use calcium ions as intracellular messengers, signaling environmental stresses and inducing changes in gene expression. Calcium is also an important structural component of cells, present in cell walls and membranes, and is a counterion for anions in cell vacuoles. The presence of too much calcium is, however, problematic and organisms have evolved ways to manage excess calcium. For example, earthworms contain calciferous glands that excrete calcium carbonate when too much calcium has been ingested and certain tree species (e.g., Norway spruce (*Picea abies* (L.) Karst.)) are thought to incorporate excess calcium into extracellular calcium oxalate crystals in their foliage.

Calcium is categorized as a major mineral nutrient for plants, and deficiencies of calcium affect plant health. Studies have demonstrated correlations between calcium availability and susceptibility of trees to insect, drought, frost damage, and disease. Tree species, such as sugar maples (*Acer saccharum* Marsh.), with greater requirements for calcium are more readily damaged due to low calcium availability. Accordingly, Ca deficiency in ecosystems can lead to shifts in plant species composition that may in turn have effects on the entire food web. Generally speaking, while both monocots (e.g., grasses, corn, and other grains) and dicots require calcium, monocots need less of it than dicots. Legumes, on the other hand, need roughly twice as much calcium as grasses. Plants can be classified into two groups, calcifuges (e.g., rhododendrons, heaths, and azaleas), which grow in acid soils with low calcium, and calcicoles (e.g., the Brassicaceae family including cabbage, broccoli, and kale), which require calcium-rich soils.

Invertebrates such as snails and mollusks use calcium to build their shells. Wood lice and millipedes prefer Ca-rich soils and serve as sources of calcium for their predators. Freshwater crayfish require at least a month of exposure to calcium-rich water after molting or their exoskeletons and claws fail to harden. Predatory birds can be affected if their prey and main calcium source becomes scarce due to inadequate levels of calcium; specifically, studies have demonstrated that egg shells become more fragile as prey (caterpillars, snails, arthropods) populations decline in areas of calcium depletion. Higher organisms, such as birds and humans, require calcium for more cellular structures and biochemical processes than the aforementioned; most of their calcium is contained within bones, but it is also used in nerve impulses, muscle contractions (e.g., heart contractions), DNA transcription, and blood clotting.

The calcium needs of organisms within lakes are the same as any other organism. The most basic difference between land and water terrestrial ecosystems is that organisms in lakes are submerged in water and the calcium content is linked with lakewater pH and, therefore, the survivability of an organism. Bodies of water with low relative concentrations of calcium are usually oligotrophic and can be dystrophic. Dystrophic lakes often have high concentrations of decaying organic matter, high concentrations of organic acids, and a low pH (e.g., bog lakes). The low abundance of calcium arises because the lakes are in an area with Ca-poor rocks or the dissolved organic matter has reacted with all of the available calcium (or both).

Marine Ecosystems

In addition to playing critical roles in the biochemistry of living cells, the major role calcium plays within marine ecosystems is as a major component of biominerals such as calcium carbonate ($CaCO_3$). Corals, mollusks, including the pelagic pteropods, and the green alga, *Halimeda*, produce aragonite, the more soluble polymorph of calcium carbonate. Coccolithophores, most foraminiferans, coralline algae, many crustaceans, and echinoderms produce the more stable polymorph, calcite.

Calcium-biomineralizing organisms play no small role in ocean ecosystems. Coccolithophores are one of the main types of phytoplankton in the ocean and their production of calcium carbonate significantly diminishes the effectiveness of the biological pump for sequestering carbon dioxide (CO_2) in the deep ocean. Pteropods are a major source of food for carnivorous zooplankton, fishes such as cod, salmon, and herring, and baleen whales. Corals and *Halimeda* and other calcareous algae together form tropical reefs which serve as habitat and food for a diverse community of microbes, invertebrates, and fish. Echinoderms such as sea urchins, brittle stars, and sea stars are important predators, grazers, and scavengers in benthic ecosystems from tropics to poles, shallow waters to deep.

The Global Calcium Cycle

Sources of Calcium to Terrestrial and Marine Ecosystems

Ultimately, calcium inputs to global ecosystems come from the chemical weathering of calcium-containing minerals (**Figure 1**). Weathering of silicate, carbonate, phosphate, and sulfate minerals in rocks, sediments, and

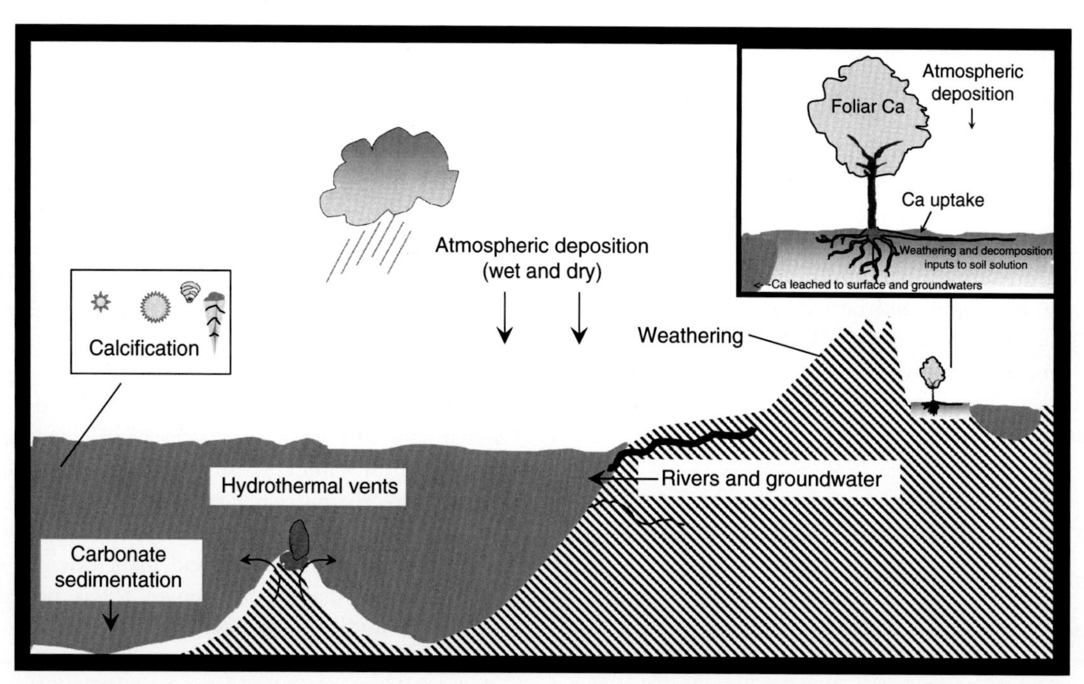

Figure 1 Schematic representation of the calcium cycle. Calcium is liberated in continents by the weathering of calcium-containing minerals, common constituents of most rocks. Calcium pools in terrestrial ecosystems include reservoirs in soils and soil minerals, organisms, and decomposing organic matter. Riverine and groundwater inputs transfer calcium in its ionic form from continents to the oceans. Hydrothermal vents also serve as calcium inputs in the oceans. In marine ecosystems, calcium ions are abundant in seawater. Marine organisms use calcium to make shells and hard parts. Calcium is removed from the oceans primarily by the sedimentation of these calcified organisms. Atmospheric deposition, both wet and dry, contributes calcium throughout both terrestrial and marine ecosystems. Anthropogenic effects including acid deposition, changes in land use (e.g., desertification), harvesting, and increased atmospheric carbon dioxide concentrations influence this natural calcium cycle.

soils releases calcium ions into solution. The two most important of these weathering reactions can be generalized, for calcium-bearing silicates, as

$$3H_2O + 2CO_2 + CaSiO_3 \rightarrow Ca^{2+} + 2HCO_3^- + Si(OH)_4$$

and for carbonates as

$$CaCO_3 + H_2O + CO_2 \rightarrow Ca^{2+} + 2HCO_3^-$$

The activities of plants are known to increase mineral weathering rates, and studies have suggested that in soils, plants involved in symbiotic relationships with mycorrhizal fungi might also directly access mineral-bound calcium before it enters the soil solution.

Calcium is also deposited into Earth surface ecosystems via wet and dry deposition. Calcium from sea salt and soil dust present as atmospheric particulate matter can be deposited by both these mechanisms. Anthropogenic contributions of calcium to atmospheric deposition come from biomass and fuel burning and the manufacturing of cement. The overall amount and relative contribution from each atmospheric source to an area varies seasonally and with factors such as the proximity of the area to the ocean.

Calcium is stored in biological materials, and calcium can be recycled in ecosystems via the breakdown of organic matter and biominerals containing calcium. Large amounts of calcium are generally exported from terrestrial ecosystems into the ocean through groundwater and surface waters. These ground- and surface waters supply $2–3 \times 10^{13}$ mol Ca yr^{-1} to the oceans versus the 0.3×10^{13} mol Ca yr^{-1} supplied strictly to the oceans through deep-sea hydrothermal vents.

The Production and Solubility of Calcium Biominerals in the Ocean

Approximately balancing the input of $2–3 \times 10^{13}$ mol Ca yr^{-1} to the ocean is the output of calcium as calcium biominerals formed by marine organisms. In shallow waters, aragonitic corals dominate the production of these minerals, providing 20% of the total output of calcium from the ocean. Calcitic foraminiferans and coccolithophores and aragonitic pteropods form the majority of the output sedimenting to the deep sea, with foraminiferans comprising somewhere between 20% and 60% of the total calcium output from the oceans and pteropods contributing only a few percent of it.

Despite the heavy biological usage of calcium, the relatively high abundance of Ca^{2+} in seawater means that Ca^{2+} concentrations are relatively invariant and never limiting to the growth of organisms. However, the concentration of Ca^{2+} together with the carbonate ion (CO_3^{2-}) concentration defines the saturation state of seawater with respect to calcite and aragonite. The saturation state of seawater is an integral part of the global Ca cycle and has profound implications for biota producing Ca biominerals.

Both the production of calcium carbonate biominerals and their dissolution can be summarized by the simple, reversible reaction

$$Ca^{2+} + CO_3^{2-} \leftrightarrow CaCO_3$$

The product of the calcium ion and carbonate ion concentrations in seawater (i.e., $[Ca^{2+}] \times [CO_3^{2-}]$) determines whether conditions are supersaturated or undersaturated with respect to the carbonate mineral in question. When the ion activity product is higher than the saturating value, conditions are favorable for mineral formation and dissolution does not occur. The lower the ion activity product below the saturating value, the more difficult it is to precipitate the minerals and the quicker the dissolution of the mineral.

Because the concentration of Ca^{2+} is relatively invariant in the oceans, it is the variability in the carbonate ion concentration in seawater (along with temperature) that affects biological calcification and the dissolution of carbonate minerals. Carbonate ion concentrations drop with pH, as the increasing H^+ concentration favors the protonation of CO_3^{2-} to form bicarbonate ion (HCO_3^-). Aragonite and calcite also become increasingly more soluble at cooler temperatures. Thus warm, tropical surface waters with their relatively low pH are supersaturated with respect to both aragonite and calcite. Cooler waters require greater carbonate ion concentrations to sustain saturating conditions, making polar waters less saturated than tropical waters. The pH of cold deep waters is relatively low due to the addition of CO_2 from the decay of sinking organic matter, and so these waters contain lower concentrations of carbonate ion and may be undersaturated with respect to both calcite and aragonite. As a result, production of massive, shallow water reefs by corals occurs only in the tropics, polar waters favor noncalcareous phytoplankton, and carbonate sediments do not accumulate below depths of several thousand meters.

Ecological Consequences of Anthropogenic Perturbations to the Calcium Cycle

Acid Deposition

Sulfate (SO_4^{2-}) and NO_x, anthropogenically emitted to the atmosphere, are oxidized and hydrolyzed to form sulfuric (H_2SO_4) and nitric (HNO_3) acids which are then introduced to ecosystems through precipitation or condensation of water vapor on foliage. These acids drive terrestrial ecosystems toward or into calcium limitation. Contact between acid fog and foliage leaches calcium directly out of leaf membranes, causing tissue damage and calcium depletion in plants. The deposition of extra anions (SO_4^{2-} and NO_3^-) causes the increased formation of calcium salts, such as $CaSO_4$, in soils and these neutral compounds are easily leached from soils. Acid deposition also increases the concentration of hydrogen ions (H^+) in soil solutions. These excess protons, in turn, compete with other cations (especially Ca^{2+}) for space on the negatively charged surfaces (cation exchange sites) of soil particles. Calcium ions that are not bound to particles are easily removed from the soil and lost to the ecosystem by leaching.

As the concentration of H^+ rises in a soil, weathering rates of aluminosilicates and other minerals increase, inflating concentrations of total soluble Al and soluble free Al^{3+}. This leads to the increasing displacement of Ca^{2+} and other cations from the cation-exchange sites of soils. Eventually, the concentration of H^+ reaches a point where the protons outcompete Al^{3+}. The aluminum ions then become mobile in the soil solution where they may directly interact with plants. This compounds the problem of the depletion of soil Ca^{2+}, because, by damaging root tissue and displacing Ca^{2+} from exchange sites on the xylem walls of plants, Al^{3+} diminishes the ability of plants to take up calcium.

Acid deposition also affects terrestrial bodies of water. The main source of calcium for streams and lakes is runoff and groundwater; as the minerals and soils surrounding these bodies of water become increasingly depleted in Ca^{2+}, so does the water. A lowering of calcium ion concentrations affects the acid-neutralizing capacity (ANC) of waters, resulting in greater pH fluctuations with changes in proton fluxes and increasing the effects of acidic deposition. The runoff will also contain increasing concentrations of Al^{3+}, potentially delivering lethal doses for some aquatic organisms and severely impacting or destroying local food webs.

Anthropogenic Impacts on the Atmospheric Dry Deposition of Calcium

As noted above, calcium is added to both terrestrial and marine ecosystems via the deposition of particulate matter from the atmosphere. Humans make significant contribution to these fluxes via industry (e.g., the manufacturing of cement) and biomass and fuel burning. An equally large, and in some localities much larger, anthropogenic influence is found in the abundance of dust. Land-use changes that result in desertification and dry soils at construction sites increase the amount of Ca-containing dust in the atmosphere. These inputs may sometimes temporarily offset the effects of acid deposition, but when these inputs are stopped, calcium depletion resumes.

Effects of Increased Carbon Dioxide on Terrestrial Ecosystems

Higher atmospheric concentrations of carbon dioxide eventually lead to higher mineral weathering rates, resulting in the enhanced leaching of calcium and other elements from soils. Such changes in weathering rates are not balanced by the natural recharge rate of soil calcium. Increased carbon dioxide concentrations also increase plant growth rates. Relieved of a carbon dioxide limitation, the plants will grow until they are limited by some other nutrient, often calcium. The increased plant growth could translate into a faster biological cycling of calcium, a process that has unpredictable results.

Carbon Dioxide, Calcium Biominerals, and Marine Ecology

Anthropogenic effects on the marine calcium cycle primarily occur through the acidification of seawater by carbon dioxide. The resultant lowering of pH and carbonate ion concentrations decreases the saturation state of seawater with respect to calcite and aragonite, thus making it more difficult for marine organisms to produce and maintain calcium biominerals.

Since the beginning of the industrial revolution, CO_2 concentrations in the atmosphere have increased by 90 μatm (i.e., by more than a third) and show no signs of slowing down. The CO_2 added to atmosphere will eventually be absorbed by the ocean, acidifying it. The pH of surface waters has already dropped by 0.1 units (a significant amount), and within 40 years carbonate ion concentrations below aragonite saturation will begin to occur in polar waters, spreading eventually into lower latitudes. If CO_2 emissions continue unabated, the pH of the ocean will eventually sink to levels lower than it has been for hundreds of millions of years.

A drop in ocean pH has implications for the existence and ecology of coral reefs because a decrease in the saturation state of seawater with respect to calcium biominerals will have a corresponding decrease in the rate of calcification of coral reefs. Calcification rates in the tropics have already dropped by 10% since the beginning of the industrial revolution and a doubling of atmospheric CO_2 from pre-industrial levels could diminish coral calcification rates by as much as 50%. Such decrease in calcification rates will result in reefs shrinking in size and structural integrity, as reef size and strength result from the balance struck between calcium carbonate production and erosion. A decrease in the areal extent of coral reefs in turn, diminishes the habitat and food available for the hundreds of thousands of species of organisms that dwell within coral reef ecosystems. This is true for both the familiar warm water coral reefs of the tropics whose productive ecosystems are an important resource for subsistence fishers and billion-dollar tourism economies alike, and the deeper-dwelling, cold-water coral reef ecosystems that provide habitat and nursing grounds for numerous species including commercially important fish like rockfish and orange roughy.

The reduced saturation state of seawater with respect to calcium biominerals may also affect the production and maintenance of shells and exoskeletons of organisms like mollusks, echinoderms, and crustaceans. The first impacts will be seen in polar ecosystems whose cold waters may become undersaturated with respect to aragonite at the doubling of pre-industrial CO_2 expected by 2050. Experiments and material collected in sediment traps have shown that the shells of the aragonitic pteropod mollusks become rapidly pitted and begin to dissolve upon exposure to undersaturated waters. Even the shells of live pteropods begin to dissolve under conditions equivalent to those expected for polar waters at the end of this century. Pteropods should not survive if they cannot maintain their shells, and their disappearance from polar waters would have a significant impact on polar ecosystems. Pteropods are important prey for many zooplankton, fish, and baleen whales and their fecal pellets and mucous feeding webs are important vectors for the sinking of organic matter to deep-sea ecosystems and sediments.

The lowering of the saturation state of seawater for calcium carbonate minerals will also make it thermodynamically less favorable for the calcitic plankton, foraminiferans, and coccolithophores to biomineralize. If the lowering of carbonate ion concentrations means that the high internal pHs required for calcite precipitation take more energy to maintain, these organisms will have a lesser portion of their total energy budget available for growth and reproduction. Although, experimentally, the response of coccolithophore species to increased acidification is mixed, an acidification-driven shift in the phytoplankton toward noncalcareous forms such as diatoms would have an impact on the cycling of CO_2, nutrients, and alkalinity in the ocean by altering efficiency of the biological pumping of particulate organic matter and biominerals into the deep sea. If the ratio of particulate organic carbon to calcium carbonate sinking into the deep sea were to increase due to the lesser production of calcium biominerals by foraminiferans and coccolithophores, the biological pump would be more effective at sequestering CO_2 in the deep sea, lowering atmospheric concentrations at the expense of more quickly lowering the pH of the deep sea.

Tracking Calcium Cycling in Biogeochemical Systems

Changes in the calcium cycle due to pollution and other anthropogenic influences and their impact on ecosystems make unraveling and quantifying the fluxes of calcium through ecosystems a pressing concern. Three types of tools have been employed, trace element ratios, such as Sr/Ca, ratios of nonradioactive isotopes of strontium or calcium (e.g., $^{87}Sr/^{86}Sr$, $^{44}Ca/^{42}Ca$, $^{44}Ca/^{40}Ca$), and, less commonly, studies of artificially enriched stable (e.g., ^{48}Ca) and radioactive (e.g., ^{45}Ca) isotopes. Such tracers reflect sources of Ca to ecosystems and, when reconstructed from the wood of long-lived trees, may serve as a means of reconstructing the acidification of environments over the past century or so.

The use of Sr/Ca in terrestrial ecosystems takes advantage of the fact that different minerals serving as sources of Ca^{2+} contain different Sr/Ca signatures. Studies employing this method rely on the assumption that Sr^{2+} and Ca^{2+} ions are not fractionated as they cycle through ecosystems because of their comparable charge and size. The Sr/Ca of different calcium reservoirs (e.g., soil waters, soils, and vegetation) should thus reflect it of their sources of calcium. Complicating the use of Sr/Ca, however, are data suggesting that Sr/Ca is fractionated during uptake by and internal cycling within plants and the fact that all Sr/Ca inputs to ecosystems (specifically, mineral pools) have not been identified.

Strontium and calcium isotopic signatures may also usefully identify the sources of Ca to terrestrial environments. Strontium isotopes, which have the advantage of not being biologically fractionated, are used based on the assumption that Sr and Ca in terrestrial environments have been derived from the same sources. The source of calcium to plants, for example, is identified from their $^{87}Sr/^{86}Sr$ because it reflects the bulk $^{87}Sr/^{86}Sr$ of the materials from which the strontium came. Calcium isotopes provide a more direct way of investigating Ca cycling through ecosystems. Solutions enriched in the stable calcium isotope with the lowest natural abundance (^{48}Ca) or a radioactive isotope of calcium (^{45}Ca) have been released to study the movement of calcium through the environment. Natural abundances of Ca isotopes provide a way to directly study the calcium cycle in ecosystems. Such work is in its infancy and studies are underway to characterize the Ca isotopic composition of minerals, natural waters, and vegetation and to define Ca isotopic fractionation during weathering, soft tissue formation, biomineralization, and between different plant tissues. Such studies pave the way for this new tracer to be universally applied.

In marine systems, such trace elements and isotopic systems are not as useful for tracking anthropogenic changes to the calcium cycle as direct measurements of pH, alkalinity, and calcification rates. Reconstructions of the depth distributions of calcite sediments and the calcium isotopic composition of marine sediments, however, help to identify past perturbations in the calcium cycle and their links to climatic and ecological events.

The Future

Quantifying the effects of anthropogenic perturbations on calcium cycling and ecosystems is challenging because the effects are not instantaneous. Outcomes, such as deteriorating tree health (or die-offs) and declining bird populations due to calcium-depleted eggshells, may only be obvious after years of cumulative damage to the environment. On a hopeful note, grave and large-scale impacts such as these can inspire shifts in industrial practices; in response to the problems acidification was causing to terrestrial ecosystems, care has been taken in recent years by industrialized nations to lower emissions of sulfate and NO_x (although emissions have not ceased entirely). Unfortunately, decreases in calcium-containing emissions have diminished the unintended anthropogenic amelioration of the calcium depletion caused by acidification.

Land-use changes also have mixed effects. Re-vegetation of areas may decrease their production of dust, but reforesting an area after repeated harvesting of crops accelerates the calcium depletion of the area as the calcium contained in the removed biomass has been lost. Accordingly, attempts at environmental remediation have been made through the application of calcium-rich compounds, such as lime or wollastonite. Although forest-scale manipulations have been set up to assess the effectiveness of these applications, these experiments are ongoing and, thus, conclusions about the effectiveness of these treatments cannot yet be made. Ecosystem models have also been employed to understand and predict impacts of these ecological manipulations on the terrestrial calcium cycle, but as our understanding of the complexities of the terrestrial calcium cycle is currently limited, these models primarily serve to provide broad guesses of future impacts.

As in terrestrial ecosystems, although the impacts of anthropogenic perturbations of the calcium cycle on marine ecosystems have been predicted and modeled, the extent of the impact of ocean acidification on ocean ecosystems remains uncertain. On one hand there are undoubtedly effects and variables that have not been considered, and on the other hand it is an open question to what extent calcifying organisms will adapt to and cope with the lower pHs, lower carbonate ion concentrations, and lesser degrees of saturation with respect to calcium carbonate minerals. Already stressed by pollution and overfishing, ecosystems centered around calcareous organisms like corals may collapse under the additional burden of acidification. Alternatively, although the rate of pH change is occurring at an unprecedentedly rapid timescale relative to evolution, genetic adaptation to the more acidic conditions could occur before widespread alteration of the ecosystems occurs. At the moment no experiments on appropriately long time frames have been conducted, in terms of either pollution or remediation, for the outcomes to be clear. Even where we immediately cease to perturb the calcium cycle, we have still made significant changes to the global calcium cycle and it will take some while for the full consequences of our inadvertent global-scale experiment on the calcium cycle to unfold.

See also: Anthropospheric and Antropogenic Impact on the Biosphere; Climate Change 2: Long-Term Dynamics; Deforestation; Global Change Impacts on the Biosphere.

Further Reading

Berner EK and Berner RA (1996) *Global Environment: Water, Air and Geochemical Cycles*, 376pp. Upper Saddle River, NJ: Prentice Hall, Inc

Bullen TD and Bailey SW (2005) Identifying calcium sources at an acid deposition-impacted spruce forest: A strontium isotope, alkaline earth element multi-tracer approach. *Biogeochemistry* 74: 63–99.

De La Rocha CL and DePaolo DJ (2000) Isotopic evidence for variations in the marine Ca cycle over the Cenozoic. *Science* 289: 1176–1178.

Driscoll CT, Lawrence GB, Bulger AJ, *et al.* (2001) Acidic deposition in the northeastern United States: Sources and inputs, ecosystem effects, and management strategies. *Bioscience* 51: 180–198.

Hoff CJ, Bryce JG, Hobbie EA, Colpaert JV, and Bullen TD (2005) Quantification of calcium isotope fractionation in ectomycorrhizal trees. *Eos: Transactions of the American Geophysical Union* 86: Fall Meeting Supplement, Abstract B53B-05.

Huntington T (2000) The potential for calcium depletion in forest ecosystems of southeastern United States: Review and analysis. *Global Biogeochemical Cycles* 14: 623–638.

Langdon C, Takahashi T, Sweeney C, *et al.* (2000) Effect of calcium carbonate saturation state on the calcification rate of an experimental reef. *Global Biogeochemical Cycles* 14: 639–654.

Lawrence GB and Huntington TG (1999) Soil-calcium depletion linked to acid rain and forest growth in the eastern United States. USGS WRIR 98-4267.

Likens GE, Driscoll CT, Buso DC, *et al.* (1998) The biogeochemistry of calcium at Hubbard Brook. *Biogeochemistry* 41: 89–173.

McLaughlin SB and Wimmer R (1999) Calcium physiology and terrestrial ecosystem processes. *New Phytologist* 142: 373–417.

Orr JC (2005) Anthropogenic ocean acidification over the twenty-first century and its impact on calcifying organisms. *Nature* 437: 681–686.

Raven J, Caldeira K, Elderfield H, *et al.* (2005) *Ocean Acidification Due to Increasing Atmospheric Carbon Dioxide. Policy Document 12/05 of The Royal Society*. London: The Royal Society.

Schaberg PG, DeHayes DH, and Hawley GJ (2001) Anthropogenic calcium depletion: A unique threat to forest ecosystem health? *Ecosystem Health* 7: 214–228.

White PJ and Broadley MR (2003) Calcium in plants. *Annals of Botany* 92: 487–511.

Yanai RD, Blum JD, Hamburg SP, *et al.* (2005) New insights into calcium depletion in northeastern forests. *Journal of Forestry* 103: 14–20.

Carbon Cycle

V N Bashkin, VNIIGAZ/Gazprom, Moscow, Russia

I V Priputina, Institute of Physico-Chemical and Biological Problems of Soil Science RAS, Moscow, Russia

Introduction

Cyclic processes of exchange of carbon mass are of particular importance for the global biosphere, both in terrestrial and oceanic ecosystems especially owing to the close connections to the global climate changes.

This element is distributed in the atmosphere, water, and land as follows. According to existing data there are 6160×10^9 t or 1.4×10^{16} mol of CO_2 in the atmosphere (1680×10^9 t of C). A major source of atmospheric carbon dioxide is respiration, combustion, and decay, compared with oxygen, whose main source is photosynthesis. In its turn, an important sink of CO_2 is photosynthesis (about $66 \times 10^9\ \mathrm{t\,yr^{-1}}$ or $1.5 \times 10^{15}\ \mathrm{mol\,yr^{-1}}$). Since carbon dioxide is somewhat soluble in water ($K_H = 3.4 \times 10^{-2}\ \mathrm{mol\,l^{-1}\,atm^{-1}}$), exchange with the global ocean must also be considered. The approximate global balance of atmosphere–ocean water exchange is $7 \times 10^{15}\ \mathrm{mol\,yr^{-1}}$ ($308 \times 10^9\ \mathrm{t\,yr^{-1}}$) being taken up and $6 \times 10^{15}\ \mathrm{mol\,yr^{-1}}$ ($264 \times 10^9\ \mathrm{t\,yr^{-1}}$) being released in different parts of the oceanic ecosystem. The residence time of CO_2 in atmosphere is about 2 years, which makes the atmospheric air quite well mixed with respect to this gas. However, a more recent analysis shows that the terrestrial ecosystems have much stronger sinks of carbon dioxide uptake.

In the global ocean, along with occurrence in living organisms, carbon is present in two major forms: as a constituent of organic matter (in solution and partly in suspension) and as a constituent of exchangeable inorganic ions HCO_3^-, CO_3^{2-}, and CO_2:

$$\begin{aligned} CO_2(g) &\leftrightarrow H_2CO_3(aq) \leftrightarrow -H^+/+H^+ \leftrightarrow HCO_3^-(aq) \\ &\leftrightarrow -H^+/+H^+ \leftrightarrow CO_3^{2-} \leftrightarrow +Ca^{2+} \leftrightarrow CaCO_3(s) \end{aligned}$$

The amount of CO_2(aq) in the oceans is 60 times that of CO_2(g) in the Earth's air, suggesting that the oceans might absorb most of the additional carbon dioxide being injected at present into the atmosphere. However, there are some drawbacks restricting this process. First of all, CO_2 uptake into surface oceanic waters (0–100 m) is relatively slow ($t_{1/2} = 1.3$ years). Second, these surface waters mix with deeper waters very slowly ($t_{1/2} = 35$ years). Consequently, the surface oceanic waters have the capacity to remove only a fraction of any increase in the anthropogenic CO_2 loading (**Figure 1**).

The known analytical monitoring data obtained over many years at the Mauna Loa Observatory in Hawaii, a location far from any anthropogenic sources of carbon dioxide pollution, show a pronounced 1-year cycle of CO_2 content (**Figure 2**).

One can see the peak about April and then through around October each year. These data indicate that the content of carbon dioxide in the Earth's atmosphere is not perfectly homogeneous. Some explanations would be of interest to understand this figure.

Hawaii is in the Northern Hemisphere where the photosynthetic activity of vegetation is maximal in summertime (May–September). In this period CO_2 is

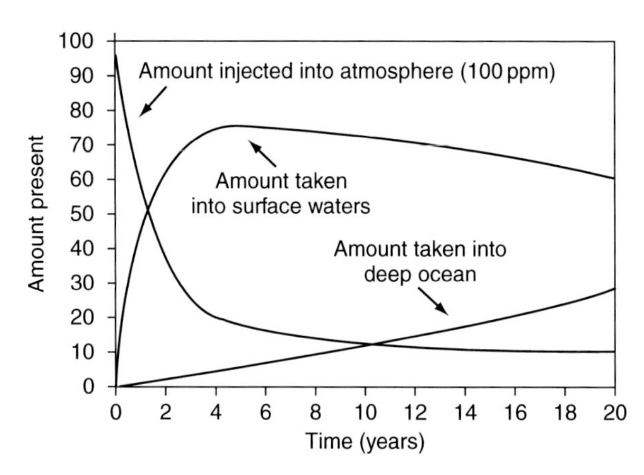

Figure 1 Calculated uptake of CO_2 from atmosphere to the surface and deep oceanic waters.

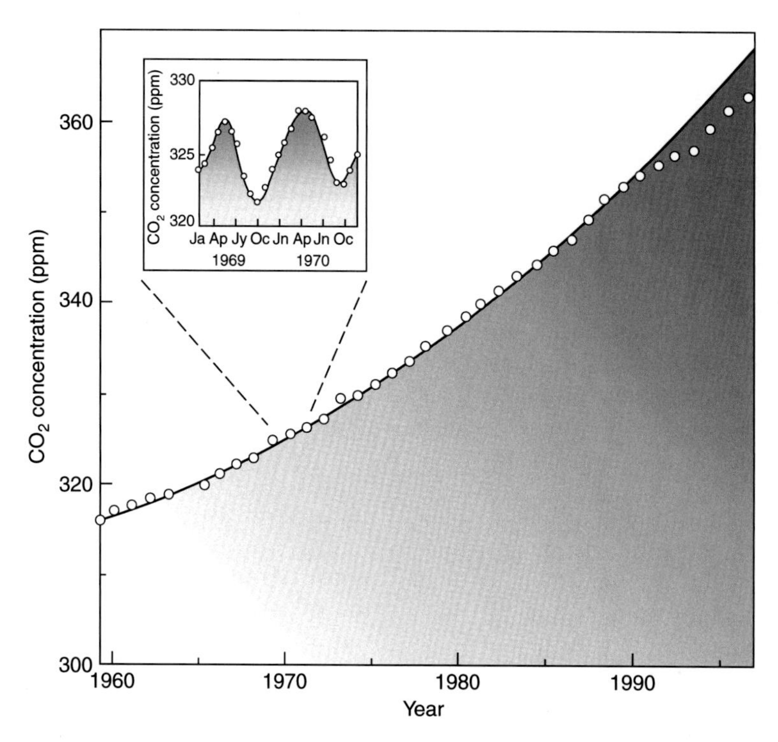

Figure 2 Observations of CO_2 concentration at the Mauna Loa Observatory for the period of 1958–99.

removed from the air a little bit faster than it is added. The reverse situation occurs during winter. This is a reasonable explanation and accordingly the monitoring stations in the South Hemisphere show the highest concentration of CO_2 in October, and the lowest in April (see http://www.mlo.noaa.gov).

A gradual increase in the partial pressure of carbon dioxide over the last decades is clearly pointed out in **Figure 2**. The value of $p(CO_2)$ was *c.* 315 ppmv in 1958, it had reached 350 ppmv in 1988, and >370 ppmv in the beginning of twenty-first century. Accordingly, this trend can give a doubling of carbon dioxide content in the Earth's atmosphere sometime during the end of the twenty-first century and this seems a reasonable prediction.

Here we should refer to the opinion of some other authors who have argued that increased CO_2 levels in the atmosphere may be a consequence of atmospheric warming, rather than the cause. The statistical analysis of various authors led to the conclusion that, although there is a correlation between $p(CO_2)$ and global temperatures, the changes in $p(CO_2)$ appear to lag behind the temperature change by *c.* 5 months. A possible explanation, if this trend is proved correct, would be that natural climatic variability like the solar activity alters the temperature of the global ocean, which contains about 90% of total CO_2 mass. In turn, this leads to increase of CO_2 flux from the warmer oceanic water to the atmosphere in accordance with Henry's law.

Turnover of Carbon in the Biosphere

As has been pointed out earlier, terrestrial ecosystems are the main sink of carbon dioxide due to the photosynthesis process. The present bulk of living organisms is confined to land and their mass (on dry basis) amounts to 1880×10^9 t. The average carbon concentration in the dry matter of terrestrial vegetation is 46% and, consequently, the carbon mass in the land vegetation is about 865×10^9 t.

In accordance with various estimates, the oceanic biomass of photosynthetic organisms contains 1.7×10^9 t of organic carbon, C_o. In addition, we have to include a large number of consumers. This gives 2.3×10^9 t of C_o. Totally, the oceanic organic carbon is equal to 4.0×10^9 t or about 0.5% from that in land biomass.

Moreover, a substantial amount of dead organic matter as humus, litterfall and peat is also present in the terrestrial soil cover. The mass of forest litter is close to 200×10^9 t, mass of peat is around 500×10^9 t, and that of humus is 2400×10^9 t. Recalculation of this value for organic carbon amounts to 1550×10^9 t.

However, the greatest amount of carbon in the form of hydrocarbonate, HCO_3^- ($38\,600 \times 10^9$ t) is contained in the ocean, 10 times higher than the total carbon in living matter, atmosphere, and soils.

Thus, in the terrestrial ecosystems the least amount of carbon is monitored in living biomass, followed by dead biomass and atmosphere.

Table 1 Mass distribution of carbon in the Earth's crust

Earth's compartments	*Mass (10^{18} t)*	*Average concentration (%)*			*Mass (10^{15} t)*				
		CO_2	*C_c*	*C_o*	*CO_2*	*C_c*	*C_o*	*$C_c + C_o$*	*Ratio of C_c/C_o*
Total Earth's crust	28.5	1.44	0.38	0.07	409	108	20	128	5.4
Continental type including:	18.1	1.48	0.40	0.08	267	72	14	86	5.1
Sedimentary layer	1.8	9.57	2.61	0.50	177	48	9	57	5.3
Granite layer	6.8	0.81	0.22	0.05	55	15	3	18	5.0
Basalt layer	9.4	0.37	0.10	0.02	35	9.4	1.9	11	5.0
Subcontinental type	4.3	1.37	0.36	0.07	58	16	3	19	5.3
Oceanic type	6.1	1.35	0.36	0.05	82	21	3	24	7.0
Earth's sedimentary shell	2.4	12.4	3.37	0.62	297	81	15	96	5.4
Phanerozoic sedimentary deposits	1.3	15.0	4.08	0.56	194	53	7	60	7.5

The mass distribution of carbon in the Earth's crust is of interest for understanding of the global biogeochemistry of this element. These values are shown in **Table 1**. One can see that carbon from carbonates (C_c) is the major form. The C_c/C_o ratio is about 5 for the whole Earth's crust as well as for its main layers (sedimentary, granite, and basalt) and crustal types: continental, subcontinental, and oceanic. However, for the latter this ratio is higher.

The sedimentary layer of the Earth's crust is the main carbon reservoir. The C_c and C_o concentrations in the sedimentary layer are by an order of magnitude higher than in granite and basalt layers of lithosphere. The volume of sedimentary shell is about 0.10 from the crust volume; however, this shell accounts for 75% of both carbonate and organic carbon. Dispersed organic matter (kerogen) contains most of the C_o mass. Localized accumulation of C_o in oil, gas, and coal deposits are of secondary importance. It has been estimated that the oil/gas fields amount to 200×10^9 t of carbon, and the coal deposits contain 600×10^9 t, totally 800×10^9 t. This is by three orders of magnitude less than the carbon mass of dispersed organic matter in the sedimentary shell. The general carbon distribution between reservoirs is shown in **Table 2**.

Thus, there are two major reservoirs of carbon in the Earth: carbonate and organic compounds. It should be stressed that both are of biotic origin. Nonbiotic carbonates, for instance, from volcanoes, are the rare exception of the rule. A connecting link between the carbonate and organic species is CO_2, which serves as an essential starting material for both the photosynthesis of organic matter and the microbial formation of carbonates.

Atmospheric CO_2 provides a link between biological, physical, and anthropogenic processes. Carbon is exchanged between the atmosphere, the ocean, the terrestrial biosphere, and, more slowly, with sediments and sedimentary rocks. The faster components of the cycle are shown in **Figure 3**.

The component cycles (**Figure 3**) are simplified and subject to considerable uncertainty (cf. **Table 2**, for example). In addition, this figure presents average values. The riverine flux, particularly the anthropogenic portion, is currently very poorly qualified and is not shown here. While the surface sediment storage is approximately 150×10^9 t, the amount of sediment in the bioturbated and potentially active layer is of order of 400×10^9 t. Evidence is accumulating that many of the key fluxes can fluctuate significantly from year to year (e.g., in the terrestrial sink and storage). In contrast to the static view conveyed by figures such as this one, the carbon system is clearly dynamic and coupled to the climate system on seasonal, interannual, and decadal timescale.

Thus, the obvious discrepancies between data shown in **Tables 2** and **3** as well as in **Figure 3** and above-mentioned discussion in the text are related to both the uncertainties in data sources and different authors' speculations on the topic. At the state of the art of present knowledge, one cannot make more precise estimates of carbon fluxes and pools at the global scale.

The carbonate formation and photosynthesis have to be considered as two general processes in the global activity of living matter over geological history of the Earth. The C_c-to-C_o mass ratio may specify the 'growth limit' of living matter at sequential stages of Earth's geological history over the period of 3.5–3.8 billion years. This ratio tends to decrease regularly with the last 1.6 billion years. The C_c/C_o ratio was 18 in the sedimentary layers of the Upper Proterozoic period (1600–750 million years); that of the Paleozoic (570–400 million years), 11; of the Mesozoic (235–66 million years), 5.2; and of the Cainozoic (66 million years to the present), 2.9. The never interrupted increase in the relative content of organic matter in the ancient stream loss provides evidence for a progressively increasing productivity of terrestrial photosynthetic organisms. This provides also the proof for the growing importance of global terrestrial ecosystems in the fixation of CO_2. Apparently, the increasing productivity of land vegetation would be the major sink of CO_2 under the increasing content of this green-house gas in the

Table 2 The major global carbon reservoirs

Reservoirs	*C (10^9 t)*
Atmosphere, CO_2	1680
Global land	
Vegetable biomass prior to human activity (estimates)	1150
Present natural vegetable biomass	900
Soil cover	
Forest litterfall	100
Peat	250
Humus	1200
Total	1550
Ocean	
Photosynthetic organisms	1.7
Consumers	2.3
Soluble and dispersed organic matter	2100
Hydrocarbonate ions in solution	38 539
Total	40 643
Earth's crust	
Sedimentary shell, C_o	15 000 000
Sedimentary shell, C_c	81 000 000
Continental granite layer, C_o	4 000 000
Continental granite layer, C_c	18 000 000
Total	118 000 000
Total present global C mass	118 044 773

atmosphere; however, the role of increasing input of nitrogen, for instance, with atmospheric deposition, has to be considered. Moreover, both carbonate formation and the photosynthesis of organic matter share in the common tendency for removal from the atmosphere of CO_2 continually supplied from the mantle. Consequently, these processes take part in the global mechanisms for maintaining the present low concentration of carbon dioxide in the Earth's gas shield, which is an essential parameter in the greenhouse effect.

Carbon Fluxes in Terrestrial Ecosystems

All three CO_2-controlling processes (ocean soaking, photosynthesis, and carbonate formation) play an important role in maintaining equilibrium in the biosphere–atmosphere–hydrosphere system. The photosynthetic process is of great importance for living plants and microorganisms. The difference between total photosynthesis and respiration processes is defined as 'net primary production', NPP. The global NPP distribution in the Earth's major ecological zones is shown in **Table 3**.

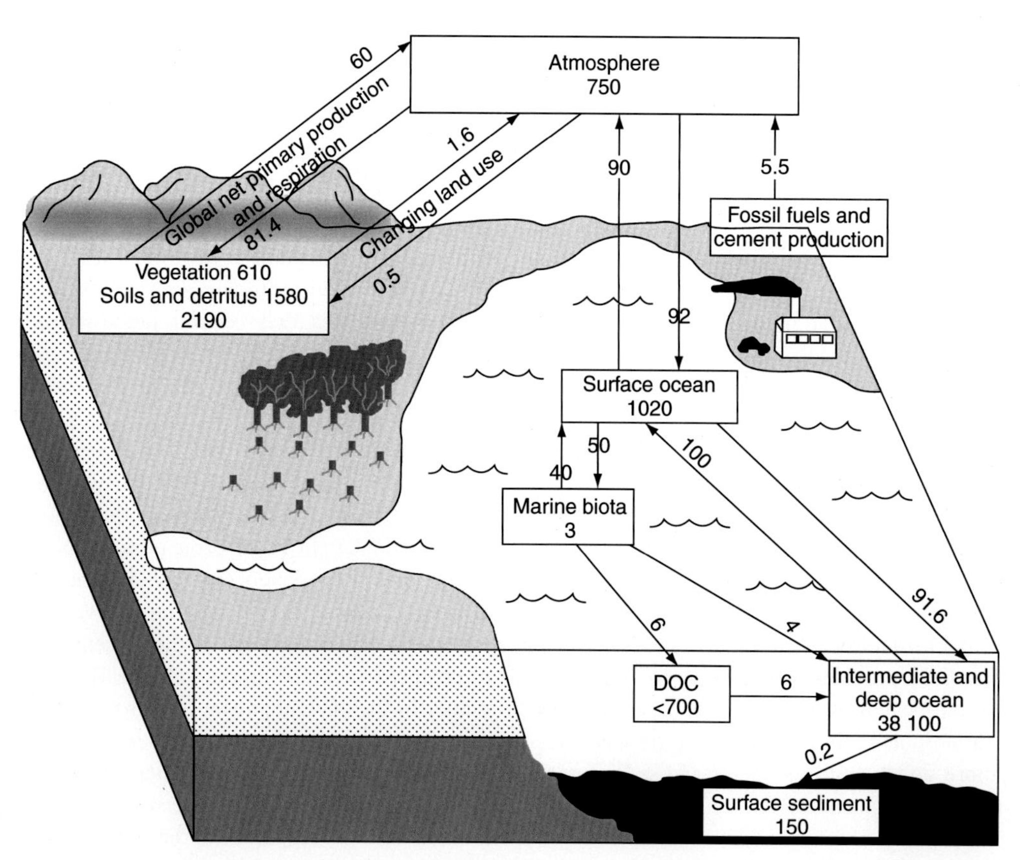

Figure 3 The global carbon cycle, showing the reservoirs (in 10^9 t yr^{-1}) relevant to the anthropogenic perturbation as annual averages over the period 1980–89.

Table 3 Net primary production of the Earth's major ecosystems

Global ecosystem zone	*Area (10^6 km^2)*	*Plant mass (10^9 t)*	*C-NPP (10^9 t)*
Polar	8.1	13.8	1.3
Coniferous forests	23.2	439.1	15.2
Temperate	22.5	278.7	18.0
Subtropical	24.3	323.9	34.6
Tropical	55.9	1347.1	102.5
Total land	133.9	2402.1	171.6
Lakes and rivers	2.0	0.04	1.0
Glaciers	13.9	0	0
Total continents	149.3	2402.5	172.6
Oceans	361.0	0.2	60.0
Earth total	510.3	2402.7	232.6

Oceans, despite their much larger surface area, contribute much less than half of the global NPP. The reason is related to high nutrient deficiency in surface waters, which limits the photosynthesis process. Oceanic production is mainly concentrated in coastal zones, especially where upwelling of deep water brings the nutrients (P and N, of major interest) into the surface layer, 0–100 m. On land the photosynthetic process is also often limited by nutrient deficit; however, the influence of water storage and low temperature plays a more important role. That is why subtropical and tropical ecosystems contribute much more to global NPP than their proportional share.

The amount of annually decaying organic matter is the subject of speculation. However, some estimates might be done. For instance, in terrestrial ecosystems only, the humus accumulation of carbon in soils is about 70% of the total accumulation of CO_2 in the atmosphere. We may presume therefore that the stable long-lived humic compounds acquire some 30% of carbon annually from the dead organs of plants, and the complete renewal of humus in soils extends over the period of $(0.3–1.0) \times 10^3$ years. The variance depends on the moisture and temperature conditions in the region of question.

Terrestrial biomass is divided into a number of subreservoirs with different turnover times. Forest ecosystems contain 90% of all carbon in living matter on land but their NPP is only 60% of the total. About half of the primary production in forest ecosystems is in the form of twigs, leaves, shrubs, and herbs that only make up 10% of the biomass. Carbon in wood has a turnover time of the order of 50 years, whereas these times for carbon in leaves, flowers, fruits, and rootlets are less than a few years. When plant material becomes detached from the living plant, carbon is moved from phytomass reservoir to litter. 'Litter or litterfall' can refer to a layer of dead plant material on the soil surface. A litter layer can be a continuous zone without sharp boundaries between the obvious plant structures and a soil layer containing amorphous organic carbon. Decomposing roots are a kind of litter that seldom receives separate treatment due to difficulties in distinguishing between living and dead roots. Total litter is estimated as 60×10^9 C and total litterfall as 40×10^9 t C yr^{-1}. The average turnover time for carbon in litter is thus about 1.5 years, although for tropical ecosystems with mean temperature above 30 °C, the litter decomposition rate is greater than the supply rate and so storage is impossible. For colder climates, NPP exceeds the rate of decomposition in the soil and organic matter in the form of peat is accumulated. The total global amount of peat might be estimated at 165×10^9 t C. Average temperature at which there is a balance between production and decomposition is about 25 °C.

Humus is a type of organic matter in terrestrial ecosystems that is not readily decomposed and therefore makes up the carbon reservoir with a long turnover time (300–1000 years). An assessment of the various carbon pools for a temperate grassland soil is presented in **Figure 4**.

The undecomposed litter (4% of the soil carbon) has a turnover time measured in tens of years, and the 22% of the soil carbon in the form of fulvic acids is intermediate with turnover times of hundreds of years. The largest part (74%) of the soil organic carbon (humic acids and humins) also has the longest turnover time (thousands of years).

Comparison of Carbon Biogeochemical Processes in Terrestrial and Aquatic Ecosystems

The synthesis and degradation of organic matter in the ocean are significantly distinct from those in terrestrial ecosystems. The phytoplankton provides for a large part of photosynthetic organic matter. The dry mass of phytoplankton is three orders of magnitude less than global terrestrial mass, whereas the annual production is only about 3 times smaller. This can be related to the much

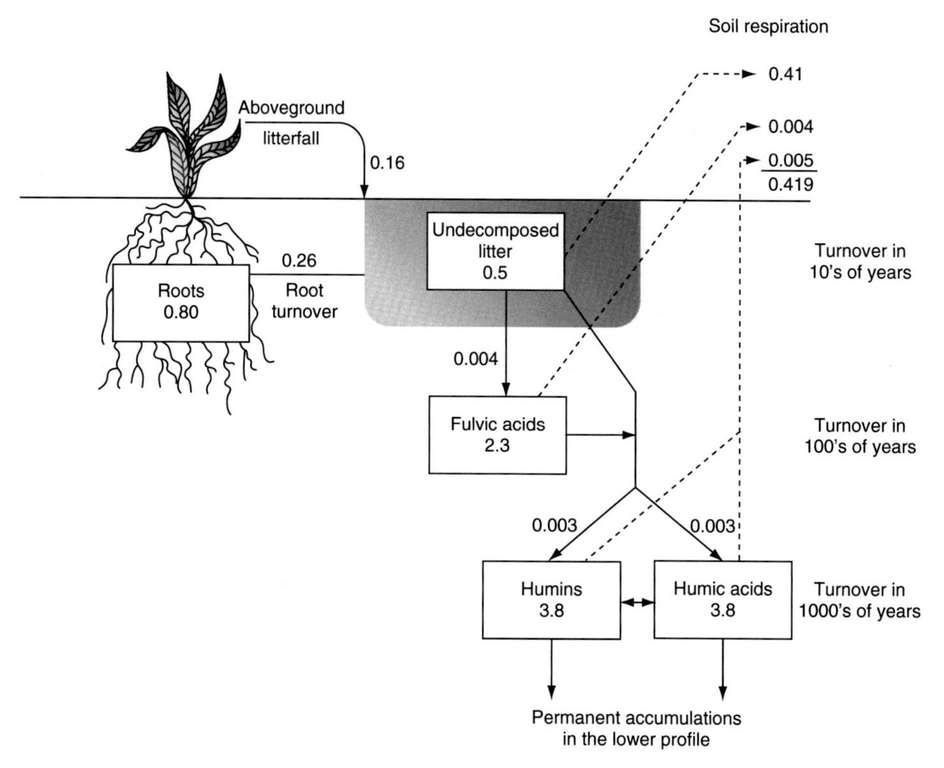

Figure 4 Detrital carbon dynamics for the 0–20 cm layer of chernozem grassland soil. Carbon pool (kgC m^{-2}) and annual transfers (kgC m^{-2} yr^{-1}) are shown. Total profile content down to 20 cm is 10.4 kgC m^{-2}.

faster life cycles of plankton organisms in comparison with the terrestrial vegetation.

Let us consider the renewal of terrestrial and oceanic organic matter. The terrestrial biomass might be assessed as $(2400–2500) \times 10^9$ t of dry organic matter and annual production as $(170–175) \times 10^9$ t. These values present a period of 13–15 years for complete renewal of organic matter. In the oceans, the problem is much more complicated. The various authors give eight- to tenfold discrepancy in the existing estimates of phytoplankton productivity and biomass. It is estimated also that phytoplankton mass cycle takes 1–2 days to be completed. Taking this into account, one can reasonably consider that the renewal of the total biomass in the global ocean takes about 1 month. Based on modern assessments, the annual production of photosynthesis varies from $(20–30) \times 10^9$ to 100×10^9 t of organic carbon and the average values are $(50–60) \times 10^9$ t C_o. Furthermore, one can hypothesize that the plankton-synthesized organic matter is almost completely assimilated in subsequent upper food webs. Thus, the organic precipitation would not exceed 0.1×10^9 t. These calculations present the annual uptake of terrestrial and oceanic living organisms of about 440×10^9 t CO_2 or 120×10^9 t C_o. Most of this amount recycles into the ocean and atmosphere.

Carbon Dioxide Interactions in Air–Sea Water System

The interaction between carbon dioxide in the atmosphere and the hydrosphere is the principal factor for understanding large carbon biogeochemical cycles. As it has been mentioned above, the gases of the troposphere and the surface layer of the ocean persist in a state of kinetic equilibrium.

Compared with the atmosphere, where most carbon is presented by CO_2, oceanic carbon is mainly present in four forms: dissolved inorganic carbon (DIC), dissolved organic carbon (DOC), particulate organic carbon (POC), and the marine biota itself.

DIC concentrations have been monitored extensively since the appearance of precise analytical techniques. When CO_2 dissolves in water it may hydrate to form H_2CO_3(aq), which, in turn, dissociates to HCO_3^- and CO_3^{2-}. This process depends on pH and specification is shown in **Figure 5**.

The conjugate pairs responsible for most of the pH buffer capacity in marine water are HCO_3^-/CO_3^{2-} and $B(OH)_3/B(OH)_4^-$. Although the predominance of HCO_3^- at the oceanic pH of 8.2 actually places the carbonate system close to a pH buffer minimum, its importance is maintained by the high DIC concentration (~2 mm).

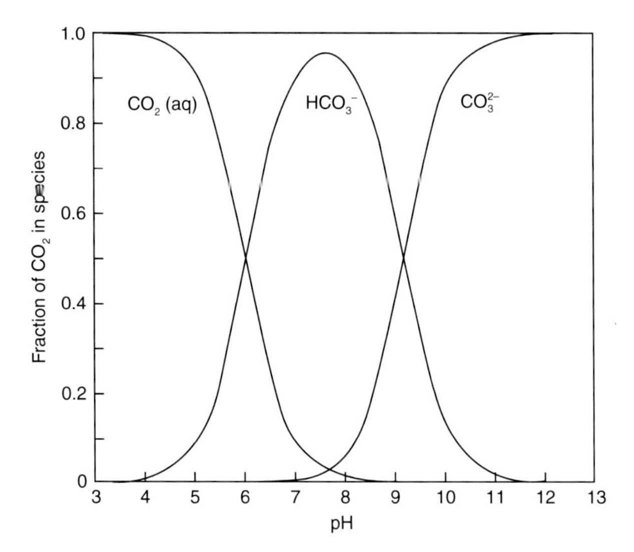

Figure 5 Distribution of dissolved carbon species in seawater as a function of pH at 15 °C and a salinity of 35. Average oceanic pH is about 8.2.

Ocean water in contact with the atmosphere will, if the air–sea gas exchange rate is short compared to the mixing time with deeper water, reach equilibrium according to Henry's law. At the pH of oceanic water around 8.2, most of the DIC is in the form of HCO_3^- and CO_3^{2-} with a very small proportion of H_2CO_3. Although H_2CO_3 changes in proportion to CO_2 (g), the ionic form changes little as a result of various acid–base equilibrium.

From chemical aqueous carbon specification, the alkalinity, Alk, representing the acid-neutralizing capacity of the solution, is given by the following equation:

$$\mathrm{Alk} = [OH^-] - [H^+] + [B(OH)_4^-] + [B(OH)_3] + 2[CO_3^{2-}]$$

Average DIC and Alk concentrations for the world's oceans are shown in **Figure 6**.

With an average DIC of 2.35 mmol kg^{-1} seawater and the world oceanic volume of 1370×10^6 km^3, the DIC carbon reservoir is estimated to be $37\,900 \times 10^9$ t C. The surface waters of the world's oceans contain a minor part of DIC, $\sim 700 \times 10^9$ t C. However, these waters play an important role in air-deep water exchange (see above).

Oceanic surface water is supersaturated everywhere with respect to the two solid calcium carbonate species calcite and aragonite. Nevertheless, calcium precipitation is exclusively controlled by biological processes, specifically the formation of hard parts (shells, skeletal parts, etc.). The very few existing amounts of spontaneous inorganic precipitation of $CaCO_3$(s) come from the Bahamas region of the Caribbean.

The detrital rain of carbon-containing particles can be divided into two groups: the hard parts comprised of calcite and aragonite, and the soft tissue containing

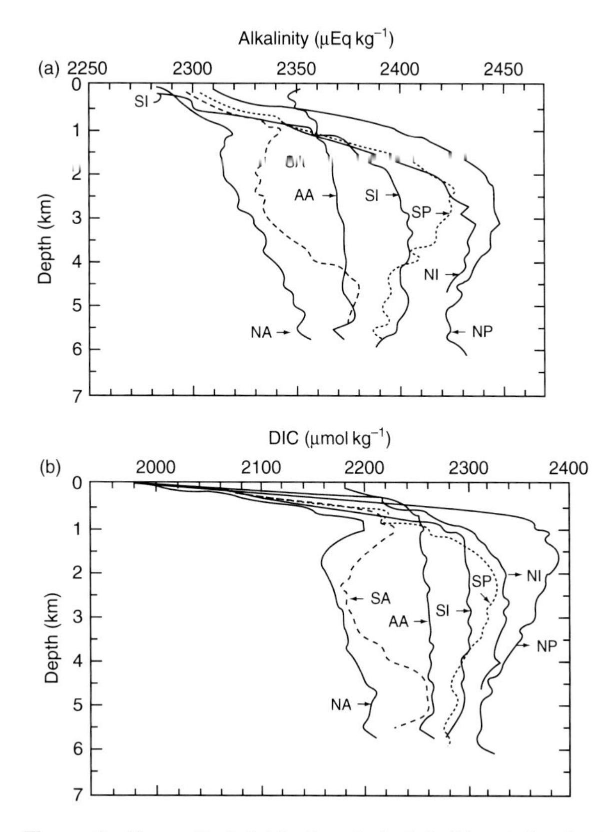

Figure 6 The vertical distribution of alkalinity (a) and dissolved inorganic carbon in the world's ocean (b). Ocean regions are shown as NA (North Atlantic), SA (South Atlantic), AA (Antarctic), SI (South Indian), NI (North Indian), SP (South Pacific), and NP (North Pacific).

organic carbon. The composition of the soft tissue shows the average ratio of biophils as P:N:C:Ca:S = 1:15:131:26:50, with C_c:C_o ratio as 1:4.

The estimation of C_c and C_o mass annually eliminated from the biogeochemical cycles in ocean is a very uncertain task (see above). The carbonate–hydrocarbonate system includes the precipitation of calcium carbonate as a deposit:

Atmosphere CO_2

$\uparrow\downarrow$

Surface ocean layer $H_2O \leftrightarrow H_2CO_3 \leftrightarrow H^+ + HCO_3^- \leftrightarrow H^+ + CO_3^{2-} + Ca^{2+}$

$\uparrow\downarrow$

Deep ocean water $CaCO_3$

The binding of carbon into carbonates is related to the activity of living organisms. However, the surface runoff of Ca^{2+} ions from the land determines the formation of carbonate deposits to a significant degree. The Ca^{2+} ion stream is roughly 0.53×10^9 t yr^{-1}, which can provide for a $CaCO_3$ precipitation rate of $1.33 \times 10^9\,t\,yr^{-1}$. This would correspond to the loss of $0.57 \times 10^9\,t\,CO_2$, or $0.16 \times 10^9\,t\,C$ from the carbonate–hydrocarbonate system.

The surface runoff from the world's land plays an important role in the global carbon mass exchange. The continental runoff supply of HCO_3^- is $2.4 \times 10^9\,t\,yr^{-1}$, that is, $0.47 \times 10^9\,t\,yr^{-1}$ for carbon. Besides, stream waters contain dissolved organic matter at $6.9\,mg\,l^{-1}$, which make up to an annual loss of $0.28 \times 10^9\,t\,yr^{-1}$. The average carbon concentration of suspended insoluble organic matter in the stream discharge is $5\,mg\,l^{-1}$, which gives the loss of about $0.2 \times 10^9\,t\,yr^{-1}$. Most of this mass fails to reach the open ocean and becomes deposited in the shelf and the estuarine delta of rivers. One can see that equal amounts of C_c and C_o (0.5×10^9 t for each) are annually lost from the world's land surface.

The formation of carbonates and the accumulation of organic matter are not confined solely to oceans; these processes occur also on land. The mass of carbonates annually produced in the soils of arid landscapes appears to be high enough.

Table 4 Fluxes of carbon in the biosphere

Fluxes	*C ($10^9\,t\,yr^{-1}$)*
World's ocean	
Turnover of planktonic photosynthesis organisms	50
CO_2 uptake by ocean	30
CO_2 release by ocean	30
C_o deposited in precipitation	0.08
C_c deposited in precipitation	0.16
World's land	
Biological cycle (photosynthesis–degradation of organic matter)	85
HCO_3^- ion mass exchange between land and troposphere	
Supply to troposphere	0.136
Rainfall washout from troposphere	0.139
Stream loss of	
DIC	0.47
DOC	0.28
POC	0.20
Transport of oceanic airborne HCO_3^- ions to land	0.003

Global Carbon Fluxes

Two large cycles determine global dynamics of carbon mass transport in the biosphere. The first of these is provided for by the assimilation of CO_2 and decomposition of H_2O through photosynthesis of organic matter followed by its degradation to yield CO_2. The second cycle involves the uptake–release of carbon dioxide by natural waters via chemical reactions of CO_2 and H_2O leading to buildup of a carbonate–hydrocarbonate system. The cycles are intimately related to the activity of living matter. The living matter of the biosphere, the global water cycle, and carbonate–hydrocarbonate system regulate the cyclic mass exchange of carbon between atmosphere, land, and ocean. These global carbon fluxes are shown in **Table 4**.

A specific feature of these two major biogeochemical cycles of carbon is their openness, which is related to the permanent removal of some carbon from the turnover as dead organic matter and carbonates. The carbon burial in the sea deposits is of great importance for biosphere development.

There is a suggestion that the alteration of glacial and interglacial periods in the Pleistocene was mainly due to fluctuations of CO_2 in the atmosphere. It may be hypothesized that the spread of land ice and the drastic reduction of forest areas with their typically high biomass were favorable for an elevated content of carbon dioxide in the atmosphere and the subsequent climatic warming up. In its turn, the resulting contraction of glacial areas and reforestation was attended by an increased CO_2 uptake from the atmosphere and by its binding to the biomass and soil organic matter. The resulting effect was a gradual cooling and the onset of a new glaciation followed by reduction of forest areas and a repetition of the whole cycle.

The role of carbon dioxide in the Earth's historical radiation budget merits modern interest in raising atmospheric CO_2. There are however other changes of importance. The atmospheric methane concentration is increasing, probably as a result of increasing cattle population, rice production, losses during natural gas exploration and transportation, and biomass burning. Increasing methane concentrations are important because of the role they play in stratospheric and tropospheric chemistry. Methane as a greenhouse gas (GHG) species is also important to the radiation budget of our planet.

Analyses of ice cores from Vostok, Antarctica, have provided new data on natural variations of CO_2 and CH_4 levels over the last 220 000 years. The records show a marked correlation between Antarctic temperature, as deduced from isotopic composition of the ice and the CO_2/CH_4 profiles (**Figure 7**).

Clear correlations between CO_2 and global mean temperature are evident in much of the glacial–interglacial paleo-record. This relationship of CO_2 concentration and temperature may carry forward into the future, possibly causing significant positive climatic feedback on CO_2 fluxes.

Global Climate Changes and Critical Loads of Sulfur and Nitrogen in the European Ecosystems

Global biogeochemical cycle of carbon and its alterations attracted great attention in the mass media due to CO_2 increase in the atmosphere being closely related to the

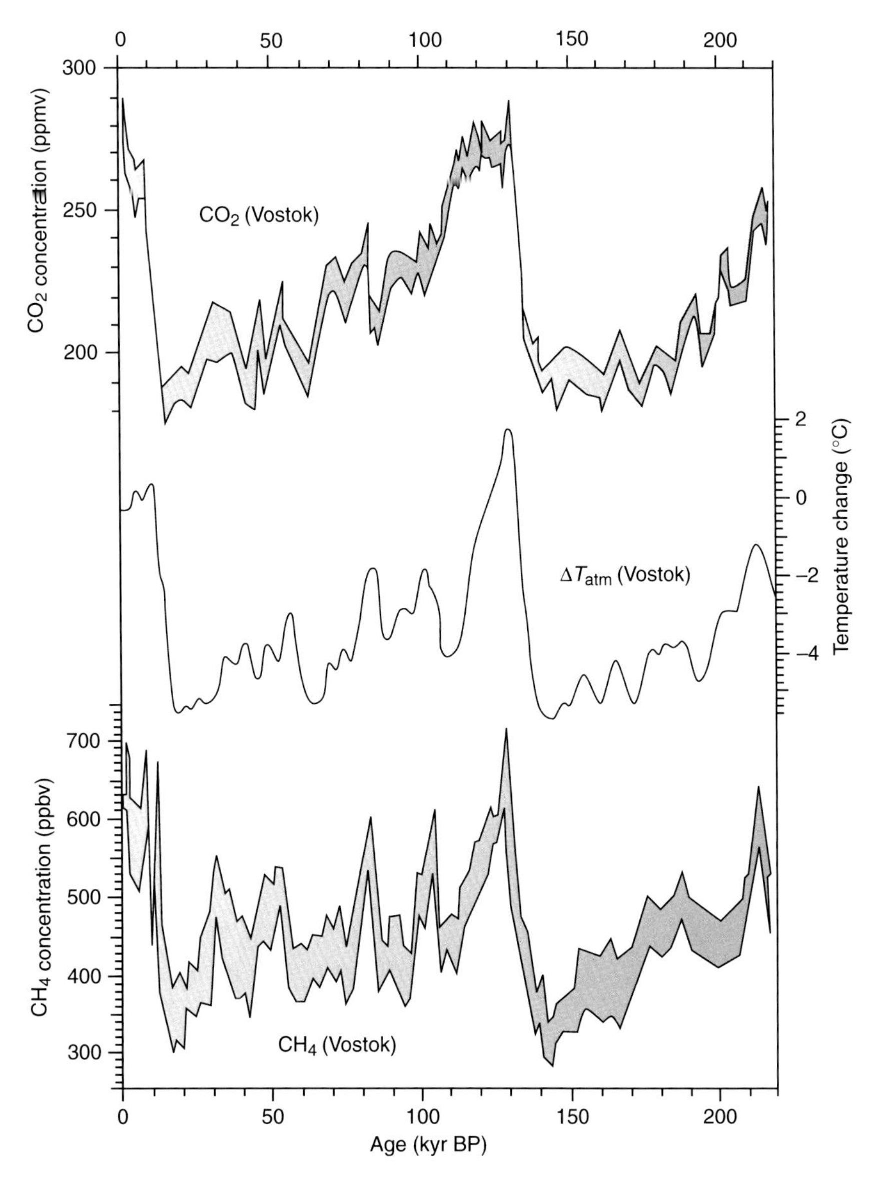

Figure 7 Temperature anomalies and methane and carbon dioxide concentrations over the past 220 000 years as derived from the ice-core records at Vostok, Antarctica.

various changes in the Earth biosphere. These changes include both climate changes and environmental pollution. Here one should mention a project on the integrated assessment of regional air pollution and climate change in Europe, AIR-CLIM project, which *inter alia* examined whether climate change will alter the effectiveness of agreed-upon or future policies to reduce regional air pollution in Europe. Climate changes and emission abatement strategy can be estimated using the calculations of pollutants' critical loads and their exceedances.

A critical load has been defined as the maximum input of pollutants (sulfur, nitrogen, heavy metals, POPs, etc.), which will not introduce harmful alterations in biogeochemical structure and function of ecosystems in the long term, that is, 50–100 years according to present knowledge. Starting from this general definition, methodologies have been developed during the 1990s by the working group on effects (WGE) under the long-range transboundary air pollution (LRTAP) convention for calculating and mapping critical loads in Europe. This has been used by European countries to calculate critical loads of pollutants for various ecosystems (forests, surface waters, and seminatural vegetation).

Transboundary air pollution by sulfur, nitrogen, heavy metals, and persistent organic species is not the only environmental problem calling for internationally agreed abatement policies. In developed countries, it is the issue of climate change, which currently attracts most attention and resources, and negotiations under the framework convention on climate change (FCCC) are trying to come up with equitable mitigation policies. To date, climate change policies are mostly discussed in isolation. However, any measures taken (or not taken) to slow down global warming are likely to have an impact on other environmental problems.

Eight scenarios for different combinations of future GHG, sulfur, and nitrogen emissions, covering the years 1990–2100, were developed during the AIR-CLIM project (**Table 5**).

To assess the risk of ecosystem damage due to a given scenario, critical loads have to be compared with the resulting deposition patterns. Within the integrated assessment framework of AIR-CLIM, deposition fields due to emission scenarios are computed with the source-receptor matrices (SRMs) derived from the EMEP long-range atmospheric transport model. The SRMs derived for the meteorological years 1985–96 were averaged to minimize the effects of interannual variability. With the aid of these SRMs, the sulfur and nitrogen ($NO_x + NH_3$) emissions of the European countries and the respective depositions in every grid cell are computed. If the depositions are greater than critical loads, we say the critical loads are exceeded. While in the case of a single pollutant the exceedance can be defined in an obvious manner, for example, Ex(Ndep) = Ndep − CLnut(N), there is no unique exceedance (i.e., amount of deposition to be reduced to reach nonexceedance) in the case of acidifying N and S.

Figure 8 depicts the temporal development of the percentage of forest area for which critical loads of acidity and nutrient nitrogen are exceeded under the eight AIR-CLIM scenarios. The area for which critical loads are exceeded declines under all scenarios, starting from 41% for acidity critical loads and 75% for nutrient N in 1990. The speed decrease after 2010, however, differs between the two sets of scenarios, with larger decreases in the B1-set. In all the cases, the A1-P and the B1-450-A scenarios are the least and most stringent ones, respectively, with the other scenarios giving intermediate results. The most striking conclusion is that acidification (almost) ceases to be a problem, with exceedance percentages in 2100 between 4.7% (A1-P) and 0.7% (B1-450-A). In drawing this conclusion it has to be borne in mind that considering the in-grid variability of deposition (e.g., by reducing the grid size) would certainly lead to higher exceedances. Furthermore, areas that cease to be exceeded at some point in time are not at once without the risk of adverse effects. The recovery of the chemical, and especially the biological status of the soil is delayed due to finite buffers, which have to equilibrate with the lower deposition. Only dynamic models can provide estimates of the times needed for a full recovery.

Eutrophication, on the other hand, continues to be a widespread problem, even under the most stringent scenario, which brings the exceedance hardly down to 15% of the forest area. This confirms the conclusion that nitrogen is the main pollutant in need of future mitigation.

Summary

The changes in the global biogeochemical carbon cycle become more and more obvious; however, the relative contributions of natural and anthropogenic activities are still uncertain. However, the role of carbon dioxide in the Earth's historical radiation budget merits modern interest in raising atmospheric CO_2. The relative changes include both climate changes and environmental pollution.

Critical loads have been widely used to formulate European emission reduction policies for sulfur and nitrogen. The critical load values depend on the ecosystem characteristics that might be altered due to climate changes. An investigation of the impact of

Table 5 Overview of AIR-CLIM scenarios

Scenario	*Greenhouse gas policies*	*SO_2/NO_x policies*
A1-P	None	Present policies
A1-A	None	Advanced policies
A1-550-P	To achieve 550 ppm CO_2 stabilization	Present policies
A1-550-A	To achieve 550 ppm CO_2 stabilization	Advanced policies
B1-P	None	Present policies
B1-A	None	Advanced policies
B1-450-P	To achieve 450 ppm CO_2 stabilization	Present policies
B1-450-A	To achieve 450 ppm CO_2 stabilization	Advanced policies

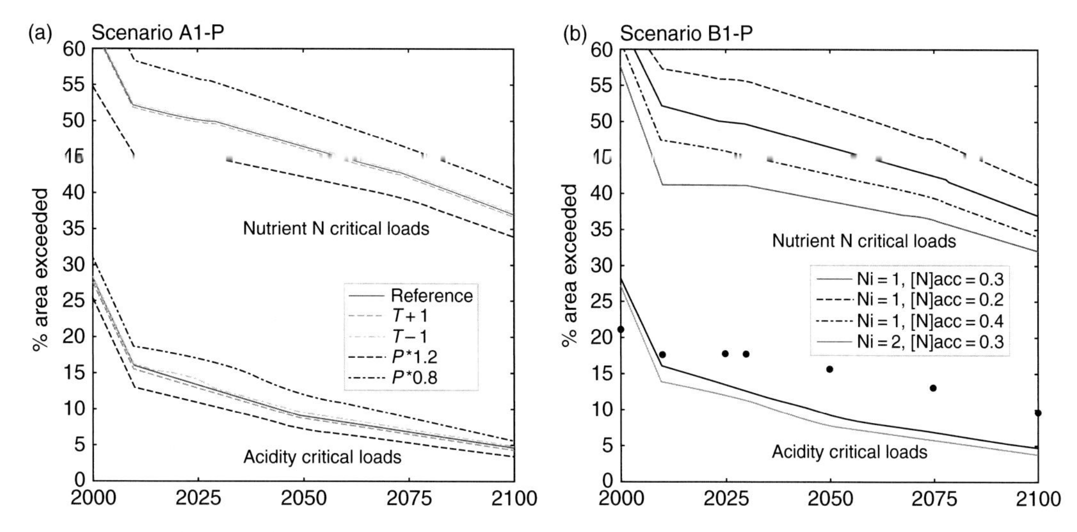

Figure 8 Temporal development of the percentage of forest area for which the critical loads of acidity and nutrient nitrogen are exceeded for the four scenarios in the A1-set (left) and for corresponding four scenarios in the B1-set (right).

different scenarios of climate change on critical loads and their exceedances is of both scientific and political interests.

Recent estimates have shown that the acidity critical loads will be exceeded only in small parts of Europe under all scenarios. It should be borne in mind, however, that: (1) nonexceedance does not mean immediate recovery, and (2) higher-resolution deposition fields, capturing some of their small-scale variability, would certainly lead to more widespread exceedances.

Eutrophication, on the other hand, will continue to be a problem even under the most stringent scenario. This confirms the important and increasing role nitrogen plays in environmental problems, both in its oxidized and reduced forms. Thus, research should focus on the effects of nitrogen in the environment, especially under conditions of climate change, whereas policies should concentrate on further reductions of nitrogen emissions. This not only reduces acidification and eutrophication, but also helps curbing the formation of tropospheric ozone.

Furthermore, the global overview on the carbon biogeochemical cycle should point out its interaction with other cycles such as that of nitrogen and sulfur both for the respective and perspective trends.

See also: Climate Change 1: Short-Term Dynamics.

Further Reading

Barnola J-M, Pimienta P, Raynaud D, and Korotkevich TS (1991) CO_2–climate relationship as deduced from Vostok ice core: A re-examination based on new measurement and on a re-evaluation of the air dating. *Tellus* 43B: 83–90.

Barrett K and Berge E (eds.) (1996) *Transboundary Air Pollution in Europe,* EMEP/MSC-W Report 1/1996. Oslo: Norwegian Meteorological Institute.

Bashkin VN (2002) *Modern Biogeochemistry*, 572pp. Dordrecht: Kluwer Academic Publishers. (in cooperation with Howarth RW).

Bashkin VN and Park S-U (eds.) (1998) *Acid Deposition and Ecosystem Sensitivity in East Asia*, 427pp. New York: Nova Science Publishers.

Bunce N (1994) *Environmental Chemistry*, 2nd edn. Winnipeg: Wuerz Publishing.

Dobrovolsky VV (1994) *Biogeochemistry of the World's Land*, 362pp. London: Mir Publishers.

Holsten K (1992) The global carbon cycle. In: Butcher SS, Charlson RJ, Orians GH, and Wolfe GV (eds.) *Global Biogeochemical Cycles*, pp. 239–316. London: Academic Press.

Jouzel J, Barkov NI, Barnola J-M, *et al.* (1993) Extending the Vostok ice-core record of paleoclimate to the penultimate glacial period. *Nature* 364: 407–412.

Kuo C, Lindberg C, and Thompton DJ (1990) Coherence established between atmosphere carbon dioxide and global temperature. *Nature* 343: 709–713.

Mayerhofer P, De Vries B, Den Elzen M, *et al.* (2002) Integrated scenarios of emissions, climate change and regional air pollution in Europe. *Environmental Science & Policy* 5: 273–305.

Posch M (2002) Impacts of climate change on critical loads and their exceedances in Europe. *Environmental Science and Policy* 5: 307–317.

Schimel D, Enting IG, Heimann M, *et al.* (2000) CO_2 and the carbon cycle. In: Wigley TML and Schimel DS (eds.) *Carbon Cycle*, pp. 7–36. Cambridge: Cambridge University Press.

Schlesinger WH (1997) *Biogeochemistry. An Analysis of Global Changes*, 443pp. New York: Academic Press.

Energy Balance

A Kleidon, Max-Planck-Institut für Biogeochemie, Jena, Germany

Introduction

All ecosystems are affected by and interact with their environment. At the global scale, the Earth's environment is characterized by the global energy balance, the balance of all heating and cooling terms that shape the climatic variations in space and time, especially with respect to surface temperature, precipitation, and light. From an energy balance viewpoint, the interrelationships between ecosystems and their environment are threefold: (1) ecosystems utilize energy sources from their environment, and thereby are a part – though small – of the energy balance; (2) ecosystem processes are affected by environmental conditions that are directly or indirectly connected to the energy balance (e.g., precipitation affects the levels of water limitation of terrestrial productivity); and (3) the form and functioning of ecosystems affect energy balance terms. This article reviews the basics of the global energy balance, how it is reflected in the seasonal and geographic distribution of mean climatic properties, and how it interacts with life through ecosystem functioning.

Global Energy Balance

At the planetary scale, the energy balance is driven by the absorption of sunlight and the emission of radiation to space. Planetary properties and the global energy balance give a first impression of the relevant processes that shape the environmental conditions at the surface and how habitable these are to life.

Planetary Energy Balance

The planetary energy balance is driven by the absorption of about 240 W m^{-2} of solar radiation, which is then re-emitted into space as long-wave radiation. The planetary energy balance is approximately at a steady state when the amount of absorbed radiation is balanced by the emission of radiation. In this case, the planetary energy balance is

$$I_{0,\mathrm{mean}}(1-a_\mathrm{P}) = \sigma T_\mathrm{R}^4$$

where the amount of absorbed solar radiation is expressed by the mean incident amount of solar radiation at the Earth's orbit $I_{0,\mathrm{mean}} = 342\ \mathrm{W\,m^{-2}}$ and the Earth's planetary albedo (or reflectivity, see below) $a_\mathrm{P} = 0.30$. The amount of emitted radiation is expressed by the Stefan–Boltzmann radiation law, with $\sigma = 5.67 \times 10^{-8}\ \mathrm{W\,m^{-2}\,K^{-4}}$ and T_R being the radiative temperature. These numbers yield a value of $T_\mathrm{R} = 255\ \mathrm{K}$ for present-day Earth. The cooling of the Earth's core adds less than 0.1 W m^{-2}, which is very small in comparison to the amount of absorbed solar radiation, and can therefore be neglected for Earth's energy balance consideration.

The observed global mean surface temperature of $T_\mathrm{s} = 288\ \mathrm{K}$ is notably higher by 33 K than the radiative temperature. This additional warming of the surface is due to the atmospheric greenhouse effect. It results from the absorption of long-wave radiation by greenhouse gases in the atmosphere that were emitted from the surface (**Figure 1**). The absorbed radiation is re-emitted to space, but also back to the surface, thereby providing additional heating to the surface. The comparison of Earth's planetary characteristics to those of the planetary neighbors shows the importance of a well-balanced greenhouse effect in providing a habitable environment (**Table 1**).

Surface Energy Balance

At the surface, the energy balance is written as

$$Q_\mathrm{g} = Q_\mathrm{sw,down}(1-a_\mathrm{s}) - Q_\mathrm{lw,net} - Q_\mathrm{sh} - Q_\mathrm{lh} - Q_\mathrm{trans}$$

where Q_g is the ground heat flux (positive adds heat to the ground), $Q_\mathrm{sw,down}$ is the downwelling flux of solar radiation at the surface, a_s is the surface albedo, $Q_\mathrm{lw,net}$ is the net emission of terrestrial radiation from the surface, Q_sh and Q_lh are the turbulent fluxes of sensible and latent heat, respectively, and Q_trans the horizontal transport of heat (relevant for oceans, but not for land).

The surface energy balance directly links the change in surface temperature T_s with the heating and cooling

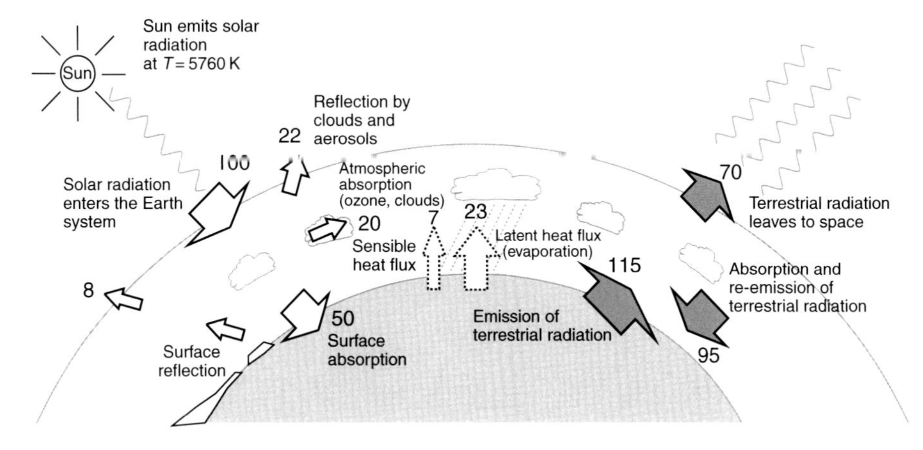

Figure 1 Earth's global energy balance. The dominant energy fluxes and their brief description within the Earth's climate system, expressed as percentage of the average amount of incoming solar radiation of 342 W m^{-2}.

Table 1 The Earth in comparison to its planetary neighbors. Orbital charateristics, atmospheric composition, albedo, absorbed radiation, radiative temperature, surface temperature, and the strength of the atmospheric greenhouse effect for selected inner planets and the Earth's moon

	Earth	*Venus*	*Mars*	*Moon*
Orbital characteristics				
Distance to Sun	150×10^6 km	108×10^6 km	228×10^6 km	*b*
Obliquity	23.45°	<3°[a]	25.2°	6.7°
Eccentricity	0.017	0.007	0.094	0.055
Length of day	24 h	2802 h	24.7 h	708.7 h
Length of year	365.2 days	224.7 days	687 days	27.3 days
Atmospheric compositon				
Surface pressure	1 bar	92 bar	6.4 mbar	3×10^{-15} bar
Carbon dioxide (CO_2)	360 ppm	96.5%	95%	
Nitrogen (N_2)	78%	3.5%	2.7%	
Oxygen (O_2)	21%	0	0.13%	
Climatic properties				
Planetary albedo	0.30	0.71	0.16	0.11
Absorbed solar radiation	239 W m^{-2}	190 W m^{-2}	124 W m^{-2}	304 W m^{-2}
Radiative temperature	255 K	233 K	210 K	275 K
Surface temperature	288 K	737 K	210 K	100–400 K
Greenhouse effect	+33 K	+504 K	+0 K	

[a]Venus rotates in the opposite sense than Earth.
[b]Distance Moon–Earth is 0.378×10^6 km.

terms at the surface. Important factors that determine surface temperature change are:

1. the ground heat flux Q_g: it depends on the conductivity and specific heat capacity of the ground surface. The conductivity describes how well heat is conducted by the material, while the specific heat capacity measures by how much the temperature changes with a certain change in heating. For instance, water surfaces have a much higher heat capacity; therefore, they show a much reduced temperature response than dry soil or air for the same amount of heating or cooling (**Table 2**).
2. the amount of heating by absorption of solar radiation: it combines the effects of incoming solar radiation and the reflectivity of the surface.
3. the partitioning of net radiative heating (solar and terrestrial) into sensible and latent heat (evaporation).
4. the amount of heat transport at the surface: through the oceanic circulation, warm water is removed from the tropics and transported toward the poles. The heat transport by the oceanic circulation averages to zero at the global scale.

Global Energy Balance Components

In the global climatic mean, $\mathrm{d}T_s/\mathrm{d}t = 0$ and $\sum Q_{trans} = 0$. The estimates of each of the flux components are shown in **Figure 1**. Of the incoming 342 W m^{-2} of solar radiation ($I_{0,mean}$) at the top of the atmosphere, 22% is reflected by

Table 2 Specific heat capacity of selected substances. Substances change their temperature by differing amounts for a given amount of heat, depending on their specific heat capacity and their density. The last column, the product of the former two quantities, describes the amount of heat that is necessary to raise the temperature of 1 m^3 of a given substance by 1 K

	Specific heat ($J kg^{-1} K^{-1}$)	*Density ($kg m^{-3}$)*	*Heat capacity ($J m^{-3} K^{-1}$)*
Water[a]	4182	1000	4.18×10^6
Sandy soil, saturated	1480	2000	2.96×10^6
Sandy soil, dry	800	1600	1.28×10^6
Soil, inorganic	733	2600	1.91×10^6
Soil, organic	1921	1300	2.50×10^6
Peat soil, saturated	3650	1100	4.02×10^6
Peat soil, dry	1920	300	0.58×10^6
Snow, fresh	2090	100	0.21×10^6
Snow, old	2090	480	1.00×10^6
Ice	2100	920	1.93×10^6
Air[a]	1004	1.2	0.001×10^6

[a]Density depends on temperature. Values given are for 293 K.

clouds and aerosols in the atmosphere and another 8% by the surface. These two numbers add up to the planetary albedo of about $a_P = 0.30$. The remaining radiation is absorbed in the atmosphere (20% of $I_{0,mean}$, by ozone in the stratosphere and by clouds) and at the surface (50% of $I_{0,mean}$). Additional surface heating is provided by the atmospheric greenhouse effect (gray arrows in **Figure 1**), which adds almost twice as much energy to the surface than solar radiation. These heating terms are balanced by cooling through emission of terrestrial radiation and turbulent fluxes (the sum of sensible and latent heat flux). The atmosphere is heated by the absorption of solar radiation (20% of $I_{0,mean}$), absorption of terrestrial radiation emitted by the surface (115% of $I_{0,mean}$), turbulent fluxes (30% of $I_{0,mean}$), and cooled by the emission of terrestrial radiation to space (70% of $I_{0,mean}$) and to the surface (95% of $I_{0,mean}$, the greenhouse effect).

Global Entropy Budget

An important aspect that is not captured by the global energy balance is that many of the involved energy conversions are irreversible in their nature, that is, they proceed only in a certain direction. For instance, heat is transported from warm to cold regions, but not in the other direction. This aspect is captured by the global entropy budget, where the rates of entropy production tell us about the strength of energy degradation of the different processes that convert energy.

Irreversibility of Processes

The entropy budget quantifies the irreversible nature of Earth system processes, that is, that these processes proceed in a unique direction. The entropy budget is given by

$$dS/dt = \sigma + \text{div}(F_e)$$

where dS/dt is the change in local entropy S with time t, σ is the production of entropy within the system, and $\text{div}(F_e)$ is the divergence of entropy fluxes F_e resulting from the exchange of energy and mass. The entropy S of the system characterizes its organization, with a lower value representing more organization, and a maximum value representing thermodynamic equilibrium. The rate of entropy production measures the degree of irreversibility (with $\sigma = 0$ characterizing a reversible process).

Planetary Entropy Budget

A steady state of the entropy budget is defined by $dS/dt = 0$. Note that this is a different, less common definition of the climatic steady state that includes a representation of internal organization within the system. The entropy budget of the Earth can be estimated from this steady-state condition since the rate of entropy production σ is then balanced by the difference of entropy fluxes across the Earth–space boundary. The entropy production by the flux of energy from a warm to a cold point can be expressed by

$$\sigma = Q(1/T_c - 1/T_w)$$

where Q is the flux of energy, and T_c and T_w are the temperatures. For instance, at the planetary level in steady state, solar radiation that was emitted at $T_{sun} = 5760$ K is absorbed and emitted by the Earth at a much lower temperature of $T_R = 255$ K, resulting in an estimate for the planetary entropy production of

$$\sigma_{earth} = Q_{sw,abs}(1/T_R - 1/T_{sun}) = 892\,mW\,m^{-2}\,K^{-1}$$

with the average amount of absorbed radiation of $Q_{sw,abs} = 238\,W\,m^{-2}$. Note that mass fluxes do not contribute to the global entropy balance since the mass of the Earth is approximately conserved.

Entropy Production by Earth System Processes

The planetary rate of entropy production results from various irreversible processes (**Table 3**):

- *Scattering of solar radiation.* Scattering of solar radiation within the atmosphere and at the surface results in the irreversible conversion of a direct, focused beam with small solid angle to radiation being distributed over a wide solid angle. The resulting amount of entropy production is $\sigma_{scatter} = 26\,mW\,m^{-2}\,K^{-1}$.
- *Absorption of solar radiation.* Absorption of solar radiation at a temperature lower than the Sun's emission temperature is irreversible (i.e., it cannot be re-emitted as shortwave radiation). On Earth, absorption occurs within the atmosphere (e.g., ozone in the stratosphere) and at the surface. The two associated rates of entropy production are estimated to be: (1) atmospheric absorption of $Q_{sw,abs,atm} = 68\,W\,m^{-2}$ at a stratospheric temperature of about $T_{strat} = 252\,K$ results in $\sigma_{sw,abs,atm} = 258\,mW\,m^{-2}\,K^{-1}$; (2) surface absorption of $170\,W\,m^{-2}$ at a global mean temperature of about $T_s = 288\,K$ results in $\sigma_{sw,abs,srf} = 561\,mW\,m^{-2}\,K^{-1}$.
- *Absorption of terrestrial radiation.* Of the net transfer of energy of $68\,W\,m^{-2}$ from the surface to the atmosphere by terrestrial radiation, *c.* $40\,W\,m^{-2}$ escape to space without absorption, while the remaining $28\,W\,m^{-2}$ is absorbed at a lower temperature of about $T_a = 252\,K$. This results in $\sigma_{lw,abs} = 14\,mW\,m^{-2}\,K^{-1}$.
- *Moist convection.* Water is evaporated at the surface at $T_s = 288\,K$ and subsequently condenses within the atmosphere at a lower temperature of about $T_c = 266\,K$ (i.e., evaporation into an unsaturated atmosphere is irreversible). The associated global mean latent heat flux of $Q_{lh} = 79\,W\,m^{-2}$ yields an estimate of the overall entropy production of $\sigma_{moist} = 23\,mW\,m^{-2}\,K^{-1}$. This term includes various irreversible processes associated with moist convection, such as the phase transitions liquid–gas, the mixing of air masses of different humidity, and dissipation of kinetic energy of falling raindrops.
- *Dry convection.* Under dry conditions, the sensible heat flux reflects the dominant form of heat transport by turbulent motion from the surface into the convective boundary layer. Using a typical temperature of $T_{bl} = 280\,K$ and the global mean value of $Q_{sh} = 24\,W\,m^{-2}$ results in an entropy production of $\sigma_{dry} = 2\,mW\,m^{-2}\,K^{-1}$.
- *Frictional dissipation of large-scale motion.* Motion in the atmosphere and ocean are generated from density differences. The associated physical work W performed to accelerate the atmosphere and ocean is balanced on average by the amount of friction dissipation D. Through the motion, heat is transported and counteracts the density differences. With an average amount of atmospheric heat transport of $Q_{ht} = 10\,W\,m^{-2}$ and typical temperatures of the tropics, $T_{trop} = 300\,K$, and of the poles, $T_{pole} = 255\,K$, this yields an estimate of $\sigma_{ht} = 6\,mW\,m^{-2}\,K^{-1}$.
- *Biotic activity.* On the global scale, the biosphere with a gross primary production of $200\,GtC\,yr^{-1}$ converts approximately $8\,W\,m^{-2}$ of solar radiation into organic carbon compounds that are eventually respired into heat at $T_s = 288\,K$. The resulting entropy production is $\sigma_{bio} = 5\,mW\,m^{-2}\,K^{-1}$. Note that this term is already included in the above estimate of entropy production by absorption of solar radiation at the surface. Not included in this estimate is the additional work done by transpiring vegetation (included in the estimate of moist convection).

Table 3 Global entropy budget. The global entropy budget characterizes the irreversibility of various Earth system processes. The columns give typical values of the heat flux Q (see global energy balance) and the temperatures T_{cold} and T_{warm} at which the energy is being transformed. The entropy production σ is then estimated by $\sigma = Q\,(1/T_{cold} - 1/T_{warm})$ using the steady-state assumption

	Heat flux (*$W\,m^{-2}$*)	*T_{cold}* (*K*)	*T_{warm}* (*K*)	*σ* (*$mW\,m^{-2}\,K^{-1}$*)
Scattering of solar radiation	103	n/a[a]	n/a[a]	26
Atmospheric absorption of solar radiation	68	252	5760	258
Surface absorption of solar radiation	170	288	5760	561
Atmospheric absorption of terrestrial radiation	28	252	288	14
Moist convection (evaporation–precipitation)	79	266	288	23
Dry convection (sensible heat into boundary layer)	24	280	288	2
Frictional dissipation of large-scale circulation	10	255	300	6
Biotic activity	8	288	5760	5[b]
Planetary	235	255	5760	881[c]

[a]Entropy produced by scattering originates from broadening of the solid angle, not from temperature differences.
[b]Term included in surface absorption of solar radiation.
[c]Total does not balance individual contributions due to estimated nature of the budget.

There are various other irreversible processes, such as seasonal freeze–thaw associated with sea ice and snow cover, seasonal storage and release of heat, wetting and drying of soils, etc., that are not included here. Yet these simple estimates provide an important additional component of the workings of the global energy balance.

Radiative Exchange

Absorption, reflection, and emission of radiation are critical processes that shape the global energy balance and its regional variations. To understand these variations, the nature of radiation, the processes that reflect and absorb it, as well as the resulting latitudinal variation of radiative fluxes for the present-day climate are explained in the following.

Electromagnetic Radiation

Electromagnetic radiation is characterized by its wavelength λ, or alternatively by its frequency v. The two variables are related by $\lambda \times v = c$, with c being the speed of light ($c = 3 \times 10^8\,\text{km s}^{-1}$ in vacuum). Climatically relevant are mainly the following wavelength ranges: (1) ultraviolet radiation, corresponding to wavelengths of less than 400 nm; (2) visible light, ranging from 400 nm (blue light) to 750 nm (red light); and (3) infrared radiation, referring to wavelengths longer than 750 nm. Radiation with shorter wavelengths is generally referred to as more energetic.

The peak of emission of solar radiation is about 550 nm (green light), while the Earth with its much lower emission temperature has its peak emission at about 11 μm (infrared). The peak of emission is described by Wien's law ($\lambda_{\text{peak}} = 0.2898 \times 10^{-3}\,\text{m K}/T_{\text{R}}$). Since these peak wavelengths and the associated distributions are well separated, electromagnetic radiation in climatology is generally classified into two types: solar (or shortwave) radiation that is emitted by the Sun, and terrestrial (or longwave) radiation associated with emission of radiation within the Earth system.

Solar Radiation

In order to understand spatial and temporal variations in temperature, one needs to consider the causes of variability in the heating and cooling terms of the energy balance. The main factor in this variaibility is the variations of solar radiation that can result from three aspects:

1. *The amount of emitted radiation by the Sun (solar luminosity L_0).* The typical value of the solar luminosity is $L_0 = 3.9 \times 10^{26}$ W, corresponding to a surface emission temperature of = 5760 K. The actual value of L_0 varies, for instance, on decadal timescales through the sunspot cycle (11 years, by less than 1.5 W m^{-2}), and increases over geologic time (L_0 was about 70% of the present-day value 4.5 billion years ago).

2. *The distance d_{Earth} of the Earth to the Sun.* The flux of solar energy remains constant through any surface around the Sun, but the density decreases quadratically with distance, so that at the mean distance of the Earth's orbit of about $d_{\text{Earth}} = 150 \times 10^6$ km, an average amount of $I_0 = L_0/4\pi d_{\text{Earth}}^2 = 1367\,\text{W m}^{-2}$ illuminates the Earth. The value of I_0 is referred to as the solar constant. Considering that the Sun illuminates the Earth's cross section of size $\pi d_{\text{Earth}}{}^2$, but the surface area of the Earth is $4\pi d_{\text{Earth}}^2$, the mean solar radiation used above is obtained by $I_{0,\text{mean}} = I_0/4 = 342\,\text{W m}^{-2}$. The mean distance of the Earth varies between 147×10^6 and 152×10^6 km throughout the year, due to Earth's slightly eccentric orbit (**Figure 2**). The location of the Earth's path that is closest (farthest) to the Sun is called the perihelion (aphelion). The perihelion currently occurs in early January, so that the Earth in total receives about 7% more sunlight in January than it does in July. These orbital parameters – perihelion, eccentricity, and obliquity (or tilt) – vary on longer timescales and relate to the timing of ice ages. Even though the direct impacts of solar radiation are well understood, the indirect effects and feedbacks that amplify the Earth system response to these orbital changes are not yet fully understood.

3. the orientation of the surface toward the Sun, as characterized by the solar zenith angle θ: The amount of incident solar radiation for a given region depends on latitude and time within the year. It is calculated from the zenith angle θ, which measures the position of the Sun to the vertical, and the declination angle δ, which characterizes the relation of the Earth's tilt to the direction of sunlight (**Figure 3**). Integration yields a global mean solar radiation $I_{0,\text{mean}} = I_0/4$.

Reflection of Radiation

Reflection of radiation is mainly due to scattering. The size of the scattering particle plays an important role and affects the amount of radiation of a certain wavelength λ that is scattered, resulting in three forms of scattering:

1. Rayleigh scattering applies to very small particles with diameters of 0.1–1 nm, such as electrons. The intensity of scattering varies with λ^{-4}, therefore affecting primarily radiation of short wavelength. This process of scattering results in blue skies since blue light is scattered much more strongly than red light.
2. Mie scattering involves particles with diameters of 0.01–1 μm, such as aerosols. The intensity of scattering varies with λ^{-1}, so that the intensity of scattering is more evenly spread across wavelengths. This scattering process results, for example, in hazy skies at a

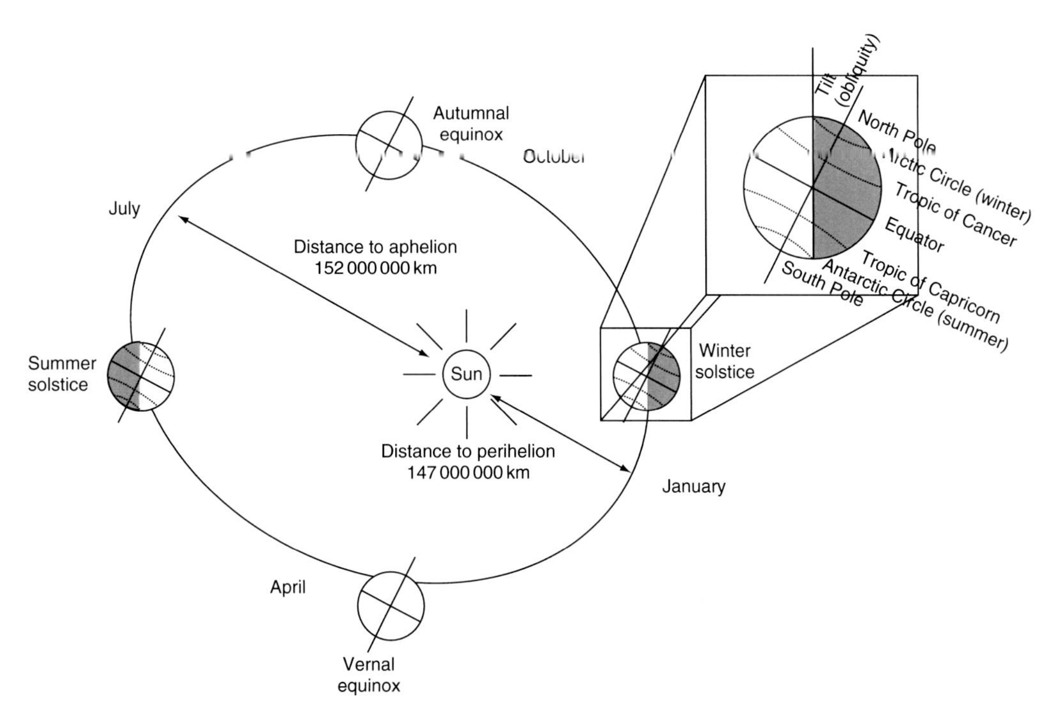

Figure 2 The orbit of the Earth around the Sun and its relation to seasons. The orbit of the tilted Earth around the Sun results in the seasons, as indicated for the Northern Hemisphere (NH). In the NH winter, the Earth's axis of rotation is pointed away from the Sun, resulting in less incident solar radiation and the polar night at latitudes above the Arctic Circle. This situation is reversed in the summer.

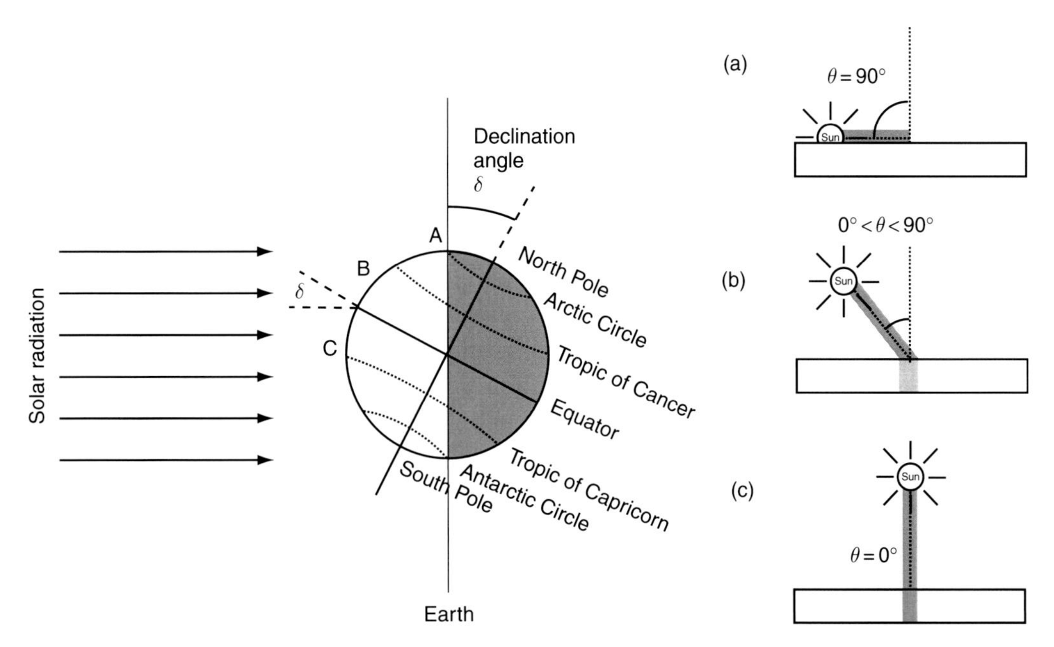

Figure 3 Effects of the orientation of the Earth's surfaces toward the Sun on the amount of incident solar radiation at different locations of the Earth's surface. Left: The amount of solar radiation that reaches the surface at a given latitude depends on the declination angle δ. The declination angle measures the angle between the Earth's axis of rotation and the vertical plane of the orbit, or, alternatively, the angle between the direction of solar radiation and the Earth's equator. The declination angle defines Earth's major regions: the tropics (latitudes $-\delta$ to $+\delta$) and the polar regions ($90° - \delta$ to pole). Earth's declination angle is currently at 23.45°. Right: At a given location on Earth, the zenith angle measures the angle between the vertical and the Sun. It depends on hour, latitude, and time of year. In the situation shown on the left (Northern Hemisphere winter solstice), the zenith angle at location A at noon is 90°, that is, the Sun does not rise above the horizon and no solar radiation is incident at the surface. At location B, the zenith angle is in between 0° and 90°. At location C at the tropic of Capricorn, the zenith angle is 0° at noon, and the incoming solar radiation is vertical to the surface. In sloped terrain, a correction needs to be applied for the calculation of incident radiation to correct for the slope.

windy day at the beach due to sea spray, or over cities due to air pollution (i.e., aerosol production by traffic).
3. Geometric scattering applies to large particles with sizes ranging from 10 to 100 μm, such as cloud droplets. Intensity of scattering does not vary with wavelength. This form of scattering makes clouds appear white.

Typical values of albedo (or reflectivity) of different surfaces are summarized in **Table 4**. The reflectivity also depends on other factors, such as the zenith angle, and wavelength. For instance, vegetated surfaces are generally much more reflective (30–50%) in the near infrared (at wavelengths of 0.8–1.0 μm) but absorbent in the red part of the spectrum at about 0.6 μm, with a low reflectivity of around 5%. This difference in absorptive characteristics is used for the remote sensing of vegetation greenness.

Absorption of Radiation

Radiation is absorbed by different processes and at different intensities, depending on material characteristics and the wavelength of incident radiation λ:

1. Photoionization refers to highly energetic radiation with wavelengths of less than 100 nm; it can remove electrons from atoms, resulting in ionized atoms. This process can be found in the higher atmosphere at heights of 100 km and above in the so-called ionosphere.

2. Photodissociation applies to highly energetic radiation of short wavelengths, where the energy of the radiation is absorbed by breaking up molecular bonds. This process occurs in the atmosphere mainly for wavelengths shorter than visible light. An example is the absorption of ultraviolet radiation by molecular oxygen and ozone in the stratosphere.

3. Electronic absorption is relevant to the absorption of visible light. Here, radiation is absorbed by raising electrons into excited states. While this form of absorption has little relevance in atmospheric absorption, it is essential for photosynthesis, where electronic absorption is used to separate hydrogen ions from the water molecule.

4. Molecules can absorb radiation if the electronic charge is unevenly distributed in the molecule, causing a dipole moment. The absorption results in rotation or vibration of the molecule. This mechanism of absorption is relevant for radiation with low energy and long wavelengths (near infrared and longer).

Table 4 Range and typical values of reflectivity of clouds and surface albedo for different surfaces

	Range (%)	*Typical value* (%)
Atmosphere		
Cirrus clouds		21
Cumulus clouds		48
Stratus clouds		69
Ice, snow, and water		
Deep water, small zenith angle	3–10	7
Deep water, large zenith angle	10–100	
Sea ice	30–45	30
Snow, fresh	70–95	80
Snow, old	35–65	50
Snow, forested	11–35	25
Bare land surfaces		
Sand, wet	20–30	25
Sand, dry	30–45	35
Clay, wet	10–20	15
Clay, dry	20–40	30
Humus soil, moist	5–15	10
Desert	20–45	30
Concrete	15–35	20
Asphalt pavement	5–10	7
Vegetated surfaces		
Tundra	18–25	15
Grassland	16–26	19
Coniferous forest	5–15	12
Deciduous forest	10–20	17
Evergreen forest	12–25	13
Cropland		18

In the atmosphere, water vapor (H_2O) absorbs very well by rotational and vibrational modes due to the architecture of the molecule, where the oxygen atom attracts the electrons more than the two hydrogen atoms. Other climatically relevant gases that absorb by this mechanism are carbon dioxide (CO_2), methane (CH_4), and nitrous oxide (N_2O).

Because the Earth's surface emits radiation mainly in the infrared, gases that absorb in these wavelengths are called greenhouse gases. Water vapor and clouds are by far the most important contributors to the strength of the present-day greenhouse effect. The special role of carbon dioxide originates from two facts: (1) water vapor absorbs poorly at the peak of the Earth's surface emission at about 11 μm, where the CO_2 molecule has a dominant absorption peak nearby at 15 μm and therefore absorbs very well; and (2) the concentration of water vapor in the atmosphere is constrained to at or below its saturation level, which in turn depends on the ambient air temperature. Hence, the concentration of water vapor reacts to other prevailing conditions and by itself does not act as a driver for change. For instance, cold air is unable to hold large amounts of water vapor, and consequently water vapor plays a less important role, for example, in cold regions of the atmosphere, and in winter seasons in polar regions.

Zonal Distribution of Radiative Fluxes

The zonal distribution of solar and terrestrial radiation for the present day are shown in **Figure 4** for the top of the atmosphere and at the surface. The imbalance of net

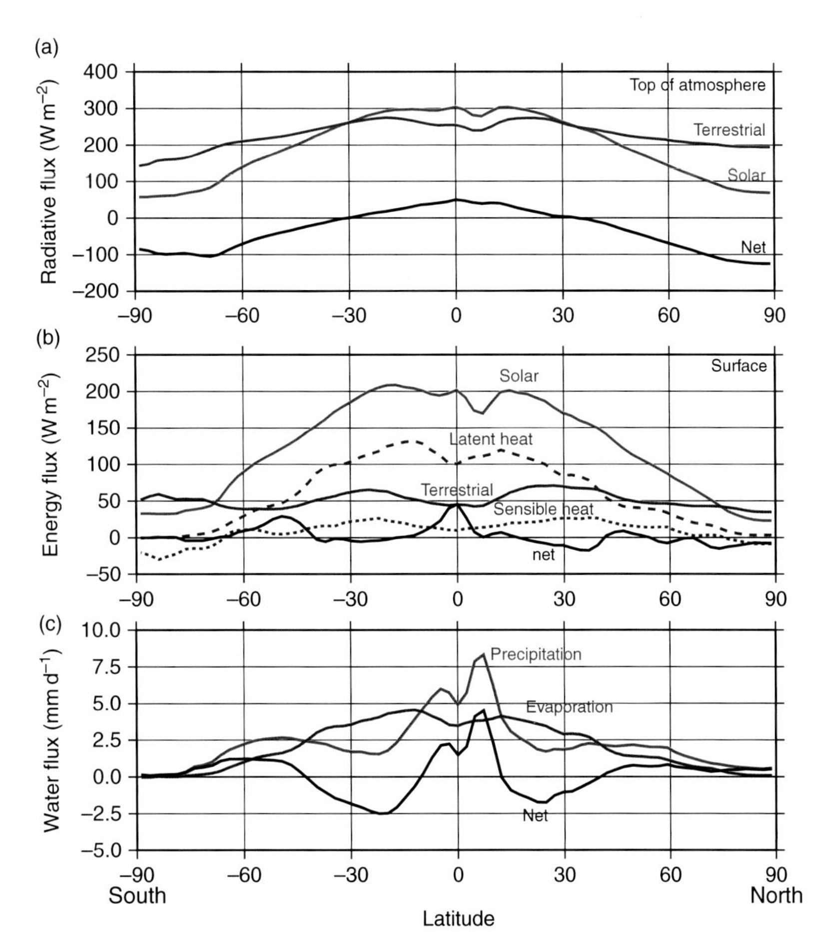

Figure 4 Zonally and annually averaged components of the energy and water balance at the top of the atmosphere and the surface from the time period 1980 to 1990. (a) The graph shows the fluxes of net solar radiation (incoming minus reflected, red line, 'solar'), terrestrial radiation (outgoing long-wave radiation, blue line, 'terrestrial'), and the difference between both (black line labeled 'net') at the top of the atmosphere. The positive values of net radiation in the Tropics (i.e., more solar radiation is absorbed than terrestrial radiation emitted to space) indicates that heat is transported by the atmosphere and ocean systems toward the polar regions, where net radiation is negative. (b) Surface energy balance components of absorbed solar radiation (red line, 'solar'), net emission of terrestrial radiation (blue solid line, 'terrestrial'), latent heat flux associated with evaporation (blue dashed line, 'latent heat'), sensible heat flux (blue dotted line, 'sensible heat'), and the residual (black line, 'net'). The residual consists of the effects of ocean heat transport and heat fluxes due to freeze/thaw of sea ice. (c) The atmospheric water budget, reflected by annual mean precipitation (red line), evaporation (blue line), and the difference ('net', black line). Regions where evaporation exceeds precipitation ('net' is negative) are regions where the atmosphere gains moisture, which is transported by the atmospheric circulation to regions where precipitation exceeds evaporation ('net' is positive). The plots were created using the European Centre for Medium-Range Weather Forecasts (ECMWF) reanalysis data sets. Data sets have been obtained from the ECMWF data server.

fluxes at the top of the atmosphere, where the tropics absorb more solar radiation than is emitted as terrestrial radiation, reflect overall heat distribution within the climate system that is governed by atmospheric and ocean dynamics.

Dynamics

Atmospheric and oceanic motions redistribute heat and mass at the hemispheric scale, thereby playing a critical role in the surface energy and water balance. Motion results from differences in heating, but motion also interacts with these forcings, as seen in the radiative imbalance at the top of the atmosphere (**Figure 4**).

Atmospheric Motion

Atmospheric dynamics are driven by density differences that are caused by differential heating and cooling. As the surface absorbs solar radiation and the atmosphere aloft cools by emitting terrestrial radiation, a difference in density is generated that drives vertical convection. The heat carried by convection in form of sensible and latent heat cancels out these differences in heating and cooling. The same applies for large-scale horizontal motion that is caused by the radiative heating imbalance due to the zonal variation of solar radiation.

The atmospheric circulation has important consequences as it provides the driving force for the hydrologic cycle (**Figure 4**), shaping the large-scale patterns of precipitation (see section titled 'Global energy balance and climate').

Oceanic Motion

Oceanic motion is set into motion by the same principles, except that differences in salinity also result in density differences, with saltier water being more heavy than freshwater at the same temperature. The resulting circulation is therefore known as the thermohaline circulation. This provides an important link to the hydrologic cycle as it sets the freshwater balance of the oceans.

However, compared to the atmospheric heat transport of $= 5 \times 10^{15}$ W, recent estimates of oceanic heat transport of 1–1.3 $\times 10^{15}$ W make it noticeably smaller. This is evident in **Figure 4**, which shows that the top of atmosphere net imbalance – showing the combined effect of atmospheric and oceanic heat transport – is significantly larger than the net imbalance at the surface, which reflects mainly the effect of oceanic heat transport.

Hypothesis of Maximum Entropy Production

The radiative imbalance at the top of the atmosphere exemplifies the importance of atmospheric heat transport and is further explained in **Figure 5**. The emergent magnitude of poleward heat transport can be understood by the hypothesis of maximum entropy production (MEP).

Different amounts of heat transport are associated with different top-of-atmosphere imbalances, surface temperature gradients, and magnitudes of entropy production associated with the atmospheric circulation. A maximum in entropy production results from the tradeoff between flux and force: a greater poleward heat flux results in a lower temperature gradient and smaller force that is needed to maintain the circulation. This maximum in entropy production is associated with maximum dissipation of kinetic energy and corresponds to an atmospheric circulation at maximum strength.

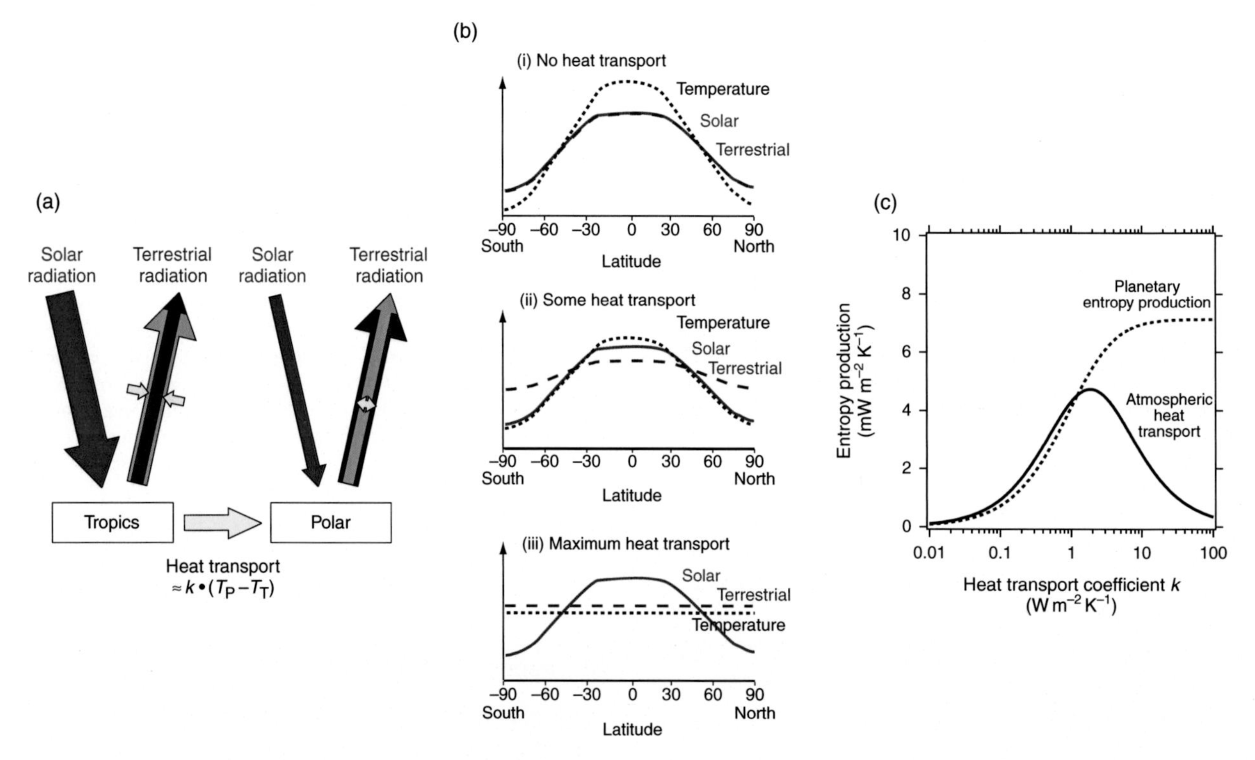

Figure 5 Heat transport, its effects on the radiative balance at the top of the atmosphere, and its effects on entropy production. (a) Conceptual diagram to illustrate the effect of poleward heat transport on the radiation balance at the top of the atmosphere. With heat transport, less terrestrial radiation is emitted to space in the tropics, but more is emitted in the polar regions. This effect is indicated by the yellow arrows. (b) Sketches of how solar and terrestrial radiation and surface temperature would vary for no, some, and maximum amount of heat transport. (c) Conceptual model results that demonstrate the existence of a maximum in entropy production associated with poleward heat transport, that is, a state where the atmosphere works and dissipates kinetic energy as much as possible.

The hypothesis that the atmospheric circulation is maintained at a state of MEP has been confirmed with more detailed calculations with atmospheric circulation models and may have wider-ranging applications beyond turbulent processes within the atmosphere.

Global Energy Balance and Climate

Variations in solar radiation and the energy balance components strongly shape how the climatic conditions near the surface vary in space and time. This linkage is important to understand for the present-day climate, but also how it reacts and interacts with change.

Geographic and Temporal Variations

The geographic variation of the main climatic variables, surface temperature and precipitation, and their seasonal variation are strongly connected to the energy balance (**Figure 6**).

The annual mean temperature distribution largely follows the variation of solar radiation with latitude. Some deviations in this pattern can be found at west coasts in the subtropics where cold ocean currents affect temperature, and in high altitudes, such as the Andes and the Tibetan plateau. The seasonality in temperature is much stronger over land than over oceans. This reflects the differences in heat storage (**Table 2**) and the lack of oceanic heat transport on land.

Mean precipitation is at its peak in the tropics and is associated with general upward motion of the atmospheric circulation. The subtropics are dominated by a lack of rainfall, which is associated with large-scale sinking motion, which prevents air from cooling and saturating. This shapes the large-scale distribution of deserts. Seasonality is strongest in the tropics, which is associated with the seasonal change of solar radiation.

The availability and seasonality of precipitation on land has large impacts on the surface energy and water balance as it limits the amount of water than can be evaporated or transpired by the vegetative cover, thereby affecting the latent heat flux.

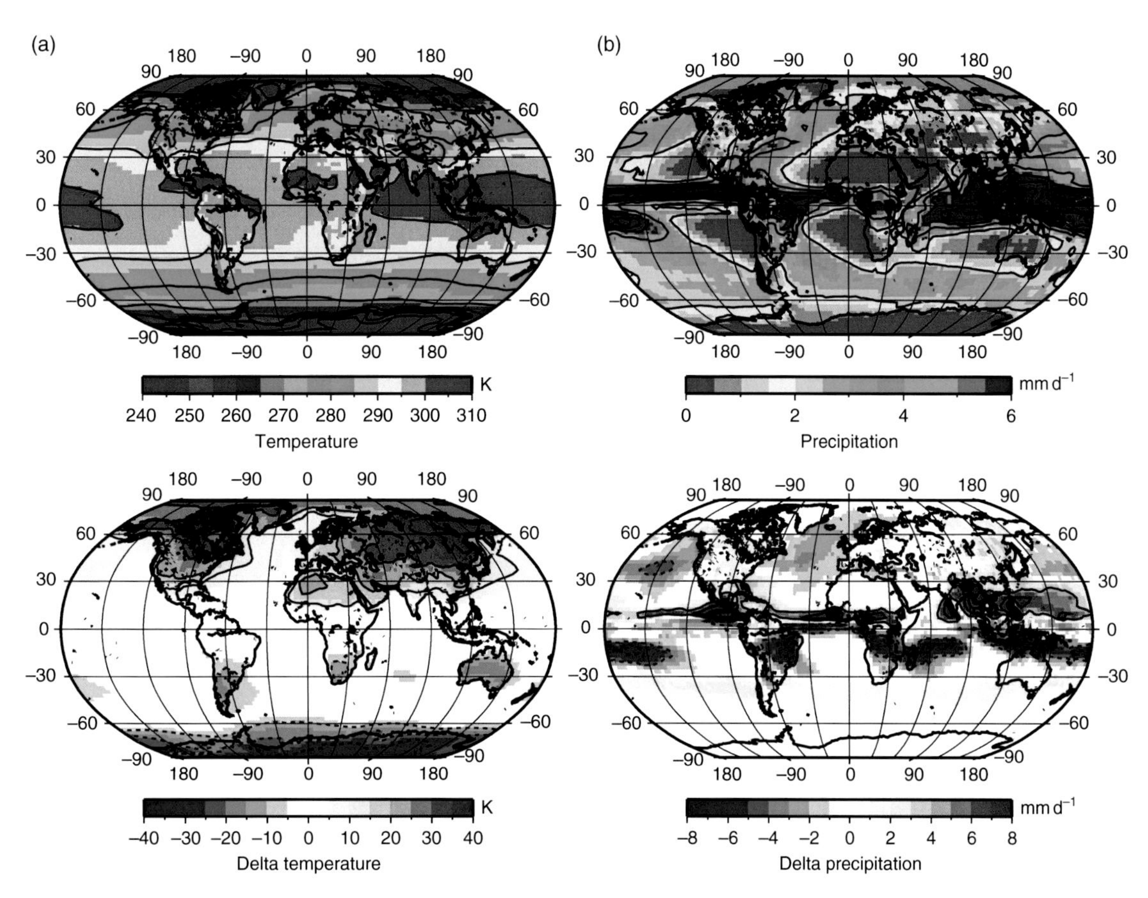

Figure 6 Annual mean climate during the period 1980–90. (a) Annual mean near-surface air temperature and its seasonal variation (June–August average minus December–February average). (b) The same, but for annual mean precipitation and its seasonal variation. The plots were created using the European Centre for Medium-Range Weather Forecasts (ECMWF) reanalysis data sets. Data sets have been obtained from the ECMWF data server.

Feedbacks

Global changes (e.g., orbital parameters, atmospheric concentration of carbon dioxide) affect energy balance components, which in turn alter climate. These changes can be amplified or reduced due to feedbacks. Feedbacks characterize the response of the global energy balance to a perturbation or external forcing. They formalize the nonlinearities and interactions in the climate system. Feedbacks are classified into positive and negative feedbacks. Positive feedbacks enhance the response of a chosen variable (mostly temperature) to the external forcing, while negative feedbacks stabilize the system, making it less responsive. In other words, a positive feedback makes a positive (negative) external change more positive (negative). Examples for important climate system feedbacks are given in **Figure 7**.

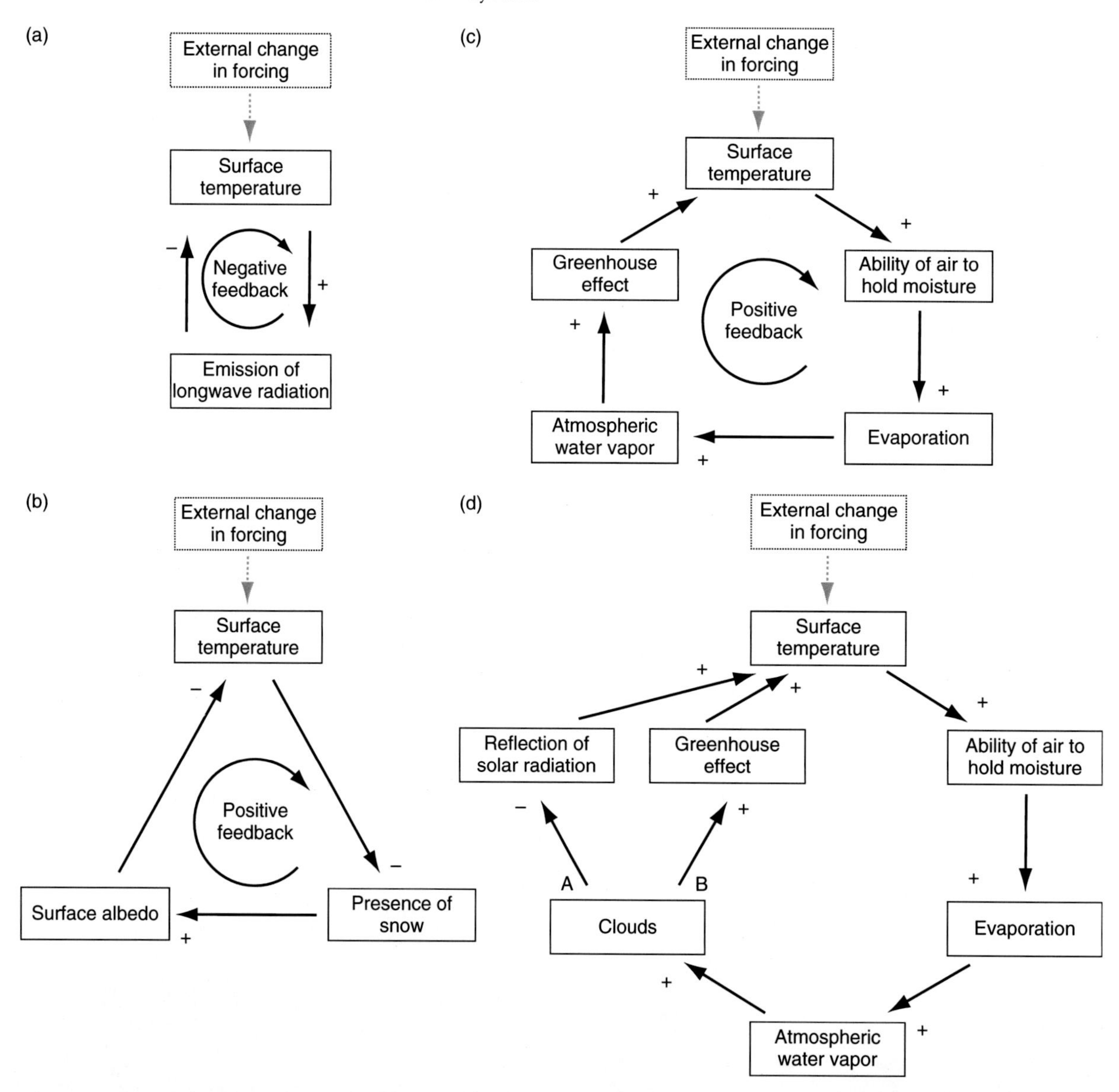

Figure 7 Dominant feedback processes that shape the response of the global energy balance and surface temperature to external change. The diagrams show the variables involved in four important feedback processes. The (+/−) signs at the arrows indicate positive/negative influences. (a) The thermal radiation feedback. An external change in forcing that would increase surface temperature would also increase the emission of long-wave radiation (a '+' influence). An increased emission would result in a lower surface temperature (a '−' influence). The enhanced emission of long-wave radiation therefore counteracts the initial change, resulting in a negative feedback loop. The same line of reasoning also applies for an external change that would reduce surface temperature. (b) The snow/ice albedo feedback. An external change that warms the surface reduces the presence of snow, lowers the surface albedo, thereby amplifying the warming (a positive feedback). (c) The water vapor feedback. An external change that warms the surface heats the lower atmosphere. Since warmer air can hold more moisture, this enhances surface evaporation and the amount of water vapor in the atmosphere. More water vapor results in a stronger atmospheric greenhouse effect, thereby amplifying the initial change (a positive feedback). (d) Two types of cloud feedbacks. Continuing from the water vapor feedback, more water vapor in the atmosphere can result in more clouds. Depending on the balance of increased cloud cover on shortwave reflection (path A) or increased greenhouse forcing (path B), cloud feedbacks can form both positive and negative feedback loops on surface temperature.

Ecosystems and the Global Energy Balance

Ecosystems affect the global energy balance directly by utilizing solar radiation by photosynthesis, and also indirectly by altering components of the surface energy balance such as surface albedo and evapotranspiration rates, and by strongly affecting biogeochemical fluxes and atmospheric composition.

Direct Biotic Effects

Photosynthesis utilizes about $8\ W\ m^{-2}$ of solar radiation (**Table 3**) Since most of the carbohydrates are respired within relatively short time at the same location, most of the energy is released as heat by respiration. Hence, the energy fluxes associated with photosynthesis and respiration are generally neglected in considerations of the surface energy balance.

Indirect Biotic Effects and Feedbacks

Ecosystems interact with their physical and geochemical environment. The effect of ecosystems on the energy balance are categorized into two types of effects:

1. Biogeophysical effects modify components of the surface energy balance and the physical functioning of the climate system. These effects are strongest for terrestrial vegetation, which affects the energy balance over land in various ways: (a) the surface albedo of vegetated surfaces is generally darker than bare surfaces (**Table 4**), thereby enhancing absorption of solar radiation; (b) a heterogeneous canopy cover enhances the aerodynamic roughness of the surface, which enhances turbulent fluxes and frictional dissipation; (c) vegetation root systems enhance the ability to recycle soil moisture through transpiration, thereby affecting the latent heat flux; and (d) vegetated surfaces modulate the partitioning of sensible and latent heat through stomatal functioning. These effects have considerable effects on the surface energy and water balance and the overlying atmosphere (**Figure 8**) and result in two biogeophysical feedback processes (**Figure 9**).
2. Biogeochemical effects modify chemical cycling of elements and the atmospheric composition (**Table 1**). Some of these have important consequences for the global energy balance and atmospheric dynamics, such as (a) carbon cycling affects the strength of the greenhouse effect, (b) oxygen concentrations affect stratospheric absorption of sunlight, and (c) the production of some compounds, such as dimethyl sulfide by marine algae, act as cloud condensation nuclei, thereby modifying cloud cover.

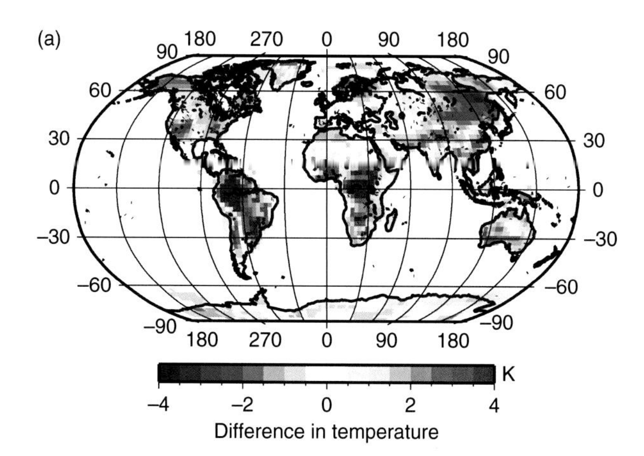

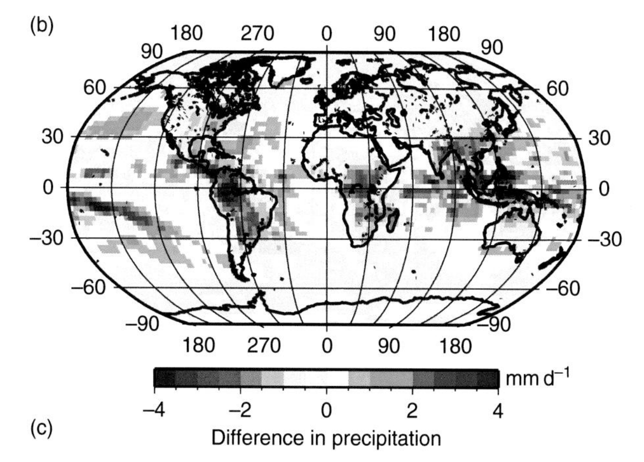

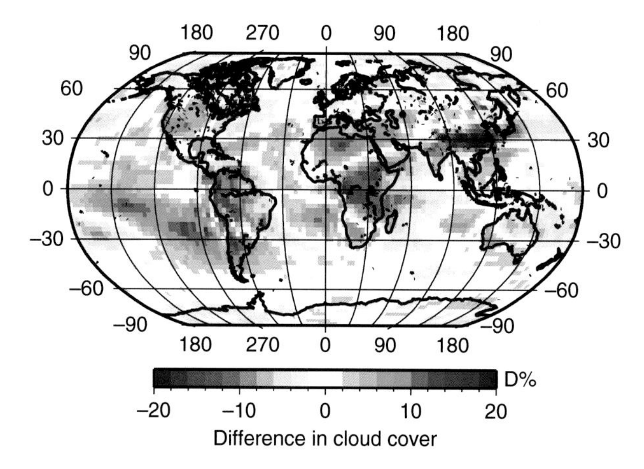

Figure 8 Climatic differences of a 'desert world'. Annual mean differences in (a) near-surface air temperature, (b) precipitation, and (c) cloudiness between the simulated climate of a 'desert world' void of terrestrial vegetation and the simulated present-day climate. These climatic differences result from the effect of vegetation on surface albedo, aerodynamic surface roughness, and the depth of the rooting zone.

Gaia Hypothesis

In the extreme form, strong negative biotic feedbacks on temperature can regulate the global energy balance into a state of homeostasis (i.e., no temperature sensitivity to

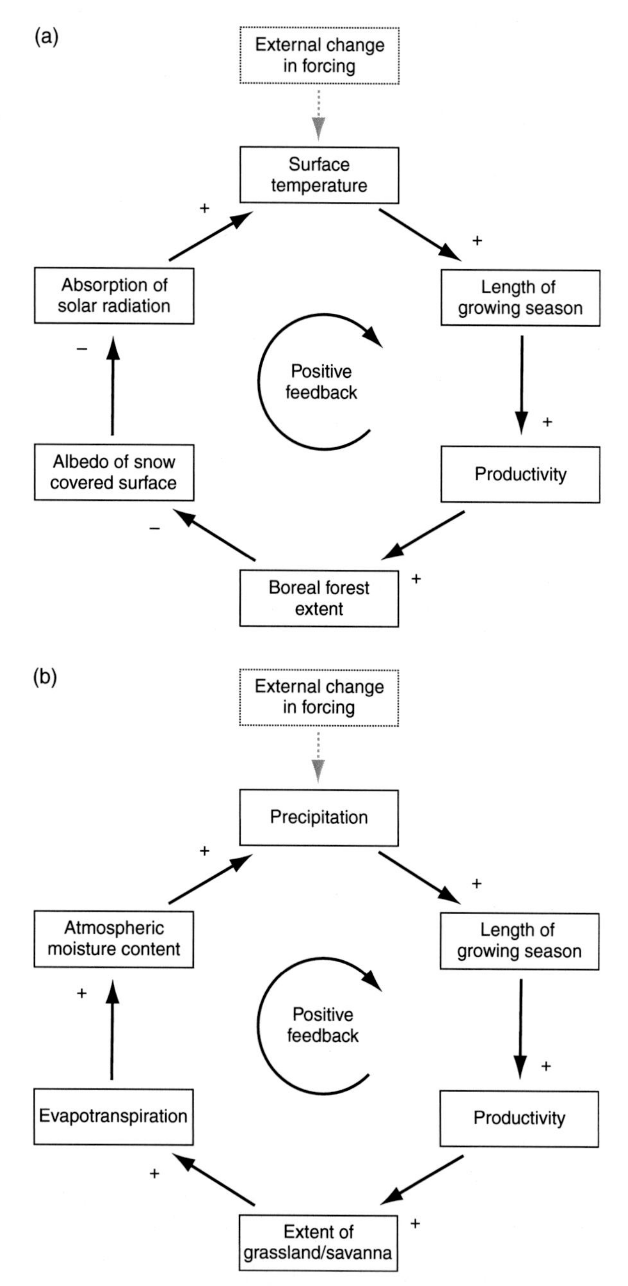

Figure 9 Vegetation feedbacks on the surface energy balance. The diagrams show the two major feedback loops by which vegetation directly affects the physical functioning of the surface energy balance. (a) The snow-masking feedback. An external change in forcing that would increase surface temperature in regions where temperature limits terrestrial productivity (such as the Arctic) increases the length of the growing season. A longer growing season would result in higher productivity, which extends the boreal forest cover in temperature-limited regions. Enhanced boreal forest cover masks the presence of snow at the surface, thereby lowering the surface albedo. This results in enhanced absorption of solar radiation, which amplifies the initial change, resulting in a positive feedback loop. (b) The water cycling feedback. An external change that results in enhanced precipitation in regions where water limits productivity (such as the semiarid tropics) increases the length of the growing season, resulting in higher productivity. This extends vegetative cover, and thereby evapotranspiration, atmospheric moisture content, resulting in more precipitation. The initial change is hence amplified, resulting in a positive feedback.

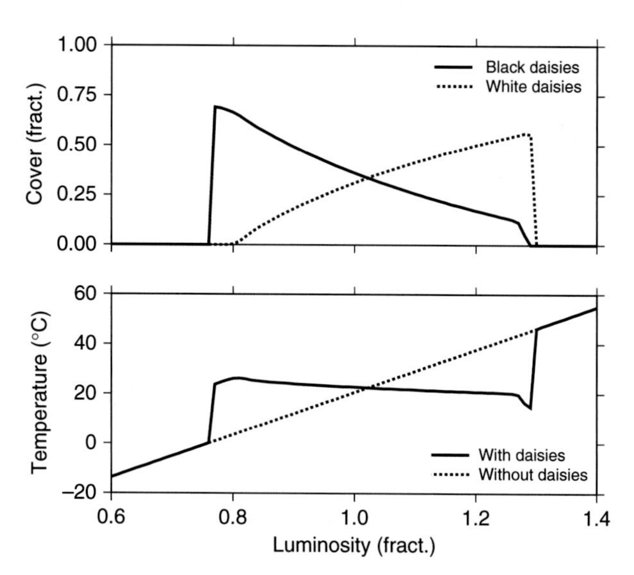

Figure 10 Emergence of temperature regulation in the conceptual 'Daisyworld' model. The 'Daisyworld' model is a conceptual model of a virtual world in which the planetary albedo is regulated by the population dynamics of black and white daisies. (a) The fractional cover of black and white daisies for different values of solar luminosity, expressed as the fraction of its present-day value. (b) The different proportions of daisies result in an overall planetary albedo that results in constant temperature conditions over a wide range of solar luminosity values.

external forcing). This has been suggested by the Gaia hypothesis of James Lovelock, stating that the atmosphere is regulated by and for the biosphere. This notion was originally motivated by the observation that the Earth's atmospheric composition is far from thermodynamic equilibrium, and is maintained in that state by the photosynthesizing biota.

The conceptual 'Daisyworld' model was developed to demonstrate the possibility of global homeostasis. This model describes a world where the planetary albedo is determined by the fractions of black and white daisies, and of bare ground. Using equations of population dynamics and a temperature-dependent growth parametrization, 'Daisyworld' demonstrates that homeostasis is a possible outcome of population dynamics coupled to the global energy balance (**Figure 10**).

However, the notions of Gaia and Daisyworld remain controversial. Surface temperatures in Earth's history have been far from constant, and the representation of biospheric dynamics in Daisyworld is highly simplistic. Yet the challenge to find general organizing principles that can explain the interactions of the biosphere with the global energy balance remains an active research topic.

See also: Carbon Cycle; Climate Change 3: History and Current State; Coevolution of the Biosphere and Climate; Energy Flows in the Biosphere; Entropy and Entropy Flows in the Biosphere; Gaia Hypothesis; Radiation

Balance and Solar Radiation Spectrum; Temperature Patterns; Water Cycle.

Further Reading

Aoki I (1988) Entropy flows and entropy productions in the Earth's surface and in the Earth's atmosphere. *Journal of the Physical Society of Japan* 57: 3262–3269.
Bonan GB (2001) *Ecological Climatology*. Cambridge: Cambridge University Press.
Hartmann DL (1994) *Global Physical Climatology*. San Diego, CA: Academic Press.
Kiehl JT and Trenberth KE (1997) Earth's annual global mean energy budget. *Bulletin of the American Meteorological Society* 78: 197–208.
Kleidon A, Fraedrich K, and Heimann M (2000) A green planet versus a desert world: Estimating the maximum effect of vegetation on land surface climate. *Climatic Change* 44: 471–493.
Kleidon A and Lorenz RD (eds.) (2005) *Non-Equilibrium Thermodynamics and the Production of Entropy: Life, Earth, and Beyond*. Heidelberg: Springer.
Lovelock JE and Margulis L (1974) Atmospheric homeostasis by and for the biosphere: The Gaia hypothesis. *Tellus* 26: 2–9.
Oke TR (1987) *Boundary Layer Climates*, 2nd edn. New York: Halsted.
Peixoto JP and Oort AH (1992) *Physics of Climate*. New York: American Institute of Physics.
Schneider SH, Miller JR, Crist E, and Boston PJ (2004) *Scientists Debate Gaia: The Next Century*. Cambridge, USA: MIT Press.
Walker D (1992) *Energy, Plants, and Man*. Brighton, UK: Oxygraphics Ltd.
Watson AJ and Lovelock JE (1983) Biological homeostasis of the global environment: The parable of Daisyworld. *Tellus* 35B: 284–289.
Wilson MF and Henderson-Sellers A (1985) A global archive of land cover and soils data for use in general circulation climate models. *Journal of Climatology* 5: 119–143.

Relevant Website

http://nssdc.gsfc.nasa.gov – Lunar and Planetary Science at the National Space Science Data Center (NSSDC).

Energy Flows in the Biosphere

Y M Svirezhev, Potsdam Institute for Climate Impact Research, Potsdam, Germany

Introduction

Life on Earth is a product of so-called 'photon's mill', which has started to function when our solar system was in the form of matter's 'clots' embedded into the ocean of 'cold' photons with temperature $T = 2.7$ K. Evolution and self-organization of planets (including life on our planet) is a result of this mill's functioning, which is happening due to the fact that Sun's surface irradiates the 'hot photons' at $T = 5800$ K, and these photons reach the cold planet's surfaces. Then they cool down to the temperature of the surface and irradiate back into the space. For the Earth, this temperature is equal to 253 K; it is the temperature, which could be measured by an observer at the top of atmosphere. Formally, the photon's mill is a typical 'heat machine' that is functioning by Carnot cycle, but its working body is the photon gas (for details see Entropy and Entropy Flows in the Biosphere).

Incoming and Outgoing Radiations and the Planetary Energy Balance

A parallel flow of solar radiation at the Earth's mean distance from the Sun is equal to 1368 W m^{-2}; this is an irradiation of blackbody with $T = 5800$ K (see **Figure 1**). Since the Sun's flow is parallel its effective section is πr_{Erth}^2, where r_{Erth} is the Earth's radius. An area of the Earth's surface is $4\pi r_{\mathrm{Erth}}^2$, that is, four times larger than the section's area; therefore, only one-fourth of the total flow, 342 W m^{-2}, is coming to the area unit of the upper boundary of the 'Earth + atmosphere system' (EAS). It is obvious that specific flow is varying from point to point on the globe and in time within 1-year interval, so that these and other local values connected with energy flows are averaged over the globe and the year.

Incoming radiation is described by the energy spectrum $E^{\mathrm{in}}(\lambda, x, y, t)$, where λ is a wavelength, x and y are

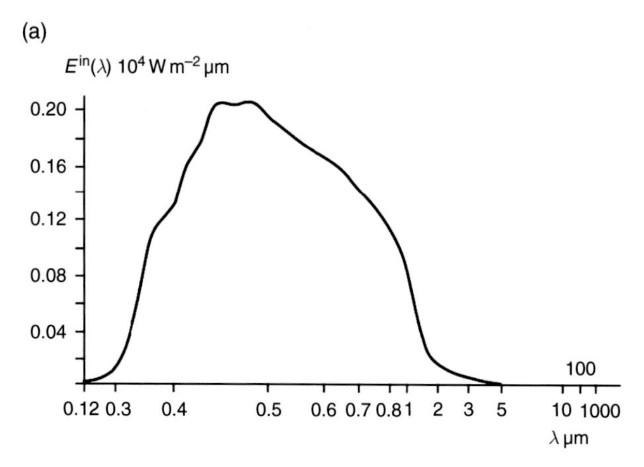

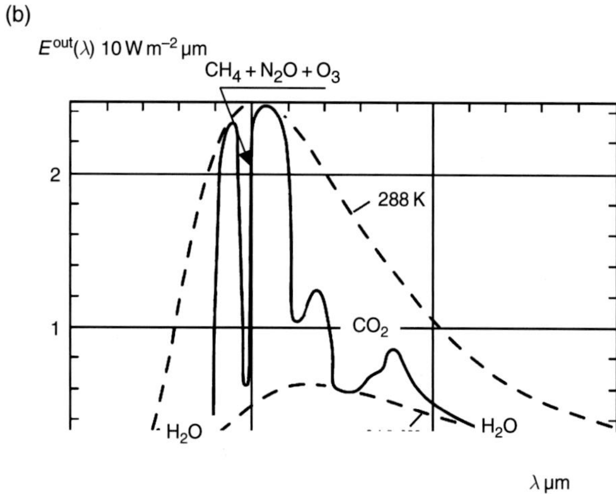

Figure 1 Spectra of incoming and outgoing solar radiation. (a) Standard spectrum of solar radiation at the top of the atmosphere. (b) Typical spectrum of the Earth's thermal radiation.

geographic coordinates, and t is a time. By interacting with a surface, it is transformed onto the energy spectrum of outgoing radiation, $E^{\mathrm{out}}(\lambda, x, y, t)$:

$$E^{\mathrm{in}}(\lambda, x, y, t) \overset{F(\lambda,x,y,t)}{\Rightarrow} E^{\mathrm{out}}(\lambda, x, y, t) \qquad [1]$$

where $F(\lambda, x, y, t)$ is the transition operator. The spectrum of outgoing radiation is close to the blackbody spectrum with $T = 253$ K.

Really, we have information only about these two spectra measured sufficiently frequently, sufficiently densely, in sufficiently big number of spectral bands (today up to 120), in the course of sufficiently long time. Outside of the EAS, satellites are carrying out these measurements today. At the level of the Earth's surface (the ground), it is performed also by satellites, and when they are corrected by data of the ground measurements. Note that outgoing radiation contains a lot of information about a surface, interacting with incoming radiation. Spectra of incoming and outgoing radiations are shown in **Figure 1**.

The simplest form of $F(\lambda, x, y, t)$ is a shift operator $R(\lambda, x, y, t) = E^{\mathrm{in}}(\lambda, x, y, t) - E^{\mathrm{out}}(\lambda, x, y, t)$; a convolution $\bar{R}(x, y, t) = \int_\Omega R(\lambda, x, y, t)\mathrm{d}\lambda$ over all wavelengths is named a local 'radiative (radiation) balance'. Convolutions $\bar{E}^{\mathrm{in}}(x, y, t) = \int_\Omega E^{\mathrm{in}}(\lambda, x, y, t)\mathrm{d}\lambda$ and $\bar{E}^{\mathrm{out}}(x, y, t) = \int_\Omega E^{\mathrm{out}}(\lambda, x, y, t)\mathrm{d}\lambda$ are the total energy of incoming and outgoing radiation at the given point and in the given moment. If we average $\bar{R}(x, y, t)$ over the EAS surface, S_{EAS}, and 1-year interval, t_1, we get the 'annual planetary radiative balance'

$$\hat{R} = \frac{1}{S_{\mathrm{EAS}} \cdot t_1} \int_G \int_T R(x, y, t)\mathrm{d}x\,\mathrm{d}y\,\mathrm{d}t \qquad [2]$$

is equal to zero. This is a typical 'empirical generalization'.

The wavelengths λ is usually measured in the band Ω: (0.2, 50 μm), which contains almost 100% of the total energy of incoming and outgoing radiations. Its most part (~99%) is a shortwave radiation (SWR) with wavelengths λ lying within the spectral band S: (0.2, 5.0 μm), where 53.5% constitutes a radiation with $\lambda \in (0.4, 0.7\ \mu\mathrm{m})$, so-called photosynthetically active radiation (PAR). This spectral band is called 'visible'. The radiation with $\lambda \in L$ (5, 50 μm), the long-wave radiation, LWR, constitutes only 0.45% of the total radiation. About 0.5% constitutes an ultraviolet radiation, $\lambda < 0.2$ μm, that is, fortunately, almost completely detained by ozone layer. Note that other divisions of the total spectral band are often used, for instance, S: (0.3, 3.0 μm), etc.

Later on we shall operate with values of energy integrated over these two spectral bands and averaged on the total EAS' surface and 1-year interval: $\hat{E}_S^{\mathrm{in}}$, $\hat{E}_L^{\mathrm{in}}$ and $\hat{E}_S^{\mathrm{out}}$, $\hat{E}_L^{\mathrm{out}}$. In accordance with observed data we have with sufficient accuracy that for the EAS: $(\hat{E}_S^{\mathrm{in}})_{\mathrm{EAS}} = 340\ \mathrm{W\,m^{-2}}$, $(\hat{E}_L^{\mathrm{in}})_{\mathrm{EAS}} = 0$, and $(\hat{E}_S^{\mathrm{out}})_{\mathrm{EAS}} = 102\ \mathrm{W\,m^{-2}}$, $(\hat{E}_L^{\mathrm{out}})_{\mathrm{EAS}} = 238\ \mathrm{W\,m^{-2}}$, so that $(\hat{E}^{\mathrm{in}})_{\mathrm{EAS}} = (\hat{E}^{\mathrm{out}})_{\mathrm{EAS}}$. Earth on the whole gets $238\ \mathrm{W\,m^{-2}} \times 5 \times 10^{14}\ \mathrm{m^2} = 1.2 \times 10^{17}$ W of the SWR, and irradiates as much LWR again.

The relation between these values can also be presented as

$$\hat{R}_{\mathrm{EAS}} = \left(\hat{E}_S^{\mathrm{in}}\right)_{\mathrm{EAS}}(1 - \alpha_{\mathrm{EAS}}) - \left(\hat{E}_L^{\mathrm{out}}\right)_{\mathrm{EAS}} \qquad [3]$$

where $\alpha_{\mathrm{EAS}} = (\hat{E}_S^{\mathrm{out}})_{\mathrm{EAS}} / \left(\hat{E}_S^{\mathrm{in}}\right)_{\mathrm{EAS}} = 0.3$ is so-called planetary 'albedo' (from Latin 'whiteness'), that is, a coefficient of reflectance of the EAS with respect to incoming radiation.

Transformation of Solar Energy Inside the EAS

What is the fate of 340 W m^{-2} of SWR coming into the top of the atmosphere? Clouds reflect about 68 W m^{-2} of the total incoming radiation. Molecules of atmospheric gases and aerosols still scatter $16\,\mathrm{W\,m^{-2}}$ by forming so-called diffuse radiation, half of which finally goes out to the space, and other half reaches the ground. The $26\,\mathrm{W\,m^{-2}}$ of SWR is reflected by the ground. So, the $68+8+26=102\,\mathrm{J\,m^{-2}}$ of energy in the form of SWR leaves our planet every second. The rest, $238\,\mathrm{W\,m^{-2}}$, is consumed and used by the EAS to maintain the work of the Earth's climate machine: evaporation and 'precipitation', 'oceanic currents', atmospheric circulation, etc.

About $78\,\mathrm{W\,m^{-2}}$ is absorbed by the atmosphere and then used in different phase transitions: clouds formation, precipitation, formation of dew, etc. Another part, $160\,\mathrm{W\,m^{-2}}$, is absorbed by the ground. The measurements show that the spectrum of radiation coming into the Earth's surface is a mixture of SWR, $(\hat{E}_S^{in})_G = 186\,\mathrm{W\,m^{-2}}$, and LWR (a counter-radiation from the warmed atmosphere), $(\hat{E}_L^{in})_G = 102\,\mathrm{W\,m^{-2}}$. The ground irradiates into the atmosphere $(\hat{E}_S^{out})_G = 26\,\mathrm{W\,m^{-2}}$ of SWR that corresponds to the mean albedo $\alpha_G = (\hat{E}_S^{out})_G/(\hat{E}_S^{in})_G \approx 0.14$ and $(\hat{E}_L^{out})_G = 160\,\mathrm{W\,m^{-2}}$ of LWR. The radiation absorbed by the ground, $160\,\mathrm{W\,m^{-2}}$, is transformed to heat. The depth of heat penetration depends on the properties of underlying surface. For the land, the depth depends on the properties of soil (for instance, soil moisture) and equals a few meters. As to the ocean, oceanic waves intermix effectively the surface layer, so that the depth is about 100–200 m. If we take into account that the mean depth of the ocean is 3800 m, then the depth of heat penetration is negligibly small in comparison with the mean depth. Therefore, the warmed body is a very thin film, which is not able to warm the 'lithosphere' (their masses are not commensurable), but it is warming up the atmosphere. Seasonal and diurnal oscillations of the temperature at the levels lying below the film stopped. The film is considered as a low boundary (basement) for the EAS, and it is namely identified with the Earth's surface. A heat flow across the low boundary is negligibly small. For instance, the thermal flow from the Earth's upper mantle is maximally $0.2\,\mathrm{W\,m^{-2}}$. So, one can say that the EAS consists of the atmosphere and the upper layers of the 'hydrosphere and lithosphere'.

Greenhouse Effect

If Earth would not have the atmosphere, then all absorbed $160\,\mathrm{W\,m^{-2}}$ has to be irradiated into space in the form of LWR, since the temperature of the Earth's surface does not change. However, different atmospheric gases, transparent for SWR, may be weakly transparent for the LWR. For instance, the water vapor strongly absorbs LWR (in the cloudless atmosphere) in the spectral band (5–7.5 μm) and the carbon dioxide in the band (13–17 μm); absorption in the band (9–12 μm) is relatively small (see **Figure 1**). As a result, a part of the ground infrared radiation is detained, the atmosphere is warmed, and becomes in turn a source of LWR. Appearing here as a 'counter-radiation', $E_{count} = (\hat{E}_L^{in})_G = 102\,\mathrm{W\,m^{-2}}$, it compensates a significant part of the ground LWR. A difference between the ground LWR and the counter-radiation is called an 'effective radiation' of the Earth's surface, $E_{eff} = (\hat{E}_L^{out})_G - (\hat{E}_L^{in})_G = 58\,\mathrm{W\,m^{-2}}$. Its spectrum is close to the blackbody one with $T = 288\,\mathrm{K}$, and namely this amount of LWR is irradiated by the Earth's surface into the space, the rest $180\,\mathrm{W\,m^{-2}}$ is irradiated by the atmosphere. This is an essence of the 'greenhouse effect'.

Albedo

Theoretically, the albedo's value may change from 0 of a blackbody until 1 of a 'white body' completely reflecting the solar radiation. It is natural that an albedo depends on spectrum of the radiation, since different surfaces reflect differently in different spectral bands.

Albedos of typical underlying surfaces in visible light range from 0.04 for charcoal, one of the darkest substances, up to 0.90–0.95 for fresh snow. Albedo of salt and sand deserts are 0.45–0.5, while the albedo of coniferous forest is 0.1. Note that the maximal albedo of a surface, covered by vegetation (meadow), is 0.25. Albedo of wet soils is usually less than the albedo of dry ones, for instance, the albedo of chernozem (~0.15) is reduced to 0.05 under moistening conditions.

The classic examples of albedo's effect are the snow–, vegetation–, and moisture–temperature feedbacks. If a snow-covered area warms and the snow melts, the albedo decreases down to 0.4–0.5, more sunlight is absorbed, and the temperature tends to increase. If a desert is covered by vegetation, then the albedo decreases, and the temperature has to increase. While the increase in plant biomass tends to slow down in the carbon dioxide concentration in the atmosphere that, in turn, tends to decrease the temperature, so that the balance of these feedbacks may become very complex. Similar considerations are valid also for 'soil moisture– temperature' feedback.

Albedo of water bodies differs from albedo of land surfaces, since the reflection of SWR from water depends on the angle of incidence. At small angles, most part of radiation is reflected from the surface, not penetrating deeply into water body. As a result the albedo increases up by a few tenths, while the albedo at the great angles, that is, when the Sun elevation is high, is equal to a few hundredths. For instance, if the angle of incidence $\beta < 10°$, then $\alpha > 0.22$; if $\beta > 45°$, then

$\alpha < 0.05$. Albedo of scattered radiation does not really depend on the angle of incidence, and it is almost constant, about 0.10.

The significant part of incoming radiation is reflected by clouds; their albedo, depending on the thickness of cloudiness, is equal on average to 0.4–0.5, so that the mean albedo of the Earth is about 0.29. This is far higher than for the ocean primarily.

The Earth's surface albedo is regularly estimated via 'Earth observation satellite sensors' such as NASA's MODIS instruments onboard the Terra and Aqua satellites.

Equations of Radiative Balance

Due to the greenhouse effect, the Earth's surface gets the 102 W m^{-2} of radiative heat additionally. An amount of 80 W m^{-2} of this heat is used in the process of evaporation and transpiration of water by plants, evapotranspiration, and 20 W m^{-2} are transported into the atmosphere by a turbulent (sensible) heat flow, E_{turb}, caused by a difference in the temperatures of the ground and the atmosphere. The first term is named a 'latent' flow; it is equal to $L \cdot Q$, where $L = 2453\,\text{J g}^{-1}$ is the specific enthalpy of evaporation (heat content) and Q is the flux of water, evaporated from the surface of waterbodies, soils, and plants, and also water, condensed on these surfaces. To close the balance, we add the value of $E_{\text{mech}} = 2$ W m^{-2} that is a dissipated mechanical energy (friction). The corresponding equation of radiative balance for the Earth's surface is

$$\hat{R}_G = \left(\hat{E}_S^{\text{in}}\right)_G (1-\alpha_G) - E_{\text{eff}} \qquad [4]$$

which is positive, $\hat{R}_G = 102$ W m^{-2}.

Since the radiative balance of the EAS is equal to zero, the radiative balance of the atmosphere

$$\hat{R}_{\text{a}} = \hat{R}_{\text{EAS}} - \hat{R}_G = \left(\hat{E}_S^{\text{in}}\right)_{\text{EAS}} (1-\alpha_{\text{EAS}}) - \left(\hat{E}_S^{\text{in}}\right)_G (1-\alpha_G) - \left[\left(\hat{E}_L^{\text{out}}\right)_{\text{EAS}} - E_{\text{eff}}\right] \qquad [5]$$

has to be negative. The negativeness is compensated by the latent and turbulent heat flows.

The main carriers of heat between the ground and the atmosphere are precipitation and water vapor. Then the radiative balance for the EAS can be represented as

$$\hat{R}_{\text{EAS}} = F_s + L(Q - P) \qquad [6]$$

where the term F_s is the sum of the heat inflows and outflows across the vertical walls of the EAS column with unit basement, the term $L(Q-P)$ is a difference between the flows of latent heat $L \cdot Q$ and heat brought by precipitation, $L \cdot P$, where P is the sum of all precipitations. Since for the globe and 1-year interval $Q = P$ and $F_s = 0$, then the equation of energy (heat) balance for the EAS has a simple form:

$$\hat{R}_G = 0 \qquad [7]$$

The balance equations for the atmosphere and the Earth's surface are

$$\begin{aligned} \hat{R}_{\text{a}} &= -L \cdot P - E_{\text{turb}} - E_{\text{mech}} \text{(atmosphere)} \\ \hat{R}_G &= L \cdot Q + E_{\text{turb}} + E_{\text{mech}} \text{(Earth's surface)} \end{aligned} \qquad [8]$$

A generalized scheme of energy flows and their transformation is shown in **Figure 2**.

At least, we can estimate the internal energy of the atmosphere, which is equal to 8.6×10^{23} J (1.7×10^{9} J m^{-2}), the storage of latent heat 3×10^{22} J (6×10^{7} J m^{-2}), and the storage of mechanical energy 2.5×10^{15} (5×10^{5} J m^{-2}). About 40% of the total atmospheric internal energy constitutes a potential energy (0.7×10^{9} J m^{-2}), but only 4 W m^{-2} is necessary to maintain turbulent flows.

Parallel to the vertical redistribution of solar energy, there are powerful energy flows redistributing it over the Earth's surface. All of them form the complex system of atmospheric circulation and oceanic currents that provides to transport heat from the low latitudes to the high latitudes by 'softening' the Earth's climate.

Energetics of Photosynthesis and Vegetation

We described above the main processes of the transformation of solar energy, which are the principal components of the global energy balance, forming in essence the thermostat for the biosphere. However, there are other processes, which do not really influence the energy balance. Their heat flows are very small (mentioned above as the dissipation of mechanical energy, dew condensation, etc.). Some of them, nevertheless, play the principal role in the biosphere, for instance, photosynthesis.

Green plants (autotrophs) convert solar energy into the chemical energy of new living biomass in the process of photosynthesis. The process uses energy of visible light, which is absorbed by the chlorophyll molecules of plants to convert carbon dioxide and water into carbohydrates and oxygen. Note that the presence of oxygen in the Earth's atmosphere is a result of photosynthesis. Proteins, fats, nucleic acids, and other compounds are also synthesized during the process, as long as elements such as nitrogen, sulfur, and phosphorus are available.

Then the stored chemical energy flows into herbivores, carnivores (predators), parasites, decomposers, and all other forms of life. Photosynthesis produces the living biomass of vegetation, constituting more than 95% of the global biomass and being a main agent in the 'global biogeochemical cycle of carbon'.

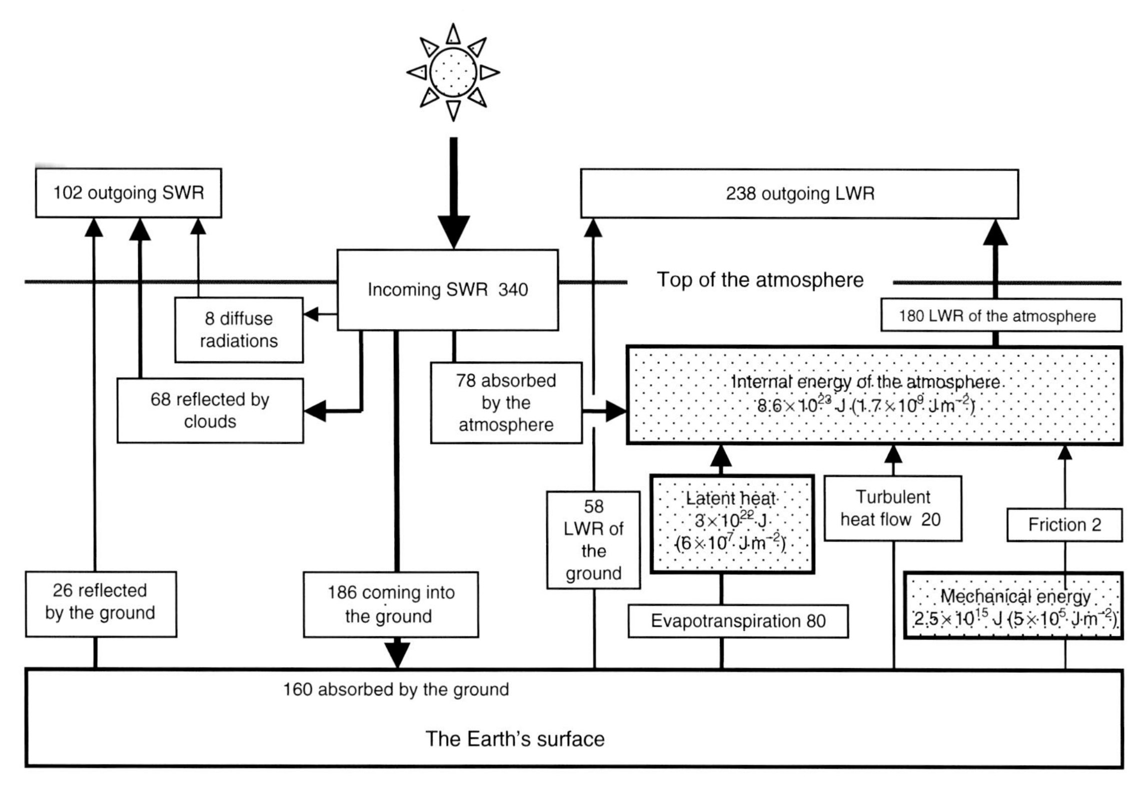

Figure 2 Energy flows in the system 'the Earth's surface + atmosphere'.

Photosynthesis

The basic equation of photosynthesis is

$$CO_2 + H_2O + h\nu \rightarrow (CH_2O) + O_2 + 470\,\text{kJ mol}^{-1}$$

where $h\nu$ is a photon energy and (CH_2O) is a fragment of carbohydrate molecule, releasing $470\,\text{kJ mol}^{-1}$ of energy (that equals an increase in enthalpy, ΔH). Since the change of free energy is equal to $\Delta G = 504\,\text{kJ mol}^{-1}$, and $G\Delta = \Delta H - T\Delta S$, the change of entropy, ΔS, is equal to $(470 - 504)/273 = 116\,\text{J K}^{-1}\,\text{mol}^{-1}$ ($T = 293$ K), that is, photosynthesis is an antientropic process.

Efficiency of photosynthesis is defined in different ways. Its theoretically maximal value is the ratio of ΔG to the total energy of eight photons ($E_{ph} = 1470\,\text{kJ mol}^{-1}$), which are necessary to get one molecule of O_2, $\eta_{max} = 504/1470 = 34\%$. On the other hand, since the 'useful' work, which can be performed by photosynthesis, is 'exergy', $\text{Ex} = -T\Delta S = 34\,\text{kJ mol}^{-1}$, then efficiency is defined as $\eta_{ex} = \text{Ex}/E_{ph} = 34/1470 = 2.3\%$.

Efficiency of Vegetation

If the working process creating a new biomass is photosynthesis, then the working machine is plant. Therefore, it is natural to say about efficiency of plant (efficiency of vegetation) or 'green leaf' than about efficiency of photosynthesis. The rate of photosynthesis depends on the amount of light reaching the leaves, the temperature of surrounding air, and the availability of water and other nutrients such as nitrogen and phosphorus. One of these factors ('limiting factors') already limits the rate, so that the real efficiency of vegetation is lower than its theoretical value, 34%, and what is more, this efficiency should not exceed the 'exergic' efficiency, 2.3%.

From the thermodynamic point of view, a 'green leaf' is a heat machine with photosynthesis as the working process, and molecules of chlorophyll, adenosine triphosphate (ATP) etc., transferring energy of photons into leaves as the working body. Efficiency of the heat machine is $\eta_{leaf} = (T_{leaf} - T_{air})/T_{leaf}$, where T_{leaf} and T_{air} are the mean daily temperatures of leaves and surrounding air; since the reaction of photosynthesis is exogenous, the leaf is warmed, $T_{leaf} > T_{air}$. Under summer conditions in temperate forest, $T \sim 5\,°C$ and $T_{air} \sim 20\,°C$ on average; therefore, $\eta_{leaf} = 5/298 = 1.7\%$.

It is known that about 98–99% of solar energy, reaching the Earth's surface, is reflected from leaves and other surfaces and absorbed by other molecules, which convert it to heat. Thus, vegetation is available to catch about 1–2% of incident solar energy, that is, these numbers constitute its efficiency.

The rate at which plants convert PAR (or inorganic chemical energy) to the chemical energy of organic matter is named gross primary production (productivity)

(GPP). This value (as well as biomass) is often reported in grams or metric tons of either dry weight or carbon (the latter is about one-half of the first). Since enthalpy of 1 of carbon is equal to ~42 kJ g^{-1}, then production and biomass can be also reported in joules.

Fifteen to sixty percent of the energy assimilated by plants immediately is spent in cellular respiration, when carbohydrates, proteins, and fats are broken down, or oxidized, to provide energy (in the form of ATP) for the cell's metabolic needs. The residual (40–85%) is stored in biomass as net primary production (NPP). The highest annual NPP, 2000 gC m^{-2} yr^{-1}, occurs in swamps, marshes, and tropical rainforests; the lowest, 20 gC m^{-2} yr^{-1}, occurs in deserts. The mean NPP for terrestrial ecosystem is about 400 gC m^{-2} yr^{-1}. Among aquatic ecosystems, the highest NPP, 2000 gC m^{-2} yr^{-1}, occurs in estuaries; the mean NPP in the ocean is 75 gC m^{-2} yr^{-1}, so that the ocean is a desert (see **Table 1**).

Efficiency of solar energy utilization by vegetation can be defined as the ratio of enthalpy, contained in the NPP, to the solar radiation, reaching to the Earth's surface and integrated over the vegetation period. The corresponding values for continents and for land overall are shown in **Table 1**.

One square meter of the terrestrial vegetation on average utilizes in the course of 1 year about 17 million joules of solar energy, but this gigantic number constitutes only 0.37% of the total solar energy that comes into the Earth's surface.

The total annual production of terrestrial vegetation is about 60 gigatons (Gt, 1 Gt = 10^9 t) of carbon, while for the ocean this value is estimated as 25 Gt with NPP = 75 gC m^{-2} yr^{-1}. Thus, the mean global NPP is 186 gC m^{-2} yr^{-1}, and the mean efficiency of global vegetation is about 0.1%. However, it is necessary to take into account that much energy is consumed in the process of forming and maintaining of the thermostat for vegetation. It is very similar to the situation with greenhouse, where the most part of energy is used for its heating.

Table 1 The NPP and the efficiency of utilization

Continents	*NPP ($gC\,m^{-2}yr^{-1}$)*	*Efficiency (%)*
Europe	365	0.54
Asia	421	0.38
North America	353	0.40
South America	899	0.49
Africa	443	0.25
Australia with Oceania	370	0.19
Land on average	408	0.37

Energy Transfers, Trophic Chains, and Trophic Networks

If we look at a global pattern of pathways on which the solar energy stored in biomass is flowing within the gigantic (and unique) ecosystem (often associated with the biosphere), we see the network entangling the Globe. It is named a 'trophic network or a food web', and as a rule subdivided on local networks. The trophic network is described by an oriented graph with vertices corresponding to species that constitute the ecosystem, and links indicating trophic interaction between them (their directions show the energy flowing, for instance, prey → predator). In the network structure the 'trophic levels' are naturally distinguished, that is, groups of species having no direct trophic interactions; however, species of one level usually either compete for life resource or cooperate in its utilization. It is natural that some part of energy is spent (and later on dissipated as heat) in such kind of interactions – this is a payment by means of energy for stability of the network structure. Another significant part of consumed energy (from 30% to 70%) is spent for maintaining life in the process of 'metabolism' (respiration).

In any trophic network a structure, in which every two adjacent species form a prey–predator pair and which is described by a linear graph, can be distinguished. It is called a 'trophic (food) chain'; their interlacing and branching form a trophic network. Since the energy dissipation along the chains is very high they are usually short (their length measured in the number of links is about 4–6).

The basic trophic species of chain usually are 'producers' (plants, autotrophic organisms that accumulate the solar energy and 'nutrients' – carbon, nitrogen, phosphorus, etc.); the next species are 'primary consumers' (herbivorous, heterotrophic organisms), and 'secondary consumers' (carnivores, predators, preying on herbivorous). Really, the chain may be longer. It is not necessary that the chain be originated by an autotroph: it may be any species considered as a resource for consequent ones. For instance, if a resource is 'detritus' (faeces, dead organic matter) then a special 'detritus chain' can be considered. At last, trophic chains could be 'open' and 'closed'; as a rule they are open in relation to the energy flowing through an ecosystem and carbon that is accumulated in the process of photosynthesis and spent in the respiration, and closed in relation to nutrients turning in the ecosystem.

In order to start a 'biogeochemical machine' we have to 'close' the chain by species named 'decomposers' (protozoa, bacteria, fungi, scavengers, and carrion eaters), which in the course of their vital activity split complex organic compounds into simpler mineral substrates (nutrients) for autotrophs. A principal scheme of such kind of 'biogeochemical machine' is shown in **Figure 3**.

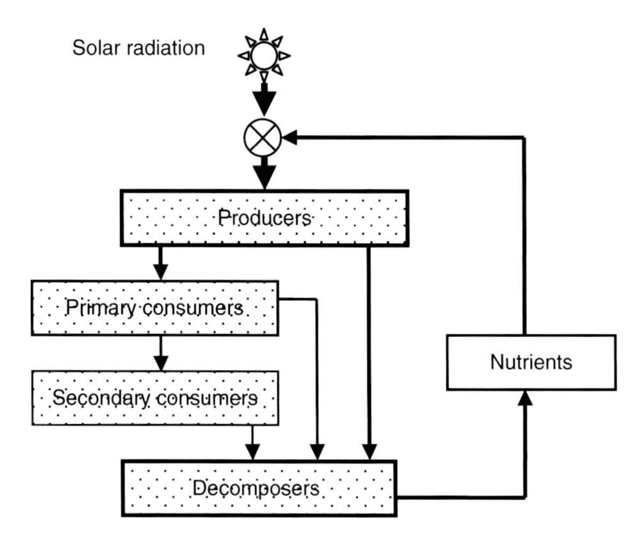

Figure 3 Flows of mass and energy in an elementary biogeochemical machine.

Really, the closure is not complete: about 1% of dead organic matter is deposited in the deep ground ('kerogen'), and has accumulated over long periods of geologic time (oil and coal repositories).

A small amount of the energy passes from one trophic level to another; for instance, only from 5% to 25% of plant biomass is consumed by herbivores, the rest falls out and becomes a resource for decomposers (detritophages). Efficiency of this passing is called 'ecological efficiency'; on average, it equals 10%. The rate at which these consumers use the chemical energy of their food for growth and reproduction is called 'assimilation efficiency'. For instance, assimilation efficiency of herbivores lies in the interval from 15% to 80%, while the interval for carnivores is from 60% to 90%.

It is easy to estimate that such a predator as *Homo sapiens* (the third trophic level) gets only 1% of solar energy stored by plants. Unfortunately, the situation has not improved; if he would be a vegetarian, by winning in the ecological efficiency, he would lose in the assimilation efficiency.

The result of such kind of consequent energetic transitions is a pyramid of energy, with most energy concentrated by autotrophs at the bottom of trophic chain and less energy at each higher trophic level. As an example the trophic chain and the pyramid of biomass of the concrete ecosystem of warm Silver Springs in Florida are presented in **Figure 4**. Note that it is a classic object that has been studied by H. T. Odum. The ecosystem has four trophic levels: (1) producers (phytoplankton), (2) herbivores (zooplankton), (3) carnivores (fish), (4) higher predators (predacious fish), and one special level, decomposers, with biomass equal to 105 kJ m^{-2}. Since the system is through-flowing, that is, described in the terms

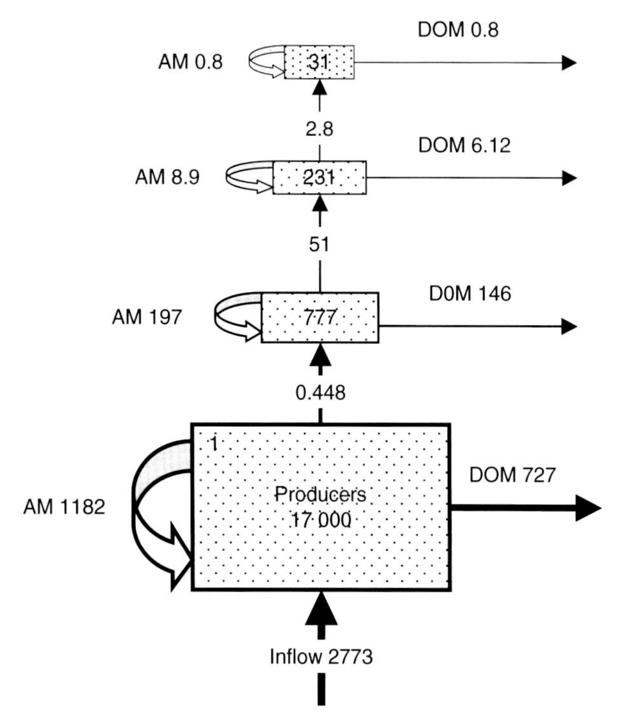

Figure 4 Trophic chain of the ecosystem of Silver Springs. Energy flows and biomasses are measured in mW m^{-2} and kJ m^{-2}, respectively. DOM, dead organic matter; AM, assimilation and metabolism.

of energy flows, therefore the chain may be considered open, without decomposers.

Conclusion

As described above biosphere machines from the anthropocentric point of view are badly made, with very low efficiency. They dissipate the solar energy by heating the environment more than perform some useful work. Nevertheless, they are significantly reliable. Since it is necessary to pay for their reliability and stability, they pay by high dissipation of energy that in turn decreases their efficiency.

See also: Radiation Balance and Solar Radiation Spectrum.

Further Reading

Budyko MI (2001) *Evolution of the Biosphere (Atmospheric and Oceanographic Sciences Library)*, 444p. Berlin: Springer.
Jørgensen SE and Svirezhev YuM (2004) *Towards a Thermodynamic Theory for Ecological Systems*, 366p. Amsterdam: Elsevier.
Morowitz HJ (1978) *Foundations of Bioenergetics*. New York: Academic Press.
Smile V (2002) *The Earth's Biosphere: Evolution, Dynamics, and Change*. Cambridge, MA: MIT Press.

Entropy and Entropy Flows in the Biosphere

Y M Svirezhev, Potsdam Institute for Climate Impact Research, Potsdam, Germany

Introduction

From the thermodynamic point of view, Earth is a closed system, which gets 1.2×10^{17} J of energy from the Sun every second in the form of a short-wave radiation, which corresponds to the density of energy flow in $238\ \mathrm{W\,m^{-2}}$. The same amount of energy is irradiated into the space in the form of a long-wave (infrared) radiation (see Energy Flows in the Biosphere, Ecological Network Analysis, Environ Analysis, and Energy Balance). We assume that the Earth's total mass and its mean temperature (more precisely, the temperature of the Earth–Atmosphere System (EAS)) are not changing in the course of rather long time ($\sim 10^3$ years). The latter means that the planetary radiative balance is constant. These are plausible hypotheses, which can be considered as 'empirical generalizations'.

Carriers of energy are 'hot' photons with temperature $T_S = 5800$ K of the Sun's surface, and the energy is carried away by 'cooled' photons at $T_E = 253$ K. This is the so-called 'photon mill'; evolution and self-organization of planets (including life on Earth) is a result of its work.

Formally, the photon mill is a typical 'heat machine' functioning as a Carnot cycle, but its working body is the photon gas, whose 'molecules' have no mass, so that in this case it becomes slightly incorrect to talk about a heat machine (although Gibbs has indicated it). Later on, Prigogine stated that such a classic thermodynamic concept as the heat machine is also applicable to the photon gas. Note that this 'roughness' is not necessarily present, if the concept of 'exergy' is used.

Let $d_i\sigma/dt$ be the internal production of entropy by the EAS, and $d_e\sigma/dt$ be the exchange flow of entropy between the Sun and the EAS, then the change in the total entropy of the EAS is

$$\frac{d\sigma}{dt} = \frac{d_e\sigma}{dt} + \frac{d_i\sigma}{dt} \qquad [1]$$

The value of $d_e\sigma/dt$ can be estimated as the algebraic sum of the entropy flow from Sun to Earth, $q_{SE} = (4/3)(238\ \mathrm{W\,m^{-2}})(1/T_S)$, and the entropy flow from the EAS to space, $q_{ES} = -(4/3)(238\ \mathrm{W\,m^{-2}})(1/T_E)$:

$$\frac{d_e\sigma}{dt} = \frac{4}{3}(238\mathrm{W\,m^{-2}})\left(\frac{1}{5800} - \frac{1}{253}\right) \approx -1.2\ \mathrm{W\,K^{-1}\,m^{-2}} \qquad [2]$$

where factor 4/3 is the so-called Planck's form factor. The annual entropic balance for the globe overall is equal to $-2 \times 10^{22}\ \mathrm{J\,K^{-1}\,yr^{-1}}$.

We assume here implicitly that the irradiation of the EAS is the blackbody irradiation with $T_A = T_E = 253$ K. Indeed, the irradiation is a sum of the blackbody irradiations with the temperatures from 215 to 288 K, so that this estimation is a zero approximation.

In accordance with Prigogine's theorem, at the dynamic equilibrium the system's entropy must be constant, that is, $d\sigma/dt = 0$; whence

$$-\frac{d_e\sigma}{dt} = \frac{d_i\sigma}{dt} \qquad [3]$$

The value of $d_e\sigma/dt$ is known; if we could estimate the value of $d_i\sigma/dt$, and if equality [3] holds, or, in other words, if the internal production of entropy is balanced by its export into the environment, we would prove one important statement: the EAS is at the dynamic equilibrium with its environment, the space. Note that equality [3] has to hold overall for the EAS, but each of its subsystems may be in nonequilibrium, so that the main statement of Prigogine's theorem (equality [3]) in relation to each subsystem does not necessarily hold.

Entropy Flows in the EAS

Let us look at the simplified scheme of the energy flows in the EAS shown in **Figure 1** (see Energy Flows in the Biosphere).

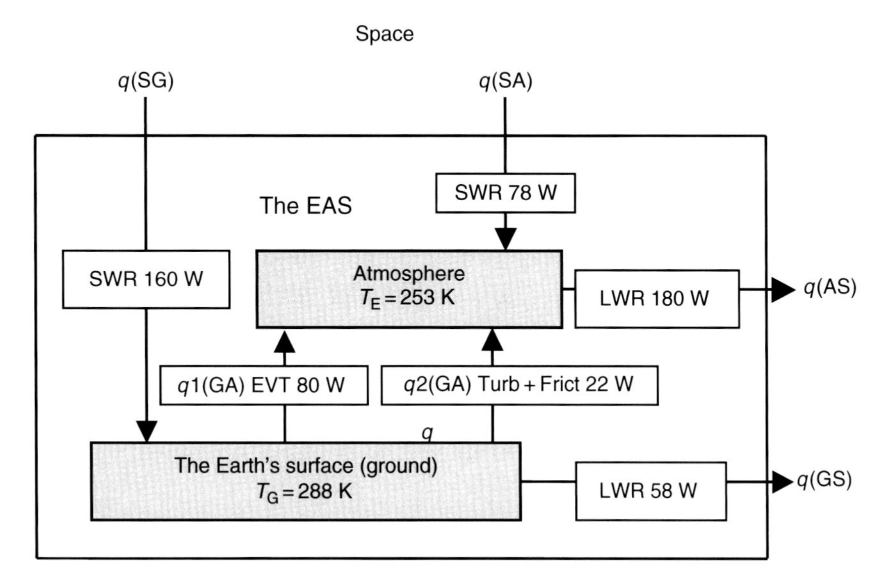

Figure 1 Energy flows in the EAS. SWR is the flow of short-wave radiation with $T_S = 5800$ K, LWR is the flow of long-wave radiation with the temperatures of the Earth's surface (ground), $T_G = 288$ K, or the atmosphere $T_E = 253$ K; EVT is the flow of latent heat (evapotranspiration), and Turb + Frict is the flow of turbulent heat (sensible flow) + the flow of heat discharged in mechanical movements (friction).

Using **Figure 1** we can write equations of entropic balance for the EAS (atmosphere + ground) in a more detailed way:

$$\frac{d\sigma_G}{dt} = \frac{4}{3}\frac{q(SG)}{T_S} - \frac{q_1(GA)}{T_G} - \frac{q_2(GA)}{T_G} - \frac{4}{3}\frac{q(GS)}{T_G} + \frac{d_i\sigma_G}{dt} \approx -0.586\ \mathrm{W\,K^{-1}\,m^{-2}} + \frac{d_i\sigma_G}{dt} \quad [4]$$

$$\frac{d\sigma_A}{dt} = \frac{4}{3}\frac{q(SA)}{T_S} + \frac{q_1(GA)}{T_G} + \frac{q_2(GA)}{T_G} - \frac{4}{3}\frac{q(AS)}{T_E} + \frac{d_i\sigma_A}{dt} \approx -0.576\ \mathrm{W\,K^{-1}\,m^{-2}} + \frac{d_i\sigma_A}{dt} \quad [5]$$

It is interesting that the exchange entropic flows, $d_e\sigma_G/dt \approx -0.586\ \mathrm{W\,K^{-1}\,m^{-2}}$ for the ground and $d_e\sigma_A/dt \approx -0.576\ \mathrm{W\,K^{-1}\,m^{-2}}$ for the atmosphere, are almost equal, and their sum is equal to $d_e\sigma/dt \approx -1.16\ \mathrm{W\,K^{-1}\,m^{-2}}$, that almost coincides with the value given by eqn [2]. In accordance with Prigogine's theorem,

$$-\frac{d_e\sigma}{dt} = \frac{d_i\sigma}{dt} = \frac{d_i\sigma_A}{dt} + \frac{d_i\sigma_B}{dt} \approx 1.16\ \mathrm{W\,K^{-1}\,m^{-2}} \quad [6]$$

Entropy Storage of the Biota

The EAS is divided into four subsystems: atmosphere (A), hydrosphere (H), pedosphere (P), and biota (B). The atmosphere is a mixture of different gases: mainly nitrogen and oxygen; in lesser concentrations, carbon dioxide, water vapour, argon, etc., which determine the thermal regime of our planet. The hydrosphere's mass is a mass of all water (including salt dilutions and excluding polar ice and glaciers). The pedosphere is soils. All these are exchanging energy and matter with each other, and in turn the EAS is exchanging, however, only energy with space. In particular, the matter exchange is realized by means of the global biogeochemical cycles (see also Matter and Matter Flows in the Biosphere).

The atmosphere, hydrosphere, and pedosphere have stored gigantic amounts of entropy. For instance, the entropy storage of atmosphere is $3.5 \times 10^{22}\ \mathrm{J\,K^{-1}}$ that in general is close to the global entropy balance; the storages of other subsystems are significantly larger. The exchange entropy flows that bound them with the biota are relatively weak with respect to their entropy storages, and do not really change their state, but they are able to change the state of biota. Thus, the latter is important for us.

Since the atmospheric CO_2 is one of the 'life-forming' gases, it is interesting to estimate its entropy, which is equal to $1 \times 10^{19}\ \mathrm{J\,K^{-1}}$.

The biota is defined as all of the Earth's living matter. Apparently, this is one of the reasons why the term 'biosphere' is often used (especially in Anglo-Saxon literature) in the sense of 'biota'. The present bulk of living organisms are confined to land, and their mass (on dry basis) amounts to 1.88×10^{18} g. For instance, the oceanic biomass is about 0.5% of that in land. Since the terrestrial vegetation constitutes the most part of biota, mainly contributing to its dynamics, the biota is identified with the terrestrial vegetation.

So, biota is the terrestrial phytomass, put into a thermostat with the mean annual temperature of the Earth's surface, $T_B = 15\ °C$. The total phytomass is known; hence, if only the specific entropy of living matter is also known, there is no problem in calculating the total entropy of biota. However, here we deal with a strongly nonequilibrium system, and it is unknown how to define the entropy in this case. What can be done here is to calculate the entropy of dead organic matter (DOM; in dry weight), $s(\mathrm{DOM}) = h(\mathrm{DOM})/T_B$, where $h(\mathrm{DOM}) = (16.4 - 18.4)\,\mathrm{kJ\,g^{-1}}$ is its specific enthalpy. Therefore, $s(\mathrm{DOM}) = 60.4\,\mathrm{J\,K^{-1}\,g^{-1}}$, and the total entropy of 'dead biota' is $S(\mathrm{DOM}) \approx 1.1 \times 10^{20}\,\mathrm{J\,K^{-1}}$, which is less by two orders of magnitude than the atmosphere entropy.

Change of Entropy in the Terrestrial Biota

Let $d_i S_B/dt = \dot{S}_B^i$ be the annual internal production of entropy by the global biota, and $d_e S_{jB}/dt = \dot{S}_{jB}$, $j = \mathrm{S, P, H, A}$ be the flows of entropy from jth subsystem into biota, then the rate of its entropy change is

$$dS_B/dt = \left(\dot{S}_{SB} + \dot{S}_{PB} + \dot{S}_{HB} + \dot{S}_{AB}\right) + \dot{S}_B^i \quad [7]$$

Here we make a very important assumption: although we do not know what the specific entropy of living matter is, we can still speak about the change of entropy. For instance, the value of $(-T_B \cdot \Delta S)$ can be interpreted as an ability of a living system to perform the work, which in turn can be measured. This ability is named 'exergy'.

For the Earth's surface, using the data given above, $d_e\sigma_G/dt \approx -0.586\,\mathrm{W\,K^{-1}\,m^{-2}}$. Since the area of globe is $5.1 \times 10^{14}\,\mathrm{m^2}$, $d_e S_G/dt \approx (-0.586\,\mathrm{W\,K^{-1}\,m^{-2}})(5.1 \times 10^{14})(3.15 \times 10^7) = -9.41 \times 10^{21}\,\mathrm{J\,K^{-1}\,yr^{-1}}$.

Exchange between Space and Biota

We assume that plants use only the solar short-wave radiation, absorbed by the Earth's surface, $160\,\mathrm{W\,m^{-2}}$. Since plants absorb only 53.5% of this energy (photosynthetically active radiation), $85.6\,\mathrm{W\,m^{-2}}$, and vegetation covers 72.5% of land area, $A_{veg} \approx 1.1 \times 10^{14}\,\mathrm{m^2}$, the biota gets annually

$$\begin{aligned}\dot{S}_{SB} &= \frac{4}{3}\left(85.6\,\mathrm{W\,K^{-1}\,m^{-2}}\right)\frac{A_{veg}}{T_S}\left(3.15 \times 10^7\right)\\ &= 0.53 \times 10^{20}\,\mathrm{J\,K^{-1}\,yr^{-1}}\end{aligned} \quad [8]$$

Here $T_S = 5800\,\mathrm{K}$.

Exchange between Pedosphere and Biota

This flow is mainly determined by the flow of DOM from terrestrial biota into pedosphere q_{DOM}. Assume that annual flow of DOM is equal to the net primary production (NPP), 140 Gt of dry matter, then $q_{DOM} = -1.4 \times 10^{17}\,\mathrm{g}$ d.w. of DOM per year. Since a living biomass contains about 65% (on average) of water, then it is natural to assume that standing dead vegetation contains the same percentage of water, and the flow of dry DOM, $-1.4 \times 10^{17}\,\mathrm{g}$, has to be accompanied by the water flow, $-2.6 \times 10^{17}\,\mathrm{g}\ H_2O$. By taking into account that specific entropy of H_2O is $3.89\,\mathrm{J\,K^{-1}\,g^{-1}}$, the total entropy flow $\dot{S}_{PB} = -(60.4 \times 1.4 + 3.89 \times 2.6) \times 10^{17} = -0.947 \times 10^{19}\,\mathrm{J\,K^{-1}\,yr^{-1}}$.

There is also a reversible flow of minerals (the nutrients: nitrogen, phosphorus, potassium, etc.), which are used in the process of creation of new biomass. All these substances come into the biota in the form of water solutions, entropy of which is the sum of the water entropy and exactly these elements. Note that their contribution constitutes less than 1% of the contribution of water.

Exchange between Hydrosphere and Biota

This flow is defined as $\dot{S}_{HB} = s(H_2O) \cdot q_{HB}(H_2O)$, where $q_{HB}(H_2O)$ is the annual flow of water consumed by biota, and $s(H_2O) = 3.89\,\mathrm{J\,K^{-1}\,g^{-1}}$ is the specific entropy of liquid water at $T = 288\,\mathrm{K}$. We assume that $q_{HB}(H_2O)$ is equal to the annual transpiration of global vegetation, $4.8 \times 10^{19}\,\mathrm{g}\ H_2O$. Then $\dot{S}_{HB} = 3.89 \times 4.8 \times 10^{19} = 1.87 \times 10^{20}\,\mathrm{J\,K^{-1}\,yr^{-1}}$.

Exchange between Atmosphere and Biota

This flow, $\dot{S}_{AB}$, is a sum of the following flows:

1. entropy flow caused by diffusion of CO_2 through stomata into leaves, $\dot{S}_{AB}(CO_2)$;
2. entropy flow caused by diffusion of O_2 through stomata into the atmosphere, $\dot{S}_{BA}(O_2)$; and
3. entropy flow caused by the transpiration of water, $\dot{S}_{BA}(H_2O)$.

The first and second flows are defined as $\dot{S}_{AB}(CO)_2) = s(CO_2)_{AB} \cdot q_{AB}(CO_2)$ and $\dot{S}_{BA}(O_2) = -s(O_2) \cdot q_{BA}(O_2)$, where $q_{AB}(CO_2) = \mathrm{NPP[gC]} \cdot (44/12) = (6.6 \times 10^{16}\,\mathrm{gC})(44/12) = 2.42 \times 10^{17}\,\mathrm{gCO_2\,yr^{-1}}$ and $q_{BA}(O_2) = -(6.6 \times 10^{16}\,\mathrm{gC})(32/12) = -1.76 \times 10^{17}\,\mathrm{gO_2\,yr^{-1}}$ are the rates of consumption and release of carbon dioxide and oxygen by plants in the process of photosynthesis. The specific entropies are: $s(CO_2) = 4.86\,\mathrm{J\,K^{-1}\,g^{-1}}$ and $s(O_2) = 6.41\,\mathrm{J\,K^{-1}\,g^{-1}}$. Then $\dot{S}_{AB}(CO_2) = 1.18 \times 10^{18}\,\mathrm{J\,K^{-1}\,yr^{-1}}$ and $\dot{S}_{BA}(O_2) = -1.13 \times 10^{18}\,\mathrm{J\,K^{-1}\,yr^{-1}}$. The summation of these flows gives $\dot{S}_{AB}(CO_2) + \dot{S}_{BA}(O_2) = (1.18 - 1.13) \times 10^{18} = 5 \times 10^{16}\,\mathrm{J\,K^{-1}\,yr^{-1}}$, that is, the exchange flows of entropy related to CO_2 and O_2 are almost balanced by each other, so that their sum is reduced by two orders of magnitude.

The entropy flow $\dot{S}_{BA}(H_2O) = s(\mathrm{WV}) \cdot q_{BA}(H_2O)$, where $q_{BA}(H_2O) = -q_{HB}(H_2O) = -4.8 \times 10^{19}\,\mathrm{g}\ H_2O\,\mathrm{yr^{-1}}$ is the annual transpiration through stomata (we assume that all consumed water is transpired), and $s(\mathrm{WV})$ is the specific entropy of water vapor at $T_B = 288\,\mathrm{K}$ and 1 atm. We see that maximal entropy flows are associated with water in liquid and vapor forms, that is, with the global water cycle. Their total balance $\dot{S}(H_2O) = \dot{S}_{HB} + \dot{S}_{BA}(H_2O) = q_{BA}(H_2O) \cdot (h_{ev}/T_B)$, where

$h_{ev} = 2462$ J per g H_2O is the specific enthalpy of evaporation, and is equal to a jump of entropy caused by the phase transition 'liquid water → water vapor', $\dot{S}(H_2O) = -4.1 \times 10^{20}\,\mathrm{J\,K^{-1}\,yr^{-1}}$.

The internal production of entropy, $\dot{S}_B^i$, can be represented as a sum of two terms: $\dot{S}_B^i = \dot{S}_B^{DOM} + \dot{S}_B^{Work}$. The first is mainly connected with chemical reactions forming structural molecules of biomass (cellulose, proteins, carbohydrates, lipids, etc.). Organic compounds containing phosphorus take an active part in such type of reactions. All these processes are associated mainly with the carbon, nitrogen, and phosphorus biochemical cycles, the entropy flows of which are less than in the water cycle approximately by two (and less) orders of magnitude. Certainly, knowing the chemical composition of living matter, we can calculate its chemical entropy as a sum of corresponding specific entropies weighted proportionally to their percentages. However, since the dead matter has the same composition, then the specific chemical entropies of living and dead matter do not differ from each other. Hence, we can assume that the processes of forming of the new biomass and falling off the DOM with respect to their chemical composition are mutually reversible, that is, $\dot{S}_B^{DOM} + \dot{S}_{PB} = 0$.

The second term, $\dot{S}_B^{Work}$, is the entropy produced by the biota during its working cycle (see details below).

By summing all these flows, we get

$$\begin{aligned} dS_B/dt &\approx \left[(0.53-4.11) \times 10^{20} \right. \\ &= \left. -3.58 \times 10^{20}\,\mathrm{J\,K^{-1}\,yr^{-1}}\right] + \dot{S}_B^{Work} \end{aligned} \quad [9]$$

that is, from the thermodynamic point of view, biota is a strongly nonequilibrium system. The structure of the exchange entropic flows for the biota is shown in **Figure 2**.

If we compare $d_eS_B/dt \approx -3.57 \times 10^{20}\,\mathrm{J\,K^{-1}\,yr^{-1}}$ and $d_eS_G/dt \approx -9.41 \times 10^{21}\,\mathrm{J\,K^{-1}yr^{-1}}$, then it is easy to see that these values differ by 26 times. Even if we take into account that in the case of biota we deal with the land area (covered by vegetation), which is less by almost fivefolds than the globe area, then we have almost five-multiple excess. Nevertheless, if we now compare the biota and the Earth's surface with respect to the energy obtained (the first obtains less than 1% in comparison with the second), then we can conclude that the biota is one of the main actors on the entropic scene.

Biota Performs the Work

We have an argument carrying the concept of self-organization of the biosphere: the very existence of the living biota is in necessary disequilibrium with the nonliving part of the biosphere. The main consequence of this disequilibrium is that the biota is able to perform some useful work, and a measure of this work is 'exergy'. By performing work, the biota produces additional entropy; namely, this entropy 'closes' its balance. This in turn allows us to say that the biota is in dynamic equilibrium at least in the course of last millenaries. What is this work? This is mainly the chemical work of the biogeochemical cycles, the work forcing to move the matter flows, that is, forcing the 'wheels' of the 'biosphere machine' to be turned, and then to move evolution. In particular, all the work to produce and maintain the gigantic overproduction of offspring (namely this is one of the main ideas of Darwin, while the concept of natural selection dates back to antiquity) is the work of biota.

So, this work (more correctly the ability to perform the work, i.e., 'exergy') is equal to

$$\begin{aligned} \mathrm{Ex} &= -T_B(\text{change of entropy}) = -T_B(d_eS_B/dt) \\ &\approx 1 \times 10^{23}\,\mathrm{J\,yr^{-1}} \end{aligned} \quad [10]$$

Exergy is also defined as $\mathrm{Ex} = \beta\, h_{DOM} \cdot \mathrm{NPP}$, where $h_{DOM} = 17.4\,\mathrm{kJ\,g^{-1}}$ is the specific enthalpy of DOM, $\mathrm{NPP} = 1.4 \times 10^{17}$ g d.w. is the annual NPP, and the factor β is some specific genetic characteristic of a living organism defined by a number of nonrepetitive genes in its genome. For plants, this value lies in the range 29–87. In our case $\beta \approx 42$, which is close to the mean of this interval, $\hat{\beta} = 58$.

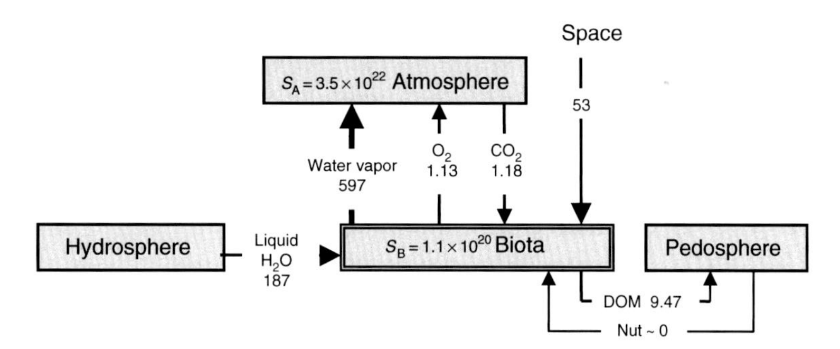

Figure 2 Exchange entropic flows for the biota (all the flows are shown in $10^{18}\,\mathrm{J\,K^{-1}\,yr^{-1}}$). DOM is the entropy flow determined by falling off the dead organic matter, Nut is the entropy flow of nutrients (nitrogen, phosphorus, potassium, etc.).

It is not a secret that all these estimations for the total biomass of vegetation, its annual production, the volume of water transpired by plants, etc., are rather conditional and strongly varying. Nevertheless, the corresponding values of β in most cases get into the interval from 29 to 87.

Humans and Biota

The annual production of artificial energy consumed by the anthroposphere constitutes about $3\times10^{20}\,\mathrm{J\,yr^{-1}}$, which is ~12% of the global terrestrial NPP. If all the energy is transformed into heat, then the annual production of entropy is $\dot{S}^{\mathrm{Art}}\approx1\times10^{18}\,\mathrm{J\,K^{-1}\,yr^{-1}}$, which is comparable with some entropic flows in the biota (e.g., with flows caused by uptake of CO_2 and release of O_2).

Now the biosphere and anthroposphere are in the state of strong competition for common resources, such as land area and fresh water. Contamination of the environment and reduction of the biota diversity are the consequences of the competition.

Since the biosphere (considered as an open thermodynamic system) is in dynamic equilibrium, then all entropy flows have to be balanced too. Therefore, the entropy excess, which is created by the anthroposphere, has to be compensated by means of two processes: (1) reduction of the biota and degradation of the biosphere, and (2) change in the work of the Earth's climate machine (in particular, an increase in the Earth's mean temperature). Note that in any case it is desirable to include the entropic flow $\dot{S}^{\mathrm{Art}}$ into the total balance of entropy for both, the atmosphere and biota, but we shall assume that the anthropogenic impact concentrates only on the biota. Then $\dot{S}_{\mathrm{B}}^{\mathrm{i}}=\dot{S}_{\mathrm{B}}^{\mathrm{DOM}}+\dot{S}_{\mathrm{B}}^{\mathrm{Work}}+\dot{S}^{\mathrm{Art}}$. By assuming that equalities $\dot{S}_{\mathrm{AB}}(\mathrm{CO_2})+\dot{S}_{\mathrm{BA}}(\mathrm{O_2})\approx0$ and $\dot{S}_{\mathrm{B}}^{\mathrm{DOM}}+\dot{S}_{\mathrm{PB}}=0$ hold in this case also, we get the following simplified equation:

$$\mathrm{d}S_{\mathrm{B}}/\mathrm{d}t\approx\dot{S}(\mathrm{H_2O})+\dot{S}_{\mathrm{B}}^{\mathrm{Work}}+\dot{S}^{\mathrm{Art}}\qquad[11]$$

The total flow of transpiration can be represented as $|q_{\mathrm{BA}}(\mathrm{H_2O})|=bB$, where B is the total mass of biota (in d.w. of DOM) and $b=|q_{\mathrm{BA}}(\mathrm{H_2O})|/B$ is the specific intensity of transpiration (in g $\mathrm{H_2O\,g^{-1}}$ d.w.), which is constant. We implicitly assume here that the power of transpiration 'pump' is proportional to the biomass of plant.

On the other hand, since this value of water is necessary to transpire in order to create P units of a new biomass (P=NPP in d.w.), then $|q_{\mathrm{BA}}(\mathrm{H_2O})|=pP$, where p is the amount of transpired water, which is necessary for creating 1 g of biomass. Therefore, the coefficient $P/B=b/p$. It is known that the P/B coefficient is a biome-specific value; apparently, we can let it be a constant. Since $|q_{\mathrm{BA}}(\mathrm{H_2O})|=4.8\times10^{19}$ g $\mathrm{H_2O\,yr^{-1}}$, $B=1.86\times10^{18}$ g d.w. and $P=1.4\times10^{17}$ g d.w. $\mathrm{yr^{-1}}$, then $b=25.8\,\mathrm{g\,H_2O\,g^{-1}}$ d.w. per year, $p=343$ g $\mathrm{H_2O\,g^{-1}}$ d.w., and $P/B=0.075\,\mathrm{yr^{-1}}$. So, $\dot{S}(\mathrm{H_2O})=-0.221\times10^3B\,\mathrm{J\,K^{-1}\,yr^{-1}}$ or $\dot{S}(\mathrm{H_2O})=-2.93\times10^3P\,\mathrm{J\,K^{-1}\,yr^{-1}}$.

Let us consider the entropic flow $\dot{S}_{\mathrm{SB}}$, which is proportional to area A_{veg} covered by vegetation. Since vegetation covers the globe by a relatively thin layer, then the equality $\dot{S}_{\mathrm{SB}}=aB$ or $\dot{S}_{\mathrm{SB}}=[a/(P/B)]P$ are rather plausible hypotheses. The value of a is easily found from eqn [8]: $a=28.5\,\mathrm{J\,K^{-1}}$ per g d.w. per year.

The entropic flow $S_{\mathrm{B}}^{\mathrm{Work}}=[\beta s(\mathrm{DOM})]P=[\beta s(\mathrm{DOM})(P/B)]B$, so that $S_{\mathrm{B}}^{\mathrm{Work}}=60.4\beta P=4.53\beta B$.

Finally, eqn [11] is rewritten as

$$\begin{aligned}\mathrm{d}S_{\mathrm{B}}/\mathrm{d}t&\approx(-192+4.53\beta)B+\dot{S}^{\mathrm{Art}}\\&\approx(-25.5+0.604\beta)\times10^2P+\dot{S}^{\mathrm{Art}}\end{aligned}\qquad[12]$$

This equation allows us to estimate different critical bounds of the impact of humankind on the biosphere. The impact may be manifested through: (a) increase in energy, $E=T_{\mathrm{B}}\dot{S}^{\mathrm{Art}}$ leads to decrease in the total biomass, B; (b) increase in energy inhibits the NPP, that is, $B=B(E)$, $\partial B/\partial E\le0$ and $P=P(E)$, $\partial P/\partial E\le0$. The simplest form of these functions may be linear, $B=B_{\mathrm{nat}}(1-E/E_{\mathrm{crit}}^{\mathrm{B}})$ and $P=P_{\mathrm{nat}}(1-E/E_{\mathrm{crit}}^{\mathrm{P}})$, where $B_{\mathrm{nat}}=1.86\times10^{18}$ g and $P_{\mathrm{nat}}=1.4\times10^{17}$ g are natural (without anthropogenic impact) values of biomass and NPP, $E_{\mathrm{crit}}^{\mathrm{B}}$ and $E_{\mathrm{crit}}^{\mathrm{P}}$ are critical values of energy with respect to the biomass and NPP (they vanish at these values).

The biota is living if $\mathrm{d}S_{\mathrm{B}}/\mathrm{d}t<0$; therefore, the upper bound for human energy production, E^* is

$$E^*=\frac{E_{\mathrm{crit}}^{\mathrm{B}}\xi^{\mathrm{B}}}{E_{\mathrm{crit}}^{\mathrm{B}}+\xi^{\mathrm{B}}}=\frac{E_{\mathrm{crit}}^{\mathrm{P}}\xi^{\mathrm{B}}}{E_{\mathrm{crit}}^{\mathrm{B}}+\xi^{\mathrm{B}}}\qquad[13]$$

where $\xi^{\mathrm{B}}=T_{\mathrm{B}}(192-4.53\beta)B_{\mathrm{nat}}$, $\xi^{\mathrm{P}}=T_{\mathrm{B}}(25.5-0.604\beta)\times10^2P_{\mathrm{nat}}$. In the previous section, we gave some meaningful interpretation to the parameter β, but here β is regarded as a free parameter.

Let us consider two simple examples.

Example 1. The lower bound of β is $\beta_*=29$, which is equivalent to full disappearance of the biosphere of vascular plants. From eqn [13] we get, for $\beta=29$: $\xi^{\mathrm{B}}\approx3.2\times10^{22}\,\mathrm{J\,yr^{-1}}$. In order to estimate $E_{\mathrm{crit}}^{\mathrm{B}}$ we assume that this value is equal to the full enthalpy of biota, that is, $E_{\mathrm{crit}}^{\mathrm{B}}\approx3.2\times10^{22}\,\mathrm{J\,yr^{-1}}$, then $E^*\approx(1/2)E_{\mathrm{crit}}^{\mathrm{B}}=1.62\times10^{22}\,\mathrm{J\,yr^{-1}}$. Today humans are consuming about 3.24×10^{20} J annually. If humans would be doubling their energy consumption by every decade, then they would reach and exceed this bound during the next 70 years.

Example 2. The work performed annually by the biota is $W_B = T_B \dot{S}_B^{Work} = T_B h_{DOM} \beta P$. Since $P = P_{nat} \left(1 - E/E_{crit}^P\right)$ where $P_{nat} = 1.4 \times 10^{17}$ g, then $W_B = W_B^{nat}\left(1 - E/E_{crit}^P\right)$, where $W_B^{nat} = h_{DOM} \beta P_{nat} \approx 1 \times 10^{23}$ J yr^{-1} is the work of the 'natural' biota. Therefore, the relative work corresponding to the bound E^* is $W_B^*/W_B^{nat} = E_{crit}^P / \left(E_{crit}^P + \xi^P\right)$. One of the possible estimations of $W_B^*/W_B^{nat} \approx 0.95$, that is, only 5% of the potential work of the biosphere can be used to maintain its structure (in particular, animals) and its evolution; the rest is spent to turn the 'wheels' of the global biogeochemical cycles, so that $\xi^P(\beta) = 0.0526 E_{crit}^P$. By substituting this value into eqn [13], we get $E^* = 0.05 E_{crit}^P$. We assume that E_{crit}^P is equal to the total enthalpy of the NPP $\sim 2.44 \times 10^{21}$ J yr^{-1}, then $E^* = 1.22 \times 10^{20}$ J yr^{-1}. By comparing this value with the current energy uptake, 3×10^{20} J yr^{-1}, we see that we already have serious problems today.

Entropy Balance in Elementary Ecosystems

From the thermodynamic point of view, any ecosystem is an open system. An ecosystem being in a 'climax' state corresponds to a dynamic equilibrium, in which the internal production of entropy is balanced by the entropic outflow to the environment.

An 'elementary ecosystem' is the area unit of land, covered by some type of vegetation, which is properly the main part of any terrestrial ecosystem, and upper layer of soil with litter, in which DOM is decomposed. We neglect horizontal exchange flows of matter, energy, and entropy between this and other ecosystems.

The equation of energy balance for this area is $R = h_{evp} q_W + Q_{turb} + h_{DOM}$GPP (see also Energy Flows in the Biosphere). Here $h_{evp} = 2462$ J g^{-1} H_2O is the specific enthalpy of evaporation, q_W is the flow of evapotranspiration, Q is the turbulent heat flow, transporting heat from the surface into the atmosphere, $h_{DOM} = 17.4$ kJ g^{-1} is the specific enthalpy of DOM, and GPP is the gross primary production (in g d.w.). Oxidation of biomass (respiration and decomposition of DOM) gives an additional source of heat, therefore the left side of the balance equation has to be $R + (Q_{met} + Q_{dec})$, where Q_{met} is a metabolic heat and Q_{dec} is heat releasing in the process of decomposition.

Let us group items of the radiative balance into two classes (in square brackets), which differ by values of their elements: $[R - h_{ev} q_W - Q_{turb}] + [Q_{met} + Q_{dec} - \text{GPP}] = 0$, where the difference may constitute a few orders. For instance, the energy acting in the process of evapotranspiration is higher by two orders of magnitude than the energy of photosynthesis. Then we can equate each of the brackets to zero (it is the so-called 'asymptotic splitting': $[R - h_{ev} q_W - Q_{turb}] = 0$ and $[Q_{met} + Q_{dec} - \text{GPP}] = 0$.

We assume that the fulfilment of the first equality provides the existence of some 'thermostat', which should be called the 'environment'. Then the fulfilment of the second equality is determined by a consistency of the processes of production, on the one hand, and metabolism of plants and decomposition of DOM in litter and soil, on the other.

In accordance with a standard definition, the internal production of entropy is equal to $d_i S/dt \approx Q_{ox}/T$, where T is the system temperature, and Q_{ox} is the heat generated by the system. The total heat production is a result of two processes: metabolism or respiration (Q_{met}) and decomposition of DOM (Q_{dec}). Since these processes can be considered as a burning of corresponding amount of organic matter, then the values of Q_{met} and Q_{dec} can be also expressed in enthalpy's units. Thus, $d_i S/dt \approx (Q_{met} + Q_{dec})/T$. The mean annual temperature at the surface of given site is the system temperature.

Since the equality $[Q_{met} + Q_{dec} - \text{GPP}] = 0$ must hold, then $d_i S/dt \approx \text{GPP}/T$. At the dynamic equilibrium the internal entropy production must be compensated by the entropy export from the system, so that

$$\frac{d_i S}{dt} = \left|\frac{d_e S}{dt}\right| = \frac{\text{GPP}}{T} \qquad [14]$$

where $|d_e S/dt|$ is so-called 'entropy pump', which 'sucks' the redundant entropy (that is existing in the system for a long time), out of the ecosystem. We assume the local climatic, hydrological, soil, and other environmental conditions are adjusted in such a way that only one natural ecosystem corresponding specifically to these conditions can exist at this site and be in dynamic equilibrium. This is a concept of 'entropy pump'.

Any natural ecosystem is in dynamic equilibrium if and only if the internal entropy production within the system is balanced by an entropic outflow from the system to its environment (the 'entropy pump' is working). Suppose that additional inflows of artificial energy (energy load, W_{ae}) and chemical substances (chemical load, W_{ch}) start entering into the system. This is a typical impact of industry (or, in a broader sense, technological civilization) and industrialized agriculture on the environment. The internal production of entropy by the 'disturbed' ecosystem is given by

$$\frac{d_i S}{dt} = \frac{1}{T}[W + \text{GPP}(W)] \qquad [15]$$

where $W = W_{ae} + W_{ch}$ is the total anthropogenic impact. Since a certain part of the entropy is released by the 'entropy pump' with power $|d_e S/dt| = \text{GPP}_0/T$, where GPP_0 is the gross primary production of undisturbed 'wild' ecosystem located at a given point, then the total entropy balance is given by

$$\frac{dS}{dt} = \sigma = \frac{1}{T}[W + GPP(W) - GPP_0] \qquad [16]$$

Under the anthropogenic pressure, the system moves toward a new state, gaining the ability to perform some work, then it returns to the initial state, performing the work and producing the entropy. This is a typical two-time working cycle of a thermodynamic machine called an 'elementary ecosystem'.

If this system tends to some stable dynamic equilibrium with respect to W ($W_{eq} = W^*$) and, in addition, satisfies to Prigogine's theorem, then $GPP(W^*) + W^* = GPP'_0$ and $GPP(W) + W \to \min_W$ at $W^* \neq 0$. Here GPP'_0 is a new value of the power of 'entropy pump', corresponding to a new equilibrium, which is established in the process of succession from natural to 'anthropogenic' ecosystem. Unfortunately, the proper time of this transition is rather long, and often the transition is not successfully finished (e.g., the 'old field' succession recovers a structure of pre-anthropogenic natural ecosystem, and does it very fast).

As a rule, the decrease in entropy, obtained at the first stage of the cycle, does not compensate its increase at the second stage. The further destiny of this 'superfluous' entropy could be different: (1) it is accumulated by the system, the system (in particular, its environment) degrades, and after a while, dies; (2) entropy may be exported from the system, the initial state is reestablished, and the system is again ready for the next cycle. The latter strategy may be realized by means of an import of additional low-entropy energy that could be used for the system restoration: soil reclamation, pollution control, or generally speaking, ecological technologies, etc. In other words, this refers to the so-called 'ecological management'. Using such entropy calculation, we can estimate the necessary investments (in energy units).

Unfortunately, there is a 'third alternative' to restore the initial state: to divide the system on two parts – a proper biological community and its abiotic environment, pumping over the superfluous entropy from one to another. In other words, we try to resolve the problem at the expense of environmental degradation. Note that the value of entropy excess σ could be used as a measure of the latter, or, as the entropy fee which has to be paid by society (actually suffering from the degradation of environment) for modern industrial technologies.

From the thermodynamic point of view, the environmental degradation leading to a decrease in the GPP is a typical system's reaction tending the internal entropy production to decrease (Prigogine's theorem and Le Chatelier's principle), while it may be considered as a disaster from the anthropocentric position. Thus, in order to avoid the anthropogenic disaster, we have to compensate the positive increment of entropy at each working cycle of this machine.

All these concepts are visibly illustrated in the case of agroecosystems.

Agricultural (Agro-) Ecosystems

What concerns agroecosystems, which are typical representatives in the class of anthropogenic ecosystems exploited by *Homo sapiens*, it is obvious that by increasing the input of artificial energy we increase their (agricultural) production. Note that the increase does not have an upper boundary and can continue infinitely. However, this is not the case, and there are certain limits, determined by the second law of thermodynamics. In other words, we pay the cost for increasing of agricultural productivity, which is a degradation of the physical environment, in particular, soil degradation. As an example, we shall analyze, as a case study, the maize production in Hungary of 1980s.

To start with, we apply the previous results to the case of agroecosystems. By taking into account that only some fraction of the GPP, $(1 - k)(1 - r)$GPP, participates in the local production of entropy, another fraction, $y = k(1 - r)$GPP, is exported from the system as a crop yield. Here r is the respiration coefficient and k is the fraction of biomass corresponding to the crop yield y. Note also that the latter and the flow of artificial energy is usually bounded by some linear relation, $y = \eta W$, where η is the so-called Pimentel's coefficient. Then instead of eqn [16] we write

$$\frac{dS}{dt} = \sigma = \frac{1}{T}\left[y\left(\frac{1}{\eta} + \frac{1}{s} - 1\right) - GPP_0\right] = \frac{1}{T}\left[W\left(1 - \eta + \frac{\eta}{s}\right) - GPP_0\right], \quad s = k(1 - r) \qquad [17]$$

The agroecosystem will exist for an infinitely long time without degradation if the annual overproduction of entropy will be equal to zero ($\sigma = 0$). This is a typical situation of the local sustainability.

Therefore, eqn [17] under the condition $\sigma = 0$ gives us the value of 'limit energy load':

$$W_{sust} = W_{sust} = \frac{GPP_0}{1 - \eta + \eta/s} \qquad [18a]$$

which provides sustainability of the agroecosystem, if $W \leq W_{sust}$. Using another form of eqn [18] we get

$$y_{sust} = \frac{GPP_0}{1/s + 1/\eta - 1} \qquad [18b]$$

This is an evaluation of some sustainable yield, that is, the maximal crop production, which could be obtained without a degradation of agroecosystem, in other words, in a sustainable manner.

In our case $W = 27\ GJ\ ha^{-1}$, $y = 4.9$ ton d.m. per hectare $= 73.5\ GJ\ ha^{-1}$, $\eta = 2.7$, $r = 0.4$, $k = 0.5$, $s = 0.3$. It is natural to take the Hungarian steppe as a reference natural ecosystem with $GPP_0 = 118\,GJ\,ha^{-1}$. By substituting these values into [17] we get $\sigma T = 81\,GJ\,ha^{-1}$; therefore, to compensate for the environmental degradation we must increase the energy input by three times, when two thirds of it is used only for soil reclamation, pollution control, etc., with no increase in the crop production.

Using eqns [18a] and [18b] we get $W_{sust} = 16\,GJ\,ha^{-1}$ and $y_{sust} = 2.9$ ton d. m. per hectare. It is interesting that the first value is very close to different estimations of the 'limit energy load', 14–15 $GJ\,ha^{-1}$, derived from economical considerations or empirically. It is the maximal value of the total anthropogenic impact (including tillage, fertilization, irrigation, pest control, harvesting, grain transportation and drying, etc.) on 1 ha of agricultural land; and if the anthropogenic impact exceeds this limit, an agroecosystem is destroyed (soil acidification and erosion, chemical contamination, etc.).

As to the second value, let us now keep in mind that the contemporary maize yield in the USA is equal to 3 ton; and also after 'black storms of 1930s', the modern agricultural technologies allow us to avoid the strong soil erosion.

Entropy (more correctly the dissipative function, $\sigma_{er}T$) corresponding to the destruction of one ton of soil in the Hungarian case is $\sigma_{er}T = 2.54\,Gt\,ha^{-1}$; then the annual loss of soil per one hectare is $\sigma T/\sigma_{er}T \approx 32$. Therefore, the high maize production would cost us 32 ton of soil loss annually. It is obvious that the value of 32 ton per hectare is an extreme value: the actual losses are less, approximately 13–15 ton. This means that also other degradation processes take place, such as environmental pollution, soil acidification (the latter is very significant for Hungary), etc.

Myth of Sustainable Development

Thanks to the Brundtland Commission book *Our Common Future. From One Earth to One World*, the concept of sustainability has become rather 'fashionable' today. Unfortunately, the sustainable development runs counter the second law of thermodynamics. What kind of arguments could be used to prove this thesis?

Our technological civilization: (a) uses nonbiospheric, nonrenewable sources of energy (fossil fuels and nuclear energy); (b) applies technological processes, which increase concentrations of chemical elements in comparison with their concentrations in the biosphere (metallurgy, chemical industry, etc.); (c) disperses chemical elements decreasing their concentrations in comparison with their biotic concentrations. All these processes produce redundant entropy, which is not sucked out by the biosphere's entropy pump, which is tuned in natural conditions. Thus, degradation of the environment is the only way to compensate for the entropy overproduction. Of course, we can avoid the degradation by applying ecological technologies, but they are rather expensive. Therefore, another way is often used by TNC. What is this way?

Since the overproduction is spatially heterogeneous, the redundant entropy naturally overflows from one site with high entropy to others with lower entropy, or it is artificially transported. If in the first case the process manifests as spreading of different pollutants by natural agents (wind, rivers, etc.), then in the second case this is either a purposeful export of industrial waste and polluting technologies to other regions, or import of low-entropy energy (e.g., fossil fuels) from other regions. Finally, we formulate the following thesis: sustainable development is possible only locally, in selective areas of the planet, and only at the expense of creating 'entropy dumps' elsewhere.

Note that in order to 'save' the sustainability concept, in the sustainability literature one talks about the so-called 'strong' sustainability, which is impossible due to the second law, and then 'weak' sustainability where losses are replaced by other gains. For instance, our technological civilization is generally using nonrenewable energy resources and materials that inevitably will lead to a loss of sustainability, but if we develop new technologies based on renewable sources of energy and materials, we are still doing well with respect to weak sustainability. However, in this case we shall deal with some slow movement of the biosphere from its contemporary equilibrium to some new unknown one. Certainly, the equilibrium might either be more suitable and comfortable for *Homo sapiens*, or might not be – that we do not know. There is one more rock in this slow movement: the small changes are accumulated without some visible effect, but sooner or later it could result in a disaster. This behavior is typical for nonlinear system such as the biosphere.

Conclusion

The author would like to complete the article by quoting the British physicist Robert Emden:

> When I have been a student, I have read with pleasure F. Wald's small book under the title "The Queen of the World and her Shadow". Energy and entropy were kept in mind. Now, when I understand these concepts deeper, I think that their positions should be interchanged. In the giant factory of natural processes, the entropy law is a director who controls and manages all the business, while the energy conservation law is only an accountant who is keeping a balance between debit and credit. (Emden, 1938).

See also: Energy Balance; Energy Flows in the Biosphere; Matter and Matter Flows in the Biosphere; Oxygen Cycle.

Further Reading

Aoki I (1995) Entropy production in living systems: From organisms to ecosystems. *Thermochimica Acta* 250: 359–370.
Ebeling W, Engel A, and Feistel R (1990) *Physik der Evolutionsprozesse*, 374pp. Berlin: Akademie Verlag.
Jørgensen SE and Svirezhev Yu M (2004) *Towards a Thermodynamic Theory for Ecological Systems*, 366pp. Amsterdam: Elsevier.
Kleidon A and Lorenz RD (eds.) (2005) *Non-Equilibrium Thermodynamics and the Production of Entropy. Life, Earth, and Beyond, Series: Understanding Complex Systems, XIX.*, 260pp. New York: Springer.
Morowitz HJ (1970) *Entropy for Biologists: An Introduction to Thermodynamics*, 195pp. New York: Academic Press.
Morowitz HJ (1978) *Foundations of Bioenergetics*. New York: Academic Press.
Svirezhev Yu M (2005) Application of thermodynamic indices to agro-ecosystems. In: Jørgensen S-E, Costanza R, and Xu F-L (eds.) *Handbook of Ecological Indicators for Assessment of Ecosystem Health*, pp. 249–277. New York: CRS, Lewis Publishers.

Information and Information Flows in the Biosphere

P J Georgievich, Russian Academy of Sciences, Moscow, Russia

Information is a concept intuitively clear to everybody and quite correctly associated with knowledge or – which is similar in meaning – with eliminating uncertainties. Knowledge is naturally considered to be useful as it increases efficiency of person's activities, ensures his better adaptability to changing environments, and therefore enhances his vital capacity and sustainability. In fact, however, that is not always so and excessive knowledge may be dangerous. Whether the knowledge appears to be constructive or destructive for an entity (subject) depends mostly on the subject's own state or own structure. When turning our own attention to ourselves, we could confidently assert that information (knowledge) obtained is capable of making changes in organization (structure, order, regulations) of our thoughts, organization of technological processes, engineering structures, social communities, etc., the latter gaining in efficiency and stability in the process.

C. Shannon worked out the law of the rate of information transfer from transmitter to receiver over a channel of any physical nature with noise. The law has been derived from a random process model and has direct analogy with models of thermostatics and therefore with models of diversity.

The capacity of a channel of band W perturbed by white thermal noise power N when the average transmitter power is limited to P is given by

$$C = w\log\left(1 + \frac{P}{N}\right) \qquad [1]$$

where $N = wN_0$ (N_0 is the power of noise on unit of a band of frequencies).

It is easily seen that this law is nothing else than a logarithmical form of allometric relationship which widely occurs both in living organisms and in inorganic nature. By conditions, the law of information transfer gives rise to a fractal set. In linguistics, the frequency band is related with the alphabet length, and the signal power with the length of a word. For biological systems, the frequency band is associated with the specialization level (the narrower is the band, the more specialized is the system). The signal power (dispersion) depends on the environment strength (energy) and diversity. There is a linear dependence between noise and frequency band – the narrower is the band, the less are errors, though communication channel capacity decreases accordingly. Assuming that there is some probability of errors becoming lethal on accumulation, the individual stability of the receiver (in terms of error-free operation) would increase with narrowing of the frequency band (increasing specialization). Taking improvement of stability to be a target function of evolution, we come to a conclusion that specialization is a natural way to this target. A specialized system, however, has a lower channel capacity and therefore lesser resistance to fluctuations in environments. If we assume, for example, that a population of organisms should have a certain minimum of diversity, it is easy to see that the most specialized and least fertile organisms are likely to inhabit environments of tropical rainforests, while the least specialized organisms of maximum fertility would be found under conditions of cold climate of taiga and tundra, and partly in deserts. This dependence is a matter of common knowledge. It follows from the law of communication channel capacity that there exists a limit of information amount that can be transmitted per unit of frequency band equal to 1.443 natural units. All the limitations bring us to the conclusion that no supersystem can exist that could receive information within an arbitrary large frequency band; a number of receiving systems (mutually complementary by alphabet) are necessary for effective transformation of the information. Hence it immediately follows that a diversity is necessary in the receiving system, and the more powerful is the transmitter, the greater is the number of various receivers needed for complete transformation of information. Under certain simple assumptions, it may be inferred from eqn [1] that the diversity is given approximately by

$$\text{Number of species } S = aN^b$$

where $b < 1$ and N is sample size and $N = f(\text{area, habitat capacity})$. This is identical to the relationships derived from thermostatics. A connection between the quantitative information model and thermostatic model is determined by their common mathematical basis: information is defined as the inverse of the entropy.

If a noisy channel is fed by a source, there are two statistical processes at work: the source and the noise. Thus there are a number of entropies that can be calculated. First there is the entropy $H(x)$ of the source or of the input to the channel (these will be equal if the transmitter is nonsingular). The entropy of the output of the channel, that is, the received signal, will be denoted by $H(y)$. In the noiseless case $H(y) = H(x)$. The joint entropy of input and output will be $H(xy)$. Finally, there are two conditional entropies, $H_x(y)$ and $H_y(x)$ – the entropy of the output when the input is known and conversely. Among these quantities, we have the relations $H(x,y) = H(x) + H_x(y) = H(y) + H_y(x)$ All of these entropies can be measured on a per-second or a per-symbol basis. The rate of transmission I can be written in two other forms due to the identities noted above. We have $I = H(x) - H_y(x) = H(y) - H_x(y) = H(x) + H(y) - H(x,y)$.

Entropy differs from diversity in that a quantity of information within a closed system transmitter increases, and not decreases, with time, as uncertainty of the transmitter decreases with time and its behavior may be more reliably predicted by the receiver. If, however, the transmitter is an open system, its uncertainties do not depend generally on the duration of transmission, and its behavior keeps up at a constant level of unpredictability. As the model of the communication channel capacity is homologous to the thermostatic model, information may be considered a phenomenon of universal occurrence. It is the transmission of information from environment to any object within a certain frequency band that controls the existing order or structure of the object. Even in case of ceasing external action or transfer of information from outside, the structure appears steady for a long time in the existing environments. In that case, there is a good reason to speak of stored information.

The rate of information flow received by the system within a certain time interval may be estimated in terms of difference in diversity at the moments of time under comparison. Information may also be measured by Kulback entropy in comparison with a diversity under conditions of equilibrium or steady (stationary) state. There is practically no study aimed at measurement of the quantitative information flows. There was a rather keen interest in information theory as applied to natural sciences, and to biology in particular, in the 1950s and the 1960s. One of the sections of information theory (i.e., theory of coding) made a considerable contribution to solving problems of genetic code and molecular synthesis. Limited possibilities for measurements and inadequate equipment hindered fruitful application of information theory for ecological research. Though a connection between information theory and thermodynamics was evident as early as the 1950s, a real integration of the two branches became possible only on a basis of developed theory of nonequilibrium thermodynamics and synergetics. All the above accounts for an exponential growth of published papers dealing with the considered problem during the last 10–15 years. The studies are mainly focused on explaining the evolution of both living matter and human society.

Evidently, the law of quantitative transfer of information does not cover all the aspects of what we instinctively associate with knowledge. A signal received may be meaningless in the receiver's perception and would not change its state, and, vice versa, a signal of negligible strength may induce drastic changes. Accordingly, information includes both quantitative and semantic components. In the simplest case, the latter may be dealt with in terms of decoding of signals coming from transmitter to receiver. It implies that there is an outside observer who establishes rules of decoding, and records signal characteristics at the input and consequent changes in the receiver state. Formally, it is a problem of statistic analysis aimed at a search for invariants with respect to signal receiver toward the transmitter. This important and by no means trivial problem of biosemiotics is related to partial interaction analysis and is potentially capable of simulation of all possible partial relations. However, its solution does not necessarily give an insight into the problem at the macroscopic level.

It should be stressed that, as follows unambiguously from our experience, an interaction between two systems may produce some new systems, and structure and properties of the latter may appear completely unpredictable, even if we have a complete knowledge of the initially interacting systems (emergence). Generally, it is impossible even to define a set of possible outcomes, that is, expected uncertainty. Therefore, the appearance (emergence) of a new, earlier unknown structure may be defined as origination of new information. The only condition for it is some energy input to the system. On the other hand, any former locally steady structure may disappear, together with related information. It seems conceivable, therefore, that conservation laws do not apply to information at the macroscopic level. Being a measure of order, information arises from chaos and returns to it; the evolution based on memory (selection of locally stable structures) proceeds by progressive retrieval of order and accumulation of information. Open macrosystem receives energy in various forms from conventionally separated environment and generates flow of information and its continuous increase.

Actually, it is this phenomenon that brings us to change our understanding of entropy as a measure of disorder; it makes us revise the classic thermodynamic model (that admits only mechanical forms of energy conversion), thus eliminating discrepancy between the observed evolution and the second principle of thermodynamics. There are numerous researches dealing with this problem. More than 60 monographs have been recently published by Springer-Verlag publishers. S. D. Khaitun, in particular, gives a meticulous review of existing opinions on thermodynamic irreversibility and concludes on advisability of coming back to wording of the second principle as stated by W. Thomson (Lord Kelvin); according to the latter, mechanical energy dissipates (depreciates) in the course of irreversible processes – its amount decreases when passing into other kinds of energy. It is a mechanical approach, where all the processes in the system are described by movement of constituent particles and its state is exhaustingly characterized in terms of coordinates and impulses so that the energy appears to be their function. Mechanical energy differs from nonmechanical in that its movement may be completely described by a set of coordinates and impulses; in other words, the energy is described by the Hamiltonian function. Nonmechanical energy is related to entropy information, because a part of the energy is spent for new structure synthesis and maintenance and for synthesis of new information. When considered together with the law of the information transfer rate over a communication channel, the results bring us to a conclusion that power (energy) of any external action is spent partly for synthesis of some elements of known type and partly for creation of new structures, with unknown characteristics; those enlarge the band (where the external actions are reproduced) and reduce the noise level in every individual case of the information reception.

S. E. Jorgensen and Yu. M. Svirezhev introduce information into a biological system through Kulback entropy, the latter being a measure of the system deviation from the stationary state. In their model, the system evolution is governed by consumed energy and inner order generated by the system itself and controlling the exergy (useful work). The evolution is aimed at increase in exergy, that is, at a synthesis of structures far from equilibrium or stationary state. Demonstrating a fact of information synthesis, A. M. Khasen supplemented the nonequilibrium thermodynamics model developed by I. Prigogine. He considered entropy information as a function of complex variables, which permitted to recognize it in two constituents, namely basic information and semantic information. The expanded model generates new structures and increases entropy information within the self-developing and self-organizing system. It would be natural to suggest that a constant analogous to Boltzmann constant (length of a word or width of frequency band) appears as a function of self-development and creates a hierarchy (of the word–phrase–paragraph type). The hierarchy arises from a limited transmission capacity at a currently accepted level of energy transformation. Increase of the transmissivity is due to self-organization of the synthesized systems into systems of the next (higher) level, with narrower frequency band. Accordingly, the number of hierarchic levels increases with total signal strength, while diversity decreases at every higher level.

The chosen descriptive (qualitative) models of information synthesis and transformation, in common with other analogous models, predict an exponential growth of information in biosphere and therefore 'cancel' a danger of the 'heat death' imminent according to the second principle of thermodynamics. Within the frame of those models, the biosphere is considered a system of a practically unlimited growth of information complexity. That does not mean that individual elements cannot fail; but every lost element would be replaced by two to four new ones, so that the rate of diversity synthesis grows progressively. It should be noted, however, that there is no universally accepted model of information processes in the biosphere; at present, we can only speak about a search for an adequate theory.

Summary

At present, there is no general information theory. In the theory of quantitative information, it is described as elimination of uncertainty. This definition implies existence of a finite set of possible states or relationships and their prior probabilities. In a more comprehensive sense, information is understood as the appearance (emergence) of order or structure with unknown characteristics from the chaos. In that case, there is no closed set of states or their prior probabilities. The information may be measured post factum, for example, in terms of distance between the emerged structure and its stationary analog, by some other means. There exists a distinct trend toward inclusion of information into thermostatic model as a missing variable which controls evolution and its irreversibility. The very fact of living matter evolution (including evolution of human beings) demonstrates that in the course of time the set of its stages gains in power and new locally stabilized systems appear; they become more and more complicated and require increasing flow of energy for their maintenance. A growth of the consumed energy flow is compensated by enhanced total transmission capacity. A great problem in synthesis of new structures consists of balance between the memory controlling admissible variants of new structures (targeted evolution) and environmental influence either through selection or by way of direct or indirect perception of its properties by the evolving object. Under actual conditions, the information flows are measurable, though an experience in such measurements is rather scarce.

See also: Matter and Matter Flows in the Biosphere.

Further Reading

Ashby WR (1956) *An Introduction to Cybernetics*. London: Chapman and Hall.

Jorgensen SE and Svirezhev Yu M (2004) *Towards a Thermodynamic Theory for Ecological Systems*, 366pp. Amsterdam: Elsevier Science.

Khaitun SD (1996) *Mechanika I Neobratimost* (Russ.) (Mechanics and Irreversibility), 445pp. Moscow: Janus.

Khazen AM (2000) *Razum Prirodi I Razum Cheloveka* (Russ.) (Nature's Intelligence and Intelligence of Man), 608pp. Moscow: Mosobluprpoligrafizdat (ISBN 5-7953-0044-6).

Shannon CE (1948) The mathematical theory of communication. *Bell Systems Technology Journal* 27: 379–423, 623–656.

Shannon CE (1949) Communication in the presence of noise. *Proceedings of the Institute of Radio Engineers* 37: 10–21.

Iron Cycle

K A Hunter and R Strzepek, University of Otago, Dunedin, New Zealand

Introduction

This article presents a scientific overview of the biogeochemical cycling of iron in the ocean, focusing in particular on what is currently known about the importance of this element as a micronutrient for the growth of oceanic phytoplankton. The first section focuses on the basic biochemistry of iron in phytoplankton metabolism, followed by consideration of the biogeochemistry of this element and how this affects its chemical speciation and bioavailability. The remaining sections deal with large-scale experiments involving iron enrichment in the ocean and the mechanisms that phytoplankton have developed to acquire iron for metabolic processes.

Why Is Iron Important to Phytoplankton?

It has long been known that iron is an essential element for the metabolism of many organisms, including humans. Iron is one of the most abundant elements in the Earth's crust, and the Fe(II)–Fe(III) redox couple provides for facile electron-transfer reactions:

$$Fe^{3+} + e^- \rightarrow Fe^{2+}$$

$$E^0 = 0.77\,V$$

As a result, iron-containing enzyme proteins are among the most common electron-transfer catalysts (**Table 1**).

Of particular importance to photosynthesizing algae are the photosystem proteins which are involved in the splitting of water to form O_2 and which contain a number of Fe redox centers. Prokaryotic phytoplankton (microbes lacking a cell nucleus), which evolved early on in the evolution of the ocean during the Archean period at least 2.5 billion years ago, used iron as the

Table 1 Some iron-containing enzyme proteins and their functions

Cytochromes	*Photosynthetic and respiratory e^- transfer*
Cytochrome oxidase	$O_2 + 4H^+ + 4e^- \rightarrow 2H_2O$
Fe-superoxide dismutase	$O_2^- + 2H^+ \rightarrow H_2O + O_2$
Catalase	$2H_2O_2 \rightarrow 2H_2O + O_2$
Peroxidase	$R(OH)_2 + H_2O_2 \rightarrow RO_2 + 2H_2O$
Ferredoxin	e^- to $NADP^+$, NO_3^-, SO_2, N_2, thioredoxin
Succinate dehydrogenase	$FAD + succinate \rightarrow FADH_2 + fumarate$
Nitrate reductase	$NO_3^- + 2H^+ + 2e^- \rightarrow NO_2^- + H_2O$
Nitrite reductase	$NO_2^- + 8H^+ + 6e^- \rightarrow NH_4^+ + 2H_2O$
Nitrogenase	$N_2 + 8H^+ + 6e^- \rightarrow 2NH_4{}^+$

basis of these redox catalysts because of the high abundance of iron in the early ocean. In the absence of free O_2, iron forms the quite soluble form Fe(II) and probably had concentrations on the order of 1 mM during Archean times.

Ironically, photosynthesis began to win out over other biological strategies, and free O_2 became increasingly available during Proterozoic times. This change in the oxygen status of the ocean had serious consequences for the photosynthesizing algae, for it led to the oxidation of Fe(II) to the much less soluble form Fe(III). During the latter stages of the Archean, this gave rise to immense deposits of Fe(III) oxides on the seafloor known as the 'banded iron formations' (BIFs). The banded nature of BIFs resulted from periodic cycles of 'boom and bust' as the algae coped with the episodic delivery of Fe(II) through upwelling from the anaerobic deep ocean and its subsequent oxidation by the O_2 generated by the algae. This process resembles a giant titration of the Earth's Fe(II), both in the ocean and on the land, by the algal waste product O_2.

Abundance and Sources of Fe in the Ocean

Eventually, the modern ocean evolved about 1.2 billion years ago in which O_2 of photosynthetic origin permeated almost all of the ocean's depth, not to mention the atmosphere, laying the foundation for the rich and complex biological systems we know today. Under these conditions, iron has become a very rare trace element in the oceans, having concentrations in most regions of the order of 1 nM or less, except in coastal and estuarine regions under the influence of terrestrial runoff.

In surface waters, where Fe is needed for phytoplankton growth, the lowest Fe concentrations are found in those oceanic regions most remote from land. More specifically, away from the direct runoff of Fe in rivers, the main external source of Fe entering oceanic surface waters in the modern Earth system is soil-derived dust transported over great distances from the arid areas of the Earth's surface (**Table 2**). Of particular importance are the Sahara and Sahel desert regions which deliver Fe to the equatorial and North Atlantic Ocean, and the Asian deserts which are a major source for the western North Pacific Ocean. The Southern Ocean contains no major dust sources other than the desert regions of Australia and Patagonia well to the north. Not surprisingly, this region turns out to be particularly depleted in iron.

Table 2 Annual flux of dust delivered to different ocean basins

Region	*Dust flux (Tg yr^{-1})*
N. Pacific	480
S. Pacific	39
N. Atlantic	220
S. Atlantic	24
N. Indian	100
S. Indian	44
Global	910

Iron Limitation and Iron-Enrichment Experiments

Less than two decades ago it became clear that the low abundance of iron in certain remote areas of the surface ocean represented an important limitation to phytoplankton growth. Over most of the temperate and tropical latitudes of the world's oceans, the main factor controlling phytoplankton growth rates is thought to be the availability of the nutrients nitrate and phosphate. The concentrations of both of these nutrients are extremely low in such waters, having been efficiently consumed by plankton. Indeed, as originally postulated by Arthur Redfield, the molar ratio of nitrate to phosphate in the global ocean is remarkably constant at about 15:1, almost exactly the same as the requirements of phytoplankton for these elements. This constant ratio is probably maintained by a balance between phosphate availability and the more biochemical alternative of nitrogen fixation that is available to nitrogen-fixing plankton.

However, in certain areas of the ocean, these nutrients remain at high residual concentrations, suggesting that another factor has come into play as a limitation on growth. These regions, which are characterized by high nutrient but low chlorophyll (HNLC), became strikingly obvious once both detailed surface maps of nutrients became available (starting with the pioneering GEOSCS Program in the 1960s and later through the programs JGOFS and WOCE) (e.g., **Figure 1**) and also the ability to map surface water chlorophyll concentrations using satellites such as the Coastal Zone Color Scanner (**Figure 2**). A comparison of these figures indicates that relatively low chlorophyll concentrations are found in regions of high nutrients, especially the Southern Ocean HNLC region.

The late John Martin, of Moss Landing Marine Laboratory in California, first made the suggestion that a lack of iron inhibited phytoplankton growth in these HNLC waters. He did this using incubation experiments in which seawater samples were inoculated with small additions (1–2 nM) of iron. Several days after inoculation, considerable increases in chlorophyll and plankton growth rate were observed compared to controls. A key to the success of these experiments was the ability to collect and handle seawater samples under scrupulously clean conditions that minimized the influence of dust contamination introduced by the experimenter. From these results, Martin speculated that iron was the

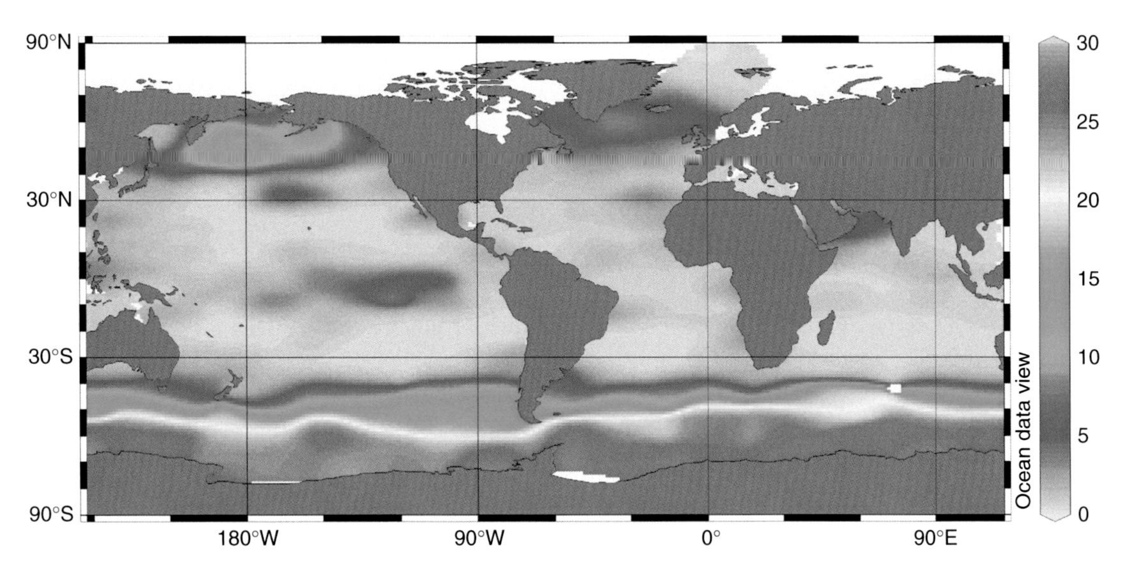

Figure 1 Map showing the annual mean concentration of nitrate in ocean surface waters. Drawn by the authors using data collected during the World Ocean Circulation Experiment (WOCE).

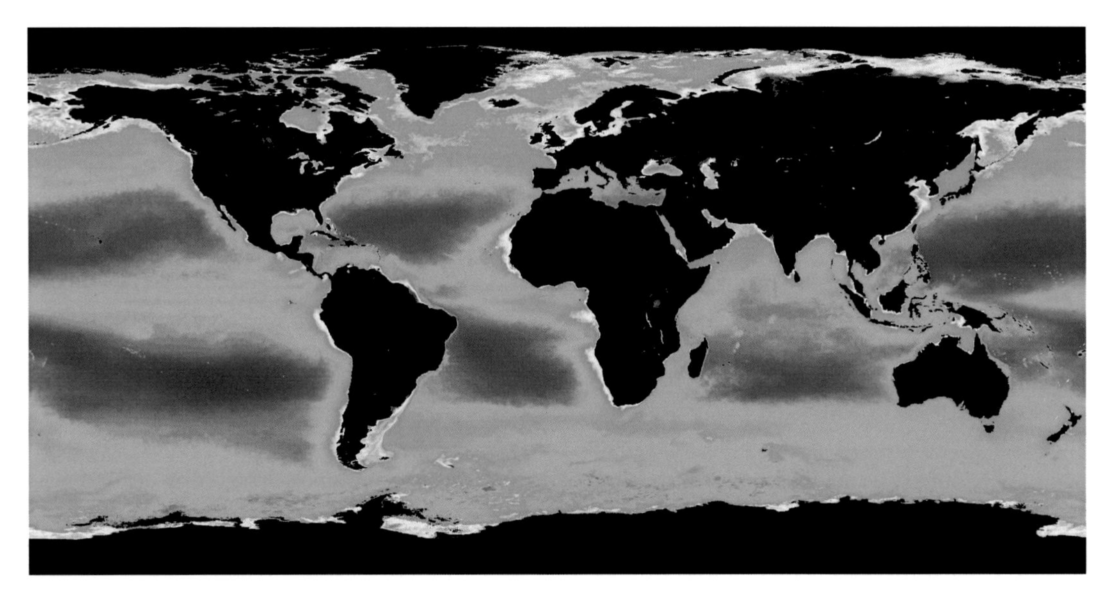

Figure 2 Satellite map showing the annual mean chlorophyll concentration of ocean surface waters (blue indicates low values; red indicates high values). Provided by the SeaWiFS Project, NASA/Goddard Space Flight Center, and ORBIMAGE.

growth-limiting factor in HNLC waters. He also went on to claim that periods of enhanced growth during glacial times might have been a result of enhanced dust input during more arid glacial climates. Periodic inputs of such dust are recorded in the polar ice core record, and seem to correlate well with periods of low atmospheric CO_2, consistent with enhanced plankton growth.

In spite of these convincing arguments, there were many skeptics. A major criticism centered on the artificiality of the small bottle incubation experiments. Grazing is also an important controlling factor on phytoplankton populations, and small bottles would not contain a sufficient population of the larger grazers. This criticism was settled by several mesoscale iron-enrichment experiments initiated in the mid-1990s. In these, a large area typically $8 \times 8\ km^2$ was fertilized with several tonnes of iron (as $FeSO_4$) along with an inert tracer SF_6 to mark the patch of iron-fertilized water. The first two experiments, IronEx I and II, took place in the equatorial Pacific Ocean, which is mildly HNLC. However, in 2002 a group of NZ and British scientists conducted the Southern Ocean Iron Enrichment Experiment (SOIREE) in the HNLC waters of the Southern Ocean south of Tasmania.

In these experiments a dramatic increase in chlorophyll as a result of a phytoplankton bloom was observed several days after the initial infusion of iron. This was accompanied by a decrease in the CO_2 equilibrium partial pressure in the water, indicating biological uptake of CO_2 by plankton. More detailed examination showed that the main beneficiaries of the added iron, and thus the main source of the new chlorophyll, were large pennate diatoms such as *Fragilieria kergulensis*. These are not the dominant organisms under normal, low-Fe conditions. All other things being equal, the best strategy for surviving under limited iron conditions is to have as small a cell as possible. The SOIREE iron-induced bloom was particularly intense, and evidence was still visible in chlorophyll satellite images up to 55 days after the initial iron infusion.

The Geritol Fix

These iron-enrichment experiments engendered considerable interest both inside and outside the scientific community. They raised the possibility of bioengineering of the oceanic ecosystem as a palliative measure against rising levels of fossil fuel CO_2 by the controlled addition of iron to HNLC areas of the ocean. This became known as the 'Geritol fix', named after a popular iron-containing tonic for 'tired blood' that was popular many decades ago.

Initially, this idea looked promising. Calculations based on the biological response in these enrichment experiments indicated that a single Fe atom could theoretically initiate the uptake of many thousands of CO_2 molecules. This means that to sequester billions of tonnes of fossil fuel carbon (the current global input from fossil fuels) might only require a few million tonnes per year of iron. This is a very small fraction of the total amount of iron smelted each year. Iron is, after all, an extremely abundant element in the Earth's crust. It has even been reported that John Martin quipped "Give me half a tanker of iron and I'll give you an ice age."

However, things are not that simple. It is not sufficient for iron enrichment to stimulate new CO_2 uptake through additional phytoplankton growth. It is also necessary for the biological carbon sequestered in this way to survive respiration long enough to sink out of the mixed layer, thus removing the sequestered carbon from the ocean–atmosphere system. Once 'pumped' into deep water in this manner, the sequestered carbon will not return to equilibrate with the atmosphere for 1000–2000 years, the turnover time of the deep water circulation system.

Thus, experiments were conducted to measure the flux of biological carbon sinking into deep water as a result of iron enrichment, and here things began to look less promising. In the initial IronEx experiment, some sinking of carbon into deep water was observed, but this may have merely been a result of subduction of the water itself. In the remaining experiments, especially during SOIREE, no increase in the flux of biological carbon to deep water was observed. This was in spite of the fact that the majority of organisms that bloomed were relatively large diatoms.

In reality, the sinking of biological carbon into deep water is a much more complex process because an entire food web is involved. Much of the carbon flux is mediated by grazing zooplankton which produce large, rapidly settling fecal pellets. During artificial experiments like SOIREE, the principal grazers of the large diatoms are probably not very abundant before iron infusion stimulates the rapid growth in numbers of their prey, and the predominant grazers may have been too small to take advantage of a bloom of very large phytoplankton. However, if iron infusions were carried out on a semicontinuous basis, who knows what permanent changes to the food web might be induced? Although this offers tantalizing benefits for mitigating climate change, it does seem to be a very dangerous experiment.

Speciation and the Bioavailability Conundrum

Not surprisingly, the discovery of the importance of iron in regulating plankton productivity in HNLC areas of the ocean stimulated a renaissance of interest in the marine chemistry of this element. Very quickly, new knowledge began to emerge that made out understanding of this complex situation even more difficult. As already mentioned, iron is very difficult to measure accurately at the very low concentrations observed in seawater, and even now there is no universal agreement on its distribution in ocean waters. This is in spite of some carefully designed intercalibration experiments that have attempted to sort out the best experimental methods for sample collection, handling, and analysis. Nonetheless, some features are now clear.

As mentioned, in the modern ocean iron is present mostly as Fe(III); this oxidation state is very insoluble in seawater at its normal pH of about 8 because of the very insoluble hydroxide $Fe(OH)_3$. Careful laboratory measurements using purely inorganic salt solutions suggest that at this pH the solubility of $Fe(OH)_3$ is about 0.2 nM. Yet the so-called 'dissolved' Fe concentrations, measured using filtered samples of seawater, are invariably up to 3–4 times higher, even in remote regions. One reason for this discrepancy is that Fe(III) readily forms colloidal particles of $Fe(OH)_3$ which are small enough to pass through most filters, thus masquerading as 'dissolved' Fe. However, very small ultrafilters can be used to eliminate a lot of the colloidal fraction, but even then the concentrations of the apparently soluble fraction still exceeds the theoretical solubility limit of 0.2 nM.

We now know that this is a result of the interaction of Fe(III) with natural organic matter (NOM) dissolved in seawater which form coordination complexes with NOM ligands. A number of very sensitive techniques are now available to probe the nature of these NOM complexes, and while there is some variation in the reported results, some general trends are clear. Seawater appears to universally contain an excess of NOM ligands that bind Fe(III), some of which are extremely strong in a thermodynamic sense (large equilibrium constant for formation). In surface waters, there is mounting evidence that the main NOM ligands are of direct biological origin, similar to the 'siderophore' compounds known to be produced by certain terrestrial microorganisms as a mechanism to sequester iron in, for example, soil waters. Iron-binding NOM persists throughout the oceanic water column, and it has been estimated that as a result of their presence, the total oceanic inventory of Fe(III) is raised by a factor of at least 4 over the solubility limit. Clearly this is very important, especially for phytoplankton growing in HNLC areas such as the Southern Ocean, where the main Fe supply may well be the upwelling of deep waters rich (relatively speaking) in NOM-bound iron.

Increasing the solubility of dissolved Fe is advantageous only if the Fe bound to NOM can then be rapidly taken up and released as inorganic Fe inside the cell. This may not be a problem for marine prokaryotes. Heterotrophic bacteria and cyanobacteria isolated from marine habitats also produce siderophores when Fe-limited, some of which have been isolated and chemically characterized. Moreover, marine bacteria transport Fe bound to siderophores regardless of whether or not they produce their own. Little is known of the mechanism by which marine bacteria obtain siderophore-bound Fe, but there is evidence that its fundamental features resemble those of terrestrial bacteria, which possess outer-membrane receptors that transport a wide range of intact Fe(III)–siderophore complexes through the cell wall.

However, the binding of Fe by NOM ligands generates a puzzling conundrum for marine eukaryotic phytoplankton. For these organisms, the principal effect of the formation of a coordination complex by a metal ion with NOM ligands is considered to be a 'reduction' in bioavailability. In this paradigm, in order for a metal ion to become available for cellular uptake, it must first dissociate from the NOM complex and become converted into a kinetically available inorganic form such as the free ion Fe^{3+} or its complexes formed with simple ligands such as OH^- or Cl^-. Only these forms are considered kinetically accessible to ion-uptake mechanisms on the cell wall.

This is why the chelator ethylenediamine tetraacetic acid (EDTA) is added to many culture media. Without it, the metal ions present as impurities in the salts used to prepare the media would be far too toxic for any phytoplankton to grow. Similarly, chelators like EDTA are used to strip metal ions like Pb^{2+} when people suffer from lead poisoning.

The conundrum is that Fe, a biologically essential element in drastically short supply in HNLC areas, appears to be bound up by NOM that ought to make it unavailable to much of the phytoplankton community. Worse still, the NOM appears to be of biological origin. So what is really going on with Fe(III) and the NOM complexes it forms? It does not make sense that phytoplankton, already struggling with a lack of iron supply, should synthesize iron-binding compounds like siderophores unless the formation of Fe(III) complexes by these materials actually assists them in acquiring iron. That implies that they have some specific mechanisms on the cell surface for unlocking Fe bound by NOM. In support of this, some very elegant culture experiments using radiolabeled Fe conducted on board ship made it clear that oceanic plankton from HNLC areas were able to take up iron much faster than it could possibly dissociate from NOM complexes to form readily available inorganic forms of Fe(III).

However, at the time of writing, we have no clear idea how this works. One possibility is that photochemistry may play a role. Fe(III)-containing complexes can be photochemically reduced to Fe(II) in seawater, in which form the Fe is much more biologically available. However, although recent work has shown that Fe(II) is generated during daylight hours in seawater, the amount of Fe(II) produced does not seem to be enough to support much plankton growth.

Biologically mediated reduction of Fe may be an alternative means to increase the biological availability of Fe bound to NOM. Experiments conducted on marine diatoms have shown that Fe(III)–NOM complexes can be accessed through use of a cell membrane Fe(III) reductase, similar to systems found in some vascular plants and other eukaryotes. Under Fe deficiency the activity of the reductase is enhanced, enabling these diatoms to acquire Fe bound to a number of natural and synthetic Fe chelators and to grow rapidly. In this type of non-ligand-specific system, reduction of organically bound Fe(III) results in dissociation of the complex, allowing uptake as inorganic Fe(II) or as Fe(III) after reoxidation.

An interesting twist in the reductive uptake process of Fe NOM complexes is the possible involvement of copper. There is evidence from a marine diatom that Fe acquisition involves two consecutive redox transformations of Fe. First Fe(III) is enzymatically reduced to Fe(II) by cell membrane reductases, then Fe is taken up by a protein complex containing a multicopper oxidase, which oxidizes Fe(II) back to Fe(III) during the membrane transport step. Even though the oxidation of Fe(II) occurs spontaneously and rapidly in oxygenated seawater, a multicopper oxidase may be important in order to acquire Fe before it diffuses away from the cell. This Fe transport pathway is highly analogous to that identified in common yeast, and some fungi and green algae. Genes homologous to those that encode for the proteins of this pathway have been identified in the recently sequenced genome of the diatom *Thalassiosira pseudonana*.

Further Reading

Boyd P, Watson AJ, Law CS, *et al.* (2000) A mesoscale phytoplankton bloom in the polar Southern Ocean stimulated by iron fertilization. *Nature* 407: 695–702.

Hunter KA and Turner D (eds.) (2001) *The Biogeochemistry of Iron in the Ocean*. New York: Wiley.

Jickells TD, An ZS, Anderson KK, *et al.* (2005) Global iron connections between desert dust, ocean biogeochemistry and climate. *Science* 308: 67–71.

Saito M, Sigman D, and Morel FMM (2003) The bioinorganic chemistry of the ancient ocean: The co-evolution of cyanobacterial metal requirements and biogeochemical cycles at the Archean/Proterozoic boundary? *Inorganica Chimica Acta* 356: 308–318.

Matter and Matter Flows in the Biosphere

S V Chernyshenko, Dnipropetrovsk National University, Dnipropetrovsk, Ukraine

Introduction

Cyclic structure is typical for living systems of all levels of organization. On the top levels (the levels of ecosystems and the biosphere as a whole) the main cyclic process is a turnover of the matter (a composition of chemical elements) through different components of ecosystems. The matter turnover is peculiar to the Earth life from the first steps of its development; and it is quite probable that the elemental cycling is initial and forming core of the life origin process (see Phenomenon of Life: General Aspects).

The cyclic character of stable natural systems can be explained by a number of considerations from philosophy, systems theory, chemistry, etc. Thus, it is logically evident that any long-term dynamics in bounded environment should have a periodical character. It is understandable that in stable chemical processes a dynamic balance between synthesis and destruction has to be observed. In accordance with cybernetic principles, formulated by N. Wiener in 1948, for stability of systems, their structure should include feedback loops (extreme importance of this principle for biological systems was grounded by L. von Bertalanfy in 1964). It is demonstrative that the wheel is a symbol of the World in many philosophical and religious systems. The list of theoretical reasons, pointing to the importance of cyclic processes, can be expanded.

Let us consider peculiarities of the natural matter cycling. We shall start from the consideration of global geochemical cycles, which are the basis of biogeochemical ones. In the next two sections, pure biological effects and the structure of global cycling (including cycling of separate biogenic elements) will be analyzed. Finally, the questions about driving forces, main dynamical properties and evolution of biogeochemical cycling, will be discussed.

Global Geochemical Cycling

The global matter cycling consists of a more or less periodical process of migration and transformation of chemical agents in nature. In part, it is initiated by abiotic factors and proceeds without participation of biological objects. Such a pure geochemical cycle is sometimes called as the big or geological turnover and characterized by global scale and extremely long course. It demonstrates abilities of even abiotic matter to primitive self-organization.

Calcium Cycling

For example, the global abiotic cycle of calcium includes the following main stages: natural destruction (denudation, erosion, etc.) of limestone; generation of soluble calcium salts (bicarbonates, etc.) and their dissolution in water; transportation of the salts by rivers to oceans (present discharge is about 5×10^{11} kg of calcium per year); sedimentation of calcium-containing substances; metamorphization of sediments, limestone formation; sea regression, return of limestone to the land. In this way the cycle is terminated, but the repetition is not complete: the type of limestone can essentially change; for example, Paleozoic limestone is richer in carbonic magnesium than younger rocks. Biological components can hasten some steps of the cycle: rock destruction and, especially, accumulation of calcium from oceanic water and its sedimentation. However, even with regard to biotic effects, the period of this type of calcium circulation is estimated to be 10^8 years.

Water Cycling

Much more intensive abiotic cycling is typical for water. Contrary to the calcium circulation, where the main work is done by the radiogenic energy of the Earth (sea regression) and only partially by the solar energy (rock destruction), for the water turnover the solar energy is the constitutive driving force.

Abiotic water cycling is extremely important for forming the global biogeochemical turnover, playing the role of its background (see Water Cycle). The main flows, forming the water cycling, are represented in **Figure 1**. Theoretically the cycle can be divided into five separate cyclic processes: I – water circulation under the ocean; II – circulation under the land; III – water interchange between the ocean and the land; IV – groundwater cycling in the lithosphere; and V – the big water cycling through the Earth mantle.

Cycle I is initiated by evaporation from ocean surface (flow 1). The intensity of the flows in **Figure 1** is represented in $10^{15}\,\mathrm{g\,yr^{-1}}$ ($10^{9}\,\mathrm{t\,yr^{-1}}$; flow 1 has a maximum value – 384. Then the main part of the water (347) returns back to the ocean with precipitations (flow 2). The last value ($347 \times 10^{18}\,\mathrm{g\,yr^{-1}}$) can be used as a characteristic of the total intensity of cycle II. In the cycle an additional loop (flow 3) can be noted. Arctic and Antarctic ice can accumulate part of incoming water and then return it to the cycling in corresponding conditions. This loop can be considered as a damping structure, which reflects self-organizational abilities of abiotic systems.

The intensity of cycle II is much smaller, mainly because the aquatic surface in the land (lakes, rivers, marshes, etc.) is not so big. The intensity of precipitations (flow 6) and evaporation (flow 4) is equal to 102 and 17, respectively. In **Figure 1**, one biogenic flow (flow 5) is also represented. It illustrates the importance of terrestrial plants even for the mainly abiotic water cycling. It is the process of respiration, in the course of which plants accumulate from soil and give back to the atmosphere some part of the precipitation. The intensity of flow 5 (which equals 48) is much bigger than the intensity of the passive

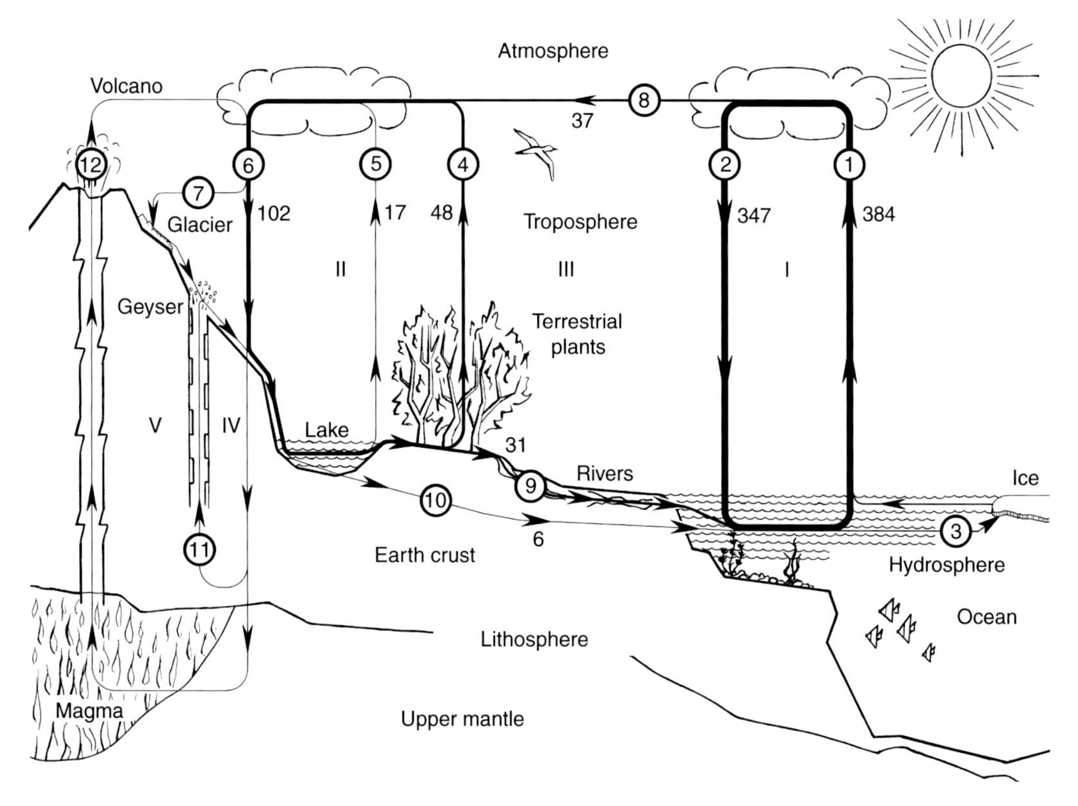

Figure 1 The global water cycling.

evaporation. Totally the intensity of cycle II can be estimated as $55 \times 10^{18}\,g\,yr^{-1}$. In the terrestrial water cycling, the role of damping device is played by mountainous glaciers (flow 7).

Between the ocean and land water circulation, there are essential links, which form cycle III. An essential part of evaporated water is transported by winds from the ocean to the land and in the opposite direction. Because the evaporation of the ocean is essentially bigger, we can generally talk about flow 8, directed from the ocean to the land. Then the water returns to the oceans with rivers (flow 9) and groundwater (flow 10), extracting solubles from lithospheric rocks on the way. The general intensity of cycle III is about $37 \times 10^{18}\,g\,yr^{-1}$.

Some part of the precipitated water can be involved in more large-scale processes. Cycle IV is formed by groundwater, which leaks to the lower stratum of the Earth crust, concentrates under the influence of pressure and temperature, and returns to the surface as thermal waters (flow 11).

Cycle V through the upper mantle is called big or geological cycling in the narrow sense of this term. Water can be bound by sedimentary rocks, migrate together with them to deeper in the Earth, accumulate to the magma, and return to the surface in the course of volcanic eruptions (flow 12). The intensity of cycles IV and V is very low in accordance with other cycles mentioned.

Although the water cycling is the fastest abiotic turnover, its influence on general geochemical situation is very gradual. Thus, the period of total abiotic renewal of oceanic water is estimated to be 10^6 years (whereas the biological renewal lasts about 2800 years).

Biological Cycling

Geochemical turnovers can circulate without participation of biological compartments, but their rate is much higher because of such participation. For calcium circulation it is especially right, since mainly the sedimentation of calcium salts is a result of calcium absorption and accumulation in animal and plant bodies (skeletons of corals and higher animals, shells of mollusks, etc.), which are concentrated on the seabed after the organisms' death. The influence of living organisms is not so important for the water cycling, but higher plants essentially accelerate it. They play the role of peculiar 'pumps', sucking out water from soil and returning it to the atmosphere in the course of transpiration.

Historically, the biological component is embedded into the abiotic cycles. It was based on peculiarities of these preceding cycles, changed them, and, finally, formed new biochemical 'nutrient' cycles, which are much faster and localized. Sometimes they are called small or biological cycles. Now all matter flows within the Earth proceed under significant influence of biological compartments. From the Cambrian period, biota became an important geological factor, which has formed the modern face of the Earth.

Both big (geological) cycling and small (biological) cycling function under the influence of the external source of solar energy.

Biosphere

The totality of living organisms, unified by participation in the biogeochemical cycles, is called biosphere. This term was proposed, in a slightly different sense, by J.-B. Lamarck in 1802. The modern meaning of the word was proposed by E. Suess in 1875. The theory of biosphere was essentially developed by V. I. Vernadsky (during 1927–44), who especially studied links between biotic and abiotic components and considered the role of humanity in the current biosphere functioning.

The borders of the biosphere in space are determined by the location of biogenic elements, involving biogeochemical cycles. It covers all the territory of the Earth, and it is limited from the top by the upper board of the troposphere (8–16 km of altitude) and from the bottom by the lower board of sedimentary rocks in the lithosphere (2–3 km of depth) and deepest oceanic depressions (11 km of depth). Thus, geometrically, it is really a sphere of thickness of about 20–30 km. Most of the organisms concentrate in the much thinner geographical shell, which includes the lowest layer of the troposphere and the upper layers of the lithosphere and hydrosphere (not deeper than 200 m). The biochemical cycling functions mainly in this shell of Earth.

Biogeocoenose

The biosphere is the biggest object, which can be called an ecological system. The next, lower level of ecological systems is represented by big territorial and functional units with more or less homogeneous structure. The systems include, besides biological populations, abiotic compartments: 'soil–ground' (edaphotop) and atmosphere (clymatop). For the biological part of the object, a special term 'biocoenose' was introduced by T. Mobius in 1877. Later the term was translated into English as ecosystem by A. G. Tansley in 1935, who also modernized the concept by emphasizing of systems nature of biological societies. Concurrently, the concept of biocoenose was developed by V. N. Sykachev in 1940, who proposed the term 'biogeocoenose' for the system, including not only biological objects, but also its direct abiotic environment. The last term is more preferable for description of territorial biosphere units, whereas the ecosystem can be understood as an ecological system of any scale.

Biogeocoenose is a system of many populations of different species (biocoenose) acting in a relatively homogeneous abiotic environment (biotop) and characterized by relatively stable biogeochemical cycling. Biogeocoenoses (and corresponding types of cycling) can be classified by the type of natural conditions for forests, grasslands, marshes, lakes, aquatic ecosystems, etc.

Extraction of next levels of ecological organization, associated with separate biogeochemical cycles, is not evident. There are many concepts realizing the functional (consortium, coenoelement, etc.) or territorial (parcella, gap, locus, etc.) approach. In any case, the existence of elementary cycles, formed by dominant population of plant–autotroph species and satellite species of plants, animals, and microorganisms, is considered as an ascertained fact. For such an abstract elementary cycle, the term 'coenome' can be used (for more details, see later).

Food Webs

Although biological cycling of each biogenic element is characterized by its own properties (see Carbon Cycle, Oxygen Cycle, Nitrogen Cycle, Phosphorus Cycle, Calcium Cycle, and Sulfur Cycle), all of the elements include migration of biomass in food webs. Transfer of the matter in the course of the cycling involves the following main steps: absorption and accumulation by living organisms of elements from abiotic environment; distribution of the matter among organisms as a result of herbivory, predation, and parasitism; territorial migration of organisms; formation of dead organic matter (DOM or mortmass) as a result of excretion and death of organisms; decomposition of the mortmass and return of the elements to the abiotic environment.

In accordance with the place, occupied by species in the food webs, they are usually divided into three main groups: producers (which use external energy, solar or inorganic chemical, and realize biosynthesis: generate organic matter), consumers (which use chemical energy of living tissue of other organisms), and reducers (which use chemical energy of mortmass and do its biodegradation: decomposition to simple inorganic agents). A classification of such organisms was initially proposed by A. L. Lavoisier in 1792 and then, in another form, was developed by W. Pfeffer in 1886.

The main players in the nutrient cycling are producers and reducers. The former are an 'engine' of the cycling; they involve elements from the abiotic environment in the turnover and send them further in the composition of permanently generating high-energetic organic matter.

The reducers 'close' the cycling; they return the elements to the abiotic environment, where they can be used by producers again. Abiotic decomposition takes place, but its intensity is very low. Without producers available elements would concentrate in the biomass and leave the environment; cycling would stop, and life development would end. The simplest artificial stable ecosystems, functioning in closed flasks, included populations of producers (unicellular algae) and reducers (bacteria and fungus).

Theoretically it is possible to envision producers, independently realizing the function of reducing with respect to their own biomass. But in reality this possibility is not realized. This fact can be explained by the absence of evolutionary reasons of forming 'self-sufficient' organisms, if the hypothesis about the origin of heterotrophs (consumers and reducers) before producers is correct. Stable biological cycles could form gradually, during the process of co-adaptation of producers and reducers. Close symbiosis and species peculiarity are typical for relations between producers and reducers.

The most common flows of matter in the biosphere, including food chains and abiotic topical ways, are presented in **Figure 2**. There are four ecosystems of different nature in the **Figure 2**: terrestrial; shelf; open sea; deep-sea black geyser. The core cycle for each ecosystem is cycling 1: 'producers–mortmass–reducers–inorganic salts–producers'. For the terrestrial ecosystem it can be written: 'producers–litter–reducers–soil–producers'; for the water one: 'producers–sediments–reducers–water–producers'. The cycles are initiated by the producer block, which transform the solar energy to the chemical one in the course of the photosynthetic process. It is the most rapid and mainframe cycle in the biosphere; its dynamic properties are considered below.

The amount of different biogenic elements in cycling is determined mainly by features of the environment. The intensity of cycling depends on properties of the producer block. The additional loop 2 reflects the role of the consumer block. The energy and matter flow through it are approximately 10 times less than directly from the producers to the mortmass, but the consumers, influencing the producer block, 'bootstrap' the turnover. Similarly, loop 3, involving the block of consumers of second order and also 10 times less intensive than loop 2, contributes to increasing intensity of the total cycling. In **Figure 2** for the marine ecosystems the blocks of producers, reducers, and consumers are marked by the first letters P, R, and C, respectively.

Elements' Pathways: Terrestrial Ecosystems

The terrestrial biogeocoenose, in addition to the blocks, involved in the described main cycle, includes the atmospheric block. Flows 1–6, connecting this block with others, are internal flows of the biogeocoenose. The

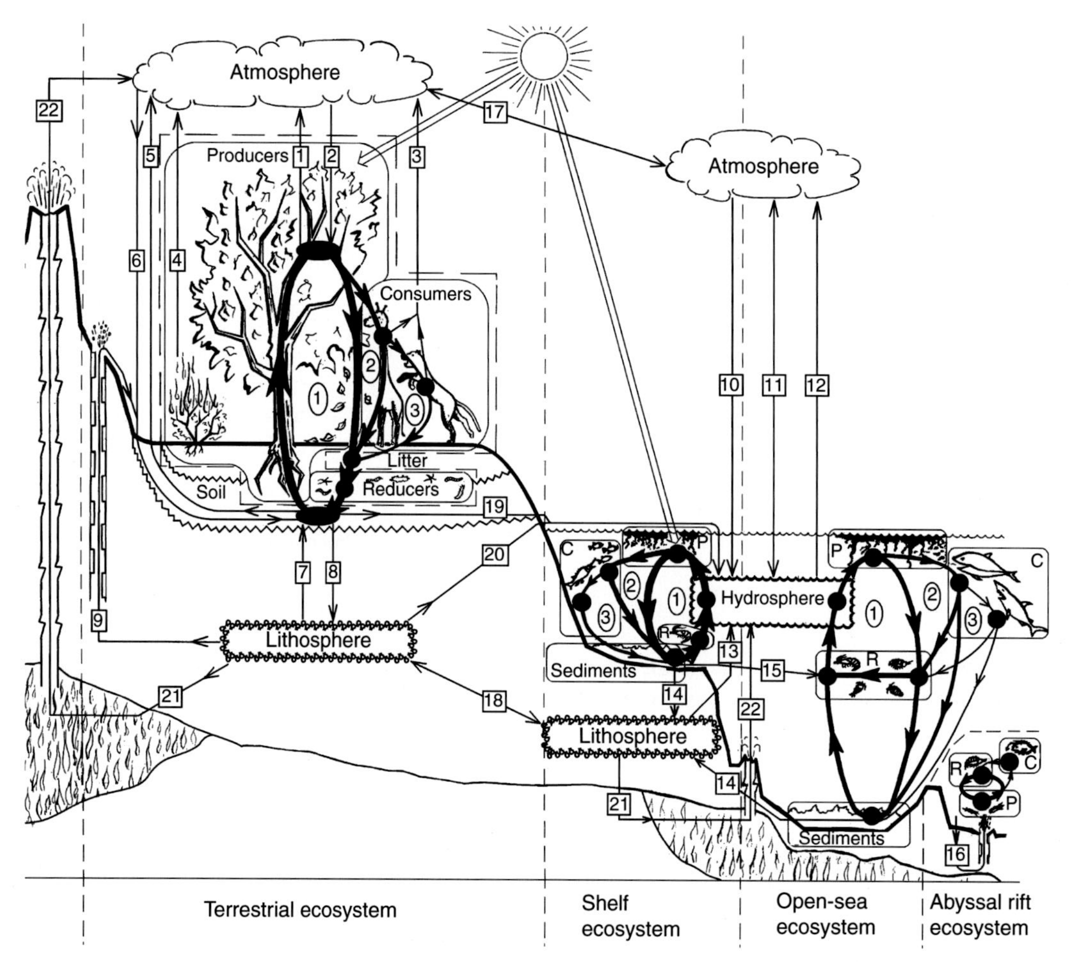

Figure 2 Main matter flows in the biosphere.

importance of the cycle formed by flows 1 and 2 is close to importance of the central cycle 1. The main source of carbon for photosynthetic plants is carbonic gas from atmosphere. In contrast, in the course of respiration, plants send off carbonic gas and consume oxygen. Also, as shown in **Figure 1**, plants play important role in passing water from soil to atmosphere (the process of transpiration).

Flow 3 describes the process of respiration of other members of biocoenose: consumers and reducers. Another way of sending carbonic gas (and other chemical substances) from plant body to atmosphere is fire, produced by such natural factors as thunderbolt (flow 4). Connections between soil and atmosphere are reflected in flows 5 and 6: precipitation, soil respiration, and diffusion.

Interaction between soil and upper lithosphere (which is considered as a part of the biosphere, but not of biogeocoenoses) is described by flows 7 (lixiviation), 8 (leakage, mineralization, fossilization), and 9 (thermal water circulation).

Elements' Pathways: Marine Ecosystems

Aquatic biogeocoenoses have the same principal 'cybernetic' structure, but with special role of the water environment. The latter plays the functional role of soil (or, more precisely, soil solution), which stores ions and 'feeds' producers, and, partially, the role of atmosphere, which delivers carbonic gas and oxygen to living organisms. Connections between water environment and real atmosphere are pictured by flows 10 (precipitations), 11 (diffusion), and 12 (evaporation). The hydrosphere is also connected with the upper lithosphere by flows 13 (dissolution) and 14 (fossilization of sediments).

It is important to stress that total biomass of marine ecosystems is much less (approximately 800 times) of the terrestrial one (about 5×10^9 t contrary to 4000×10^9 t). At the same time the primary production of marine ecosystem is only 4 times less. It is explained by much more intensive matter cycling in the ocean. Correspondingly, the matter involved in marine and terrestrial cycles is similar (about 215×10^9 and 270×10^9 t correspondingly). Apart from similarity of the primary production (170×10^9 and 60×10^9 t), it is explained by intensive migration of gas substances among the atmosphere and the ocean. Biomass of marine ecosystems is renewed many times during a year.

The marine ecosystems can be divided into shelf or sublittoral (functioning on the shelf, shallow coastal part of the sea) and open sea (connected with open deep-sea territories). The former are much richer in energy and biomass, and are characterized by much more intensive nutrient cycling. This fact is explained by possibility of photosynthesis in the sea floor (the bottom is above the photosynthetic horizon, which lies at a depth of about 200 m) and easiness of involving the mortmass, concentrating in the seabed, in the cycling.

Open-sea ecosystems are characterized by two types of nutrient cycling. The first is connected with surface water, which includes not only producers, but also reducers, 'intercepting' diving particles of mortmass. This cycle, naturally, is not closed, because an essential part of the mortmass falls to the bottom. Return of this matter to the ecosystem is difficult because of (1) bottom conditions, which are adverse for the life and, particularly, for reducer's activity; and (2) the usual absence of ways of transportation of biogenic elements from the bottom to the layers under the photosynthetic horizon. As a result, the biogenic elements are concentrated in the bottom sediments, and the process of fossilization (flow 14) is much more intensive in the open sea than in the shelf. It leads to much more essential losses of biogenic elements in this type of biological cycles.

The upper cycle, in its part, usually can be divided into two forms. The first, in a depth of 25–40 m, is characterized by the best conditions for photosynthesis, but limited by deficiency in accessible biogenic elements, which are mainly results of local reducers' activity. The second form of cycling is connected with producers' activity in a depth of 70–90 m, where photosynthesis cannot be so intensive, but the reducers are much better provided by nutrition, which come in from above, from the upper cycle, and from below, as a result of turbulent diffusion.

The richest marine ecosystems are observed in the territories where cold sea currents lift deep water to the surface or where local conditions promote mixing of water layers (e.g., near coral reefs). In other words, open-sea ecosystems function better, if the second, bigger cycle, including deep-water sediments, is closed.

The shelf and open-sea ecosystems are unified, naturally, by the water; these are mixed by currents and winds. Besides, there is a transfer of biogenic elements from the shelf to open sea (flow 15), where they are utilized by the upper-water reducers or accumulated by bottom sediments.

Abyssal Rift Ecosystems

A special type of marine nutrient cycling is typical for the so-called rift life concentrations or 'oases'. They were discovered in 1977 and are located around abyssal rifts where deep-sea thermal springs occur. Such abyssal rift ecosystems are based on populations of microbial chemosynthetic producers, which use chemical energy, not solar, for organic matter production . Flow 16 represents the movement (with lifting hot water) of high-energy substances (mainly hydrogen sulfide) from lithosphere to ocean water. The ecosystems can include many organisms, consumers and reducers, adapted to these special conditions: worms, pogonophors, mollusks, fishes, etc. Some of these organisms are in symbiotrophic relations with chemosynthetic bacteria, which live in their bodies.

Probably, rift ecosystems can function without contacts with other parts of biosphere, although they are not really isolated and, particularly, can participate in enrichment of seawater by organic substance. There is an opinion that biogeochemical cycling, characterized for the rift ecosystems, has some common properties with the first cycling, formed in the Earth on the stage of hydrogen–helium atmosphere.

Mathematical Models of Biological Cycling

A basic element for formal description of biological cycling is a model of elementary turnover. A special term 'coenome' can be proposed for such elementary biogeocoenotic cyclic element. The coenome can include one or several populations of producers (as a source of energy and a kernel of the association); reducers, that close biogeochemical cycles; and consumers that stimulate energetic processes in the system. According to V. N. Sukachev, producers can be subdivided into three general groups: edificators (determinators), co-edificators, and assectators. Coenome species are characterized by some level of co-adaptation. The main features of cross-population relations in a coenome are the following: a low level of competition, as a result of effective separation of niches; high-developed mutualism, especially, in pairs 'producer–reducer'; and an optimal (from energetic reasons) level of trophic relationship. Naturally, the areas of ecological optimum of coenome species must have a common part.

Biogeocoenose may be characterized by one dominant coenome, but usually it is also possible to recognize some minor coenomes in its structure. If biogeocoenose is characterized by two or more dominant coenomes, it is amphicoenoses. In general, most natural ecosystems are superposition of several coenomes as a result of spatial heterogeneity, exogenic factors, etc.

The principal scheme of matter and energy flows in coenome (without consumers) is represented in **Figure 3**.

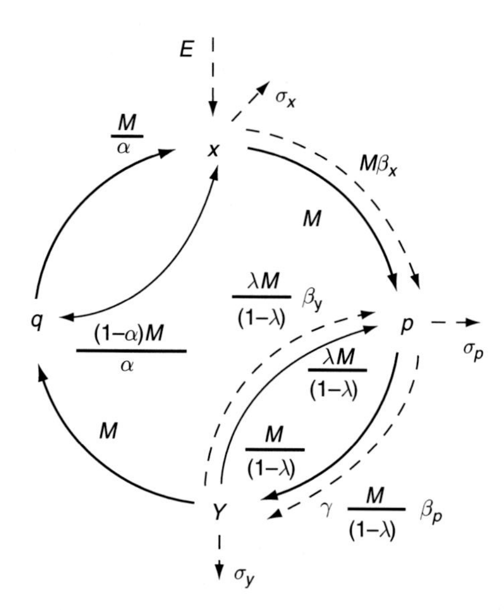

Figure 3 Matter and energy flows in a coenome.

The symbols x, y, p, q denote contents of some chemical element, respectively, in the biomass of producers, in the biomass of reducers, in the mortmass, and in the inorganic matter, accessible for producers. Characteristic intensity of the element cycling is symbolized by M. The coefficient α $(0<\alpha<1)$ determines what part of the element does escape the populations of producers as organic matter, and the coefficient λ $(0<\lambda<1)$ determines the same for populations of reducers.

The cyclic structure of a coenome illustrates the importance of positive feedback in food webs which exists simultaneously with negative feedback circuits in usual competition and trophic relations. This nontrivial 'cybernetic' structure of ecosystem permits its capacity of fast development in appropriate conditions, concurrently with the capacity of homeostasis.

The simplest closed model of matter cycling in a coenome (the firm lines in **Figure 3**) can be designed on the basis of Lotka–Volterra models. Without describing effects of saturation and self-limitation of populations, it can be written as follows:

$$\begin{aligned}
\mathrm{d}x/\mathrm{d}t &= aqx - bx\\
\mathrm{d}p/\mathrm{d}t &= \alpha bx + \lambda sy - ryp\\
\mathrm{d}y/\mathrm{d}t &= rpy - sy\\
\mathrm{d}q/\mathrm{d}t &= (1-\alpha)bx + (1-\lambda)sy - aqx
\end{aligned} \qquad [1]$$

The coefficients b and s are the death rates of producers and reducers; a and r estimate intensity of use, respectively, of inorganic matter by producers and mortmass by consumers. The model does not determine the stable values x and y, but only their ratio: $x/y=(1-\lambda)s/\alpha b$. It is explicable both from the mathematical point of view (the sum $(x+y+p+q)$ does not change during the process) and by biological reasons (the model describes closed circulation of the element) that the element amount in the cycle is determined by some external factors. Under natural assumptions that $a>b$, $r>s$, system of equations in [1] is stable, but not asymptotically.

A more realistic model can be obtained by the use of the equations with Michaelis–Menten functions in the right-hand members:

$$\begin{aligned}
\mathrm{d}x/\mathrm{d}t &= A(cq/(x+cq) - b/A)x\\
\mathrm{d}p/\mathrm{d}t &= \alpha bx + \lambda sy - Rhpy/(y+hp)\\
\mathrm{d}y/\mathrm{d}t &= R(hp/(y+hp) - s/R)y\\
\mathrm{d}q/\mathrm{d}t &= (1-\alpha)bx + (1-\lambda)sy - Acqx/(x+cq)
\end{aligned} \qquad [2]$$

System [2] dictates the same parameter $M=\alpha bx$ and the same ratio between biomasses of populations, as in model [1]. However, in this case, the quantities p and q depend on population sizes.

For the energetic flows in coenome (the dotted lines in **Figure 3**), it is possible to write, similarly to the model [2], the following system:

$$\begin{aligned}
\mathrm{d}e_x/\mathrm{d}t &= (E-\alpha b-\sigma_x)e_x\\
\mathrm{d}e_p/\mathrm{d}t &= \alpha be_x + \lambda se_y - Rhe_pe_y/(e_y+he_p) - \sigma_pe_p\\
\mathrm{d}e_y/\mathrm{d}t &= (\gamma Rhe_p/(e_y+he_p) - \lambda s - \sigma_y)e_y
\end{aligned} \qquad [3]$$

where the variables e_x, e_y, e_p estimate energy in the biomass of producers, reducers, and mortmass, respectively; σ_xe_x, σ_ye_y, σ_pe_p are energy losses for each group; $E\ e_x$ is the intensity of the energy flow into populations of producers; γ $(0<\gamma<1)$ is the part of the dead biomass energy, which passes to the energy of the reducers' biomass.

The first equation of the system [3] does not depend on others. The condition of existing steady nonzero value e_x (which depends on initial conditions) is $\sigma_x = E-\alpha b$. The conditions of positivity of the steady state are $\gamma > \lambda s/R$, $\sigma_y < \gamma R - \lambda S$.

The problem of interrelation between the matter and energy flows in the coenome can be considered with the use of the variables β_x, β_y, β_p, which are the measures of energy per biomass unit of producers, reducers, and mortmass, respectively. The variables can be described by the equation

$$\begin{aligned}
\mathrm{d}\beta_x/\mathrm{d}t &= (E-\alpha b-\sigma_x)\beta_x\\
\mathrm{d}\beta_p/\mathrm{d}t &= h(R-s)(1-\lambda)\beta_x + h(R-s)\frac{\lambda - R\beta_p}{(R-s)\beta_y + s\beta_p}\beta_y\\
\mathrm{d}\beta_y/\mathrm{d}t &= \left(\frac{\gamma Rs\beta_p}{(R-s)\beta_y + s\beta_p} - \lambda s - \sigma_y\right)\beta_y
\end{aligned} \qquad [4]$$

The variables β_x, β_p, β_y can be used as measures of the level of organization (or thermodynamic instability) of population biomass. It is possible to interpret system [4] as the model of information transformation in the coenome, if the term 'information' is used as a synonym of negentropy.

The flows of matter, energy, and information in the models [1]–[4] are ultimately determined by producer activity. In some way, the variables x, e_x, and β_x are external parameters of the models, which should be described by some additional equations. A possibility to estimate the steady values of these coordinates has to be studied on the basis of other reasons; for example, one can use the Liebig law. Concentration of noncritical ions in soil does not influence productivity, and even inhibits it in the case of too high concentration. A scale of cycling is determined not only by the amount of accessible bioelements, but also by the biochemical properties of soil.

Global Biogeochemical Cycling

Biosphere-wide biogeochemical cycling is formed by matter transfer between the land and the ocean. There are several ways of this transfer (**Figure 2**). Particularly, water steam and other substances in the atmosphere are spread over the Earth by winds (flow 17). Tectonic movement leads to interchange by matter of different parts of the lithosphere; or some rocks can become a land from an ocean floor, and vice versa, as a result of sea regression or transgression (flow 18). Water drain (rivers and groundwaters, flow 19) transports solute mineral matter from the land to the ocean. The matter is to get to the drain from the lithosphere as a result of erosion and ablation processes (flow 20).

A slow cycling of biogenic elements is connected with the big geological turnover. In the course of lowering lithospheric layers, some rocks transform to magma (flow 21) and return to the biosphere as a result of volcanic eruptions (flow 22).

Structure

When one talks about global biogeochemical cycling, it does not mean that there is only one cycle. As is evident from **Figure 2**, the real structure is closer to a web with many cyclic cells. A global property of the web, which allows considering it as a cycle, is the necessity of balance between inputs and outputs for every acting subject or subsystem. For example, in accordance with **Figure 2**, for terrestrial biogeocoenose the input of each chemical element (the total intensity of flows 2, 6, 7, 9, 22) should be equal to its output (the total intensity of flows 1, 3, 4, 5, 8, 19). Although, as ecosystems are never in equilibrium, all these balances do not absolutely hold true.

The other important peculiarity of biological cycles is their essential spatial heterogeneity. In different natural conditions, biosphere is represented by different biogeocoenoses (forests, steppes, deserts, etc.), which are characterized by special types of biogeochemical cycling. Data on net primary production of main kinds of ecosystems are represented in **Table 1**. The primary production reflects the amount of energy (and, indirectly, matter) put into cycling every year. Finally, this amount estimates the intensity of biogeochemical cycling in corresponding ecosystems.

In **Figure 2**, flows connecting biogeochemical cycling with external environment are not shown. For example, the biosphere loses hydrogen and helium, dissipating to the outer space, and, probably, permanently needs carbonic gas from the mantle.

General Characteristics

The biosphere as a whole can be interpreted as a peculiar heat engine, which uses, for its operation, solar (and, unessentially, radioactive geothermal) energy. In addition to artificial mechanical engines, the global biological machine is based on periodical processes and has its 'work cycle': the global biogeochemical cycling. Contrary to human-made engines, the biogeochemical one has no external destination. Finally, nearly all accepted energy is converted into the thermal energy of the environment, although only a small part is spent for biosphere self-organization.

An important characteristic of the biological turnover is intensity of the matter or energy flow in the cycle. It is necessary to note that in a stable state this intensity (measured in $kg\,s^{-1}$ or, for the energy flow in watts) should be the same in all sections of the cycle (except very long term effects of biosphere development). Particularly, the amount of producing organic matter must be equal to the amount of the reducing one. Another important fact is the relative independence of two values: the flow intensity through a cycle's component and its saturation by mass or energy. The concept of

Table 1 Indices of biogeochemical cycles' intensity (the net primary production) of the main types of ecosystems

Types of ecosystems	*1*	*2*	*3*	*4*	*5*	*6*	*7*	*8*	*9*	*10*	*11*	*12*	*13*	*14*
Net primary production ($10^6\,J\,m^{-2}\,yr^{-1}$)	2.5	15	25	11	9	0.6	13.5	37.5	37	9	37	11.5	7	1.7
Total world net primary production ($10^{12}\,J\,yr^{-1}$)	25	200	440	75	85	40	210	750	60	25	60	175	190	800

1, Tundra; 2, coniferous forests; 3, deciduous forests; 4, forest steppes; 5, steppes; 6, deserts; 7, savanna; 8, tropic forest; 9, bogs and marshes; 10, lakes and rivers; 11, reaches; 12, agricultural lands; 13, continental shelf; 14, open ocean.

the trophic pyramid is based on the idea about a direct relation between these values, but it is known that the concept does not work in some cases. For example, in aquatic ecosystems, autotrophic algae communities can have relatively small biomass (less than the accompanied herbivores), but they 'pump' through themselves much more matter than the herbivorous block.

Other global turnover characteristics are its period, 'length', 'width', and 'amplitude' (amount of involved matter). In general, the amplitude equals to the product of the intensity and period or, otherwise, the length and width. It is important to take into account that the real turnover is not a circle, but a network with a lot of paths and loops; and for any portion of matter, each element has its own way in the cycle with unique values of period, length, etc. The general characteristics are the average values, corresponding to an abstract average cycle of the pure circle form.

Average terms, needed for general renewal of biosphere components, can be used as integral indices of the matter turnover. Thus, the total living matter is renewable on average during 8 years; the marine one circulates much quicker, during 33 days. Corresponding values for plant biomass are even more contrasting: 14 years and 1 day. In the mean, all water of the hydrosphere does its cycling during 2800 years, and it passes through photosynthetic decomposition during 5 millions years only.

Energy Aspects

The intensity of cycling is determined by two main factors. The first key factor is the amount of available biogenic elements, which will be considered below. The second and the main one is the power of the producers block. An amount of energy that can be put into the cycle per unit time by producers determines both the cycle intensity and length. The energetic concept of A. Lotka and H. Odum, based on the consideration of energy as a peculiar ecological 'currency' is very useful for understanding functioning of biogeochemical cycling.

In contrast to the element cycling, it is not correct to consider energy cycling. Energy does not circulate; it is fixed by producers from the solar irradiation and then step-by-step degrades in food webs, transforming to thermal energy and dissipating in the outer space. The role of energy in biogeochemical cycling is compared sometimes with the role of water in water-mill functioning. A millwheel, as well as the elements in the course of biogeochemical cycling, really circulates; all its components permanently repeat their positions. At the same time water, rotating the wheel, does not return back; each of its portions leaves the mill forever. It is reasonable to talk about energy flow, but not cycling.

Similarly to energy flows through ecosystems, it is possible to consider information flows. If we associate information with negentropy, as it is often done, we can talk about the creation of information by producers and its gradual destruction by consumers. Although two consumers can use the same amount of energy, their production of entropy can be different. Some consumers can keep information better and transmit it to the next level. More careful use of energy gives essential evolution advantages to species. It is especially important for consumers of high levels, which deal with a relatively poor flow of energy; but energetic effectiveness also demands careful use of information. These considerations explain partly why intelligence of species usually increases in the direction of the top of trophic pyramid.

Cycles of the Main Biogenic Elements

All chemical elements of the Earth crust are involved in the biogeochemical cycling, but the intensity of their turnover and their importance for biosphere are quite different things. Each element is characterized by its own paths in the biosphere; importance of different matter flows (1–22 in **Figure 2**) for different elements is quite changeable. Some characteristics of the most important elements cycling are presented in **Table 2**.

The elements can be divided by different criteria. By their importance for living organism functioning (and representative in their bodies) they are classified for obligatory biogenic elements (the first six rows in **Table 2**: oxygen, hydrogen, carbon, nitrogen, sulfur, and phosphorus). Because water is the main part of biomass (60% of terrestrial organisms, 80% of water ones), the latter is formed essentially by oxygen and hydrogen. Both of these elements are important components of organic substances, together with carbon (the master biogenic element). Nitrogen, sulfur, and phosphorus are not represented in biomass in such big amounts, but they are absolutely necessary for forming such key organic substances as proteins, DNA, etc. Hydrogen is the only important element (the other one is helium), which has principally unclosed cycling, since its molecules permanently leave the Earth because of their low molecular weight.

Elements of the next group (potassium, calcium, magnesium, sodium, chlorine, silicon) together with the obligatory elements form the group of macroelements. Their concentrations in the total biomass are more than 0.01%. Two other elements from **Table 2** (iron and manganese) represent the group of microelements, which embraces practically all stable elements from the periodical table.

Based on how elements involve in biogeochemical cycling they can be divided into products of Earth mantle

Table 2 Participation of the main biogenic elements in the global biogeochemical cycling

Element	*Part in terrestrial biomass (%)*	*Part in marine biomass (%)*	*Mass, involved in the terrestrial cycling (10^9 t)*	*Mass, involved in the marine cycling (10^9 t)*	*Element migration from land to ocean (10^9 t)*	*Flows of migration (in accordance with Figure 2)*
O	69	74.1	170	125	6	1, 3, 5, 6, 8, 12, 14, 16, 17, 19–22
H	10.2	12.4	20	15		1–3, 5–10, 12, 14, 16, 17, 19, 21, 22
C	18	9.4	70	50	0.8	1–5, 7–9, 11, 13–15, 18–22
N	0.75	1.6	3.4	6	0.05	4–6, 9, 11, 13–16, 18–22
S	0.19	0.38	0.6	1.32	0.16	6–11, 13–16, 18–22
P	0.08	0.1	0.35	1.2	0.025	7–9, 13–15, 18–20
K	0.45	0.6	1.8	1.2	0.34	5–11, 13, 14, 18–20
Ca	0.6	0.2	2.3	1.1	0.9	5, 7–9, 13, 14, 17–20
Mg	0.13	0.1	0.5	0.8	0.7	7–9, 13, 14, 18–20
Na	0.05	0.4	0.2	2.8	1.3	7–9, 11, 13, 14, 17–20
Cl	0.08	0.09	0.3	4.4	2.3	7–9, 11, 13, 14, 17–22
Si	0.2	0.45	0.86	5.5	0.2	5, 7, 9, 10, 14, 18–20
Fe	0.01		0.034	0.047	0.9	7–9, 13, 14, 18–20
Mn	0.01		0.035	0.001	0.02	7–9, 13, 14, 18–20

degasification (O, H, C, N, S, Cl) and Earth crust lixiviation (P, K, Ca, Mg, Na, Si). These processes took place during Earth formation in earlier geological epochs and still are typical for the present big geological cycling of corresponding elements (flows 7, 13 or 21, 22).

Water (and, correspondingly, oxygen and hydrogen) follows the standard ways (see **Figure 1**), for example, by flows 1, 3, 6. Carbon, as a component of carbonic gas, is also involved in the circulation through the atmosphere (flows 1–5). Some other biogenic elements (such as nitrogen and phosphorus) are characterized by relative stable circulation in biological cycles 1–3.

Chemical Elements' Presentation in Cycling

The general terrestrial (flows 1–9) and marine (flows 10–16) cycles are more or less close. Amounts of different elements participating in these cycling are represented in **Table 2**. The numbers are very rough approximations of real values, because the current state of ecology as a science does not allow integrating data within all type of ecosystems and, additionally, the situation can essentially change from year to year.

The element representations in the cycles are determined by both presenting of elements in environment and their necessity for organisms of ecosystems. The last two factors are under mutual influence: organisms adapt to deficiency or excess of some elements and, conversely, can change the environment in desirable direction (e.g., soil bacteria radically increase the amount of available nitrogen in environment).

Rates between different elements, involved in cycling, are more or less proportional to their presentation in biomass. There are the so-called stoichiometric rates. For example, the rates of Redfield describe relations between amplitudes of cycles of carbon, nitrogen, and phosphorus.

In general, the amplitude of the cycling is determined by the total mass of available carbon, but in separate links it can be limited by the amount of other bioelements, playing an important role in internal organization of these special links. This fact is reflected in the well-known law of Liebig.

Usually, the limiting elements in terrestrial and especially aquatic ecosystems are nitrogen and phosphorus. As in accordance with the rates of Redfield the amount of carbon is strictly connected with amounts of these limiting elements, the amplitude of carbon cycle can be determined by their availability. Carbon buffer in soil humus, peat, and 'ocean humus' (dissolved organic substances in ocean water) includes the most part of cycling carbon and can be used very quickly for stabilization of global biogeochemical cycling.

The close character of the cycles determines the relative stability of ecosystems, although the permanent migration of many elements from the terrestrial cycle to the marine one (flows 19, 20) takes place. Corresponding estimations are also represented in **Table 2**. In marine ecosystems the surplus of elements is compensated by the process of sedimentation (flow 14); in other words, the elements are directed to the big biogeochemical cycling. Correspondingly, terrestrial ecosystems mainly compensate the losses of elements from the big cycles (flows 7, 18, 21). The unclosed character of some biogeochemical cycles by some elements, however, produces long-term processes of ecosystems' development, adaptation, and self-organization. In many cases, the biological evolution is a result of instability of biogeochemical cycling and, in its turn, produces such instability. One of the brightest manifestations of life on Earth is its embedding in the geochemical cycling and radical transformation of the latter.

Evolution of Biogeochemical Cycling

Structural Peculiarities of Biogeochemical Cycles' Development

Although ecological systems can be considered as the next level of matter self-organization after the biological one, it is quite probable that biogeochemical cycling preceded the origin of separate organisms. First, the cyclic chemical reactions, including both synthetic and decomposing processes, were formed in the ocean. Then their main links were shaped as separate self-reproduced organisms. The formation of the biological cycling was not the result of unification and cooperation of organisms or populations that existed before; all other forms of biological units arose in the course of structurization of biological cycles (as a result of special processes of discretization or corpuscularization). Such a view on forming ecosystems is analogous to the model of M. Eigen (his famous hypercycle) of first organisms' origin.

Another important point is the fact that evolution of the organic world never totally destroyed the cycles that previously existed, but only supplemented and transformed them. All new forms are forced to adapt to existing conditions and are not 'interested' in their destruction. Besides, the biosphere is characterized by extremely various conditions; in some cases 'old-fashioned' cycles still were the most effective. Thus, biogeochemical evolution is not a process of the change of old cycles by new ones; it is a process of cycles 'layering', development of complex dynamical networks. Such a multilevel, diverse character of cycling provides unique vitality to the biosphere, which successfully develops in spite of various natural cataclysms, which took place during the existence of life.

The stability of biosphere is explained substantially by a complex dialectic interaction between its living and nonliving components. Biota is under strong influence of abiotic environment and should adapt to it, but at the same time actively transforms it, making life more convenient. Such important environment parameters, as chemical composition of air, water, and lithosphere, climate, character of solar radiation on the surface, etc., are under essential control of the biosphere.

Although such external factors as volcanism, ocean regressions, transgressions, etc., cannot be influenced by biota, they also contribute to the life development. Abrupt or continuous changes in external conditions play the role of oscillations in the process of 'sifting' of living forms. The most impressive advances of the organic world took place after essential changes in the environment; only by radical disturbance of a system it can pass from a local optimum to other, more optimal steady state.

Forming of Biogeochemical Cycling

The last micropaleontological research has shown that the life and, correspondingly, biogeochemical cycling took place from very early stages of the Earth's history. They originated during its first billion years; the age of first primitive prokaryotic organisms is about 4 billion years. Probably, first cycles were based mostly on chemotrophic organisms, which used for biomass synthesis chemical energy of inorganic or simple organic substances (by way of their degradation with the use of oxygen or sulfur). Ecosystems of the black geysers can give an estimate of the first biological cycles. First chemotrophs lived in the age of forming the Earth crust and essentially influenced its composition. In the initial biosphere, as well as in the present one, the processes of decomposition of silicates, sedimentation of silicon, iron, phosphorus, manganese, cycling of sulfur, etc., took place.

The first cycles operated in condition of high temperature and dense hydrogen and helium atmosphere. The last elements permanently dissipated in the interplanetary space, and about 2.7 billion years ago the initial atmosphere disappeared. It was the first ecological catastrophe, which caused essential degradation and further reformation of biogeochemical cycling.

Dynamics of Oxygen and Carbonic Gas Concentration in the Atmosphere

During next 400 000 years, forming and intensive development of phototrophs took place. Correspondingly, the atmosphere became more and more oxygenic. The accumulation of oxygen was promoted by methanogenic bacteria, which consolidated free hydrogen. The main pathway of hydrogen (originated from water dissolution) led to its dissipation in the outer space through methane, which migrated into upper atmosphere layers and was destructed under the influence of ultraviolet radiation. The development of oxygenic atmosphere caused the next ecological catastrophe (sometimes called 'oxygen revolution'), because free oxygen was toxic for organisms of those ages. Some of them evolved to aerobic forms, others remained in anaerobic conditions, but the majority died out.

From 2.3 to 0.3 billions years ago the oxygen concentration in the atmosphere permanently increased. As a result, the character of photosynthesis was changed. According to Gaffron, the first phototrophs used energy of ultraviolet radiation. Since free atmospheric oxygen produced the ozone layer protecting the surface from ultraviolet radiation, phototrophs evolved to use the visible light, forming the chlorophyll photosynthesis and recent type of biogeochemical cycling.

The formation of the ozone screen about 700 million years ago (with oxygen concentration corresponding to Pasteur point) gave a possibility of starting the terrestrial life and terrestrial biogeochemical cycling.

Decreasing of carbonic gas concentration in the atmosphere accompanied the increasing of oxygen one. The control of carbonic gas is one of the most evident functions of the biosphere; without this control the gas concentration can reach 98%, as it takes place, for example, in the abiotic atmosphere of Venus. Probably, the antagonism between carbonic gas and oxygen in the atmosphere (determined by biotic factors) produces long-term oscillation of their concentrations. The stage of relative abundance of carbonic gas in carbon (generated by high geological activity and led to further conservation of essential amount of carbon in the form of coal in the Earth crust) was changed by the stage of high concentration of oxygen in Mesozoic era (stimulated development of huge forms of animals).

The abundance of carbon gas is quite important for development of vegetation. It is the main biomass-generating substance, a peculiar ecological 'currency', for which, similarly to the real one, little inflation should take place. Volcanic activity permanently adds the gas to the cycling and creates such inflation. One of the pessimistic ecological forecasts is connected with gradual decrease of geological activity. When the carbonic gas entry to the biosphere finishes, 'stagnation' of the biogeochemical cycling is inevitable.

At the same time, ecological problems of the recent moment are connected mostly with increase of carbonic gas concentration in the atmosphere, accompanied by the notorious greenhouse effect. The problem is a result of large-scale activity of human society, which is deeply involved in the current biogeochemical cycling.

Mankind as a Biogeochemical Factor

The origin of the sentient life about 500 000 years ago and the civilization about 10 000 years ago determined a principally new stage of the global cycles' development. The fact that people are very effective consumers with high-level abilities on information processing is not so significant in this context. Much more important factors are essential for changing of matter pathways as a result of human impact. The invention of fire led not only to efficiency of consuming biomass use (because of its use in cooking) and areal broadening (because of heating), but for people it created the possibility to play the role of reducers, promptly transforming useless biomass to mineral substances. The slash–burn clearing produced a new kind of intensive biological cycling.

Agricultural and industrial activity of man is accompanied by more and more large-scale involvement of new substances into the cycling. The humanity has intensified water cycling (by creation of water reservoirs, artesian wells, etc.), and added to the turnover a big amount of different elements, including toxic ones. This impact is not always negative for the biosphere (if it is possible to talk about the use of interference into the nature). Enrichment of elements in biogeochemical cycling can hasten it. For example, in the course of the so-called eutrophication, a natural process of lakes transformation into bogs and then to meadows can run much quicker. But such process of cycling acceleration is considered, from human point of view, as a negative impact.

Involvement of a huge amount of carbon as a result of use of coal and oil deposits has led to an increase of carbon gas in the atmosphere and, correspondingly, to the greenhouse effect. At the same time, plant nutrition improves; it can partially compensate forest destruction. Besides, the greenhouse effect can compensate the already-mentioned global tendencies to the climate getting colder.

Human impact on forming biological cycles became determinative during the last centuries. It has a geological and planetary character. The new stage of biosphere development was called as the noosphere by V. I. Vernadsky in 1944, who used the term, proposed by E. Le Roy with a 1927, with a somewhat different meaning.

Humanity should coordinate its activity with global biogeochemical cycling. According to D. H. Meadows, three main principles of sustainable development are closely connected with integrity of global cycling: the rate of use of renewable resources should correspond to the rate of their regeneration; rate of use of nonrenewable resources should correspond to the rate of their change by renewable resources; the rate of production of pollutants should correspond to the rate of their decomposition in the environment.

Both in agriculture and industry, humanity was forced to embed elements of biogeochemical cycles as a part of general technologies. The recent ecological situation and tendencies of its change demand another approach: to embed technologies in the global biogeochemical cycling. Conscious control of the cycling, creation of real noosphere (sphere of intellect), is the only one way for humanity to survive.

General Tendencies of Biogeochemical Cycles' Developments

The question about direction, driving forces, and 'purport' of biosphere evolution is quite complex and vexed. An answer can be based on different philosophical concepts. Let us consider the energetic approach, which was actively developed by A. Lotka, V. I. Vernadsky, and H. Odum. From this point of view the main direction of biogeochemical cycles' development is permanent increase of the energy flow through biosphere.

Energetic efficiency is a key parameter of species in the process of competitive selection. Ability of better assimilation of solar energy or energy collected by other organisms is a prior evolutional advantage. More and more effective populations are involved into global cycling, increasing its intensity. As a result, the biosphere power (consumed energy per unit time) permanently grows. V. I. Vernadsky formulated three main tendencies of biosphere evolution: biogenic migration tends to maximum manifestation; biological evolution causes intensification of biogenic migration; covering the Earth by life is maximally possible for current abilities of the biosphere.

The intensification of cycling is mainly a result of competition of producers, which are forced to maximize production for keeping place in ecosystem. Another extremely important factor is activity of consumers. They withdraw producers' biomass and additionally intensify cycling. Probably, the global role of consumers in the biosphere consists exactly in the spin-up of ecological cycles.

Successions

The process of increasing biosphere power includes not only improvement of existing cycles, but very often it is replacement of less effective cycles by more effective ones (and, not so often, forming new type of cycling). Since the type of cycling geographically corresponds to biogeocoenoses, one can talk about perfection of cycling in the course of biogeocoenoses change or, in other words, during successions. The latter can take place after either change of external conditions or disturbance of ecosystem steady state (climax) as a result of external impact (or, rarely, of ecosystem elements evolution).

It is important to take into consideration that change in biogeocoenoses is a result of 'struggle' (or competition) between biological cycles of different types. Such a struggle geographically takes place on territories between different types of biogeocoenoses (e.g., between forest and grassland). The corresponding ecosystem including two or more types of cycles is called ecotone or amphicoenose. By the definition of A. L. Belgard, amphicoenose is a system of several antagonistic biological cycles. Intermediate conditions do not allow one cycle to displace another; it is a case of 'glitching' succession. On the other hand, amphicoenoses ecosystems, including two struggling cycles, are natural intermediate stages of any succession. The amphicoenotic character of ecosystems dynamics is one of the main sources of their biodiversity.

A model of amphicoenose can be designed as a combination of models [1] or [2] of coenome. The scheme of amphicoenose structure for the two-dimensional case is represented in **Figure 4**. The two coenomes are unified by the soil block. The model correctly describes the main dynamic properties of amphicoenoses: competitive exclusion of plant associations, crossing of reducer populations between associations, etc. The result of computer simulation of amphicoenose dynamics (which was considered as a struggle of biological cycles of different types) is shown in **Figure 5**. The use of the cyclic models determines much more complicated dynamics than using the standard Volterra-type models of successions.

Summary

Global biogeochemical cycling is a superposition of big geological cycles and biological nutrient ones. Probably, this superposition started its development from the very early geological ages. Biogeochemical cycling can be interpreted as a stage of the general process of matter self-organization, as a manifestation of principle of maximization of energy flow through the open system. It is realized in the course of competition between biological

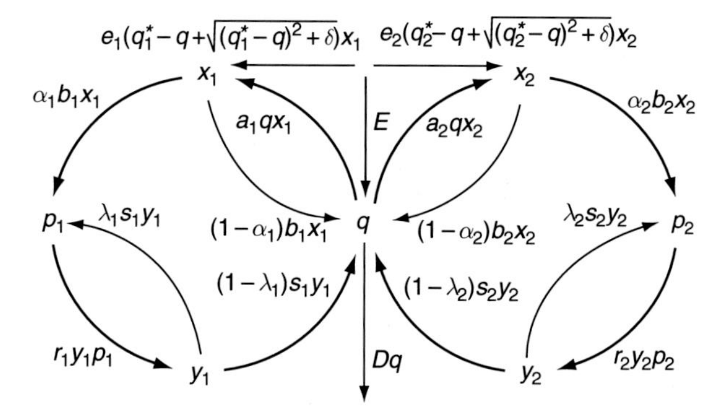

Figure 4 Matter flows in an amphicoenose.

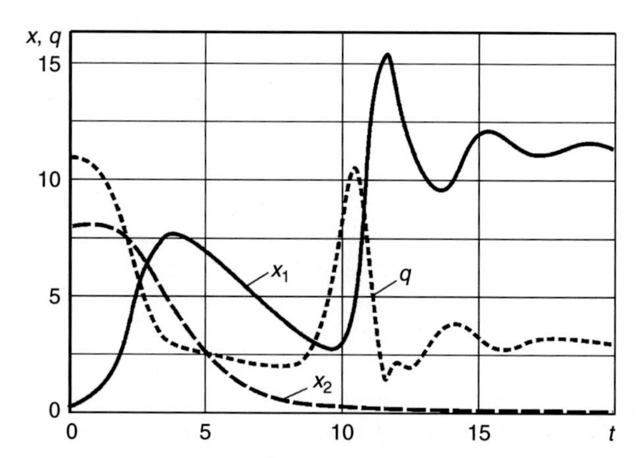

Figure 5 Results of simulation of the amphicoenose dynamics; x_1, x_2 – contents of the element in the producer blocks, q – the same for the soil block.

cycles and their components. Improvement and intensification of cycling occur in the form of successions, that is, consecutive structural changes of ecosystems.

Circulation of different chemical elements can be different, but there are a number of standard pathways. Nontrivial interaction of biotic and abiotic components creates driving force of biosphere's progress and a guarantee of its stability. During the last centuries, the mankind became a new important participant of the process. Organic inclusion of civilization into existing global cycles is the prerequisite of survival of the humanity.

See also: Biogeocoenosis as an Elementary Unit of Biogeochemical Work in the Biosphere; Biosphere. Vernadsky's Concept; Energy Flows in the Biosphere; Structure and History of Life.

Further Reading

Ågren GI and Bosatta E (1996) *Theoretical Ecosystem Ecology: Understanding Element Cycles*. Cambridge: Cambridge University Press.
Brimblecombe P and Lein AY (eds.) (1989) *Evolution of the Global Biogeochemical Sulphur Cycle*. London: Wiley.
Chapin FC, III, Matson PA, and Mooney HA (2002) *Principles of Terrestrial Ecosystem Ecology*. New York: Springer Science.
Chernyshenko SV (1996) Dynamic processes in biogeocoenoses: Mathematical modelling on the base of M. Eigen's hypercycle. *Journal of Ecology of Industrial Regions* 2: 77–90.
DeAngelis DL (1992) *Dynamics of Nutrient Cycling and Food Webs*. New York: Chapman and Hall.
Field CB and Raupach M (eds.) (2004) *The Global Carbon Cycle: Integrating Humans, Climate, and the Natural World*. New York: Island Press.
Gorham E (1991) Biogeochemistry: Its origins and development. *Biogeochemistry* 13: 199–239.
Jones EW (1968) Fundamentals of forest biogeocoenology by V. Sukachev and N. Dylis. *Nature* 220: 199–203.
Jørgensen SE (1988) Use of models as experimental tool to show that structural changes are accompanied by increased exergy. *Ecological Modelling* 41: 117–128.
Melillo JM, Field CB, and Moldan B (eds.) (2003) *Interactions of the Major Biogeochemical Cycles: Global Change and Human Impacts*. New York: Island Press.
Odum HT (1994) *Ecological and General Systems and Introduction to Systems Ecology*. Niwot: University of Colorado Press.
Pickett STA (1976) Succession: An evolutionary interpretation. *American Naturalist* 110: 107–119.
Pokarzhevskii AD and Van Straalen NM (1996) A multi-element view on heavy metal biomigration. *Applied Soil Ecology* 3: 95–98.
Reeburgh WS (1997) Figures summarizing the global cycles of biogeochemically important elements. *Bulletin of the Ecological Society of America* 78: 260–267.
Rodin LE and Basilevich NI (1967) *Production and Mineral Cycling in Terrestrial Ecosystems*. Edinburgh: Oliver and Boyd.
Schlesinger WH (1997) Biogeochemistry: An Analysis of Global Change. San Diego, CA: Academic Press.
Svirezhev Yu M (2000) Thermodynamics and ecology. *Ecological Modelling* 132: 11–22.
Tiessen H (ed.) (1997) *Phosphorus in the Global Environment: Transfer, Cycles and Management*. New York: Wiley.
Walter H and Box E (1976) Global classification of natural terrestrial ecosystem. *Vegetatic* 32: 231–238.
Wiens JA, Stenseth NC, Van Horne B, and Ims RA (1993) Ecological mechanisms and landscape ecology. *Oikos* 66: 369–380.

Microbial Cycles

G A Zavarzin, Russian Academy of Sciences, Moscow, Russia

Introduction

Microorganisms, particularly bacteria (=prokaryotes), represent the first sustainable system in the biosphere into which all other living beings are superimposed and included. Sustainability of the system depends on the catalytic role of bacteria in the cycles of biogenic elements and their mediating role in the transformation of other elements. Development of cooperative microbial community led to the biogeochemical succession, the most prominent result being oxygenation of the atmosphere around 2.4 billion years ago with interconnected changes for chemical compounds. The role of bacteria in the biosphere depends on their functional diversity and formation of cooperative trophic systems, scaling up from the local ecosystems to the biosphere as a whole. The limits of life are delineated by the topic adaptability of bacteria, while all other living beings remain within these frames. The number of bacteria exceeds 10^{28} in the active layers with rapid turnover and might be 10^{30} in total. It makes them by far the most important group in the conceptual structure of the sustainable biosphere. The trophic structure of the microbial system makes the framework of the biosphere. The interconnection of biogeochemical cycles makes the functional role of microorganisms in the biosphere most fundamental.

Biogeochemical cycles represent the main system by which the energy of the Sun is transformed into energy of the chemical compounds by living beings and products of their activity. The cyclic arrangement is the main principle of sustainability in the Earth system. It means that the compound involved in the process after sequential transformations is regenerated as its end product. Cycles are regarded as the cycles of the elements. Stepwise reactions of the cycles are catalyzed by specific groups of microorganisms. The system of higher organisms is superimposed into the initial cooperative system constructed by bacteria.

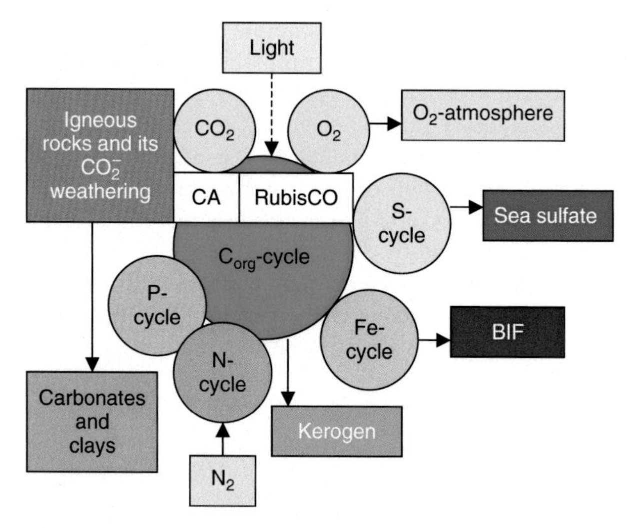

Figure 1 Interlink between the cycles of the main biogenic elements. Cycle of C_{org} makes the main driving force of machinery coupled to the cycles of biomass constituents N_{org} and P_{org} and catabolic cycles of oxygen, sulfur, and iron. Cycles of these elements are coupled to reservoirs of inorganic matter in the geosphere. For each time period approximate material balance should be sustained. Misbalance leads to the biogeochemical succession on the large time scale. Modified from Zavarzin GA (2004) *Lekcii po Prirodovedcheskoi Mikrobiologii* (*Lectures in Environmental Microbiology*). Moscow: Nauka.

C_{org}-Cycle

The driving force of the system of interlinked cycles is the cycle of organic carbon (C_{org}). The cycle involves two steps: production and destruction. During production CO_2 is assimilated in the biomass; during destruction dead biomass is decomposed into CO_2. Composition of biomass includes in addition to C_{org}, as the main components N_{org} and P_{org} in approximate molar ratio 106:16:1. This ratio calculated for marine phytoplankton is quoted as 'Redfield ratio'. H and O are included in the biomass in water in the ratio 2:1, making the reductive level of C_{org} close to $[CH_2O]$. Strong deviations from Redfield ratio are known for terrestrial biomass with organic supportive structures as in trees with C_{org}:N_{org} about 500; minor deviations are caused by storage products. There are other elements included in the biomass such as S_{org}, and a number of essential 'mineral' elements beginning with K, Fe, Mg, Ca, and microelements. Composition of the living biomass might be considered as invariable with minor deviations (**Figure 1**).

Transformation of CO_2 into C_{org} is performed by autotrophic organisms by the metabolic pathways where the Calvin cycle is quantitatively dominating; other autotrophic reactions seem not important quantitatively on the global scale. The key enzyme is ribulose-1,5-bisphosphate carboxylase/oxygenase (RubisCO) which carboxylates phosphopentose regenerated in the cyclic metabolic pathway (Calvin cycle). Due to the discrimination of ^{13}C during autotrophic assimilation isotopically lighter carbon with $\delta\,^{13}C_{org} \sim -25‰$ is produced, which is considered as the isotopic signature of a biotic source. Careful interpretation of isotope fractionation data is strongly recommended since they depend both on biotic pathways and inorganic diffusion. Assimilation depends on the source of energy, light of the Sun for photosynthesis, or oxido-reductive reaction of inorganic compounds for chemosynthesis (chemolithotrophy is the later synonym) in subterranean systems. In photosynthesis the overall reaction $CO_2 + H_2O \rightarrow [CH_2O] + O_2$ takes place. It makes a coupled cycle with an equimolar ratio of CO_2/O_2. The quantity of O_2 liberated is equivalent to the total C_{org} extracted from the system into the biomass and the reduced products of its decomposition. The link between reservoirs of inorganic and organic carbon is performed by enzyme carboanhydrase (CA) in the reaction $CO_2 + H_2O + CA \leftrightarrow H_2O\text{–}CA\text{–}CO_2 \leftrightarrow HCO_3^- + H^+ + CA$. In cyanobacteria, CA and RubisCO are integrated into the structural unit carboxysome. In eukaryotes, intracellular localization of enzymes is different. CA is responsible for CO_2 evolution during respiration. The production is measured either by O_2 production in water systems or by ^{14}C-bicarbonate assimilation. The cycle of C_{org} is linked to the reservoir of C_{inorg} with a strong influence of the calcium cycle.

Photosynthesis is the dominating process in production. Primary production is proportional to the illuminated surface or more precisely to the density of chlorophyll, with an approximate ratio of annual assimilation 145 kg C_{org} per kg of chlorophyll in terrestrial boreal ecosystems. Formation of C_{org}-pool occurs through several steps. The first one is gross primary production (GPP), counterbalanced by photorespiration in which approximately half of carbon is lost. Netto primary production (NPP) is calculated on the annual basis of the growing season for a plant. Evidently for algae with a short life cycle, the concept is different. C_{org} balance in the ecosystem is different and includes losses by the respiration of decomposers; it is referred to as Netto ecosystem production (NEP). Optimal conditions for photosynthesis and destruction are different: destruction has higher optimal temperature than photosynthesis, and different dependence on water, being suppressed by the excess of water, causing anaerobiosis. This causes zonal variance for biomes. The accumulation of C_{org} on the decadal scale is designated as Netto biome production (NBP) for which accumulation of nondecomposed C_{org} as humic substances and peat is the main parameter. For marine ecosystems, 'dissolved C_{org}' substitutes soil humus. The recalcitrant C_{org} of humic substances has a residence time of about millennia. It is converted into carbon of sedimentary rocks known as 'kerogen', which makes the main reservoir of reduced carbon on the planetary scale with a residence time of more than millions of years, depending on geological recycle. The reservoir of kerogen is sufficient to balance oxygen in marine sulfates and iron oxides deposits. Only 5% of the total oxygen produced remains in the transitionary reservoir of the atmosphere.

From a brief description of C_{org}-cycle, it is evident that the residence time in reservoirs is to be included into consideration. Seasonal variations in CO_2 fluxes are illustrated by the annual oscillations of atmospheric CO_2 in the continental Northern Hemisphere with an amplitude of about 20 ppm (parts per million) in Hawai and increasing in higher latitudes. In the oceanic Southern Hemisphere, oscillations are smoothened by the carbonate/bicarbonate system of the ocean.

Destructive pathways begin by decomposition of the dead biomass. Transition from living to dead biomass is accompanied by autolysis, which liberates part of the organic matter. For cyanobacteria and algae, lysis induced by viruses or phagi is quite important. Density of population is important, and below 10^5 cells per milliliter, phagolysis is ineffective. Lysis produces two components: dissolved organic compounds (DOCs) and particulated organic compounds (POCs), which consist mainly of structural components of the cell. Osmotrophic microorganisms can immediately use DOC; the threshold depends on the dilution with 1–10 mg l^{-1} still utilizable depending on the inflow. POC is to be hydrolyzed by hydrolytic exoenzymes before osmotrophic organisms can utilize it – bacteria in the sea or fungi in terrestrial ecosystems. Destructive pathways are formed by organotrophic organisms, which traditionally are named heterotrophs. This term is imprecise since it refers to the assimilative pathway leading at the end to the secondary production. There are three main metabolic pathways for C_{org}: proteolytic, saccharolytic, and lypolytic, according to the composition of biomass. The Winogradsky rule (1896) says that each natural compound has its specific microbial decomposer. The number of species of prokaryotes exceeds 5000 of cultivated and 2×10^4 clones of noncultivated. That gives sufficient functional diversity to perform biogeochemical essential reactions. As a result, specific trophic groups of organisms characterized by the utilizable substrate (e.g., cellulolytic, or lypolytic, or lignin-decomposing fungi), appear. The set of these organisms makes the functional biodiversity in the trophic system, which should make a complete community for each habitat. In the terrestrial environment, mycelial fungi are most important. Wood consists of 20–30% lignin, which is decomposed by fungi and that gives the lower limit of their involvement in terrestrial C_{org}-cycle as 1/3–1/4 of CO_2 producers. Adding cellulose decomposition would at least double their contribution.

In the presence of O_2, aerobic organisms regenerate approximately one-third of C_{org} in secondary production with CO_2 as the product of respiration. Consumption of O_2 in the dark is by the usual estimation of respiration by the so-called biochemical oxygen demand (BOD) test.

In an anoxic environment, a cascade of reactions begins with fermentation, which is also the main pathway for many hydrolytic decomposers. As products, a mixture of organic acids and H_2 appears, and this is the reason why this stage is designated as acidogenic or hydrogen producing. Organic acids as nonfermentable compounds can be utilized only with an external oxidant, such as nitrate or ferric iron, sulfate, or CO_2. A cascade of anaerobic reactions makes a community function as an entity with an integrated trophic network.

Without an external oxidant, anaerobic decomposition is completed by methanogenesis, a process which dominates in terrestrial mires and lake mud. In the sea, it takes place in deep layers of sediments, when sulfate is exhausted from interstitial water. Methanogens make a group of Euryarchaeota usually named as *Methano-*.... There are three pathways for methanogenesis: either hydrogenotrophic with $H_2 + CO_2$, or acetoclastic, or methylotrophic for C-1 compounds. Acetoclastic pathway dominates in C_{org}-abundant environment, for instance, in methane tanks. Methylotrophic methanogens develop noncompetitive pathways in saline environment, while they do not compete with sulfate reducers. Hydrogenotrophic methanogens can use endogenous H_2 formed by reaction of water with superheated rocks and belong to hyperthermophiles, for example, *Methanobacterium fervidus*, which develops at temperatures over 100 °C. More important is the role of hydrogenotrophic methanogens in community, where they act as H_2-sink. They establish H_2-concentration below 10^5 ppm and this allows us to oxidize acetate and other nonfermentable substrates in cooperative action with H_2-producing syntrophic organisms. Biogenic methane is identified by its isotopically light composition. Most of methane is ^{13}C-depleted.

Methane either remains in the sediments or escapes into the oxic zone where it is oxidized with O_2 by a specific group of methanotrophs. Under geologically favorable conditions, methane is stored in sedimentary rocks. Another possibility is the formation of crystallohydrates, which at appropriate hydrostatic pressure and low temperature make an ice-like cover for deep methane. At present, about 500 Mt yr^{-1} of methane comes into the atmosphere, where it is oxidized photochemically. Many times more than this quantity is oxidized by methanotrophs, which form an oxidative filter on the path of CH_4 to the atmosphere. The genera of methanotrophs are designated as *Methylo-*.... Oxidation of CH_4 includes its enzymatic transformation in C-1 compounds in the cell by a special metabolic pathway and thus methanotrophs represent a specialized group of the Proteobacteria. In the ocean, methane is oxidized by anaerobic consortia of methanogens with sulfate-reducing or denitrifying bacteria. The microbial cycle of CH_4 is most important for the biosphere.

Involvement of oxidized N, Fe, and S compounds as oxidants conjugates C_{org}-cycle with cycles of other elements. Transition from oxic to anoxic zone favors retainment of nondecomposed organic matter and leads to formation of oil and gas deposits in aquatic environments and coal in terrestrial ecosystems.

Particulated components including bodies of bacteria are consumed by phagotrophic Protists or/and by zootrophic multicellular animals. The trophic chains of animals are arranged into a trophic pyramid with a number of levels, including herbivorous and carnivorous. The size of the prey determines the nutritional pyramid. Animals that use filtration for nutrition are important in aquatic environment, keeping the density of microorganisms on the threshold level of about 10^5 cells ml^{-1}. The total amount of bacteria in the active zone of the ocean and soil is at least of the order 10^{28}, with the biomass of each cell about 10^{-12} g.

Microbial Nitrogen Cycle

C_{org}-cycle is coupled with N_{org}-cycle. Nitrogen cycle begins by nitrogen fixation. Nitrogenase enzyme is present in prokaryotes exclusively and distributed among different groups. N_2-fixation is an energetically expensive process. It occurs only on severe limitation on the bound nitrogen. It is facilitated in a reductive environment. The main groups of nitrogen-fixing bacteria are cyanobacteria and anaerobic bacteria. Aerobic nitrogen-fixing bacteria are either plant symbionts or organisms within the community supplied in excess by the nitrogen-free organic matter. C_{org}:N_{org} ratio over 20 stimulates N_2-fixation. Fixed N_2 is included into the biomass. Then the regenerative cycle begins. In the sea, its zone is just below the photic zone. Nitrogen is liberated in the proteolytic pathway with ammonification as summation of the process. NH_4^+ is either re-assimilated, or is metabolized by the two-step conversion into nitrate by chemosynthetic nitrifiers in the presence of O_2. Nitrate is assimilated by phytoplankton or plants. Part of it escapes from the productive zone. In the anoxic zone, in the presence of available organic matter, various denitrifiers use nitrate and nitrite as oxidants and reduce it to N_2 closing the cycle. As a variant, nitrate reduction to ammonia occurs but this is a less important pathway. Denitrification is the closing step in the nitrogen cycle. Nitrate makes a reservoir of bound nitrogen in the deep cold ocean of ~20Eg. It is noticeable how important denitrification is in the marine environment for decomposition of organic compounds including hydrocarbons. Limitation of the availability of bound nitrogen is the major problem for plant productivity. In the sea, seasonal exhaust of nitrates definitely determines algal development as it was demonstrated for the Northern Sea. For the Pacific, a peculiar change in plankton composition indicates an N- and P-cycle interrelation: in nitrate-limited conditions cyanobacteria dominate when there is enough phosphate, whereas in phosphate-limited conditions algae are the main group in phytoplankton.

Phosphorus

Phosphorus comes into the ecosystem due to the weathering of rocks. The productivity of the lakes is proportional to the phosphate availability. For instance, soda lakes are often eutrophic in spite of extreme environment and limitation by other elements. Phosphorus is mobilized from its minerals by many acid-producing microorganisms, which make dissolution zones on the plates with an enamel of phosphate containing minerals. Phosphorus of the sea has terrestrial origin. Assimilated phosphate is included into the nucleic acids and phospholipids of the biomass. Regenerative cycle of phosphorus includes liberation of phosphate by the action of phosphatases. Phosphorus escapes from the cycle by binding into insoluble compounds of phosphorites on reaction with Ca and F. It should be noted that deposits of micritic phosphorites were formed by cyano-bacterial mat; phosphatisized microfossils of cyanobacteria are clearly visible in the scanning electron microscope. Cyanobacteria store phosphate as intracellular polyphosphate, which is the transitional source for rapid phosphatization. The iron pump demonstrates liberation of phosphate in anoxic environment: ferric iron binds phosphate in an insoluble compound but reduction to ferrous state liberates phosphates. On a large scale, phosphorus is the limiting element for primary production, which depends on its availability.

Sulfur Cycle

Sulfur cycle is the most important cycle conjugated to C_{org}. Assimilation of sulfate into S_{org} is quantitatively of minor importance in spite of the fact that it is the main source of dimethylsulfide – a volatile compound contributing to the source of S in the atmosphere. Its photochemical oxidation leads to the formation of aerosol in the stratosphere and is most important for the climate. In the destructive pathway coupled to the C_{org}-cycle, sulfate is reduced to H_2S by sulfate-reducing bacteria (SRBs), which by now are taxonomically numerous but functionally uniform. There are the following trophic groups: H_2-utilizers, SRBs with incomplete oxidation of organic acids (*Desulfovibrio*-type) producing acetate, and complete oxidizers (*Desulfobacter*-type), which use various unfermentable products of fermentations. Hydrogenotrophic SRBs are H_2-scavengers, which allow them to serve as intermediary oxidants to H_2-producing syntrophic bacteria and thus oxidize a variety of organic compounds. Most interesting is their interaction with methanogens in anaerobic methane oxidation in marine sediments. Methane is oxidized by reversed methanogenesis with the formation of isotopically light carbonates and evolution of H_2S by SRBs. H_2S, if not bound by iron into pyrite, escapes to the surface of the mud where it is oxidized by pelophilic sulfur bacteria, which can either use intracellular S_0 for oxidation into sulfate, if O_2 or NO_3^- is available, or use it as an oxidant in sulfur reduction. Magnificent benthic mats of trichomic sulfur bacteria are found on the shelf close to Chile and West Africa. Here, the so-called thiobios is formed by sulfur bacteria of *Thioploca*-type. Large filamentous bacteria cross the ox–red boundary and receive H_2S from the anaerobic layer and the current near the surface of the mud brings oxidant as nitrate or O_2, which is used for chemosynthesis. H_2S escaping in the water mass in the bodies of water with limited circulation makes a chemocline with the reductive zone below the oxic zone; Black Sea is a conventional example. The same occurs in stratified lakes. It is supposed that Mid-Proterozoic stratified ocean had the same structure. If H_2S zone comes to the photic zone, anoxic sulfur phototrophs develop. There are a variety of anoxygenic phototrophic bacteria, which belong to phylogenetically distant phyla. Purple layers of phototrophs make a remarkable landscape when they come up to the surface on the beach or in the soda lakes. In oxygenated photic zone, H_2S is oxidized into sulfate by various thionic bacteria. It is noteworthy that the appearance of sulfates in the palaeocean correlates with the oxygenation of the atmosphere around 2.4 Ga, and before 1 Ga its composition became close to the present one. It might be speculated that sulfates of the sea are biogenic in their origin. What was the initial source of mineral S? If the source was massive sulfides, then for their mobilization oxidative step was needed by aerobic acidothiobacteria used now in biohydrometallurgy in the general reaction, $FeS + O_2 \rightarrow Fe^{3+} + SO_4^{2-}$. The reaction strongly depends on the availability of O_2. Oxidation of sulfides leads to the formation of extreme acid conditions. It is most spectacular on volcanic thermal fields with sulfur exhalations, so-called solfataras. When A. Humboldt visited Vesuvius before its eruption, he noted that hot vapors were neutral in spite of possible SO_2 production in the heat, while cold walls of the crater were strongly acidic for Lakmus paper. Now it is known that oxidation of sulfur occurs mainly by acidophilic thionic bacteria and only in outlets of fumaroles, extremely thermophilic archaea are active. In the deeper parts of thermal fields, S_0 is used as an oxidant by anaerobic archaea with H_2S production. Short cycle $S_0 \leftrightarrow H_2S$ works also in microbial mats where white sulfur is deposited from H_2S by microaerobic sulfur bacteria and reduced when oxidant is not available. A large variety of microorganisms are involved in the cycle. Sulfur cycle closes destruction of organic matter in anoxic zone with sulfate regeneration either by anaerobic phototrophs or by aerobic sulfur oxidizers. Its function strongly depends on the transport processes across chemocline. The outcome from the cycle depends on availability of iron, which forms insoluble sulfides first as hydrotroillit and then pyrite.

Iron Cycle

The production of H_2S is environmentally incompatible with dissolved iron because of the formation of sulfides. Thus in terrestrial wetlands, where sulfate is limiting, iron cycle develops. Bacterial Fe-cycle takes place now in swamps. It includes oxidation of Fe^{2+}-bicarbonate under O_2-limited conditions with formation of $Fe(OH)_3$ ferrihydrite. Energy of oxidation might be used for chemosynthesis by *Gallionella* with precipitation of Fe(III) on the slimy stalks. Precipitation of Fe(III) on slimy structures is well known for the number of so-called 'iron bacteria', among which *Leptothrix ochracea* is best known for large deposits of ochre in slow-flowing water. Historically, that was the first example of geological activity of microorganisms described by Ehrenberg. Ochre-forming deposits were used as a 'swamp-ore' in the beginning of the Iron Age. However, two processes should be distinguished: chemosynthetic oxidation of iron and precipitation of iron hydroxides on mucous polysaccharides. Both processes are geologically significant. Ferrihydrite is readily reduced by iron-reducing bacteria, which use H_2, acetate, and a number of other C_{org}-

compounds as electron donors. There are two possible end products: siderite $FeCO_3$ is formed in excess of organic matter and magnetite Fe_3O_4 under more restricted conditions. Iron-reducing bacteria substitute nitrate reducers in moderately reductive habitats. There are also thermophilic iron reducers. For formation of ferrihydrite in anoxic environment, there are two possible pathways, both phototrophic: one possibility is oxidation of Fe^{2+} by cyanobacteria but it is unclear if it is direct or indirect and caused by O_2 production; and the other is definitely direct and is performed by nonsulfur purple bacteria such as *Rhodomicrobium*. Oxidation of Fe^{2+} by anoxygenic phototrophic nonsulfur bacteria was described only recently. Product of oxidation in the light is ferrihydrite. This process closes the iron cycle in anoxic environment. Large deposits of layered silicified iron oxides composed of alternating layers of hematite and magnetite known as banded-iron formations (BIFs) were formed during the Early Proterozoic 1.8 billion years ago. Their origin remains unclarified. Fe is of hydrothermal origin. Total amount of iron oxides contains about 40% of O_2 evolved corresponding to C_{org} of kerogen. Iron migrated in the ancient ocean most probably as bicarbonate. Period of BIF is clearly incompatible with sulfate reduction.

Oxidation of sulfides, first of all pyrite, involves both cycles of iron and sulfur. Oxidation involves two functions. At low pH Fe^{2+} is stable in the air. Oxidation of sulfide produces sulfuric acid with a drop to $pH<2$. Some pyrite-oxidizing chemosynthetic bacteria such as *Acidothiobacillus ferrooxidans* use energy of both sulfur and iron oxidation. However in nature, these two functions are often divided between iron-oxidizing *Leptospirillum* and sulfur-oxidizing *Acidothiobacteria* working in concert. There are also other examples of these bacteria, especially thermophilic, which are most important in bacterial hydrometallurgy because they are able to dissolve various sulfides, and copper, gold, and other metals that also come into solution.

In addition to the cycles of major elements used by chemosynthetic bacteria, it is worth mentioning cycles of arsenic, manganese, and selenium, where both oxidative and reductive pathways operate. The general rule is, chemosynthetic microbes use oxido-reductive reaction and develop in the thermodynamic field of stability of the product of this reaction. Energy generated in the reaction must be sufficient to support the formation of ATP.

Biologically Mediated Reactions

The so-called biologically mediated reactions are also very significant in the involvement of microorganisms in the biogeochemical cycles. In these reactions, microbes form an environment in which certain forms of minerals are stable according to the fields of thermodynamic stability. It is most important for rare elements, whose amount is insufficient to ensure the existence of certain specific groups. These trace elements act as indicators of the environment. Uranium is one of the examples. Another example is the deposition of metals in the zone of sulfate reduction. Such microbially mediated pathways form many sedimentary deposits. The so-called biogeochemical barriers represent the sites of drastic changes in the environment, which cause precipitation of minerals. Biogeochemical barriers are sustained by countercurrents of solutes and by the activity of microorganisms. There are different kinds of barriers: oxido-reductive, alkaline, sulfidic, etc. Chemocline in the lakes is an example. Change in the state of environment often causes transition from migrating chemical species to insoluble one.

The cycle of calcium belongs to biologically mediated reactions. It begins by CO_2 weathering of rocks. The essential point here is the concentration of carbon in the biomass and concentrated release of CO_2 during decomposition. Formation of active products of decomposition as acids or chelators is also important. Released Ca^{2+} comes into waterways as $Ca(HCO_3)_2$ and migrates to the ocean. Here it might be used by calcareous eukaryotes for the formation of skeleton and release of CO_2. The main part of $CaCO_3$ at present is biogenic by origin. Another possibility is the release of CO_2 in warm shallow water when its solubility decreases and the reaction, $Ca(HCO_3)_2 \rightarrow \downarrow CaCO_3 + \uparrow CO_2 + H_2O$ develops. The surface of precipitated $CaCO_3$ is covered by a microbial biofilm. It gives to the precipitate a laminated texture due to the slime produced. If the microbial biofilm is formed by cyanobacteria, then the additional sink of CO_2 assimilated in C_{org} might result in a drop of pH and precipitation of carbonate. The layered structures of precipitated carbonates are known as stromatolites. They are recorded for Proterozoic as the most important deposits. Their abundance indicates that they represent significant deposits of CO_2. Stromatolites correlate with deposits of dolomites, but sometimes pieces of black chert are included and in these cherts, silicified microfossils of cyanobacteria are observed. Preservation is excellent and one can identify taxa, with the aid of books on systematics, of extant cyanobacteria to approximately 2.4 billion years ago. The scale of stromatolite formation delineating ancient warm shallow water environment is of the order of millions of square kilometers. Height of such deposits is up to hundreds of meters. Mass development of stromatolites ended with the end of bacterial exclusive domination in the biosphere. Calcium cycle contributes to the neutrophilic conditions on the Earth.

Trophic Organization of Microbial Communities

The organization of the prokaryotic community is most clearly demonstrated by the cyano-bacterial mat. Sign (-) denotes here two components: cyanos as prime producers and bacteria as decomposers in regenerative cycle. In the mat, distinct layers are found: the upper illuminated level is occupied by cyanobacteria; below is the white layer of sulfur bacteria, followed by the purple layer of anoxygenic phototrophs, and then the black layer of sulfide-producing bacteria. Still below are the layers of dead bacteria. The whole system has dimensions of 2–4 mm. It is called in German 'Streiffarbsandwatt'. The architecture of cyano-bacterial mats is similar in hypersaline lagoons, soda lakes, thermal springs, etc. The structure of mat is maintained by exopolysaccharides produced by cyanobacteria, which are edificators (from 'edifice') for the community. The main factor is illumination and self-shadowing by the upper layers of cyanobacteria, which move to the optimal illumination. Minor differences in the composition of mats are caused, for instance, by the absence of purple bacteria in thermal habitats or strong development of planktonic forms in soda lakes. The 'Winogradsky column' illustrates stratified planktonic microbial community: cylinder with water from the site supplemented by mud with organic debris and gypsum at the bottom. Blooming microbes in the column produce alternating black, purple, and green layers. The column might sustain for years.

Trophic links in the microbial community are organized into the trophic network of a cascade of degradative reactions. The rule is that each step should be sufficient to support the species performing the transformation. Degradation begins with hydrolysis of biopolymers, the most resistant being structural components of the cell walls. Aerobic and anaerobic dissipotrophic bacteria in the cascade of reactions utilize low-molecular-weight compounds, dissipating from the sites of hydrolysis. The final result should be complete decomposition of organic matter, so-called 'mineralization'. In fact, decomposition is not complete and recalcitrant substances are formed in minor part, giving rise to humic substances, and dispersed organic matter. It should be noted that the physical environment strongly contributes to the trapping of undecomposed organic matter, preventing microbial activity.

This is a brief description of biogeochemical cycles catalyzed by bacteria. The main conclusion is that bacteria act as a cooperative community with the interlinked metabolic pathways of the main elements; only the cooperative community is autonomous due to the links between productive and regenerative cycles. Each step of catalysis is performed by a functional or trophic group of specialized bacteria. Links provide the trophic network. Cyclic pathways make such a community autonomous, depending mainly on the energy for photosynthesis. Cooperative community is an operational unit for the ecosystem at the landscape level. However, biogeochemical cycles are not entirely closed: there is formation of products, which escape recycle. The changes in the community composition known as succession are caused by the accumulation of products as well as exhaust of substrates. In the microbial community, it is the development from fast-growing copiotrophs to a climacteric community with well-balanced interactions. In fact, the microbial community exists all the time in a transitional state.

When we consider a larger temporal scale, the most important concept of biogeochemical succession arises. It may be illustrated by the composition of the atmosphere, which is formed by microbial activity, since the main components of the atmosphere are metabolized by bacteria: CO_2, O_2, CH_4, CO, H_2, N_2, NO_x, NH_3, and sulfur species. Due to the accumulation of the waste products – oxygen is the most evident example – biosphere becomes uncomfortable to the community, here the initial anoxic microbial community. Less evident but more important is accumulation of C_{org} in sedimentary rocks, which rolls cycles with the passage of time. As a result, the atmosphere moves from the neutral to the oxygenated state. It changed conditions on the Earth's surface. Biosphere overturned: anaerobic pockets remain under the shield of aerobic O_2 consumers. Much the same occurred with the ocean (or hydrosphere), which is in equilibrium with the atmosphere. Three main steps seem to be identified: the first before approximately 2.4 billion years with the domination of iron cycle; beginning of the biosphere with O_2 in the atmosphere and pronounced S-cycle; and present-day biosphere with O_2-atmosphere where biogenic O_2 substituted part of CO_2. Pathways in the community were reoriented to the new environment. Consider that any sustainable system should be able to support its own existence by effective feedbacks otherwise it is not sustainable. S. Winogradsky in 1896 suggested the qualitative concept of cycle of life as a 'huge organism' (or the goal-oriented system) with microorganisms acting as the main catalysts. Later V. Vernadsky in *The Biosphere* (1926) introduced the quantitative approach, considering biogeochemical cycles as the main mechanisms. The Geospheric–Biospheric Program and Global Change concept represent the contemporary approach to the problem. The expediency of the links in the biosphere leads to its interpretation as 'Gaia'.

However, biosphere was always within the geographic envelope of the Earth. This means that there was always a mosaic of landscapes arranged in climatic zones. Landscapes give the possibility of lateral interaction and formation of geochemical barriers. The mosaic of landscapes furnishes refugia, places for survival of particular communities. Landscapes on geological timescale are dependent on the tectonic. Weathering–sedimentation pathway leads to equilibrium if no metamorphism and geological cycle occurs.

Since bacteria catalyzed main cycles and established the primary biogeochemical system, they form the dynamic environment, into which Protists, Metaphyta, and Metazoa were incorporated in the course of evolution. The system was developed by the substitution of prime producers by algae, kelps, and plants. The terrestrial system changed with the appearance, about 300 Ma ago, of vascular plants, which changed the atmospheric hydrological cycle by the involvement of deeper layers of ground water and producing a new illuminated surface within the leaves for derivates of cyanobacteria converted into chloroplasts. Plant cover significantly changed the terrestrial environment. However, microbes remain as the main catalysts in the system of biogeochemical cycles.

See also: Nitrogen Cycle; Oxygen Cycle.

Further Reading

Brock TD and Madigan MT (1991) *Biology of Microorganisms*, 6th edn., 874pp. Englewood Cliffs: Prentice-Hall.
Lengeler JW, Drews G, and Schlegel HG (eds.) (1999) *Biology of the Prokaryotes*, 955pp. Stuttgart: Thieme.
Schlesinger WD (1997) *Biogeochemistry: An Analysis of Global Change*, 2nd edn., 588pp. San Diego: Academic Press.
Zavarzin GA (2004) *Lekcii po Prirodovedcheskoi Mikrobiologii (Lectures in Environmental Microbiology)*. Moscow: Nauka.

Nitrogen Cycle

P E Widdison and T P Burt, Durham University, Durham, UK

Introduction

The nitrogen cycle is arguably the second most important cycle, after the carbon cycle, to living organisms. Nitrogen is essential to plant growth, and therefore is a significant contributor to the human food chain, but its presence in the environment is strongly influenced by anthropogenic activities.

Here, we will describe the global nitrogen cycle; we then examine the long-term trends at national and global scales for both terrestrial and aquatic ecosystems; next we describe how nitrogen is transported at local and long-distance scales; finally we consider how public policy for environmental protection aims to mitigate against pollution effects.

Nitrogen was discovered in 1772 by Daniel Rutherford, who called the gas 'noxious air'. During the late eighteenth century other chemists, such as Scheele, Cavandish, Priestly, and Lavoisier were also studying 'dephlogisticated' air, the term then used for air without oxygen. By the late nineteenth century its vital role as a plant nutrient was understood and by the early twentieth century, the Haber–Bosch process was able to 'fix' nitrogen from the atmosphere on an industrial scale. Nitrogen fixation influences the amount of food present within an ecosystem. Prior to the industrial process of N production, crop growth was sustained by recycling crop residues and manures on the same land where food was grown. Any 'new' N was created by growing rice and legumes, or by mining guano and nitrate deposits. However, as the human population increased so has the demand for food and with that the dependence on inorganic fertilizers to sustain agriculture. This trend has affected the nitrogen cycle at global, national, and local scales.

The Nitrogen Cycle

Nitrogen comprises approximately 79% of the Earth's atmosphere in the form of biologically unavailable dinitrogen (N_2) gas. This reservoir is estimated to be in the order of 3.8×10^9 kg N, approximately 90% of the global reservoir. Crustal reservoirs comprise the remaining 10% (**Figure 1**). By comparison the amount of N stored in the biomass (terrestrial and oceanic) and soil is small, but this, of course, is the vital component as far as living organisms are concerned.

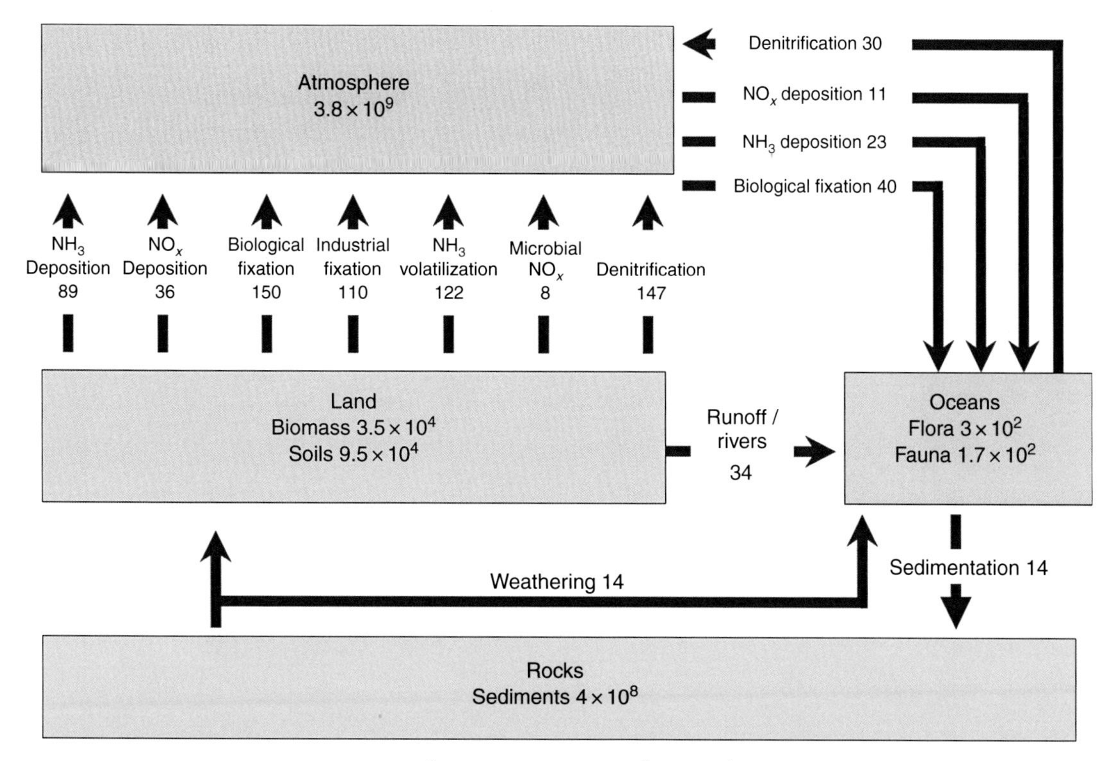

Figure 1 Global nitrogen reservoirs (units kg N yr^{-1}) and fluxes (units × 10^9 kg N yr^{-1}).

The global nitrogen cycle (**Figure 2**) is driven by biological and physical processes, which depend on a variety of environmental factors such as solar energy, precipitation, temperature, soil texture, soil moisture, the presence of other nutrients, and atmospheric CO_2 concentrations.

These factors control N fluxes into and out of soils and vegetation, thereby influencing the mass of N in these compartments, and therefore its availability. **Figure 3** illustrates the global distribution of nitrogen in soil and vegetation. Tropical forest soils show the least amount of storage because of high decomposition rates; but vegetation in temperate and tropical forests have higher N storage due to the higher production rates. In general terms, human activity has tended to accelerate nitrogen cycling, increasing flux rates from one store to another.

In order for nitrogen to be used for plant growth, it must be available in inorganic formal ammonia (NH_3), ammonium (NH_4), nitrite, (NO_2), or nitrate (NO_3). In the terrestrial nitrogen cycle (**Figure 4**), soil nitrogen cycling processes dominate, with surface application (fertilizer and manure) providing most of the nitrogen inputs. Microbes break down organic matter to produce much of the available nitrogen in soils. Mineralization/immobilization, nitrification, nitrate leaching, denitrification, and plant uptake can then occur. Nitrate is completely soluble in water and since it is not adsorbed to clay particles, it is vulnerable to being leached out of the soil by percolating rainfall or irrigation water. Generally, the movement of nitrogen can occur in one of three directions: (1) upward – crop uptake and gaseous loss, (2) downward – as leaching to groundwater, and (3) lateral – via surface and subsurface flow to surface waters.

The nitrogen cycle is strongly influenced by anthropogenic activities. During the twentieth century land-use changes, such as intensive agriculture, over-fertilization, deforestation, biomass burning, combustion of fossil fuels, industrial activities, and energy production, have significantly disturbed 'natural' N biogeochemical cycling. In natural ecosystems plant growth rates are low and annual uptake of N is relatively small. Cultivated crops are much more demanding with nutrient uptake ranging from about 100 kg N ha yr^{-1} for wheat and up to 450 kg N ha yr^{-1} for sugar cane. Improved grasslands for livestock rearing typically require 250 kg N ha yr^{-1}. The mineralization capacity of soils is almost always insufficient to maintain optimum growth; therefore, chemical fertilizers and manures are required to supply N for intensive agriculture. This has resulted in changes to the long-term trends within the N cycle at global, regional, and local scales.

Long-Term Global and Regional Trends in the Nitrogen Cycle

Globally, nitrogen is found in the terrestrial ecosystem as dead organic matter (89.5%), with live biomass accounting for 4% and inorganic nitrogen 6.5% of this

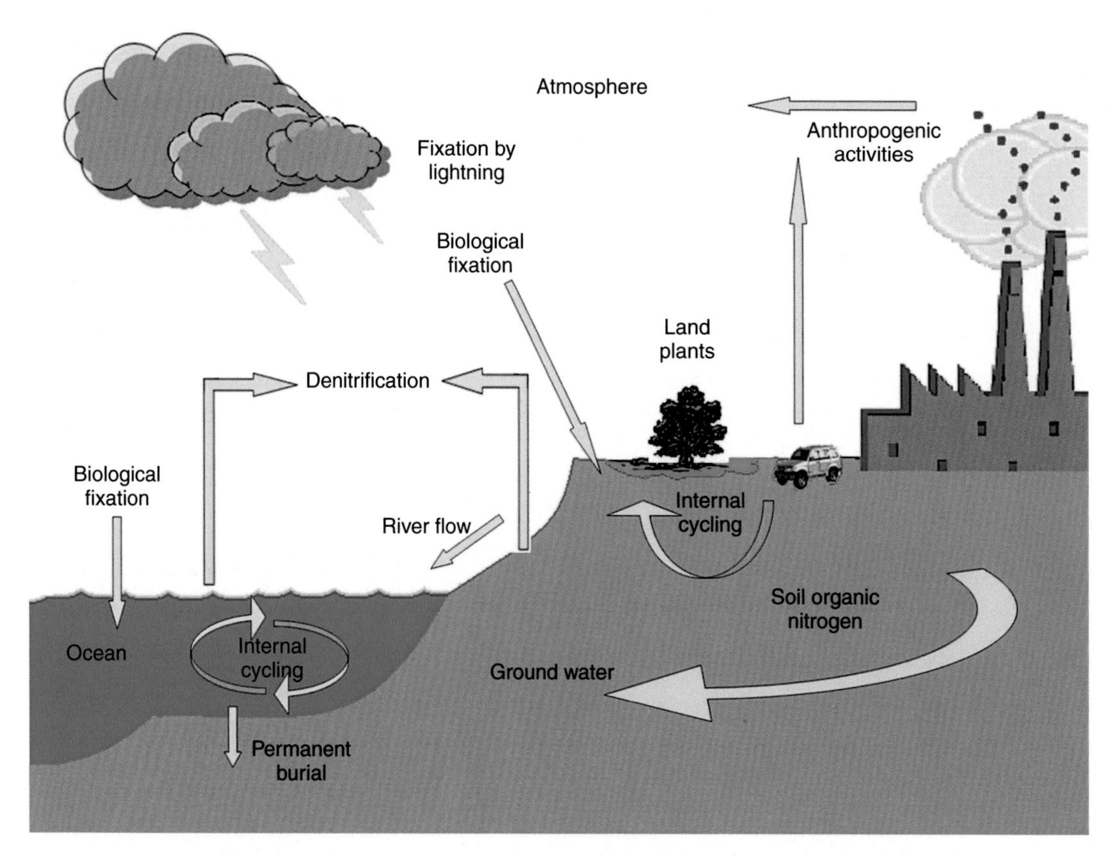

Figure 2 Global nitrogen cycle.

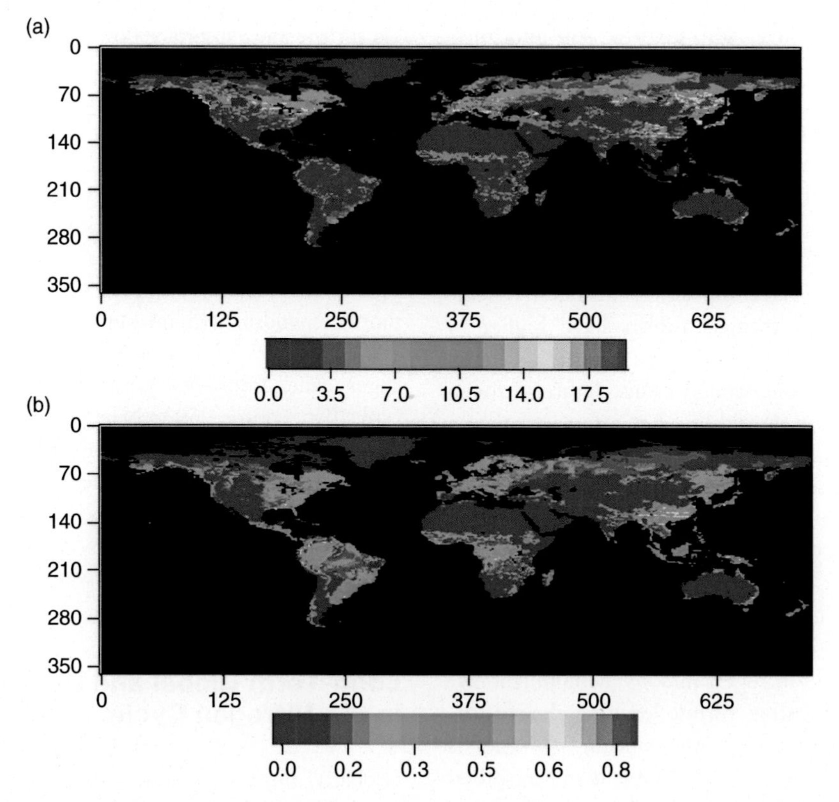

Figure 3 Global distribution of nitrogen storage (kg m^{-2}) in soil (a) and vegetation (b). Reproduced from Bin-Le Lin, Sakoda A, Shibasaki R, Gato N, and Suzuki M (2007) Modelling a global biochemical nitrogen cycle model in terrestrial ecosystems. *Ecological Modelling* 135(1): 89–110, with permission from Elsevier.

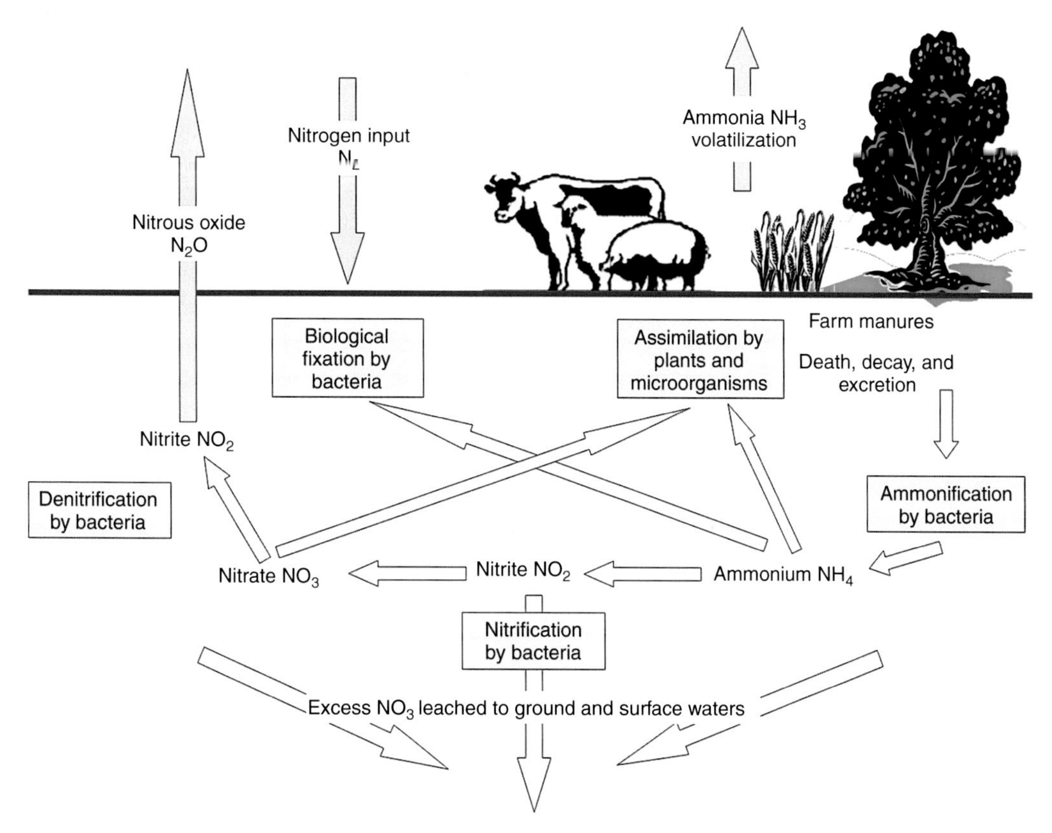

Figure 4 The terrestrial nitrogen cycle.

Table 1 Nitrogen production ($10^9\,kg\,N\,yr^{-1}$)

Nitrogen production ($10^9\,kg\,N\,yr^{-1}$)	*1890*	*1990*
Anthropogenic sources	15.0	140.6
Terrestrial ecosystem	100.0	89.0
Marine ecosystems	140.0	140.0
Fixation by lightning	5.0	5.0
Total	~260	~374

source. Natural sources of nitrogen have seen a small decline since 1890 (**Table 1**). Losses of biomass due to large-scale burning and forest clearances during the late twentieth century have contributed to the decline of this reservoir.

Natural reservoirs now cannot provide nitrogen in the quantity required for global food production. In 1890, total anthropogenic N production was approximately $15 \times 10^9\,kg\,N\,yr^{-1}$, but by 1990 this had risen by an order of magnitude to approximately $140 \times 10^9\,kg\,N\,yr^{-1}$.

In the terrestrial ecosystem, globally, nitrogen production is generally driven by the use of fertilizers for intensive agriculture, with cultivation and combustion contributing approximately 38% to all anthropogenic sources. However, this is not evenly distributed across the world regions. Asia produces almost half the world's nitrogen fertilizers, followed by Europe then North America (**Table 2**). Africa, Latin America, and Oceania combined contribute less than 12% of global nitrogen production.

Significant changes to the nitrogen cycle have been apparent since the 1960s. This is closely linked to expanding human populations and an increasing demand for food and energy. Creation of anthropogenic nitrogen in Asia increased from $\sim 14.4 \times 10^9\,kg\,N\,yr^{-1}$ in 1961 to $\sim 68 \times 10^9\,kg\,N\,yr^{-1}$ by 2000, and is set to increase to $105 \times 10^9\,kg\,N\,yr^{-1}$ by 2030. North America doubled its N production between 1961 and 1997, with most of the increase occurring during the 1960s and 1970s. Although the largest increase was in use of inorganic N fertilizer, emissions of NO_x from fossil fuel combustion also increased substantially. By 1997, even though N fixation had increased, fertilizer use and NO_x emissions had increased more rapidly and two-thirds of reactive N inputs were denitrified or stored in soils and biota, while one-third was exported, the largest export being in riverine flux to coastal oceans, followed by export in food and feeds, and atmospheric advection to the oceans. The consumption of meat protein is a major driver behind N use in agriculture in North America. Without changes in diet or agricultural practices, fertilizer use will increase over the next 30 years, and fluxes to coastal oceans may increase by another 30%.

Similar trends are mirrored in the European N budget (**Table 3**). By 1990, N inputs are approximately

Table 2 Global anthropogenic nitrogen production 1990 ($\times 10^9$ kg N yr^{-1})

World region	*Fertilizer production*	*Cultivation*	*Combustion*	*Total*
Africa	2.5	1.8	0.8	5.1
Asia	40.1	13.7	6.4	60.2
Europe and Former Soviet Union	21.6	3.9	6.6	32.1
Latin America	3.2	5.0	1.4	9.6
North America	18.3	6.0	7.4	31.7
Oceania	0.4	1.1	0.4	1.9
Total	~86	~31	~23	140.6

Table 3 European N Budget 1990

N input	$\times 10^9$ *kg N yr*$^{-1}$	*N output*	$\times 10^9$ *kg N yr*$^{-1}$
N-fertilizer production	14.0	Denitrification	13.8
Combustion and industry	3.3	Emissions of NH_3 and NO_x	7.8
Biological N fixation	2.2	Sewage and industry	2.6
Deposition	7.3	Riverine flux to oceans	4.0
Imported products	7.6	Exported products	6.3
Total	34.5	*Total*	34.5

34.5×10^9 kg N yr^{-1}. The major process of N fixation being fertilizer production at $\sim 14.0 \times 10^9$ kg N yr^{-1}, with industry and combustion accounting for a further $\sim 3.3 \times 10^9$ kg N yr^{-1}. Imported N from products such as animals, animal feeds, food, fertilizers, forestry products, exceeds the amount of N exported outside Europe. Furthermore, exports in riverine flux to oceans accounts for $\sim 4.0 \times 10^9$ kg N yr^{-1}.

Nitrogen Export by Rivers

Water is a carrier of N from pollution source to river outlet. The fraction that ultimately reaches the outlet depends on amount of runoff and distribution between different runoff components. Time delay between inputs at the soil surface and inputs to surface water additionally depends on groundwater residence times. The natural water quality of a river will be determined primarily by the catchment soil type and underlying geology to which water, falling on the catchment as rain, is exposed as it drains to the river. Climate provides an important context for nitrogen cycling by controlling the propensity for carbon and nitrogen to be stored within the catchment; thus in the UK, upland soils tend to conserve organic matter as peat, whereas organic matter tends to decompose much more readily in lowland soils. Deviations from this baseline water quality are generally caused by the influence of human activities through point and diffuse pollution sources. Up to 40% of total nitrogen reaches the aquatic system through direct surface runoff or subsurface flow. Nitrogen delivery to surface waters is further controlled by (1) soil structure and type, (2) rainfall, (3) the amount of nitrate supplied by fertilizers, and (4) plant cover and root activity.

In pristine river systems, the average level of nitrate is about 0.1 mg l^{-1} as nitrogen (mg l^{-1} N). However, in Western Europe, high atmospheric nitrogen deposition results in nitrogen levels of relatively unpolluted rivers to range from 0.1 to 0.5 mg l^{-1}. In recent years, nitrate concentrations in European rivers have been rising and progress still needs to be made in reducing the concentration of nitrate in Europe's rivers. High rates of nitrogen input to rivers and coastal waters are not confined to Europe. In USA as late as 1998, more than one-third of all river miles, lakes (excluding the Great Lakes), and estuaries did not support the uses for which they were designated under the Clean Water Act (1987). For example, **Table 4** illustrates the extent of N inputs to rivers and coasts in areas of America, Africa, and Asia. These

Table 4 Nitrogen inputs to rivers and coastal waters

River	*N inputs to rivers (kg yr*$^{-1}$*)*	*N exports to coastal waters (kg yr*$^{-1}$*)*
Mississippi	7489	597
Amazon	3034	692
Nile	3601	268
Zaire	3427	632
Zambezi	3175	330
Rhine	13941	2795
Po	9060	1840
Ganges	9366	1269
Chang Jiang	11823	2237
Juang He	5159	214

trends are cause for concern as seasonal hypoxia develops during the summer months, resulting in a depletion of sea bed vegetation and changes in fish stocks.

It is now widely acknowledged that agriculture is the main source of N pollution in surface waters and groundwater in rural areas of Western Europe and USA. The UK House of Lords' report Nitrate in Water (1989) commented on the conflicts that can arise when the use of land for farming comes into conflict with the use of land for water supply. Concern for this initially focused on the alleged links between high nitrate concentrations in drinking water and two health problems in humans: the 'blue-baby' syndrome (methaemoglobinaemia) and gastric cancer. Now, there are also major concerns about environmental degradation. Nutrient enrichment in water bodies encourages the growth of aquatic plants (see **Figure 5**).

Reed beds and other marginal plants may be attractive on a small scale, but when these and, particularly, underwater plant growth are excessive, this can cause a narrowing of waterways, and become a nuisance to recreational users of rivers and lakes. Furthermore, eutrophication (a group of effects caused by nutrient enrichment of water bodies) can adversely affect the aquatic ecosystem. An algal bloom may cut out light to the subsurface, and when it dies, decomposition uses the oxygen supply needed by other species. Some algae are toxic to fish, while others, for example, cyanobacterial species, are toxic to mammals including domestic pets. Studies in Asia have demonstrated the link between increasing use of fertilizers and increasing incidence of algal blooms. For example in some Chinese provinces, fertilizer application is greater than 400 kg ha^{-1}. This is usually applied as a single application and with crop utilization efficiency as little as 30–40%, a high proportion is lost to rivers, lakes, and coastal waters. The environmental impact at the regional level is the incidence of red tides (algal blooms). During the 1960s less than 10 red tides per year were recorded, but in the late 1990s over 300 per year were being recorded.

Figure 5 Choked watercourse, River Skerne, UK. Source: P. Widdison.

Land-Use Controls to Reduce N Enrichment to Surface Waters

The popular misconception that the nitrate problem is caused by farmers applying too much nitrate fertilizer is too simplistic. Nevertheless, there is now little doubt that the high concentrations of nitrate in freshwaters noted in recent years have mainly resulted from runoff from agricultural land and that the progressive intensification of agricultural practices, with increasing reliance on the use of nitrogenous fertilizer, has contributed significantly to this problem. Since 1945, agriculture in the industrialized world has become much more intensive. Fields are ploughed more frequently; more land is devoted to arable crops, most of which demand large amounts of fertilizer; grassland too receives large applications of fertilizer to ensure a high-quality silage for winter feed; stocking densities in general are higher leading to increased inputs of manure on grassland and problems of disposal of stored slurry; cattle often have direct access to water courses resulting in soil and bank erosion and direct contamination from animal waste; many low-lying fields are now underdrained, encouraging more productive use of the land and speeding the transport of leached nitrate to surface water courses. Lowland rivers close to urban areas may receive larger quantities of nitrogen from sewage effluent, but budgetting studies confirm that agriculture is the main source of nitrate in river water, except in the most urbanized river basins.

In mainland Britain mapped nitrate concentrations demonstrate a marked northwest to southeast gradient, reflecting relief, climatic conditions, and agricultural activity. Upland areas in the north and west are usually characterized by nitrate concentrations below 1 mg NO_3-N l^{-1}. This reflects the high rainfall and low temperatures of such areas: upland soils tend to conserve organic matter and mineralization rates are low. In contrast, a decreasing ratio of runoff to rainfall and an increasing intensity of agricultural land use toward the south and east of Britain results in higher mean concentrations of nitrate in river water. Many of the lowland rivers are characterized by concentrations above 5 mg NO_3-N l^{-1}; in East Anglia and parts of the Thames basin, mean nitrate concentrations in rivers are close to the European Union limit of 11.3 mg NO_3-N l^{-1}, a level exceeded in some spring waters especially in the Jurassic limestone of the Cotswold's and Lincolnshire Wolds.

The changing pattern of British lowland agriculture since 1945 is reflected in long-term records of nitrate for surface and groundwaters (**Figure 6**). Such graphs confirm the accelerated nitrogen cycling in recent decades and increasing fluxes from the terrestrial to aquatic compartments of the N cycle.

For both large and small rivers, there has been a relatively steady upward trend in nitrate concentrations, often of the order of 0.1–0.2 mg NO_3-N l^{-1} a^{-1}. Analyses for relatively short time series of just a few years have shown that the upward trend may be interrupted, either because of climatic variability (drier years are associated with lower nitrate concentrations) or because of land-use change. Nevertheless, statistical analysis of long time series shows that the main effect is a steady increase in nitrate levels over time which is independent of climate. If trends continue, the mean nitrate concentration of many rivers in Europe will soon be above the EU limit; in many cases this level is already exceeded during the winter when nitrate concentrations reach their maximum. In catchments where groundwater is the dominant discharge source, this long-term trend may be prolonged since it may take years for nitrate to percolate down to the saturated zone. In such basins, nitrate pollution may remain a problem for decades to come. In recent years, a number of options have been considered as a means of halting the upward trend.

Trends in water management in Europe include moves toward catchment-level management, improved intersectoral coordination and cooperation, and frameworks facilitating stakeholder participation. This approach is developed by the European Union in its Water Framework Directive (2000/60/EC), which sets targets for good ecological status for all types of surface water bodies and good quantitative status for groundwater. More localized schemes, like the UK Nitrate Vulnerable Zones, involve greater restrictions on farming practice, such as restricting the amount and timing of organic and inorganic fertilizer application. The EU Common Agricultural Policy is to change the way payments are made to farmers. Single-farm payments will encourage farming in a more environmentally friendly way. Financial payments may be available to farmers for loss of income or for changing farming practice such as improving slurry storage and fencing off watercourses to restrict livestock access. Much interest currently focuses on the use of riparian land as nitrate buffer zones.

The terrestrial–aquatic ecotone occupies the boundary zone between the hillslope and the river channel, usually coinciding with the floodplain. Given their position, near-stream ecotones can potentially function as natural sinks for sediment and nutrients emanating from farmland. Observed denitrification rates in floodplain sediments may be sufficient to remove all nitrate from groundwater flowing under a riparian woodland, with a floodplain width of 30 m. Saturated, anoxic soils, rich in carbon, are exposed to nitrate-rich groundwater. Rates of denitrification are high within this zone since the nutrients required by denitrifying bacteria are abundant. Wetlands and wet meadows (defined as areas where the water table is at or above land surface for long enough each year to promote the formation of hydric soils and support the growth of aquatic vegetation) also have potential as nitrogen sinks. High production rates by wetland vegetation result in an abundance of carbon providing an organic substrate for bacterial processes. Wetland plants transport oxygen into anaerobic sediments which can enhance denitrification leading to losses of nitrogen as N_2O or N_2 from wetland sediments.

The type of vegetation found on the floodplain controlling the efficiency of nitrate absorption is the subject of much debate. Several studies have argued the presence of trees is crucial, yet others state the role of surface

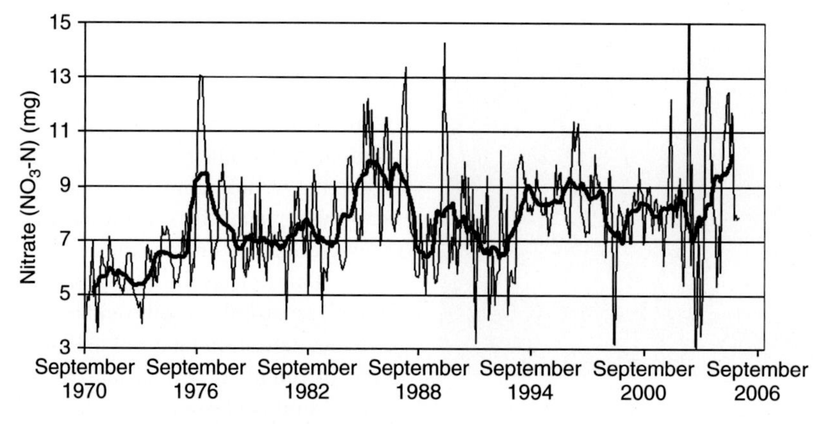

Figure 6 Long-term nitrate record 1970–2005: Slapton Wood catchment (UK).

vegetation is secondary to presence of saturated conditions together with a carbon-rich sediment. Denitrifying bacteria operate best at the junction anaerobic/aerobic zones where both carbon and nitrate are abundant. It is clear that nitrate losses may be reduced by creating a nutrient-retention zone between the farmland and the river. Given that many floodplains around the world are part of an intensive agricultural system, creating permanently vegetated buffer strips between field and water courses is an idea that should be actively promoted. However, buffer strips will only be successful nutrient sinks if they are managed in an appropriate way. Underlying artificial drainage should be broken or blocked up to prevent a direct route to the watercourse for solutes and grassland strips need maintenance to prevent them becoming choked with sediment and losing their sediment retention potential.

Solving the problem of nutrient enrichment of surface waters cannot be seen in the short-term. Long-term land-use change is needed. Taking farm land immediately adjacent to water courses out of production is one option that could go some way to allow modern agriculture and water supply to coexist in the same basin. Such proposals inevitably raise questions about who pays for them – farmers, water supply companies, or the taxpayers.

See also: Anthropospheric and Antropogenic Impact on the Biosphere.

Further Reading

Addiscott TM (1996) Fertilizers and nitrate leaching. In: Hester RE and Harrison RM (eds.) *Agricultural Chemicals and the Environment. Issues in Environmental Science and Technology*. Cambridge, UK: Royal Society of Chemists.

Betton C, Webb BW, and Walling DE (1991) *Recent Trends in NO_3-N Concentration and Load in British Rivers. IAHS Publication 203*, pp. 169–180. Wallingford: IH Press.

Bin-Le Lin, Sakoda A, Shibasaki R, Goto N, and Suzuki M (2000) Modelling a global biogeochemical nitrogen cycle model in terrestrial ecosystems. *Ecological Modelling* 135(1): 89–110.

Burt TP, Heathwaite AL, and Trudgill ST (eds.) (1993) *Nitrate: Processes, Patterns and Management*. Oxford: Wiley.

Burt TP and Johnes PJ (1997) Managing water quality in agricultural catchments. *Transactions of the Institute of British Geographers* 22(1): 61–68.

Butcher SS, Charlson RJ, Orians GH, and Wolfe GV (eds.) (1992) *Global Biogeochemical Cycles*, 379pp. London: Academic Press.

De Wit M, Behrendt H, Bendoricchio G, Bleuten W, and van Gaans P (2002) The contribution of agriculture to nutrient pollution in three European rivers, with reference to the European Nitrates Directive. *European Water Management Online*.

Eckerberg K and Forsberg B (1996) Policy strategies to reduce nutrient leaching from agriculture and forestry and their local implementation: A case study of Laholm Bay, Sweden. *Journal of Environmental Planning and Management* 39(2): 223–242.

Haycock NE, Burt TP, Goulding KWT, and Pinay G (eds.) (1997) *Buffer Zones: Their Processes and Potential in Water Protection*. Harpenden: Quest Environmental.

Hem JD (1970) *Study and Interpretation of the Chemical Characteristics of Natural Water*, 363pp. Washington: United States Government Printing Office.

Kessler E (ed) (2002) Special Report. *Ambio* 31(2).

Norse D (2003) Fertilisers and world food demand. Implications for environmental stress, IFA-FAO Agriculture Conference, Rome. http://www.fertilizer.org/ifa/publicat/PDF/2003_rome_norse.pdf

Ribaudo M (2001) Non-point source pollution control policy in the USA. In: Shortle JS and Abler DG (eds.) *Environmental Policies for Agricultural Pollution Control*. Oxford: CAB International.

Sprent JI (1987) *The Ecology of the Nitrogen Cycles*, 151pp. Cambridge: Cambridge University Press.

White RE (1987) *Introduction to the Principles and Practice of Soil Science*, 244pp. New York: Blackwell.

Oxygen Cycle

D J Wuebbles, University of Illinois at Urbana-Champaign, Urbana, IL, USA

Introduction

Without atmospheric oxygen, life on Earth would be extremely different. The increase in atmospheric oxygen to present levels, and the corresponding increase in atmospheric levels of ozone, with their ability to absorb biologically harmful high energy radiation from the Sun, allowed life to evolve and to emerge from the oceans to the land, undergoing an amazing evolution to its present diversity.

Oxygen is one of the most abundant elements on our planet. It is the most abundant element by mass in the

Earth's crust and is found in most rocks. Oxygen also accounts for 89% of the mass of the oceans. After molecular nitrogen, oxygen is the second most abundant element in the Earth's atmosphere, with molecular oxygen accounting for 20.95% of the atmospheric content. There is nearly uniform mixing of molecular oxygen in the atmosphere until above the mesosphere, roughly 80 km above the Earth's surface. Therefore, because of the nearly exponential decrease in pressure with altitude, the bulk of molecular oxygen is found in the first few kilometers above the Earth's surface. However, although the turbulent mixing in the lower atmosphere keeps the molecular oxygen nearly in constant mixing ratio (relative to the total air density), measurements indicate that there are seasonal latitudinal variations of as much as 15 ppm (parts per million molecules of air). These seasonal variations are most pronounced at high latitudes in the Northern Hemisphere, where the seasonal cyclic variations in photosynthesis and respiration are most strongly felt.

While there are many gases and particles in the atmosphere containing oxygen, the other gas that needs to be discussed as part of the oxygen cycle because of its great importance both to the atmosphere and to life on Earth is ozone. While atmospheric concentrations for ozone are much smaller (ppm levels) than that for molecular oxygen, ozone is important for several reasons: (1) it absorbs biologically harmful levels of ultraviolet (UV) radiation, keeping this radiation from reaching the Earth's surface; (2) it is a 'greenhouse' gas that influences the Earth's climate; and (3) direct contact with ozone pollution in the lower atmosphere can be harmful to plants, animals, and humans.

The following sections discuss the oxygen cycle in more detail, focusing on the processes affecting molecular oxygen and ozone in the atmosphere, the historical changes in the amounts of these important gases, and their projections for the future.

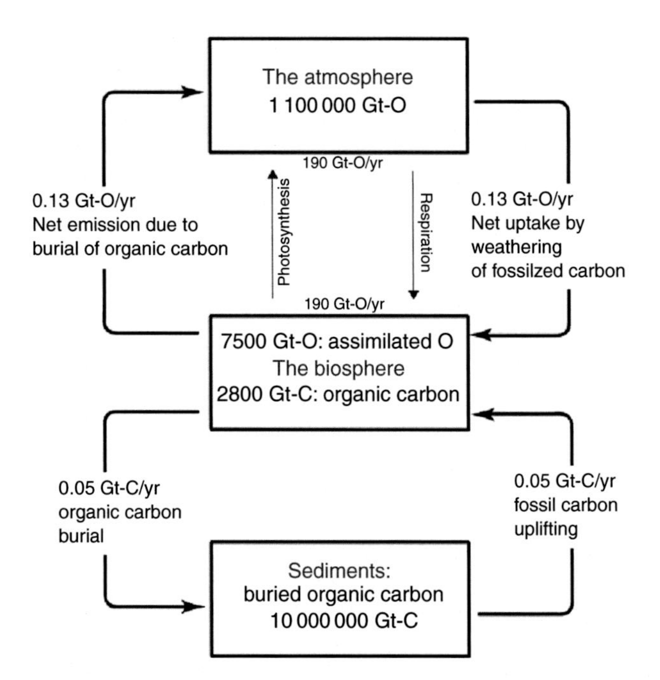

Figure 1 A simple model of the oxygen cycle, and its relationship to the carbon cycle, that considers the three major reservoirs: the atmosphere, the biosphere, and the sediments. Redrawn from graph presented by http://atoc.colorado.edu/~fasullo/pjw_class/images/oxygencycle.gif.

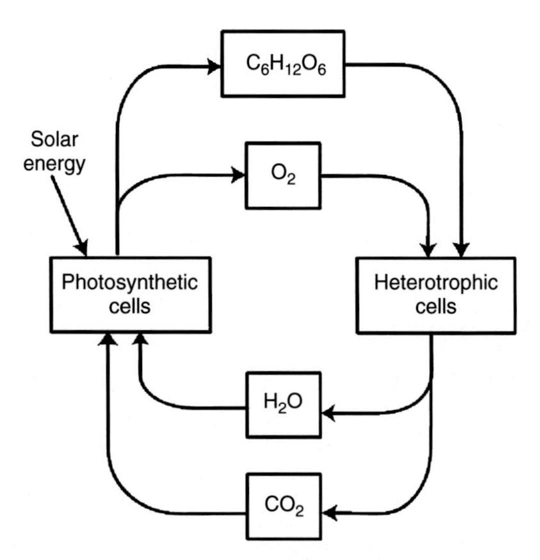

Figure 2 A representation of the basic exchange of atmospheric oxygen between biospheric systems.

The Oxygen Budget

Atmospheric molecular oxygen is generated and consumed by a wide range of processes in the Earth system. Biological, chemical, and physical processes both on and beneath the Earth's surface all contribute to the budget for oxygen in the Earth's atmosphere. **Figure 1** shows a simple representation of the budget for oxygen. This figure also shows how the oxygen budget is closely linked to the carbon cycle. While it is difficult to measure the annual fluxes shown in **Figure 1**, oxygen isotopes (^{17}O, ^{18}O) do provide constraints leading to the approximate values shown.

The major mechanism by which molecular oxygen is produced on our planet is through photosynthesis. As seen in **Figures 2** and **3**, the net reaction of photosynthesis is to convert carbon dioxide (CO_2) and water to molecular oxygen. This large annual flux is counteracted by the equally large annual flux from the effects of respiration in removing oxygen from the atmosphere. Thus, as suggested by **Figures 1** and **2**, the current atmospheric equilibrium is maintained by this cycle between photosynthesis and respiration. As indicated in **Figure 3**, there are several slower removal processes in addition to this more rapid cycling.

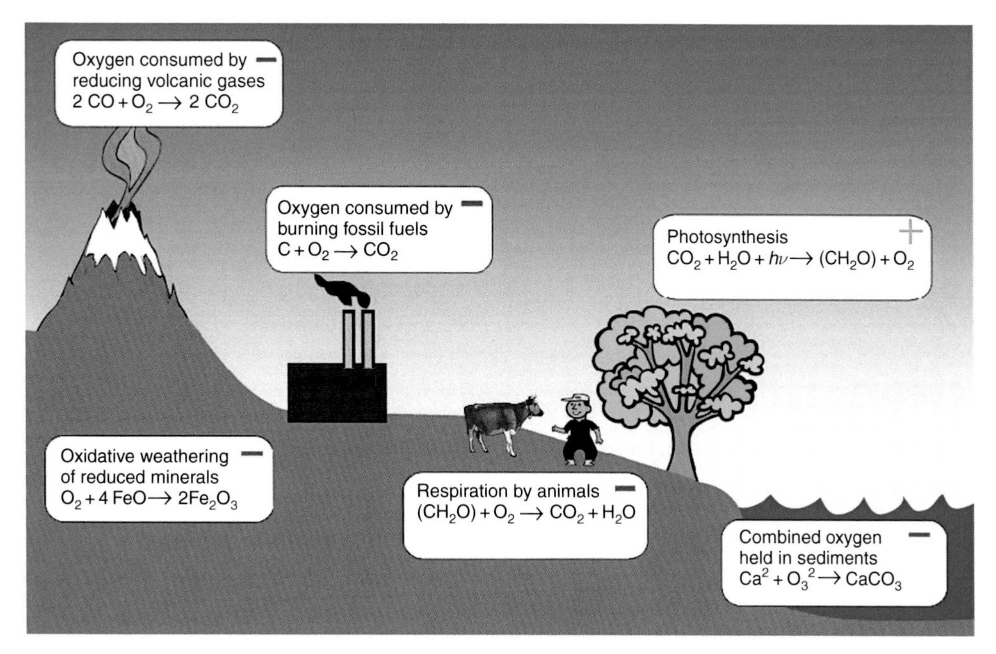

Figure 3 A simple representation of the key chemical reactions controlling the fluxes of oxygen to and from the atmosphere.

The mean residence time of molecular oxygen in the atmosphere is roughly 4000 years. With the approximate balance between photolysis and respiration, this residence time is largely controlled by the long-term burial of reduced carbon in ocean sediments. This burial of organic matter results is initiated via the photosynthesis from organisms in the upper levels of the ocean, whose carbon then is deposited to deeper levels of the ocean. The actual rate of burial depends on the area of the ocean floor subject to anoxic conditions, which in turn varies inversely with the concentration of atmospheric oxygen. The balance between the burial of organic matter and its oxidation thus plays a significant role in the maintenance of the atmospheric molecular oxygen at 20.95% of the atmospheric density.

A large amount of oxygen has also been consumed by weathering of reduced crustal materials, such as those containing iron and sulfur, through the geologic history. However, the residence time of atmospheric oxygen at the current rate of exposure would be roughly 2 million years (My) due to removal mechanism.

The History of Atmospheric Oxygen

Primitive life began in the absence of free oxygen. However, in the first 400 My of the Earth, bacteria-like organisms developed that could take advantage of the light energy from the Sun to initiate photosynthesis, although early production of oxygen likely oxidized crustal materials instead of building up in the atmosphere. Geochemical evidence suggests there was little oxygen in the atmosphere during the Achaean (~2.5–4 billion years ago), but near the beginning of the Proterozoic, about 2.5 billion years ago, cyanobacteria – photosynthetic prokaryotes also known as blue-green algae – started the process that eventually led to a massive increase in the concentration of atmospheric oxygen. The definitive worldwide change that signaled the appearance of a significant increase in free oxygen was the appearance of red beds, stratified layers of sedimentary rocks, characterized by abundant red oxides of iron (ferric oxides), which occurred about 2.0 billion years ago. When the oxygen began to accumulate, the resulting oxidation of other atmospheric constituents totally changed the nature of the atmosphere and the oxygen 'poisoning' was devastating to many of the existing life forms then found on Earth and its oceans. Some bacteria however were able to endure the oxygen atmosphere. A symbiosis between bacteria and the formerly free-living mitochondria enabled eukaryotes to eventually evolve in response to the crisis. Oxygen-based metabolism came into being. The environment changed and life evolved.

Figure 4 provides an estimate of the growth of free oxygen in the Earth's atmosphere, along with the corresponding evolution in the types of various life forms found on our planet. With the increasing proliferation of life, the amount of photosynthesis increased and atmospheric oxygen also continued to increase. The oxygen buildup eventually resulted in the appearance of the first higher order cells with nuclei, called eukaryotes or eukaryotic cells (some analyses suggest this occurred earlier than shown in the figure). These cells depend on atmospheric oxygen for the formation of complex energy-producing compounds. Such cells provide the

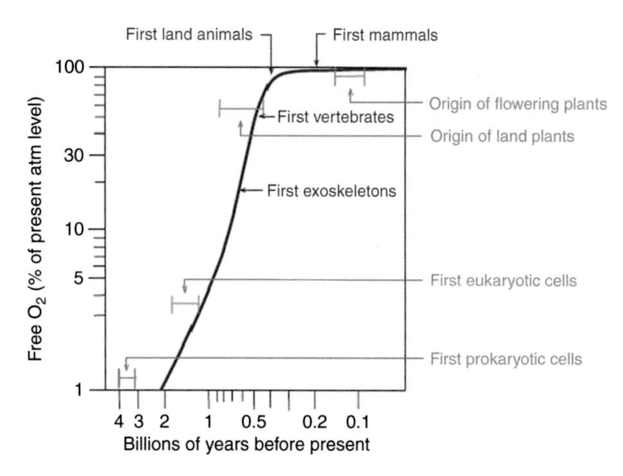

Figure 4 Estimated growth of free oxygen in the Earth's atmosphere. Based on graph from http://www.ldeo.columbia.edu/edu/dees/U4735/projections/free_o.html.

makeup of all nonbacterial (or nonbacteria-like) forms of like on Earth.

Atmospheric oxygen grew to its present levels of roughly 21% of the atmospheric content about 400 million years ago (Ma). While it was likely slowed in reaching this state by consumption in terrestrial weathering processes, the amount of atmospheric oxygen, as suggested by the paleontologic record, appears to have varied little since then. The variations that have occurred (while we have little evidence of actual variations, concentrations may have varied from as little as 15% to as much as 30% during the last 500 My) are likely related to periods when there were strong variations in the deposition of organic matter.

The Recent Decline in Atmospheric Oxygen

Since 1989, when Ralph Keeling of the Scripps Institution of Oceanography began extremely high precision measurements, there has been a steady decline of molecular oxygen in our atmosphere. The decrease is small, just 0.03% in the past 20 years or an annual rate of about 2 ppm (out of 210 000 ppm of atmospheric oxygen). Measurements of the bubbles of air trapped in ice cores from Greenland and Antarctica suggest that this oxygen decline started in the late eighteenth century, at the beginning of the industrial revolution, when fossil fuel burning increased dramatically.

This decrease in atmospheric oxygen is not unexpected, for the combustion of fossil fuels, while in the process of producing carbon dioxide, destroys O_2. For every 100 atoms of fossil-fuel carbon burned, it has been estimated that roughly 140 molecules of O_2 are consumed. Since every molecule of additional carbon dioxide locks up two oxygen atoms, the free oxygen decline is greater than the carbon dioxide increase. However, the rate of decline in molecular oxygen is only about two-thirds of that expected from fossil fuel combustion. While not fully verified, the difference may be explained by the increase in biomass known to occur as a result of the increase in CO_2. Plants grow a bit faster than before, leading to a greater storage of carbon in tree wood and soil humus. For each atom of extra carbon stored in this way, roughly one molecule of extra oxygen accumulates in the atmosphere.

Although the oxygen decrease is unlikely to get to be large enough an effect to be of major concern, the oxygen decrease is expected to continue in the future as fossil burning continues and concentrations of atmospheric carbon dioxide continue to increase.

The Role of Ozone

As oxygen increased in the Earth's atmosphere, so did the levels of ozone, O_3. The formation of the ozone layer in the upper atmosphere is generally believed to have played an important role in the development of life on Earth, particular in the development of life on land. The accumulation of oxygen molecules in the atmosphere allowed for the production of ozone. Gradually, the increasing levels of ozone led to the formation of the stratosphere, a region of the upper atmosphere where temperature increases with altitude largely as a result of the absorption of solar radiation by ozone. The resulting screening of lethal levels of solar UV radiation by the ozone layer is thought to have been important to allowing life to migrate from the oceans onto land.

Ozone, O_3, is composed of three oxygen atoms and is a gas at atmospheric pressures and temperatures. Approximately 90% of the atmospheric ozone is in the stratosphere. Most of the remaining ozone is in the troposphere, the lower region of the atmosphere extending from the Earth's surface up to roughly 10 km at mid-latitudes and 16 km in the tropics. At mid-latitudes, the peak concentrations of ozone occur at altitudes between 20 and 30 km. At high latitudes, the peak occurs at lower altitudes, largely as the result of atmospheric transport processes and the lower height of the tropopause (the transition region between the troposphere and stratosphere).

Ozone in the stratosphere is often called 'good' ozone because it protects life on Earth from harmful levels of UV radiation from the Sun. Therefore, a decrease in the amount of ozone would allow an increase in the amount of the UV radiation from the Sun to reach the Earth's surface. Corresponding to an increase in UV are projected significant impacts on ecosystems and human health, including increases in incidences of skin cancer, eye cataracts, damage to genetic DNA, and suppression of the efficiency of the immune system. It is the concerns about increased biologically harmful levels of UV from the decreasing levels of ozone that has largely been the driver for policy actions to protect the ozone layer.

On the other hand, ozone near the Earth's surface is called 'bad' ozone because of its direct effects on plants, ecosystems, and humans. Ozone pollution is a concern during the summer months because strong sunlight and hot weather result in harmful ozone concentrations in the air we breathe. Many urban and suburban areas throughout the world have high levels of 'bad' ozone during summer months, and the effects of winds can carry these high ozone levels to rural areas.

Breathing ozone can trigger a variety of health problems including chest pain, coughing, throat irritation, and congestion. It can worsen bronchitis, emphysema, and asthma. 'Bad' ozone also can reduce lung function and inflame the linings of the lungs. Repeated exposure may permanently scar lung tissue. Ground-level ozone also damages vegetation and ecosystems. It leads to reduced agricultural crop and commercial forest yields, reduced growth and survivability of tree seedlings, and increased susceptibility to diseases, pests, and other stresses such as severe weather.

Ozone can also radiatively affect the Earth's climate. Ozone absorbs solar radiation but it also is a so-called greenhouse gas that can absorb infrared radiation from the Earth that otherwise would be emitted to space. It is the balance between the solar and infrared radiative processes that determines the net effect of ozone on climate. Decreases in ozone in the stratosphere above about 30 km (roughly 18 miles above the Earth's surface) tend to increase the surface temperature as a result of the increased absorption of solar radiation, effectively increasing the solar energy that warms the Earth's surface. Below about 30 km, decreases in ozone tend to cool the surface temperature, as the infrared greenhouse effect dominates in this region. Scientific analyses have shown that the decrease in stratospheric ozone over recent decades have had a cooling effect, counteracting a fraction of the warming effect over this time from increasing concentrations of carbon dioxide and other greenhouse gases.

The Production and Destruction of Ozone

Without human intervention, the stratospheric ozone layer would be produced and destroyed through natural processes. The amounts of ozone in the stratosphere vary naturally throughout the year as a result of production and destruction processes, and as a result of winds and other transport processes that move the ozone molecules around the planet. In addition, changes in ozone occur associated with changes in the solar radiation reaching the Earth during the 11-year solar cycle and with various events such as large explosive volcanic eruptions.

Production of ozone in the stratosphere results primarily from photodissociation of oxygen, O_2, molecules. The breaking of the molecular bond by high energy solar photons at wavelengths less than 242 nm results in oxygen atoms that generally react rapidly with an oxygen molecule to form ozone. The sequence of reactions to form ozone is represented as:

$$O_2 + h\nu\,(\lambda < 242\ \text{nm}) \rightarrow O + O \qquad [1]$$

$$O + O_2 + M \rightarrow O_3 + M \qquad [2]$$

where $h\nu$ represents a photon, λ is wavelength, and M is a third atmospheric gas, normally N_2 or O_2, the primary components of air.

Since the atmosphere is uniformly filled with oxygen molecules, most of the ozone is generated where there is a balance between the decreasing atmospheric density with altitude and where there is available UV solar radiation in the wavelength region less than 242 nm. This primarily occurs in the upper stratosphere.

The high-energy solar radiation needed to produce ozone is largely absorbed in the stratosphere resulting in the production of the ozone layer. Too little UV radiation reaches the troposphere for it to be a major cause of ozone production in this region. In contrast, the lesser amount of ozone in the troposphere is largely formed through a series of chemical reactions, generally referred to as smog reactions. The primary source of ozone in the troposphere (and in the smog in urban areas) comes through the conversion of nitric oxide, NO, to nitrogen dioxide, NO_2, which then photolyzes at visible wavelengths to release an oxygen atom that produces ozone through reaction (see eqn [2]). The transport of stratospheric ozone to the troposphere is also important to the budget of tropospheric ozone.

If there was no natural ozone destruction, most of the oxygen in the stratosphere, and perhaps throughout the atmosphere, would eventually be converted to ozone. Such concentrations of ozone would be intolerable to many forms of life on Earth, both because of the direct toxicity of ozone and because of the resultant elimination of any UV radiation reaching the Earth's surface.

Ozone photodissociates at UV and visible wavelengths to produce O and O_2. However, because the oxygen atom will generally react to reform ozone, this mechanism produces no net change in the amount of odd oxygen. The actual destruction of ozone and odd oxygen in the stratosphere occurs mainly through catalytic reactions with other gases. In the stratosphere, important catalysts for ozone destruction include nitric oxide (NO), hydroxyl (OH), chlorine (Cl), and bromine (Br). Such gases can be directly substituted for X in the following catalytic mechanism,

$$X + O_3 \rightarrow XO + O_2 \qquad [3]$$

$$XO + O \rightarrow X + O_2 \qquad [4]$$

which results in the net reaction of O + O_3 being converted to two oxygen molecules. As gas X is recycled through these reactions, it continues to destroy ozone until some other reaction converts X to a less reactive form, such as HNO_3 or HCl. In addition to the mechanism described by reactions [3] and [4], there are a number of other catalytic mechanisms also affecting ozone. In this way, a single chlorine atom can lead to the net destruction of thousands of ozone molecules. Because of such cyclic reaction mechanisms, relatively small concentrations of reactive chlorine in the stratosphere can have a significant impact on the amount and distribution of stratospheric ozone.

Human Effects on Ozone

Human activities have had a devastating effect on the concentration and distribution of stratospheric ozone over the last few decades. Measurements of ozone by satellite and ground-based measurements over the last several decades indicate that stratospheric ozone levels have decreased. Atmospheric ozone has decreased globally by more than 5% since 1970 (see **Figure 5**). Atmospheric measurements have also shown that the depleted levels of ozone have indeed increased the amount of UV at the surface. The significant global decrease in stratospheric ozone since the 1970s is well correlated with increasing amounts of chlorine and bromine in the stratosphere. The sources of this chlorine and bromine are chlorofluorocarbons (CFCs) and other halocarbons produced industrially for a variety of uses such as refrigerants in refrigerators, air conditioners, and large chillers, as propellants for aerosol cans, as blowing agents for making plastic foams, and as solvents for dry-cleaning and for degreasing of materials. Atmospheric measurements have clearly corroborated theoretical studies showing that the chlorine and bromine released from the destruction of these halocarbons in the stratosphere is reacting to destroy ozone. **Figure 6** shows the excellent comparison between the observed trend in ozone based on measurements from the solar backscatter ultraviolet (SBUV) satellite instrument averaged over latitudes from 50° N to 50° S and results from the University of Illinois zonally averaged model that is based on the changes in atmospheric gases and particles. In addition to the human related emissions, natural forcings on ozone from the effects of the 1991 Mount Pinatubo volcanic eruption (due to sulfur emissions) and from the effects of solar flux variations during the 11-year sunspot cycle also contribute to the observed trends.

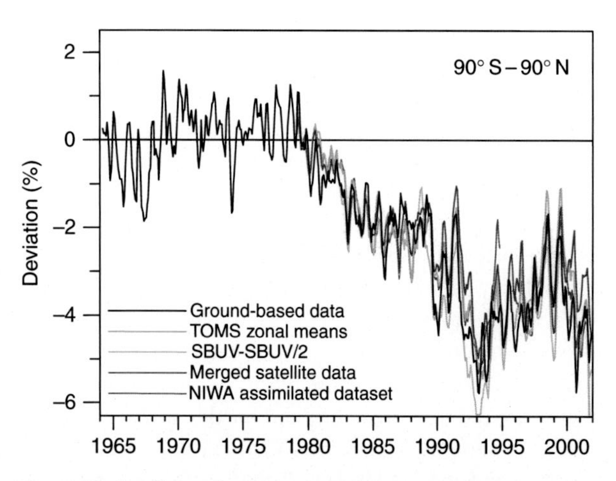

Figure 5 Deviations in total ozone with time relative to January 1979 from various ground-based and satellite measurements (TOMS and SBUV). The data are area weighted over 90° S–90° N. Graph provided by Vitali Fioletov as an update to earlier analyses presented in Fioletov VE, Bodeker GE, Miller AJ, McPeters RD, and Stolarski R (2002) Global and zonal total ozone variations estimated from ground-based and satellite measurements: 1964–2000. *Journal of Geophysical Research* 107: 4647 (doi: 10.1029/2001JD001350).

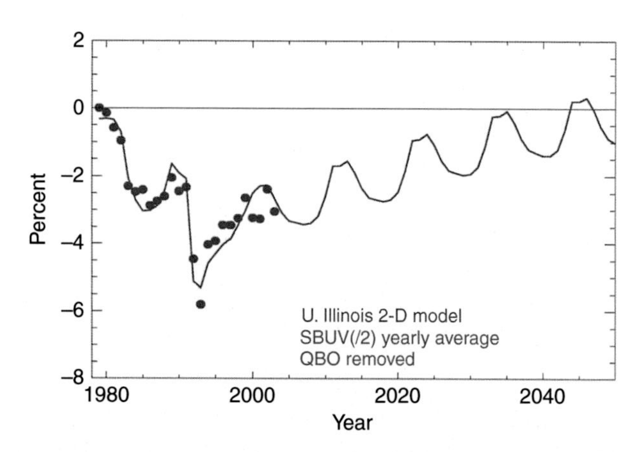

Figure 6 Comparison of trend in annually averaged, observed total ozone (red dots) from 50° N–50° S based on measurements from the solar backscatter ultraviolet (SBUV) satellite instrument and the derived trend from the University of Illinois zonally averaged chemistry-transport model of the global atmosphere. The effects of the natural cycle due to the quasi-biennial oscillation have been removed from the satellite data. While the overall decline in ozone results from increases in concentrations of stratospheric chlorine and bromine from human-related emissions of various halocarbons, the cyclic variations in the observations and model results are due to the effects of the 11-year solar sunspot cycle, while the extra deep minimum in ozone in the early 1990s is due to the effects of emissions from the Mount Pinatubo volcanic eruption in 1991. The model results also show the projected recovery of ozone if the Montreal Protocol reduces halocarbon emissions and atmospheric concentrations as expected.

Beginning in the late 1970s, a special phenomenon began to occur in the springtime over Antarctica, referred to as the Antarctic ozone 'hole'. A large decrease in total ozone, now over 60% relative to pre-hole levels, has been observed in the springtime (September to November) over Antarctica. Dr. Joseph Farman and colleagues first documented this rapid springtime decrease in Antarctic ozone over their British Antarctic Survey (BAS) station at Halley Bay, Antarctica. These analyses attracted the attention of the scientific community, who soon found that decreases in the total ozone column were greater than 50% compared with historical values observed by both ground-based and satellite techniques. At the time, there was no expectation that such a phenomenon would be discovered. As a result of the Farman paper, a number of hypotheses arose attempting to explain the large ozone destruction in the springtime over Antarctica. It was initially proposed that the chlorine catalytic cycle might explain the observed ozone decrease, but this did not match the expected ozone decrease possible from the reactive chlorine available at the high latitudes. A special measurement campaign in 1987, as well as later measurements, proved that chlorine and bromine chemistry indeed was indeed responsible for the ozone 'hole' but because of heterogeneous reactions occurring on polar stratospheric clouds in the lower stratosphere.

The air over the Antarctic becomes extremely cold during the winter as a result of the lack of sunlight over the polar region and because of greatly reduced mixing of the lower stratospheric air over this region with air outside this region. During the winter, a circumpolar vortex, also called the polar winter vortex, forms which isolates the air in the polar region from that outside of the region as a result of a stratospheric jet of wind circulating between approximately between 50° S and 65° S. The extremely cold temperatures inside the vortex lead to the formation of clouds in the lower stratosphere (from roughly 12 to 22 km), called 'polar stratospheric clouds'. Heterogeneous reactions occur on these particles that convert less reactive forms of chlorine to much more reactive ones with ozone. When daylight starts to occur over Antarctica in the early spring, this chlorine is available to react with and destroy ozone. Bromine compounds and nitrogen oxides can also react heterogeneously on the particles of these clouds. The ozone destruction continues until the polar vortex breaks up, usually in November.

In the late 1980s, it was generally thought that the Arctic lower stratosphere did not get cold enough to lead to decreases in ozone during the winter and springtime like those found in the Antarctic. The polar vortex is not generally as strong in the Northern Hemisphere, and, although polar stratospheric clouds would form, they would not likely last long enough for extensive decreases in ozone. However, since 1990, ozone decreases of as much as 30% have been found in the Arctic in those years when lower-stratospheric temperatures in the Arctic vortex have been sufficiently low to lead to ozone destruction processes similar to those found in the Antarctic ozone 'hole'. As with Antarctica, large increases in concentrations in reactive chlorine have been measured in the regions where the large ozone destruction is occurring.

The recognition of the harmful effect of chlorine and bromine on ozone spawned international action to restrict the production and use of CFCs and halons and protect stratospheric ozone. These included the 1987 Montreal Protocol on substances that deplete the ozone layer, the subsequent 1990 London Amendment, the 1992 Copenhagen Amendment, and the 1997 Montreal Amendment. In the Montreal Protocol and its Amendments, there is a distinction between the control measures in developed and developing countries. These agreements initially called for reduction of CFC consumption in developed countries. A November 1992 meeting of the United Nations Environment Program held in Copenhagen resulted in substantial modifications to the protocol because of the large observed decrease in ozone, and called for the phase-out of CFCs, carbon tetrachloride (CCl_4), and methyl chloroform (CH_3CCl_3) by 1996 in developed countries. As part of this, the United States, through the Clean Air Act, has eliminated production and import of these chemicals. Production of these compounds is to be totally phased out in developing countries by 2006, while production of halons in developed countries was stopped in 1994. Human-related production and emissions of methyl bromide were not to increase after 1994 in developed countries, and should slowly decline with total elimination by 2005. Hydrochlorofluorocarbons (HCFCs), many of which have been used as replacements for the CFCs, still contain chlorine that can destroy ozone, are to be phased out in the developed countries by 2030.

Worldwide compliance with the international agreements to protect ozone is resulting in significant reductions in the emissions of the CFCs, halons, and other halocarbons having the largest effects on ozone; as a result, levels of stratospheric ozone should slowly begin to recover over the coming decades as the reactive chlorine and bromine in the stratosphere declines. This recovery will be gradual, primarily because of the long time it takes for CFCs and halons to be removed from the atmosphere. Atmospheric model results, such as that shown in **Figure 6**, suggest that the effects of halocarbons on ozone should return to 1980 ozone levels by 2040–50.

While ozone slowly recovers, scientists and policymakers will need to work together to ensure that new problems do not develop as a result of the introduction of new chemicals into the marketplace. They will also need to interact with industry and governments to ensure that the potential effects of increasing concentrations of other gases changing as a result of human activities, such as methane and nitrous oxide, do not produce their own significant impacts on ozone. Further understanding is also needed towards evaluating the potential effects on ozone from various natural events, such as volcanic eruptions and solar events, and other possible human activities, including nuclear explosions.

See also: Carbon Cycle; Climate Change 2: Long-Term Dynamics; Matter and Matter Flows in the Biosphere; Radionuclides: Their Biochemical Cycles and the Impacts on the Biosphere.

Further Reading

Fioletov VE, Bodeker GE, Miller AJ, McPeters RD, and Stolarski R (2002) Global and zonal total ozone variations estimated from ground-based and satellite measurements 1964–2000. *Journal of Geophysical Research* 107: 4647 (doi:10.1029/2001JD001350).

Heimann M (2001) The cycle of atmospheric molecular oxygen and its isotopes. In: Schultze ED, Heimann M, Harrison S, *et al.* (eds.) *Global Biogeochemical Cycles in the Climate System*, pp. 235–244. San Diego: Academic Press.

Holland HD and Turekian KK (eds.) (2005) *Treatise on Geochemistry, Vol. 8: Biogeochemistry*. Oxford: Elsevier – Pergamon.

Jacobson MZ (2002) *Atmospheric Pollution: History, Science, and Regulation*. New York: Cambridge University Press.

Keeling RF (1995) The atmospheric oxygen cycle: The oxygen isotopes of atmospheric CO and O and the O/N ratio (U.S. National Report to IUGG, 1991–1994). *Reviews of Geophysics* 33(supplement): 95RG00438.

Margulis L and Schwartz KV (1988) *Five Kingdoms*, 2nd edn. New York: W. H. Freeman.

Schlesinger WH (1997) *Biogeochemistry: An Analysis of Global Change*. San Diego: Academic Press.

Turco RP (2002) *Earth Under Siege: From Air Pollution to Global Change*, 2nd edn. New York: Oxford University Press.

Turekian KK (1996) *Global Environmental Change: Past, Present, and Future*. Upper Saddle River, NJ: Prentice Hall.

Volk T (1998) *Gaia's Body: Toward a Physiology of Earth*. New York: Copernicus and Springer.

Phosphorus Cycle

Y Liu and J Chen, Tsinghua University, Beijing, People's Republic of China

Introduction

Phosphorus (P) is important because it is an essential ingredient of the energy metabolism of all forms of life. It is one of the three macronutrients needed by all crops (together with N and K). Human activity has quadrupled the mobilization of phosphorus and although the availability of this nonrenewable resource does not seem to pose a problem at the moment, there are other aspects of our phosphorus metabolism which do require our attention, namely the wastes (water and soil sinks) and how we affect the normal cycle of phosphorus on Earth. Throughout the metabolism of phosphorus in our economy, there are large amounts of wastes and emissions as will be shown later. P sources coming from industry, farmland, animal feed, and household consumption are all main contributors to over-nutrient water bodies, causing eutrophication. For the soil sink, P is accumulated in both agricultural and natural soils due to fertilizer application exceeding crop assimilation, and due to dumped industrial, agricultural, and animal wastes which slowly leach into the soil. Thus a huge amount of P is immobilized in soils, which results in deterioration of farmlands and the inefficient use of P resource.

The Human-Intensified Phosphorus Cycles

Inorganic Cycle

Phosphorus circulates through the environment in three natural cycles. The first of these is the inorganic cycle, which refers to phosphorus in the crust of the Earth. Through millions of years, phosphorus has moved slowly through the inorganic cycle, starting with the rocks which slowly weather to form soil, from which the phosphorus is gradually leached from the land into rivers and onward to the sea, where it eventually forms insoluble calcium phosphate and sinks to the seafloor as sediment. There it remains until it is converted to new, so-called sedimentary rocks as a result of geological pressure. On a timescale of hundreds of millions of years, these sediments are uplifted to form new dry land and the rocks are subject to weathering, completing the global cycle. In addition, some phosphorus can be transferred back from the ocean to the land by fish-eating birds whose droppings have built up sizable deposits of phosphate as guano on Pacific coastal regions and islands, and by ocean currents that convey phosphorus from the seawater to these regions. A simplified schematic of the global phosphorus cycle is presented in **Figure 1**.

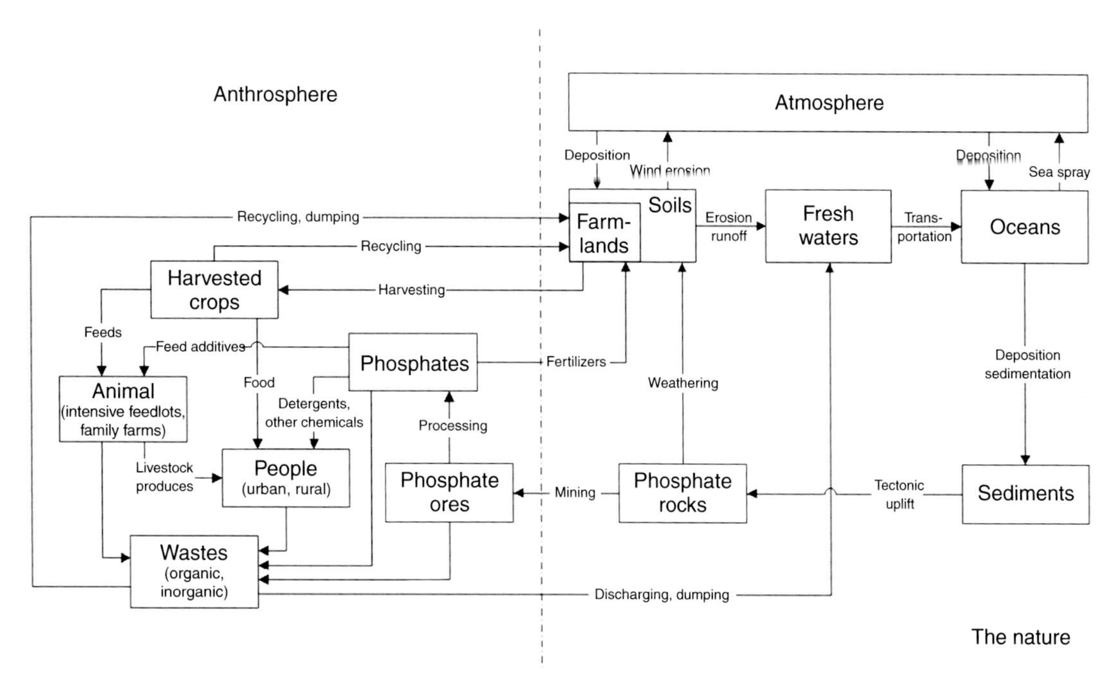

Figure 1 The human-intensified global phosphorus cycles.

The global cycle of phosphorus is unique among the cycles of the major biogeochemical elements in having no significant gaseous compounds. The biospheric phosphorus flows have no atmospheric link from ocean to land. A little phosphorus does get into the atmosphere as dust or sea spray, accounting for 4.3 million metric tons of phosphorus per year (MMT P yr^{-1}) and 0.3 MMT P yr^{-1}, respectively, but the amounts are several orders less important than other transfers in the global phosphorus cycle. The amount 4.6 MMT P yr^{-1} of atmospheric phosphorus deposition, being balanced by the phosphorus carried by the wind and the sea spray, cannot offset the endless drain of this element from the land due to erosion and river transportation. Fortunately, increased anthropogenic mobilization of the element has no direct atmospheric consequences.

Nearly all the phosphorus on land is originally derived from the weathering of calcium phosphate minerals, especially apatite [$Ca_5(PO_4)_3OH$]. Around 13 MMT P yr^{-1} of this is released to form soils each year. However, this amount cannot offset the annual losses of phosphorus from the land. Taking into account all four forms of phosphorus (dissolved and particulate, organic and inorganic), the total amount of annual phosphorus losses from the lithosphere into freshwaters is estimated at 18.7–31.4 MMT P yr^{-1}.

The uncertainty in the estimate is mainly due to a lack of knowledge on the biogeochemical processes of the particulate inorganic phosphorus (PIP), which constitutes the major component in the total loss. Not all of the eroding phosphorus can eventually reach the ocean. About 3.0 MMT P yr^{-1} is carried away by wind into atmosphere, and at least 25% of that is redeposited on adjacent cropland and grassland or on more distant alluvia. Consequently, the amount of phosphorus transported by freshwaters into the ocean is probably in the range 12–21 MMT P yr^{-1}. This result agrees with the most likely value of 17–22 MMT P yr^{-1} given by some previous estimates. The riverborne transport of phosphorus constitutes the main flux of the global phosphorus cycle. The loss, as a result of erosion, pollution, and fertilizer runoff, must be considerably higher than it was in prehuman times. It can be argued that the human-intensified phosphorus flux caused by wind and water erosion is at least 2 or even 3 times its prehistoric level.

Organic Cycles

Imposed on the inorganic cycle are two organic cycles which move phosphorus through living organisms as part of the food chain. These are a land-based phosphorus cycle which transfers it from soil to plants, to animals, and back to soil again; and a water-based organic cycle which circulates it among the creatures living in rivers, lakes, and seas. The land-based cycle takes a year on average and the water-based cycle organic cycle only weeks. It is the amount of phosphorus in these two cycles that governs the biomass of living forms that land and sea can sustain.

The amount of phosphorus in the world's soils is roughly 90–200 $\times 10^3$ MMT P according to various estimates. While the total phosphorus content of soils is large, only a small fraction is available to biota in most soils.

This constitutes an available phosphorus pool containing 1805–3000 MMT P, most likely 2000–2600 MMT P. A larger amount, in the range $27–840 \times 10^6$ MMT P, can be found in the oceans. The seawater contains $80–120 \times 10^3$ MMT P and the rest is accumulated in sediments.

The ocean water loses phosphorus continually in a steady drizzle of detritus to the bottom, where it builds up in the sediments as insoluble calcium phosphate. Despite the geological remobilization, there is a net annual loss of millions of tons of phosphate a year from the marine biosphere. Thus the ocean sediments are by far the largest stock in the biogeochemical cycles of phosphorus.

Global Phosphate Consumption

The phosphate rock is initially converted to phosphoric acid (P_2O_5) by reaction with sulfuric acid. The phosphoric acid is further processed to produce fertilizers, food-grade and feed-grade additives, and detergents. Other marginal applications include metal surface treatment, corrosion inhibition, flame retardants, water treatment, and ceramic production. Despite such widespread use, the latter applications represented only ~3% of the total consumption of various phosphates in the 1990s.

The global consumption of all phosphate fertilizers surpassed 1 MMT P yr^{-1} during the late 1930s. After reaching 14 MMT P yr^{-1} in 1980, the world consumption of phosphate fertilizers has been relatively stable. It was 14.8 MMT P (34 MMT P_2O_5) in statistical year 2002–03, and slightly decreased to 13.8 MMT P (31.5 MMT P_2O_5) in statistical year 2003–04, roughly accounting for 72.8% of the global extraction of phosphate rock. The top three economies, including China, the United States, and India, accounted for one-half of the world consumption. The area of the world current crop land is about 1.4 billion ha, implying that the global fertilizer application intensity averages 10 kg P ha^{-1}. The application rate varies significantly among continents, ranging from about 3 kg P ha^{-1} in Africa to over 25 kg P ha^{-1} in Europe. Among Western Europe countries, application levels range from 8.7 kg P ha^{-1} in Denmark and Sweden to 34 kg P ha^{-1} in Ireland.

Crop Harvests

The use of phosphates to nourish agricultural soils aims to replenish the removal of phosphorus from soil by harvests and erosion losses. Adopting the average phosphorus contents in crops and the harvest index, the global crop production harvested 12.7 MMT P from soils in 2005 as shown in **Table 1**. A study of Chinese phosphorus flows suggested that the national harvest in 2000 removed 3.4 MMT P from croplands, based on a set of 'domestic' data for the phosphorus contents and the harvest index. These two estimates agree that (1) cereals accounted for a major part of the harvested phosphorus, that is, 70% at the global level, and/or 68% in China; and (2) about two-thirds of the annually harvested phosphorus is contained in grains, and rest is contained in straw and other agricultural waste.

Since natural weathering and atmospheric deposition, as discussed elsewhere, cannot compensate the amount of phosphorus uptake from soils, application of phosphates, in both inorganic and organic forms, becomes essential to sustain today's harvests. There are several means of organic phosphorus reuse. The most direct means is to recycle crop residues *in situ*. Assuming roughly half of the annual output of crop residues (mostly cereal straw) is not removed from fields, the amount of the direct reuse of crop residues is about 2.2 MMT P yr^{-1}.

Table 1 Allocation of phosphorus in world harvest in 2005

	Harvested crops			*Crop residues*		
	Fresh weight (MMT)	*Dry matter (MMT)*	*P in grains (MMT P)*	*Residues (MMT)*	*P in straws (MMT P)*	*Total P uptake (MMT P)*
Cereals	2239	1968	5.9	2947	2.9	8.9
Suger crops	1534	476	0.5	370	0.7	1.2
Roots and tubers	713	143	0.1	219	0.2	0.4
Vegetables	882	88	0.1	147	0.1	0.2
Fruit	505	76	0.1	126	0.1	0.2
Pulses	61	58	0.3	61	0.1	0.4
Oilcrops	139	102	0.1	92	0.1	0.2
Other	100	80	0.1	200	0.2	0.3
Forages		500	1.0	0	0.0	1.0
Total	6173	3491	8.2	4163	4.5	12.7

Based on FAO (2006) FAOSTAT. Statistics Division, Food and Agriculture Organization of the United Nations.

Livestock and Animal Wastes

Animal wastes have been applied as organic manure in traditional farming and remain a relative large source of recyclable phosphorus in modern agriculture. According to the latest national survey data from the United States, beef cattle, dairy cattle, swine, and poultry produced 1.7 MMT P contained in animal manures in 1997, of which about one-half was produced by confined animals. The livestock population in the United States accounts for 7% of the world total in 1997 and the proportion has remained fairly constant. On this basis, the global production of animal wastes would be 24.0 MMT P yr^{-1}. However, the real figure may be somewhat smaller, because the animals in the United States are exceptionally well-fed. For this reason, the estimate of global production of 16–20 MMT P yr^{-1} in animal wastes, applying an average concentration of 0.8–1% of phosphorus for both confined and unconfined animal wastes, is probably more accurate.

Only the phosphorus in confined animal wastes is considered to be recyclable for croplands, while unconfined animal wastes mainly return to pastures. Assuming that animal biomass remains relatively constant, the amount of phosphorus in animal wastes is equal to the consumption of phosphorus contained in all kinds of feeds. In 2003, livestock consumed 36% of the harvested cereals (excluding the amount of cereals processed for beer), 21% of the harvested starchy roots, and 20% of the harvested pulses. Consequently, the annual livestock consumption of phosphorus in the harvests is accounted as about 2.9 MMT P yr^{-1}.

Some part of crop residues is used as animal fodder. However, the reuse ratio of crop residues as fodder considerably varies globally. For instance, it was reported that the percentage of crop residues – mostly the straws of rice, wheat, and corn (maize) – used as fodder ranged from 3.6% in Shanghai to 42.8% in Gansu Province in 2000 in China, depending on crop and livestock species, farming and feeding traditions, and local economic profiles, and averaged 22.6% across the nation. Since over 70% of world livestock are raised in developing countries where commercial feeds are less used, the global recycling rate of crop straws as fodder is probably about 25%, leading to an absolute quantity of 1.1 MMT P yr^{-1}.

Another major source of animal daily phosphorus intake is via feed additives. Around 6% of the global yield of phosphoric acid has been processed as animal feed-grade additives since 2000. This constitutes an annual phosphorus flux of 1.0 MMT P yr^{-1} input to livestock husbandry.

Adding all above three sources, the global livestock consumption of phosphorus amounts to *c.* 5.0 MMT P yr^{-1}. Taking into account the recycling of various industrial by-products and kitchen organic wastes (which is prevalent in rural family-based farms in developing countries), this figure could be as much as 20% higher, resulting in a total of 6.0 MMT P yr^{-1}. Of course, the phosphorus flux to livestock of 6.0 MMT P yr^{-1} is mainly consumed by animals in confined facilities, while the world's cultivated and natural pastures provide a major source of phosphorus for unconfined animals. Assuming that 1.0 MMT P yr^{-1} goes to unconfined animals in pastures, about 5.0 MMT P yr^{-1} consumed by confined animals gives an approximation for the maximum potential of recoverable phosphorus for croplands. If one-half of the organic phosphorus in confined animal wastes is subject to recycling, animal manure is responsible for about 2.5 MMT P yr^{-1} returns to global croplands.

Food Consumption and Human Wastes

The third source of organic phosphorus available, in principle, for cropland is human waste. Assuming the world human body mass averaging 45 kg per capita (reflecting a higher proportion of children in the total population of low-income countries), and phosphorus content in human body averaging 470 g P per capita, implies a global anthropomass contains approximately 3.0 MMT P. The typical daily consumption is about 1500 mg P per capita for adults. This is well above the dietary reference intake (DRI), the amount human individual should take each day, as recommended by the Food and Nutrition Board, Institute of Medicine, US National Academy of Science. The US recommended intakes are 700 mg per capita for adults over 18 years of age, 1250 mg per capita for young adults between 9 and 18 years of age, and 500 mg per capita for children.

A similar estimate for China is derived from a previous study: the individual daily intake of phosphorus was 1400 mg P per capita for urban residents and 1470 mg P per capita for rural residents in 2000. This exceeds the DRI of 1000 mg per capita recommended by the Chinese Nutrient Society. In addition, livestock products provided 30% of daily phosphorus intake for Chinese urban residents, and 14% of that for rural population.

Although the phosphorus content of anthropomass is marginal in comparison with that of soil biota or ocean biota, the ecological consequences can still be significant. Given the annual world population growth of 78.4 million since 2000, the net accumulation of phosphorus in the global anthropomass is around 0.04 MMT P yr^{-1}. Compared with the global consumption of phosphorus in foods, the slight increase of phosphorus in the anthropomass stock implies a low assimilation rate of about 0.5%. Assuming a global average dietary consumption of 1400 mg P per capita, human excreta must have contained about 3.3 MMT P yr^{-1}, of which urban and rural population generated 1.6 and 1.7 MMT P yr^{-1}, respectively.

Application of human excreta as organic fertilizer is common both in Asia and in Europe, but less prevalent elsewhere in the world. The nutrient linkage between farmers and croplands has been relatively stable, but the human wastes in urban areas are less recycled than in rural areas. For instance, less than 30% of human wastes in urban areas were recycled for agricultural uses in the late 1990s in China. This percentage dramatically decreased from 90% in 1980. In contrast, about 94% of human wastes in rural areas were returned to croplands in the 1990s. In European countries, the recycling rate of urban sewage averaged about 50% over the 1990s. Globally, it could be appropriate to assume that about 20% of urban human wastes and about 70% of rural human wastes are recycled at present. Therefore, recycled human wastes amount to 1.5 MMT $P\,yr^{-1}$.

Adding the quantities of the phosphorus recycled as crop residues, animal manures, and human wastes, the total organic fertilizers applied to croplands amounts to 6.2 MMT $P\,yr^{-1}$. This is equivalent to 45% of the applied amount of inorganic fertilizers. Thus, the global input of phosphorus to croplands is probably 20 MMT $P\,yr^{-1}$ in total, or 1.6 times the amount of the phosphorus removed from the soil by harvesting. This leads to a net accumulation of 7.3 MMT $P\,yr^{-1}$ or 4.7 kg $P\,ha^{-1}$ in global soils, disregarding erosion and runoff losses.

Phosphates in Soil

The distribution, dynamics, and availability of phosphorus in soil are controlled by a combination of biological, chemical, and physical processes. These processes deserve special attention, as a considerable proportion of the applied phosphate is transformed into insoluble calcium, iron, or aluminum phosphates. On average, only a small proportion, perhaps 15–20% of the total amount of phosphorus in the plant, comes directly from the fertilizer applied to the crop. The remainder comes from soil reserves. For most of the twentieth century, farmers in Western countries were advised to add more than double the amount of phosphate required by a crop, because these immobilized calcium, iron, and aluminum phosphates had been assumed to be permanently unavailable to plants.

The primary source of phosphorus taken up by plants and microorganisms is dissolved in water (soil solution). The equilibrium concentration of phosphate present in soil solution is commonly very low, below 5 μmol. At any given time, soil water contains only about 1% of the phosphorus required to sustain normal plant growth for a season. Thus, phosphates removed by plant and microbial uptake must be continually replenished from the inorganic, organic, and microbial phosphorus pools in the soil. These continuous processes dominate contemporary agricultural production to remove about 8.2 kg $P\,ha^{-1}$ from cropland each year on a world average (based on our own estimate), and commonly 30 kg $P\,ha^{-1}$ from the US and European fertile agricultural soils.

Each phosphate mineral has a characteristic solubility under defined conditions. The solubility of many compounds is a function of acidity (pH; **Figure 2**). An increase in pH can release sediment-bound phosphorus by increasing the charge of iron and aluminum hydrous oxides and therefore increasing the competition between hydroxide and phosphate anions for sorption sites. Also, organic acids can inhibit the crystallization of aluminum and iron hydrous oxides, reducing the rate of phosphorus occlusion. The production and release of oxalic acid by fungi explains their importance in maintaining and supplying phosphorus to plants.

The relative sizes of the sources and stocks of phosphate in soil change as a function of soil development (**Figure 3**). The buildup of organic phosphates in the soil is the most dramatic change. As time goes on, this becomes the chief reservoir of reserve phosphate in the soil. In most soils, organic phosphates range from 30% to 65% of total phosphorus, and it may account for as high as 90%, especially in tropical soils. The reasons for this are their insolubility and chemical stability. It has been noted that acid soils tend to accumulate more total organic phosphorus than do alkaline soils. This is almost certainly because organic phosphates react with iron and aluminum under acid conditions and become insoluble. Being the salts or metal complexes of phosphate esters they release their phosphate by hydrolysis, but only very slowly. Phosphate esters can have half-lives of hydrolysis of hundreds of years. This process can be greatly speeded up by

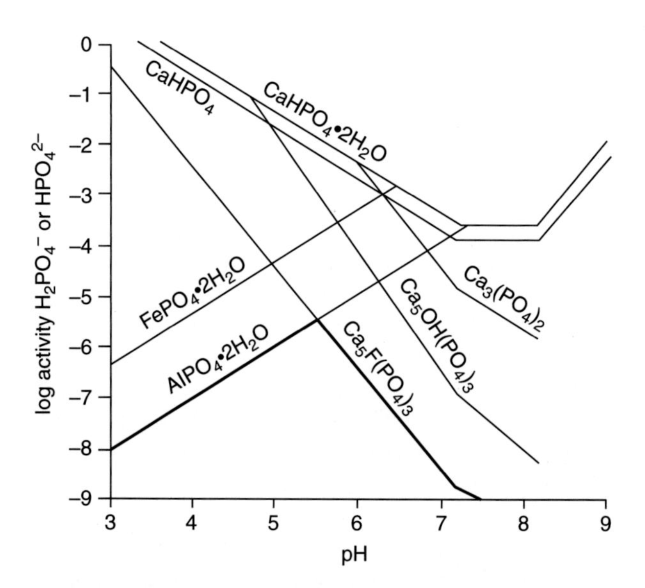

Figure 2 The solubility of phosphorus in the soil solution as a function of pH. Adapted from Schlesinger WH (1991) *Biogeochemistry: An Analysis of Global Change*. San Diego, CA: Academic Press.

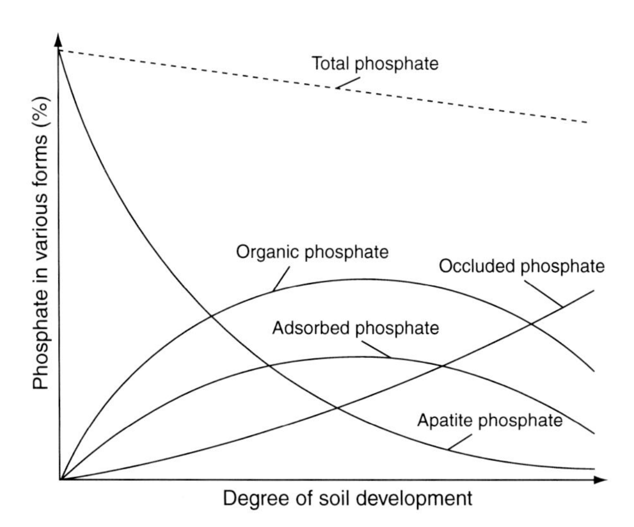

Figure 3 Phosphates in the soil vary with soil development. Adapted from Emsley J (1980) The phosphorus cycle. In: Hutzinger O (ed.) *The Handbook of Environmental Chemistry: The Natural Environment and the Biogeochemical Cycles*, pp. 147–167. Heidelberg: Springer.

the action of phosphatase enzymes in the soil whose function is to facilitate reaction by catalyzing it. At later stages of soil development, phosphorus is progressively transformed into less-soluble iron- and aluminum-associated forms, and organic phosphorus contents of the soil decline. At this stage almost all available phosphorus is found in a biogeochemical cycle in the upper soil profile, while phosphorus in lower depths is primarily involved in geochemical reactions with secondary sediments.

When the supply of dissolved phosphates to growing biomass is abundant, a net immobilization of inorganic phosphorus into organic forms will occur. Vice versa, inadequate inorganic phosphorus supply will stimulate the production of phosphatases and the mineralization of labile organic forms of phosphorus for microbial uptake. A continuous drain on the soil phosphorus pools by cultivation and crop removal will rapidly deplete both labile inorganic and organic phosphorus in soils.

Allowing soil reserves of readily available phosphorus to fall below a critical value, determined by field experiments, can result in a loss of yield. The turnover of available phosphates by plants in soil solution is determined by rates of releases of phosphorus from insoluble forms to soluble phosphates. All kinds of soil particles can contribute to this process and in some cases it is not only chemical balance that maintains the supply but also the action of microbes and enzymes that release phosphate from organic debris in the soil. It is believed that the biogeochemical control of phosphorus availability by symbiotic fungi is a precursor to the successful establishment of plants on land.

However, our existing knowledge – briefly discussed above – cannot yet provide a comprehensive understanding of the complex movements and transformations of phosphorus, especially of its organic forms. This hampers the efficient application of phosphate fertilizers, and the efficient control of phosphorus losses.

Phosphorus Losses

Phosphorus is lost from croplands via erosion or runoff. Quantifying phosphorus losses in eroding agricultural soils is particularly uncertain, as erosion rates vary widely even within a single field. It is also because that few nations have comprehensive, periodic inventories of their soil erosion.

The phosphorus loss from croplands can be roughly estimated based on the amount of topsoil erosion and average phosphorus content. A crop takes up the majority of the nutrients it requires from topsoil. The topsoil is often identified as the 'plough layer', that is, the 20–30 cm depth of soil which is turned over before seedbed preparation. The volume of topsoil in the plough layer is around $2500\,m^3\,ha^{-1}$and weighs approximately 2000 t. A ton of fertile topsoil can contain 0.6–3.0 kg of phosphorus, based mainly on US and European agricultural practices.

It has been estimated that the annual soil erosion from agricultural systems of the United States, China, and India was 3.0, 5.5, and $6.6\,Gt\,yr^{-1}$, respectively. Since these three countries hold 23% of global agricultural lands, the global soil erosion could reach $66\,Gt\,yr^{-1}$. Some of the most serious soil erosion takes place in the agricultural systems of Southeast Asia, Africa, and South America. Hence, the real global soil erosion loss could be even higher, perhaps as much as $75\,Gt\,yr^{-1}$.

The erosion intensity from croplands varies a lot among countries, ranging from 0.5 to $400\,t\,ha^{-1}\,yr^{-1}$. Worldwide, soil erosion is highest in the Asia, Africa, and South America, averaging $30–40\,t\,ha^{-1}\,yr^{-1}$ of soil loss. It was suggested that the global average erosion rate is at least $20\,t\,ha^{-1}\,yr^{-1}$. The lowest erosion rates occur in the United States and Europe where they average about $10\,t\,ha^{-1}$ each year. It is evident that soil erosion in the US has been reduced by soil conservation policies; a national survey showed that the total soil erosion between 1982 and 1992 decreased by 32%. The annual sheet and rill erosion rate in the US fell from an average of $10\,t\,ha^{-1}$ in 1982 to $7.7\,t\,ha^{-1}$ in 1992, and the wind erosion rate fell from an average of $8.1\,t\,ha^{-1}\,yr^{-1}$ to $5.9\,t\,ha^{-1}\,yr^{-1}$ over the same period. Assuming global erosion rate averaging $25\,t\,ha^{-1}$ gives the soil loss of $38.5\,Gt\,yr^{-1}$ from cropland (cf. **Table 1**). Furthermore, most of the loss is permanent and may not be replenished by weathering. For instance, the excessive soil loss, a rate that would impair long-term crop productivity, is estimated at about $25.4\,Gt\,yr^{-1}$ in around 1980.

Erosion from pastures is commonly less intensive than that from ploughed fields. However, soil losses have been greatly increased by overgrazing, which now affects more than half (i.e., at least 1720×10^6 ha) of the world's permanent pastures with a high erosion rate of 15 t ha^{-1} each year. This leads to 25.8 Gt yr^{-1} of the soil loss from overgrazed pastures. Together with the amount of soil loss from cultivated grassland, the world's permanent pastures lose their topsoil at an annual rate of 34.4 Gt yr^{-1}. Adding the losses from cropland and pastures, the world soil erosion from agricultural areas amounts to 72.9 Gt yr^{-1} in total, or 15 t ha^{-1} yr^{-1} on average. This is similar to previous estimates as discussed above.

Allowing for the poor condition of topsoil in developing countries, it might be appropriate to assume that the global phosphorus content in topsoil averages about 0.5 kg P t^{-1}, or 1.0 t P ha^{-1}. This gives the world phosphorus losses at 19.3 and 17.2 MMT P yr^{-1} from cropland and pastures, respectively, as shown in **Table 2**.

The surface runoff loss of applied inorganic phosphate fertilizer varies significantly with a number of agronomic factors. Typical runoff rates of phosphorus in European countries range from 0.2% to 6.7%, an average of 3.5%. Worldwide, the maximum rate can reach 10% under certain soil characteristics and climatic condition. Roughly, the world total phosphate fertilizer application can lead to a loss of 0.5 MMT P yr^{-1} in surface runoff.

Phosphate Balance in Cropland

The national phosphorus balance varies significantly from one country to another, due to differences in the use of mineral fertilizers and manure, and differences in animal husbandry practices. Broadly speaking, in developing countries, soils tend to be deficient in phosphorus, while in developed countries the phosphorus content of the soils is adequate or even excessive. Taking into account applications of mineral fertilizers and manure, the balance for some West European countries is positive, particularly in the Netherlands where it exceeds 39 kg P ha^{-1} each year. For the majority of Western European countries, the phosphorus balance ranges from 8.7 to 17.5 kg P ha^{-1} annually. China, one of the largest agricultural systems in transition, also achieved a positive balance around 1980 at the national level, in parallel with increasing application of synthetic fertilizers. In 2000, the national surplus of phosphorus in Chinese soils was estimated at an average of 16 kg P ha^{-1}.

The world phosphorus budget for cropland is summarized in **Table 3**. To balance the phosphorus budget for the world's cropland, two natural inflows of phosphorus to croplands should be taken into account. Based on the ratio of cropland area to world total land areas, weathering and atmospheric deposition contribute 2.0 MMT P yr^{-1} as inputs to world croplands. Although the magnitudes of recycling of animal wastes and soil erosion need further verification, the budget provides a comprehensive overview on the global phosphorus flows associated with the farming sector, which is the most intensive and complicated subsystem of the anthropogenic phosphorus cycle.

Table 2 Global soil erosion and phosphorus losses from agricultural land (2003)

		Permanent pasture	
	Cropland	*Overgrazed*	*Ordinary*
Total area (million ha)	1540	1720	1720
Soil erosion			
Erosion rate (t ha^{-1} yr^{-1})	25	15	5
Erosion quantity (Gt yr^{-1})	38.5	25.8	8.6
Phosphorus loss			
P content in topsoil (kg P ha^{-1})	0.5	0.5	0.5
P loss (MMT P yr^{-1})	19.3	12.9	4.3

Ecological Impacts of Phosphorus Use

The phosphorus-related ecological issues fall into a broad range. Some of them are caused by inappropriate use of the material, some are not. Eutrophication, being regarded as the most immediate environmental consequences of extensive phosphorus usage in contemporary societies, has received wide attention. However, it is not

Table 3 Phosphorus budget for the world's cropland in recent year (2004)

Flows	*Annual fluxes (MMT P)*
Inputs	*22.0*
Weathering	1.6
Atmospheric deposition	0.4
Synthetic fertilizers	13.8
Organic recycling	6.2
Crop residues	2.2
Animal wastes	2.5
Human wastes	1.5
Removals	*12.7*
Crops	8.2
Crop residues	4.5
Losses	*19.8*
Erosion	19.3
Runoff	0.5
Balance	*−10.5*
Input shares	
Fertilizer application	69%
Organic recycling	31%
Uptake efficiency	64%

the whole story, and other issues deserve to be taken into consideration.

Mineral Conservation

Keeping the annual mining rate of about 19.5 MMT P yr^{-1} constant, the world's known phosphate reserves could be exhausted in about 120 years. Moreover, it has been projected that the utilization trend is unlikely to decline in next 30 years. It will instead probably increase at the rate of 0.7–1.3% annually. This strongly suggests that phosphate rock, as a finite nonrenewable resource, may be exhausted in a much shorter time. It has been shown that the global average phosphorus content in raw ores dropped to 29.5% in 1996 from 32.7% in 1980, and that global reserves can sustain the current mining intensity for only another 80 years. Some phosphorus-rich deposits around the world can be exploited much sooner. China's phosphorus reserves, for instance, constitute 26% of the world's total reserve base, second only to Morocco and the Western Sahara. With a high intensive extraction activity as well as losses incurred during mining, the basic reserve of the nation's phosphorus resources, that is, 4054 million tons with average P_2O_5 content of 17–22%, could be exhausted in 64–83 years. Certainly, the larger reserve base and probably more reserves to be discovered in the future guarantee a longer lifespan of the extraction. Even so, the deposits of phosphorus in the lithosphere will inevitably be depleted before new igneous or sedimentary rocks to be formed via the biogeochemical process at the timescale of millions of years.

Soil Erosion

One of the most important results derived from the phosphorus budget (cf. **Table 3**) is that the world cropland has lost phosphorus at a surprising rate of 10.5 MMT P yr^{-1}. This massive loss from croplands is mainly caused by wind and water erosion of topsoil. Soil erosion has been recognized as one of the most serious environmental crisis suffering the world. It is estimated that 10 million ha of cropland is abandoned each year worldwide due to lack of productivity caused by the soil erosion. Nearly 60% of present soil erosion is induced by human activity, increasing 17% since the early 1900s.

In contrast to the erosion loss, a huge amount of phosphorus has been mobilized in cultivated soils. Contemporary scientific knowledge cannot fully explain the complex transportation of phosphorus between plant roots, soil waters, and soil particulates. More complete understanding of these processes might suggest a possibility of controllable remobilization of soil phosphorus that would benefit the environment via reducing both the input of fertilizers and the loss of phosphorus from soils.

Animal Wastes

Livestock husbandry, in particular large intensive feedlots, has become a major problem both for recycling of organic phosphorus and for emission of phosphorus pollutants. Worldwide, the structure of animal agriculture has changed as livestock are concentrated in fewer but larger operations. In the United States, in spite of losing nearly a fourth of the livestock operations between 1982 and 1997, the total number of animal units (an equivalent that converts various kinds of animals into cattle based on individual nutrient excretion) has remained fairly constant at *c.* 91–95 million. In China, a large number of intensive feedlots appeared in suburbs and rural areas during the last decade. According to a national investigation, the output of hogs, meat chicken (broilers), and egg chicken (layers) produced by intensive feedlots and farms accounted for 23%, 48%, and 44% of the national total in 1999, respectively.

As livestock operations have become fewer, larger, and more spatially concentrated in specific areas, animal wastes have also become more concentrated in those regions. This leads to a considerable phosphorus surplus in manure, as the amount of manure nutrients relative to the assimilative capacity of land available on farms for application has grown, especially in specific high-production areas. Consequently, off-farm manure export requirements are increasing.

But because of its bulk, uneven distribution, and prohibitive cost of transport beyond a limited radius, a large proportion of manure phosphorus is now subject to disposal instead of recycling. If construction of necessary infrastructures for appropriate disposal of manure lags behind, animal wastes become a major source of phosphorus loads in surface waters. Uncontrolled phosphorus emission from intensive feedlots and farms in China has escalated in parallel with the gradual growth in total animal feeding operations and the rapid shift in breeding structure. The emission of China's livestock was estimated at 36% of its national phosphorus load to aquatic environments in 2000. Thus, livestock husbandry is the most significant source of phosphorus flux to surface waters in China, similar to the situation in European countries in the early 1990s.

Sewage Treatment

In the 1960s, many developed countries began to alleviate the pollution in surface water by constructing municipal sewage infrastructures and implementing phosphorus discharge restrictions on production sectors. The giant infrastructure of centralized wastewater treatment has drastically reshaped the phosphorus cycle within modern cities. Despite high economic costs, its environmental benefits with regard to removal of phosphorus from wastewater are far less than satisfactory worldwide. However, some progress has been achieved in European countries and the United States. As the centralized control strategy just removes 'pollutants' into sewage sludge rather than promotes a recovery and recycling of resources, including phosphorus, it does not really solve the long-term ecological problem. The costly and rigid infrastructures have significantly reduced agricultural reuse of urban human excreta and contributed to a disconnect of nutrient cycles between urban areas and croplands. Unfortunately, no available technologies for stable recovery and recycling of phosphorus are likely to be successfully commercialized in the near future. Hence, most of the phosphorus in urban human wastes is not subject to efficient recycling and is permanently lost from the land.

Proposals for recovery of phosphorus via decentralized source-separated strategies have received increasing attention since the mid-1990s. This decentralized and downsized sanitation concept, focused on ecologically sustainable and economically feasible closed-loop systems rather than on expensive end-of-pipe technologies, advances a new philosophy. It departs from the one-way flow of excreta from terrestrial to aquatic environments, as introduced by the conventional flush-and-discharge sewage system. The new alternative separates nutrients and domestic used water at source and handles both components individually based on material flows approaches. Thus, it avoids the disadvantages of conventional wastewater solutions and enables and facilitates nutrient recycling. Although the reinvention and transition of urban wastewater systems poses a major challenge, it does provide a promising prospect for future phosphorus recovery and recycling in an ecological and economic efficient way. (Detailed studies are essential as a first step, *inter alia*, of technological, organizational, economic, and social aspects. In addition, the involvement of multi-stake-holders, such as residents, building owners, farmers, politicians, officials, and other interested parties, from the start seems essential. All these problems cannot be solved overnight, as it requires nothing less than a paradigmatic change of a large sociotechnical system.)

Detergents Use

The use of sodium tripolyphosphate ($Na_5P_3O_{10}$, STPP), the most widely used detergent additive, has been identified as a significant contributor to eutrophication. STPP was first introduced in the US in 1946. After reaching a peak in the 1960s, global production has finally fallen down to one-half of the peak level, *c.* 1.0 MMT P yr^{-1}, mainly due to bans on phosphorus-containing detergents in developed countries. The total quantity of STPP production was estimated at 0.865 MMT P in 2004. In the late 1990s, phosphorus-free detergents accounted for 45%, 97%, and 100%, respectively, in the United States, Japan, and the European countries.

There has been a controversy on the environmental impacts of STPP since the mid-1980s. Today, it is acknowledged that limiting or banning household consumption of phosphorus-containing detergents would not lead to a significant or a perceivable improvement of eutrophication. It would have little impact on water and human health compared to other substitute chemicals (sodium carbonates, sodium silicates or zeolites A, and sodium nitrilotriacetate), both from an ecological and an economic perspective. In parallel with these discoveries, some Nordic countries ecolabeled STPP as an environmentally friendly component of detergents in 1997, and have repromoted the production and consumption of STPP since then.

Eutrophication

Eutrophication is an unwanted explosion of living aquatic-based organisms in lakes and estuaries which results in oxygen depletion that can destroy an aquatic ecosystem. It has been regarded as the most important environmental problem caused by phosphorus losses. Significant eutrophication took place in the 1950s in the Great Lakes of North America and has been prevalent in many lakes and estuaries around the world. Phosphorus is often the limiting factor responsible for eutrophication, since nitrogen fluxes to water bodies are relative large.

Phosphorus losses from industries, croplands, animal farms, and households constitute the main sources. **Figure 4** illustrates a cross-country comparison of the phosphorus loads in European countries and in China to their domestic aquatic environments.

The results show that the phosphorus loads range from 0.2 kg P ha^{-1} in Sweden to 2.5 kg P ha^{-1} in Belgium at the national level. China lies between Germany and Northern Ireland in turns of the load per unit land area. At the basin level, the comparison of phosphorus loads shows a similar figure. These results suggest that the Chinese economy is in general processing phosphorus 'wastes' less efficiently than developed countries. However, regarding phosphorus control strategy, it is nearly impossible to determine a common benchmark (of a desired phosphorus load) to prevent water bodies from eutrophication. This is because the complex interrelations between the amount of aquatic biomass and the phosphorus load are affected by a number of hydrological, meteorological, and biochemical factors that remain unclear under current knowledge. Further improvement of our current understanding of phosphorus movement and transformation within aquatic and territorial ecology, across socioeconomic and ecological boundaries, is desired.

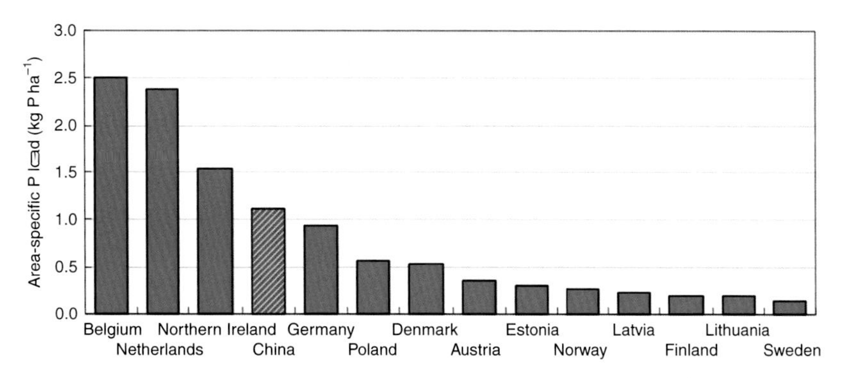

Figure 4 Comparison of phosphorus loads to aquatic environments by country (unit: kg P ha^{-1}). Based on EEA (2005) *Source Apportionment of Nitrogen and Phosphorus Inputs into the Aquatic Environment*, No.7/2005. Copenhagen: European Environment Agency; Liu Y (2005) *Phosphorus Flows in China: Physical Profiles and Environmental Regulation*. PhD Thesis, Environmental Policy Group, Wageningen University, Wageningen, The Netherlands.

Regulating the Societal Phosphorus Flows

Phosphorus (P), intensively extracted from the natural sink in the lithosphere and processed through various production–consumption cycles, ultimately deposits in soil or reaches the water body by different pathways. The societal P flows are characterized by complicated physical interconnections in high intensities among a number of production and consumption sectors. Hence, it is vitally important to connect the P flows with environmental regulations introduced to intervene in social practices and human behaviors.

Instead of continuously trying to limit the growth of P (like the bans of detergent P), there is a great need to reconstruct the physical structure of P flows, in particular by redirecting the crucial P flows with highly negative environmental impacts. The ecological restructuring of the current one-through mode of societal P metabolism is thus desired, leading to a structural shift in the societal production and consumption of P flows. The ecologically rational switch can contribute to a substantial decline of P outflows by minimizing P input and maximizing P recycling. Since ecologizing the P flows only succeeds when measures are institutionalized into the economy and society as a whole, this process will most likely be a gradual one rather than a radical revolution.

See also: Nitrogen Cycle.

Further Reading

Compton JS, Mallinson DJ, Glenn CR, *et al.* (2000) Variations in the global phosphorus cycle. In: Glenn CR (ed.) *Marine Authigenesis: From Global to Microbial*, pp. 21–33. Tulsa, OK: Society for Sedimentary Geology (SEPM).

Driver J (1998) Phosphates recovery for recycling from sewage and animal wastes. *Phosphorus and Potassium* 216: 17–21.

EEA (2005) *Source Apportionment of Nitrogen and Phosphorus Inputs into the Aquatic Environment*, No.7/2005. Copenhagen: European Environment Agency.

Emsley J (1980) The phosphorus cycle. In: Hutzinger O (ed.) *The Handbook of Environmental Chemistry: The Natural Environment and the Biogeochemical Cycles*, pp. 147–167. Heidelberg: Springer.

FAO (2006) FAOSTAT. Statistics Division, Food and Agriculture Organization of the United Nations.

Filippelli GM (2002) The global phosphorus cycle. *Reviews in Mineralogy and Geochemistry* 48: 391–425.

Grove TL (1992) Phosphorus, biogeochemistry. In: Nierenberg WA (ed.) *Encyclopedia of Earth System Science*, pp. 579–587. London: Academic Press.

Hart MR, Quin BF, and Nguyen ML (2004) Phosphorus runoff from agricultural land and direct fertilizer effects: A review. *Journal of Environ Quality* 33(6): 1954–1972.

Litke DW (1999) Review of phosphorus control measures in the United States and their effects on water quality. *Water-Resources Investigations Report 99-4007*. Denver, CO: US Geological Survey.

Liu Y (2005) *Phosphorus Flows in China: Physical Profiles and Environmental Regulation*. PhD Thesis, Environmental Policy Group, Wageningen University, Wageningen, The Netherlands.

Pimentel D (2006) Soil erosion: A food and environmental threat. *Environment, Development and Sustainability* (8): 119–137.

Richey JE (1983) The phosphorus cycle. In: Bolin B and Cook RB (eds.) *The Major Biogeochemical Cycles and Their Interactions*, pp. 51–56. New York: Wiley.

Schlesinger WH (1991) *Biogeochemistry: An Analysis of Global Change*. San Diego, CA: Academic Press.

Smil V (2000) Phosphorus in the environment: Natural flows and human interferences. *Annual Review of Energy and the Environment* 25: 53–88.

Turner BL, Frossard E, and Baldwin DS (eds.) (2005) *Organic Phosphorus in the Environment*. Wallingford, UK: CABI Publishing.

Valsami-Jones E (ed.) (2004) *Integrated Environmental Technology Series: Phosphorus in Environmental Technology: Principles and Applications*. Cornwall, UK: IWA Publishing.

Radiation Balance and Solar Radiation Spectrum

I N Sokolik, Georgia Institute of Technology, Atlanta, GA, USA

Introduction

Global radiation balance ultimately controls the climate of the Earth and thus plays an important role in the functioning of all ecosystems. The two key components of the global radiation balance are solar radiation, which is emitted by the Sun, and terrestrial radiation, which is emitted by the Earth's surface and atmosphere. Besides regulating climate, solar radiation is vital for sustaining life on our planet, given that almost all living organisms need sunlight for their well-being. This article discusses the global radiation balance and its main components, including solar radiation and its spectrum.

Solar Constant and Solar Spectrum

The Sun is a typical G2 star in the universe powered by nuclear reactions, mainly the nuclear fusion of hydrogen atoms to form helium. Because the Sun is the closest star to our planet with a mean distance of about 1.496×10^{11} m, it is the major external source of energy for the Earth. In its absence, the Earth would be a cold and lifeless planet. Solar radiation, or electromagnetic radiation emitted by the Sun, corresponds to a blackbody emission temperature of about 5780 K. According to the Stefan–Boltzmann law, the radiant energy emitted by the blackbody at temperature T is

$$F = \sigma_b T^4$$

where F is the radiative flux (energy per unit time across unit surface area) and σ_b is the Stefan–Boltzmann constant, equal to $5.671 \times 10^{-8}\,\mathrm{W\,m^{-2}\,K^{-4}}$. This gives a solar flux of $6.329 \times 10^7\,\mathrm{W\,m^{-2}}$. Then the solar flux reaching the Earth can be estimated from the energy conservation law as

$$F \times 4\pi a_s^2 = S_0 \times 4\pi r^2$$

where a_s is the radius of the Sun, and S_0 is the solar constant, defined as the solar flux corresponding to the mean distance, r, between the Earth and the Sun incident on a surface at the top of the Earth's atmosphere normal to the direction of propagation. Taking $a_s = 6.96 \times 10^5$ km, one finds an approximate value of the solar constant of $1370\,\mathrm{W\,m^{-2}}$. Given the importance of solar energy, knowledge of the exact value of the solar constant is of fundamental interest. Historically, the first measurements of the solar constant have been done with the ground-based instruments. Unfortunately, these observations have a limited accuracy because of the atmospheric contamination. Over the past 30 years, a number of sophisticated sensors flown on the satellites have provided a reach data set. Based on this data, a mean value of $1366\,\mathrm{W\,m^{-2}}$ with an uncertainty of $\pm 3\,\mathrm{W\,m^{-2}}$ has been suggested.

Although it is called the constant, the value of the solar constant does vary with time, mainly due to the Sun's activities. Periodic changes in the solar constant are related to the 11-year sunspot cycle. The sunspots are the relatively cold regions on the surface of the Sun with an average temperature of about 4000 K. Also, the solar constant varies due to the varying distance of the Earth from the Sun over the course of the year. Although overall changes in S_0 are relatively small, they are of great importance to understanding the natural climate variability.

The distribution of solar radiation as a function of wavelength is called the solar spectrum. The emission of a blackbody having temperature of the Sun provides a good approximation to the solar spectrum. This emission is given by the so-called Planck function which relates the emitted energy to the wavelength and the temperature of the blackbody:

$$B_\lambda(T) = \frac{2hc^2}{\lambda^5 (e^{hc/k_B T\lambda} - 1)}$$

where λ is the wavelength, h is the Planck's constant ($=6.63 \times 10^{-34}$ J s), k_B is Boltzmann's constant ($=1.38 \times 10^{-23}\,\mathrm{J\,K^{-1}}$), c is the velocity of light, and T is the absolute temperature (in kelvins) of the blackbody.

As a matter of convention, the electromagnetic spectrum is subdivided into discrete spectral bands with assigned names. The approximate wavelength boundaries and commonly used names are given in **Table 1**. Spanning the ultraviolet (UV), visible, and near-infrared (near-IR) bands, the region between about 0.01 and 4 μm contains practically

Table 1 The electromagnetic spectrum

Name of the spectral region	*Wavelength region* (μm)
X-ray	<0.01
Ultraviolet (UV)	0.01–0.4
Visible	0.4–0.7
Near-infrared (near-IR)	0.7–4
Thermal	4–1000
Microwave	10^3–10^6
Radio	>10^6

all solar energy that is of relevance to the Earth's radiation balance. Radiation that falls in this region is called solar radiation or shortwave radiation.

Figure 1 shows the spectrum of solar radiation reaching the top of the atmosphere and the spectrum of the blackbody at $T = 6000$ K. The percentage of solar energy in the UV, visible, and near-IR bands is also shown. Although the fraction of solar radiation at the top of the atmosphere that falls in the UV region is only about 8%, this radiation plays a major role in atmospheric photochemistry and heats the stratosphere and mesosphere. The visible region extending from about 0.4 to 0.7 μm contains about 40% of solar radiation. This is the only part of the spectrum that is visible to the human eye. Not only is it important to the radiation budget, the solar radiation in this spectral region is vital to life on our planet. It is needed to aid terrestrial plants in the process of photosynthesis; hence this radiation is called photosynthetic active radiation (PAR).

The solar spectrum is fundamental to understanding the fate of solar radiation in the Earth's surface–atmosphere system because interactions of electromagnetic radiation with matter depend on wavelength. A wavelength of about 4 μm separates the spectrum containing most solar radiation from that containing most terrestrial (emitted by the Earth) radiation. Therefore, it is common to consider the fate of solar and terrestrial radiation independently and to add them together when the radiant energy balance is of interest.

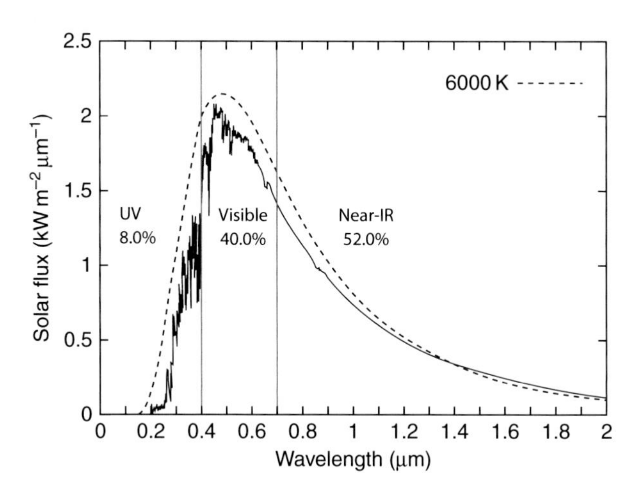

Figure 1 The solar spectrum at the top of the Earth's atmosphere (solid curve) and the spectrum emitted by a blackbody at $T = 6000$ K (dashed curves). Also shown are the positions of the UV, visible, and near-IR bands along with the approximate percentage of solar energy in each band.

Solar Radiation Flow in the Earth's Surface–Atmosphere System

As it propagates through the atmosphere, solar radiation undergoes scattering and absorption by gases, aerosol, and clouds. The fraction of solar radiation that survives and reaches the surface is partly reflected back to the atmosphere. The remainder is absorbed by the surface. The scattered (or diffused) radiation can undergo many acts of scattering and reflection until it is either reflected back to the space or absorbed by the Earth's surface–atmosphere system. Absorption of solar radiation, a process by which the energy transported by electromagnetic waves is converted to other forms, is the sole significant source of heat that ultimately supports the climate and life on our planet. Unlike absorption, scattering is a process that conserves the radiant energy but redirects the energy from the incident wave in all directions.

Atmospheric gases scatter solar radiation at all wavelengths but they absorb only in selected wavelength bands. In the upper atmosphere, molecular oxygen (O_2) and ozone (O_3) absorb almost completely the solar UV radiation with wavelengths less than about 0.3 μm. In the troposphere, water vapor (H_2O) is the chief absorbing gas in the visible band. Water vapor also absorbs in the near-IR band along with carbon dioxide (CO_2). The main gases that can absorb solar radiation are present only in small amounts compared to nitrogen (N_2) and oxygen (O_2) that are responsible for >99% of the total mass of the atmosphere. Molecular (Rayleigh) scattering controlled by N_2 and O_2 is the largest near the surface and decreases with altitude as air density decreases. The characteristic feature of molecular scattering is the inverse proportionality to the fourth power of the wavelength. This causes more blue light to be scattered than green, yellow, and red, so the sky appears blue on clear days.

Scattering and absorption by particles (aerosols and cloud drops) depend on their size and composition. In particular, the amount of scattered energy and its directional distribution strongly depend on the ratio of the incident wavelength and particle size. The larger the ratio, the larger the amount of radiation scattered in the forward direction. Because of larger sizes of cloud drops, scattering by clouds is much greater than molecular (Rayleigh) scattering and scattering by aerosol particles. For this reason the presence of clouds is a main factor controlling the amount of solar radiation scattered back to space. Clouds appear white because, unlike molecules, they scatter all visible wavelengths equally. The cloud albedo increases with cloud water path (a total mass of cloud water in a vertical column per unit surface area). Clouds also absorb some solar radiation in the near-IR. Both cloud albedo and absorption of solar radiation are sensitive to sizes of cloud particles. For the fixed cloud water path, clouds consisting of smaller drops tend to have larger albedos. Clouds strongly vary in space and time, but on average they cover *c.* 62% of the entire planet.

Aerosols, liquid, solid, or mixed-phase particles suspended in the air, all can scatter solar radiation. Whether they can absorb solar radiation or not depends on their chemical compositions. In the troposphere, the common aerosol types are sulfates, nitrates, carbonaceous (organic and black carbon), mineral dust, and sea salt. Of those, black carbon and mineral dust absorb solar radiation. Some organic aerosols can absorb in the UV, but this absorbed energy is too small to be important in the global radiation balance. In the stratosphere, aerosol particles originating mainly from volcanic eruptions do not absorb sunlight. The amount of solar radiation scattered and absorbed by aerosols also depends on aerosol particle concentration. Both concentration and aerosol composition vary greatly with time and location. Thus aerosol scattering adds to the amount of solar energy that is reflected back to space, whereas absorption of solar radiation by black carbon and dust particles when present contributes to the radiant energy that stays in the system. Collectively, atmospheric gases, aerosols, and clouds absorb only about 20% of solar radiation.

The Earth's surface also absorbs some of the solar radiation that survived passing through the atmosphere. The remainder is reflected back to the atmosphere. The fraction of solar radiation reflected by a surface is called surface albedo. Surfaces with low albedo reflect a small amount of sunlight; those with high albedo reflect a large amount. The larger the albedo, the lower the amount of the solar energy absorbed by the surface. Different types of vegetated and bare surfaces have different albedos. Although it is a function of wavelength, in the context of the energy balance the surface albedo averaged over the solar spectrum is of interest. **Table 2** gives some examples for common natural surfaces. The water surfaces have the lowest values while ice- and snow-covered surfaces have the highest. Forests typically have lower albedo than deserts. The albedo of vegetated surfaces varies temporally (e.g., seasonal changes) more than that of bare soils.

Table 2 Albedo of various surfaces

Surface	*Albedo* (%)
Grass	15–25
Bare soil	10–25
Water	<10
Fresh snow	70–90
Old snow	35–65
Sand, desert	25–40
Deciduous forest	15–25
Coniferous forest	5–15

Ultimately, the reflection from the surface and the atmosphere controls the amount of the solar energy returning back to space, which can be expressed in terms of the albedo of the Earth as a whole, called planetary albedo. The Earth's planetary albedo is about 0.3, that is to say that about 30% of solar radiation is reflected and the remainder 70% is absorbed by the Earth's surface–atmosphere system.

The Nature of Terrestrial Radiation

In addition to solar radiation reflected back to space, our planet also loses the energy via thermal emission. The emission from the Earth's surface corresponds to the emission of a blackbody at 288 K. Because the thermal emission strongly depends on temperature and Earth's temperatures are much smaller than those of the Sun, electromagnetic radiation emitted by the Earth's surface–atmosphere system occurs at the longer wavelengths, with its maximum at about 10 versus 0.5 μm for the Sun. About 99% of the radiant energy emitted by the Earth's surface and atmosphere is found in the band of 4–100 μm. This radiation is called terrestrial radiation. Other commonly used names are longwave radiation and thermal IR radiation. Emission at wavelengths larger than 100 μm is very small and practically irrelevant for the global energy budget, although it is actively used in various remote sensing applications.

Most land and water surfaces are very efficient emitters at thermal IR wavelengths. Their emissivities, the degree to which an object behaves like a perfect blackbody, are between ~0.9 and 1. The longwave radiation emitted from the surface is absorbed and re-emitted by gases, clouds, and, to a lesser extent, aerosols throughout the atmosphere. The atmospheric gases that can absorb IR radiation and are important to the radiation balance are water vapor, carbon dioxide, ozone, methane (CH_4), and nitrous oxide (N_2O), with H_2O being the chief IR absorber. Collectively, they all are called the greenhouse gases because they trap the fraction of the radiant energy emitted by the surface which otherwise will escape to space. Gases absorb IR radiation selectively, that is, they absorb not all but certain wavelength bands in the IR. Some bands are more important to the global radiation balance than others, depending on their position in the surface emission spectrum.

The cloud-free atmosphere is the most transparent to IR radiation between about 8 and 12 μm, the so-called atmospheric window. Practically all energy emitted by the surface in this band escapes to space. Outside of this band, the atmosphere is largely opaque.

Clouds have a profound effect on longwave radiation. In general, they work in the same way as greenhouse gases do: they trap the emission from the surface and re-emit some energy back to the system. Efficiency of clouds to absorb and emit longwave radiation depends on several factors including their amount and makeup (water drops or ice crystals) as well as height of cloud tops. Most water clouds emit as blackbodies. Vertically extended cloud systems with the high tops at temperatures much lower than the surface temperature emit little longwave radiation to space while they absorb essentially all surface emission. However, these clouds have offsetting effects on solar and terrestrial radiation in terms of the energy balance of the planet. By reducing the thermal emission they contribute to a heating, but they result in a cooling by reducing the amount of absorbed solar radiation due to a generally higher albedo than the underlying surface. Unlike water clouds, cirrus (ice) clouds are partially transparent to thermal IR radiation, and their albedo at the short-wave band is smaller than that of water clouds. As a result, cirrus and water clouds can cause an opposite effect. Satellite observations indicate that overall on the global mean, clouds reduce the radiant energy of the planet.

Annual Global Mean Radiation Balance

For the Earth to be in equilibrium, solar radiation incident on the planet should be equal to the radiation lost by the planet. The radiant energy remaining in the Earth's surface–atmosphere system is then available to support a climate conducive to life. The energy balance should be achieved at the top of the atmosphere and at the surface. In the former case, the balance involves the exchange of radiative fluxes only, while in the latter sensible and latent heat fluxes must be added to radiative fluxes to maintain the energy balance, that is to say that the gain of radiant energy by the surface is countered by the transport of sensible and latent heat fluxes out of the surface. Sensible heat is the energy transferred between the surface and the atmosphere when there is a difference in temperature between them. Latent heat is the energy required to evaporate water.

Figure 2 shows the estimates of the annual global mean energy budget based on modeling results and observations. At the top of the atmosphere, incident solar radiation is about 342 W m^{-2}, or one-fourth of the solar constant. This is because the cross-section area at which the Earth intercepts the solar flux is πa_e^2, but the area of the spherical planet is $4\pi a_e^2$, where a_e is the radius of the Earth. Taking the planetary albedo of 30%, *c.* 235 W m^{-2} (or 70%) is available on average to warm the atmosphere and the surface. Of this, about 50% is absorbed by the surface. In turn, the longwave radiation emitted by the Earth's surface is about 390 W m^{-2} that is far larger than the amount of solar radiation absorbed by the surface (168 W m^{-2}). This stresses the importance of the atmosphere as a thermal blanket that keeps the surface warm. The major fraction of the surface longwave radiation is absorbed and re-emitted by the greenhouse gases and clouds (and to a lesser extent by aerosols) at lower temperature. This emission returns *c.* 324 W m^{-2} back to the surface, whereas *c.* 235 W m^{-2} of terrestrial radiation is lost to the space. Only a small fraction (40 W m^{-2}) of surface-emitted radiation is transmitted through the atmospheric window directly to space, considering the global cloud cover of *c.*

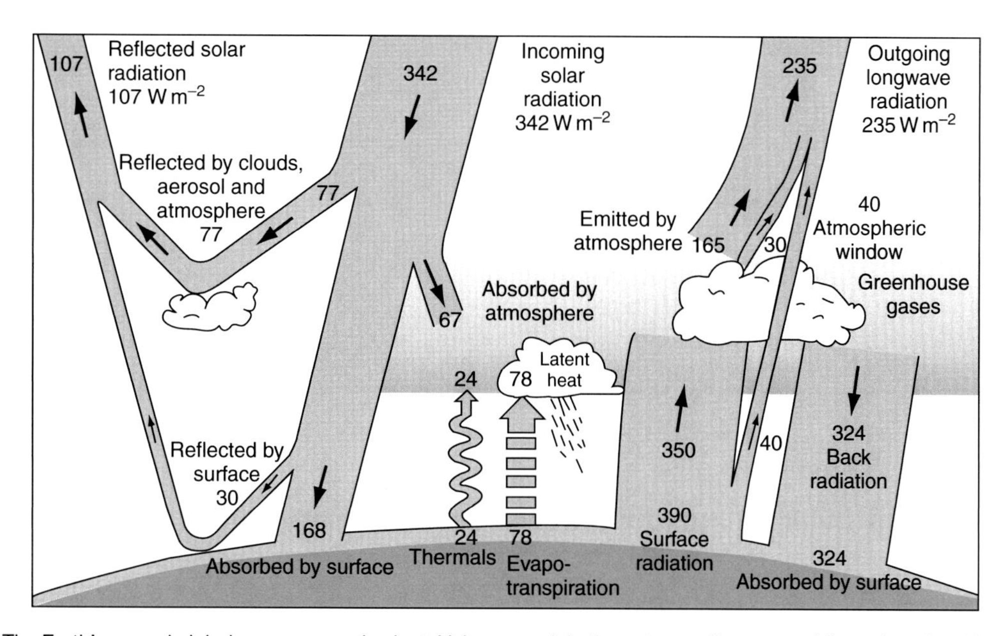

Figure 2 The Earth's annual global mean energy budget. Values are globally and annually averaged fluxes in units of W m^{-2}. The left side of the figure shows the fate of solar radiation in the Earth's surface–atmosphere system, while the right side depicts the processes (emission and absorption) governing the longwave (IR) radiation. Thermals and evapotranspiration illustrate the contribution from sensible and latent heat fluxes, respectively. From Kiehl JK and Trenberth KE (1997) Earth's annual global mean energy budget. *Bulletin of American Meteorological Society* 78: 197–208.

62%. Thus while the absorbed solar radiation provides the source of energy, the thermal IR plays a major role in the redistribution of the radiant energy within the atmosphere and between the surface and the atmosphere.

Globally, latent heat releases from the surface about three-fourths (or 78 W m^{-2}), and sensible heat one-fourth (or 24 W m^{-2}), of the annual mean radiant (shortwave and longwave) energy absorbed by the surface. Ecosystems exert a major influence on the partitioning of the energy between sensible and latent heat. The presence of vegetation tends to increase latent heat fluxes relative to sensible heat fluxes. If there were no latent and sensible heat transfer, the Earth's surface temperature would be much higher than the observed value of 288 K.

Examination of the global radiation balance reveals a vital role of the atmospheric composition and the surface in controlling the climate of our planet. Any changes (natural or human induced) in gases, aerosols, clouds, or the surface has the potential to alter the radiation balance. Anthropogenic changes in CO_2 concentrations and in aerosols (especially, carbonaceous and sulfates) have been suggested as the important global climate change drivers.

Summary

The global mean radiation balance represents the fundamental state of the Earth's climate system. The inflow of solar energy that is absorbed by the Earth's surface–atmosphere system must be balanced by the radiant energy emitted by this system to the space over the long time periods and over the entire globe. The atmospheric gases, aerosols, clouds, and the Earth's surface are the key factors that control the fate of solar and terrestrial radiation. Clouds have a profound influence on both solar and terrestrial radiation. Aerosol particles are more important in the solar band than in the IR. Atmospheric gases that absorb in the thermal IR region play a key role in reducing the thermal emission to space, called greenhouse effect. The surface absorbs a large fraction of incident solar radiation, emits/absorbs thermal radiation, and controls the latent and sensible heat fluxes.

See also: Climate Change 3: History and Current State; Coevolution of the Biosphere and Climate; Energy Balance; Energy Flows in the Biosphere.

Further Reading

Kiehl JK and Trenberth KE (1997) Earth's annual global mean energy budget. *Bulletin of American Meteorological Society* 78: 197–208.
Lean J and Rind D (1998) Climate forcing by changing solar radiation. *Journal of Climate* 11: 3069–3094.
Liou K-N (2002) *Introduction to Atmospheric Radiation*. San Diego, CA: Academic Press.
Wilelicki BA, Barkstrom BR, Harrison EF, *et al.* (1996) Clouds and the Earth's Radiant Energy System (CERES): An Earth observing system experiment. *Bulletin of American Meteorological Society* 77: 853–868.

Radionuclides: Their Biochemical Cycles and the Impacts on the Biosphere

H N Lee, US Department of Homeland Security, New York, NY, USA

Introduction

In recent years, there have been great concerns about the fate of radioactivity released from the detonations of nuclear weapons tests and the impacts on health and ecology caused by the global dispersion of radioactivity. The transport of radioactive dust and its deposition can have significant impacts of radioactivity levels in the ecological environment. Many processes govern the movement of radionuclides through multiple media, including air, water, land, and biota, to human, known as biogeochemical cycles. One way to assess and predict the movement of radionulcides is to understand the processes in each medium. We focus on the processes used in modeling transport and the movement of radionuclides related to the biogeochemical cycles for assessing and predicting the radioactivity level in the atmosphere and in terrestrial and aquatic ecosystems. This article begins to present the primary sources for releasing the radionuclides. The first section is an overview of historical background of the nuclear weapons tests as a source of emitting non-natural radionuclides and the collection of global fallout of radionuclides resulting from the test. The subsequent sections focus on the transport processes and the global pathways (i.e., atmospheric, terrestrial, and aquatic) of radionuclides as collected and measured from the series of nuclear weapons tests. We then discuss the biogeochemical cycles of long-lived radionuclides. Finally, the radioecological effects from a nuclear power accident at the Chernobyl site in Ukraine are also presented.

Nuclear Weapons Tests as a Source

Significant amounts of atmospheric testing of nuclear weapons took place from 1945 to 1980. The first nuclear weapons test, TRINITY, was conducted on a steel tower at Alamogordo in the south-central New Mexico on 16 July 1945. The nuclear explosion creates a radioactive cloud that usually takes the form of a huge mushroom. Explosion converts a small atomic mass into an enormous amount of energy through nuclear fission or fusion. Fission releases energy by splitting uranium or plutonium atoms into radioactive elements. Fusion, triggered by a fission explosion that forces tritium or deuterium atoms to combine into larger atoms, produces more powerful explosive yields than fission. The nuclear weapons test resulted in the release of substantial quantities of radioactive debris to the environment. The debris spread over large areas downwind of test sites, depending on the heights of bursts, the yields, and the meteorological conditions of temperature, precipitation, wind speed, and direction that vary with altitude. Usually, large particles settle locally, whereas small particles and gases may travel a long distance. There was evidence of long-range transport and fallout of debris from the test when the beta activity was first picked up by the film packaging material at mills of Kodak Research Laboratories in Indiana and Iowa in the summer of 1945. On the other hand, large atmospheric explosions may inject radioactive material into the stratosphere, 10 km or more above the ground, where it could remain in the atmosphere for years and subsequently be distributed globally and eventually deposited into the ground (i.e., global fallout) thereby contaminating the radioactivity level in the ecological environment.

After the TRINITY test, the Soviet Union conducted its first nuclear weapons test at a site near Semipalatinsk, Kazakhstan in 1949. The fallout from the series of detonations at the Nevada Test Site in 1951 had resulted in long-range transport of radioactive dust collected at the Eastman Kodak Company in Rochester, New York. The United States, the Soviet Union, and the United Kingdom continued nuclear weapons tests in the atmosphere until a limited test ban treaty was signed in 1963, except France and China. France undertook atmospheric testing from 1960 through 1974 and China from 1964 through 1980. Altogether, over 500 weapons tests were conducted in the atmosphere at a number of locations around the world, yielding the equivalent explosive power of 440 Mt of trinitrotoluene (TNT; i.e., 1 kt of TNT = 4.184 TJ energy released) which was estimated as 189 Mt of fission yield and 251 Mt of fusion yield, as shown in **Figure 1**. That is equivalent to 29 333 Hiroshima-size bombs, which was 15 kt of TNT on 6 August 1945. This would have been equivalent to exploding a Hiroshima-size bomb in the atmosphere every 11 h for the 36 years between 1945 and 1980. Clearly, significant amount of radionuclides was dumped into the atmosphere in these 36 years. That would certainly increase the level of ambient radioactivity to cause extreme harm to the ecosystems.

Fallout Collection and Sampling

A number of national and international monitoring programs were established to collect the long-term fallout from the atmosphere. Many monitoring techniques and methods have been established for collecting samples at the surface and/or in the troposphere and stratosphere. The simple and robust method for collecting fallout at the surface was using an ion-exchange column collector. The column collector consists of a funnel, an ion-exchange column, and a leveling device constructed of polyethylene. The ion-exchange column is packed with a paper pulp filter and an anion-exchange resin. At the end of sampling, the column is shipped to a central laboratory for analysis.

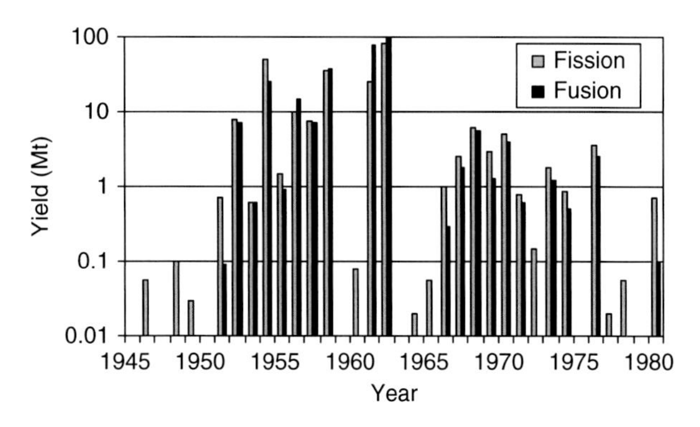

Figure 1 Fission and fusion yields for each year during nuclear weapons tests. From Beck H and Bennett BG (2002) Historical overview of atmospheric nuclear weapons testing and estimates of fallout in the continental United States. *Health Physics* 82: 591–608.

Other sampler for collecting wet or dry fallout was to automatically expose a 'wet collector' during rainfall and a 'dry collector' during dry days. The unit for measuring the activity of fallout deposited on the ground or other surface is becquerels (Bq), defined as the number of radioactive disintegrations per second. The activity of each radionuclide per square meter of ground is important for calculating both external and internal doses. The unit for deposition is usually becquerels per square meter per month or year.

The common technique for sampling in the troposphere/stratosphere was to use high-altitude balloons equipped with large-volume filter samplers. The balloons were made of polyethylene sheet and helium was used as the buoyant gas. The size of the balloon and the amount of helium varied depending on the float altitude needed for sampling. The balloon can be floated up to 40 km height and beyond. Another sampling platforms used were aircraft. The direct-flow impactors were mounted in the front probe of the aircraft. The flow rate was governed by the altitude and aircraft air speed. Aircraft sampling provides useful information because the aircraft flies at various locations and altitudes to collect samples, unlike balloon sampling, which is fixed at a specific location.

Pathway of Radionuclides and Transport Processes in the Atmosphere

Many radionuclides can be created from a nuclear weapons test, by the processes through nuclear fission, nuclear fusion, and neutron activation. The neutrons produced in fission and fusion can induce nuclear reactions that produce radioactive isotopes. One example of this neutron activation process is the reaction with atmospheric nitrogen producing ^{14}C (carbon-14) with a radioactive half-life, $t_{1/2} = 5730$ years. The physical and chemical form of radionuclides may vary depending on the conditions of release and transport in addition to the element's properties. They can be gases, aerosols, and particulate material. The fate of radionuclides that are emitted into the atmosphere is determined by various physical processes of transport, removal via dry deposition or wet deposition, and turbulent mixing that govern the atmospheric flow and diffusion. Radionuclides that are deposited on the land surface can be resuspended into the atmosphere through the processes of resuspension that depend on the surface conditions and atmospheric wind over lands and on seaspray if the radionuclides are deposited in the sea.

Atmospheric transport and deposition models were generally used to study the transport processes that control the distribution of radionuclides in the atmosphere. These models were constructed to satisfy the condition of mass conservation and boundary and initial conditions and are governed by the following basic equation of atmospheric transport–diffusion of mean concentration $[C_i = C_i(x, y, z, t)]$ for a specific radionuclide, i, in the vector form of three-dimension (x, y, z):

$$\frac{\partial C_i}{\partial t} + \nabla \cdot (\mathbf{V} C_i) = \nabla \cdot (\mathbf{K} \cdot \nabla C_i) - \lambda_i C_i - D + S \qquad [1]$$

where $\mathbf{V}(x, y, z, t)$ is a vector of the mean wind velocity, $\mathbf{K}(K_x, K_y, K_z)$ is a diagonal matrix of the turbulent eddy diffusivity, λ_i is the decay coefficient of the ith radionuclide, $D(x, y, z, t)$ is the deposition due to dry and wet removal, and $S(x, y, z, t)$ is the source term including resuspension. Under the assumptions of homogeneous turbulence over a flat terrain in a large diffusion time, the above equation can be simplified to have a Gaussian solution, which serves as a basis for the Gaussian plume model. This Gaussian plume model requires only horizontal wind speed and direction at the release location along with the estimates of atmospheric stability and source term and can be quickly performed to have results of distributions of radioactivity in a gross view. Generally, having a more realistic view, the above equation is numerically solved by difference equation that can take into account the various effects of terrain and spatially varying turbulence and meteorology, including varied wind, temperature, and precipitation. The model obtained is then used for evaluating the consequences and predicting the distributions of an atmospheric release of radioactivity. But the model requires validation before it can be used. The common approach for model validation is to use the natural radionuclides, as described below.

Natural Radionuclides Used as Tracers for Global Model Validation

The presence in the atmosphere of radioactive debris, particularly of ^{90}Sr and ^{137}Cs from nuclear weapons tests, provided a unique opportunity to study the atmospheric transport. Currently, the depositions of these radionuclides have been significantly reduced to extremely low level (see **Figure 5**). On the other hand, the natural sources of atmospheric radioactivity used by scientists for improving the understanding of transport processes are ^{222}Rn (radon-222), ^{210}Pb (lead-210), and ^{7}Be (beryllium-7). Because the global distributions of the source–sink terms of these natural radionuclides by latitude, longitude, and altitude are reasonably well known, their radioactivities can be easily measured to produce many data that are useful and available. The atmospheric ^{210}Pb, which has a radioactive half-life of $t_{1/2} = 22.26$ years, is produced in the lower troposphere from the decay of ^{222}Rn gas ($t_{1/2} = 3.8$ days) that is naturally emitted from the Earth's land surface as a result of uranium decay in soil. The atmospheric ^{7}Be, which has a radioactive $t_{1/2} = 53.44$ days, is produced naturally by spallation reactions in the upper troposphere and the lower stratosphere.

Atoms of ^{7}Be and ^{210}Pb attach themselves to nonreactive submicron-size aerosol particles and, therefore, act as aerosol-borne tracers. ^{222}Rn acts as gas tracer. These tracers are used for assessing the characteristics of airflow and the transport of aerosols in the large- and global-scale atmospheric models. For instance, relatively high ^{7}Be concentrations accompanied by low ^{210}Pb concentrations could indicate subsidence of airflow from upper altitudes and vertical air movement. Atmospheric scientists have found these natural radionuclides to be useful tracers for validating transport models and studying atmospheric circulation, mixing processes, deposition or removal processes, air pollutant transport and ozone sources, and variability related to climate changes. For example, for validating global models scientists use ^{222}Rn as a model input to simulate the global deposition of ^{210}Pb, as shown in **Figure 2**. The results of simulation are then compared with the measurements. The model comparisons with the measurements help us improve our understanding of the atmospheric transport and the transport processes involved in atmospheric models.

The extensive database on ^{222}Rn, ^{210}Pb, and ^{7}Be continues to provide the scientific community with tracer data used to verify global model simulations. The simulations of distributions of ^{222}Rn, ^{210}Pb, and ^{7}Be radionuclides might establish the standards for how well a global model can represent the air concentration, and the aerosol concentration, and its deposition for air and an aerosol, respectively, passing through a monitoring site.

Sampling Method of Radionuclides in the Atmosphere

The technique for collecting radionuclides is to use an air sampler fitted with a filter that is connected to an air pump to collect radionuclide samples. The typical flow rate through an air filter using a 1 hp pump is about 1 m^3 min^{-1}. A pressure gauge is usually installed in the air sampler for measuring air flow as a function of air pressure drop across the filter. The filter is changed daily or weekly depending on the radioactivity levels in the atmosphere. Typically the filter is made of a $20.3 \times 25.4\,cm^2$ rectangle with an effective exposure area of about 407 cm^2 and is composed of three layers of 100% polypropylene fibers. Increasing the flow rate of the sampler is the easiest way to increase the sensitivity for isotopes with short half-lives or low fission yields. The filter samples collected are then brought back to laboratory for counting and analysis.

Pathways of Radionuclides and Transport Processes in the Terrestrial and Aquatic Ecosystems

Radionuclides that are injected into the Earth's atmosphere eventually deposit through gravitation, dry deposition, or precipitation (i.e., wet deposition) processes and migrate toward terrestrial and aquatic ecosystems via surface waters or groundwaters into deeper soil layers and

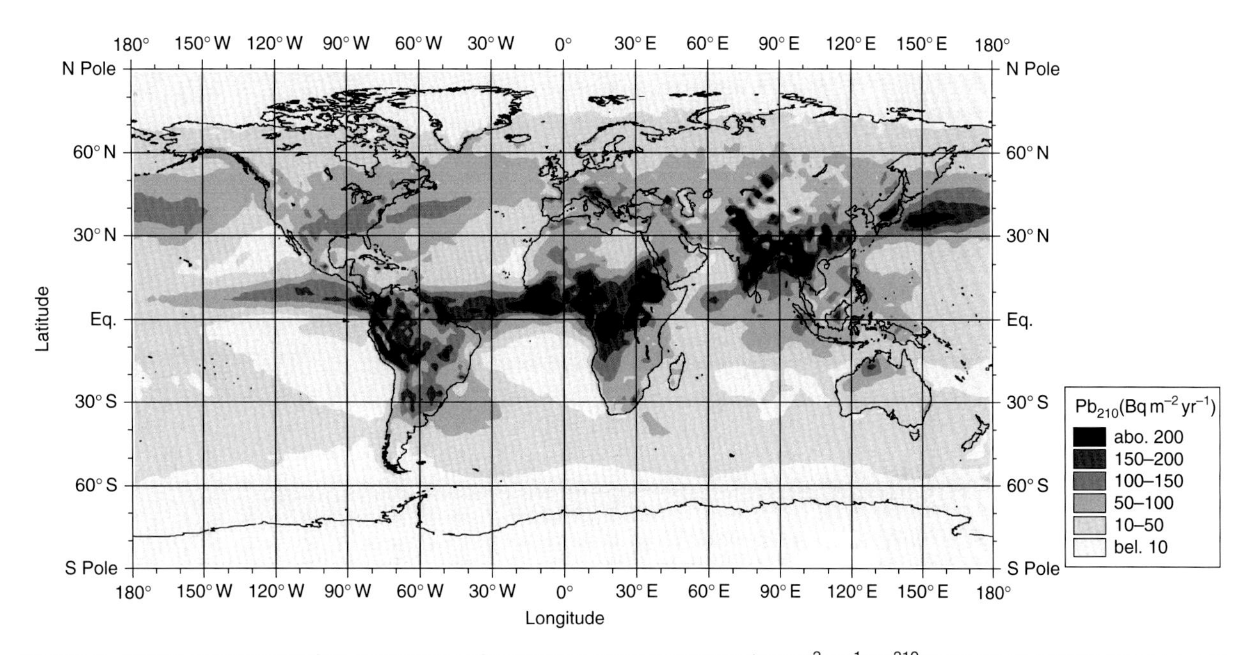

Figure 2 Model-calculated global distributions of yearly mean total deposition (Bq m^{-2} yr^{-1}) of ^{210}Pb for the year 2002. From Lee HN, Wan G, Zheng X, *et al.* (2004) Measurements of ^{210}Pb and ^{7}Be in China and their analysis accompanied with global model calculations of ^{210}Pb. *Journal of Geophysical Research (Atmospheres)* 109: D22203 (doi: 10.1029/2004JD005061).

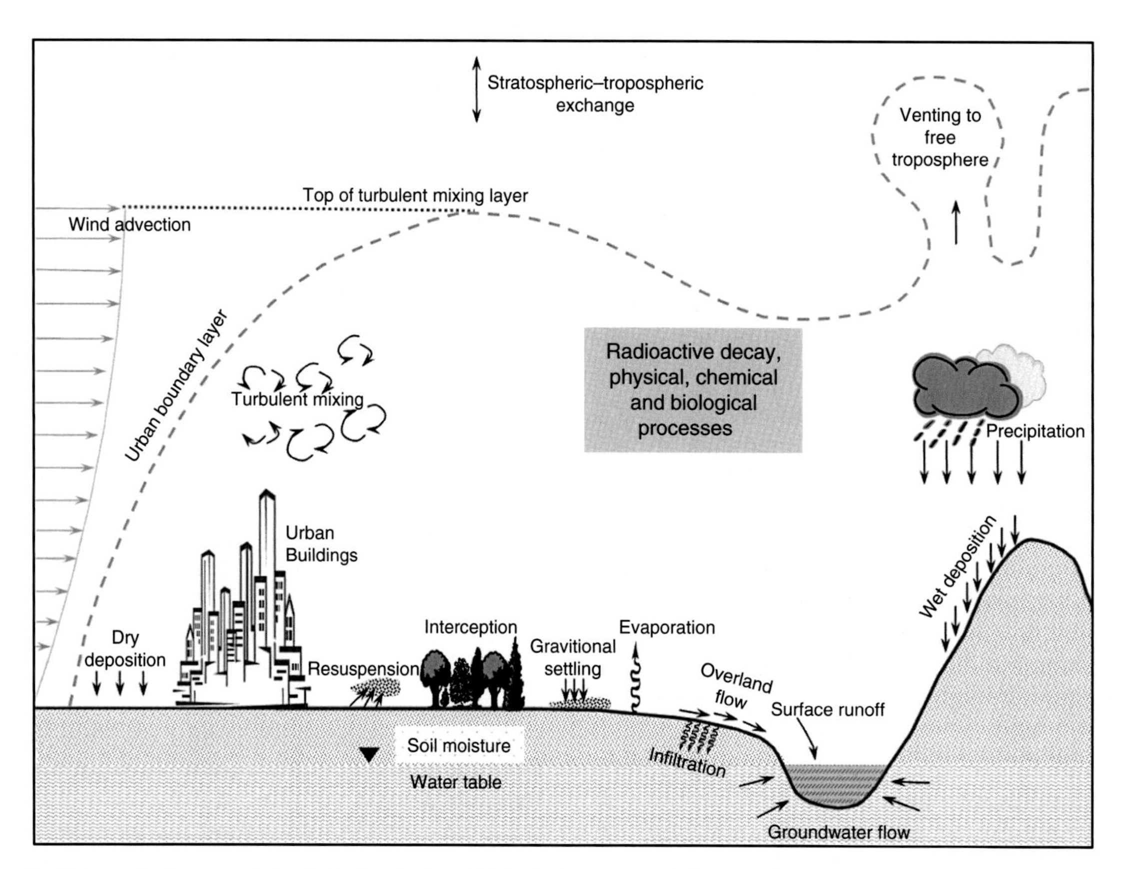

Figure 3 Schematic diagram of physical, chemical, and biological processes influencing the biogeochemical cycles of radionuclides.

reservoirs, as shown in **Figure 3**, potentially increasing risks to public health and the ecological environment.

The mechanisms of transport in the terrestrial and aquatic ecosystems involve a few main processes, such as the physical processes (i.e., interception, runoff, soil infiltration, resuspension, and underground water transport) that are independent of the radionuclides, the biological processes (i.e., uptakes by plants and animals), and the chemical processes (i.e., ion-exchange) that are strongly dependent on the element and its chemical form. Each process is complicated and poorly understood, requiring further investigation and detailed study. They are briefly described below.

Interception Processes

The transport and deposition of radionuclides in the atmosphere can be intercepted by trees and other vegetation. Also, a large amount of rain precipitation can be caught by trees and vegetation as interception. Therefore, the simple process of modeling deposition due to the wet removal ($D_w = D_w(x, y, z, t)$) by taking into account the effect of leaf interception in the terrestrial ecosystem can be described below for a specific radionuclide, i, with mean concentration $C_i = C_i(x, y, z, t)$:

$$D_w = \frac{k_1 L[1 - \exp(-k_2 R)]}{R} C_i \qquad [2]$$

where L is the leaf area index, $R(x, y, z, t)$ is the amount of rain precipitation (mm), k_1 and k_2 are the constants. The wet deposition process can be refined by employing the following improved equation:

$$D_w = \frac{FL[1 - \exp(\beta T_c)]}{\Delta t} C_i \qquad [3]$$

where D_w depends not only on the amount of rain precipitation but also on the frequency and duration of the precipitation event. $F(x, y, z, t)$ is the fraction of the cloud in which the precipitation occurred. β is the frequency with which cloud water is converted to rainwater and then removed from the cloud. $T_c(x, y, z, t)$ is the duration of the precipitation event within the time step (Δt).

Runoff Processes

Runoff is usually considered as a streamflow that involves surface runoff and groundwater flow that reaches the streams. Surface runoff is a function of intensity, duration, and distribution of rain precipitation; permeability of ground surface; surface coverage (i.e., arid or semiarid); geometry of stream channel; depth of water table; and the slope of the land surface. Surface runoff is commonly represented in the form of a hydrograph, which is a time record of stream-surface elevation or stream discharge at a given cross section of the stream. Surface runoff and groundwater flow are determined by the rainfall intensity and duration, rate of infiltration, volume of infiltrated water, soil moisture deficiency, and other characteristics of the basin.

Soil Infiltration Processes

Soluble radionuclides can infiltrate through the surface waters and penetrate into the deep soil layer. The infiltration depends on the surface conditions of soil (i.e., dry or wet, and the amounts of foliage and plants at the surface soil), the morphology of the surface, the soil porosity, and the duration, intensity, and volume of rainfall. Generally, soil infiltration can be calculated from the time-varying depth of water fluctuation in the streams during a storm event. The stream water depth is based on the runoff hydrographs. The obtained infiltration information and runoff hydrographs are then used as loadings for calculating contaminant concentrations in the ground.

Resuspension Processes

Because a large amount of radionuclides deposit on the ground as the result of global fallout, resuspension of radionuclides attached to surface soil particles provides another mechanism of deposition on other surfaces. Historically, the concern with resuspension has been with isotopes of plutonium, which have a very long half-life. But, the resuspensions of ^{90}Sr and ^{137}Cs have raised the major concern of current radioactivity in the ecological environment. Resuspension has generally been treated by means of an empirical resuspension factor (RF) defined as the ratio of a resuspended radionuclide concentration in air (Bq m^{-3}) to the total ground deposition density (i.e., activity per unit area, Bq m^{-2}) of a radionuclide. Many empirical models for studying resuspension processes have been proposed but do not capture details of the mechanisms because the processes, which depend on the surface conditions (i.e., wet or dry soil, arid or semiarid) and need to take into account the soil particle size and density, air density, surface wind and atmospheric turbulence, are complex. There has not been universal agreement that resuspension is an important pathway, but it is now generally accepted that there are a few situations in which this pathway could be the dominant one. For example, the surface observations of ^{137}Cs/^{90}Sr activity ratios have shown the indication of long-range transport of resuspended surface dust carrying ^{90}Sr and ^{137}Cs during the dust storm. The approach is to compare the measured concentrations of ^{90}Sr and ^{137}Cs and their activity ratio from the surface dust collected at remote site with the activity ratio and concentrations measured at the local site that is proved to be a receptor to the remote site during the transport. A group of scientists in Japan has used this approach for studying the resuspension of ^{90}Sr and ^{137}Cs attached to surface dust in a dust storm event.

Underground Water Transport Processes

Radionuclide behavior in the aquatic environment is determined by transport of the liquid and solid phases as well as the chemical interactions between phases and their biological cycling. Once the radioactive debris released by a nuclear weapons test or from a nuclear power plant accident in the atmosphere enters the terrestrial ecosystem, the debris can infiltrate into the deep soil layer to contaminate the groundwater in the aquatic ecosystem. Various radionuclides, whether they are naturally or non-naturally made or leaked from a nuclear reactor, that have appeared in the groundwater have been of most concern. Mobility of radionuclides in the groundwater involves several processes: precipitation, dissolution, adsorption, desorption, and ion exchange. Propagation of a radioactive plume through groundwater is a dynamic event in which all of these processes may occur simultaneously.

To evaluate potential risks of radioactive contamination in soil and groundwater, fate and transport modeling were used to calculate and predict the migration of site contaminants through the ecosystem. Commonly used fate and transport models in the groundwater transport are inherently simplistic because of the complexities of the biogeochemical processes. Most common models assume an equilibrium state, that is, linear sorption isotherm, which assumes a reversible adsorption of masses between the solid and liquid phases in the ground such that a constant distribution coefficient (K_d), which is the ratio of the amount of a solute sorbed onto solid to the concentration of the solute in the liquid solution, is used. But, variations in K_d are likely to occur not only because of possible biological and colloidal effects but also due to changing solution and sediment chemistries becoming a nonequilibrium state. In many cases, the injection/extraction of groundwater can cause mass transfer processes among the phases to be in a nonequilibrium state, which is an irreversible process.

It has been shown that the ion-exchange process including chemisorption has a profound impact on the model calculations of underground contaminants, such as uranium plume in the aquifer. The processes of uranium sorption to iron oxides and iron oxyhydroxides are not completely reversible. These oxide phases act as irreversible sinks for uranium in soil and groundwater. This irreversible process leads to attenuation of the solute.

The nonequilibrium model of groundwater transport for the liquid and solid phase of a radionuclide resolved in contaminated water can be described as follows:

$$\frac{\partial C}{\partial t} + \nabla\cdot(\mathbf{v}C) = \nabla\cdot(\mathbf{k}\cdot\nabla C) - \frac{\partial(\rho_b S)}{\partial t} + B$$
$$\frac{\partial(\rho_b S)}{\partial t} = Q_1 C\left(1-\frac{S}{S_m}\right) - Q_2\rho_b S\left(1-\frac{C}{C_0}\right) - \frac{\partial(\rho_b S_c)}{\partial t} \quad [4]$$
$$\frac{\partial(\rho_b S_c)}{\partial t} = Q_3\rho_b S_c$$

where $C(x, y, z, t)$ is the mean concentration of a radionuclide in water in the liquid phase, $\mathbf{v}(x, y, z, t)$ is a vector of the mean groundwater velocity, $\mathbf{k}(k_x, k_y, k_z)$ is a diagonal matrix of the dispersion coefficient, $S(x, y, z, t)$ is the mean concentration of sorbing radionuclide in the solid phase, S_m represents the maximum amount of radionuclide that can be absorbed in the solid phase, C_0 represents the solubility limit of radionuclide, $S_c(x, y, z, t)$ is the mean concentration of chemisorbed radionuclide, ρ_b is the density of adsorbed radionuclide in the solid phase, Q_1 is the adsorption rate, Q_2 is the desorption rate, Q_3 is the chemisorption rate, and $B(x, y, z, t)$ indicates the biological and colloidal or chemical processes other than the processes of precipitation, dissolution, adsorption, desorption, and chemisorption. The sorption rates Q_1, Q_2, and Q_3 can be determined by performing a sorption experiment of soil samples in the laboratory.

Nonequilibrium sorption for mass transfer between liquid and solid phases has been examined and applied to studies of the transport of the underground uranium plume at various nuclear production sites where the soil has been contaminated by leaks and spills from processing activities. The nonequilibrium sorption model is still required for more examination of quantifying the various sorption rates to improve the accuracy of the calculation in reality.

Uptake Processes

The uptake routes of radionuclides by plants are foliar absorption and root absorption, and by animals the route is daily intake of foods that have been contaminated with radioactivity as a result of global fallout or a nuclear power plant accident involving various radionuclides. There are many reports that the cycles of transfer of radionuclides from air to soil, soil to plant, and plant to animal can occur. One example of the uptake process is the observation of global fallout radioactivity of ^{137}Cs in lichens in arctic regions. Observations have indicated a long-term buildup in response to weapons testing of ^{137}Cs in lichens, reindeer, and caribou, and the people who consumed these animals as a major food source. A few of many other possible examples are provided in the next section.

Biogeochemical Cycles of Long-Lived Radionuclides

The radioactive debris injected into the stratosphere from nuclear weapons tests takes years to deposit, during which time the shorter-lived radionuclides largely disappear through substantially radioactive decay and gravitational settling, whereas the longer-lived radionuclides (such as ^{90}Sr (strontium-90, $t_{1/2} = 28.8$ years), ^{137}Cs (cesium-137, $t_{1/2} = 30$ years), and the plutonium isotopes ^{238}Pu ($t_{1/2} = 87.7$ years), ^{239}Pu ($t_{1/2} = 2.4 \times 10^4$ years), and ^{240}Pu ($t_{1/2} = 6.6 \times 10^3$ years)) remain in the atmosphere. The study of long-lived radionuclides has yielded useful information for understanding global biogeochemical cycling and various physical, chemical, and biological processes in terrestrial and aquatic ecosystems. ^{137}Cs and Pu isotopes are chemically reactive, and ^{90}Sr is less reactive. Generally, ^{90}Sr migrates more rapidly than ^{137}Cs in the soil layer. In other words, in the soil ^{90}Sr is more mobile than ^{137}Cs, which is strongly adsorbed on clay and is essentially nonexchangeable. But the migration of ^{137}Cs depends largely on soil characteristics. Hence, unexpectedly high levels of ^{137}Cs in milk, vegetation, and animal tissues were found in a region of the southeastern United States characterized by very sandy soils in which ^{137}Cs could rapidly migrate into vegetation through uptake and eventually to the entire food chain. This is in sharp contrast to the behavior of ^{137}Cs in most ecosystems in the United States, where ^{137}Cs is relatively quickly and nearly irreversibly bound to clay in soil. The uptake of ^{137}Cs from a sandy soil is about five times higher than from a clay soil; for a loam soil, this factor is about 2. High organic matter content in soil is also believed to play a major role for the mobility of cesium. Another important factor that influences uptake is pH. Uptake increases with decreasing pH values.

The weapons testing fallout of ^{90}Sr and ^{137}Cs that has affected terrestrial and aquatic ecosystems has been reported extensively in the literature. A number of studies on ^{90}Sr, ^{137}Cs, and Pu isotopes have shown the continually downward movement of these radionuclides into the soil. These radionuclides may remain in the soil for many years, resulting in uptake by plants into vegetation and the entire food chain through biogeochemical cycles, as mentioned earlier, in the southeastern United States. As another example, the island environment of the atolls in the Marshall Islands represents a unique ecosystem where radionuclides that were introduced between 1946 and 1958 have had nearly 50 years to equilibrate. Bikini and Enewetak atolls were the sites of 66 atmospheric nuclear weapons tests. These atolls are composed of coral limestone. The composition of the atoll soil produces dramatic differences in the uptake of ^{137}Cs and ^{90}Sr at the Marshall Islands compared with the uptake rates discussed in the literature, which are based primarily on silica-clay-type soils. For instance, the concentration ratio (CR), defined as the activity concentration of the radionuclide per gram of wet plant soil divided by the activity concentration per gram of dry soil, is about 0.1 for ^{137}Cs, and about 1.0 for ^{90}Sr in silicate soils. However, in the coral soils in the Marshall Islands the CR for ^{137}Cs is about 5, and that for ^{90}Sr is about 0.0001. This is in stark contrast to the very different CRs observed for ^{137}Cs and ^{90}Sr in different soil systems. Also, the atoll ecosystem is ideal for evaluating the root uptake of $^{239+240}Pu$ by plants. It was found that the CRs for plutonium generally agreed with pot culture studies in glasshouses. The general magnitude of uptake of plutonium seems to be about the same over a wide range of soil types, with the coral soils being at one extreme with high pH and nearly pure $CaCO_3$ plus organic material. It indicates the importance of a fundamental understanding of biogeochemical cycles and pathways of radionuclides in different ecosystems when the prediction of impacts of radionuclide contamination is made.

Radioecological Effects of Radionuclides after the Chernobyl Accident

The presence of radioactivity in the environment has been indirectly affected by nuclear weapons tests, but significant accidents from nuclear reactor plants, such as at Three Mile Island in the United States in 1979 and at Chernobyl in Ukraine in 1986, can also greatly increase radioactivity in the atmosphere. The Chernobyl accident is the most serious to have occurred in the history of nuclear reactor operation. The Chernobyl accident occurred at 01:23:48 hours local time on 26 April 1986 at Unit No. 4 of the plant. The reactor continued to burn for several days. After about 10 days, the fire was effectively smothered by the large quantities of sand and other materials dropped on the reactor. Various radionuclides were released to the environment in a period of 10 days, which resulted in a wide dispersion of radionuclides over the globe.

Right after the Chernobyl accident, the shorter-lived radionuclides settled on the ground in a 30-km zone through gravitational settling, whereas the longer-lived ^{90}Sr, ^{137}Cs, and plutonium isotopes remained in the atmosphere and were transported over wider areas through the turbulent transfer and the large-scale air flows. A variety of air sampling methods were used to measure air concentrations of radionuclides, and a network of air monitoring stations were maintained in many nations. Many sites throughout the world had also been established to monitor the deposition of radionuclides on the Earth's surface. A wide range of atmospheric transport and deposition models as well as soil and biological models were used to investigate the various processes that distributed the radioactive plume released from the Chernobyl site.

The Chernobyl accident transported a significant fraction of the radioactivity that spread in the atmosphere of the whole globe after 10 days as shown in **Figure 4**. Clearly, the radioactivity was detected by many countries. **Figure 5** shows the distinguished spikes of ^{137}Cs and ^{90}Sr depositions detected and measured at Japan in 1986 from the accident.

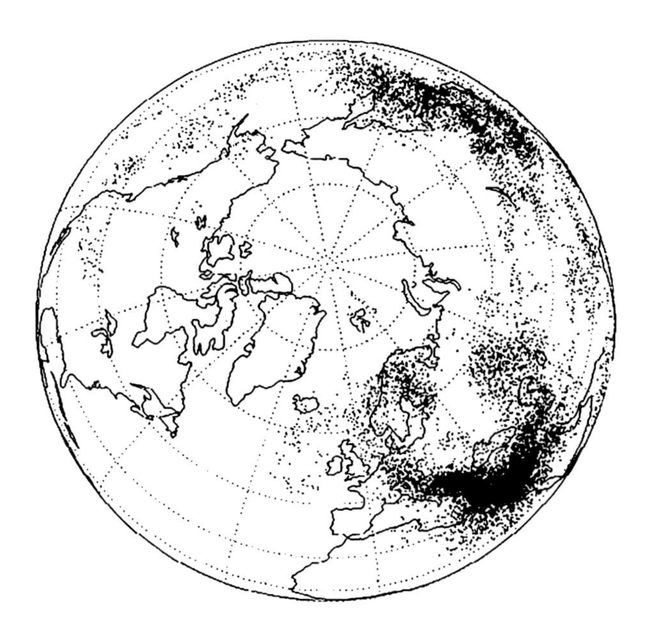

Figure 4 Model-calculated spatial distribution of radioactivity over the Northern Hemisphere 10 days after the Chernobyl accident. From Warner F and Harrison RM (1993) *Radioecology after Chernobyl: Biogeochemical Pathways of Artificial Radionuclides*, SCOPE 50. Chichester: Wiley.

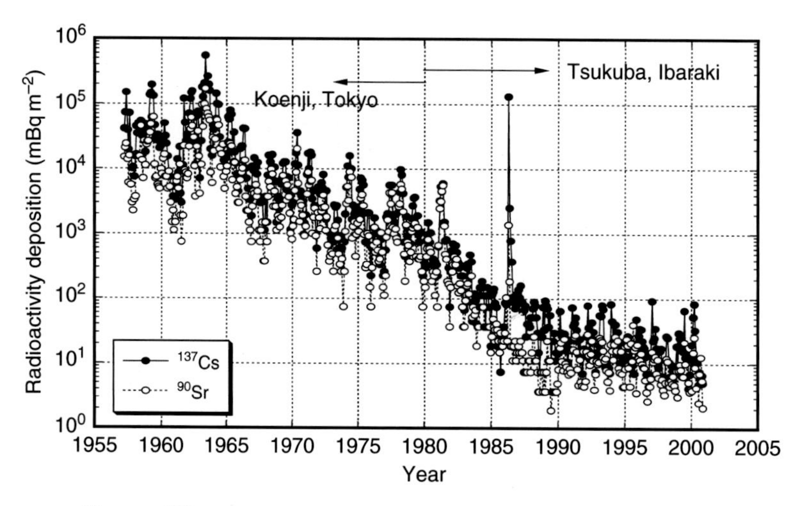

Figure 5 Monthly variations of ^{90}Sr and ^{137}Cs depositions observed at Tokyo and Tsukuba areas in Japan. From Yasuhito I, Michio A, Katsumi H, *et al.* (2003) Resuspension: Decadal monitoring time series of the anthropogenic radioactivity deposition in Japan. *Journal of Radiation Research* 44: 319–328.

The accident has produced new data for us to examine the processes in detail. For example, the soil samples collected in 1993 at Pogonnoe, Belarus, a location 20 km north of the Chernobyl nuclear power plant, have been used for measuring the chemical properties of radionuclides. The results have indicated that ^{129}I, ^{137}Cs, and $^{239+240}Pu$ have migrated into soil with identical rates of distribution in depth of soil. It seems that the process of penetration into soil is physical, rather than chemical, for radionuclides. However, the penetration process is complicated if the radionuclides deposit over a long period of time, such as from the global fallout during nuclear weapons tests. The uptake of radionuclides by plants through roots has to be taken into account during the penetration process over a long term. Therefore, the studies at the atolls in the Marshall Islands mentioned earlier are unique, as far as the global fallout is concerned. The impacts of Chernobyl accident to the ecological environment still remain in some countries, for example in the upland of United Kingdom, after initial occurrence of accident 20 years ago.

Summary and Conclusions

This article reviewed and discussed the present state of knowledge with respect to various important processes that govern transport and movement of radionuclides in atmosphere and in terrestrial and aquatic ecosystems. Each process is complicated and more data are needed to refine the transport models to assess and predict the fate of radionuclides and their impacts on ecology.

Radionuclides entering the atmosphere and terrestrial and aquatic ecosystems through various processes of biogeochemical cycles might potentially increase risks to public health because the cycles of transfer of radionuclides from air to soil, soil to plants, and plants to animals can occur and were addressed previously. The transfer is a long-term process. Therefore, it is essential to conduct a longer-term study of the changes of biogeochemical cycles of radionuclides and their impacts on ecosystems.

In conclusion, the health effects that have resulted from exposures received as a result of nuclear weapons tests include thyroid disease exposed to ^{131}I (iodine-131, $t_{1/2} = 8$ days) which concentrates in the thyroid gland as well as leukemia and solid cancers from low-dose rates of external and internal exposure. An obvious example is the 1945 atomic bombings of Hiroshima and Nagasaki in Japan more than 60 years ago. The bombings were tragic experiences for the cities and for many survivors who sustained severe radiation exposures and injuries from the blasted air shock waves. Radiation-associated deaths from leukemia (46% of all leukemia deaths) peaked within 10 years of the bombings. Many of these radiation-related cancer deaths continue to occur. It is a lifetime of suffering to the survivors. The Chernobyl accident produced similar results. To date, the United Nations reports that 4000 people developed thyroid cancer following that accident, and 56 people, mostly rescue workers, have died from radiation released during the accident.

See also: Pedosphere.

Further Reading

Anspaugh LR, Simon SL, Gordeev K, *et al.* (2002) Movement of radionuclides in terrestrial ecosystems by physical processes. *Health Physics* 82: 669–679.

Beck H and Bennett BG (2002) Historical overview of atmospheric nuclear weapons testing and estimates of fallout in the continental United States. *Health Physics* 82: 591–608.

Bell JNB and Shaw G (2005) Ecological lessons from the Chernobyl accident. *Environment International* 31: 771–777.
Bennett BG (2002) Worldwide dispersion and deposition of radionuclides produced in atmospheric tests. *Health Physics* 82: 644–655.
Fetter CW (1993) *Contaminant Hydrogeology*. Upper Saddle River, NJ: Prentice-Hall.
Hanson D (2006) Chernobyl's aftermath. *Chemical and Engineering News* 18(37): 11.
Igarashi Y, Aoyama M, Hirose K, *et al.* (2003) Resuspension: Decadal monitoring time series of the anthropogenic radioactivity deposition in Japan. *Journal of Radiation Research* 44: 319–328.
Lee HN (2001) An approach for estimating kinetic mass transfer rate parameters in modeling groundwater transport at Fernald, Ohio. *The 2001 Containment Proceedings of International Containment & Remediation Technology Conference and Exhibition*, Florida State University, Orlando, Florida. Orlando, FL: Institute for International Cooperative Environmental Research; CD-ROM Index ID, 077.
Lee HN (2004) Issues and challenges of using natural radionuclides as tracers for atmospheric studies. In: *First International Expert Meeting: Workshop on Sources and Measurements of Natural Radionuclides Applied to Climate and Air Quality Studies, GAW Report 155/WMO TD 1201*, pp. 34–38. Gif-sur-Yvette, France/ Geneva, Switzerland: Global Atmosphere Watch (GAW)/World Meteorological Organization (WMO).
Lee HN (2006) Overview of radionuclides transport. *Proceedings of International Symposium on Environmental Modeling and Radioecology*, Institute for Environmental Sciences, Rokkasho, Aomori: Japan,18–20 October 2006.
Lee HN and Feichter J (1995) An intercomparison of wet precipitation scavenging schemes and the emission rates of ^{222}Rn for simulation of global transport and deposition of ^{210}Pb. *Journal of Geophysical Research* 100: 253–270.
Lee HN, Tositti L, Zheng X, and Bonasoni P (2006) Analyses and comparisons of variations of ^{7}Be, ^{210}Pb, and $^{7}Be/^{210}Pb$ with ozone observations at two Global Atmosphere Watch stations from high mountains. *Journal of Geophysical Research (Atmospheres)* 112: D05303 (doi: 10.1029/2006JD007421).
Lee HN, Wan G, Zheng X, *et al.* (2004) Measurements of ^{210}Pb and ^{7}Be in China and their analysis accompanied with global model calculations of ^{210}Pb. *Journal of Geophysical Research (Atmospheres)* 109: D22203 (doi:10.1029/2004JD005061).
Miller K and Larsen R (2002) The development of field-based measurement methods for radioactive fallout assessment. *Health Physics* 82: 609–625.
Shaw G, Venter A, Avila R, *et al.* (2005) Radionuclide migration in forest ecosystems – Results of a model validation study. *Journal of Environmental Radioactivity* 84: 285–296.
Simon SL, Bouville A, and Land CE (2006) Fallout from nuclear weapons tests and cancer risks. *American Scientist* January–February issue: 48–57.
United Nations Scientific Committee on the Effects of Atomic Radiation (UNSCEAR) (2000) *Sources and Effects of Ionizing Radiation, 2000 Report to the General Assembly with Scientific Annexes.* New York: United Nations; Publications E.00.IX.3 (vol. I) and E.00.IX.4 (vol. II).
Warner F and Harrison RM (1993) *Radioecology after Chernobyl: Biogeochemical Pathways of Artificial Radionuclides*, SCOPE 50. Chichester: Wiley.
Whicker FW and Pinder JE (2002) Food chains and biogeochemical pathways: Contributions of fallout and other radiotracers. *Health Physics* 82: 680–689.
Yasuhito I, Michio A, Katsumi H, *et al.* (2003) Resuspension: Decadal monitoring time series of the anthropogenic radioactivity deposition in Japan. *Journal of Radiation Research* 44: 319–328.

Sulphur Cycle

P A Loka Bharathi, National Institute of Oceanography, Panaji, India

Sulfur Cycle

Most elemental cycles are operative in both oxidative and reductive mode, each fueling the other, either in a dynamic instantaneous manner in space or sequentially over time. Sulfur and its species are important geochemical agents. While the element sulfur is the fourteenth most abundant element on Earth, sulfate ion is the second most abundant ion next only to chloride in seawater and carbonate in freshwater. Elemental sulfur is produced hydrothermally and also by oxidation of sulfide by weathering. The element is also formed as an intermediate of sulfide oxidation or sulfate reduction. Sulfides exist in a variety of forms, most of which are solids. However, dissolved sulfide can occur as bisulfide (HS^-) at neutral pH, sulfide ions (S^{2-}) at alkaline pH, and H_2S at acidic pH, which is volatile and has a rotten egg smell.

Sulfur transformations govern the compositions of the oceans, and the redox balance on the Earth's surface. It is complex due to a variety of oxidation states. Besides, some transformations occur at significant rates both bacteriologically as well as chemically.

The sulfur cycle involves eight electron oxidation/ reduction reactions between the most reduced H_2S (−2) to the most oxidized SO_4^{2-} (+6). It acts as either electron donor or acceptor in many bacterially mediated reactions. The oxidation states of key sulfur compounds are given in the following table:

Organic	S (R–SH)	−2
Sulfide	(H_2S)	−2
Elemental S	(S^0)	0
Thiosulfate	($S_2O_3{}^{2-}$)	+2 (av./S)
Tetrathionate	($S_4O_6{}^{2-}$)	+2 (av./S)
Sulfur dioxide	(SO_2)	+4
Sulfite	($SO_3{}^{2-}$)	+4
Sulfur trioxide	(SO_3)	+6
Sulfate	($SO_4{}^{2-}$)	+8

Bacteria can mediate these oxidations. While bacterial oxidation of sulfur at the expense of oxygen or nitrate generally leads to chemosynthetic carbon fixation, the reductive S cycle is respiratory.

Sulfur Oxidation

Although a variety of oxidation states exist, only three forms are important, namely the sulfhydryl and elemental form besides the sulfate radical.

Reduced sulfur compounds can be used either by colorless sulfur bacteria or the colored photosynthetic bacteria. These bacteria notably belong to β purple bacteria group, namely the colored *Chromatium* sp. or the colorless *Thiobacillus* sp. The others include *Thiosphaera, Thiomicrospira, Thermothrix, Beggiatoa,* and the archaean *Sulfolobulus.* The final product sulfate and different amounts of energy are available depending on the oxidation state of sulfur used as the electron donor:

$$H_2S + 2O_2 \rightarrow SO_4^{2-} + 2H^+ \quad -798.2\,kJ$$

$$HS^- + 1/2O_2 + H^+ \rightarrow S^0 + H_2O \quad -209.4\,kJ$$

$$S^0 + H_2O + 1\,1/2O_2 \rightarrow SO_4^{2-} + 2H^+ \quad -587.1\,kJ$$

$$S_2O_3^{2-} + H_2O + 2O_2 \rightarrow SO_4^{2-} + 2H^+ \quad -822.6\,kJ$$

Those forms that can oxidize sulfur under acid conditions are also able to oxidize iron. Yet others grow at neutral pH.

$$S_2O_3^{2-} \rightarrow S_4O_6^{2-} + 2e^-$$

The oxidation of sulfur involves the reaction of sulfhydryl groups of the cell, like glutathione, with the formation of sulfide–sulfhydryl complex. The enzyme sulfide oxidase oxidizes the sulfide to sulfite.

Chemosynthetic Sulfur Oxidation

The chemotrophic pathways are involved in the oxidation of reduced sulfur compounds, H_2S, S, and $S_2O_3^{2-}$. The oxidation of $S_2O_3^{2-}$ to tetrathionates $S_4O_6^{2-}$, trithionate $S_3O_6^{2-}$, pentathionate $S_5O_6^{2-}$, and elemental S^0 depends on environmental factors like oxygen and pH. The trithionate and pentathionate are formed from tetrathionate with sulfite and thiosulfite, respectively:

$$S_4O_6^{2-} + SO_3^{2-} \leftrightarrow S_3O_6^{2-} + S_2O_3^{2-}$$

$$S_4O_6^{2-} + S_2O_3^{2-} \leftrightarrow S_5O_6^{2-} + SO_3^{2-}$$

The chemotrophic bacteria have to compete with the spontaneous oxidation of sulfide. However, this chemical oxidation rates are speeded up by bacterial intervention – for example, bacterial oxidation of sulfide ores by thiobacilli. Thiobacilli-like bacteria play an important role in thiosulfate oxidation and sulfide oxidation, sometimes at the expense of nitrate as in *Thiobacillus denitrificans*:

$$5S_2O_3^{2-} + 8NO_3^- + H_2O \leftrightarrow 10\,SO_4^{2-} + 4N_2 + 2H^+$$

Sulfide oxidation coupled to nitrate reduction could be an important process in some coastal ecosystems where sediment-produced sulfide encounters nitrate-rich overlying waters that have been depleted in oxygen.

In other ecosystems like deep-sea hydrothermal vents, sulfur/sulfide oxidation is one of the main processes that bacteria utilize for chemosynthetic production of organic matter. Gigantic tubeworms in and around the vent fields harbor symbiotic sulfide-oxidizing bacteria. Special hemoglobins that bind H_2S as well as O_2 transport both substrates to the trophosome where they are released to the bacterial symbiont, thus preventing sulfide poisoning of the host.

Sulfide concentration in some vents measured as the sum of H_2S, HS^-, and S_2 can be related to prevailing temperature, but where biological uptake is rapid, the relationship can become nonlinear. The proportion of these species is strictly determined by pH of the surrounding. At the near-neutral pH from 7 to 7.9 of low-temperature fluids, HS^- is the prevailing species. Sulfide exposed to oxygen could be inorganically oxidized with a half-life of *c.* 380 h at 2 °C, and pH 7.8 and 110 μm O_2 and 10 μm H_2S. Biological oxidation of sulfide by macroorganisms and microorganisms at Galapagos vents could be 4–5 orders of magnitude greater than spontaneous sulfide oxidation in the laboratory. Generally in nature, the transition zones where the anaerobic zone meet the aerobic, the sulfide-oxidizing bacteria can form a sumptuous source of food to protozoans, microzooplanktons, and other higher forms of life. The oxidation of reduced sulfur compounds generates organic compounds (CH_2O) from inorganic substrates and is akin to primary production. In the marine microbial sulfur cycle, there is no net gain of organic material (**Figure 1**). This is because organic material must be oxidized to generate the sulfide that is required for chemosynthetic production of organic carbon. However, at geothermal vents sulfide is released from the geochemical interaction of seawater and hot rock deep

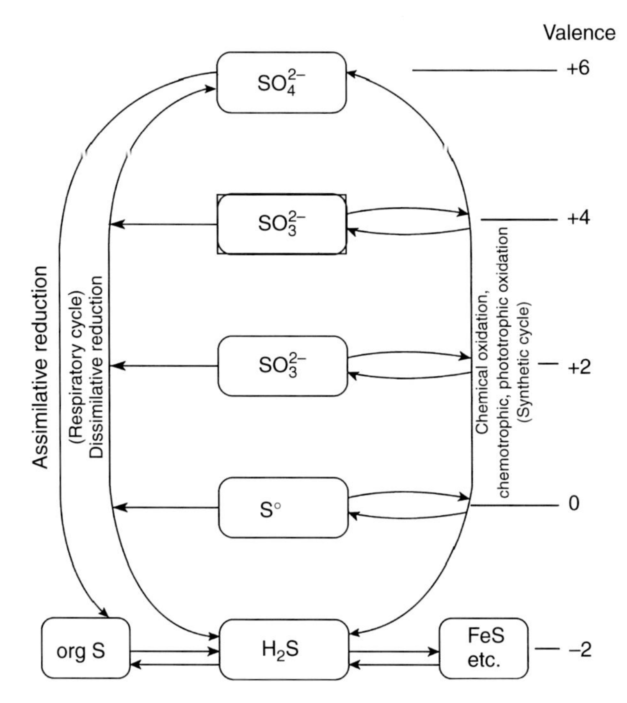

Figure 1 The microbial sulfur cycle. Modified from Fenchel T and Blackburn TH (1979) *Bacteria and Mineral Cycling*. London: Academic Press.

within the Earth crust. Under these conditions there is a net gain of organic material through the oxidation of sulfide and production of new biomass. The actual biochemical transformations are complex with light or chemical energy used to generate reducing power like NADH that is coupled to CO_2 fixation generally through Calvin–Benson cycle. Though chemosynthesis was described more than a century ago by Winogradsky, it was only with the discovery of hydrothermal vents that its significant quantitative role got established.

Geochemical Implications of Sulfur Oxidation

The sulfur-oxidizing microbes act as competitors and sinks for inorganic sulfur along with other reduced compounds as producers of organic biomass as food for zooplankton and a variety of benthic organisms.

Just as dissolved sulfides support chemosynthetic production, particulate sulfides too can support autotrophic growth. Metal sulfides, including pyrite, FeS_2, pyrrhotite, Fe_5S_6-$Fe_{16}S_{17}$, chalcopyrite, $CuFeS_2$, and sphalerite, ZnS, form widespread active and relict hydrothermal sites. Though sulfide oxidation of metal sulfides generally takes place at low pH, nonacidophiles from vent's sulfides are capable of autotrophic growth. Thus massive sulfide deposits on the seafloor may serve as potential source of electrons for autotrophic growth even when the vents are extinct. Such systems are also known to support high biomass of invertebrates.

Chemosynthetic systems in extreme environments like hydrothermal vents are likened to extreme environment systems on other planets or other extraterrestrial bodies. The search is on for chemoautotrophic forms on Mars. Such systems are also suspected to occur on Europa, a moon of Jupiter. They could represent the type of nonequilibrium systems which are thought to have been important in the origin of life on our planet.

Applications

These microbes evoke deep scientific interest because of their metabolic diversity and their adaptability to extreme environments. These characteristics can be gainfully harnessed by biotechnological firms for their enzymes or other metabolic products.

Sulfur-oxidizing bacteria could be judiciously used to contain 'crude oil souring' due to excess sulfide production in oil wells.

Sulfur Bacteria in Symbiosis

The association of thioautotrophic (autotrophic at the expense of reduced sulfur compounds) symbiotic bacteria could be very specific. Most of them are related to subdivision of Gamma proteobacteria. Mariculture of symbiont-bearing invertebrate bivalves has also been suggested as a means of treating industrial sulfur waste.

Photosynthetic Sulfur Oxidation

Anoxygenic photosynthesis is also responsible for converting reduced sulfur to sulfate, thus forming habitats called 'sulfureta'. The purple sulfur bacteria like *Chromatium* can oxidize sulfide internally through elemental sulfur to sulfate. The green sulfur bacteria do so externally. Consequently, the former are less tolerant to sulfide (−0.8 to 4 M) as compared to the latter (4–8 mM). The purple nonsulfur bacteria are least tolerant (0.4–2 mM).

When an anoxygenic bacteria grows on CO_2 as sole source of carbon, besides the formation of ATP, reducing-power NADPH must also be made available so that CO_2 can be reduced to form cell material. The source of reducing power is not water as in photosynthetic plants but reduced sulfur compounds like H_2S, S°, or thiosulfate. Hydrogen or organic compounds, such as lactate, succinate, butyrate, and malate, can also donate electrons for the activity:

$$6CO_2 + 12H_2S \rightarrow C_6H_{12}O_6 + 6H_2O + 12S^\circ$$

Habitats in which oxygenic photosynthesis is important is relatively limited in distribution and is therefore restricted to shallow coastal sediments or coral lagoon or sediments on beaches. These bacteria are at the end of anaerobic food chain that operates within microbial level. Heterotrophs feed sulfate-reducing bacteria (SRBs) and SRBs in turn feed photosynthetic bacteria, which can act as food sources for protozoans or microzooplankton.

Mixotrophy

Sulfur oxidation is carried out not only by chemolithotrophs but also by other groups like (1) mixotrophs (capable of autotrophic and heterotrophic growth); (2) chemolithotrophic heterotrophs; (3) heterotrophs which do not gain energy but derive benefits; (4) heterotrophs which gain nothing from the oxidation. Most pseudomonads are capable of growing mixotrophically on organic compound and reduced inorganic sulfur. Both marine and freshwater pseudomonads are capable of growing on thiosulfate and oxidizing it to tetrathionate.

The metabolic capabilities of a microbe sometimes cannot be too specific. It is argued that many microbes could be facultative, autotrophic at times, and heterotrophic at other, assimilating simple organic substrates that are available. Thus microbes like *Pseudomonas* sp. and *Alcaligenes* sp. have also been implicated in the heterotrophic sulfide oxidation. They could also behave like *Thiobacillus denitrificans*-like organisms (TDLOs) oxidizing reduced sulfide at the expense of nitrate and fixing carbon dioxide in the process. Such metabolic flexibility increases their competitive edge.

Sulfate Reduction

Assimilatory Sulfate Reduction

Sulfate-reducing activity (SRA) that takes place for the incorporation of sulfide radical for biosynthetic cycle is referred to as assimilatory sulfate reduction (ASR). Sulfate is reduced to sulfide in the assimilatory cycle which combines with serine to form cysteine. This in turn can be converted to methionine. These two amino acids are the main constituents of sulfur-containing molecules in the cells. Sulfur content can vary from 0.3% in eel grass to 3.3% in marine algae. As the N:S ratio in land plants is only 30:1, the reductive assimilation of sulfate is less important than nitrate. Assimilatory reduction is common among organisms and does not lead to the production of sulfide.

The eight-electron reduction of sulfate to sulfide proceeds in different stages. As the ion is stable it needs to be activated with ATP. The enzyme ATP sulfurylase catalyzes the attachment of sulfate ion to phosphate of ATP to form adenosine phosphosulfate (APS). Another P is added to APS to form phosphoadenosine phosphosulfate (PAPS) before it gets reduced to form sulfite.

Dissimilatory Reduction

The SRA that takes place in anaerobic respiration is termed as dissimilatory sulfate reduction (DSR). Here sulfate is used as terminal electron acceptor leading to the production of sulfide. Here too the sulfate ion is activated by ATP to form APS. However, in this case, the sulfate moiety of APS is reduced directly to form sulfite with the release of AMP. Thus the first product of ASR and DSR is sulfite.

Dissimilatory SRBs act as agents of synergy in sulfur cycle and bring about syntrophic associations. The end products from organic substrate oxidation and sulfate reduction lead to the formation of sulfide and carbon dioxide. SRA can account for nearly 80% of organic carbon mineralization in marine environment, especially in coastal regimes where nearly 5×10^{12} kg yr^{-1} of sulfate gets reduced.

SRA follows zero-order kinetics in marine sediments with respect to sulfate up to a concentration of *c.* 2 mM . The rate of SRA depends on both quantity and type of organic matter and the sulfate ion available which is generally not limiting in the marine environment. There is in general 2:1 molar relationship between the labile carbon utilized and the sulfate reduced. Sulfate-reducing rates (SRRs) can span several orders of magnitude: 50–500 nm cm^{-3} day^{-1} in coastal zones; 2785 nm g^{-1} day^{-1} in salt pans; 4000 nm ml^{-1} day^{-1} in salt marshes; and up to 14 000 nm ml^{-1} day^{-1} in microbial mats.

The other environmental parameter that affects SRA is temperature. Though the rates of activity can vary by factors ranging from <5 in winter to >0.30 in summer in the temperate region, in the tropics it is not very marked.

SRA predominates in marine sediments but the accumulation of the end products of sulfate respiration is pH dependent. Both metal sulfide formation and rapid biological oxidation are responsible for controlling the amount of sulfide that eventually escapes any system. Sediments harboring vegetation tend to emit less sulfide due to rapid oxidation by oxygen emitted from roots.

Though SRA is largely anaerobic there have been observations of its occurrence in the surficial oxic layers of microbial mats. Sometimes the rate measured in these oxic layers can equal or exceed the SRA of the deeper anoxic layers.

Abundance, Physiological Groups, and Taxonomic Diversity

Though SRBs are high in activity, they are low in abundance contributing to a maximum of 5–6% of total counts of bacteria. While culturable forms retrieved as colony-forming units (CFUs) range from 10^2 to 10^4 l^{-1} in agar shake tubes, most probable numbers (MPNs) methods yield 10^6–10^8 ml^{-1} and fluorescent *in situ* hybridization method (FISH) up to 10^7 ml^{-1}.

SRBs form two main groups based on their ability to utilize carbon sources completely or incompletely. In incomplete organic carbon oxidation, SRBs utilize a variety of organic substrates and oxidize it to acetate:

$$2(\text{lactate}) + SO_4^{2-} \rightarrow 2(\text{acetate}) + CO_2$$

In complete organic carbon oxidation, the organic substrate is totally oxidized to carbon dioxide, water, and sulfide:

$$\text{Acetate} + SO_4^{2-} \rightarrow 2CO_2 + S^{2-} + H_2O$$

Though the above two groups of SRBs are physiologically distinct they can coexist with the latter using the metabolic end products of the former.

SRBs also fall in the following major groups, namely Gram-negative mesophilic, Gram-positive spore-forming thermophilic bacteria belonging to δ subgroup of Proteobacteria and Archaea. The former includes two main families Desulfovibrionaceae and Desulfobacteriaceae. Desulfovibrionaceae includes the genera Desulfovibrio and Desulfomicrobium. Desulfobacteriaceae includes at least 20 genera, most of which are complete oxidizers of organic acids. The Gram-positive group comprises of *Desulfotomaculum*.

Desulfobacteriaceae are metabolically more versatile and are highly adapted to environments that undergo drastic redox changes as in salt marsh sediments or intertidal regions. Rhizosphere habitats are replete with *Desulfobulbus* species which are capable of sulfur disproportionation.

Sulfur-reducing activity

Bacteria like *Desulfuromonas acetoxidans* are capable of reducing sulfur at the expense of acetate. Some SRBs and iron-reducing bacteria are also capable of reducing sulfur. Many of these bacteria are able to generate ATP during sulfur reduction. These groups can also use organic disulfide molecules like cysteine or glutathione. Though sulfur and sulfate reducers can coexist, the latter can produce more sulfide. Most of these bacteria belong to Archaea. Methanogenic thermophilic Archaea reduce sulfur to sulfide while methane generation gets retarded. The process of sulfur reduction is an ancient process, as is suggestive from their presence in the deep branches of the phylogenetic tree. Though some sulfur reducers phylogenetically belong to δ subclass of Proteobacteria, they show affinity to other unrelated classes as well. Metabolic flexibility assures ecological competitiveness.

Sulfur disproportionation

Inorganic fermentation or disproportionation of sulfur and sulfur compounds, thiosulfate, and sulfite has been frequently encountered in SRB:

$$S_2O_3^{2-} + H_2O \rightarrow SO_4^{2-} + HS^- + H^+$$

$$4S^\circ + 4H_2O \rightarrow SO_4^{2-} + 3HS^- + 5H^+$$

$$4SO_3^{2-} + H^+ \rightarrow 3SO_4^{2-} + HS^-$$

Disproportionation seems to be very important and therefore widespread. Thiosulfate is an important intermediate as it can act as an electron acceptor or donor and thus mediate both oxidative and reductive cycle. Thus the thiosulfate shunt provides for complete anaerobic sulfur cycling. Though this is not energetically very viable, the bacteria are able to grow in the presence of metal oxides which can scavenge sulfide.

Use of Heavy Isotopes in Ecology

Sulfur exists primarily as two stable forms of isotopes: ^{32}S and to a certain extent as ^{34}S. Heavier isotopes are discriminated against; that is, most biochemical reactions prefer the lighter isotope and this preference is useful in elucidating microbial interactions. Thus sulfide production by bacterial reduction is much lighter than sulfide of strictly chemical or geothermal origin. Also the biological oxidation of sulfide to sulfur or sulfate either aerobically or anaerobically shows a preference for the lighter isotope. However, this fraction is not as great as that occurring through sulfate reduction as respiratory rates are faster and higher than synthetic ones.

Geochemical Implications of SRBs and SRA

Though the abundance of SRBs is generally low, their high respiratory activity mediates many other activities. Sulfate reduction sets other geochemical reactions in pace. The sulfide formed is responsible for the precipitation of metal sulfides which are available for autotrophic sulfide-oxidizing bacteria. It has been argued that about 90% of the sulfide produced by SRA is recycled back to sulfate to complete the cycle. The rest gets buried to form FeS_2. However, the energy gain from dissimilatory SRA is relatively low: at an energy yield $\Delta G'$ of $-128\,kJ$ with lactate and $-48\,kJ$ with acetate. Nevertheless, the sulfide produced by these bacteria can act as an energy source for other autotrophic bacteria.

The SRBs are also capable of using a variety of inorganic sulfur compounds as electron acceptors. These include dithionite, tetrathionite, thiosulfate, sulfite, bisulfite, metabisulfite, sulfur, sulfur dioxide, and dimethyl sulfoxide.

Sulfate reducers can use other electron acceptors like nitrate; group 4 oxyanions like molybdate and selenate; and even metals like uranium, chromium, technetium, gold, iron, and manganese(IV).

SRA Implications on Climate

SRBs can not only mediate synergistic reactions locally but also impact the climate on a wider scale. They are known to participate not only in the degradation of dimethylsulfoniopropionate (DMSP) but also in the flux of degradation product, dimethyl sulfide (DMS).

The SRBs are involved in the demethylation of DMSP to yield methylmercaptopropionate (MMPA), carbonate, and sulfide or oxidation of DMS to yield bicarbonate and sulfide.

DMSP demethylation

$$3/4\,SO_4^{2-} + DMSP \rightarrow MMPA + HCO_3^- + 3/4\,HS^- + 5/4\,H^+$$

DMS oxidation

$$3/4\,SO_4^{2-} + DMS \rightarrow HCO_3^- + 5/2\,HS^- + 3/2\,H^+$$

Intertidal sediments harbor algal osmolyte DMSP. DMSP could release DMS by the intervention of SRB. This could have countereffect on global warming. This reaction also decreases the effect of the potent greenhouse gas methane.

DMS emissions form an important bulk of sulfur that enters the atmosphere and affects the climate. It accounts for 90% of the biogenic sulfur emissions from the marine ecosystem. DMS produced from DMSP breakdown which reaches the atmosphere serves to decrease warming by radiative backscatter from aerosols. It also reflects radiation from increased cloud cover. Both methanogens and SRBs compete for DMS but methanogens outcompete SRBs when DMS concentrations are high.

Similarly, the breakdown of the osmoregulant glycine betaine in marine sediments releases acetate and trimethylamine. The former is a preferred substrate for SRBs and the latter for methanogens. Other sulfur compounds like carbonyl sulfide (OCS) and carbon disulfide (CS_2) species could be formed photochemically or biologically for bacterial consumption.

Thus, the gaseous products of sulfur cycle interlink land, water, and atmosphere. These include hydrogen sulfide, DMS, methane thiol, carbonyl sulfide, and carbon disulfide. These volatile sulfur compounds get photochemically oxidized to produce acid rain or aerosol sulfate particles that decrease the incoming solar radiation and lead to cloud condensation nuclei. These processes influence the global radiative balance and consequently the climate.

Applications

The activity of SRB could be deleterious to all underground constructions because of their involvement in corrosion. The sulfide they produce is responsible for anodic corrosion, and their propensity to scavenge hydrogen generated in underwater metal structures could cause cathodic corrosion. However, some of these activities could be used in metal recovery from wastewater treatment as metal sulfides. The synergy existing between SRB and other microbes could be effectively used in bioremediation and ecosystem management. This trait could be exploited to contain mercury and other heavy metal contamination in water bodies.

Fluxes of the Global Biogeochemical Sulfur Cycle

The sulfur fluxes, both natural and anthropogenic, have been derived from various studies. A summary diagram of the global sulfur cycle with quantitative estimates of the sulfur fluxes is given in **Figure 2**. The numbers near the arrows designate the total sulfur flux in Tg S yr^{-1} for all compounds. The contributions from anthropogenic activities are indicated by numbers in parentheses. About 120 Tg S are extracted annually by man from the lithosphere in fossil fuels and sulfur-containing raw materials for the chemical industry of which about 58% (70 TgS) gets emitted to the atmosphere. About half of the remaining 50 Tg S directly enters rivers through sewage and residual waters, and another part from fertilizers to agricultural land. Simultaneously, volcanic gases contribute markedly to the atmospheric sulfur cycle over continents amounting to 29 Tg yr^{-1}. The major transfer of sulfur from continents to the ocean by river runoff amounts to 224 Tg of which anthropogenic contribution is about 109 Tg. The total flux of various sulfur forms, that is, organic, sulfate, and pyrite from oceanic water to sediments and further to the lithosphere, amounts to 130 Tg yr^{-1}. Thus the estimates suggest that the anthropogenic sulfur fluxes to the atmosphere and hydrosphere have reached a level comparable with that of natural fluxes. The natural sulfur flux from the lithosphere, its main reservoir, is compensated by the reverse flux of sulfur compounds to the lithospheric sediments of the ocean. Further, there is also indication that by the end of this century the anthropogenic sulfur fluxes could notably increase all over the world.

All of the main reactions of the sulfur cycle involving living organisms are closely related to the carbon cycle. The amount of carbon involved in the fluxes of the sulfur cycle through biogenic processes varies depending on the type of organisms undertaking the metabolism of the sulfur compounds. In the processes of bacterial chemosynthesis, which are characterized by low amounts of energy utilized for the CO_2 assimilation, only relatively small amounts of carbon are transformed into organic matter. In anaerobic bacterial photoassimilation of CO_2 where sulfur compounds are used as electron donors, the amounts of oxidized sulfur and assimilated carbon are comparable. In anaerobic sulfate reduction, 24 g of organic carbon is

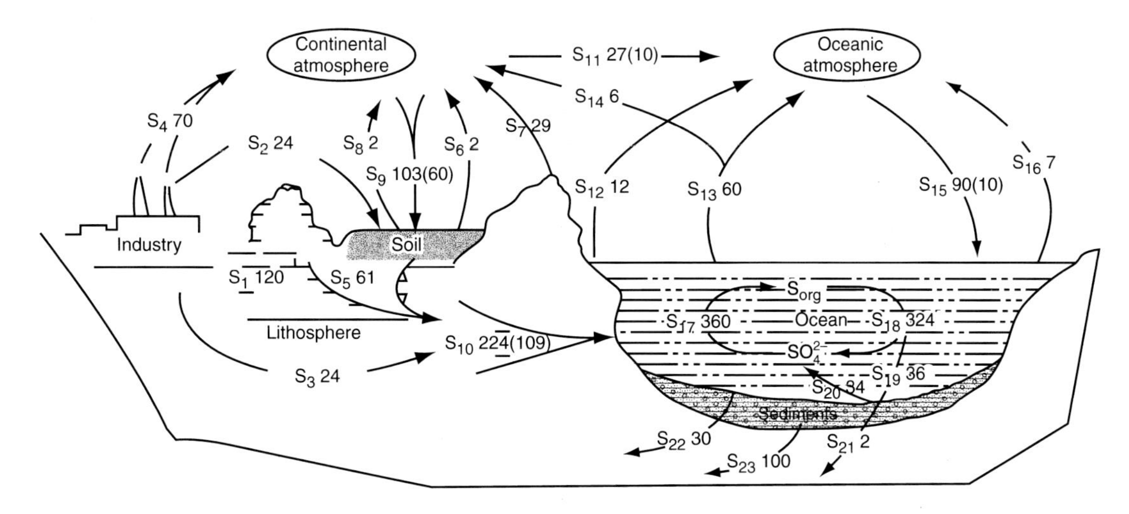

Sulfur flux with anthropogenic contributions in parenthesis

P1	Mining from lithosphere	P2	Fertilizers from soil
P3	Industrial sewage	P4	Anthropogenic sulfur to atmosphere
P5	Erosion	P6	Biogenic sulfur
P7	Volcanic	P8	Dust emission
P9	To land from atmospheric precipitation	P10	From river runoff
P11	Anthropogenic and natural flux from continent to oceans	P12	Biogenic H_2S from shallow coastal sediments
P13	Marine sulfur from sea spray	P14	Marine sulfur to continents
P15	Ocean atmosphere to ocean	P16	Reduced sulfure mission from ocean
P17	Biomass from marine plants	P18	Mineralized sulfur from dead marine organisms
P19	Organic sulfur to sea bottom	P20	Sulfate oxidized from organic sulfur returns to sea
P21	Organic sulfur buried in marine sediment	P22	Sulfate buried in marine sediments
P23	Reduced sulfur buried in marine sediments		

Figure 2 Fluxes of the global biogeochemical sulfur cycle. Modified from Ivanov MV (1981) Global biogeochemical sulfur cycle. In: G E Likens (ed.) *Some Perspectives of the Major Biogeochemical Cycles*, ch. 4. Chichester, UK: Wiley.

mineralized for each 32 g of reduced sulfate sulfur. Thus, in ecosystems with an advanced development of photoauotrophic bacteria and SRBs, both groups of microorganisms transform significant amounts of carbon compounds and, consequently, these organisms should be considered not only as participants in the sulfur cycle but also as active biogeochemical agents of the carbon cycle.

Summary

Microbes, especially bacteria, play an important role in oxidative and reductive cycle of sulfur. The oxidative part of the cycle is mediated by photosynthetic bacteria in the presence of light energy and chemosynthetic forms in the absence of light energy. At the end of the anaerobic food chain in bacteria they serve to purify the system of sulfide and other metabolic end products. In the process sulfur is returned to the system as sulfate. In transition zones from anaerobic to aerobic, photosynthetic bacteria can form a food source to protozoans and microzooplankton. Chemosynthetic sulfur-oxidizing bacteria are the dominant bacterial forms that support thriving ecosystems in hydrothermal vents. Scientists are seeking evidences from such extreme environment for similar life on other planetary bodies.

The reductive cycle on the other hand is mostly driven by the sulfate/sulfur-reducing bacteria which use sulfate as the electron acceptor in anaerobic respiration to produce sulfide. Their close association with other microbes can have profound geochemical influence. Their metabolic activity dictates the availability of trace metals to other forms of life. While sulfide gets precipitated, phosphate gets released into the systems. Nitrogen fixation by these anaerobes also adds to the nitrogen economy of the environment they inhabit. In sediments of continental shelves that hold the reserve of gas hydrates, these microbes can modulate the concentration of methane in such ecosystems. Most importantly, the interaction with DMSP, an osmolyte from phytoplankton, can have wide-ranging climatic implications.

The main reactions of the sulfur cycle involving living organisms are closely related to the carbon cycle. The amount of carbon involved in the fluxes of the sulfur cycle through biogenic processes varies depending on the type of organisms undertaking the metabolism of the sulfur compounds. The estimates suggest that the anthropogenic sulfur fluxes to the atmosphere and hydrosphere have reached a level comparable with that of natural fluxes. The natural sulfur flux from the lithosphere, its main reservoir, is compensated by the reverse flux of sulfur compounds to the lithospheric sediments of the ocean. Further, there is also indication that by the end of this century the anthropogenic sulfur fluxes could notably increase all over the world.

This is NIO Contribution No. 4296.

See also: Anthropospheric and Antropogenic Impact on the Biosphere; Carbon Cycle; Climate Change 1: Short-Term Dynamics; Energy Flows in the Biosphere; Global Change Impacts on the Biosphere; Matter and Matter Flows in the Biosphere; Methane in the Atmosphere; Microbial Cycles; Nitrogen Cycle; Oxygen Cycle.

Further Reading

Fenchel T and Blackburn TH (1979) *Bacteria and Mineral Cycling*. London: Academic Press.
Hines M (1996) Emission of sulfur gases from wetlands. In: Adams DD, Crill PM, and Seitzinger SP (eds.) *Mitteilungen der IVL, Vol. 25: Cycling of Reduced Gases in the Hydrosphere*, pp. 153–161. Stuttgart: Science Publishers.
Ivanov MV (1981) Global biogeochemical sulfur cycle. In: Likens GE (ed.) *Some Perspectives of the Major Biogeochemical Cycles*, ch. 4. Chichester, UK: Wiley.
Jørgensen BB (1988) Ecology of the sulfur cycle: Oxidative pathways in sediments. In: Cole JA and Ferguson SJ (eds.) *The Nitrogen and Sulfur Cyles*, pp. 31–63. Cambridge: Cambridge University Press.
Loka Bharathi PA (2004) Synergy in sulfur cycle: The biogeochemical significance of sulfate reducing bacteria in syntrophic associations. In: Ramaiah NN (ed.) *Marine Microbiology Facets and Opportunities*, pp. 39–51. Panaji, India: National Institute of Oceanography.
Madigan MT, Martinko JM, and Brock PJ (1997) *Brock Biology of Microorganisms*, 8th edn. Upper Saddle River, NJ: Prentice-Hall.
Van Dover CL (2000) *The Ecology of Deep-Sea Hydrothermal Vents*, Princeton, NJ: Princeton University Press, pp. 115–226.

Water Cycle

Z W Kundzewicz, RCAFE Polish Academy of Sciences, Poznań, Poland

Introduction

Water resources of the planet Earth take part in an infinitely recurrent water cycle. It is the largest movement of matter in the Earth's system. The hydrosphere, that is, all water on the Earth's surface, is interconnected with all the other 'spheres' in the Earth system, partaking in exchange of water with the atmosphere (gaseous surrounding of the Earth), the lithosphere (solid Earth), and the biosphere.

Water molecules take one of three states, with liquid state being most commonly occurring in the Earth's conditions. Water undergoes phase changes: from liquid to gaseous phase – by evaporation (evapotranspiration); from gaseous to liquid phase – by condensation; from liquid to solid state – by freezing; and from solid to liquid state – by thawing. Direct phase change between the solid and the gaseous phase is also possible, in the process of sublimation.

There is a quality dimension in the water cycle. Water is an excellent solvent, able to dissolve many chemical compounds, for example, mineral salts. It plays a substantial role in other biogeochemical cycles (of carbon, phosphorus, nitrogen) as solvent and carrier. Water interacts with both the atmosphere and the lithosphere, acquiring solutes from each. Processes partaking in the hydrological cycle transport dissolved and particulate substances. Global sediment fluxes are very high, estimated from 9.3 to $64.0\,\mathrm{Gt\,yr^{-1}}$. Water cycle contains natural purification mechanisms. Evaporation purifies (distills) salty oceanic water. Evaporate is freshwater, but salt remains in the oceans as water evaporates. Moreover, water self-purification takes place in rivers and wetlands.

Water is the basic element of the life support system of the planet and a constituent in plant and animal tissues. Water transfer plays an essential physiological role in human organisms – for example, physical workers in warm climate lose water and drink water in the same time. Due to the thermoregulation mechanism of the human body, they are sweating and, simultaneously, feeling thirst. There are closed (or almost closed) water cycles in many industrial plants, particularly in countries where water prices are high (e.g., Japan). Wastewater from industrial process is treated in the plant and fed back to the process.

Movement of Water between Stores

Water in the hydrosphere is stored in a number of reservoirs (stores), which can be defined in several ways. Stores are present in different spheres of the Earth's system: geosphere (hydrosphere proper – oceans, seas, lakes, rivers, marshes; cryosphere – ice and snow; lithosphere – groundwater, water in rocks, and Earth crust; and atmosphere – clouds) and biosphere (living organisms, flora, and fauna).

The total global water resources constitute approximately 1.385 billion km^3. Water is the most abundant substance at the Earth's surface, with 96.5% of its volume (1.338 billion km^3) contained in salty oceans, which cover nearly 71% of the Earth's surface area. Oceans, the largest water store, play an essential role in the water cycle as the main source of water in the atmosphere.

Other water stores on Earth contain much smaller volumes. Glaciers and permanent snow cover contain 24.4 million km^3 of water, that is over 50 times less than the ocean volume. The third largest global water store is groundwater (23.4 million km^3), but more than half of groundwater is not fresh. Even if frozen hydrosphere (cryosphere) is the largest reservoir of freshwater, groundwater is the largest available source of freshwater. All the lakes on Earth contain 176 400 km^3 of water, with freshwater constituting more than half of the total volume. Approximately 16 500 km^3 of water is stored in the soil (0.04% of total freshwater), while all the rivers of the world carry, on average, in any time instant about 2120 km^3 of water, being only 0.006% of total freshwater. The atmosphere itself stores approximately 13 000 km^3 (0.04% of total freshwater) and wetlands about 11 500 km^3 of water. Biological water has a global volume of 1120 km^3. Total freshwater resources are estimated to be in excess of 35 million km^3.

The most essential, and universal, law guiding the water cycle is the rule of balance (expressed by the continuity equation, also called equation of conservation of mass). It reads, for any fixed control volume:

$$\text{Inflow} - \text{Outflow} = \text{Change of storage}$$

Considering only the most essential hydrological processes, that is, the total precipitation on the basin, evaporation from the basin, runoff (river flow in a cross section terminating the basin), and change of storage in the basin (manifesting itself via surface waters – rivers, lakes, ponds, wetlands; soil moisture, groundwater, and intercepted water), one can formulate the continuity equation for a river basin as

$$\text{Precipitation} - \text{Evaporation} - \text{Runoff} = \text{Change of storage}$$

The total volume of water in the hydrosphere has been nearly constant over a longer timescale. Hydrosphere is a closed system and water takes part in recycling rather than loss and replenishment processes.

The major water fluxes are evaporation and precipitation. Every year, solar energy lifts about 500 000 km^3 of water, 86% of which (i.e., 430 000 km^3) evaporates from the oceanic surface and 14% (i.e., 70 000 km^3) from land. About 90% of the volume of water evaporating from oceans precipitates back onto oceans, while 10% is transported to areas over land, where it precipitates. About two-thirds of the latter evaporate again and one-third runs off to the ocean. By virtue of the continuity equation for stationary conditions, the global volume of precipitation is equal to that of evaporation, that is, 500 000 km^3 of water falls as atmospheric precipitation (on the ocean 390 000 km^3 and on land 110 000 km^3). The resulting imbalance – difference between precipitation on and evaporation from land surface ($110\,000 - 70\,000 = 40\,000\ km^3\ yr^{-1}$) – represents the water vapor movement from oceans to terrestrial atmosphere over continents and islands, being equal to the total runoff of Earth's rivers and direct groundwater runoff to the ocean. **Figure 1** illustrates the principle of the global water cycle, as explained above. Solid arrows represent movement of water in liquid and solid phases, while broken-line arrows represent movement of water vapor.

Since the large volumes are not easy to interpret, the global water cycle can be expressed in units of length (thickness of water layer). On average, a layer of 140 cm of water evaporates from the oceans, and 127 cm of water precipitates onto the oceans. The difference of 13 cm is very important, as it drives the continental phase of the

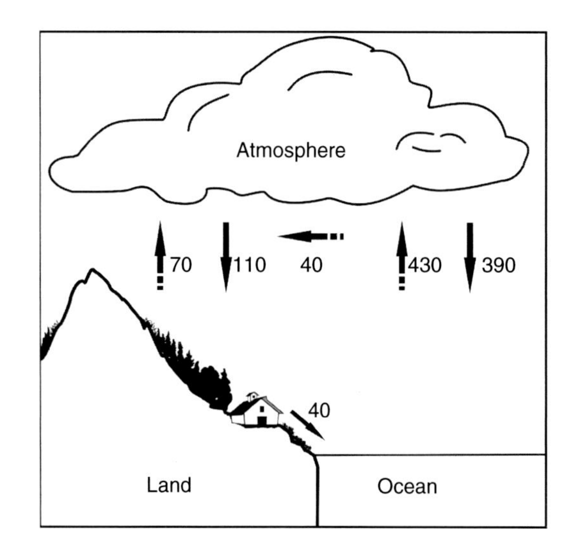

Figure 1 Principle of the global water cycle. Numbers refer to annual fluxes in thousands of cubic kilometers. The processes of vertical movement of water are precipitation and evaporation, and processes of nearly lateral movement are runoff and advection of atmospheric moisture from ocean areas to the land areas.

water cycle. Since more water evaporates from the ocean than precipitates on it, there is a surplus of moisture, which moves over land and precipitates there. In result, precipitation on the land (80 cm) is much higher than evaporation from land (48 cm).

The mean sojourn time of a water particle in different stores varies from hours to millennia. Slow turnover is typical in ocean bodies, large lakes, and deep groundwater, where mean residence time of a water particle depends on the depth, and in ice sheet and glaciers, due to their frozen, immobile nature in cold (low-energy) climates. A water particle spends, on average, about 10 000 years in underground ice in the permafrost area or the eternal snows and polar ice, 2500 years in the ocean, 1600 years in mountain glaciers, and 1400 years in groundwater. In lakes, wetlands, and the soil, the mean residence times of a water particle read 15–17, 5, and 1 year, respectively. Much faster is the turnover of stored water in rivers (16 days), atmospheric water (8–10 days), and biological water (a few hours). Since water in the atmosphere is completely replaced once every 8–10 days, one can state that the atmosphere recycles its contents about 40 times per year.

For millennia, people have not properly interpreted the hydrological cycle, even if they understood water as an indispensable condition of life, and carried out advanced water management. In the ancient times, water was treated as one of four elements (beyond fire, air, and Earth). It was easy to comprehend that the gravity force dominates in the atmospheric precipitation, overland (surface) flow, river runoff, and infiltration. It was followed in the water supply technology (aqueducts). However, it was not clear how the water got up to the source areas located in higher altitudes, against the force of gravity. It was difficult to understand how the loop of the water cycle was closed – what was the force lifting water to the atmosphere, so that it could precipitate onto the ground. It was also not clear why the sea level does not grow despite the perpetual inflow of gigantic rivers. In short, many thinkers falsely interpreted the closing of the water cycle, expecting an underground connection. A concept of the water cycle, similar to the present interpretation, was known in ancient Rome over 2000 years ago, but possibly earlier in China and Greece.

Hydrological Processes

Water cycle can be interpreted as a set of hydrological processes that transfer water between different stores (reservoirs).

The water cycle is powered by solar energy, mainly through direct vaporization. Water evaporates from the Earth's surface, vapor is lifted into the atmosphere to form clouds, and transported in the atmosphere. The heating is spatially uneven, as most of the solar energy warms tropical seas and drives evaporation there. Water vapor, which has risen to the atmosphere, is carried by winds away from the tropics, where it condenses, releasing latent heat and precipitates over oceans and land. Vertical processes of evaporation and precipitation are the (two) major fluxes in the water cycle.

Precipitation water falling down on land is the main source of the formation of land waters: rivers, lakes, groundwater, and glaciers. Precipitated water may be intercepted by vegetation, infiltrate into the ground, be stored in ponds, lakes, and depressions at the Earth's surface, or run off. A portion of atmospheric precipitation evaporates, a part infiltrates and contributes to groundwater, and the rest gets as river flow to the ocean, then evaporates, so that the process repeats again and again. A portion of global river discharge, in the drainless areas of endorheic runoff, does not reach the ocean. Water moves also in the biosphere, in oceans, seas, and lakes, in snow pack, and even in seemingly immobile glaciers.

A list of major hydrological processes partaking in the water cycle, and their short characteristics, are compiled in **Table 1**.

One can conceptually divide precipitated water into 'green' water and 'blue' water. The former is a part of precipitation that evaporates and sustains plant growth, while the latter is liquid water in surface and subsurface water bodies, which can be withdrawn for human use (e.g., for irrigation). Blue water turns to green water in the ecosystems (including agriculture).

There is a considerable movement of water within the ocean bodies. Mixing between two stratified layers (upper – warmer; deeper – colder), separated by the thermocline, is very slow. Motions (currents) of ocean's waters result from the density differences, dependent on temperature and salinity. The thermohaline circulation is intense – the Gulf Stream is a conveyor belt of heat responsible for the relatively mild climate of Europe.

Extremes

At times, volumes of water in stores and fluxes of water between stores take extremely high or extremely low values. For over a century, highest precipitation values have been recorded in meteorological stations worldwide, for various time intervals. The records range from 38 mm of precipitation in 1 min interval, through 1825 mm in 1 day (24 h), to 9.3 m in 1 month, 22.45 m in 6 months, 26.46 m in 1 year, and 40.76 m in 2 years.

Too much rainfall can cause excess runoff, or flooding, while too little rainfall leads to drought, decrease in water level, or even drying out of surface-water bodies, drop of groundwater level and soil moisture, often accompanied by failure of crops, and adverse effects related to water supply, navigation, hydropower, and wild fire.

Amounts of precipitation may considerably vary from year to year. The same areas may experience a drought in

Table 1 Major hydrological processes

Process	*Description*
Evaporation	Transformation of liquid water from the Earth's surface into the vapor state
Transpiration	Transformation of liquid water into the vapor state and transfer of water vapor to the atmosphere via plant metabolism
Evapotranspiration	Joint category embracing evaporation and transpiration. Potential evapotranspiration is the upper limit to evaporation and transpiration, assuming unlimited supply of water and full opening of the stomata (daylight hour)
Condensation	Transformation of water from vapor into (denser) liquid form in the air, producing clouds or fog
Advection	Transport of water by horizontal movement of mass of air
Precipitation	Transfer of water from the atmosphere to the Earth's surface in liquid (rainfall, fog drip) or solid (snow, graupel, hail and sleet) state
Runoff	Transfer of water across the land. This category includes surface runoff (overland flow), subsurface flow, groundwater runoff, and river flow
Sublimation	Change of water state directly from solid (snow or ice) to gaseous
Infiltration	Flow of water from the ground surface (also bottom of water body) into the ground, under the combined forces of gravity, viscosity, and capillarity. Infiltrating water contributes to soil moisture (within the vadose or aeration zone), or percolates deeper to become groundwater (in the aquifer, that is, saturated zone)
Snowmelt	Transfer of water from snow cover (solid state) to liquid state, by melting process
Interception	Storage of precipitated water (in liquid or solid state) by plant foliage. Intercepted water may evaporate back to the atmosphere or fall onto the ground. The amount of intercepted water depends on the duration of the storm, wind speed, temperature, and the density of foliage. A dense forest can intercept nearly all water from a low-intensity rainfall
Subsurface flow	Flow of water under the ground surface, in the vadose (aeration) zone or saturation zone (aquifer)
Capillary rise/ exfiltration	Movement of infiltrated water back toward the Earth surface, driven by an upward capillary potential gradient (caused by evaporation)

one year and a flood in next year (or even – drought and flood in the same season). For instance, the River Elbe flooded Dresden in summer 2002 with a record high level and featured extremely low flow in summer 2003. In 1988, an intense drought disrupted agriculture in the Midwestern United States (too little water), while in 1993 the same area was subjected to severe inundations, greatly reducing the annual harvest (too much water).

Human Impacts

The Earth's freshwater resources remain constant, but man is capable of altering the water cycle and the water resource itself, in both quantity and quality context.

Since ancient times, man has interacted with the water cycle, influencing hydrological processes, in order to accelerate the water movement (e.g., improving conveyance in open channels), or to slow it down (by damming a river and catching water in a reservoir rather than letting it flow promptly to the sea). Man has benefited from the utilitarian values of water – its kinetic or dynamic energy, and the value of water itself.

The water resources have always been distributed unevenly in space and time and man has tried to reduce this unevenness and smooth the spatial–temporal variability. Regulating flow in time to suit human needs can be achieved by storage reservoirs (capturing water when abundant and using it when it is scarce), while regulating flow in space can be achieved via water transfer.

Water transfer is an old idea. Man-made water conduits (aqueducts) date back to the ancient world. Already 6000 years ago, river water was diverted for irrigation agriculture in Mesopotamia. In Mesopotamia and Egypt, qanats – underground lateral canals – were used, through which water was led without losses by evaporation.

The rationale for a large-scale water transfer is backed by the following observations:

- the aggregate renewable freshwater resources on the Earth (river discharge) are sufficient to meet the water demands for many decades ahead;
- freshwater resources on the Earth are distributed in a very uneven way; and
- man's economic activities additionally exacerbate natural unevenness in spatial distribution of water resources.

It is therefore tempting to transfer water from the regions where it is abundant to water-scarce regions. However, effects of large-scale water transfers on the natural environment have to be thoroughly analyzed. Adverse side effects may prove to be large.

Today, water storage reservoirs serve multiple purposes – water supply for agriculture, households, industry, hydropower, navigation, recreation. Irrigated agriculture is indispensable to feed the increasing population. Hydropower developments have been enhanced by solid fuel deficit. However, beneficial impacts of reservoirs are not for free. Among adverse effects of dams and reservoirs are disturbances to ecosystems, barrier to fish, inundation of fertile land, and relocation of people. Moreover, enhanced evaporation from a large water surface reduces available water resources of the region. Hence, reservoirs should be taken into account in estimations of water consumption. Dams have been built for millennia, but most large dams have been constructed since the second half of the twentieth century. At present, the total design volume of world reservoirs exceeds 6000 km^3, and the total water surface area reaches 500 000 km^2.

Beyond water storage and water transfer schemes, the water cycle is exposed to many other human impacts: changes in land use and land cover, deforestation or aforestation, modification and compression of soil layers, field management, urbanization and agricultural activities. Water withdrawals and uptake from rivers and wells for irrigation, and municipal and industrial water usages modify water cycle significantly in both quantitative and qualitative aspects. Impacts of population growth, economic activities, and consumptive lifestyle on hydrological cycle and water withdrawals result in rising water stress. Water withdrawals increase directly with the growth of population and water usage per capita, and indirectly through the increase in food production.

Despite growing demand for water, it is necessary to allocate a share of water to maintain the functioning of freshwater-dependent ecosystems. There is a need to meet conditions of environmental flows and environmental water requirements, that is, to maintain the water regime within a river or wetland, that suits ecosystems where flows are regulated. Nowadays, this is very difficult in some areas; even large rivers in China and Central Asia may run dry due to water withdrawal.

Man has changed the quality of the world's water, creating acute problems in densely populated regions of the Earth where no efficient wastewater purification takes place. It is estimated that only 5% of the world's wastewater is treated. Important water pollution problems are caused by bacteriological and organic contamination, salinization of freshwater (groundwater, rivers, lakes), driven by water withdrawals, pollution by nutrients (nitrogen, phosphorus), remains of agricultural chemistry products, metals, and radioactive material.

Particularly important are easily accessible freshwater resources, such as surface waters and shallow groundwater, part of which is accessible to plant roots. At present, about 600–700 km^3 of annual water withdrawal stem from groundwater. A large part of this groundwater is used for irrigation and municipal needs. For areas with almost no river runoff (e.g., Arabian Peninsula, Libya), groundwater and desalinated seawater are the main water sources, but much of the groundwater is nonrenewable. The recharge took place in past climates, and after withdrawal, the groundwater resources will not be replenished.

The ongoing globalization has increased the transport and trade of 'virtual water' worldwide. The virtual water content of goods is equal to the amount of water required if the transferred goods are produced in the importing and consuming area.

Access to freshwater is now being regarded as a universal human right and extending access to safe potable water is one of the Millennium Development Goals. However, due to a number of changes in nonclimatic factors, such as population increase and rising living standards, the availability of water of appropriate quality is likely to become increasingly restricted.

Climate Change – Acceleration of Water Cycle

Climatic and freshwater systems are interconnected in a complex way, so that any change in one of these systems induces a change in the other. Climate change exerts considerable impact on the water cycle and all the hydrological processes partaking in it.

Earth's climate has always been changing, reflecting regular shifts in Earth's orbit and solar activity and radiation, and irregular volcanic eruptions. However, a large part of the ongoing climate change is due to human activity. Man has been carrying out a planetary-scale experiment, disturbing the natural composition of the atmosphere by increasing the contents of greenhouse gases, by unprecedented level of burning of fossil carbon (coal) and hydrocarbons (oil and natural gas), and large-scale deforestation (reduction of carbon sink). In consequence, the greenhouse effect becomes more intense, leading to global warming. The global mean temperature of the Earth has increased by over 0.74 °C since 1860 and further increase is projected, by up to 1.1–6.4 °C by 2100, depending on the socioeconomic (hence, emission) scenarios.

Apart from the warming, there are several further manifestations of climate change and its impacts on freshwater resources, many of which have already been observed, and further (and more pronounced) impacts have been projected. Observational evidence indicates an ongoing intensification of the water cycle, with increasing rates of evaporation and precipitation. There is more water vapor in the warmer atmosphere, and this creates potential for enhanced intense precipitation. There is a poleward shift of the belt of higher precipitation. Increase in midsummer dryness in continental interiors has been observed and further increase is projected.

Effects of future climate change on average annual river runoff across the world indicate some generally consistent patterns of change – increases in high latitudes and the wet tropics, and decreases in mid-latitudes and some parts of the dry tropics. However, the magnitude of change varies between climate models driving projections.

Changes in streamflow volume, both increases and decreases, have been recorded in many regions, but often these trends cannot be definitively attributed to changes in climate, due to existence of several other factors. The effect of climate change on streamflow, lake levels, and groundwater recharge, which varies regionally, largely follows changes in precipitation. A robust finding is that warming leads to changes in the timing of river flows where much of the winter precipitation currently falls as snow. The effect is greatest at lower elevations (where snowfall is more marginal). Winter flows increase and summer flows decrease.

Water quality can be degraded by higher water temperature, but this may be offset regionally by dilution where flows increase. Carbon dioxide enrichment improves efficiency of plant water use, reducing stomatal conductance and leaf-scale evaporation, but this is partly offset by increased plant growth. Widespread glacier retreat has been observed, and many rivers draining glaciated regions have increasing flows, due to increase of melt. The flow would dramatically fall if glaciers disappear. Model-based projections for the future, particularly related to expected changes in precipitation, are highly uncertain, hence directly unusable for credible and accurate assessment of future freshwater availability.

See also: Climate Change 3: History and Current State; Coevolution of the Biosphere and Climate; Hydrosphere; Precipitation Pattern; Matter and Matter Flows in the Biosphere.

Further Reading

Chahine MT (1992) The hydrological cycle and its influence on climate. *Nature* 359: 373–380.
Eagleson PS (1970) *Dynamic Hydrology*, 462 + xvi p. New York: McGraw-Hill.
Herschy RW and Fairbridge RW (eds.) (1998) *Encyclopedia of Hydrology and Water Resources*, 803 + xxvii p. Dordrecht, The Netherlands: Kluwer.
Jones JAA (1997) *Global Hydrology: Processes, Resources and Environmental Management*, 399 + xiv p. Harlow, UK: Longman.
Kundzewicz ZW, Mata LJ, Arnell NW, *et al.* (2007) Freshwater resources and their management. In: Parry ML, Canziani OF, Palutikof JP, Van der Linder PJ, and Hanson CE (eds.), *Climate Change 2007: Impacts, Adaptation and Vulnerability Contribution of Working Group II to the Fourth Assessment Report of the Intergovernmental Panel on Climate Change*, pp. 173–210. Cambridge: Cambridge University Press.
Shiklomanov IA and Rodda JC (eds.) (2004) *World Water Resources at the Beginning of the Twenty-First Century*. Cambridge: Cambridge University Press.

Xenobiotics Cycles

V N Bashkin, VNIIGAZ/Gazprom, Moscow, Russia

Introduction

Persistent organic pollutants, POPs, are a wide class of chemical species with different physicochemical properties and toxicology. Here we will consider the following priority list of POPs: 1,1,1-trichloro-2,2-bis (4-chlorophenyl) ethane, DDT; hexachlorocyclohexanes, HCHs; hexachlorobenzene, HCBs; polychlorinated dibenzo-*p*-dioxins and dibenzofurans, PCDD/Fs; polychlorinated biphenyls, PCBs; polycyclic aromatic hydrocarbons, PAHs. Environmental pollution by POPs is one of the global problems that is drawing attention at national and international levels. The transboundary aspects of POP transport and pollution of various environmental media require study of the relevant effects on human health and the environment, including quantification of those effects. In accordance to Protocol on POPs to the UN ECE Convention on Long-Range Trans-Boundary Air Pollution that entered into force in October 2003, the parties to the protocol shall encourage research, development, monitoring, and cooperation related, in particular, to an effects-based approach which integrates appropriate information on measured or modeled environmental levels, pathways, and risk to human health and the environment.

Evaluation of POP Deposition

Calculated fields of depositions and concentrations give the opportunity to assess the changes in atmospheric contamination and deposition of POPs and to select 'hot spots' of contamination. As an example, the spatial distribution of PCDD/Fs depositions to the EMEP region, calculated for the beginning and the end of the considered period, is given in **Figure 1**. 'Hot spots' are particular cells of the EMEP grid characterized by the highest values of PCDD/Fs deposition fluxes in both years (marked by arrows). As seen from the data presented, deposition fluxes over the European countries decreased substantially. PCDD/F deposition at one 'hot spot' near Prague (the Czech Republic) decreased more than 2 times. Such calculated fields of depositions for other considered POPs are also available on the Internet (http://www.msceast.org).

Spatial Pattern of PCDD/Fs Contents in Various Environmental Compartments

For PCDD/Fs the spatial distribution of concentrations in air in comparison with that for soil concentrations in 2001 is shown in **Figure 2**. Note that significant differences in the spatial distribution of air and soil concentrations in most European countries are observed. This fact can be explained by the long-term accumulation of PCDD/Fs in soil and relatively low degradation rates in this medium in combination with changes in the emissions during a long time period.

To take into account the effect of accumulation of POPs in different environmental compartments (soil, seawater, and vegetation) the modeling of their long-range transport was performed for a more prolonged period of time (1970–2001). In general, the trends of PCDD/F content in air and seawater followed the emission variation. The trend of PCDD/F accumulation in soil was strongly different from that of emissions. Emissions began to reduce in 1980, whereas the decrease in soil contamination started in 1990. The rate of the soil content decrease is much lower. This causes substantial PCDD/F re-emission flux from soil, which slows down the tendency for a decrease in PCDD/F content in the atmosphere.

Such pollutants as PAHs and PCBs also tend to be accumulated in the terrestrial environment but HCB and γ-HCH in the marine ones. Thus, this information gives us an idea of the POP exposure pathways to human beings.

To identify the areas and regions which were the most polluted by the considered POPs, the preliminary model results on the spatial distribution of their concentrations in different environmental media of the EMEP region were obtained. As an example, the spatial distributions of PCDD/F concentrations in soil, vegetation, and seawater with a spatial resolution of 50×50 are presented in **Figure 3**.

POP Transport in the Northern Hemisphere

To evaluate the long-range transport ability of the considered POPs, the amount of each of these pollutants emitted in Europe and transported outside the EMEP region (outflow) was estimated. For the considered POPs, the percentage ratio of outflow to annual emissions ranged from 20% to 80%. For pollutants with the highest long-range transport potential, such as PCBs, HCHs, and HCB, calculations on the hemispheric scale were made. To evaluate the importance of intercontinental transport for these pollutants, calculations of their transport from different groups of sources such as European, American, and so on were carried out. To make these calculations tentative, hemispheric emission data for these pollutants were used.

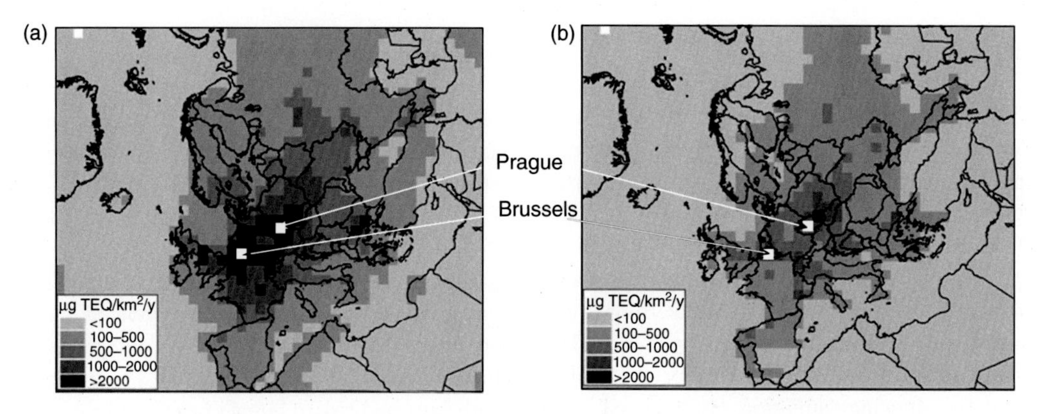

Figure 1 Spatial distribution of PCDD/Fs depositions, (a) 1990 and (b) 2001 (the arrows show 'hot spots').

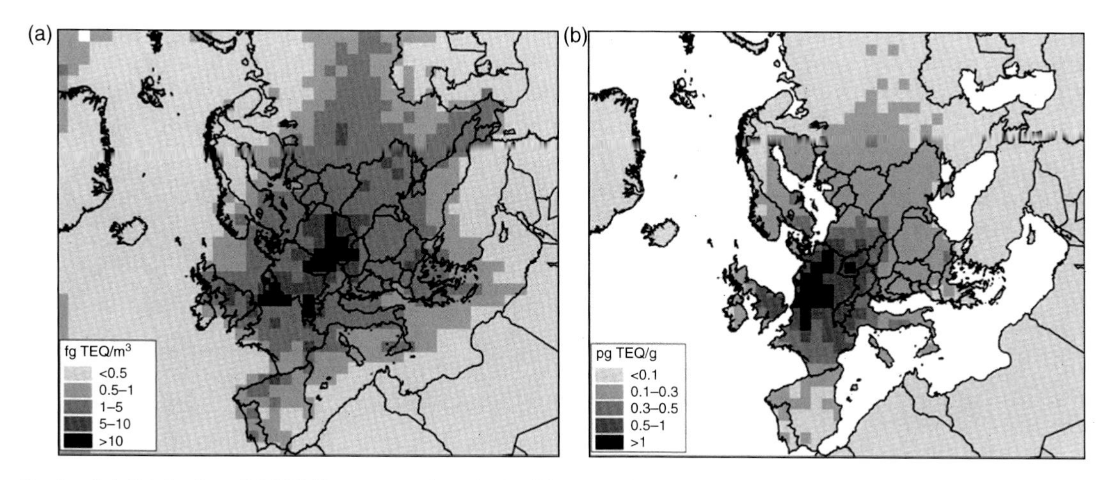

Figure 2 Spatial distribution of PCDD/F concentrations in the (a) air and (b) soils of Europe in 2001.

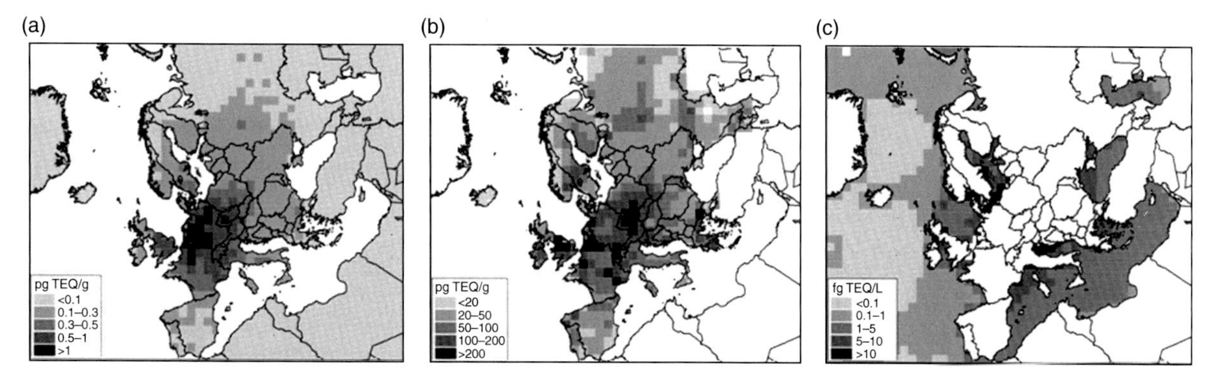

Figure 3 Spatial distribution of PCDD/F concentrations in (a) soil, (b) vegetation, and (c) seawater in 2001.

On the basis of calculations made, contributions of different groups of emission sources located in the Northern Hemisphere, for instance, to HCB depositions over Europe and the Arctic were evaluated (**Figures 4a** and **4b**). The contributions of remote source groups in the contamination of these regions are essential. Contributions of Russian emission sources to the European and Arctic contamination amount to about 19% and 31%, respectively. The relevant sum values of Canada and USA are 7% for the European domain and 17% for the Arctic.

At present evaluation of POP depositions to various types of the underlying surface are under investigations. The spatial distribution of PCB-153 depositions to areas covered with forests, soil, and seawater in 2000 is demonstrated in **Figure 5**. Depositions of this pollutant to forests, soil, and seawater were estimated using different parametrizations of dry deposition velocities for different

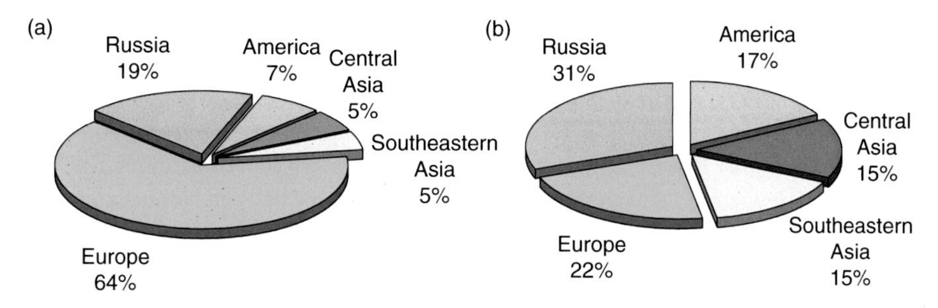

Figure 4 Contributions of emission sources located in the Northern Hemisphere to depositions of HCB over Europe (a) and the Arctic (b), 2000.

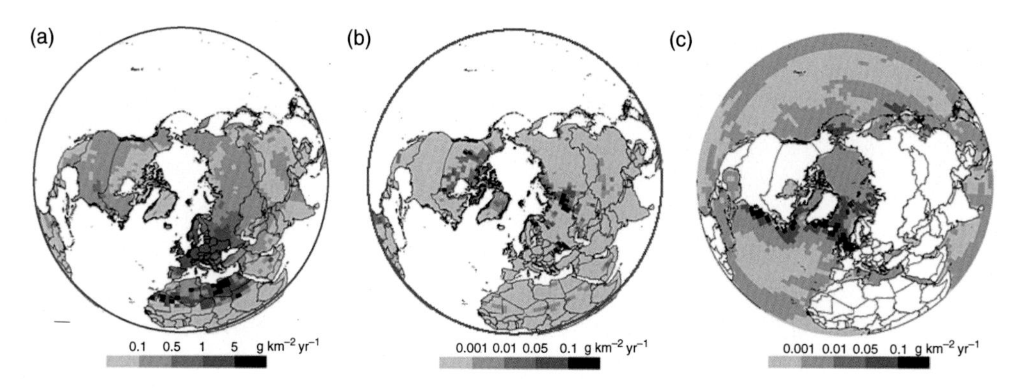

Figure 5 Spatial distribution of PCB-153 depositions on (a) forests, (b) soil, and (c) sea in 2000.

types of underlying surfaces. This resulted in considerable differences in depositions to the considered areas. As seen from the maps, the highest levels of PCB-153 depositions were characteristic of forested areas.

Exposure Pathways of Dioxins and Dioxin-Like PCBs to Human

General Description of Dioxins

In this section, we consider the exposure pathways of POPs to human beings on the example of PCDD/Fs, often called just 'dioxins'. These species consist of two groups of tricyclic aromatic compounds with similar chemical and physical properties. The number of chlorine atoms in each molecule can vary from one to eight. The number of chlorine atoms and their positions are of utmost importance for the toxicological potency of each congener. PCDD/Fs have never been produced intentionally, except for pure substances used as references in analytical and toxicological research, and have never served any useful purpose, unlike many other POPs such as PCBs and DDT. PCDD/Fs are formed as unwanted by-products in many industrial and combustion processes. They have also been shown to be formed in the environment by forest fires and volcanoes, and also via enzymatically catalyzed processes.

Primary sources of environmental contamination with PCDD/Fs in the past were the production and use of organic chemicals containing chlorine. PCDFs were formed as inadvertent by-products in the production and use of PCBs and, in combination with PCDDs, in such high-temperature processes as waste incineration, the metal industry, home heating, and other energy production processes.

PCDFs are also found in residual waste from the production of vinyl chloride and the chlor-alkali process for chlorine production. Factors favorable for the formation of PCDD/Fs are high temperatures, alkaline media, the presence of ultraviolet light, and the presence of radicals in the reaction mixture/chemical process.

Previous production of pentachlorophenol, as well as the bleaching process in pulp and paper mills, has been shown to be a major source. Changes in industrial processes have resulted in a reduction of PCDD/Fs concentration in products. Whereas in the past the chemical industry and, to a lesser extent, the pulp and paper industry were considered to be the main sources of PCDD/Fs (and also the cause of many of today's contaminated sites in several industrialized countries), today's dioxin input is mainly due to thermal processes. There is still a considerable focus on waste incineration but, owing to requirements for dioxin reduction in stack gases set by several national authorities, the importance of this category has declined during the last years. Examples can be seen especially in the European emission inventories. An overview of combustion sources known to generate and emit PCDD/Fs is presented in **Table 1**.

PCDD/PCDFs are found not only in stack gases but also in solid residues from any combustion process such as bottom ash, slag, and fly ash. With advanced technology and better burnout of the ashes and slag (characterized by a low content of organic carbon), PCDD/F concentrations have declined.

Secondary sources of PCDD/Fs, their reservoirs, are those matrices where they are already present, either in the environment or as products. Product reservoirs include PCP-treated wood, PCB-containing transformers and sewage sludge, compost and liquid manure, which can be used as fertilizers in agriculture and gardens. Reservoirs in the environment are, for example, landfills and waste dumps, contaminated soils (mainly from former chemical production or handling sites), and contaminated sediments (especially in harbors and rivers with industries discharging directly to the waterways).

Although these reservoirs may be highly contaminated with PCDD/Fs, the chemical and physical properties of these compounds imply that dioxins and furans will stay adsorbed to organic carbon in soils or other particles. On

Table 1 Sources of emission of PCDD/Fs

Stationary sources	
Waste incineration	Municipal solid waste, clinical waste, hazardous waste, sewage sludge
Steel industry	Steel mills, sintering plants, hot-strip mills
Recycling plants	Nonferrous metals (melting, foundry: Al, Cu, Pb, Zn, Sn)
Energy production	Fossil fuel power plants, wood combustion, landfill gas
Diffuse sources	
Traffic	Cars
Home heating	Coal, oil, gas, wood
Accidents	PCB fires, fires in building, forest fires, volcanic eruptions

From Fielder H (1999) Sources of PCDD/PCDF and impact on the environment. *Chemosphere* 32: 55–64.

the other hand, mobilization can occur in the presence of lipophilic solvents (leaching into deeper layers of soils and/or groundwater) or in cases of erosion or runoff from topsoil (translocation into the neighborhood). Experience has shown that transport of PCDD/Fs due to soil erosion and runoff does not play a major role in environmental contamination and human exposure.

PCBs have been used commercially since 1929 as dielectric and heat exchange fluids and in a variety of other applications. The presence of PCBs in human and wildlife tissues was first recognized in 1966. Investigations in many parts of the world have since revealed widespread distribution of PCBs in the environment, including remote areas with no PCB production or use. There is evidence that the major source of PCB exposure in the general environment is the redistribution of PCBs previously introduced into the environment. It is believed that large bodies of water, such as the Baltic Sea and the Canadian Great Lakes, may release significant amounts of PCB residues from previous uses into the atmosphere. The fact that PCB levels seem to decline in a similar way at different latitudes indicates that primary sources may still play an important role. The amount of dioxin-like PCBs might vary in the environment but the sources, transport, and distribution, as well as persistence, show similarities with the general properties of PCBs.

Potential for Long-Range Transboundary Air Pollution

PCDD/Fs are very persistent compounds; as their Kow and Koc are very high, they will intensively adsorb on to particles in air, soil, and sediment and accumulate in fat-containing tissues. The strong adsorption of PCDD/Fs and related compounds to soil and sediment particles means that their mobility in these environmental compartments is negligible. Their mobility may be increased by the simultaneous presence of organic solvents such as mineral oil. The air compartment is probably the most significant compartment for the environmental distribution and fate of these compounds.

Some of the PCDD/Fs emitted into air will be bound to particles while the rest will be in the gaseous phase, which can be subject to long-range transport (up to thousands of kilometers). In the gaseous phase, removal processes include chemical and photochemical degradation. In the particulate phase, these processes are of minor importance and the transport range of the particulate phase will primarily depend on the particle size. PCDD/Fs are extremely resistant to chemical oxidation and hydrolysis, and hence these processes are not expected to be significant in the aquatic environment. Photodegradation and microbial transformation are probably the most important degradation routes in surface water and sediment.

The number of chlorine atoms in each molecule can vary from one to eight. Among the possible 210 compounds, 17 congeners have chlorine atoms at least in the positions 2, 3, 7, and 8 of the parent molecule and these are the most toxic, bioaccumulative, and persistent ones compared to congeners lacking this configuration. All the 2,3,7,8-substituted PCDDs and PCDFs plus coplanar PCBs (with no chlorine substitution at the *ortho*-positions) show the same type of biological and toxic response.

PCDD/Fs are characterized by their lipophilicity, semivolatility, and resistance to degradation. The photodegradation of particle-bound PCCD/Fs in air was found to be negligible. These characteristics predispose these substances to long environmental persistence and to long-range transport. They are also known for their ability to bioconcentrate and biomagnify under typical environmental conditions, thereby potentially achieving toxicologically relevant concentrations. The tetra–octa PCCD/PCDFs have lower vapor pressures than PCBs and are therefore not expected to undergo long-range transport to the same extent; nevertheless, there is evidence for deposition in Arctic soils and sediments.

Persistence in water, soil, and sediment

Owing to their chemical, physical, and biological stability, PCDD/Fs are able to remain in the environment for a long time. As a consequence, dioxins from so-called 'primary sources' (formed in industrial or combustion processes) are transferred to other matrices and enter the environment. Such secondary sources are sewage sludge, compost, landfills, and other contaminated areas. PCBs and PCDD/Fs are lipophilic (lipophilicity increases with increasing chlorination) and have very low water solubility. Because of their persistent nature and lipophilicity, once PCDD/Fs enter the environment and living organisms, they will remain for a very long time, like many other halogenated aromatic compounds. As log Kow (typically 6–8) or log Koc are very high for all these compounds, they will intensively adsorb on to particles in air, soil, and sediment. The strong adsorption of PCDD/Fs and related compounds to soil and sediment particles causes their mobility in these environmental compartments to be negligible.

Their mobility may be increased by the simultaneous presence of organic solvents such as mineral oil. The half-life of TCDD in soil has been reported as 10–12 years, whereas photochemical degradation seems to be considerably faster but with a large variation that might be explained by experimental differences (solvents used, etc.). Highly chlorinated PCDD/Fs seem to be more resistant to degradation than those with just a few chlorine atoms.

Bioaccumulation

The physicochemical properties of PCBs and their metabolites enable these compounds to be absorbed readily by organisms. The high lipid solubility and the low water solubility lead to the retention of PCCD/Fs, PCBs, and their metabolites in fatty tissues. Protein binding may also contribute to their tissue retention. The rates of accumulation into organisms vary with the species, the duration and concentration of exposure, and the environmental conditions. The high retention of PCDD/Fs and PCBs, including their metabolites, implies that toxic effects can occur in organisms spatially and temporally remote from the original release.

Gastrointestinal absorption of tetrachlorodibenzo-*p*-dioxin (TCDD) in rodents has been reported to be in the range of 50–85% of the dose given. The half-life in rodents ranges from 12 to 31 days except for guinea pigs, which show slower elimination ranging from 22 to 94 days. The half-life in larger animals is much longer, being around 1 year in rhesus monkeys and 7–10 years in humans.

Monitoring

PCCD/Fs have been found to be present in Arctic air samples, for example, during the winter of 2000/2001 in weekly filter samples (particulate phase) collected at Alert in Canada. PCDD/PCDFs have been monitored since 1969 in fish and fish-eating birds from the Baltic. The levels of PCDD/Fs in guillemot eggs, expressed as TEQ, decreased from 3.3 ng/g lipids to around 1 ng/g between 1969 and 1990. Since 1990, this reduction seems to have leveled off and today it is uncertain whether there is a decrease or not. Fish (herring) show a similar picture.

Thus both physical characteristics and environmental findings support the long-range transport of PCCD/Fs and PCBs. There are differences, however, both between and within the groups regarding ability to undergo LRTAP.

Pathways of LRTAP-Derived Human Exposure

For decades, many countries and intergovernmental organizations have taken measures to prevent the formation and release of PCDD/Fs, and have also banned or severely restricted the production, use, handling, transport, and disposal of PCBs. As a consequence, release of these substances into the environment has decreased in many developed countries. Nevertheless, analysis of food and breast milk show that they are still present, although in levels lower than those measured in the 1960s and 1970s. At present, the major source of PCB exposure in the general environment appears to be the redistribution of previously introduced PCBs.

Significant sources and magnitude of human exposure

PCDD/Fs are today found in almost all compartments of the global ecosystem in at least trace amounts. They are ubiquitous in soil, sediments, and air. Excluding occupational or accidental exposures, most human background exposure to dioxins and PCBs occurs through the diet, with food of animal origin being the major source, as they are persistent in the environment and accumulate in animal fat.

Importantly, past and present human exposure to PCDD/Fs and PCBs results primarily from their transfer along the pathway: atmospheric emissions → air → deposition → terrestrial/aquatic food chains → human diet. Information from food surveys in industrialized countries indicates a daily intake of PCDD/Fs on the order of 50–200 pg I-TEQ/person per day for a 60 kg adult, or 1–3 pg I-TEQ/kg bw per day. If dioxin-like PCBs are also included, the daily total TEQ intake can be higher by a factor of 2–3. Recent studies from countries that started to implement measures to reduce dioxin emissions in the late 1980s clearly show decreasing PCDD/F and PCB levels in food and, consequently, a lower dietary intake of these compounds by almost a factor of 2 within the past 7 years.

Biota from the Baltic have, however, not shown any clear trend for dioxins or PCBs since 1990. Occupational exposures to both PCDDs and PCDFs at higher levels have occurred since the 1940s as a result of the production and use of chlorophenols and chlorophenoxy herbicides and to PCDFs in metal production and recycling. Even higher exposures to PCDDs have occurred sporadically in relation to accidents in these industries. High exposures to PCDFs have occurred in relation to accidents such as the Yusho (Japan) and Yucheng (Taiwan) incidents, involving contamination of rice oil and accidents involving electrical equipment containing PCBs.

Exposure levels in adults

PCDD/Fs accumulate in human adipose tissue, and the level reflects the history of intake by the individual. Several factors have been shown to affect adipose tissue concentrations/body burdens, notably age, the number of children and period of breastfeeding, and dietary habits. Breast milk represents the most useful matrix for evaluating time trends of dioxins and many other POPs. Several factors affect the PCDD/PCDF content of human breast milk, most notably the mother's age, the duration of breastfeeding, and the fat content of the milk. Studies should therefore ideally be performed on samples from a large number of mothers, taking these variables into account.

The WHO Regional Office for Europe carried out a series of exposure studies aimed at detecting PCBs, PCDDs, and PCDFs in human milk. The first round took place in 1987–88 and the second in 1992. In 2001–02, a third round was organized in collaboration with the WHO Global Environmental Monitoring System/Food Contamination Monitoring and Assessment Programme (GEMS Food) and the International Programme on Chemical Safety (IPCS). Results are currently available from 21 countries. **Figure 6** presents the temporal trends of levels of PCDDs and PCDFs expressed in WHO–TEQ for those countries participating in all three rounds or in the last two rounds of the WHO study. A clear decline can be seen, with the largest decline for countries originally having the highest level of dioxin-like compounds in human milk.

The general population is mainly exposed to PCBs through common food items. Fatty food of animal origin, such as meat, certain fish, and dairy products, is the major source of human exposure. Owing to considerable differences in the kinetic behavior of individual PCB congeners, human exposure to PCB from food items differs markedly in composition compared to the composition of commercial PCB mixtures.

PCB levels in fish have been decreasing in many areas since the 1970s, but the decrease has leveled off during the last couple of years. Today, the daily PCB intake is estimated to be around 10 ng/kg bw for an adult.

Exposure levels in children (including prenatal exposure)

Once in the body, PCBs and PCDD/Fs accumulate in fatty tissues and are slowly released. Lactation or

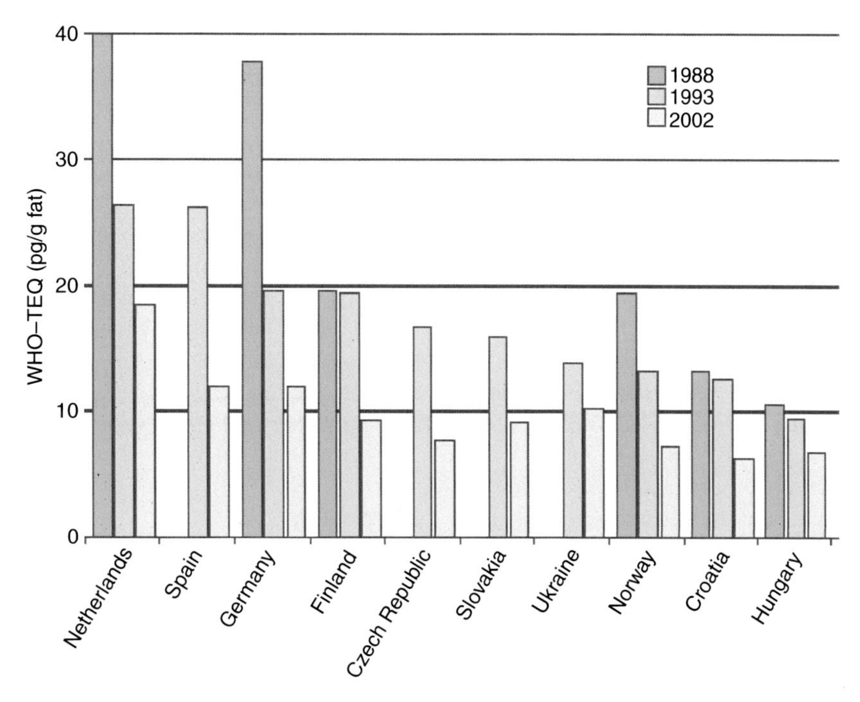

Figure 6 Temporal trends in the levels of dioxins and furans in human milk in various countries participating in consecutive rounds of the WHO exposure study. From van Leeuwen FXR and Malisch R (2002) Results of the third round of the WHO-coordinated exposure study on the levels of PCBs, PCDDs and PCDFs in human milk. *Organohalogen Compounds* 56: 311–316.

significant weight loss increases the release of the substances into the blood. PCBs can cross the placenta from mother to fetus, and are also excreted into the breast milk. PCB and PCDD/F concentrations in human milk are usually higher than in cow's milk or other infant foods. As a result, breastfed infants undergo higher dietary exposure than those who are not breastfed. This concerns particularly breastfed infants of women exposed to high levels of PCBs, including Inuit and women whose diet is mainly based on fish from highly contaminated rivers and lakes, such as the Great Lakes and the Baltic Sea. Time-trend information suggests that PCDD/F and PCB concentrations in human milk have decreased significantly since the 1970s in countries that have taken measures against these substances. However, the decrease has leveled off during the last couple of years. Therefore, current fetal and neonatal exposures continue to raise serious concerns regarding potential health effects on developing infants.

Compared to adults, the daily intake of PCDD/Fs and PCBs by breastfed babies is 1–2 orders of magnitude higher. A recent field study showed higher mean levels of PCDD/Fs and PCBs in human milk in industrialized areas (10–35 pg I-TEQ/g milk fat) and lower levels in developing countries (<10 pg I-TEQ/g milk fat). Very few studies have been performed on Arctic populations with respect to the exposure of children to these substances. It is likely, however, that the differences in exposure between children and adults demonstrated in many industrialized regions also exist in Arctic regions.

Potential for high-exposure situations

It has been shown that these substances, and especially PCBs, can occur in elevated concentration in Arctic fauna. As the diet of many Arctic populations relies to a vast extent on marine mammals that represent high trophic levels, human exposure has been shown to be considerably high compared to industrialized areas.

Health effects in humans

There are many studies on the carcinogenicity of 2,3,7,8-TCDD in accidentally exposed workers. Epidemiological studies on people exposed in connection with the accident in Seveso have generated valuable information. Excess risks were observed for ovarian and thyroid cancers and for some neoplasia of the haematopoietic tissue; these results were, however, based on small numbers. Epidemiological studies on the cohorts most highly exposed to 2,3,7,8-TCDD produced the strongest evidence of increased risks for all cancers combined, along with less strong evidence of increased risks for cancers of particular sites. The relative risk for all cancers combined in the most highly exposed and longer-latency subcohorts is 1.4.

Studies of noncancer effects in children have indicated neurodevelopmental delays and neurobehavioral effects, including neonatal hypotonia. In children in Seveso who were highly exposed to TCDD, small, transient increases in hepatic enzymes, total lymphocyte counts and subsets, complement activity, and nonpermanent chloracne were observed. Also, an alteration of the sex ratio (excess female to male) was observed in children born to parents highly exposed to TCDD.

Critical outcomes and existing reference values

During the last two decades, a number of different risk assessments of dioxins and related compounds have been performed. Since the mid-1990s, coplanar PCBs have often been included in the assessments. In 1997, WHO established an expert group on dioxins and related compounds. It proposed, based on the toxic equivalency Factor (TEF) scheme shown in **Table 2**, a TDI for dioxins and related compounds. The proposal was based on kinetic calculations of doses to body burden and vice versa. The body burden approach resulted in a reduced need for a safety factor for extrapolation between species. The WHO expert group calculated that a reliable LOAEL probably could be found in the range of 14–37 pg/kg bw per day. By applying a safety factor of 10 to this range, it proposed a TDI of 1–4 pg/kg bw. The group emphasized that the TDI represents a tolerable daily intake for lifetime exposure, and that occasional short-term excursions above the TDI would have no health consequences provided that the averaged intake over long periods was not exceeded. In addition, it recognized that certain subtle effects may be occurring in some sections of the general populations of industrialized countries at current intake levels (2–6 TEQ /kg bw per day), but found it tolerable on a provisional basis since these reported subtle effects were not considered overtly adverse and there were questions as to the contribution of non-dioxin-like compounds to the observed effects. The group therefore stressed that the upper range of the TDI of 4 pg TEQ /kg bw should be considered a maximum tolerable intake on a provisional basis, and that the ultimate goal was to reduce human intake levels to below 1 pg TEQ / kg bw per day. In 2001, the European Commission and the Scientific Committee for Food proposed a temporary TWI of 14 pg/kg bw for 2,3,7,8-PCDD/Fs and dioxin-like PCBs.

Summary

It has been demonstrated that dioxins and many PCBs resist degradation, bioaccumulate, are transported through air, water, and migratory species across international boundaries, and are finally deposited far from the place of release where they can accumulate in terrestrial and aquatic ecosystems. The clearest evidence for this long-range transport derives from the levels of PCDD/

Table 2 WHO TEF values for human risk assessment

Congener	*TEF value*	*Congener*	*TEF value*
Dibenzo-p-dioxins		*Non*-ortho-*PCB*	
2,3,7,8-TCDD	1	PCB 77	0.0001
1,2,3,7,8-PnCDD	1	PCB 81	0.0001
1,2,3,4,7,8-HxCDD	0.1	PCB 126	0.1
1,2,3,6,7,8-HxCDD	0.1	PCB 169	0.01
1,2,3,7,8,9-HxCDD	0.1		
1,2,3,4,6,7,8-HpCDD	0.01		
OCDD	0.0001		
Dibenzofurans		*Mono*-ortho-*PCB*	
2,3,7,8-TCDF	0.1	PCB 105	0.0001
1,2,3,7,8-PnCDF	0.05	PCB 114	0.0005
2,3,4,7,8-PnCDF	0.5	PCB 118	0.0001
1,2,3,4,7,8-HxCDF	0.1	PCB 123	0.0001
1,2,3,6,7,8-HxCDF	0.1	PCB 156	0.0005
1,2,3,7,8,9-HxCDF	0.1	PCB 157	0.0005
2,3,4,6,7,8-HxCDF	0.1	PCB 167	0.00001
1,2,3,4,6,7,8-HpCDF	0.01	PCB 189	0.0001
1,2,3,4,7,8,9-HpCDF	0.01		
OCDF	0.0001		

Fs and PCBs measured in the Arctic. Owing to long-range transboundary transport, these substances are nowadays ubiquitous contaminants of the ecosystem and are also present in the food chain. Therefore, most of the human population is exposed to PCDD/Fs and PCBs. Moreover, since dioxins and PCBs pass from mother to fetus through the placenta, and from mother to newborn through breastfeeding, infants are at risk of harmful effects in the most critical period of their development. There are just a few reports of dioxins in humans from Arctic regions, but there are plenty of animal samples analyzed for dioxins and PCBs that give information on human exposure through food. As many people living in the Arctic still practice hunting and fishing for an important part of their diet, their exposure to dioxins, PCBs, and other contaminants could be elevated compared to people living in industrialized parts of the world.

Further Reading

Alcock R, Bashkin V, Bisson M, *et al.* (2003) *Health Risk of Persistent Organic Pollutants from Long-Range Transboundary Air Pollution*, 252p. Bonn: WHO.

Bashkin VN (2003) *Environmental Chemistry: Asian Lessons*, 490p. Singapore: Kluwer Academic.

Bertazzi PA, Bernucci I, Brambilla G, Consonni D, and Pesatori AC (1998) The Seveso studies on early and long-term effects of dioxin exposure: A review. *Environmental Health Perspectives* 106 (supplement 2): 625–633.

Brzuzy LR and Hites RA (1996) Global mass balance for polychlorinated dibenzo-*p*-dioxins and dibenzofurans. *Environmental Science and Technology* 30: 1797–1804.

Dutchak S, Shatalov V, Mantseva E, *et al.* (2004) *Persistent Organic Pollutants in the Environment.* MSC-E and CCC. EMEP Status Report 3/2004, Jun. 2004.

Fiedler H (1999) Sources of PCDD/PCDF and impact on the environment. *Chemosphere* 32: 55–64.

Galiulin RV, Bashkin VN, and Galiulina RA (2005) Ecological risk assessment of riverine contamination in the Caspian Sea basin: A conceptual model for persistent organochlorine compounds. *Water, Air, and Soil Pollution* 163(1–4): 33–51.

Gray LE, Ostby JS, and Kelce WR (1997) A dose-response analysis of the reproductive effects of a single gestational dose of 2,3,7,8-tetrachlorodibenzo-*p*-dioxin (TCDD) in male Long Evans hooded rat offspring. *Toxicology and Applied Pharmacology* 146: 11 20.

Gray LE, Wolf C, and Ostby JS (1997) *In utero* exposure to low doses of 2,3,7,8-tetrachlorodibenzo-*p*-dioxin (TCDD) alters reproductive development in female Long Evans hooded rat offspring. *Toxicology and Applied Pharmacology* 146: 237–244.

Mackay D, Shiu WY, and Ma KC (1992) *Illustrated Handbook of Physical–Chemical Properties and Environmental Fate for Organic Chemicals*. Boca Raton FL: Lewis.

Mantseva E, Dutchak S, Rozovskaya O, and Shatalov V (2004) *EMEP Contribution to the Preparatory Work for the Review of the CLRTAP Protocol on Persistent Organic Pollutants.* EMEP MSC-E Information Note 5/2004.

Oehme M, Schlabach M, Hummert K, Luckas B, and Nordoy ES (1994) Levels of polychlorinated-*p*-dioxins, dibenzofurans, biphenyls and pesticides in harp seals from the Greenland Sea. *Organohalogen Compounds* 20: 517–522.

van den Berg M, Bimbaum L, Bosveld AT, *et al.* (1998) Toxic equivalency factors (TEFs) for PCB, PCDDs, PCDFs for humans and wildlife. *Environmental Health Perspectives* 106: 775–792.

van Leeuwen FXR and Malisch R (2002) Results of the third round of the WHO-coordinated exposure study on the levels of PCBs, PCDDs and PCDFs in human milk. *Organohalogen Compounds* 56: 311–316.

Wagrowski DM and Hites RA (2000) Insights into the global distribution of polychlorinated dibenzo-*p*-dioxins and dibenzofurans. *Environmental Science and Technology* 34: 2952–2958.

Relevant Website

http://www.msceast.org – Meteorological Synthesizing Centre – East.

PART C

Global Patterns and Processes

Agriculture

D Lyuri, Russian Academy of Sciences, Moscow, Russia

A Short History of Agriculture

Agriculture appeared about 10 000 years ago in one of the regions of the Middle East: south of Palestine, north of Syria and Mesopotamia. In the eighteenth and the nineteenth centuries, primitive long-fallow (slash-and-burn) agriculture was replaced by three-field agriculture; in the late nineteenth and the twentieth centuries, the modern industrial (high-input) agriculture appeared in many countries. The history of agriculture is, first of all, the history of restoration and improvement of soil fertility. At first, it was restored by natural processes during the fallow stage; then, due to the application of organic fertilizers; and presently, at the expense of mineral and organic fertilizers and other inputs. These changes made it possible to increase agricultural production from 25 to $145 \times 10^9 \mathrm{J\,ha^{-1}\,yr^{-1}}$ and more (**Table 1**). But this increase has been accompanied by a decrease in agricultural effectiveness (the output/input ratio). This phenomenon can be referred to as 'the ecology–energy law of agricultural development'.

In the middle of the twentieth century, the 'green revolution' – the rapid development of agricultural technologies, irrigation, the selection of new effective strains of plant and animal species, and the increasing application of fertilizers and pesticides – led to a considerable growth of agricultural productivity. For example, in 1952–72, wheat yields in Mexico increased by 3 times (from 0.88 to $2.72\,\mathrm{t\,ha^{-1}}$).

Agricultural Lands

Agricultural lands consist of three main types: (1) arable land (including cropland and fallows), (2) land under permanent crops, and (3) pastures and hayfields. The total area of agricultural lands in the world is 4973.4 million ha. They cover 33.3% of terrestrial surface (2003), including 10.3% of arable land and land under permanent crops and 23% of pastures and hayfields. The largest areas of agricultural lands are found in Asia (33.8% of the world value) and in Africa (23.0%); in Europe, the Americas, and Oceania, they occupy 9–12% of the world value (**Table 2**).

In the aggregate structure of agricultural lands, pastures and hayfields comprise 69.0% of the total agricultural area; arable lands, 28.2%; and lands under permanent crops, only 2.8%. But this structure is different in different regions: Oceania, Africa, and South America have the maximum portion of pastures and hayfields (80–88%) and minimum portion of arable land (10–18%). In Europe and North America, the portion of pastures and hayfields decreases to 38–56%, whereas the portion of arable land gains its maximum (41–59%). In Asia, the structure of agricultural land resembles that in the whole world (**Table 3**).

Throughout history, the area of agricultural lands has been continuously expanding in all parts of the world, except for short periods of wars, epidemics, crises, etc. At present, the total agricultural area continues to rise (**Figure 1**), but the rate of this process has been decreasing since 1985.

Moreover, from the middle of the twentieth century, the dynamics of agricultural lands have been different in different regions (**Figure 2**). Their areas have increased in Asia and South America (by 30% from 1960 to 2003) and Africa (by 6%), but they have decreased in Europe (by 14%), North and Central America (by 4%), and Oceania (by 9% since 1981). Arable lands have similar dynamics: they have increased in South America (by 85%), Africa (by 33%), and Asia (by 15%), and they have decreased in Europe (by 15%). In North and Central America, the area of arable land was increasing up to year 1981 (by 5% since 1960) and then began to decrease (by 5%). Therefore, at present, the dynamics of agricultural areas have dissimilar trends in different parts of the world.

Agricultural activity is the main process altering the plant canopy on Earth. As a result of agricultural development, areas of steppe and savannah biomes have decreased by 7–10 times and the areas of forest biomes, by one-third (**Figure 3**). Australia, USA, China, Brasilia, India, and Russia have the largest areas of natural ecosystems replaced by agricultural lands.

Agricultural lands, as well as other ecosystems with plant canopy, are able to produce phytomass. The net primary productivity of arable lands and lands under permanent crops lies in the range of $1\text{–}40\,\mathrm{t\,ha^{-1}\,yr^{-1}}$, with an average value of $6.5\,\mathrm{t\,ha^{-1}\,yr^{-1}}$. This is close to

Table 1 Production and effectiveness of different types of agriculture

	Collecting	*Long-fallow*	*Three-field*	*High-input*	*Hothouse*
Systems of crop farming					
Production (10^9 J ha^{-1} yr^{-1})	0.8	25	40	145	>400
Effectiveness (output/input) (J/J)	20	10	8	1.5	0.003
	Hunting	*On natural pastures*	*On artificial pastures*		*At farms*
Systems of animal husbandry					
Production (10^{9J} J ha^{-1} yr^{-1})	<0.8	17–34	50–59		92–110
Effectiveness (output/input) (J/J)	10.0	2.0–1.5	1.0–0.2		0.1–0.05
Systems of poultry farming					
Production (10^9 J ha^{-1} yr^{-1})	<0.04		67–75		92–126
Effectiveness (output/input) (J/J)	10.0		2.0–1.0		0.5–0.1

Table 2 The area of agricultural lands (million ha)

	Africa	*Asia*	*Europe*	*N. America*	*S. America*	*Oceania*	*World*
Agricultural lands	1146.1	1681.4	483.6	618.9	584.3	459.1	4973.4
%	23.0	33.8	9.7	12.4	11.7	9.2	100.0
Arable lands	199.4	506.9	284.1	255.2	107.1	49.7	1402.3
Permanent crops	25.9	64.0	16.7	14.8	13.6	3.3	138.3
Pastures, hayfields	920.8	1110.5	182.9	348.9	463.5	406.2	3432.8

Table 3 The structure of agricultural lands (%)

	Africa	*Asia*	*Europe*	*N. America*	*S. America*	*Oceania*	*World*
Arable lands	17.4	30.1	58.7	41.2	18.3	10.8	28.2
Permanent crops	2.3	3.8	3.4	2.4	2.3	0.7	2.8
Pastures and hayfields	80.3	66.0	37.8	56.4	79.3	88.5	69.0

the average net primary productivity of terrestrial ecosystems (7.8 t ha^{-1} yr^{-1}). The primary productivity of pastures and hayfields is near to that of corresponding grasslands, or somewhat lower, if the grazing pressure exceeds the critical level. As agricultural lands replace natural ecosystems with the highest primary productivity (first of all, forests, savannas, and meadows), the global biosphere production decreases by 12–15% under the impact of agricultural development. Moreover, agricultural activity leads to serious changes in landscapes and affects the global carbon (C) cycle. Cultivated soils have lost up to 30% of C (in the tropical zone, up to 70%); at present, the carbon emission from agricultural lands is estimated at 2.5 GtC yr^{-1} (the emission related to combustion of fossil fuels reaches 5.8 GtC yr^{-1}).

Agricultural Inputs

From the viewpoint of stability, all agricultural ecosystems (agroecosystems) may be divided into two groups. First are natural pastures and hayfields, which can preserve their stability without anthropogenic inputs. Second are arable lands, permanent crops, and artificial pastures and hayfields, which can only preserve their stability at the expense of anthropogenic inputs (management practices). Without them, these agroecosystems will be destroyed and replaced by different natural ecosystems.

'Energy' is a necessary resource for modern agriculture. Direct energy inputs (oil, electricity, etc.) ensure the work of tractors, harvesters, and other equipment, the initial processing of products, etc. World agriculture (2001) consumes 189 636 000 toe of direct energy (toe – metric tons of oil equivalent – measures the energy contained in a metric ton of crude oil; 1 toe is equal to 10^7 kcal, 41 868 GJ, or 11 628 GWh). But it is only 2.5% of the total energy consumption in the world (to compare, industry uses 32% and transport, 25.2%). In different parts of the world, energy consumption in agriculture is different. The 'energy pressure' (toe ha^{-1}) is maximum in Europe and Asia and minimum in Africa and Oceania (**Table 4**).

Indirect energy consumption in agriculture is associated with energy spent for the manufacturing of tractors,

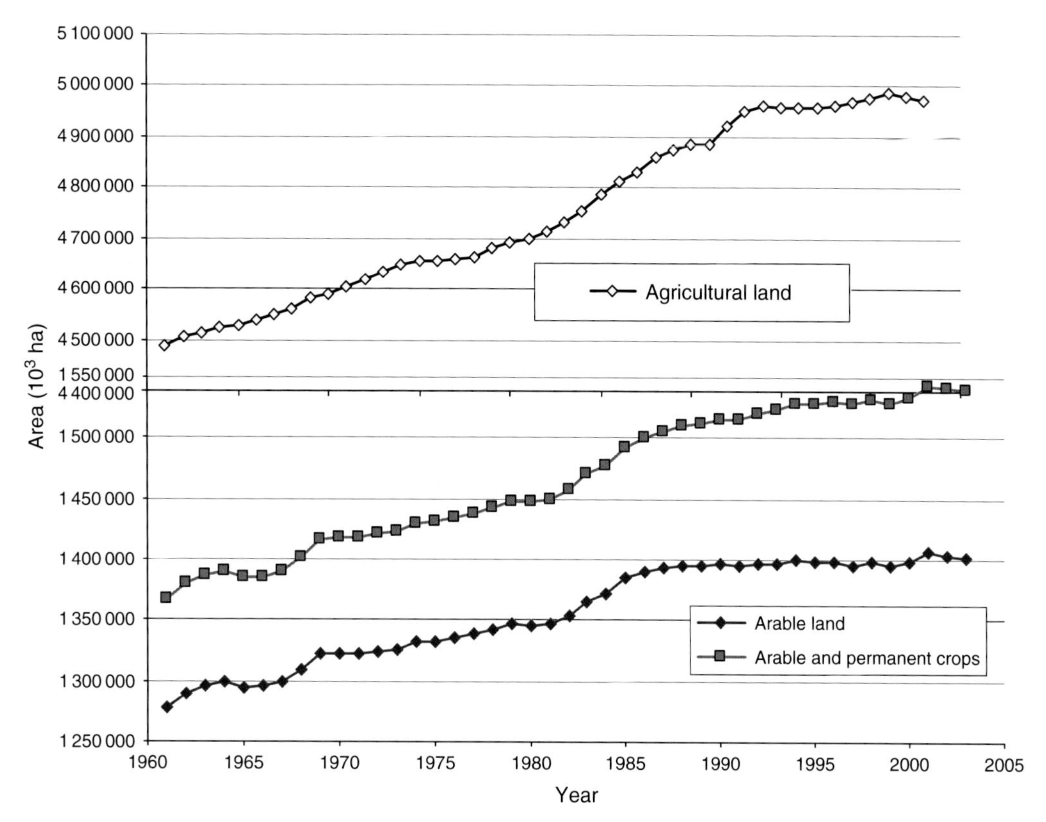

Figure 1 Dynamics of agricultural and arable lands in the world (1961–2003).

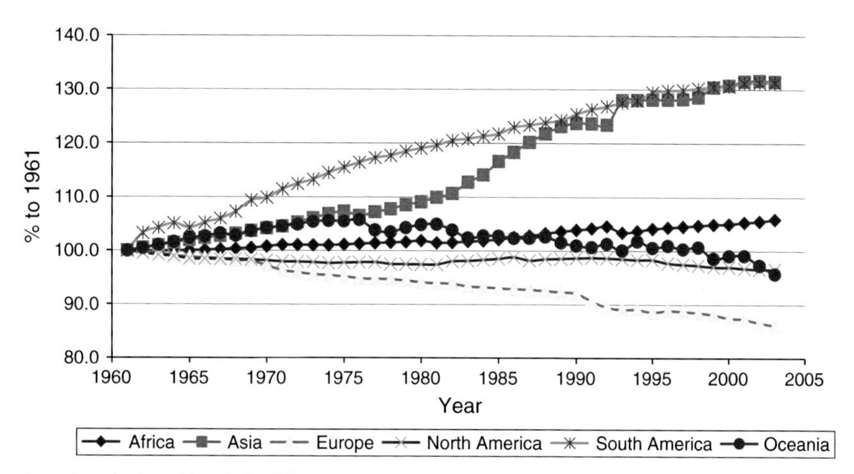

Figure 2 The dynamics of agricultural lands in different parts of the world.

equipment and fertilizers; with labor force, etc. The calculation of energy expenses makes it possible to compare different types of inputs in uniform (energy) units. In modern farms (**Table 5**), the input of direct energy averages about 40%; the input of indirect energy with fertilizers, about 32%; and with equipment and chemical weed-killers, 11–12%. The input of labor force is very small, but in traditional farms in developing countries it may reach 50% and more.

'Fertilizers' are necessary for restoring soil fertility. Organic fertilizers (manure, dung, peat) have been actively used throughout the history of agriculture. Now, their input reaches 10–40 t ha^{-1}. The application of mineral fertilizers began at the end of the nineteenth century. In 2005, their consumption reached 157 million tons annually. There are three main types of mineral fertilizers: nitrogen (93 million tons), phosphorus (39 million tons), and potassium (25 million tons); thus, they are applied in the following

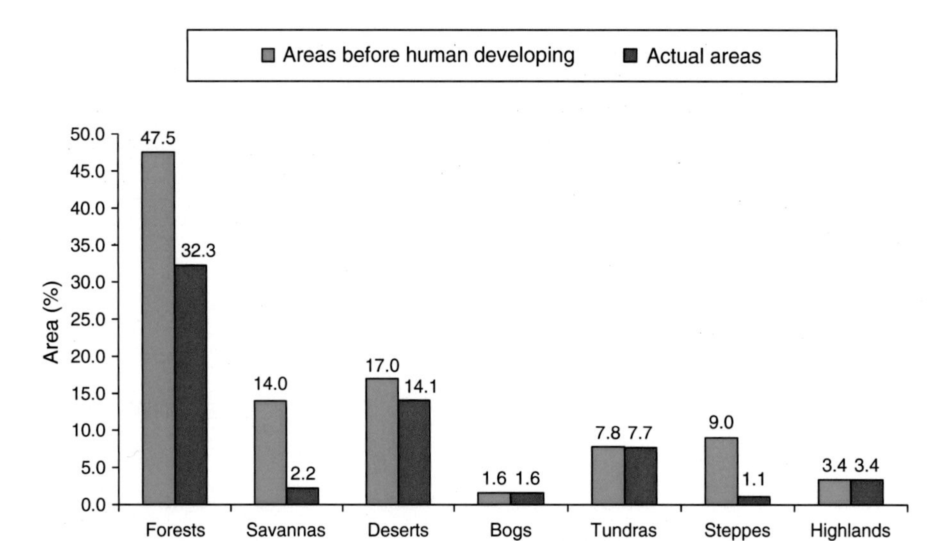

Figure 3 Changes in the areas of major biomes (percent of the Earth area) as a result of human activity.

Table 4 Consumption of direct energy in agriculture

	Africa	*Asia*	*Europe*	*N. America*	*S. America*	*Oceania*	*World*
Energy consumption (1000 toe)	12 725	63 086	50 185	22 721	13 095	1995	189 636
Energy consumption (toe ha^{-1})	0.046	0.132	0.164	0.085	0.113	0.036	0.127

Table 5 The consumption of energy in a modern farm (corn production, the USA)

Inputs	*MJ ha^{-1} yr^{-1}*	*% of total*
Gasoline and gas	7 395	21.0
Electricity	6 436	18.3
Labor force	29	0.08
Tractors and other equipment	4 158	11.8
Fertilizers	1 1355	32.2
Seeds	1 869	5.3
Herbicides and insecticides	3 762	10.7
Transport	214	0.6
Total	35 218	100

average ratios: N(55%):P(25%):K(20%). The world average input of mineral fertilizers is 91.1 kg ha^{-1}: from 32 kg ha^{-1} in Africa to 144 kg ha^{-1} in Asia (**Table 6**). In some countries, their application may be much higher (Ireland: 600 kg ha^{-1}) or smaller (Congo: 0.1 kg ha^{-1}). In many developing countries, the NPK input is less then the output of these nutrients with agricultural products, which leads to a decrease in soil fertility.

At the same time, big 'fertilizer pressure' results in contamination of food and environmental pollution. A large portion of fertilizers is washed away from croplands into rivers and lakes, which is the cause of their eutrophication (e.g., in Europe, Asia, and North America). Moreover, on the global scale, the use of fertilizers results in deep changes in the character of nitrogen, phosphorus, and potassium cycles in the biosphere. This is especially

Table 6 Different types of agricultural inputs to croplands

	Africa	*Asia*	*Europe*	*N. America*	*S. America*	*Oceania*	*World*
Labor force (men/100 ha)	80	210	10	10	20	5	90
Tractors per 1000 ha	6.9	12.4	35.9	21.7	11.1	7.2	17.6
Fertilizers (kg ha^{-1})	31.9	144	72.5	89.2	88.4	57.4	91.1

true with respect to nitrogen. The anthropogenic fixation of nitrogen via the production of nitrogen fertilizers (80 million tons per year) is only a little less than the biological fixation of nitrogen by all terrestrial ecosystems (100–130 million tons).

'Herbicides' and 'insecticides' are necessary to control weeds and pests. Their active use began in the second part of the twentieth century. Now, the average world rate of their application is 300 g ha^{-1} (from 0 in Laos to 2–3 kg ha^{-1} in the USA and Western Europe and 51 kg ha^{-1} in Costa Rica). Specialists say that these chemicals help save about one-third of the world agricultural produce. At the same time, their application results in the contamination of not only food products but also natural ecosystems on a global scale (hydrosphere, soils, etc.). Herbicide DDT was found even in Antarctica.

Without 'machineries' and 'equipment', modern intensive agriculture is impossible, because it involves many manipulations with agricultural lands, primary processing, cattle, etc. The average world number of tractors per 1000 ha is 17.6 – from 7 tractors/ha in Africa to 36 tractors/ha in Europe (**Table 6**). Note that their use results in soil compaction on croplands and additional CO_2 emission because of the utilization of fossil fuels. Therefore, agricultural lands, first of all arable lands, play the role of ecosystems with a negative carbon balance on the global scale.

In modern agriculture, 'labor force' is, foremost, the 'management block' of agricultural systems, and only in backward farming, it is the 'energy source'. At present, the average world 'labor force pressure' is 90 men/100 ha (from 5–10 men/100 ha in Oceania, Europe, and North America to 210 men/100 ha in Asia) (**Table 6**).

'Irrigation' is one of the biggest inputs to agriculture in arid zones. The world area of irrigated lands is not very large – only 277 098 thousand ha, or 18.1% of the total cropland (from 5–8% in Oceania and Europe to 35.3% in Asia) (**Table 7**). But irrigated lands are the largest consumer of water resources: 2661.2 km^3 or 70% of the total world water consumption is spent for irrigation (industry – 20%, domestic use – 10%). In some countries, the portion of irrigation water in the total water use may reach 95–99% (Mali, Cambodia, Thailand, etc.). In the case of normal irrigation, almost 50% of water is lost for evaporation; overhead (sprinkling) irrigation gives 30% losses; and only very capital expensive drip irrigation, 1–2%. Thus, irrigation is the crucial factor altering the water circulation, first of all in arid regions. For example, the disappearance of the Aral Sea is the result of irrigation of large areas in Central Asia. Moreover, intensive irrigation may result in soil salinization. Because of salinization, 0.2–0.3 million ha of irrigated land is being abandoned in the world every year.

'Drainage' is one of the types of agricultural inputs in humid zones. The world area of artificially drained lands is about 157 million ha, including 56 million ha in North and Central America, 40 million ha in Europe, and 15 million ha in the former USSR. In Finland, more than 90% of cropland is drained area; in Hungary, 74%; in the UK, 61%. This method of soil reclamation provides possibility to increase crop yields in overmoistened areas, but is also associated with the destruction of peat layers and with a decrease in soil fertility in 15–20 years. Moreover, drainage worsens the water regime of landscapes on a regional scale.

Agricultural Production

At present, global agriculture uses about 2500 species of cultivated plants and a few hundred species of cultivated animals, but only several hundred cultivated species are prevalent. The dominant groups of cultivated plant species are cereals (wheat, barley, rye, rice, corn, sorghum, etc.), pulses, roots and tubers (potatoes, yam, etc.), fruits, and vegetables. The dominant cultivated animal species are cattle, sheep, goats, pigs, equines, buffaloes, camels, chickens, and ducks.

The productivity of agriculture is the main index of its intensity: the increase in agricultural input is accompanied by the rise in agricultural output. The yield of cereals gains its maximum in the USA (5.2 t ha^{-1}) and in Western Europe (5–7 t ha^{-1}; in Belgium, 8.4 t ha^{-1}) and its minimum, in Africa (Botswana: 0.2 t/ha); North America, Middle East, and North Africa produce high yields of roots and tubers (**Table 8**).

Table 7 Irrigated lands and water input for irrigation

	Asia (excl. Middle East)	*Europe*	*Middle East and North Africa*	*Sub-Saharan Africa*	*N. America*	*S. America*	*Oceania*	*World*
Irrigated lands (% of total cropland)	35.3	8.1	28.5	3.8	10.3	8.9	4.9	18.1
Irrigation water use (million m^3)	1 739 480	132 088	279 196	99 758	199 601	111 812	18 855	2 661 624
Irrigation water (% of the total water consumption)	81	33	86	88	38	68	72	70

Table 8 The productivity of main cultivated plants (t ha^{-1}, 2003)

	Asia (excl. Middle East)	*Europe*	*Middle East and North Africa*	*Sub-Saharan Africa*	*N. America*	*S. America*	*Oceania*	*World*
Cereals	3.3	3.4	2.3	1.1	5.1	3.2	1.8	3.1
Pulses	0.7	2.1	0.9	0.5	1.5	0.8	1.0	
Roots and tubers	16.5	15.8	22.3	8.1	36.3	13.4	12.9	13.1

Table 9 Agricultural production (1000 t, 2003)

	Asia (excl. Middle East)	*Europe*	*Middle East and North Africa*	*Sub-Saharan Africa*	*N. America*	*S. America*	*Oceania*	*World*
Cereals	927 074	407 781	88 639	90 082	367 571	113 259		2 070 963
Pulses	23 137	7716	3428	8413	4344	3879	2123	55 679
Roots and tubers	283 662	134 101	16 076	170 089	25 921	47 256	3598	687 193
Meat	93 967	52 410	8427	8321	42 607	28 151	5641	246 826

Asia is the greatest producer of agricultural products: it gives about 40–45% of the global agricultural output (**Table 9**; **Figure 4**). Two countries with large population – China and India – are the main 'driving forces' of agricultural development in Asia. Only two regions can be compared to Asia: Europe (15–20% of the global agricultural output) and North America (10–15% of the global output, except for roots and tubers). Sub-Saharan Africa produces 15% of pulses and 25% of roots and tubers.

In the second half of the twentieth century, world agricultural production has been growing steadily, but with dissimilar rates in different regions. Thus, from 1961 to 2003, the world yield of cereals increased by 237%. In Oceania and South America, this increase was higher (407% and 339%, respectively). In Europe, it was lower (154%). In Africa, North America, and Asia, it was near the average value (277%, 227%, and 279%, respectively). Moreover, in Europe and North America, the production of cereals ceased to rise since the mid-1980s (**Figure 5**).

However, the huge production does not mean sufficient consumption. Because of different populations and different food allowances, the individual level of food consumption (in calories) does not coincide with agricultural production in the particular regions. The largest consumption is in North America and Europe (134% and 120% from the average world value), and the lowest consumption is in Asia and Africa (95% and 81%) (**Table 10**). Many countries can keep the food consumption at a necessary level only at the expense of import.

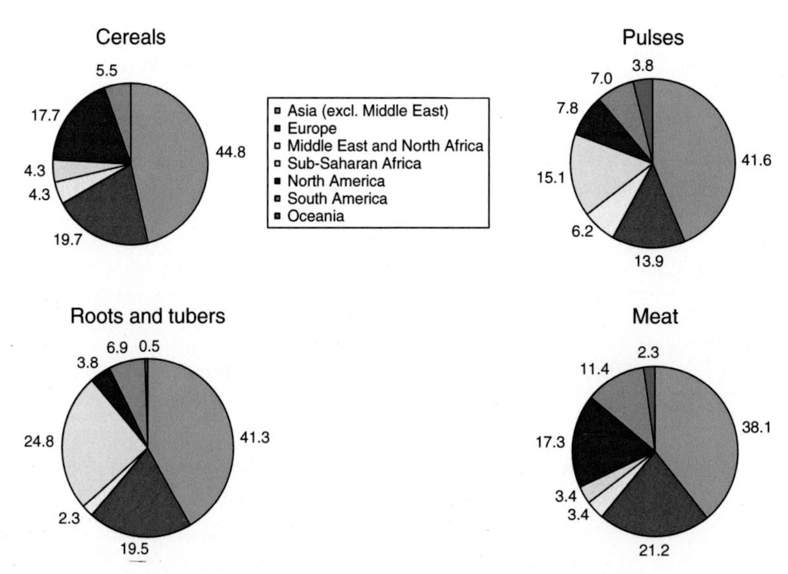

Figure 4 Agricultural production in different regions (percent of the world production, 2003).

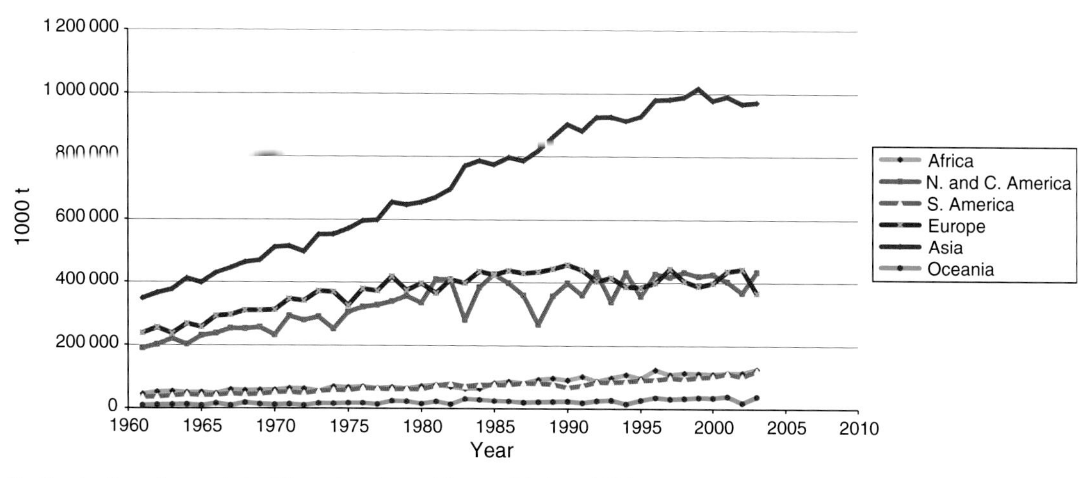

Figure 5 Production of cereals in different regions of the world since 1961.

Table 10 Calorie supply per capita in different regions of the world (2003)

	Asia (excl. Middle East)	*Europe*	*Middle East and North Africa*	*Sub-Saharan Africa*	*N. America*	*S. America*	*Oceania*	*World*
Calorie supply per capita	2681.8	3331.4	3109.8	2262.2	3755.8	2850.9	3074.3	2804.4

At present, the world market of agricultural products is the global system. Agricultural products move not only between neighboring countries but from continent to continent. The world market of cereals is the biggest (about $40 billion). The main exporters of cereals are the USA (85 Mt), France (30 Mt), Argentina (20 Mt), and Australia (20 Mt), and the main importers are Japan (26 Mt), Mexico (15 Mt), South Korea (13 Mt), China (10 Mt), and Egypt (10 Mt). All these countries are found in different parts of the world. Some countries specialize in one or two main kinds of export products and supply the whole world with them. For example, Brazil, Colombia, and Vietnam are the largest exporters of coffee, while Ecuador, Colombia, Panama, and Honduras are the largest exporters of bananas.

Abandoned Agricultural Lands

Contrary to a general increase in the world area of agricultural lands (**Figure 1**), more than 80 countries demonstrate a continuous reduction of agricultural lands. From 1961 to 2002, about 2.3 million km^2 of agricultural lands was abandoned, foremost in Russia, Australia, the USA, and West Europe (**Table 11**).

Six types of abandoning of agricultural lands can be distinguished. The first four types are typical of the countries, where the decrease in the area of agricultural lands is a result of agricultural intensification.

Table 11 Countries with maximum abandoned agricultural areas in 1961–2002

No.	*Country*	*Abandoned agricultural area (1000 km^2)*
1	Russia	617.4
2	Australia	407.7
3	United States	356.4
	West Europe Total	251.2
4	Kazakhstan	60
5	Algeria	54
6	Italy	52.4
7	France	49.8
8	Spain	30.3
9	Canada	29.5
10	United Kingdom	28.5
11	Germany	25.6
12	South Africa	23.8
13	Poland	19.7
14	Chile	19.5
15	Japan	19.3
	Other countries	553.4
	Total in the world	2297.7

1. *US type.* The decrease in agricultural area as a result of agricultural intensification is accompanied by a growing agricultural output (**Figures 6** and **7**). There are 21 countries (the USA, the UK, countries of West Europe, Australia, India, Thailand, etc.) with this type of the dynamics of agricultural lands; about 1.0 million km^2 of

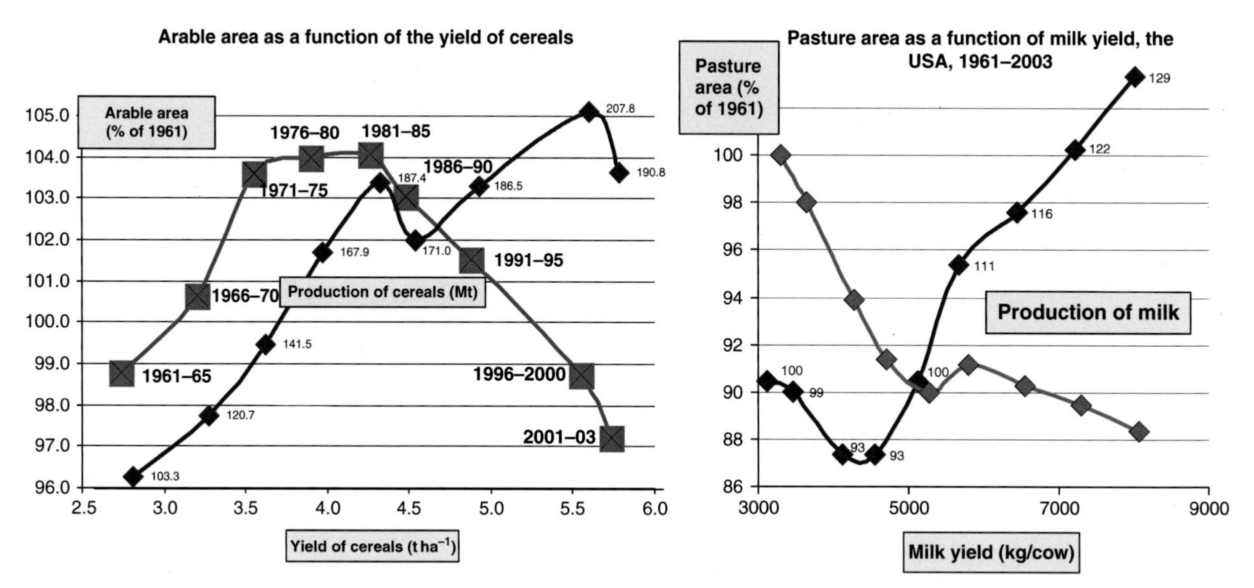

Figure 6 The decrease in agricultural area as a result of agricultural intensification with an increase in agricultural production (the USA, 1961–2003).

Figure 7 Countries with decreasing agricultural area (1961–2003).

agricultural lands has been abandoned in the recent decades (43.5% of the world value).

2. *Japan-type dynamics.* The decrease in agricultural area as a result of agricultural intensification is accompanied by the decreasing agricultural output, which signifies the orientation toward food import. Two countries – Japan and South Korea – abandoned 0.02 million km^2 of agricultural areas (<1% of the world value).

3. *France-type dynamics.* The decrease in agricultural area as a result of agricultural intensification is followed by its increase during the recent years with a general rise in the agricultural output, which signifies the orientation toward food export. There are no countries with 'clear'

France-type dynamics of agricultural lands: only a 'mix' with the US type, that is, arable land with the France type and pasture land with the US type (France, Switzerland, the Netherlands, Uruguay), or arable land with the US type and pasture land with the France type (Spain, New Zealand). Overall, 0.1 million km^2 of agricultural lands have been abandoned in these six countries (4% of the world value).

4. *Hungary-type (transitional) dynamics.* The decrease in agricultural area as a result of agricultural intensification is followed by the decrease caused by the deep economic crisis; the agricultural output increases at the first stage and decreases during the second stage. Within this type, 0.07 million km^2 of agricultural lands (3%) has been abandoned in nine countries (Poland, Hungary, and other countries of East Europe).

Two other types of the recent dynamics of agricultural lands are typical of the countries, in which the decrease in agricultural lands is, first of all, a result of crises, wars, revolutions, and others nonagricultural processes.

5. *Russian-type dynamics.* The increase in agricultural area is followed by a sharp decrease as a result of the deep economic crisis; this is accompanied by the growth and, then, reduction of the agricultural output. In 17 countries (Russia, other countries of the former USSR, Bulgaria, Romania), 0.85 million km^2 of agricultural lands (37.0% of the world value) has been abandoned.

6. *Miscellaneous-type dynamics.* The decrease in agricultural area has no relation to agricultural productivity and total agricultural output. In 21 countries with the miscellaneous-type dynamics (Bangladesh, Cameroon, Lesotho, Nigeria, Swaziland, etc.), 0.25 million km^2 of agricultural lands (10.9%) has been abandoned.

Abandoned agricultural area is replaced by two kinds of land: first – by settlements, infrastructure, industry, etc.; second – by fallows. In the latter case, the former agroecosystems are being substituted by natural ecosystems. In the forest zone, the natural succession develops toward restoration of forest ecosystems (**Figure 8**); in the steppe zone, toward restoration of grassland ecosystems.

However, the natural succession processes can be blocked. This may happen if:

- the soil or the relief of the abandoned land is destroyed by erosion or other dangerous processes;
- the abandoned agricultural land is occupied by introduced plant species, which can stop the natural succession processes;
- there are no plant species, which are necessary for succession processes, around the abandoned agricultural land or in the soil seed bank.

The Future of Agriculture

The actual forecasts of the 'dynamics of arable area' suggest that it may increase up to 3.2 billion ha in the middle of the twenty-first century, if the agricultural productivity remains at the present level, and up to 1.6 billion ha, if the productivity is raised 2 times. This increase may take place, first of all, at the expense of forest lands. But the real dynamics of arable areas in recent years (**Figure 1**) and their decrease in the countries with very intensive agriculture (**Figure 8**) indicate that these values might be overestimated.

'Agricultural technologies' will be developed along two different ways. First, their further intensification is expected in developing countries, since it is the only way to ensure food supply of their increasing population. Second, the further expansion of 'green' or 'organic' agriculture with minimum artificial inputs (mineral fertilizers, herbicides, insecticides, tillage, etc.), may take place in developed countries, where the cost of 'ecologically clean' food is very high. At present, the total area of 'organic' farmland in the world is 22.8 million ha (10.5 million ha in Australia, 5.2 million ha in Europe, 4.7 million ha in South America, and 1.4 million ha in North America), and it is steadily increasing.

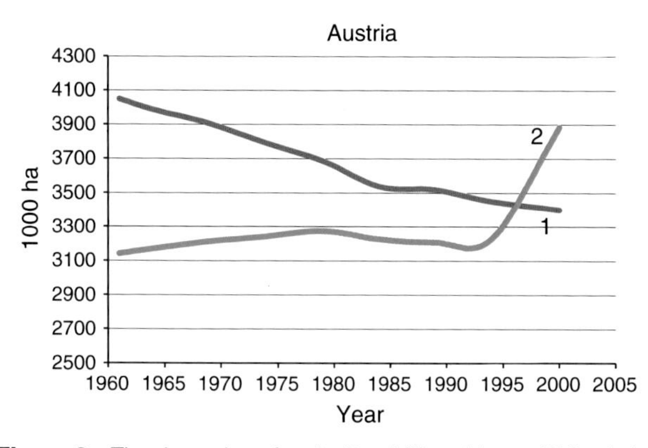

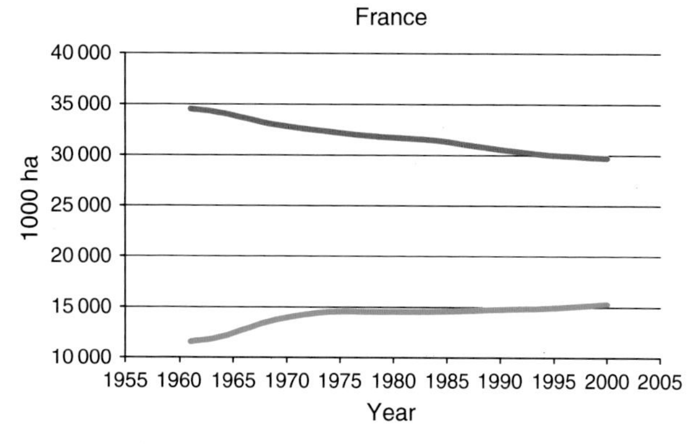

Figure 8 The dynamics of agricultural (1) and forest (2) lands in some countries of the forest zone.

Further Reading

Boehlje M, Hofing S, and Christopher R (1999) Farming in the 21st Century. West Lafayette, IN: Purdue University.
Collinson M (ed.) (2000) *A History of Farming Systems Research*. Wallingford, UK: CABI Publishing.
FAO (2004) FAOSTAT Statistical Database.
Pimentel D, Hurd LE, Belloti AC, *et al.* (1973) Food production and the energy crisis. *Science* 182(No. 4111): 443–449.
Turner BL, II (ed.) (1990) *The Earth As Transformed by Human Action*. Cambridge: Cambridge University Press.
USDA (2001) Agricultural statistics. Washington, DC: US Department of Agriculture.
USDA (2001) Food and Agricultural Policy: Taking Stock for the New Century. Washington, DC: US Department of Agriculture.
Wood S, Sebastian K, and Scherr S (2000) *Pilot Analysis of Global Ecosystems: Agroecosystems*. Washington, DC: World Resources Institute.

Material and Metal Ecology

M A Reuter, Ausmelt Ltd, Melbourne, VIC, Australia

A van Schaik, MARAS (Material Recycling and Sustainability), Den Haag, The Netherlands

Introduction

'Metals and materials' are used in a wide range of products and applications ranging from consumer products (cars, electronics, white and brown goods, etc.) to constructions (buildings, roads) and agriculture (fertilizers), etc. The social and ecological value of the materials in these applications is not only determined by the 'in-use value' of these applications such as functionality, durability, safety, reduced energy consumption, esthetics, etc., but also by the possibility of these materials to return from their original application into the 'resource cycle' after their functional lives/use at the lowest environmental impact. The design of the product determines the selection of materials to be applied in the products as well as the complexity of the material combinations and interactions within this product (e.g., welded, glued, alloyed, layered). These actions directly affect the recyclability of the materials, that is, whether the material cycle can be closed and whether one can speak of an industrial ecological system.

Figure 1 indicates that the social/environmental value of materials and metals can only be properly determined if both the resource cycle and the 'technology/design cycle' are fundamentally understood and described, but more important that tools are available to link these three inseparable disciplines. The interconnectivity between the resource cycle (i.e., the primary and secondary material cycles), the technology/design cycle (i.e., product design, recycling technology, materials processing, etc.) and the nature (social/environmental) cycle is depicted by **Figure 1**.

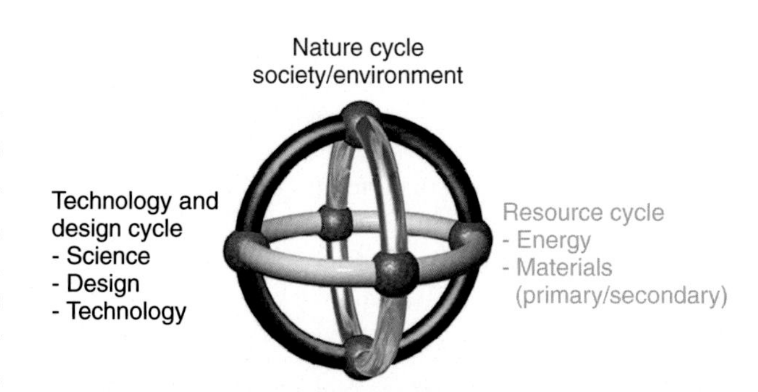

Figure 1 Philosophy: toward sustainability and material and metal ecology by linking the indicated cycles. Reproduced from van Schaik A and Reuter MA (2004) The time-varying factors influencing the recycling rate of products. *Resources, Conservation and Recycling* 40(4), pp. 301–328, with permission from Elsevier.

This section shows how complexly linked (nano-/micro-) metals (with their associated materials, plastics, etc.) that is, the resource cycle is related to products, production, manufacturing technology, and the product designer, that is, the technology and design. Ultimately these cycles intersect with nature, that is, the nature cycle, but this is disregarded from the discussion here; however, the link to geology, ores, and metal-containing minerals is considered as is the environmental impact of waste and residues from anthropogenic activity. This together composes the complex 'web of metals and materials' from an anthropogenic point of view. This web depicts the flow of metals/materials into consumer products; subsequently as consumer products into most regions of the world and finally either recycled back into consumer products and/or into nature and/or humans and/or disposed. Central to this is the metallurgical processing technology, which constitutes the 'ecological organism' in this industrial ecological system; the 'organism' that closes the anthropogenic material and metal cycles.

The web of metals and materials or 'industrial ecological metal and material system' shows how metals and associated materials flow through the resources industry as well as the consumer product and waste system. It describes this flow on the basis of the first principles of recycling technology and metallurgical processing technology theory which are closely linked to product design. This technology- and engineering-based approach provides insight into the complex web of metals and materials providing information on how to monitor, control, and improve the system (and on this basis its economics), at the same time linking this information back to product design. The best manner in which to map the complex interactions between metals is by the application of dynamic modeling, which is more advanced than material flow analysis (MFA) and provides valuable insights into the dynamic interaction and movement of metals and materials linked to consumer products. This is crucial to ensure that valuable minor elements find their way back into products, visualizing and controlling the distribution of these elements onto the surface of the Earth due to the action of consumer society and original equipment manufacturers (OEMs). Therefore, the following will be discussed providing a holistic and fundamental approach to material and metal ecology.

- Resource efficiency and future availability of materials/minor elements is of environmental, economic, and societal concern. Therefore, the recovery of material and metals within the highly connected web of metals and materials in the resource cycles of both base metals and especially environmentally relevant and valuable minor elements for automobiles, consumer electronics, miniaturization applications, and nanotechnology is of crucial importance from a sustainability perspective. Crucial here is also the effect of the closely connected materials in consumer products that cannot be separated due to their close association in end-of-life products.
- Time dependencies as well as process dynamics often have a crucial impact on the web of metals and materials and hence on the impact these metals and materials have on the environment. This is important since it takes time for consumer products and their associated metals and materials to flow through the system. Also the complex connection between the linked primary ore and secondary recycled materials chains, and rapidly changing product compositions have an important effect on determining where metals and materials report to. Social aspects such as the concentration of labor in certain parts of the world associated with metal production and recycling could be dynamically illustrated by such a dynamic visualization.
- Energy and climate change effects are directly connected to recycling as many of the environmental impacts are dominated by the energy needed and CO_2 produced to extract materials associated with products or by the prevention of this by proper recycling activities. Recycling is therefore of extreme importance to lower the consumption of energy during their production, hence directly having an impact on greenhouse gas emissions. It is shown how the web of metals and materials can be arranged to maximize energy recovery, support light-weight design, minimize toxic emissions by basing the system models on the first principles of process engineering.
- The role of product design is demonstrated by discussing the design wheel, which shows how computer-aided design (CAD) is linked to recycling, recyclate quality, metal and energy recovery, and waste creation/prevention. Often materials are connected to each other, alloyed, welded, glued, etc., which makes it impossible to consider the ecology of metals and materials only on a linear and single-element basis.

This rigorous approach provides a basis for quantifying legislation on a more technological and fundamental basis (i.e., physics, chemistry, and thermodynamics) hence providing a first-principles basis for recycling targets and a solid legal basis which OEMs can safely operate on and manufacture products (e.g., future energy-efficient recyclable light-weight cars). This dynamic and technological detailed first-principles approach supports the simplistic life cycle assessment (LCA) approach and provides the consumer with transparent information on all issues surrounding the ecological safe production and use of the product until its end-of-life phase and subsequently its recycling back into metals, materials, and energy.

The Metal Wheel – Material and Metal Ecology

Consumer products are a complex mixture of closely associated metals, plastics, chemicals, inorganic compounds, and materials. These complex connections are often difficult to separate due to the limitations of the applicable separation physics as well as incompatible thermodynamics, which sometimes render the complete recycling chain uneconomical, subject to product type. The result could be that these end-of-life products are then shipped to low-cost countries where these are hand-processed (often more efficiently than present technology permits) and/or dumped or even reused but then eventually finding their plight on an unsafe dump in an uncontrolled economic environment. The result is obviously that hazardous materials could report to the groundwater with all subsequent consequences to health by for example uncontrolled burning of these goods.

Therefore, any modeling and assessment approaches should provide valuable information to the legislator to provide a fundamental basis for global environmental legislation based on achievable technology and economics incorporating the dynamics of market flexibility and consumer behavior. Mapping of materials will inform the original equipment manufacturers (OEMs) on a technology basis where materials and elements in their products are reporting to, ensuring that a solid legal and environmental basis is maintained for marketing these products. On the other hand, the consumer can be transparently informed of all benefits and risks of using the products of the OEM, as well as providing the legislator the means to monitor the (likely) movement of end-of-life products across the globe to ensure that nature and humans do not ultimately come to harm. Thus, a first-principles modeling and simulation approach ensures that the recycling loop can be mapped and subsequently 'closed' for metals and related materials in relation to design. This approach that quantifies the material and metal flows is the only manner to ensure that sustainable resource usage will be attained.

Therefore, judicious management and understanding of the plight of valuable (and also possibly toxic) minor elements is a matter of extreme importance to OEMs, legislators, recyclers, ecologists, environmentalists as well as sociologists, general population, nongovernmental organizations (NGOs), to name but a few. Therefore, for all the issues that will be discussed below to discuss the 'ecology of materials and metals', a fundamental understanding of the size and nature of resource cycles over time is needed in order to address environmental impacts and to formulate policy, design, technological and system organizational strategies for more sustainable global resource cycles.

Figure 2 shows that each carrier commodity metal is associated in nature (geology) by a unique blend of valuable minor elements (with or without own processing infrastructure) as well as harmful elements of no economic value that are lost due to unfavorable thermodynamic and other conditions within the processing chain. The carrier element can in some cases be only the secondary material being recovered since the minor elements are of much higher economic value. Therefore, affecting the production of these carrier elements could in the end adversely affect the production of the valuable minor elements. Green processing would imply minimizing the losses of elements to the green outside band of **Figure 2**.

The intricate and unique blend of elements within each ore has led to metallurgical processing being honed to

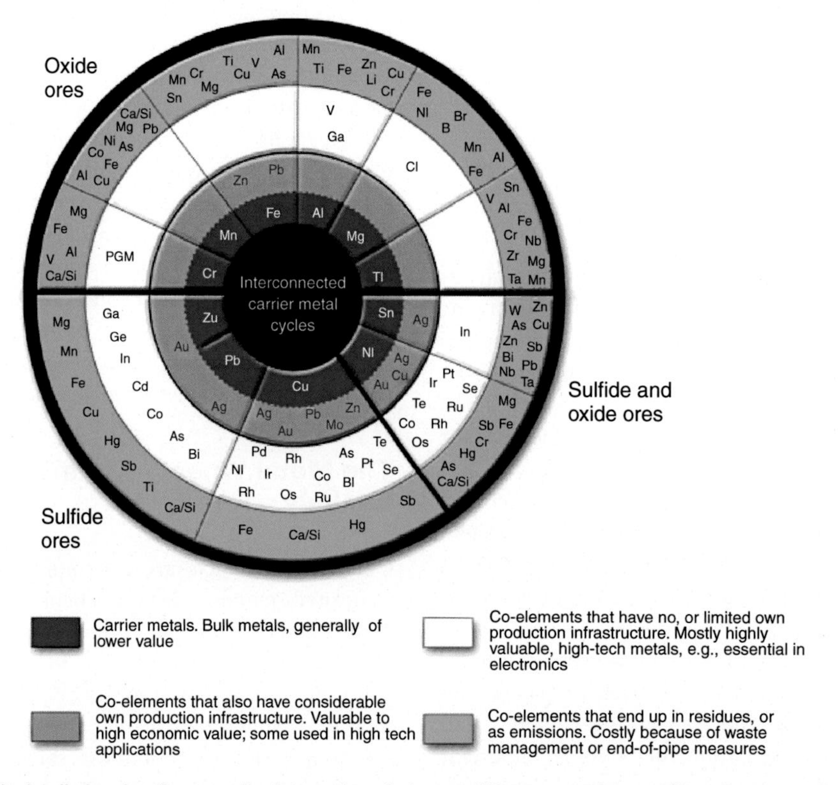

Figure 2 The 'metal wheel' showing the complex interactions between different metals as well as the economically and thermodynamically recoverability of (co-)elements. Reproduced from Reuter MA, Heiskanen K, Boin UMJ, *et al.* (2005) *The Metrics of Material and Metal Ecology*, ISBN – 13: 978-0-444-51137-9, 760pp. Amsterdam: Elsevier Science, with permission from Elsevier.

effectively recover and contain most elements economically. This complex link of materials, processing of ores, metals, and end-of-life products has led the creation of a complex web of metals and materials, in which each element has a unique position. Hence affecting the production of one element has an effect over other elements within the web.

An interesting example is indium that is used in flat TV panels. The question is how much is required, where, when, and how this changes dynamically due to the metal (especially zinc) market? For example, a country such as Japan requires around 160 t per annum but can only obtain a fraction of this via primary ore sources. Therefore, recycling has to supply the rest, but this is only achieved partially, that is, in total primary and secondary recovery only reaches around $\gtrsim$100 t.

Figure 3 shows a model that simulates this complex link between various elements and predicts their global flow over time through the complete metal and material chain,

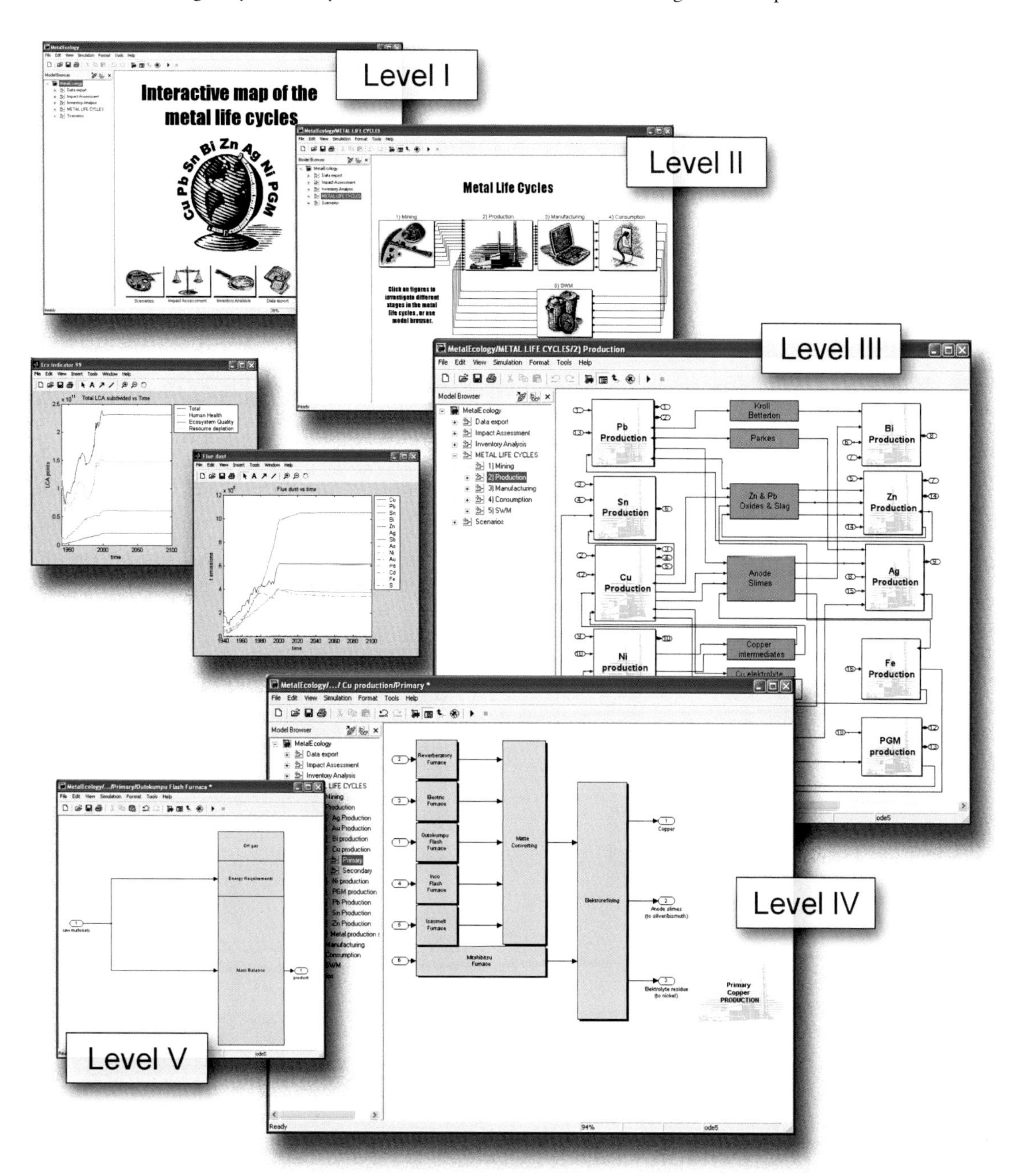

Figure 3 The various levels of the dynamic Simulink model that dynamically links the metal flow of various metals as shown in the level III slide, producing a dynamic LCA environmental score for the complete system (two small gray windows left middle). Reproduced from Reuter MA, Heiskanen K, Boin UMJ, *et al.* (2005) *The Metrics of Material and Metal Ecology*, ISBN – 13: 978-0-444-51137-9, 760pp. Amsterdam: Elsevier Science, with permission from Elsevier.

also for example the passage of the metal Indium (used in flat panel displays) through the complete material and metal system in time. Environmental indicators have been linked to the output in order to quantify the environmental performance of the complete global anthropogenic system.

Product Design and Fundamental Recycling Optimization Models

The design of a product is linked to recycling as depicted in **Figure 4**. The design does affect how materials are liberated during shredding, how efficient materials can be separated, and what the composition and quality of the recyclates will be. This determines if these recyclates can be recycled or not, therefore what the losses will be from the recycling chain. The control of the recycling chain determines what the qualities of the streams are and whether or not the recycling rate is high. This in essence determines whether materials and metals can be recycled, hence the pivot of 'industrial ecology of the materials and metals' within the car. This section will discuss the various factors influencing the 'material and metal ecology' of a consumer product, for example, the car.

Product Design, the 'Metal Wheel', and Recycling Technology

Product designers select the produced metals and materials from (primary) resources as depicted by **Figure 2** and apply them in products and applications. The product designers determine which interconnected materials are to be separated and recovered from primary ores for application in the car.

During the design of the product a range of materials/elements are once again mixed and complexly connected (gluing, welding, alloying, etc.). Modern products contain a combination of metals that are not necessarily linked in the natural resource systems as shown in **Figure 2**. As a consequence, these materials are not always compatible with the current processes in the metals production network, which was developed and optimized for the processing of primary natural resources and associated minor elements.

In general an increased complexity of recycling pyrometallurgy has arisen through the development and design of modern consumer products (such as passenger

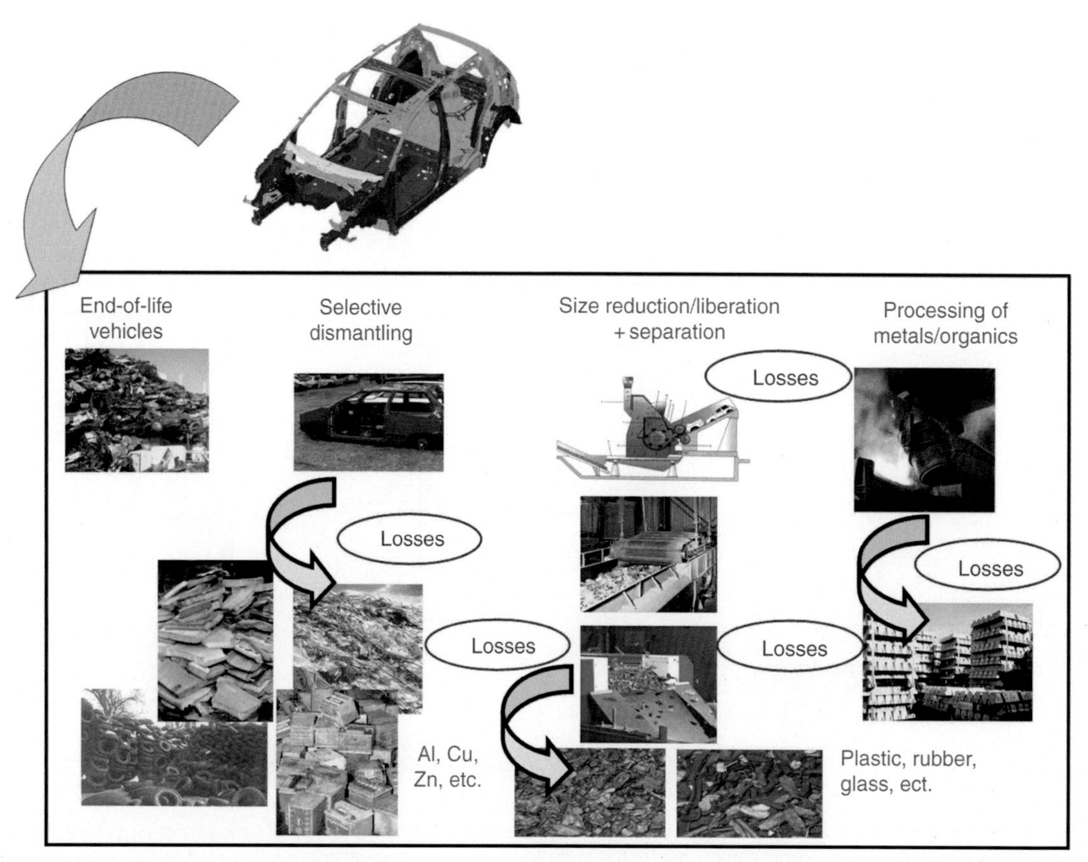

Figure 4 The link of a car design to the recycling chain, which includes selective dismantling, size reduction, and subsequent physical separation, metallurgical, and thermal treatment.

vehicles and consumer electronics). The consequence is the formation of complex residue streams or undesired harmful emissions that cannot be handled in the current system (thus the processing and recycling of those products at their end-of-life). This can be prevented by linking product design with optimized recycling technology, therefore minimizing the loss of valuable material and preventing the decrease of both quality of recyclates and recycling rates of these products.

Figure 6 The 'mineral' aluminum in its different appearances (liberation and particle-size classes) as a high-quality liberated fraction (top left) to unliberated radiator (bottom left) and various unliberated mixed fractions that cannot necessarily be recycled directly.

Recycling Optimization Models

Commercial recycling systems never create pure material streams (see **Figure 5** for materials from end-of-life-vehicles (ELVs) recycling collected from various shredder plants), never achieves 100% material recovery (recycling) during physical separation (dictated by separation physics), neither achieves 100% material recovery (recycling) during high-temperature metal production (dictated by thermodynamics), and nor achieves 100% energy recovery (dictated by thermodynamics).

The recyclability of a product is not only determined by the intrinsic property of the different materials used, but by the quality of the recycling streams (see **Figure 5**), which is determined by the mineral classes (combination of materials due to design, shredding, and separation), particle-size distribution, and degree of liberation (multimaterial particles) (see **Figure 6**). All these affect the physical separation efficiency, metallurgical and energy recovery, which all in turn determine the quality and economic value of the recycling (intermediate) products in the recycling system, which can be applied as secondary resources.

Since the quality of recyclates and the recycling rate of a product is largely affected by the design of the product (see **Figure 7**), it is required that tools are available that link CAD software and recycling models in order to predict recyclability of the car during the design phase. In addition, this predicts and determines the social and environmental value of the materials applied in the product. Fundamental knowledge of recycling processes, such as shredding, mechanical separation processes and metallurgy, and material characteristics of recycling (intermediate) products (material type, liberation, etc.) have to be combined with that of the design of the product (material combinations and connections). In order to optimize the resource cycle and maximize the recycling rate of (future) products all the parameters determining the recycling/recovery rate for each of the materials present in the multimaterial designs and applications of the present and future have to be fully understood. This should all embrace the dynamics and statistically distributed nature of the resource cycle system.

The prediction of the recyclability and recoverability of products already in the design stage requires the exploration of the limits of recycling on a fundamental

Figure 5 Impure quality materials created during physical separation of shredded ELVs (clockwise top left: steel, wires, Mg/Al/Zn/Cu/SS, steel/Cu, Mg/Al/Zn/Cu/stainless steel (SS), and plastics). Reproduced by permission of Elsevier.

Connections types	*Before shredding*	*After shredding*
Bolting/riveting		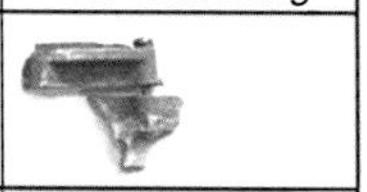
Gluing		
Insertion		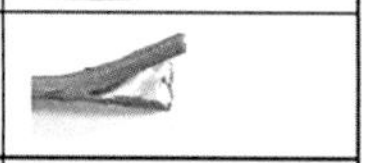
Coating/painting		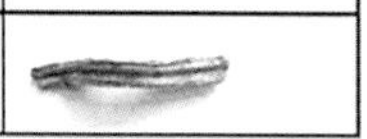

Figure 7 Possible connection types in car design with distinctive liberation behavior. From van Schaik A, Reuter MA and Richard A (2005) A comparison of the modelling and liberation in minerals processing and shredding of passenger vehicles. In: Schlesinger ME (ed.) *EDP Congress 2005*, pp. 1039–1052. TMS (The Minerals, Metals & Materials Society).

basis as has been discussed by Reuter *et al.* Recycling models have been developed by Reuter and van Schaik. These recycling models take into consideration (1) material quality (physical and chemical) and calorific values of the (intermediate) recycling streams being a function of material/mineral classes, particle-size classes, liberation classes (degree of liberation); (2) the value of intermediate streams; (3) separation physics and thermodynamics; (4) losses and emissions; (5) harmonization of plant/flowsheet architecture with changing product design; and (6) distributed and dynamic properties of present and future product designs (see **Figure 8** for a detailed flowsheet of the recycling optimization model).

Figure 9 shows how after shredding, shredded particles have different degrees of liberation, therefore creating streams of different composition (quality and economic value). This is partially caused by imperfect separation in the physical separation stage of these particles and also by the design choices as shown in **Figure 7**, affecting particle composition after shredding. The quality of recyclate streams ultimately determines in which processing steps depicted in the detail flow sheet in **Figure 8** these materials can be processed and hence how much material of sufficient quality and economic value can be recycled.

Figure 10 depicts how product design selects materials from the primary metal and material cycles and combines them into a complex multimaterial design, in which the various materials (metals and nonmetals) are complexly integrated. The combination and connection of the materials in the product design is linked (on the basis of the discussed recycling models) to the quality of recyclates as a function of the degree of liberation of the various particles after shredding. The colors in the 'design wheel' of **Figure 10** reflect the (in)compatibility of material combinations in the recyclates (either due to imperfect liberation or separation) based on the material combination matrix given in **Figure 11**, in which the (in)compatibility of material combinations is based on the thermodynamics and kinetics of metallurgical processing (see also **Figure 2**).

Figure 10 reflects the knowledge and modeling detail captured by the developed recycling optimization tools and provides feedback to the designer on desired and undesired material combinations in the design. The wheel acts as a preliminary design for recycling (DfR) tool, reflecting the complexity and detail of the developed recycling models to ensure a proper reflection of the reality of recycling system behavior and the quality and value of produced recyclates. The wheel enables DfR based on the limits and possibilities of recycling technology and recyclates quality as a function of design and separation efficiency. In summary therefore this wheel shows what can be achieved as a function of design in combination with 'best available technology' (BAT) and hence the limits of physics and chemistry as taking place in the technology.

Linking Design and Recycling

Figures 4, **8**, and **9**, respectively, depict a simple scheme for car recycling and a complex optimization model for recycling of end-of-life products such as a car. **Figure 12** depicts the (un)liberated particles after shredding (**Figures 6** and **9**) which determines its recyclability due its quality (and hence its economic value). These determine whether or not the material chain can be closed.

Table 1 explains why, if certain fractions are liberated, unliberated, whether they can be fully recycled or not. For example, copper connected to steel will dissolve in steel. Since copper is more noble (less reactive to oxygen) than steel it cannot readily be removed from the steel. This affects the steel quality negatively (e.g., its mechanical properties) and therefore it is given a red color in **Table 1**. Note, that this is dependent on the amount of the one material connected to the other (the concentration of the contaminant). In many cases, although red material combinations exist, shredding liberates the materials, which are subsequently separated during physical processing and hence they are recyclable. **Figure 11** and **Table 1** are only true if there are reasonable amounts of materials connected in the recyclates (exceeding the contamination limits), hence producing alloys outside their normal definitions. The type of models as discussed above can predict the recyclate quality and therefore link design to recycling and restrictions as indicated in **Table 1**.

The optimization model of which the flowsheet basis is depicted by **Figure 8** is far too complex to link to CAD directly, therefore fuzzy logic rule-based models that mirror the results of this complex model have been developed from the numerical results of the recycling models, and linked to CAD software and the design of the product (© MARAS). Not only can these fuzzy logic models be linked to CAD software, but they can also be integrated into LCA tools, in order to ensure that environmental models are provided with fundamental information on the end-of-life behavior of products which include (1) physics and thermodynamics of separation processes, (2) the quality and value of recyclates as a function of physical design choices (material combinations and connections), and (3) physical separation and metallurgical and thermal processing technology on a statistical basis.

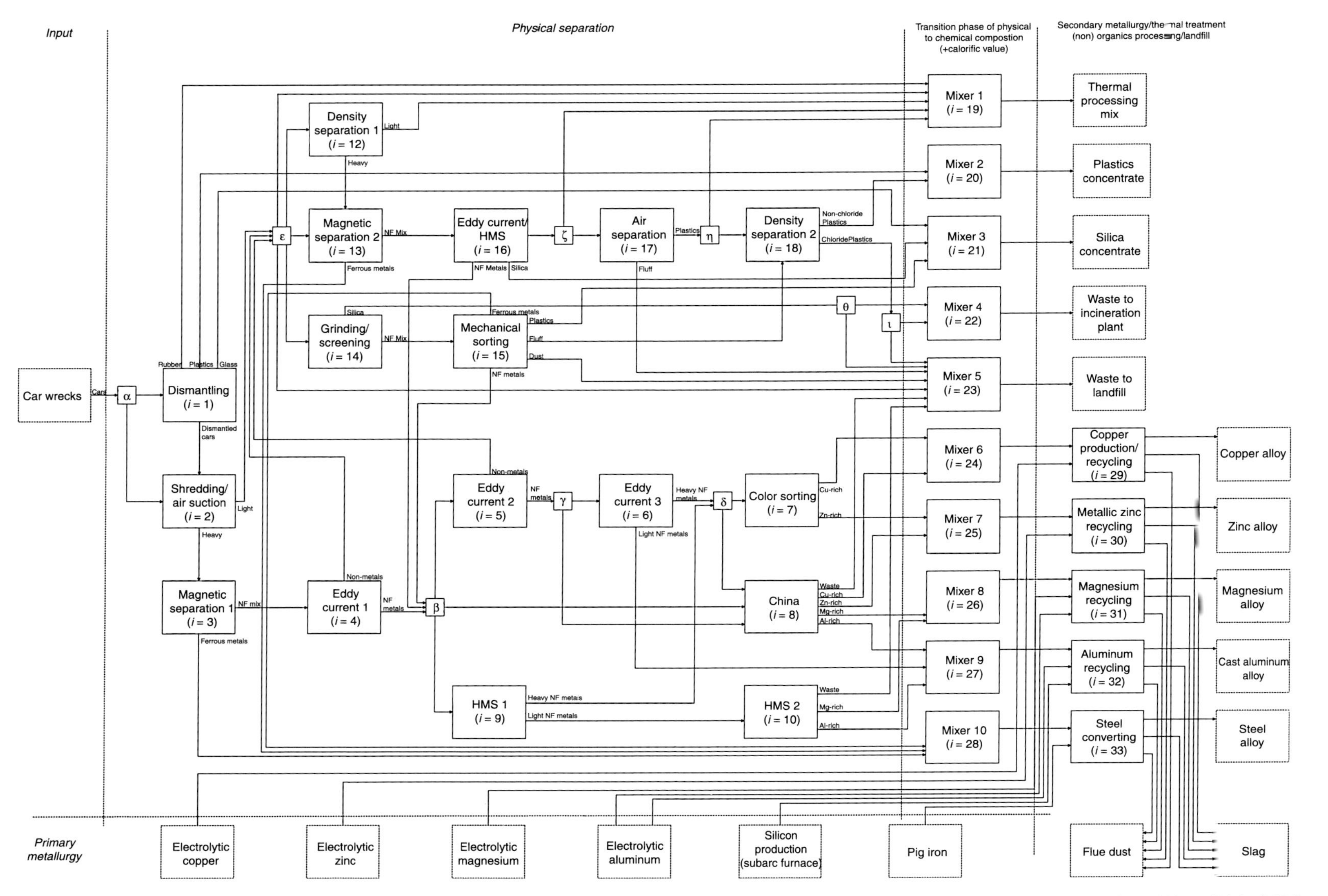

Figure 8 Flowsheet of detailed recycling system optimization model (programmed in AMPL). Reproduced from (Reuter MA, van Schaik A, Ignatenko O, and Hann GJ de (2006) Fundamental limits for the recycling of end-of-life vehicles. *Minerals Engineering* 19: 433–449, with permission from Elsevier.

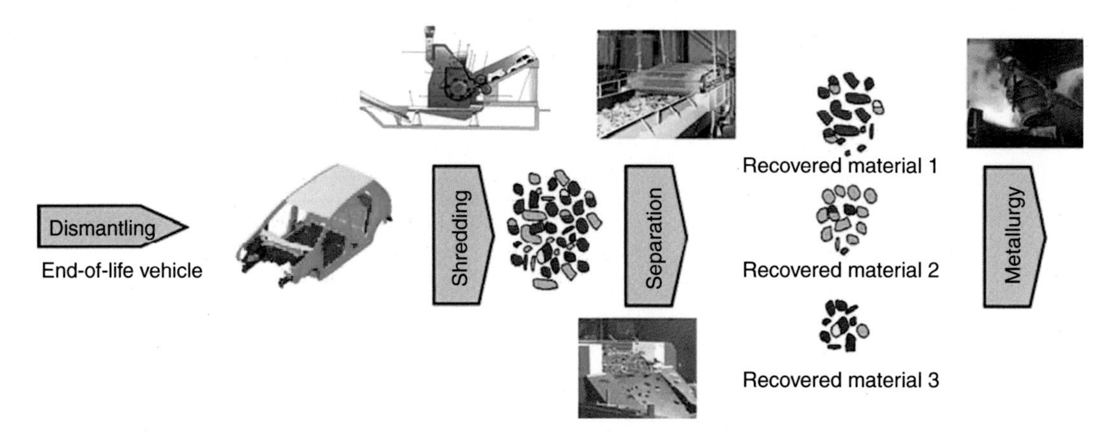

Figure 9 Shredding of a car and the creation of liberated and unliberated materials.

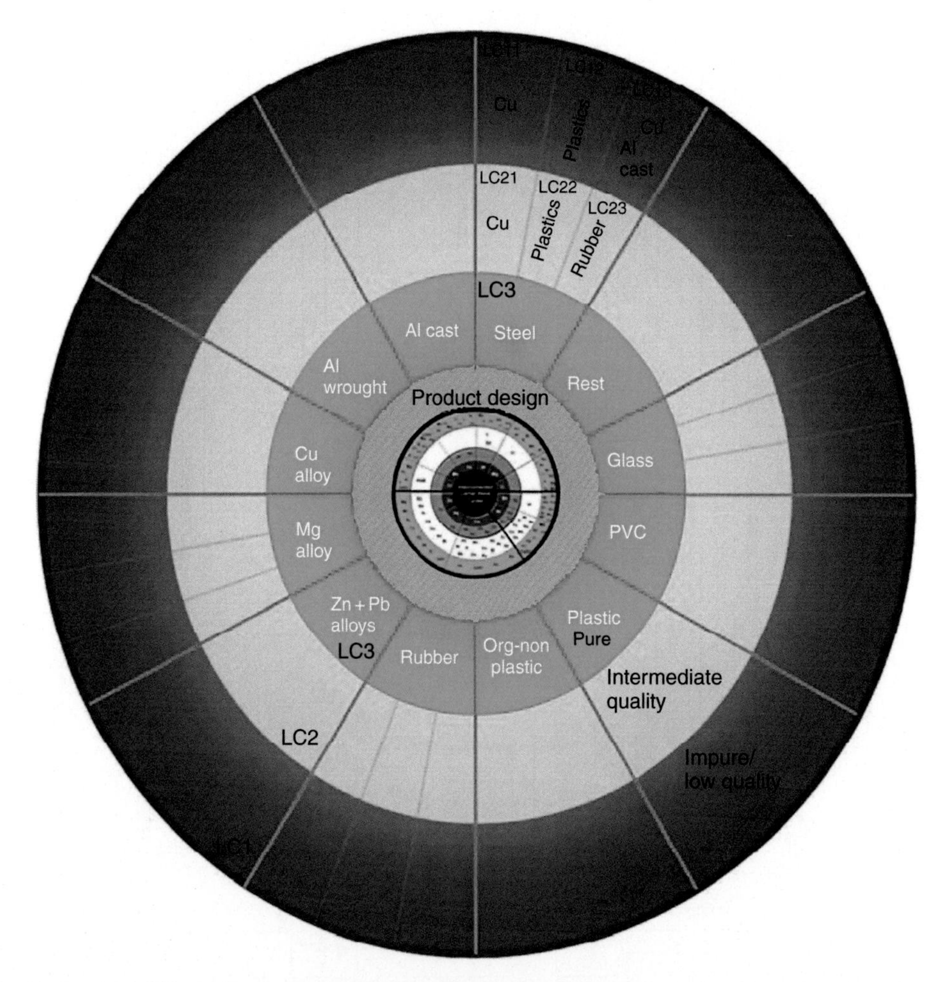

Figure 10 The 'design wheel' illustrating the underlying liberation classes that are created as a function of product design as predicted by the recycling model depicted in **Figure 8**. Reproduced from van Sehaik A and Reuter MA (2007) The use of fuzzy rule models to link product design to recycling rate calculations. *Minerals Engineering* 20: 875–890, with permission from Elsevier.

Recycling 1153 ELVs – From Theory to Practice

The developed theory as described in the previous section provides a fundamental basis for proper collection of data, supported by a good mass balance based on data reconciliation, and the corresponding statistics and how this should be performed when carrying out experiments or auditing a plant. This theory is essential to characterize and control the material and element flows in recycling

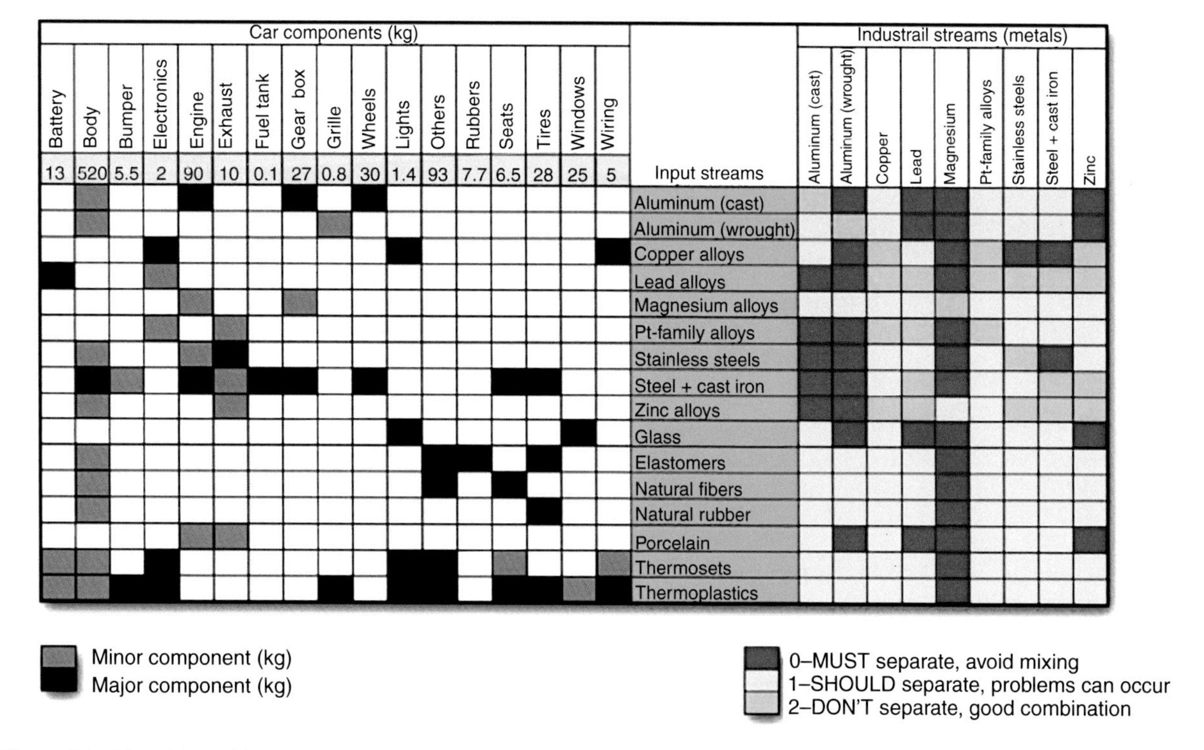

Figure 11 Material combination matrix: permitted connections and nonpermitted connections and combinations in particles after shredding and separation. Reproduced from Reuter MA, Heiskanen K, Boin UMJ, *et al.* (2005) *The Metrics of Material and Metal Ecology*, ISBN - 13: 978-0-444-51137-9, 760pp. Amsterdam: Elsevier Science, with permission from Elsevier.

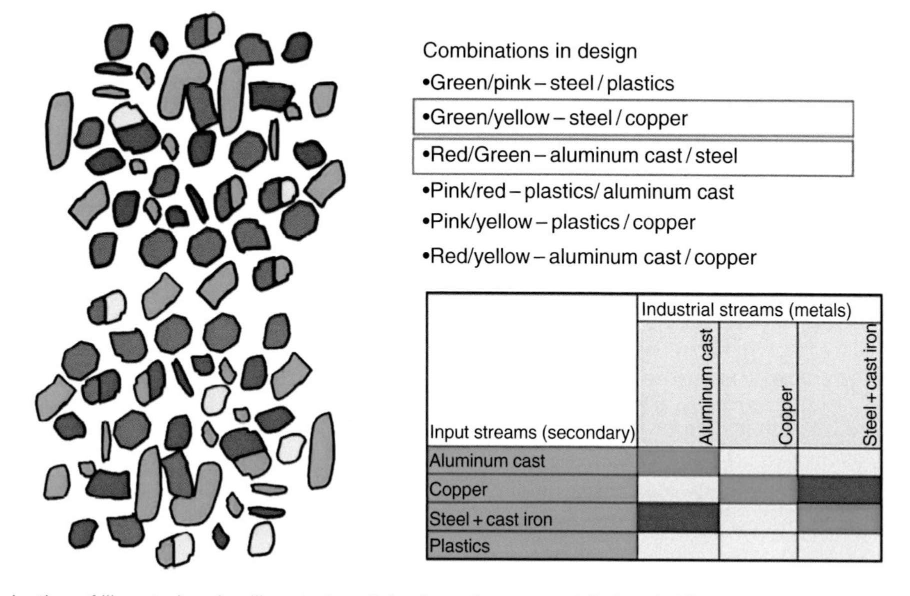

Figure 12 A selection of liberated and unliberated particles from the car model given in **Figure 4**. A section from the compatibility matrix (**Table 1** and **Figure 11** – please note that colours of particles are those of the body-in-white (BIW) in **Figures 4** and **9**).

plants and through the complete recycling system, which is extremely important for good metal/material accounting, the calculations of recycling rate on a sound statistical basis, as well as quality control of recycling streams. In fact this is the basis of any meaningful discussion on 'material and metal ecology'.

Experimental and industrial data on the composition of the car, the separation efficiency of the various processes, liberation and particle-size reduction in the shredder, the quality (or grade) of the recycling (intermediate) material streams is typical information that becomes available through a good understanding of the

Table 1 The reasons for certain materials being compatible or not (also see **Figures 11** and **12**) explained on a thermodynamic basis

Input streams (secondary) = recyclates	*Industrial streams (metals)* *Aluminum cast*	*Copper*	*Steel + cast iron*
Aluminum cast	Similar material	During copper processing Al is lost to slag	Loss of Al; Al less noble
Copper	Cu is more noble than Al cast; a certain % of Cu is allowed being one of the alloying elements for Al cast	Similar material	Cu is more noble than steel
Steel + cast iron	Steel+cast iron more noble than Al cast	Creates excessive slags, loss of steel to slag	Similar material
Plastics	Limited due to reaction of Al with C and subsequent loss of Al (Al_4C_3)	Affects energy balance of processing; fillers affect slag properties; possible dioxine creation	Affects energy balance of processing; fillers affect slag properties; possible dioxine creation

theory of recycling as discussed in the previous section. Furthermore, the collection of industrial data on recycling based on best available technology is essential to predict and calculate the recyclability of passenger vehicles, using the developed models. This is of extreme importance for a realistic definition of the type approval and end-of-life legislation of vehicles or any other consumer product. This type of data hence underpins the viability of material and metal ecology.

The theory is applied to provide a procedural basis from which the recycling rate can be calculated from an industrial experiment, in which 1153 ELVs were recycled. This experiment was executed at a large-scale industrial recycling plant and clearly illustrates how statistically sound recycling rates can and therefore should be calculated from data collected from recycling experiments based on the developed theory and classical sampling theory and statistics.

Practical Procedures for Performing Large-Scale Industrial Recycling Experiments

Mass balances of plants based on measured data mostly do not close due to inevitable weighing and sampling errors, as is also the case for the shredding and 'postshredder technology (PST)' trial as discussed here. Data reconciliation has been applied to close total and element/compound mass balances over the plant and its unit operations. A large body of data renders the mass balance more accurate and makes it possible to calculate the recovery and grade for each of the different materials over the various process steps. These data are used to calibrate the models in the optimization and dynamic models mentioned above. Classical sampling theory has been applied to calculate statistically correct sample sizes for analyses of the various material flows throughout the plant (see **Figure 4**). The mass flows and composition of the streams were measured and analyzed over all unit operations in the plant that is, on the input, intermediate, and output streams, in order to increase the amount of data available for data reconciliation, which increases the accuracy of the mass balance and its statistics.

Calculation of Recycling/Recovery Rate

Based on the mass balance and its statistics, the recycling rate of ELVs based on the discussed test could be calculated for best available technology as shown in **Figure 13**. For the first time a test was therefore concluded in which the recycling rate was calculated within a statistical framework, crucial to proving the validity of the recycling rate calculation. Ultimately the recycling rate is determined by the possibility of the market to absorb the produced output streams (either for direct application or in metallurgical or thermal processes) and is therefore determined by the quality of the recycling (intermediate) products as well as by the geographic location of the plant (due to local environmental legislation).

Statistics

Only data reported within a statistical and theoretical framework can have a legal basis and can find their way into design software for cars in order to perform 'DfR' and hence real 'material and metal ecology' on a large industrial scale. Moreover the statistics around the calculation of the recycling rate based on plant data indicates that the (calculations for the) recycling/recovery rates and requirements for type-approval of cars as imposed by legislation in Europe should also be based on a statistical basis and are meaningless if represented by a single value as is required at the moment. Any methodology to assess end-of-life systems has to take into account the statistics of design and end-of-life technology.

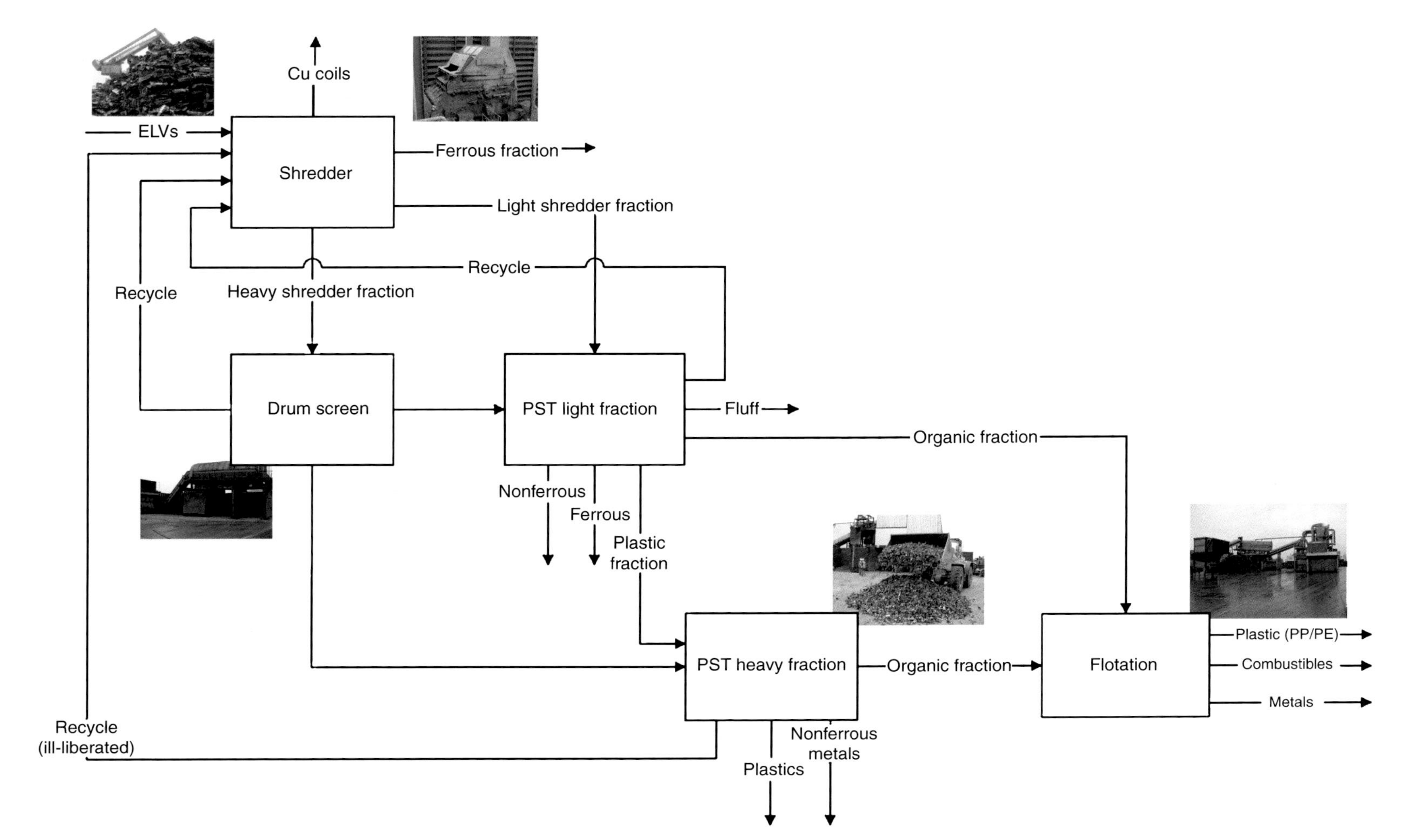

Figure 13 Simplified flowsheet of the shredding and postshredding technology (PST) plant. Reproduced from Reuter MA, Heiskanen K, Boin UMJ, *et al*. (2005) *The Metrics of Material and Metal Ecology*, ISBN – 13: 978-0-444-51137-9, 760pp. Amsterdam: Elsevier Science, with permission from Elsevier.

Material and Metal Ecology

The fundamental basis of 'material and metal ecology' is discussed in this section. The discussed aspects provide fundamentally based answers and approaches to questions of crucial importance to the industry and environmentalists such as the magnitude of the legally required recycling rate of presently designed products, emissions of processes and end-of-life products to nature, etc. This approach is being adopted by the automobile industry to ensure among others that recycling rates have the fundamental basis that challenges but also provides the legal basis for recycling legislation. It can also provide the basis for risk analyses for new car designs from a recycling point of view.

Any evaluation of the economic and/or environmental consequences and calculation of environmental scores of product and material applications can only be conducted if the interconnectivity of material/metal cycles is recognized and recycling rate calculations are based on fundamental recycling models. The model structure should be linkable to CAD product design activities, material choices, joints, etc. This is not possible with a LCA approach on its own, since it is not a simulation or predictive tool, it only represents the present and does not give technological advice about the future, how technology should be controlled and adapted, how the physical design of products have to be changed to ensure that economic recyclates can be created.

In summary the discussed approach provides a basis to 'material and metal ecology' in the following areas, which should ideally be integrated:

- Creation of a fundamental basis to define and realize 'material and metal ecology'. All the models are predictive as a function of physics, thermodynamics, and chemistry as well as time, hence they are dynamic.
- Development of fundamental recycling models for interconnected metals and materials applied in various products/applications, for example, the models are applied by the authors for the recycling of passenger vehicles, waste electric and electronic equipment (WEEE), as well as for other waste/material systems.
- Development of fuzzy logic rule-based models to link fundamental recycling models on a simplified basis to design tools and material choices, while still maintaining the detail knowledge (process, material, quality, etc.) as captured by the complex models, hence providing simple risk models for designers.
- Fundamental recycling models (calibrated with data based on a statistical basis) can be linked to environmental LCA tools/software in order to provide a fundamental technological and statistical basis for the calculations of recycling/recovery rates, prediction of quality of recyclates, process operation, recycling system architecture, etc.
- A first-principles and technological basis for 'DfR' guidelines is provided which include (1) the influence of material combinations and connections, (2) liberation, (3) particle size, and (4) physical/chemical/thermodynamical process efficiency on the quality of recyclates and the maximum achievable recycling rates.
- Support eco-design by providing fundamental knowledge on recycling systems and DfR.

Figure 14 summarizes the application of material and metal ecology, in which fundamental models link material choices in product design to recycling and interconnected material cycles. This provides fundamental knowledge and data for environmental models (LCA, MFA, etc.), therefore linking the various disciplines related to the ecological value of materials in our society. It is still required to link these material and metal ecology models to the environment so that the effect of a product design on the environment can be directly determined. This would then constitute the final objective of true material and metal ecology in the present industrialized society. **Figure 3** depicts a dynamic model that provides a basis for mapping dynamically the flow of elements, while the model depicted by **Figure 8** depicts a recycling system model. The true innovation of this work is depicted by **Figure 14** shows how **Figure 2** (metal wheel) is linked to

Figure 14 The interlinked disciplines – applying fundamental models to link design to recycling and interconnected metal cycles providing fundamental knowledge and data to environmental models (LCA, MFA, etc.).

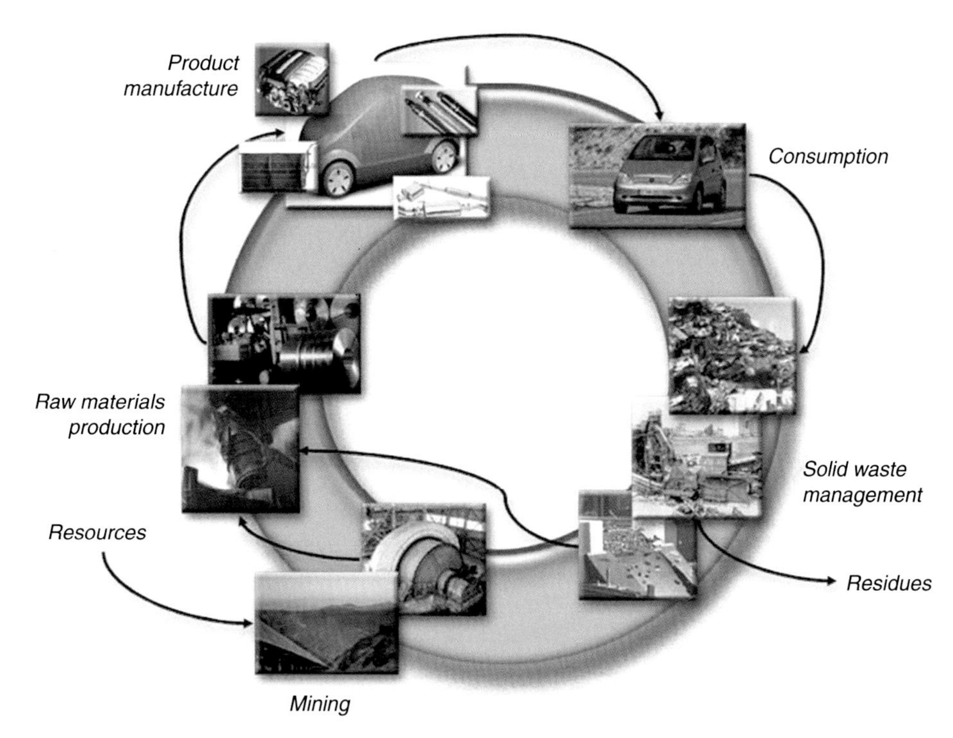

Figure 15 The anthropogenic material and metal cycle. Reproduced from Reuter MA, Heiskanen K, Boin UMJ, *et al.* (2005) *The Metrics of Material and Metal Ecology*, ISBN – 13: 978-0-444-51137-9, 760pp. Amsterdam: Elsevier Science, with permission from Elsevier.

product design (design wheel). This innovation is a key issue in controlling the anthropogenic material and metal cycle (resource cycle), ensuring that the positive interaction with the nature cycle as shown in **Figure 1** is optimized.

In order to realize a 'sustainable' material and metal ecology, the system depicted by **Figure 15** should be in balance with the material and metal cycles in nature. In summary, visualizing and describing mathematically the link between our industrialized societies with nature, as discussed above, is key to providing a measure for 'sustainability' in our present consumer society. This measure provides a tool to shape a more harmonious future!

Further Reading

Pitard FF (1993) *Pierre Gy's Sampling Theory and Sampling Practice* 2nd edn., 488pp. Boca Raton, FL: CRC Press.

Reuter MA, Heiskanen K, Boin UMJ, *et al.* (2005) *The Metrics of Material and Metal Ecology*, ISBN – 13: 978-0-444-51137-9, 760pp. Amsterdam: Elsevier Science.

Reuter MA, van Schaik A, Ignatenko O, and Hann GJ de (2006) Fundamental limits for the recycling of end-of-life vehicles. *Minerals Engineering* 19: 433–449.

van Schaik A and Reuter MA (2007) The use of fuzzy rule models to link product design to recycling rate calculations. *Minerals Engineering* 20: 875–890.

van Schaik A and Reuter MA (2004) The time-varying factors influencing the recycling rate of products. *Resources, Conservation and Recycling* 40(4): pp. 301–328.

van Schaik A and Reuter MA (2004) The effect of design on recycling rates for cars – Theory and practice. *Proceedings of the International Automobile Recycling Congress,* Geneva, Switzerland, March 10–13, 2004, 21pp.

van Schaik A, Reuter MA, and Heiskanen K (2004) The influence of particle size reduction and liberation on the recycling rate of end-of-life vehicles. *Minerals Engineering* 17(2): 331–347.

van Schaik A, Reuter MA, and Richard A (2005) A comparison of the modelling and liberation in minerals processing and shredding of passenger vehicles. In: Schlesinger ME (ed.) *EDP Congress 2005*, pp. 1039–1052. Warrendale: TMS (The Minerals, Metals & Materials Society).

Verhoef EV, Reuter MA, and Dijkema GPJ (2004) Process knowledge, system dynamics and metal ecology. *Journal of Industrial Ecology* 8(1-2): 23–43.

Ververka V and Madron F (1997) *Material and Energy Balancing in the Process Industries*. Amsterdam, The Netherlands: Elsevier.

Methane in the Atmosphere

S A Pegov, Russian Academy of Sciences, Moscow, Russia

Increase in Methane Concentration

Measured increases of methane concentration in the Earth's atmosphere began approximately 18 000 years ago, and for the period of 10 000–12 000 years ago its concentration rose from 350 up to 700 ppbv. Coming nearer our time, 200–2000 years ago, methane in the atmosphere changed from 700 to 800 ppbv. One hundred years ago, methane concentration was already at 900 ppbv. During industrialization, methane emissions and atmospheric concentrations increased quickly up to 1500 ppbv. For example, measurements for 1978–88 showed that near-ground concentration increased from 1520 to 1690 ppbv. This growth rate slowed in the late 1980s and practically stopped in 1991–92, believed to be a result of the repair of Soviet gas pipelines. Most recently, methane concentration is rising again. Taking into account significant fluctuations of methane concentration in the atmosphere measured in different places of the globe and at various times, the average rate of its modern growth was 17 ± 1 ppbv yr^{-1}, which corresponds to an increase of its contents in the atmosphere of 48 ± 3 million tons yr^{-1}. For example, over the last 40 years, methane concentration has increased approximately by 30%, and it is now between 1720 and 1770 ppbv. The current total methane in the atmosphere equals 5000 million t and could double in the next 50 years at the current rate of increase.

Methane Sources

There are three main categories of methane sources to the atmosphere: (1) biogenic methane of natural and technogenic origins; (2) abiogenic methane from lithosphere sources; and (3) anthropogenic methane from industry and agriculture.

Estimations of the total methane flux vary widely from 500 million tons yr^{-1} to not less than 3000 million tons yr^{-1} in others. The latter are based on observable annual increases of its contents, known results of measurements of distribution of its concentration at different altitudes, and conditions of balance between destruction of methane in photochemical processes and its upward transport from the Earth's surface.

Such large discrepancies can be explained by the fact that the earlier studies were limited to abiogenic methane flow from lithospheric sources, both diffuse (spreading zones, COX hydrothermal systems, etc.) and local (deposits of gas hydrates, mud volcanoes, the areas of gas, oil, and coal deposits). Latter studies included biogenic sources such as natural ecosystems – wetlands (105–300 million tons yr^{-1}), soils (10–80 million tons yr^{-1}), tundra (1.3–13 million tons yr^{-1}), termites (37–88 million tons yr^{-1}), also for agricultural activity, such as rice fields (100–350 million tons yr^{-1}), ruminants and manure (70–200 million tons yr^{-1}), as well as for a number of technogenic processes, such as biomass combustion (50–160 million tons yr^{-1}) and landfills (30–70 million tons yr^{-1}). Variations among methane emissions from biogenic sources, achieving 2–3 times, are quite satisfactory since areas of these sources are huge, and their sizes and intensities change depending on climatic conditions.

Methane leaks during extraction of fossil fuels are an important source of anthropogenic methane. More than 2000 billion m^3 of gas, over 3000 million tons of oil, and 4500 million t of coal are annually extracted in the world at present. Methane leaks in oil and gas mining occur during prospecting, extraction, transportation, storage, processing, and refilling of oil and gas products. Losses of methane in technological processes are estimated to be in a range of 30–300 million tons yr^{-1}. More precise estimation can be made through monitoring of technological processes, in particular oil and gas pipeline accidents. Estimations of methane emissions in underground and open-cast mining of coal deposits change from 35–40 million to 60–80 million tons yr^{-1}.

Special attention should be given to the lithosphere sources of abiogenic methane, which are of two main categories. The first includes deposits of fossil fuels in the Earth's crust (gas and gas condensates, oil and oil-and-gas, coal, gas hydrates). Possible pathways of natural methane into the atmosphere from fossil fuels are mainly known.

The second is due to lithospheric-scale processes (metamorphism, zones of spreading, COX systems,

Table 1 Methane content in geospheres, deposits of natural fuels, and gas hydrate deposits

Source	*The contents of methane*	
	($\times 10^6$ t)	*(m^3)*
Lithosphere	10^9	1.4×10^{18}
Hydrosphere	10^8	1.4×10^{17}
Gas fields	10^5–10^6	$(1.4–14) \times 10^{14}$
Methane in oil fields	$1–2 \times 10^4$	$(1.4–2.8) \times 10^{13}$
Methane in coal deposits	10^4–10^5	$(1.4–14) \times 10^{13}$
Sea gas hydrates	$(1.3–50) \times 10^5$	$(1.8–70) \times 10^{14}$
Gas hydrates in perma-frost	$(3–6) \times 10^4$	$(4.2–8.4) \times 10^{13}$
Atmosphere	5×10^3	7×10^{12}

Khalil MAK and Rasmussen RA (1983) Sources, sinks and seasonal cycles of atmospheric methane. *Journal of Geophysical Research* 88(C9): 5131–5144; Rasmussen RA and Khalil MAK (1991) Atmospheric methane (CH_4): Trends and seasonal cycles. *Journal of Geophysical Research* 86(C10): 9825–9832; Sokolov BA (1996) Oil-gas capacity of deposits on the Earth. *Science in Russia* (no. 6): 16–20 (in Russian).

etc.). The amount of natural methane emissions from lithosphere sources are less known due to the extensive inaccessibility, instability, and geographic dispersion necessitating territorial measurements in different regions of the world. Difficulties are also due to measurement problems, such as registration of small concentrations, passage of emissions through thick layers of water, probe poisoning, etc. At the same time, one should stress that stocks of methane in lithosphere sources and land and oceanic deposits of fossil fuels are quite large (**Table 1**).

Processes of natural methane migration and other hydrocarbons from oil and gas deposits as well as their 'catastrophic' destructions are known. Increased concentration of methane is noted in near-ground troposphere of many oilfield structures. Essential influence on methane uptake is made by seismic–tectonical processes in Earth's crust. For example, the amount of methane in near-ground troposphere exceeded its average values by several orders of magnitude during the Gasli earthquakes (1977). Various mechanisms of methane release and its huge stocks in lithosphere sources, deposits of fuels, and gas hydrate deposits are capable of providing flows of abiogenic methane over 1000 million tons yr^{-1}.

Data from literature on methane flows from the Earth's surface into the atmosphere are presented in **Table 2**, including minimal and maximal values from different publications. These estimates are coarse enough and show only orders of magnitude. We assumed that the total flow of methane into the atmosphere is not less than 3000 million tons yr^{-1}, which reflects the current growth rate of methane in the atmosphere, dynamics of its transformations in physical and chemical reactions, and distribution of its concentration on height.

Conclusions

An analysis of **Table 2** provides the following conclusions:

1. natural biogenic sources provide an average annual flow of methane about 540 million tons yr^{-1};
2. abiogenic natural sources, both lithospheric and hydrospheric, deliver about 1360 million tons yr^{-1} of methane;

Table 2 The contribution of natural and anthropogenic sources in release of methane into the atmosphere, in million tons yr^{-1}

Source	*Literary data*	*Results of generalization*
Natural		
Biogenic		
Wetlands	105–300	300
Termites	37–150	150
Soils	10–80	80
Tundra	1.3–13	13
Total	153.3–543	540 (18%)
Lithosphere and hydrosphere sources		
Oceans	3–50	50
Freshwater reservoirs	2–10	10
Volcanoes	0.2	0.2

(*Continued*)

Table 2 (Continued)

Source	*Literary data*	*Results of generalization*
Gas hydrates, metamorphism, COX, etc.	5–100	200
		500
Deposits of natural gas		500
Oil fields		10
Coal layers		100
Mud volcanoes		1–3
Total	10–160.2	1360 (45%)
Anthropogenous		
Agriculture and living activity		
Animals	70–200	200
Rice fields	100–350	350
Biomass combustion	50–160	160
Landfills	30–70	70
Total	250–780	780 (26%)
Industry and energy production		
Losses during gas and oil mining	30–90	200
Losses during coal mining	6–40	60–80
Industrial pollution and sewage waste	7.2–50	50
Motor transport	0.5–2	2
Total	43.7–182	320 (14%)
Grand total	457–1665.2	3000

3. anthropogenic sources, including methane emissions from human agricultural activity, loss of methane in organic fuel mining and its industrial emissions, provide an average annual flow of methane about 1100 million tons yr^{-1}.

Thus, the natural component of methane emissions into the atmosphere is estimated at 1900 million t yr^{-1}, which is 1.7 times more than anthropogenic emissions. Long-time regional observations and their analyses on the global scale are necessary in order to finalize quantitative estimations of methane emissions from concrete lithosphere sources.

See also: Carbon Cycle; Climate Change 1: Short-Term Dynamics; Climate Change 2: Long-Term Dynamics; Climate Change 3: History and Current State; Coevolution of the Biosphere and Climate.

Further Reading

Adushkin VV, Kudryavtsev VP, and Turuntaev SB (1998) Evaluation of the abiogenic part of the methane flow to the atmosphere. In:Global Environmental Changes, pp. 191–205. Novosibirsk: Siberian Branch of RAS NIC OIGGM (in Russian).

Blake DR and Rowland FS (1988) Continuing worldwide increase in tropospheric methane 1978 to 1987. *Science* 239: 1129–1135.

Craig H and Chou CC (1982) Methane: The record in polar ice cores. *Geophysical Research Letters* 9: 1221–1224.

Ehhalt DH, Heidt LE, Lueb RH, and Martell EA (1975) Concentration of CH_4, CH_2, H_2, H_2O, and N_2O in the upper stratosphere. *Journal of Atmospheric Sciences* 32(1): 163–169.

Etheridge DM, Pearman GL, and Fraser PJ (1992) Changes in the tropospheric methane between 1841 and 1978 from a high accumulation-rate Antarctic ice core. *Tellus* 44B: 282–294.

Fridman AI, Demidiuk LM, and Mahorin AA (1997) Emission of 'greenhouse' gases from oil – Energy plants, 90pp. M RIO RMNTK 'Nefteotdacha' (in Russian).

Khalil MAK and Rasmussen RA (1983) Sources, sinks and seasonal cycles of atmospheric methane. *Journal of Geophysical Research* 88(C9): 5131–5144.

Kriuger DV (1995) Opportunities of methane extraction and utilization from shafts methane project M, pp. 25–40 (preprint IPKON N2) (in Russian).

Nakazawa T, Machido T, Tanaka M, *et al.* (1993) Differences of the atmospheric CH_4 concentration between the Arctic and Antarctic regions in pre-industrial pre-agricultural era. *Geophysical Research Letters* 20: 973–976.

Rasmussen RA and Khalil MAK (1991) Atmospheric methane (CH_4): Trends and seasonal cycles. *Journal of Geophysical Research* 86(C10): 9825–9832.

Rinsland CP, Levin JS, and Mikes T (1985) Concentration of methane in the troposphere deduced from 1951 infrared solar spectre. *Nature* 318(6217): 245–249.

Smith AT (1995) Environmental factors affecting global methane concentrations. *Progress in Physical Geography* 20: 3–10.

Sokolov BA (1996) Oil-gas capacity of deposits on the Earth. *Science in Russia* (no. 6): 16–20 (in Russian).

Voitov GI, Starobinets IS, and Usmanov RI (1990) About density of CH_4 flow to the troposphere in oil–gas fields regions. *Doklady of Russian Academy of Sciences* 313(6): 1444–1448 (in Russian).

Zimmerman PR, Greenbery JP, Wandiga SO, and Grut-zen PJ (1982) Termites: A potentially large source of atmospheric methane, carbon dioxide and molecular hydrogen. *Science* 218(4572): 563–565.

Monitoring, Observations, and Remote Sensing - Global Dimensions*

S Unninayar and L Olsen, NASA/GSFC, Greenbelt, MD, USA

Global Ecology – Unique Perspectives from Space-Based Satellite Sensors/Instruments

Satellite remote sensing instruments provide unique global observational perspectives on the state of the biospheres occupying the land surface, coastal zones, the oceans, and the snow/ice-covered mountains and polar caps. They also provide detailed global observations of both natural and anthropogenically induced changes in land surface, atmospheric and climatic drivers that often determine ecological health, sustainability, and dispersals or dislocations. Together with more specific *in situ* observations, space-based remote sensing of the global environment enables the investigation of the interplay between the different components of the Earth/climate system and the interaction between local and global processes. These data facilitate the construction of mathematical and empirical models that strive to simulate and predict Earth system processes, especially those that can have impacts on global and regional environments and ecosystems. Time series of observations and advanced visualization tools enhance our understanding of global ecological processes, several examples of which are presented here (see **Figure 1**).

Satellite remote sensing of the Earth system may be viewed in a broad sense as 'eyes' in the sky looking down with a unique perspective defined by specific bands or channels in the electromagnetic spectrum. Depending on the particular instrument or sensor, they 'see' the Earth in the ultraviolet (UV), visible (VIS), infrared (IR), near-infrared (N-IR), and microwave (MW) wavelengths. These wavelengths include both the short-wave solar radiation bands and the long-wave bands in which the Earth emits radiation to space. The ability of a satellite with 'passive' instruments to detect an Earth surface feature depends on the spectral characteristics of the feature or object in question. Thus, different frequencies or wavelengths will capture different aspects of the atmosphere or the land surface or vegetation or coastal, marine, and ocean ecology.

Most satellites carry 'passive' sensors that collect the reflected, refracted, or emitted radiation from the Earth's atmosphere or surface. The types of Earth features that are captured by these different spectral bands are summarized in **Table 1** (see **Box 1**). Some satellites carry 'active' instruments such as radar and lidar, which generate and transmit electromagnetic signals toward the Earth. The reflected return signal or return echo carries information on the structure and composition of the atmosphere and the underlying land surface/vegetation, inland water bodies, and the vast expanses of the oceans. Currently, a number of natural hazards are routinely monitored from space. Examples include crops and droughts, dust, smoke and pollution, forest fires, floods, severe storms and hurricanes, and volcanoes. As a novel technology, a new generation of satellites also monitors the Earth's surface and subsurface geological and hydrological environments via nonphotonic measurements of the gravity anomaly field. The surface and subsurface hydrological environments can thus be monitored to provide information on the habitats that permit or limit ecosystem.

Accurate surface positioning information is now routinely obtained from the Global Positioning System (GPS) satellite constellation. Together with geographical information systems (GIS), sophisticated data-processing algorithms, and complex mathematical and empirical models, the global ecological environment and ecosystems are currently observed and analyzed to a historically unprecedented degree. A time history of remote sensing information provides crucial data on the dynamics of change in ecosystems as they respond, adapt to, and interact with the other components of the Sun–Earth system.

The space applications program of the United Nations Committee on the Peaceful Uses of Outer Space

* The views expressed herein are those of the authors and do not reflect that of any agency or program.

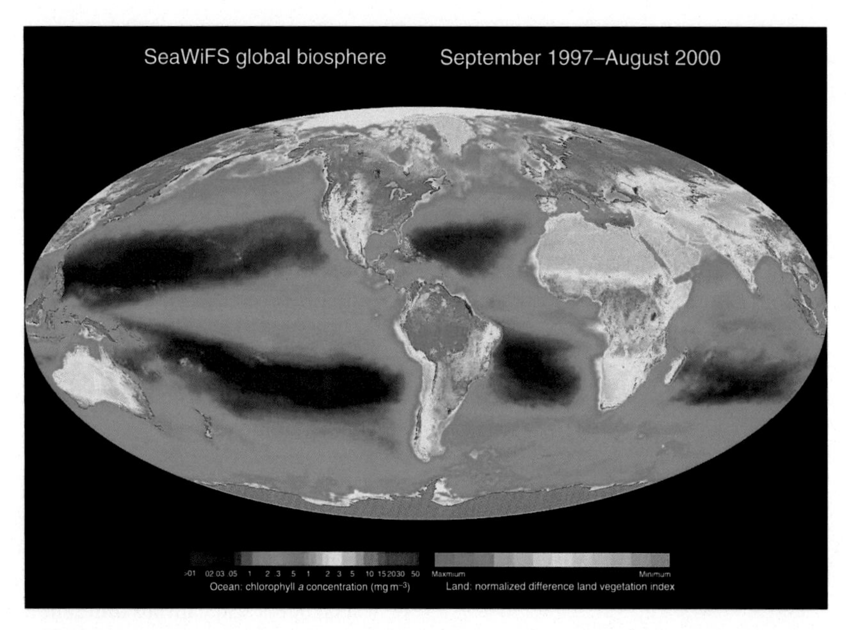

Figure 1 Plants on land and in the ocean (phytoplankton) contain chlorophyll, a green pigment used during photosynthesis. Using satellite sensors, we can measure chlorophyll concentrations on land as well as in oceans, lakes, and seas to indicate the distribution and abundance of vegetation. Since most animal life relies on vegetation for nutrition, directly or indirectly, scientists refer to these images as snapshots of Earth's biosphere. Source: David Herring, NASA. Data from SeaWiFS Project.

Table 1 EOS/TERRA-MODIS spectral bands and their key uses

Band #	*Range reflected (nm)*	*Range emitted (μm)*	*Key uses*
1	620–670		Absolute land cover transformation, vegetation chlorophyll
2	841–876		Cloud amount, vegetation land cover transformation
3	459–479		Soil/vegetation differences
4	545–565		Green vegetation
5	1230–1250		Leaf/canopy differences
6	1628–1652		Snow/cloud differences
7	2105–2155		Cloud properties, land properties
8	405–420		Chlorophyll
9	438–448		Chlorophyll
10	483–493		Chlorophyll
11	526–536		Chlorophyll
12	546–556		Sediments
13 h	662–672		Atmosphere, sediments
13 l	662–672		Atmosphere, sediments
14 h	673–683		Chlorophyll fluorescence
14 l	673–683		Chlorophyll fluorescence
15	743–753		Aerosol properties
16	862–877		Aerosol properties, atmospheric properties
17	890–920		Atmospheric properties, cloud properties
18	931–941		Atmospheric properties, cloud properties
19	915–965		Atmospheric properties, cloud properties
20		3.660–3.840	Sea surface temperature
21		3.929–3.989	Forest fires and volcanoes
22		3.929–3.989	Cloud temperature, surface temperature
23		4.020–4.080	Cloud temperature, surface temperature
24		4.433–4.498	Cloud fraction, troposphere temperature
25		4.482–4.549	Cloud fraction, troposphere temperature

(*Continued*)

Table 1 (Continued)

Band #	*Range reflected (nm)*	*Range emitted (μm)*	*Key uses*
26	1360–1390		Cloud fraction (thin cirrus), troposphere temperature
27		6.535–6.895	Mid-troposphere humidity
28		7.175–7.475	Upper troposphere humidity
29		8.400–8.700	Surface temperature
30		9.580–9.880	Total ozone
31		10.780–11.280	Cloud temperature, forest fires and volcanoes, surface temp.
32		11.770–12.270	Cloud height, forest fires and volcanoes, surface temperature
33		13.185–13.485	Cloud fraction, cloud height
34		13.485–13.785	Cloud fraction, cloud height
35		13.785–14.085	Cloud fraction, cloud height
36		14.085–14.385	Cloud fraction, cloud height

Box 1 Satellite technical details (selected examples)

MODIS (Terra and Aqua) description. Thirty-six band spectroradiometer measuring VIS and IR radiation (0.4 and 14.5 μm with a spatial resolution of 250 m, 500 m, and 1 km at nadir) for deriving products ranging from land vegetation and ocean chlorophyll fluorescence to cloud and aerosol properties, fire occurrences, snow cover on land, and sea ice in the oceans. The Terra orbit follows the Worldwide Reference System, as do the orbits of Landsat-7 (United States Geological Survey, USGS; Landsat = Land Remote Sensing Satellite), Earth Observing-1 (EO-1, NASA), and Satellite de Aplicaciones Cientificas-c (SAC-C, Argentina Comision Nacional para el Ahorro de Energis (CONAE)), all crossing the equator within 30 min of each other. These four spacecraft form the 'Morning Constellation', thus facilitating joint use of Terra data and the data from its companion missions. Terra, launched in December 1999, flies in a near-polar, Sun-synchronous orbit that descends across the equator in the morning at around 10.30 a.m. ± 5 min. The Aqua spacecraft, launched in May 2002, flies in ascending orbit with a 1.30 p.m. equatorial crossing time which enables the study of diurnal variability with the MODIS (moderate-resolution imaging spectroradiometer) and CERES (clouds and the Earth's radiant energy system) instruments onboard both Terra and Aqua. The MODIS and CERES measurements extend the measurements of their heritage sensors – the advanced very high resolution radiometer (AVHRR), the coastal zone color scanner (CZCS), and the Earth radiation budget experiment (ERBE) – but with a higher quality of calibration and characterization. Other benefits include having more spectral bands, and other collocated instruments on board, which can be used to improve the satellite retrieval algorithms and improve the atmospheric correction that often needs to be applied to obtain accurate surface features.

Data continuity from the MODIS instruments on Terra and Aqua are expected from the VIIRS (visible and infrared imaging radiometer suite) instrument on the operational National Oceanic and Atmospheric Administration (NOAA) NPOESS (National Polar Orbiting Environmental Satellite System) series to be launched beginning in the early part of the next decade. Retrospective data continuity is via the Landsat series of satellites.

SeaWiFS/Orbview. Sea-viewing wide field-of-view sensor (SeaWiFS) launched on 1 August 1997, and began taking measurements of the World Ocean in September 1997. Abbreviated SeaWiFS/Orbview, technical details: Sun-synchronous, altitude: 705 km; equatorial crossing: noon ± 20 min; inclination: 98.2°; period: 99 min; design life: 5 years; launch: 1 Aug 1997; status: operational. The eight SeaWiFS spectral bands are: Band-1: 402–422 nm; Band-2: 433–453 nm; Band-3: 480–500 nm; Band-4: 500–520 nm; Band-5: 545–565 nm; Band-6: 660–680 nm; Band-7: 745–785 nm; Band-8: 845–885 nm. Science focus areas include: carbon cycle, ecosystems, and biogeochemistry; climate variability and change; water and energy cycles. (Note the overlap between SeaWiFS and MODIS spectral bands.)

Landsat. High spatial resolution visible and infrared radiance/reflectance from terrestrial surfaces. Type: circular Sun-synchronous orbit at an altitude of 705 km with inclination 98.2°, period 98.9 min, and repeat cycle 16 days/233 orbits. The main instrument cluster is the enhanced thermatic mapper plus (ETM+) on Landsat-7 and legacy instruments (the thematic mapper, TM) on previous satellites of the Landsat. Next-generation instruments are to be found on the experimental satellite EO-1 which carries an advanced land imager (ALI), a hyperspectral instrument (hyperion), and a linear etalon imaging spectral array (LAC). ETM+ includes eight reflective spectral bands and spatial resolutions: three 30 m VIS bands; one 30 m N-IR band; two 30 m shortwave infrared (SEIR) bands; one 15 m panchromatic band; and one emissive 60 m thermal infrared (TIR) band. EO-1, launched in November 2000, is currently functional. The ALI instrument on EO-1 provides data continuity with ETM+.

Other instruments and other satellite systems.

Note 1. EOS-Terra and EOS-Aqua also carry several other instruments. They are not detailed in this article because their primary mission objectives pertain to observations of other aspects of the earth/climate system than ecology or ecosystems.

Note 2. Instruments on board Europe's Environmental Satellite, European Space Agency (ENVISAT, ESA), Japan's Advanced Earth Observation Satellite (ADEOS), and the satellites of other countries also measure various aspects of the Earth system, including ecosystems. For brevity, they are not detailed in this article either.

(UNCOPUOUS) identifies the following areas of applications that particularly benefit from remote sensing data products and monitoring (http://www.uncosa. unvienna.org/pdf/reports/IAM2006E.pdf):

- land cover and land use;
- remote and difficult-to-access areas like dense forests, glaciated areas, deserts, and swamps;
- areas undergoing rapid environmental change, including loss or fragmentation of ecosystems and related loss of biodiversity;
- wide-ranging impacts of pollution, from depletion of the ozone layer to tracing oil spills, photochemical smog, and other environmental impacts;
- identification, monitoring, and preparation of measures to cope with natural threats, such as storms, floods, droughts, forest fires, volcanic eruptions, geological faults, and mass movement;
- identification and analysis of social and physical vulnerabilities;
- disaster management; and
- areas affected by complex emergencies, such as armed conflicts.

Programs and projects of the UN system of agencies covering water management, coastal area management, disaster management, climate change, agriculture, desertification, mountain ecosystems, biodiversity, forest management, and mining are described in http://www.uncosa.unvienna.org/uncosa/en/wssd/index.html. The constellation of operational and research satellites that monitor the Earth system, coordinated by the World Meteorological Organization (WMO) with the various space agencies of nations worldwide, is summarized in http://www.wmo.int/pages/prog/sat/GOSresearch.html; also see http://www.wmo.int/pages/prog/sat/Satellites.html. In recent years, following several ministerial summits on Earth observations, an international group has been formed to coordinate global observations called the Global Observing System of Systems (GEOSS) with secretariat hosted by the WMO in Geneva, Switzerland.

This article is not intended to be a treatise on ecology or remote sensing technology. Rather, we provide a few selected examples of how remote sensing technology is used to observe and monitor global ecosystems. Substantial use is made of 'public domain' material readily available via the Internet. While extensive use is made of data from National Aeronautics and Space Administration (NASA) Earth Observing System (EOS) series of satellites and predecessor space-based platforms, it is underscored that many other space agencies of other countries and regions increasingly have advanced capabilities to monitor the global biosphere from space. Broad coverage of international satellite programs and applications are detailed on the web site of the UN Office for Outer Space Affairs which coordinates the work of the UN Committee on the Peaceful Uses of Outer Space (UNCOPUOS) and its Space Applications program (http://www.uncosa.unvienna.org/pdf/reports/IAM2006E.pdf).

Global Monitoring of Land Ecosystems

Global Vegetation

The latest quasi-operational observations of the Earth's vegetation are obtained from the moderate resolution imaging spectroradiometer (MODIS) on board NASA's EOS-Terra (launch: December 1999) and EOS-Aqua (launch: May 2002) satellites (http://eospso.gsfc.nasa.gov/eos_homepage/mission_profiles/index.php). MODIS derives from the following legacy instruments: advanced very high resolution radiometer (AVHRR), high resolution infrared radiation sounder (HIRS), Land Remote Sensing Satellite (Landsat) thematic mapper (TM), and Nimbus-7 coastal zone color scanner (CZCS). MODIS' 36-band spectroradiometer measures VIS and IR radiation with 21 spectral bands within 0.4–3.0 μm and 15 bands within 3–14.5 μm. The instrument's instantaneous field of view (FOV) at nadir is 250 m (two bands), 500 m (five bands), and 1000 m (29 bands). Derived products range from land vegetation and ocean chlorophyll fluorescence to cloud and aerosol properties, fire occurrences, surface temperatures, snow cover on land, and sea ice in the oceans. **Table 1** details the spectral bands of MODIS and their key uses. A subset of the spectral bands of MODIS is to be found on the sea-viewing wide field-of-view sensor (SeaWiFS) satellite with eight bands within 0.4–0.8 μm with a spatial resolution of about 1.13 km at nadir. We focus here on satellite platforms for which data time series are available. The best spatial resolution currently available is from the IKONOS satellite (~0.3 m).

Traditionally, for the past 25 or more years, a commonly used measure of global vegetation density or vegetation vigor has been the 'vegetation index' derived from AVHRR. Ratio transforms from visible red (VIS or R) and N-IR bands from remote sensing are widely used for studying different vegetation types and land use. The first channel from the National Oceanic and Atmospheric Administration (NOAA) AVHRR is in the VIS (red) part of the spectrum where chlorophyll absorbs most of the incoming radiation, while the second N-IR channel is in a spectrum region where spongy mesophyll leaf structure reflects most of the light. This contrast between responses of the two bands is represented by the normalized difference vegetation index (NDVI: (CH-2(NIR) − CH-1(VIS))/(CH-2 + CH-1); CH stands for channel), which is correlated with global vegetation parameters such as the fraction of absorbed photosynthetically active radiation (FPAR or fPAR), chlorophyll density, green-leaf area, and transpiration rates. The VIS (red) and N-IR detectors on the AVHRR sensors record radiance in the 0.58–0.68 and 0.725–1.1 μm wavelength regions, respectively. NDVI varies theoretically between −1.0 and +1.0, and increases from about 0.1 to 0.75 for progressively increasing amounts of vegetation and is most directly related to the fPAR absorbed by vegetation canopies, and hence to photosynthetic activity of terrestrial vegetation.

NDVI has been widely used to discriminate between vegetation types and characterize seasonal phenology. Global ecosystem models have used the AVHRR NDVI as the basis to estimate net primary production (NPP) and net ecosystem carbon flux. In most cases, the assumption is made that NDVI can be used as an accurate predictor for fPAR and therefore potential NPP, for many ecosystem types.

With the advent of Terra-MODIS, an enhanced version of the vegetation index, called EVI, has been developed, which uses the additional information obtained from MODIS' expanded range of spectral channels. This additional information enables a better characterization of vegetation in both heavily forested regions such as the Amazon, as well as in semiarid regions such as the Sudano-Sahel (**Figure 2**).

As the seasons change, the mirror effect of seasonality is seen, with vegetation alternatively blooming and fading, and one hemisphere's vegetation is high while the other is low. A 'global animation' of the seasonal change in vegetation is shown in http://earthobservatory.nasa.

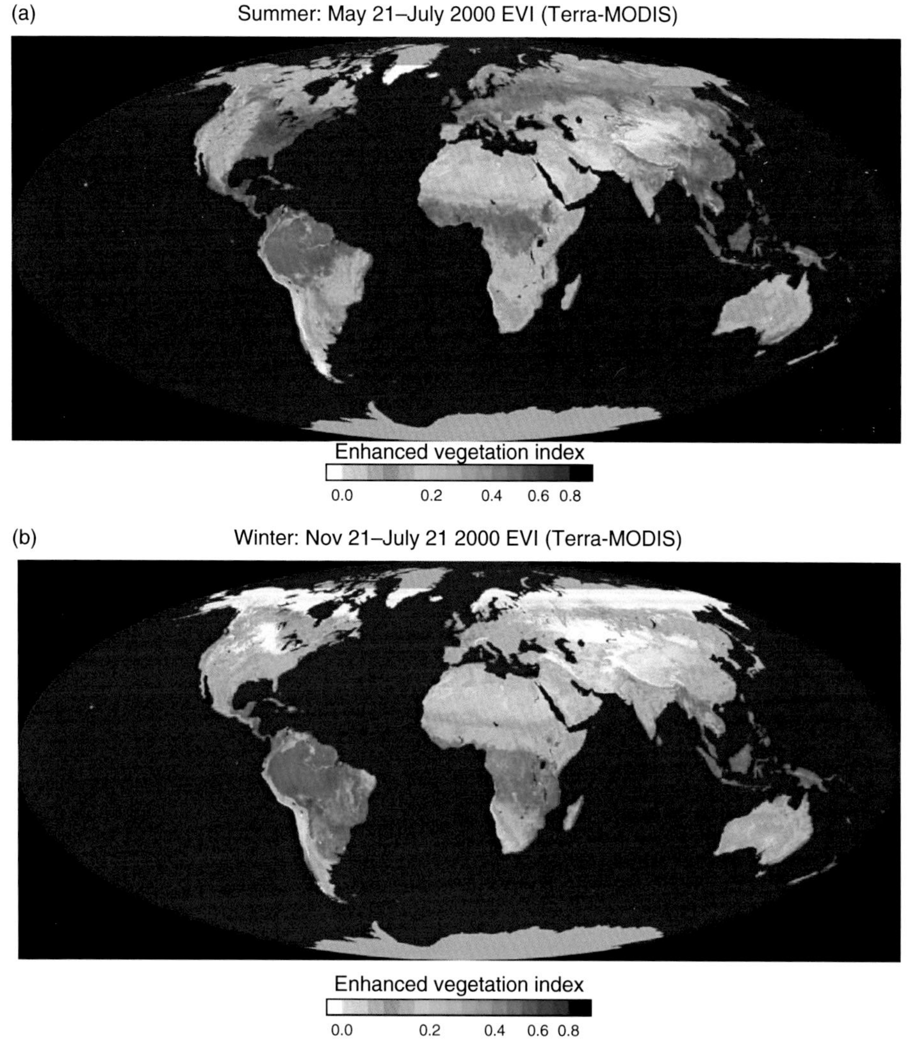

Figure 2 The images show EVI during two different seasons. Vegetation ranges from 0, indicating no vegetation, to nearly 1, indicating densest vegetation. Gray areas indicate places where observations were not collected. The EVI has increased sensitivity within very dense vegetation, and it has built-in corrections for several factors that can interfere with the satellite-based vegetation mapping, like smoke and background noise caused by light reflecting off soil (http://earthobservatory.nasa.gov/Newsroom/EVI_LAI_FPAR/). Credit: NASA/GSFC/University of Arizona.

gov/Newsroom/EVI_LAI_FPAR/Images/global_evi.mov. The biweekly and monthly vegetation index maps have wide usability by biologists, natural resources managers, and climate modelers. Naturally occurring fluctuations in vegetation, such as seasonal changes, as well as those that result from land-use change, such as deforestation, can be tracked. The EVI can also monitor changes in vegetation resulting from climate change, such as expansion of deserts or extension of growing seasons.

MODIS' observations allow scientists to track two 'vital signs' of Earth's vegetation. At Boston University, a team of researchers used MODIS data to create global estimates of the green-leaf area of Earth's vegetation, called leaf area index (LAI) and the amount of sunlight the leaves are absorbing, fPAR (http://cybele.bu.edu/modismisr/other.html). Both pieces of information are necessary for understanding how sunlight interacts with the Earth's vegetated surfaces – from the top layer, called the canopy, through the understory vegetation, and down to the ground. **Figure 3** shows an example of the representation of the vegetation by MODIS-derived LAI and fPAR.

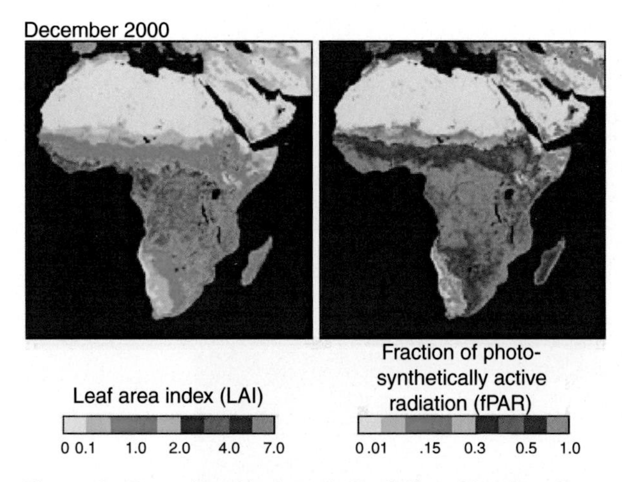

Figure 3 Examples of leaf area index (LAI) and fraction of photosynthetically active radiation (fPAR) derived from MODIS for Africa during December 2000. LAI is defined as the one-sided green-leaf area per unit ground area in broadleaf canopies, or as the projected needle-leaf area per unit ground area in needle canopies. fPAR is the fraction of photosynthetically active radiation absorbed by vegetation canopies. Color code: LAI – colors range from low LAI (0.0–0.1 is yellow) to mid-range LAI (between 2.0 and 3.0 is red) to high LAI (shades of purple); fPAR – low fPAR is in yellow (0.0–0.1, mid-range fPAR is in blues and red/brown (~0.2–0.4), and high fPAR is in shades of purple with light purple being the highest at 1.0. Images are from http://earthobservatory.nasa.gov/Newsroom/EVI_LAI_FPAR/. Original graphics credit: John Weier with design by Robert Simmon, Boston University (http://cybele.bu.edu/modismisr/other.html.) Three image sequences showing September 2000, December 2000, and April 2001 are to found in http://cybele.bu.edu/modismisr/laifpar/lai.afr.jpg and http://cybele.bu.edu/modismisr/laifpar/fpar.afr.jpg.

In Africa, rainfall is the most important factor that affects where people and animals live, and influences patterns of plant growth and ecosystem health. Animations of LAI and FPAR images can be viewed at http://earthobservatory.nasa.gov/Newsroom/EVI_LAI_FPAR/Images/LAI_wdates.mpg (LAI) and http://earthobservatory.nasa.gov/Newsroom/EVI_LAI_FPAR/Images/FPAR_wdates.mpg (fPAR).

They show the cycle of wet and dry seasons in Africa from September 2000 through May 2001 and the corresponding variation in the green-leaf area and how much sunlight the leaves are absorbing over the course of a year. The seasons in the Southern Hemisphere stand in direct opposition to those of the Northern Hemisphere, while meteorological patterns in the Northern Hemisphere roughly mirror those in the Southern Hemisphere. For example, when summer comes in the northern part of Africa in June, the winter (dry season) takes over South Africa, drying out green leaves.

NDVI time series data sets. The AVHRR and predecessor instruments have yielded long time series of NDVI data which have been used widely in many studies worldwide. Time series of NDVI data sets span several satellites and hence are prone to noise or error if not corrected for varying solar zenith angle due to orbital drift, differences in satellite sensors on board different spacecraft, sensor degradation, atmospheric absorption, equatorial crossing time, among others. Data input for atmospheric correction include aerosol optical depth, atmospheric water vapor, and ozone and other gas absorption. The physical products that are used to obtain the NDVI synthesis are also corrected for system errors such as misregistration of the different channels and calibration of all the detectors along the line-array detectors for each spectral band. An excellent comparison of the various NDVI data sets, such as AVHRR/NDVI-PAL (Pathfinder Land Program), Global Inventory Monitoring and Modeling Study (GIMMS-NDVI, and Systeme pour l'Observation de la Terre 4 (SPOT-4) VGT-NDVI), spanning from about 1981 to the present, is found elsewhere.

As an alternative to traditional approaches using predefined classification schemes with discrete numbers of cover types to describe a geographic distribution of vegetation over the Earth's land surface, Defries *et al.* applied a linear mixture model to derive global continuous fields of percentage woody vegetation, herbaceous vegetation, and bare ground from 8 km AVHRR. Linear discriminants for input into the mixture model are derived from 30 metrics representing the annual phenological cycle, using training data derived from a global network of scenes acquired by Landsat. The results suggested that the method yields reliable products that overcome apparent problems with artifacts in the multiyear AVHRR data set due to calibration problems, aerosols and other atmospheric effects, bidirectional effects, changes in equatorial crossing time, and other factors.

Land surface and vegetation classification. For global studies, the land surface and vegetation are classified into broad categories that represent large-scale aspects that can be monitored from space as well as used in land surface models that are coupled to other models of the atmospheric general circulation and climate. A typical classification would be as described by Defries *et al.* in 2002.

There are several variations to the above. For example,

Broadleaf evergreen forest and woodland	Coniferous forest and woodland
Broadleaf deciduous forest and woodland	High-altitude deciduous forest
Mixed forest and woodland	Wooded grassland
Grassland	Shrubs and bare ground
Tundra	Bare ground
Cropland	Ice

alternative versions add water bodies, cropland, urban and built-up, and barren, to render the classification system more compatible with those of the International Geosphere–Biosphere Programme (IGBP). The software and land-cover classification system developed for the Food and Agricultural Organization (FAO) and the United Nations Environment Programme (UNEP) are well described by Di Gregorio. The above also represents the typical land surface/vegetation classification system used in global climate system models. These land surface models compute the exchange of energy, water, momentum, and carbon between the biosphere and the atmosphere. They also account for the hydraulic and thermal properties of different soil types. More complex models are used to represent subpixel distributions and species composition. Conservation strategies for managing biodiversity have traditionally assumed that species distributions change relatively slowly, unless they are directly affected by human activity. However, there is increasing recognition that such strategies must include the effects or impacts of global climate change. Satellite-derived NDVI can be most useful for the development and validation of biome models. At more regional and local scales, *in situ* data are usually needed.

Global Climate Change

Satellites represent a vital observing platform to monitor the climatic environment of global ecosystems. Importantly, they provide time series information that are essential to understand the dynamics governing changes in ecosystems due to multiple stresses imposed by human activities and natural causes. Understanding the external forces that drive changes in ecosystems also helps in the understanding of how ecosystems might change in the future, due, for example, to global climate warming as projected by the international Intergovernmental Panel on Climate Change (IPCC; see http://www.ipcc.ch). The most recent assessment (Fourth Assessment Report, often abbreviated as IPCC-AR4) on the science and impacts of global climate change has just been released. The reader is referred to the IPCC web site maintained by the UK Hadley Centre for various summaries and the status of the release of the findings of the Working Groups of the IPCC: http://www.metoffice.gov.uk/research/hadleycentre/ar4/index.html.

The analysis of observed changes in the climate system reported by the IPCC are based on a combination of data from surface-based instrument networks, *in situ* observations, and a large number of operational and research satellite monitoring platforms. Selected findings of the IPCC excerpted from the summary for policymakers of IPCC Working Group I include the following:

- Eleven of the last 12 years (1995–2006) rank among the 12 warmest years in the instrumental record of global surface temperature (since 1850)....The linear warming trend over the last 50 years [0.13 °C (0.1–0.16 °C) per decade] is nearly twice that for the last 100 years.
- New analysis of balloon-borne and satellite measurements of the lower- and mid-troposphere temperature show warming rates that are similar to those of the surface temperature record.
- The average atmospheric water vapor content has increased since at least the 1980s over land and ocean as well as in the upper troposphere.
- Observations since 1960 show that the average temperature of the global ocean has increased to depths of at least 3000 m and that the ocean has been absorbing more than 80% of the heat added to the climate system. Such warming causes seawater to expand, contributing to sea level rise.
- Mountain glaciers and snow cover have declined on average in both hemispheres. Widespread decreases in glaciers and ice caps have contributed to sea level rise (ice caps do not include contributions from the Greenland and Antarctic Ice Sheets). New data (since the TAR (The Assessment Report of the IPCC)) now show that losses from the ice sheets of Greenland and the Antarctic have very likely contributed to sea level rise over 1993–2003. Flow speed has increased for some Greenland and Antarctic outlet glaciers, which drain ice from the interior of the ice sheets.
- Average Arctic temperatures increased almost twice the global average rate in the past 100 years. Satellite data since 1978 show that annual average Arctic sea ice extent

has shrunk by 2.7% [2.1–3.3%] per decade with a larger increase in summer of 7.4% [5.0–9.8%].

- Temperatures at the top of the permafrost layer have generally increased since the 1980s in the Arctic (by up to 3.0 °C). The maximum area covered by seasonally frozen ground has decreased by about 7% in the Northern Hemisphere since 1900, with a decrease in spring of up to 15%.
- Long-term trends from 1900 to 2005 have been observed in precipitation amount over many large regions. Significantly increased precipitation has been observed in eastern parts of North and South America, Northern Europe, and northern and Central Asia. Drying has been observed in the Sahel, the Mediterranean, southern Africa, and parts of southern Asia. Precipitation is highly variable spatially and temporally, and data are limited in some regions. Long-term trends have not been observed for the other large regions assessed.
- Changes in precipitation and evaporation over the oceans are suggested by freshening of mid- and high-latitude waters together with increased salinity in low-latitude waters.
- Mid-latitude westerly winds have strengthened in both hemispheres since the 1960s. More intense and longer droughts have been observed over wider areas since the 1970s, particularly in the tropics and subtropics. Increased drying linked with higher temperatures and decreased precipitation has contributed to changes in drought. Changes in sea surface temperatures, wind patterns, and decreased snowpack and snow cover have also been linked to droughts.
- The frequency of heavy precipitation events has increased over most land areas, consistent with warming and observed increases of atmospheric water vapor.
- Widespread changes in extreme temperature have been observed over the last 50 years. Cold days, cold nights, and frost have become less frequent, while hot days, hot nights, and heat waves have become more frequent.

Based on evidence, the IPCC (Working Group II) expresses high confidence that natural systems are affected. Excerpts from the IPCC reports related to effects on natural ecological and biological systems include:

- enlargement and increased numbers of glacial lakes;
- increasing ground instability in permafrost regions and rock avalanches in mountain regions;
- changes in some Arctic and Antarctic ecosystems, including those in sea-ice biomes, and also predators high in the food chain;
- increased runoff and earlier spring discharge in many glacier and snow-fed rivers;
- warming of lakes and rivers in many regions with effects on thermal structure and water quality;
- earlier timing of spring events such as leaf unfolding, bird migration, and egg laying;
- poleward and upward shifts in ranges in plant and animal species;
- shifts in ranges and changes in algal, plankton, and fish abundance in high-latitude oceans;
- increases in algal and zooplankton abundance in high-latitude and high-altitude lakes;
- range changes and earlier migrations of fish in rivers; and
- increasing oceanic acidity.

The reader is directed to the complete IPCC reports for additional detail on observed changes as well as the assessments on projected changes under various future greenhouse gas scenarios.

Monitoring Ecosystem Habitat and the Climatic Environment

Vegetation and ecosystem habitats respond to a number of climatic and environmental forcings and boundary conditions. Photosynthetic processes, fundamental to the growth of vegetative biomass, involve stomatal dynamics that control the sequestration of carbon from the atmosphere as well as plant respiration and the exchange of gases such as CO_2, O_2, and H_2O among other biochemical constituents. The primary climatic forcing parameters that vegetative growth or stress are sensitive to include temperature, precipitation/water availability, downward solar radiation at the surface and/or at the canopy level, downward long-wave radiation, relative humidity, and surface winds. These parameters affect stomatal resistance, carbon intake, and evapotranspiration among others. Solar radiation at the surface as well as long-wave radiation are modulated by cloud cover. Water availability is determined not only by local/*in situ* precipitation but also soil moisture (vadose zone) and the groundwater table which are linked to surface water flows, and subsurface recharge and water transport from distant locations in space. Snow/ice accumulation and melt introduce time lags into the dynamics and responses of such a hydroecological system. Other environmental boundaries are important for ecosystems and vegetative health and growth as well as stress and decay, for example, soil nutrient supply as well as environmental conditions that could make particular species more or less susceptible to attack by fungi and other microbial virulence. A challenge to both observing and modeling programs is to de-convolve the complexity of the vegetation/ecosystem–climate relationship so that it may later be applied to investigate the impacts of projected climate change and global warming on the biosphere.

Satellite observing systems have been deployed for over 30 years to monitor a large array of environmental parameters, including those that are critical for ecosystem function such as surface temperature, moisture, precipitation, the surface radiation balance, soil moisture, and water supply, among others. An excellent, concise summary of the various aspects and impacts already seen of global warming together with satellite video (movie loops) imagery may be found at http://www.nasa.gov/worldbook/global_warming_worldbook.html.

In a fascinating study, Balanya *et al.* linked global genetic changes to global climate warming. That climate change is altering the geographic ranges, abundances, phenologies, and biotic interactions of organisms has been demonstrated or alluded to by many researchers. Climate change may also alter the genetic composition of species, but assessments of such shifts require genetic data sampled over time. And, for most species, time series of genetic data are nonexistent or rare, especially on continental or global scales. For a few *Drosophila* species, time series comparisons of chromosome inversions are feasible because these adaptive polymorphisms were among the first genetic markers quantified in natural populations. Thus, historical records of inversion frequencies in *Drosophila* provide opportunities for evaluating genetic sensitivity to change in climate and other environmental factors. In this study, Balanya *et al.* determined the magnitude and direction shifts over time (24 years between samples on average) in chromosome inversion frequencies and in ambient temperature for populations of *Drosophila subobscura* on three continents. In 22 of 26 populations, climates warmed over the intervals, and genotypes characteristic of low latitudes (warm climates) increased in frequency in 21 of those 22 populations. Thus, they conclude, genetic change in this fly is tracking climate warming and is doing so globally.

Yet another recent study has implicated regional climate warming and its local effects on moisture, clouds, and day/night temperatures to the demise of frog varieties in Central and South America. According to the study, higher temperatures result in more water vapor in the air, which in turn forms a cloud cover that leads to cooler days and warmer nights. These conditions favor the chytrid fungus to thrive in Costa Rica and neighboring countries. The fungus which reproduces best at temperatures between 63 °F (17.2 °C) and 77 °F (25 °C) kills frogs by growing on their skin and attacking their epidermis and teeth, as well as releasing a toxin. At least 110 species of vibrantly colored amphibians once lived near streams in Central and South America but about two-thirds disappeared in the 1980s and 1990s, including the golden toad. The fate of amphibians, whose permeable skin makes them sensitive to environmental changes, is seen by scientists as a possible harbinger of global warming effects.

Numerous other studies point to the impact already seen on ecological systems due to the lengthening of the growing season and changes to temperature, precipitation, and moisture regimes. There have been shifts in plant species to higher elevations or latitude. There also have been some cases of an unusual spread of spores from distant regions carried by changing atmospheric wind circulations or the temperature of ocean currents. It is unfortunately beyond the scope of this short article on the remote sensing of global ecology to include such detail. The reader is referred to the various and excellent papers published in journals such as *Science* or *Nature*.

Large-Scale Forest Fires, Gaseous/Particulate Emissions, and Ecosystem Impacts

Choking smoke interrupted air and ship transportation in and around the islands of Sumatra and Borneo in early October 2006 as detailed in http://earthobservatory.nasa.gov/Newsroom/NewImages/images.php3?img_id=17423. Fires on the two islands were churning out a blanket of haze that mingled with clouds and reduced visibility to unsafe levels. In addition to their immediate impacts on air quality and human society, fires in tropical lowland forests affect increasingly threatened habitat for rainforest plants and animals, including the endangered orangutans. And because they release significant amounts of carbon dioxide and particle pollution, such as soot, the fires affect the global climate.

The October 2006 fires in Sumatra and Borneo had been burning for several weeks before the images in **Figure 4** were taken. During the regional dry season (roughly August–October), fires are common. Sometimes, fires are the result of slash-and-burn deforestation – clearing of rainforest for palm plantations, for example. At other times, the fires escape during brush clearing or other maintenance activities on already cleared land. Fires in the islands' low-lying forests and peat swamps generate massive amounts of smoke. Because these low-lying forests and swamp areas are inundated throughout parts of the year, the decay of dead vegetation on the ground proceeds slowly. The thick layers of dead, but undecayed, vegetation – peat – accumulate over many years. Fires burning in dry peat are very smoky and difficult to extinguish. Some can burn underground for years.

Large- and small-scale agriculture are not the only contributors to the fires. The droughts Indonesia experiences during El Niño episodes, such as the particularly severe 1997–98 event, make the forests and peat lands more likely to catch fire. Forests that have

Natural color

Shortwave- and near-infrared enhanced

Figure 4 Fires in Sumatra and Borneo in early October 2006. This pair of images from the moderate resolution imaging spectroradiometer (MODIS) on NASA's Aqua satellite from Sunday, 8 October 2006, shows the haze in the area. The top image is a photo-like image, made from MODIS' observations of visible light. Smoke appears grayish white in contrast to the bright white of clouds. Fires detected by MODIS are marked in red. The bottom image is made from a combination of VIS, short-wave IR, and N-IR light. Because smoke is more transparent in the short-wave and N-IR part of the light spectrum than it is in the VIS part, this 'false-color' type of image thins the haze and permits a view at the islands below. Smoke is transparent blue, clouds made of water droplets are white, clouds made of ice crystals are bright blue, vegetation is bright green, and the ocean is dark blue to black. Credit: NASA image created by Jesse Allen, Earth Observatory, using data provided courtesy of the MODIS Rapid Response System team (http://rapidfire.sci.gsfc.nasa.gov/).

been degraded by logging are also more likely to burn. According to a study by Page *et al.*, somewhere between 0.81 and 2.57 million tons of carbon were released by tropical lowland forest and peat land fires in Indonesia in 1997.

The recent (23 October 2007) fires in California, as captured by MODIS on NASA's Terra satellite, can be seen at http://earthobservatory.nasa.gov/Newsroom/NewImages/images.php3?img_id=17810.

Quite remarkable are the dense smoke plumes stretching over the Pacific for hundreds of kilometers. The growth and spread of the fires were fanned or 'fueled' by the powerful Santa Ana winds that whip from the high-altitude deserts of the Great Basin toward the Pacific Ocean.

Managed Ecosystems and Biofuels – A Subject of Current Interest and Concern

Some vegetation classification schemes have developed algorithms to distinguish between wooded C4 grasslands, wooded C3 grasslands, and C3 grasslands. These categories may become particularly important in the future to assess the impact of grasses and woody species being considered for biofuel production. Several grasses and woody species have been evaluated for biofuel production, with perennial rhizomatous grasses showing the most economic promise. *Arundo donax* (giant reed; native to Asia) and *Philaris arundinacea* (reed canary grass; native to temperate Europe, Asia, and North America) are two C3 grasses being considered as biofuel species that are invasive in some US ecosystems. The former threatens riparian areas and alters fire cycles; the latter invades wetlands and affects wildlife habitat. The hybrid *Miscanthus x giganteus* (native to Asia) and *Panicum virgatum* (switchgrass; native to central and eastern US) are C4 grasses being considered in Europe and the US. Several *Miscanthus* species are invasive or have invasive potential. Several traits that make these C3 grasses potentially valuable as a crop could also enhance invasiveness (ability to sprout from rhizomes, efficient photosynthetic mechanisms, and rapid growth rates). Thus, they have the potential to adversely impact local ecosystems while adding to fire susceptibility.

Importance of Land-Cover Change in Simulating Future Climates Using Global Models

Land-cover impacts on global climate can be divided into two major categories: biogeochemical and biogeophysical. Biogeochemical processes affect climate by altering the rate of biogeochemical cycles, thereby changing the chemical composition of the atmosphere. Biogeophysical processes directly affect the physical parameters that determine the absorption and disposition of energy at the Earth's surface. Albedo, or the reflective properties of the Earth's surface, alters the absorption of solar radiation and hence energy available at the Earth's surface. Surface hydrology and vegetation transpiration characteristics affect how energy received by the surface is partitioned into latent and sensible heat fluxes. Vegetation structure affects surface roughness, thereby altering momentum transport and heat transport. Summarizing the effects of land-cover change on climate has been difficult because different biogeophysical effects offset each other in terms of climate impacts and, on global and annual scales, regional impacts are often of opposite sign and are therefore not well represented in annual global average statistics. One of the methods used to separate the above-mentioned impacts is through the use of complex models together with global land cover data.

Monitoring Global Oceanic Ecosystems, Coastal Zones, and Seas

Global Distribution of Ocean Chlorophyll/ Phytoplankton Biomass

To human eyes, the ocean appears as shades of blue, sometimes blue-green. From outer space, satellite sensors can distinguish even slight variations in color to which our eyes are not sensitive. Different shades of ocean color reveal the presence of differing concentrations of sediments, organic materials, or even phytoplankton, all of which can be measured by satellites.

Due to their pigment (chlorophyll), phytoplankton preferentially absorb the red and blue portions of the light spectrum (for photosynthesis) and reflect green light. Therefore, the ocean over regions with high concentrations of phytoplankton will appear as certain shades, from blue-green to green, depending upon the type and density of the phytoplankton population there (**Figure 5**).

When considering Earth's sources of oxygen, we usually think of vast forests such as the Amazon, but about half of the oxygen we breathe comes from elsewhere; it is produced by phytoplankton. Phytoplankton are tiny, single-celled plants that live in the ocean, and they serve as the base of the oceanic food chain. Yet as important as phytoplankton are to life on Earth, their interaction with our planet has only recently been studied on a global scale. The satellite sensor that has pioneered the study of phytoplankton globally is the sea-viewing wide field-of-view sensor (SeaWiFS) based on legacy instruments such as the CZCS on NIMBUS-7. (http://earthobservatory.nasa.gov/Newsroom/NewImages/images.php3?img_id=17405).

Like their land-based relatives, phytoplankton require sunlight, water, and nutrients for growth. Because sunlight is most abundant at and near the sea surface, phytoplankton remain at or near the surface. Also like terrestrial plants, phytoplankton contain the pigment chlorophyll, which gives them their greenish color. Chlorophyll is used by plants for photosynthesis, in which sunlight is used as an energy source to fuse water molecules and carbon dioxide into carbohydrates – plant food. Phytoplankton (and land plants) use carbohydrates as 'building blocks' to grow; fish and humans consume plants to get these same carbohydrates.

The atmosphere is a rich source of carbon dioxide, and millions of tons of this gas settle into the ocean every year. However, phytoplankton still require other nutrients, such as iron, to survive. When surface waters are cold, ocean water from deeper depths upwells, bringing these essential nutrients toward the surface where the phytoplankton may use them. However, when surface waters are warm (as during an El Niño), they do not allow the colder, deeper currents to upwell and effectively block the flow of life-sustaining

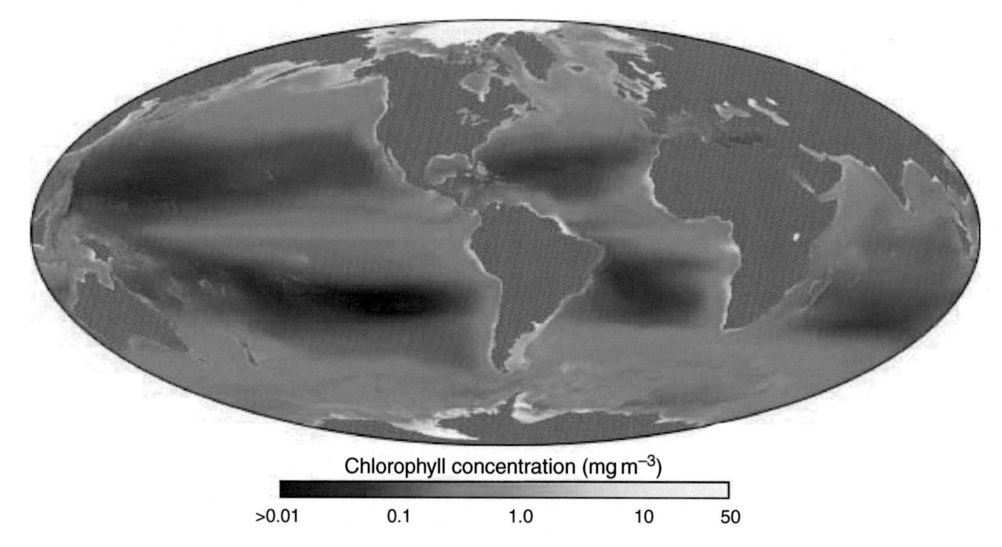

Figure 5 Nine years of ocean chlorophyll. The image shows chlorophyll measured by SeaWiFS from 18 September 1997 to 31 July 2006. Chlorophyll is shown in milligrams (a milligram is one-thousandth of a gram) per cubic meter of seawater. The greatest concentrations appear in yellow, and the sparsest appear in deep blue. Because this image shows values averaged over nearly 9 years, greater amounts of chlorophyll are observed in areas with recurring blooms. Some of the greatest concentrations appear along coastlines. Consistently high concentrations appear at the high latitudes, and medium-level concentrations appear over much of the ocean, particularly along the El Niño/La Niña route in the Pacific. Marine biologists often refer to the darkest blue areas as 'deserts', because the concentration of key nutrients in the water is usually so low that phytoplankton cannot grow. Credit: NASA image created by Jesse Allen, Earth Observatory, using data provided courtesy of the SeaWiFS Project (http://oceancolor.gsfc.nasa.gov/SeaWiFS/), NASA/Goddard Space Flight Center, and ORBIMAGE.

nutrients. (The El-Nino phenomenon is described in: http://earthobservatory.nasa.gov/Library/ElNino/.) As phytoplankton starve, so too do the fish and mammals that depend upon them for food. Even in ideal conditions an individual phytoplankton only lives for about a day or two. When it dies, it sinks to the bottom. Consequently, over geological time, the ocean has become the primary storage sink for atmospheric carbon dioxide. About 90% of the world's total carbon content has settled to the bottom of the ocean, primarily in the form of dead biomass.

Prior to the launch of SeaWiFS, scientists could only study phytoplankton on a relatively small scale. By measuring chlorophyll on a global scale over time, this sensor has been able to track how phytoplankton thrive and diminish as light and nutrient levels change. Massive phytoplankton blooms spread across the North Atlantic in the Northern Hemisphere each spring, and intense blooms also occur in the South Atlantic off the Patagonian Shelf of South America during spring in the Southern Hemisphere. Blooms fostered by changes in nutrient-rich water, though less regular, are also dramatic, especially when El Niño gives way to La Niña, and cold, nutrient-rich waters well up across the Pacific.

Hurricanes and Oceanic Phytoplankton Blooms

Analysis using SeaWiFS and MODIS data have also demonstrated that Atlantic Ocean hurricanes cause ocean deserts to bloom. The swirling hurricane rakes over the ocean surface and draws nutrient-rich water up from deeper in the ocean – fertilizing the marine desert. For 2–3 weeks following almost every storm, SeaWiFS data showed greater-than-normal phytoplankton growth, stimulated by the additional nutrients brought up to the surface. As an example, SeaWiFS took images of Hurricane Isabel on 13 and 18 September 2003. As the hurricane passed, it left behind it a trail of plankton blooms, evident by the rapid change in chlorophyll amounts. The lighter blue areas in the hurricane's wake represent higher amounts of chlorophyll. An animation of Hurricane Isabel, during 13–18 September 2003, using satellite data, is shown in http://www.nasa.gov/mpg/62507main_isabel10_320x240.mpg (credit: NASA/Orbimage).

Coastal Ecosystems and Disturbances

Coastal ecosystems are subject to effluents from neighboring land areas as well as disturbances such as hurricanes. Excessive rainfall over land and river discharge into coastal zones brings an abundance of pollutants and nutrients from agriculture that sometimes cause explosive phytoplankton blooms. Hurricane Isabel made landfall east of Cape Lookout, North Carolina, as a Category 2 (Stafford-Simpson scale) hurricane on 18 September 2003. The storm's center tracked to the northwest, passing west of Chesapeake Bay in the early morning of 19 September. Hurricane Isabel brought the highest storm surge and winds to the region since the Chesapeake-Potomac hurricane of 1933 and Hurricane Hazel in 1954. The storm surge reached a high of 2.7 m, and sustained wind speeds reached about 30 m s^{-1} with gusts of 40 m s^{-1}. Hurricane Isabel was responsible for physical and biological changes in the Chesapeake Bay on a variety of spatial and temporal scales. Short-term responses included a reduction of hypoxia by mixing, nutrient (nitrogen) inputs to the upper water column, and a large-scale phytoplankton bloom (3000 km^2), while long-term responses included an early onset of hypoxia in spring 2004, high abundance of the calanoid copepod *Eurytemora affinis* in spring 2004, and an increased recruitment of Atlantic croaker. These events highlight the importance of hurricanes in the function of this large estuarine ecosystem. The study used aircraft remote sensing technology to quantify chlorophyll (Chl a).

Example of Satellite Monitoring of the Impact of El Niño on the Productivity (Blooms) of Oceanic Phytoplankton

During normal years, when there is a steep thermocline tilt, the cold, deep currents flowing from Antarctica up the west coast of South America are allowed to upwell, bringing essential nutrients that would otherwise lie at the bottom. Phytoplankton living near the surface depend upon these nutrients for survival. In turn, fish and mammals depend upon phytoplankton as the very foundation of the marine food chain. The warm surface waters of an El Niño prevent this upwelling, effectively starving the phytoplankton population there and those animals higher up the food chain that depend upon it. Fishermen in Peru and Ecuador generally suffer heavy losses in their anchovy and sardine industries.

At Christmas Island, as a result of the sea level rise during the 1982–83 El Niño, sea birds abandoned their young and flew out over a wide expanse of ocean in a desperate search for food. Along the coast of Peru during that same time period, 25% of the adult fur seal and sea lion populations starved to death, and all of the pups in both populations died. Similar losses were experienced in many fish populations.

Meanwhile, over a 6-month period, about 100 inches of rainfall fell in Ecuador and northern Peru, ordinarily a desert region. Vegetation thrived and the region grew lush with grasslands and lakes, attracting swarms of grasshoppers and, subsequently, birds and frogs that fed on the grasshoppers. Many fish that had migrated upstream during the coastal flooding became trapped in the drying lakes and were harvested by local residents. Shrimp harvests were also very high in some of the coastal flood regions, but so too was the incidence of malaria cases due to thriving mosquito populations.

The correlation between sea surface temperature and phytoplankton productivity and blooms were well documented by the SeaWiFS satellite around the Galapagos Islands during a transition from El Niño conditions to La Niña conditions in the tropical Pacific Ocean. The images in **Figure 6** show an explosion in plankton growth as the warm El Niño waters, blamed for choking off essential ocean nutrients, are replaced by deep cold upwelled waters. The false color images of plankton concentrations between 10 May 1998 and 25 May 1998 show that life in the region to the west of the archipelago has returned in remarkable abundance associated with cooling waters.

The images in **Figure 6** are four frames from an animation in http://earthobservatory.nasa.gov/Library/ElNino/Anim/plankton_sst.mov. The animation shows sea surface temperature across the equatorial Pacific Ocean (top) during the 1997–98 El Niño; the lift-out (four frames in **Figure 6**) shows a higher-resolution image of ocean color (phytoplankton) in the region surrounding the Galapagos Islands. Notable are the transitions that occur between 10 and 25 May 1998. As the El Niño recedes, surface temperatures cool, allowing colder, nutrient-rich currents to upwell. There is a large, almost immediate bloom of phytoplankton in response to the replenished food source.

Black Water Off the Gulf Coast of Florida

Coastal ecosystems are subject to effluents from neighboring land areas, as well as changes in the oceanic circulation and disturbances such as hurricanes or severe storms. Excessive rainfall over land can cause unusual river discharges into coastal zones and bring with them an abundance of pollutants and nutrients from agriculture that sometimes cause near-explosive coastal blooms (see **Figure 7**).

Phytoplankton Blooms in the Black Sea

Many of the Europe's largest rivers, including the Danube, the Dnister, and the Dnipro (also called Dnieper), dump freshwater into the sea. The sea's only source of salty water, on the other hand, is the narrow Bosporus Strait, which connects it to the Mediterranean Sea through the Sea of Marmara. The salty water is denser than the freshwater, and so it sinks to the bottom, leaving a layer of relatively freshwater on top. The density barrier between salt- and freshwater is great enough that the two layers do not mix. As a result, when freshwater enters the sea from rivers, it only mixes with the relatively fresh water in the top 150 m of the sea. This means that fertilizers and runoff carried in the river water remain concentrated at the top of the sea

Figure 6 This sequence of SeaWiFS ocean color imagery shows the impact of a recent El Niño on the productivity of phytoplankton around the Galapagos Islands in the Pacific Ocean. The left top image (10 May 1998) was taken during the height of the 1997–98 El Niño, while the bottom right image (25 May 1998) was taken during the transition to a La Niña that followed. Note the flourishing bloom of phytoplankton as the surface waters cool, allowing the deeper, more-nutrient-rich waters to upwell. Credit: Adapted from and courtesy of Greg Shirah, NASA/Goddard Scientific Visualization Studio (http://visibleearth.nasa.gov/view_set.php?categoryID=5227), and the SeaWiFS project.

where they nourish the tiny plants (phytoplankton) that grow on or near the surface (see **Figure 8**).

In the spring of 2006, floods on the Danube River (see image captured by advanced spaceborne thermal emission and reflection radiometer (ASTER) on the Terra satellite: http://earthobservatory.nasa.gov/NaturalHazards/natural_hazards_v2.php3?img_id=13521) swept over broad stretches of farmland. The floods likely washed sediment, fertilizers, and animal waste into the Danube and the Black Sea. The extra iron, phosphates, and nitrates in the flood debris may be supporting the extensive bloom seen here. Such blooms can be both beneficial, because they provide food for fish, and dangerous, because decaying plant matter saps oxygen out of the water. If enough phytoplankton from a large bloom die and decay, the water may become so oxygen poor that fish can no longer survive in it. The result is a dead zone where little can survive.

Dead zones normally happen near the mouths of large rivers where fertilizers and agricultural waste are concentrated in the ocean. The Black Sea is one of the world's largest dead zones, though its dead zone is related to its stratification as much as to fertilizer runoff. When plants and other organic matter sink to the floor of the sea, they decay in the salty layer of water. Since the denser saltwater does not mix with the fresher water at the surface, there is no way to replenish the oxygen used during the decay process. As a result, the lower layer of the Black Sea is totally oxygen free.

Satellite Monitoring of Coral Reefs

On 26 December 2004, one of the largest earthquakes in recorded history struck offshore of the island of Sumatra, Indonesia. The ocean floor heaved in some places and sank in others, creating catastrophic tsunamis that raced across the Indian Ocean. Hundreds of thousands of people died as the waves struck coastlines from Thailand to Sri Lanka to Somalia. In addition to tsunami damage, satellite images of reefs, islands, and coastlines identified signs of permanent elevation change – sinking or uplift – along the fault between the Indo-Australia and Burma Plates. **Figure 9** shows the before and after images taken by Terra-ASTER. In the weeks and months after the earthquake, satellite images provided broad coverage of an area where ground-based observations were initially very limited. Changes in elevation were detected by a team of scientists along nearly 1600 km (994 mi) of the tectonic plate boundary. The images revealed that the earthquake rupture extended 100 km (62 mi) farther north than estimates based on seismic and GPS data suggested.

Atmosphere–Land–Ocean and Global Biogeochemical Transports

Emissions from both anthropogenic and natural sources on land and the oceans are transported to substantial altitudes in the atmosphere and many circumnavigate

Figure 7 This image of black water off the coast of Florida was acquired on 20 March 2002 by the sea-viewing wide field-of-view sensor (SeaWiFS). Scientists and local fishermen are not sure on why the coloring of the water is typically turquoise black. Amid growing concern, scientists are now trying to determine the source of the black water (http://earthobservatory.nasa.gov/NaturalHazards/natural_hazards_v2.php3?img_id=2620). Courtesy of the SeaWifs Project, NASA/Goddard Space Flight Center, and ORBIMAGE.

the globe with the air currents. During this process, various gases, aerosols, and particulates undergo chemical transformations. Some interact with cloud systems, and by changing the concentrations of cloud condensation nuclei, they affect precipitation process. They also affect the radiation budgets of the atmosphere and the surface. Aerosols can have either a heating or cooling effect depending on composition and particle size. Natural and man-made fires produce large clouds of pollution, as does industrial output. Dust from the great deserts of the world, and terpine aerosols from vegetation that produce near-surface haze, are also injected into the atmospheric circulation. Along with these aerosols and particles, also observed are the cross-oceanic transport of microbial matter that survives the journey across the oceans as encrusted spores. Moreover, the atmospheric transport of land sources of minerals also provides a pathway for nutrient resources at distant locations. As an example, it is estimated that somewhere around 250 Tg of Saharan dust is transported across the Atlantic Ocean. The dust deposited to the surface provides nutrients that feed the aquatic ecosystems of the Atlantic Ocean and the terrestrial ecosystems of South America, the Caribbean, and North America. The long-range transport of dust is, however, also a source of pollutant particulate matter. On much longer timescales, paleoclimatic and proxy records appear to suggest a 500 000 year cycle in the mineralization and fertilization of the Amazon tropical forests by desert dust from the Sahara. These cycles have been linked to solar variability and orbital forcing on the planet with resulting global climate change. Curiously, it is hypothesized that the Amazon would expire if not for the import of mineral-laden dust and aerosols from the Sahara desert. Such an event could conceivably occur if there

Figure 8 The Black Sea more closely resembled mixed paint on an artist's palette than the normally black surface of deep water when the moderate resolution imaging spectroradiometer (MODIS) on NASA's Aqua satellite captured this image on 20 June 2006. Swirls of color ranging from deep olive green to bright turquoise were created by a massive phytoplankton bloom that covered the entire surface of the sea. The sea was able to support such a large bloom largely because of its unique structure. The web site provides a more detailed explanation http://earthobservatory.nasa.gov/NaturalHazards/natural_hazards_v2.php3?img_id=13675. Courtesy Jeff Schmaltz, MODIS Land Rapid Response Team at NASA GSFC.

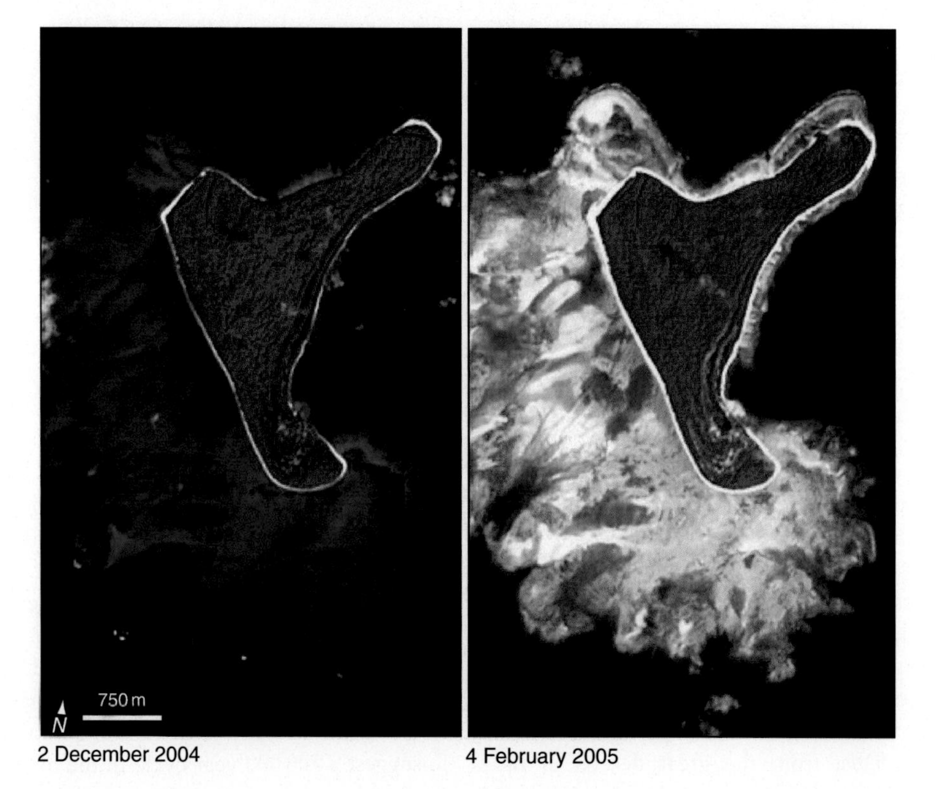

Figure 9 In places such as North Reef Island, shown in this pair of images from the advanced spaceborne thermal emission and reflection radiometer (ASTER) on NASA's Terra satellite, the quake lifted the reefs permanently out of the water. The images use VIS and IR light detected by ASTER to make different land surfaces stand out clearly from one another: water is blue, vegetation is red, coral or bare sand appears white. In the 'before' image, from 2 December 2004, the submerged reef creates a bright blue glow around the island. In the 'after' image, from 4 February 2005, the white coral stands completely up out of the water. It is even tinged with red, which suggests the exposed coral had died, and algae had colonized it. Credit: NASA images created by Jesse Allen, Earth Observatory, using data provided courtesy of the NASA/GSFC/MITI/ERSDAC/JAROS, and US/Japan ASTER Science Team (http://asterweb.jpl.nasa.gov). From http://earthobservatory.nasa.gov/Newsroom/NewImages/images.php3?img_id=17412.

were to be a strong intensification of the African monsoon systems and vegetation cover over the vast expanses of the African desert regions. Other examples include the fertilization of land vegetation by the transport and deposition of nitrous oxides, and the fertilization of the oceans by mineral iron.

Currently, satellites provide the only viable means by which the global transport of pollutants, nutrients, and minerals can be monitored. Examples include data from NASA's Terra, Aqua satellites (e.g., MODIS and multi-angle imaging spectroradiometer (MISR)), and the European Space Agency's ENVISAT satellite. Of course, in order to calibrate and validate the satellite data as also to develop the necessary algorithms for the retrieval of satellite-derived products, a large number of extensive field campaigns and experiments are typically conducted. They include instrumented balloon flights, research aircraft, research ships, and other *in situ* observing platforms and ground-based networks. An example of such a field campaign is the Intercontinental Chemical Transport Experiment-North America (INTEX-NA) to track the path of polluting gases and aerosols traveling from North America to Europe. The experiment aims at quantifying the North American import and export of ozone and associated pollutant gases, aerosols, and long-lived greenhouse gases. The INTEX-NA mission is coordinated under the International Consortium for Atmospheric Research on Transport and Transformation (ICARTT). The UK, Germany, Canada, and France will also conduct concurrent airborne campaigns (see http://www.nasa.gov/centers/goddard/earthandsun/0621_intex.html).

Global Carbon Monoxide (Air Pollution) Measurements

A short overview article by Richard Kerr (*Science*, 2007) well encapsulates the global dimensions and potential impacts of pollutant hazes and their climate-changing reach. Alluded to are conceivable changes to the atmospheric general circulation, oceanic currents, and through radiative and other feedbacks on precipitating cloud systems aerosols pollution can enhance, reduce, or delay the effects of greenhouse gas global warming. Much as an El Niño's tropical warmth can form an 'atmospheric bridge' to change the weather patterns in distant locations, the article implies that pollutants and their transport lead to global-scale teleconnections through their interaction with the global water and energy cycle. While there have been several studies (observational and modeling) of such interactions, the global dimensions of the interactions of dust, aerosols, and pollutants have only recently been conceptualized with the advent of new satellite sensors on board the current generation of satellite systems. It is to be noted, however, that several aspects of these interactions remain elusive, and more detailed analysis and modeling is required to better quantify their dynamics and energetics.

NASA's Terra spacecraft provides a complete view ever of the world's air pollution traveling through the atmosphere, across continents and oceans. For the first time, policymakers and scientists now have a way to identify the major sources of air pollution and to closely track where the pollution goes, anywhere on Earth. Carbon monoxide is a gaseous by-product from the burning of fossil fuels, in industry and automobiles, as well as burning of forests and grasslands. In the 30 April 2000 image (**Figure 10**), the levels of carbon monoxide are much higher in the Northern Hemisphere, where human population and human industry is much greater than in the Southern Hemisphere. However, in the 30 October 2000 image, immense plumes of the gas are emitted from forest and grassland fires burning in South America and Southern Africa. Researchers were surprised to discover a strong source of carbon monoxide in Southeast Asia. The air pollution plume from this region moves over the Pacific Ocean and reaches North America, frequently at fairly high concentrations. While fires are the major contributor to these carbon monoxide plumes, it is suspected that, at times, industrial sources may also be a factor.

The movements of carbon monoxide around the globe are particularly striking when viewed as a movie spanning a 10-month period. The following web sites contain animations of the images taken by MOPITT (Measurements of Pollution in the Troposphere) sensor for the Pacific, the Southern Hemisphere, and the globe, respectively: http://www.gsfc.nasa.gov/gsfc/earth/pictures/terra/pacific.mpeg, www.gsfc.nasa.gov/gsfc/earth/pictures/terra/southam.mpeg, and http://veimages.gsfc.nasa.gov/1788/mopitt_first_yeara.mpg.

The global air pollution monitor onboard Terra is the innovative MOPITT sensor, which was contributed to the Terra mission by the Canadian Space Agency. The instrument was developed by Canadian scientists at the University of Toronto and built by COM DEV International of Cambridge, Ontario. The data were processed by a team at the US National Center for Atmospheric Research (NCAR), at Boulder, CO.

Using Satellites Monitoring to Identify Ecological Niches Conducive to Disease Outbreaks

Accurate predictions of epidemics are still years away. However, in the short term, satellite monitoring could benefit public health in developing countries where resources to combat disease are limited. It is generally not feasible to send health workers everywhere, but a knowledge of where outbreaks are likely will help target those areas. Efforts can be focused where they are needed.

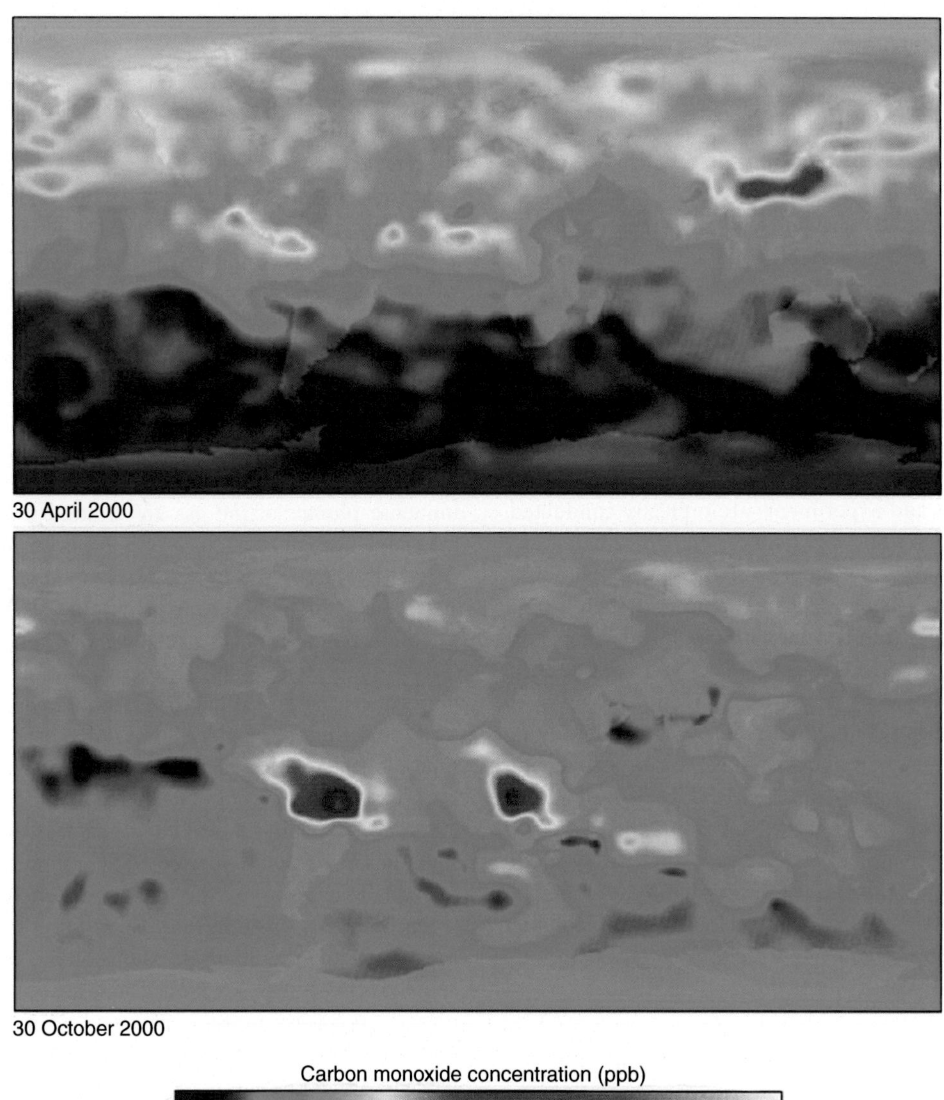

Figure 10 Images of global carbon monoxide observed by the MOPITT (measurements of pollution in the troposphere) sensor on board NASA's Terra satellite. The false colors in these images represent levels of carbon monoxide in the lower atmosphere, ranging from about 390 ppb (dark brown pixels), to 220 ppb (red pixels), to 50 ppb (blue pixels). Credit: Images and animations courtesy NASA GSFC Scientific Visualization Studio, based on data from MOPITT (Canadian Space Agency and University of Toronto). http://visibleearth.nasa.gov/view_rec.php?vev1id=8086.

Locating these vulnerable areas requires the use of satellites such as the NASA Terra satellite, to monitor vegetation on the ground. Because green vegetation cover varies with rainfall, it is a good indicator of climate variability, and therefore of conditions necessary for disease outbreaks. So far, areas of Africa that are at risk for Rift Valley fever (RVF) outbreaks have been mapped with satellite-derived information (**Figure 11**). A more detailed description than provided here and below is contained in http://eospso.gsfc.nasa.gov/newsroom/viewStory.php?id=231.

RVF outbreaks are linked to abnormally high and persistent rainfall in semiarid Africa. Ensuing flooding creates conditions necessary for breeding of mosquitoes that transmit the virus, first to domestic cattle and frequently to people as well. Though RVF causes relatively low mortality among humans (*c.* 1% of cases), it is often fatal to livestock, which can have devastating economic impacts on the countries affected.

In East Africa, animal husbandry is a major part of economy. Arab countries purchase a great deal of their meat products from East Africa. During the last RVF outbreak,

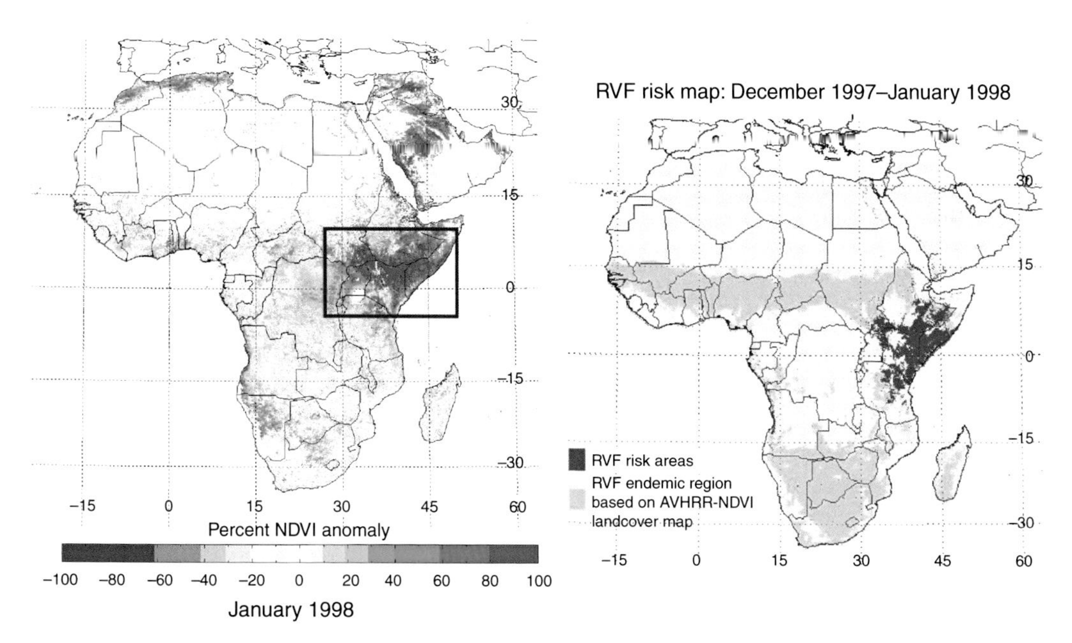

Figure 11 Left: satellite-derived normalized difference vegetation index (NDVI) showing percent deviation from mean vegetation for January 1998. Areas in the savanna lands of East Africa had an increase in vegetation vigor greater than 70% above normal, due to persistent above-normal rainfall, flooding dambo areas, that is, shallow depressions in savanna areas where mosquito eggs containing the virus are found. These are good habitats for the breeding of *Aedes* and *Culex* mosquito species, which serve as vectors for transmission of RVF. Right: this is an RVF outbreak risk map for the period December1997–February 1998. Areas shown in red (East Africa) represent areas where there was an outbreak of RVF during this period. Areas shown in green represent the savanna areas of Africa, where RVF is generally endemic and has occurred in the past. The outbreak of RVF during this period was associated with above-normal and widespread flooding during the warm El Niño–Southern Oscillation (ENSO) event of 1997–98 in the East Africa region. Credit: Compton Tucker and Assaf Anyamba, NASA Goddard Space Flight Center.

many Arab nations stopped imports from the region completely, which was catastrophic to the local economy, especially in the semiarid and arid regions of East Africa. In the late 1990s, a 'Climate and Disease Connections: Rift Valley Fever Monitor' was developed jointly by NASA/GSFC (Goddard Space Flight Center) and DoD/GEIS. The monitor, which includes climate- and satellite-derived vegetation anomalies that are associated with RVF outbreaks, is widely disseminated; it is internationally available with monthly updates at http://www.geis.fhp.osd.mil/GEIS/SurveillanceActivities/RVFWeb/indexRVF.asp.

Another example is Ebola hemorrhagic fever. Ebola is encountered in the tropical forest areas of Africa, but despite its notoriety as a highly fatal disease it remains a mystery in many respects. Though the first known Ebola epidemic occurred in Sudan in 1976, scientists have still not identified how the virus is transmitted or what animals might host it. In an effort to identify conditions under which the virus appears, scientists examined satellite data of tropical areas of Gabon and the Congo afflicted in 1994–96. They noted a sharp change from persistent dry conditions to wetter conditions over a 1–2 month period prior to the outbreaks, suggesting these dry to wet changes might be a 'trigger event'. However, it is cautioned that additional work is needed to verify the existence of the climatic trigger for Ebola. To quote Tucker (http://eospso.gsfc.nasa.gov/newsroom/viewStory.php?id=231), "It's fortunate for those affected by Ebola that we have so few outbreaks to study, but it makes [the job of associating outbreaks with specific antecedent conditions] more difficult. Drawing conclusions from a small sample is risky."

Future Challenges

Satellite remote sensing of regional and global ecology has limitations in terms of space and time resolutions and restraints on the spectral discrimination possible in regard to ecological species. Typically, remote sensing is best suited for the monitoring of relatively large spatial scales and somewhat homogeneous ecological zones. Thus, the fine features of highly heterogeneous ecological zones and complex issues concerning species composition and diversity will require detailed *in situ* observations and investigations to complement satellite observations. The challenge that faces both the scientific community of researchers, and the community of operational planners and managers who oversee and protect environmental and ecological resources, is to develop techniques to optimize the benefit obtainable from both *in situ* and remote sensing observational platforms. Beyond the scope of this article are issues involved with regard to the sustainability of the natural environment and ecosystem, a subject that entails a multitude of social, economic, cultural, and political factors or forces. Biodiversity, an important aspect of ecological studies, is also somewhat outside the capabilities of satellite remote sensing, though some features of species succession, adaptation, and movement may be captured.

We suggest here that space-based remote sensing should be viewed as an indispensable and powerful tool to observe, understand, monitor, and model global ecosystems when integrated with more detailed *in situ* observations. Another aspect for which satellite sensing technology has proved to be indispensable is in monitoring a large number of parameters that govern the geophysical, dynamic, thermodynamic, radiative, energetic, and chemical and biological environments that interact with or even determine the character of global and regional ecosystems. Space-based satellite observations, combined with *in situ* measurements, have the ability to provide a more comprehensive and holistic view of the Earth/climate system. Unprecedented advances have occurred over the past decade in satellite observing technology as well as in complex coupled models, data assimilation, and data fusion methods. The future is yet to unfold.

See also: Ocean Currents and Their Role in the Biosphere.

Further Reading

Anyamba A, Chretien JP, Formenty PBH, *et al.* (2006) Rift Valley fever potential, Arabian Peninsula. *Emerging Infectious Diseases* 12(3): 518–520.

Araujo MB and Rahbek C (2006) How does climate change affect biodiversity? *Science* 313: 1396–1397.

Bailey SW and Werdell PJ (2006) A multi-sensor approach for an on-orbit validation of ocean color satellite data products. *Remote Sensing of Environment* 102(1–2): 12–23.

Balanya J, Oller JM, Huey RB, Gilchrist GW, and Serra L (2006) Global genetic change tracks global climate warming in *Drosophila subobscura*. *Science* 313: 1773–1775.

Birk R, Miller R, and Unninayar S (2005) Taking the pulse of the planet: NASA research supports a global observing system of systems. *Earth Imaging Journal* 2(3): 16–19.

Capone DG, Popa R, Flood B, and Nealson KH (2006) Follow the nitrogen. *Science* 312: 708–709.

Chapin FS, III, Sturm M, Serreze MC, *et al.* (2005) Tipping points in the Tundra. *Science* 310: 657–660.

Cohen WB and Samuel NG (2004) Landsat's role in ecological applications of remote sensing. *Bioscience* 54(3): 535–545.

Defries RS, Bounoua L, and Colatz GJ (2002) Human modification of the landscape and surface climate in the next fifty years. *Global Change Biology* 8: 438–458.

Defries R, Field C, Fung I, *et al.* (1995) Mapping the land surface for global atmosphere–biosphere models: Toward continuous distributions of vegetation's functional properties. *Journal of Geophysical Research* 100: 20867–20882.

Defries RS, Hansen MC, and Townshend JRG (2000) Global continuous fields of vegetation characteristics: A linear mixture model applied to multi-year 8 km AVHRR data. *International Journal of Remote Sensing* 21(6–7): 1389–1414.

DiGregorio A (2005) *Land Cover Classification System: Classification Concepts and User Manual, Software Version 2*, 190pp. Rome: FAO, (original version prepared by DiGregorio A and Jansen LJM).

Feddema JJ, Oleson K, Bonan GB, *et al.* (2005) The importance of land-cover in simulating future climates. *Science* 310: 1674–1678.

Fensholt R, Nielsen TT, and Stisen S (2006) Evaluation of AVHRR PAL and GIMMS 10-day composite NDVI time series products using SPOT-4 vegetation data for the African continent. *International Journal of Remote Sensing* 27(3): 2719–2733.

Frizelle BG, Walsh SJ, Erlien CM, and Mena CF (2003) Collecting control data for remote sensing applications in the frontier environment of the Ecuadorian Amazon: The fusion of GPS technology, remote sensing methods, and social survey practices combined to generate sufficient control data for image processing and analysis in a remote and inhospitable environment. *Earth Observation Magazine* 12(7): 20–24.

Gordon HR, Clark DK, Brown JW, *et al.* (1983) Phytoplankton pigment concentrations in the Middle Atlantic Bight: Comparison of ship determinations and CZCS estimates. *Applied Optics* 22(1): 20–36.

Gordon HR and Wang M (1994) Retrieval of water-leaving radiance and aerosol optical thickness over the oceans with SeaWiFS: A preliminary algorithm. *Applied Optics* 33: 443–452.

Herrman SM, Anyamba A, and Tucker CJ (2005) Recent trends in vegetation dynamics in the African Sahel and their relationships to climate. *Global Environmental Change – Human and Policy Dimensions* 15(4): 394–404.

Holben BN (1986) Characteristics of maximum-value composite images from temporal AVHRR data. *International Journal of Remote Sensing* 7: 3473–3491.

Hooker SB, Esaias WE, Feldman GC, Gregg WW, and McClain CR (1992) An overview of SeaWiFS and ocean color. *NASA Technical Memoirs,* vol. 104566. Greenbelt, MD: National Aeronautics and Space Administration, Goddard Space Flight Center.

Iwasaki N, Kajii M, Tange Y, Miyachi Y, Tanaka Y, and Sato R (1992) Status of ADEOS mission sensors. *ACTA Astronautica* 28: 139–146.

Intergovernmental Panel on Climate Change (IPCC) (2007) *Climate Change 2007: The Physical Basis. Summary for Policy Makers. Contribution of Working Group I to the Fourth Assessment Report of the Intergovernmental Panel on Climate Change*. Geneva: IPCC. http://www.ipcc.ch (accessed December 2007).

Intergovernmental Panel on Climate Change (IPCC) (2007) *Physical Science Basis of Climate Change. Contribution of the Working Group I to the Fourth Assessment Report of the IPCC.* Solomon S, Qin D, Manning M, *et al.* (eds.), 996pp. Cambridge: Cambridge University Press. http://www.ipcc.ch (accessed December 2007).

Intergovernmental Panel on Climate Change (IPCC) (2007) *Climate Change Impacts, Adaptation, and Vulnerability. Summary for Policy Makers. A Contribution of the Working Group II to the Fourth Assessment Report of the IPCC.* Parry ML, Canziani OF, Palutikol JP, van der Linden PJ, and Hanson CE (eds.). Cambridge: Cambridge University Press. http://www.ipcc.ch (accessed December 2007).

Intergovernmental Panel on Climate Change (IPCC) (2007) Mitigation. Contribution of the Working Group III to the Fourth Assessment Report of the IPCC. Metz B, Davidson O, Bosch P, Dave R, and Meyer L (eds.). Cambridge: Cambridge University Press. http://www.ipcc.ch (accessed December 2007).

Kerr R (2006) Creatures great and small are stirring the ocean. *Science* 313: 1717.

Kerr R (2007) Pollutant hazes extend their climate-changing reach. *Science* 315: 1217.

Kunze E, Dower JF, Beveridge I, Dewey R, and Bartlett KP (2006) Observations of biologically generated turbulence in a coastal inlet. *Science* 313: 1768–1770.

Lovejoy TE and Hanna L (eds.) (2006) *Climate Change and Biodiversity*. New Haven, CT: Yale University Press.

Myneni RB, Hall FG, and Sellers PJ (1995) The interpretation of spectral vegetation indexes. *IEEE Transactions on Geoscience and Remote Sensing* 33: 481–486.

Myneni RB, Nemani RR, and Running SW (1997) Estimation of global leaf area index and absorbed PAR using radiative transfer models. *IEEE Transactions on Geoscience and Remote Sensing* 35(6): 1380–1393.
O'Reilly JE, Maritorena S, Mitchell BG, *et al.* (1998) Ocean color chlorophyll algorithms for SeaWiFS. *Journal of Geophysical Research* 103: 24937–24953.
Page SE, Siegert F, Reiley JO, Boehm HD, Jaya A, and Limin S (2002) The amount of carbon released from peat and forest fires in Indonesia during 1997. *Nature* 420: 61–65.
Parkinson CL, Ward A, and King M (eds.) (2006) Guide to NASA's Earth Science Program and Earth Observing Satellite Missions. *NASA/GSFC Earth Science Reference Handbook*, NP-2006-5-768-GSFC, 277pp. http://eospso.gsfc.nasa.gov/ftp_docs/2006ReferenceHandbook.pdf (accessed December 2007).
Pinzon JE, Wilson JM, Tucker CJ, *et al.* (2004) Trigger events: Enviroclimatic coupling of Ebola hemorrhagic fever outbreaks. *American Journal of Tropical Medicine* 71(5): 664–674.
Potter CS and Brooks V (1998) Global analysis of empirical relations between annual climate and seasonality of NDVI. *International Journal of Remote Sensing* 19: 2921–2948.
Potter CVG, Gross P, Boriah S, Steinbach M, and Kumar V (2007) Revealing land cover change in California with satellite data. *EOS Transactions, American Geophysical Union* 88(26): 269–274.
Pounds JA, Bustamante MR, Coloma LA, *et al.* (2006) Widespread amphibian extinctions from epidemic disease driven by global warming. *Nature* 439: 161–167.
Raghu S, Anderson RC, Daehler CC, *et al.* (2006) Adding biofuels to the invasive species fire? *Science* 313: 1742.
Rast M and Bezy JL (1999) The ESA medium resolution imaging spectrometer MERIS: A review of the instrument and its mission. *International Journal of Remote Sensing* 20(9): 1681–1702.
Roman MR, Adolf JE, Bichy J, Boicourt WC, *et al.* (2005) Chesapeake bay plankton and fish abundance enhanced by Hurricane Isabel. *EOS Transactions, American Geophysical Union* 86(28): 261–265.
Running SW, Loveland TR, and Pierce LL (1994) A remote sensing based vegetation classification logic for use in global biogeochemical models. *AMBIO* 23: 77–91.
Running SW and Nemani RR (1988) Relating seasonal patterns of the AVHRR vegetation index to simulated photosynthesis and transpiration of forests in different climates. *Remote Sensing of Environment* 24: 347–367.
Sellers PJ (1985) Canopy reflectance, photosynthesis and transpiration. *International Journal of Remote Sensing* 6: 1335–1372.
Tucker CJ (1979) Red and photographic infrared linear combinations for monitoring vegetation. *Remote Sensing of Environment* 8: 127–150.
Unninayar S and Schiffer RA (2002) Earth Observing Systems. In: MacCracken MC and Perry JS (eds.) *Encyclopedia of Global Environmental Change, Vol. 1: The Earth System: Physical and Chemical Dimensions of Global Environment Change*, pp. 61–80. New York: Wiley.
UN Office for Outer Space Affairs (2006) *United Nations Office for Outer Affairs*. Vienna: UNOOSA, http://www.uncosa.unvienna.org/pdf/reports/IAM2006E.pdf (accessed December 2007).
Xiao J and Moody AA (2004) Photosynthetic activity of US biomes: Response to the spatial variability and seasonality of precipitation and temperature. *Global Change Biology* 10: 437–451.
Welsch C, Swenson H, Cota SA, DeLuccia F, Haas JM, and Schueler C (2001) VIIRS (visible infrared imager radiometer suite): A next generation operational environmental sensor for NPOESS. *IGARRS* 3: 1020–1022.
Wetzel P, Maier-Reimer E, Botzet M, *et al.* (2006) Effects of ocean biology on the penetrative radiation in a coupled climate model. *Journal of Climate* 19: 3973–3987.
Wulder MA, Hall RJ, Coops NC, and Franklin SE (2004) High spatial resolution remotely sensed data for ecosystem characterization. *Bioscience* 54(6): 511–521.

Relevant Websites

http://asterweb.jpl.nasa.gov – ASTER (Advanced Spaceborne Thermal Emission and Reflection Radiometer).
http://www.geis.fhp.osd.mil – Climate and Disease Connections: Rift Valley Fever Monitor, DoD-GEIS.
http://www.esa.int – Envisat Overview, European Space Agency.
http://visibleearth.nasa.gov – Galapogos Island; First Global Carbon Monoxide (Air Pollution) Measurements, Visible Earth.
http://www.ipcc.ch – Intergovernmental Panel on Climate Change.
http://www.metoffice.gov.uk – IPCC Fourth Assessment Report, Met Office.
http://earthobservatory.nasa.gov – MODIS Instrument on NASA's Terra Satellite Improves Global Vegetation Mapping, Makes New Observations Possible, NASA News Archive; Fires on Borneo and Sumatra; Fires in Southern California; Nine Years of Ocean Chlorophyll; EO Natural Hazards: Floods in Central Europe; Earthquake Spawns Tsunamis; What is El Niño, Fact Sheet, by David Herring; La Niña Fact Sheet; Black Water off the Gulf Coast of Florida; Phytoplankton Blooms in the Black Sea; Natural Hazards, NASA Earth Observatory.
http://cybele.bu.edu – MODIS Leaf Area Index, Climate and Vegetation Research Group, Department of Geography, Boston University.
http://rapidfire.sci.gsfc.nasa.gov – MODIS Rapid Response System.
http://eospso.gsfc.nasa.gov – NASA Earth Observing System, Satellite Mission Profiles: Satellites Tracking Climate Changes and Links to Disease Outbreaks in Africa, Earth Observing System.
http://eospso.gsfc.nasa.gov – Satellites Tracking Climate Changes and Links to Disease Outbreaks in Africa, The Earth Observing System.
http://oceancolor.gsfc.nasa.gov – SeaWiFS Project: Spacecraft and Sensory Overview, Ocean Color Web, NASA.
http://www.uncosa.unvienna.org – Space Technology and Sustainable Development, United Nations Coordination of Outer Space Activities.
http://www.wmo.int – Status of Current and Future CGMS Members Satellites; The Space-Based Global Observing System, World Meteorological Organization.
http://aqua.nasa.gov – The NASA Aqua Platform.
http://modis.gsfc.nasa.gov – The NASA MODIS Instrument.
http://terra.nasa.gov – The NASA Terra Platform Website.
http://www.nasa.gov – The World Book at NASA; Global Warming; NASA Tracks Pollution across the Globe; The Ocean Chromatic: SeaWiFS Enters Its Second Decade; NASA Goddard Space Flight Center.

Ocean Currents and Their Role in the Biosphere

A Ganopolski, Potsdam Institute for Climate Impact Research, Potsdam, Germany

Introduction

This article presents an overview of the role of the ocean currents in the climate systems and the ways they affect terrestrial and marine ecosystems. The first section discusses how the ocean currents influence global and regional climate. The second section discusses paleoclimate evidences of the past instability of the ocean circulation and future modeling projections of the ocean circulation changes under global warming. The third section discusses results of model simulations and paleoclimate evidences of the impact of reorganizations of the ocean currents on terrestrial and marine ecosystems.

The Ocean Currents, Climate, and Biosphere

The Role of the Ocean Currents in the Climate System

Modern ocean circulation represents a complex three-dimensional phenomenon which is determined by the Earth's geography and spatial patterns of surface wind, and surface heat and freshwater fluxes. Surface ocean currents are directly driven by wind and the existence of large-scale oceanic gyres (**Figure 1**) is explained by prevailing westerlies in the mid-latitudes and trade winds in the tropics. The divergence of surface wind-driven currents creates upward vertical movement of water (upwelling), which plays an important role in nutrients supply to the upper ocean layer. Apart from that, winds and tidal energy are the primary sources of vertical mixing in the ocean interior. Without vertical mixing provided by wind and tides, the deep ocean would be essentially stagnant. Surface fluxes of heat and freshwater, although do not represent a direct energy source for the ocean currents, play an important role in controlling the ocean circulation by changing sea water temperature and salinity. The latter determine horizontal density gradient which drives the currents in the ocean interior.

The balance between surface heat and freshwater fluxes also determines the areas where the deep ocean water masses are formed. Currently, these deep water masses are formed in several isolated locations: in the Nordic Seas and the Labrador Sea in the North

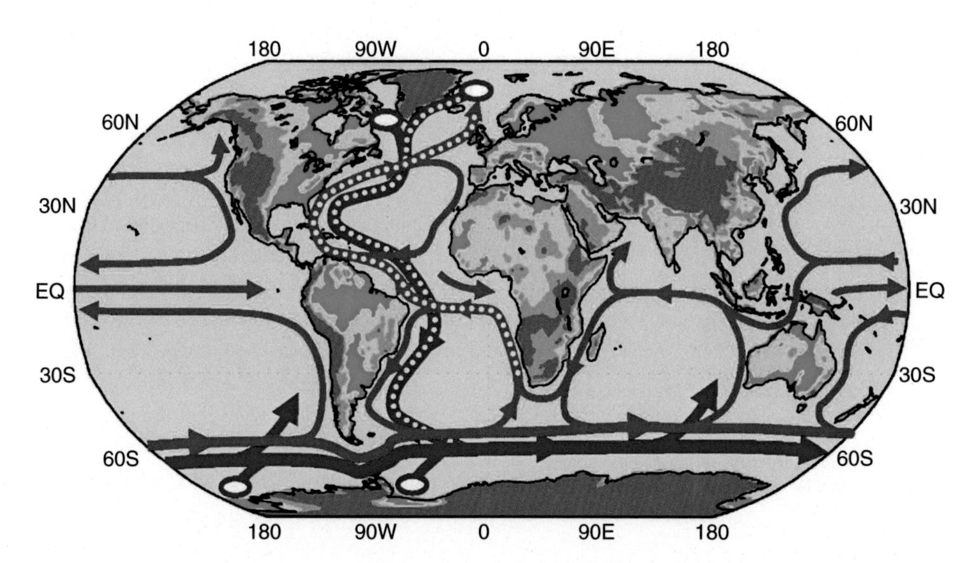

Figure 1 A simplified cartoon of the surface (red) and deep ocean currents (blue). The major areas of deep water formation are shown by ovals. Yellow dots indicate the upper branch and light blue dots indicate the lower branch of the Atlantic thermohaline circulation.

Atlantic, and around Antarctica. Although the areas of deep water formation occupy only a small fraction of the ocean, they play a fundamental role in driving of the meridional overturning circulation, also known as the ocean thermohaline circulation or 'the ocean conveyor belt'. The upper branch of the ocean conveyor is represented by the northward transport of warm water masses along the surface currents of which Gulf Stream is the most prominent one (**Figure 1**). When reaching high latitudes of the North Atlantic, surface water is cooled down by losing energy into the atmosphere and eventually reaches the high density which allows surface water to sink to the bottom of the ocean. This water then slowly moves southward along the American continental slope and reaches the Southern Ocean, where it mixes with the deep water masses formed around Antarctica. It is believed that most of deep water eventually rises to the surface in the Southern Ocean in the areas of wind-driven upwelling, thus closing the conveyor loop. The existence of the thermohaline circulation is closely related to the existence of deep water formations areas. At present, there is no deep water formation in the Pacific Ocean, and, as a result, there is no the thermohaline circulation in this ocean. The later explains very different climate conditions in the high latitudes of the Atlantic and Pacific oceans.

Although an average velocity associated with the meridional overturning circulation is rather small compared to typical velocities of surface ocean currents, the meridional overturning circulation is responsible for a large portion of the ocean meridional heat transport. Currently, about 1 PW of energy ($1\,\text{PW} = 10^{15}\,\text{W}$) is transported northward in the North Atlantic, that is about one-fifth of the total energy transport in the atmosphere–ocean system in the Northern Hemisphere. The influence of the ocean currents on climate is illustrated by **Figure 2a**. It shows deviations of local annual surface air temperature from its zonally averaged values. It is seen that annual air temperature over the northern North Atlantic, and, especially over the Nordic Seas, is much higher than average temperature for the same latitudes. Thus the main reason for mild climate conditions over most of Europe is the existence of vigorous Atlantic thermohaline circulation. Since the release of heat transported by the oceanic currents into the atmosphere occurs primary during winter, this prevents forming of the sea ice in the high latitudes, and results in a considerable reduction of the amplitude of seasonal temperature variations. As shown in **Figure 2b**, the difference between summer and winter temperatures over the Western Europe is much smaller than for the same latitudes in Asia and North America. All these factors, in combination with a stable moisture transport from warm North Atlantic, allow the existence of extended temperate and broadleaf forests over most of Europe.

Ocean Currents and Climate Change

The importance of the ocean currents for climate and climate change has been demonstrated in a number of modeling studies, which showed that the Atlantic thermohaline circulation may change rapidly in response to change in climatological conditions, such as increased freshwater flux into the North Atlantic due to massive iceberg discharge from surrounding ice sheets, as it happened many times during the glacial age, or due to intensification of atmospheric hydrological cycle and melting of the Greenland ice sheets, that may happen in the future as a result of global warming. Changes in the thermohaline circulation, in turn, lead to dramatic changes in the ocean heat transport and global climate.

Numerical experiments performed with climate models demonstrate that a complete shutdown of the Atlantic thermohaline circulation under present-day climate conditions will cause surface air temperature cooling by more than 10 °C over the Nordic Seas and northwestern Europe. The cooling is caused by cessation of the northward oceanic heat transport into high latitudes and amplified by a southward expansion of sea ice margin. The cooling is most pronounced in winter when it is almost twice stronger than in annual mean. Changes in the ocean currents not only affect local temperature but via several oceanic and atmospheric teleconnection mechanisms spread the climate change over the globe. In particular, the cooling caused by the shutdown of the thermohaline circulation is simulated over most of the Northern Hemisphere, although the magnitude of cooling in other areas is smaller than that over the northern North Atlantic. At the same time, a decrease of interhemispheric oceanic heat transport causes a warming in the Southern Hemisphere, most pronounced in the Southern Atlantic and around Antarctica. The reorganization of the thermohaline circulation also affects the hydrological cycle. In particular, cooling of surface North Atlantic reduces evaporation in this area which causes a drastic reduction of precipitation over most of Europe. Another important result of the Atlantic thermohaline circulation weakening is a southward shift in the position of the intertropical convergency zone (ITCZ), which is associated with the rain belt around the equator. Shift of ITCZ causes a considerable redistribution of precipitation in the tropics, with a decrease of precipitation north of the equator and an increase south of the equator. It also affects the strength of subtropical monsoons, with weaker Indian and African monsoons. It is also plausible that regime

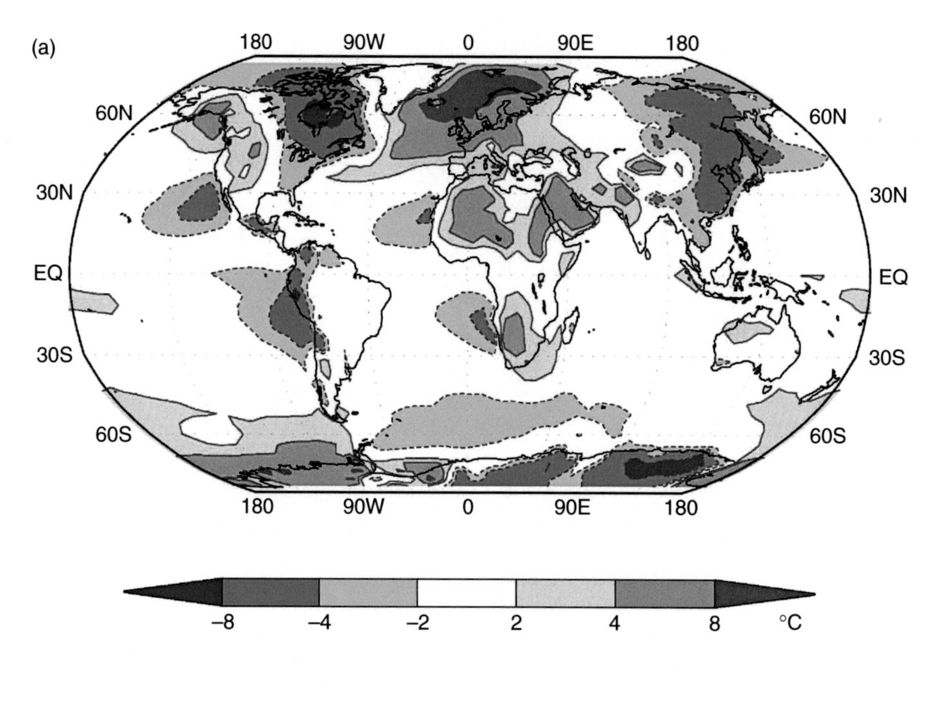

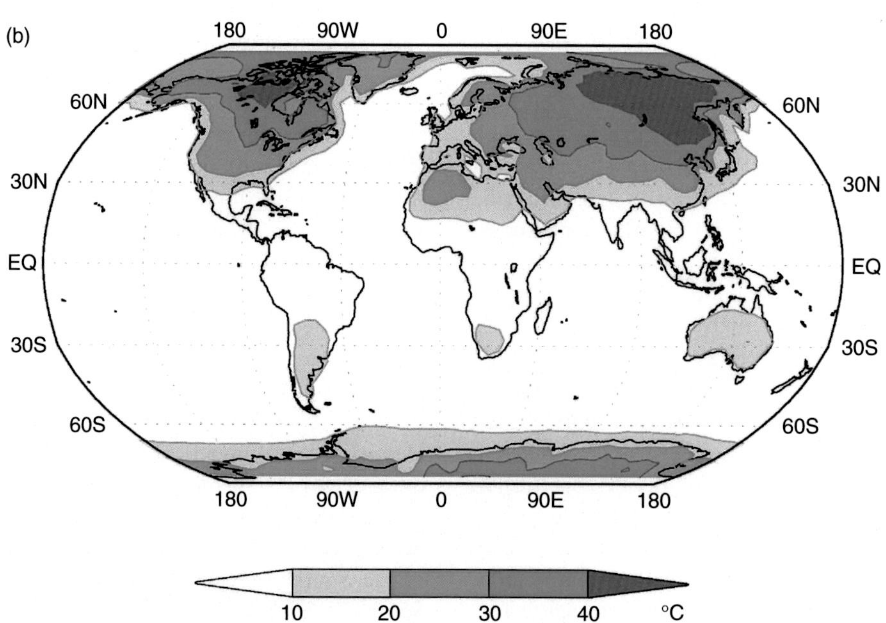

Figure 2 Deviation of the annual mean surface air temperature corrected to elevation effect from the zonal average temperatures (a), and the magnitude of seasonal variations of surface air temperature (b). The data are from Legates DR and Willmott CJ (1990) Mean seasonal and spatial variability in global surface air temperature. *Theoretical and Applied Climatology* 41: 11–21.

change of the Atlantic circulation can directly affect tropical Pacific, in particular, El Niño/Southern Oscillation cycle, which is responsible for a large portion of climate variability in the tropics and affects climate over the globe. Not all of the aforementioned processes are well understood and fundamental limitations of current climate models preclude unambiguous conclusion about possibility of the abrupt changes of the oceanic circulation in the future, but a growing body of paleoclimatological data indicates that, at least in the past, rapid and vigorous changes in the ocean circulation occurred regularly and had a widespread impact on climate and biosphere. Thereby, the possibility of abrupt climate shifts caused by changes of the ocean circulation remains one of the concerns related to anthropogenic global warming.

Past and Future Changes in Ocean Circulation

Paleoclimate Evidences for Instability of the Ocean Circulation

The issue of the stability of ocean circulation attracted a large attention after discovery of abrupt climate changes in the Greenland ice cores in the early 1990s. These records revealed that during the last glacial age, climate was rather unstable and was characterized by numerous abrupt shifts between cold and relatively warm states (**Figure 3**). The most prominent abrupt climate changes, known as Dansgaard–Oeschger events, correspond to abrupt warmings in Greenland by 10–15 °C over just several years or decades. This finding was corroborated later by numerous marine and terrestrial paleoclimate records from different locations which revealed abrupt climate changes apparently synchronous with that observed in the Greenland ice cores.

The initial idea of W. Broecker that these abrupt climate changes are related to the reorganizations of the Atlantic thermohaline circulations received in recent years a strong support from the analysis of different paleoclimate records and modeling studies. It has been shown that during the warm phases of the glacial age corresponding to Dansgaard–Oeschger events, the Atlantic thermohaline circulation was alike its present state and warm surface currents penetrated far into the high-latitude North Atlantic. During the cold periods, known also as 'stadials', although the Atlantic thermohaline circulation was still active, it was less extended to the north, and a much smaller amount of energy was transported toward the Nordic Seas. This caused a substantial southward expansion of the sea ice area and a strong cooling over the North Atlantic realm. At last, during periods of massive iceberg discharge into the North Atlantic from the North American and other Northern Hemisphere ice sheets, the Atlantic thermohaline circulation was completely shut down over centuries or even millennium causing the extreme cold climate conditions.

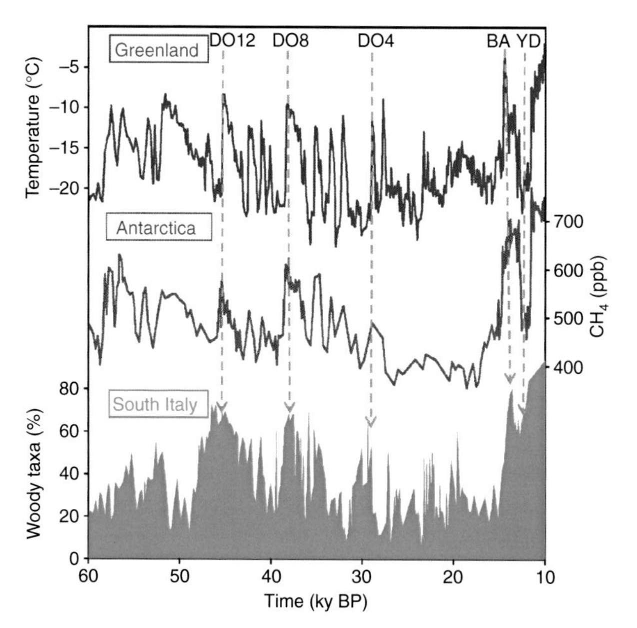

Figure 3 Paleoclimate records of Greenland temperature anomalies compared to present-day climate reconstructed from ^{18}O isotope concentration (blue), atmospheric methane concentration (red), and relative abundance of woody pollen in sediments core in the southern Italy (green). Vertical dashed lines show a probable temporal correlation between different records. DO4, DO8, and DO12 marks the Dansgaard–Oeschger events number 4, 8, and 12, respectively; BA refers to Bølling–Allerød warm event; and YD refers to Younger Dryas cold event. Greenland and Antarctic data are from Blunier T and Brook EJ (2001) Timing of millennial-scale climate change in Antarctica and Greenland during the last glacial period. *Science* 291: 109–112. Pollen data are from Watts WA, Allen JRM, and Huntley B (1996) Vegetation history and palaeoclimate of the last glacial period at Lago grande di Monticchio, southern Italy. *Quaternary Science Reviews* 15: 133–153.

There is also a growing body of paleoclimate evidences suggesting that climate impact of Dansgaard–Oeschger events was not restricted to the North Atlantic realm, and abrupt climate changes synchronous with Dansgaard–Oeschger events, recorded in Greenland, have been found in many paleoclimate records in Eurasia, tropics, and the Pacific Ocean. In the tropics, for example, abrupt climate changes are most pronounced in the paleoclimate proxies reflecting changes in hydrological conditions (precipitation) and the strength of summer and winter monsoons. This is fully consistent with results of model simulations showing a southward shift of ITCZ and weaker summer Asian monsoon for a weaker state of the Atlantic thermohaline circulation.

One of the most convincing arguments for the global-scale extent of abrupt glacial climate changes is coeval variations in methane concentration with the temperature changes in Greenland (**Figure 3**). Since the major sources of methane is the boreal and tropical wetlands, strong excursions in methane concentration comparable in the magnitude with the difference between glacial and modern conditions indicate large changes in temperature and precipitation over a large part of the globe.

Future Changes in Ocean Circulation

It is generally recognized that an increase of atmospheric concentration of carbon dioxide and other greenhouse gases due to anthropogenic activity is the primary cause of observed global warming. While the temperature rise is the most known and well-established aspect of anthropogenic climate change, the rising concentration of greenhouse gases leads to a number of other changes, such as an intensification of hydrological cycle, changes in probability of extreme weather events, gradual melting and retreat of the ice sheets and glaciers, shrinking of sea ice area, which are already supported by the analysis of observational data. It is believed that the ocean circulation, as a rather sensitive and a strongly nonlinear component of the climate system, will also undergo considerable changes in the course of anthropogenic global warming. Based on results of modeling experiments, the Atlantic thermohaline circulation is considered as the most vulnerable component of the global ocean circulation and is expected to weaken considerably in the future. Two major factors affect the ocean circulation under global warming conditions: surface warming and freshening. Both factors affect the local sea water density and meridional density gradient, the primary factor controlling the strength of the Atlantic thermohaline circulation. Surface freshening in the high latitude of North Atlantic caused by increased precipitation, enhanced river runoff, and melting of the Greenland ice sheet will lead to a substantial decrease of surface density that can hinder the formation of the North Atlantic deep water masses, the key component of the Atlantic thermohaline circulation.

In a number of numerical experiments with coupled climate models, it was shown that continuous growth of atmospheric CO_2 concentration will lead to a weakening, and, in some models, to a complete shutdown of the Atlantic thermohaline circulation. Some models, however, show only a very modest reduction of the Atlantic thermohaline circulation during the twenty-first century and do not show a complete shutdown even under a very high CO_2 concentration. The models also disagree concerning the relative role of temperature and salinity changes for the thermohaline circulation change. When assessing the results obtained with different climate models, it is important to realize that ocean models are still relatively coarse resolution, and observational data provide no constrain on sensitivity of the oceanic circulation to temperature or salinity changes, because these changes are still relatively small to be detected with confidence. Another important uncertainty in the prediction of the future of the Atlantic thermohaline circulation is related to the changes in the mass balance of the Greenland ice sheet. Model experiments indicate that under global warming conditions, an increased melting of the ice sheet will overwhelm an increase in precipitation, which implies that Greenland may become an important additional freshwater source for the North Atlantic. The latter will additionally contribute to a freshening of the area where deep water masses are formed and to a slowdown of the thermohaline circulation. However, it is still unclear whether melting of Greenland will be fast enough to cause a complete shutdown of the Atlantic thermohaline circulation.

In spite of all these uncertainties, a general consensus is that in the course of the twenty-first century, the Atlantic thermohaline circulation will weaken, but it is unlikely that abrupt (on the timescale of several years or decade) shutdown will occur. However, if the concentration of greenhouse gases will continue to rise beyond the twenty-first century, a complete shutdown of the Atalntic thermohaline circulation will become more likely. It is important to note that although weakening of the Atlantic thermohaline circulation under global warming is a common feature of many climate models, even a complete shutdown of the thermohaline circulation does not imply immediate cooling or, moreover, entering of a new ice age. Modeling results suggest that greenhouse warming will overwhelm the effect of reduced oceanic heat transport and the warming in the North Atlantic will continue even in the case of substantially reduced thermohaline circulation. This warming, however, is expected to be smaller than in other regions of the planet. At the same time, it is possible that if the thermohaline circulation weakens considerably, it will take centuries for its complete recovering.

Impact of Ocean Circulation Changes on Biosphere

Impact of the Atlantic Circulation Change on Terrestrial Ecosystems

As discussed above, changes in the ocean circulation affect temperature and precipitation worldwide. This has a direct impact on terrestrial ecosystems for which these two climate factors exert primary control on distribution of terrestrial vegetation and their productivity. Paleoclimate records reveal strong and rapid reorganizations of terrestrial ecosystems in response to climate changes. For example, **Figure 3** shows a pollen record from southern Italy during the last glacial cycle. The record reveals numerous excursions apparently coeval with the changes recorded in Greenland, with an abrupt increase of woody pollen during Greenland warm events and its almost complete disappearance of during cold periods. Detailed analysis of different pollen species in this record indicates numerous transitions between forest and cold steep ecosystems during the glacial age. Similar changes in methane concentration shown in **Figure 3** indicate abrupt and dramatic changes in the area and climate conditions of tropical and boreal wetlands synchronous with abrupt climate changes recorded in Greenland.

A strong response of terrestrial ecosystems to the reorganizations of the ocean circulation is also supported by results of modeling experiments performed both for present-day and glacial climate conditions. It was shown that a complete shutdown of the Atlantic thermohaline circulation would cause a pronounce impact on the distribution of ecosystems and their net primary production in different parts of the Earth. In the boreal latitudes of the Northern Hemisphere, the primary effect of change in the ocean circulation is a strong winter cooling and a reduction of the length of the growing season which results in a southward retreat of boreal forest area and shrinking of the area of temperate forest. In more southern locations, where the total amount of precipitation is the primary limiting factor, a southward shift of ITCZ and a weakening of summer monsoon lead to an expansion of Sahara desert, shrinking of the area of evergreen forest, and pronounced decline in productivity in certain areas. It was shown that if collapse of the Atlantic thermohaline circulation will occur under present-day climate conditions, it will result in substantial reduction of natural ecosystem biomass and net primary production. It would also have a dramatic impact on agricultural production, especially in the Western Europe and in some tropical areas.

Under the glacial climate conditions, in spite of the fact that a large portion of the most sensitive boreal zone was covered by the ice sheets, changes in the Atlantic thermohaline circulation still cause a considerable impact on ecosystem, their biomass, as well as soil carbon storage. Modeling results indicate that a reduction of terrestrial biomass might contribute to CO_2 rises observed during Heinrich events, when the coldest climate conditions prevailed over the North Atlantic due to the collapse of the Atlantic thermohaline circulations.

Ocean Currents and Marine Biota

The ocean circulation plays an important role in the ocean carbon cycle and affects marine biota. The most straightforward way of this influence is via changes in the horizontal and vertical transport of nutrients, which are the limiting factor for marine biota productivity over most of the globe. The highest productivity in the ocean apart from shelf areas is observed in the regions of strong vertical upwelling and the areas of deep convection. Both these mechanisms bring nutrients from the ocean interior to the surface layer, which is usually extremely depleted with nutrients. While strong upwelling in coastal and equatorial regions is caused primarily by surface wind, the areas of deep mixing and upwelling in the ocean interior are closely related to the thermohaline ocean circulation. This is why, changes in the ocean circulation directly, via changes in upwelling and mixing, and indirectly, via changes in surface wind, can affect marine ecosystems.

Paleoclimate data revealed strong variations in productivity in many locations, primarily in the North Atlantic, but also in Arabian Sea and coastal Pacific Ocean areas, apparently synchronous with abrupt climate changes recorded in Greenland and attributed to the reorganizations of the ocean circulation. At the same time, paleoclimate data show an enhanced productivity in Iberian and North African coastal areas during cold events associated with enhanced coastal upwelling in these regions. Model simulations confirm that shutdown of the Atlantic thermohaline circulation can cause a strong and widespread decline of the marine ecosystem productivity. In particular, in the northern part of the North Atlantic, the collapse of the Atlantic thermohaline circulation leads to a decrease of planktonic biomass by factor of 2 while the globally averaged export production decreases by 20%. The main cause for such strong reduction of the North Atlantic plankton stock is a shoaling of wintertime mixed layer which reduces supply of nutrients from relatively nutrient-rich intermediate water masses to the nutrient-depleted surface ocean layer. For the rest of the globe, changes in marine biota production are caused by changes in upwelling and surface winds. Sea ice extension and surface cooling can also affect ocean productivity in the high latitudes. Modeling experiments also show that the response time of the marine ecosystem to changes in the Atlantic circulation is the shortest in the North Atlantic, while in the Pacific and Indian oceans the response time is order of centuries.

Even though a probability of a complete shutdown of the Atlantic thermohaline circulation in the future remains uncertain, a weakening of this circulation and a shoaling of mixed layer due to surface warming and freshening is the robust feature of majority of climate model simulations. This will inevitably lead to a decline of North Alantic plankton stock, which, in turn, will have a serious consequence for the fishery in these, currently relatively productive areas and, eventually, affects the food supply to the growing population of the planet.

Summary

The ocean currents play a fundamental role in the climate system and in a number of ways affect terrestrial and marine ecosystems and global carbon cycle. Numerous paleoclimate data indicate that abrupt climate changes observed in the past were associated with changes in the ocean circulation, primarily the Atlantic thermohaline circulations. Modeling studies demonstrate that, at least on a regional scale, this effect is very important and this is supported by numerous paleoclimate records. There is a concern that anthropogenic global warming may cause a substantial reorganization of the ocean circulation, which will not only negatively affect natural ecosystem, but may cause a considerable impact on agriculture, fishery, and cause other negative socioeconomic consequences.

See also: Climate Change 1: Short-Term Dynamics; Climate Change 2: Long-Term Dynamics; Climate Change 3: History and Current State.

Further Reading

Blunier T and Brook EJ (2001) Timing of millennial-scale climate change in Antarctica and Greenland during the last glacial period. *Science* 291: 109–112.

Clark PU, Pisias NG, Stocker TF, and Weaver AJ (2002) The role of the thermohaline circulation in abrupt climate change. *Nature* 415: 863–869.

Houghton JT (ed.) (2001) *Climate Change 2001: The Scientific Basis*. Cambridge: Cambridge University Press.

Köhler P, Joss F, Gerber S, and Knutti R (2005) Simulated changes in vegetation distribution, land carbon storage, and atmospheric CO_2 in

response to a collapse of the North Atlantic thermohaline circulation. *Climate Dynamics* 25: 689–708.
Legates DR and Willmott CJ (1990) Mean seasonal and spatial variability in global surface air temperature. *Theoretical and Applied Climatology* 41: 11–21.
Rahmstorf S (2002) Ocean circulation and climate during the past 120 000 years. *Nature* 419: 207–214.
Rahmstorf S and Ganopolski A (1999) Long-term global warming scenarios computed with an efficient coupled climate model. *Climatic Change* 43: 353–367.
Schmittner A (2005) Decline of the marine ecosystem caused by a reduction in the Atlantic overturning circulation *Nature* 434: 628–633.
Watts WA, Allen JRM, and Huntley B (1996) Vegetation history and palaeoclimate of the last glacial period at Lago grande di Monticchio, southern Italy. *Quaternary Science Reviews* 15: 133–153.
Welinga M and Wood RA (2002) Global climatic impact of a collapse of the Atlantic thermohaline circulation. *Climatic Change* 54: 251–267.

Precipitation Pattern

F W Gerstengarbe and P C Werner, Potsdam Institute for Climate Impact Research, Potsdam, Germany

Introduction

In meteorology, the term precipitation means water in either a liquid or solid state which falls from the atmosphere to the Earth.

Water vapor enters the atmosphere through evaporation from the oceans, inland water bodies, and the moisture in plants and soils. Droplets or ice crystals condense on condensation nuclei, giving rise to various cloud types. The droplets or ice crystals can agglomerate, thus becoming so heavy that they fall from the clouds and reach the Earth (falling precipitation in the form of rain or snow, for instance). If condensation or sublimation of the water vapor occurs directly at or near to the surface of the Earth, it is the deposited precipitation such as dew or hoar frost.

Measurements, however, are generally confined to falling precipitation. They are given in units of volume per area ($\mathrm{l\,m^{-2}}$) or as depth of precipitation (mm): $1\,\mathrm{l\,m^{-2}}$ is equal to 1 mm precipitation depth.

The gauge prescribed by the World Meteorological Organization consists of a cylindrical container with an area of $200\,\mathrm{cm^2}$. Additional methods of measurement are the use of radar and satellite technology. There is a relatively high level of error in measuring precipitation. Moreover, the great variability of precipitation causes problems in recording it in time and space.

It is estimated that $495\,000\,\mathrm{km^3}$ of precipitation falls per annum, $385\,000\,\mathrm{km^3}$ over the oceans, and $110\,000\,\mathrm{km^3}$ over land. This amount of precipitation over land corresponds to a water column of about 700 mm, but this is very unevenly distributed. The reasons for this are the atmospheric and oceanic circulation, the distribution of land and water masses, the location of mountainous areas, vegetation, and the differing water vapor content of the atmosphere as a function of temperature. (Since there is no sufficient knowledge about the distribution of precipitation over the oceans, we do not attempt to describe it here.)

Mean Global Precipitation Patterns

The geographical distribution of precipitation is determined by a series of factors. These are given in the following.

Air Temperature

The higher the temperature, more the water vapor the air can contain, from which either a greater or lesser amount of precipitation can form. Absorption capacity increases exponentially with temperature, which is why there is much more precipitation at the equator than near the poles (see **Figure 1**).

Warming of the Lower Air Layers

The greater the angle of incidence of sunlight, the more the underlying ground surface and the air above it can warm up, and more evaporation occurs. Both increase the instability of the atmospheric layering, which in turn stimulates convection (vertical movement) and thus the formation of clouds. This process is clearly visible in the

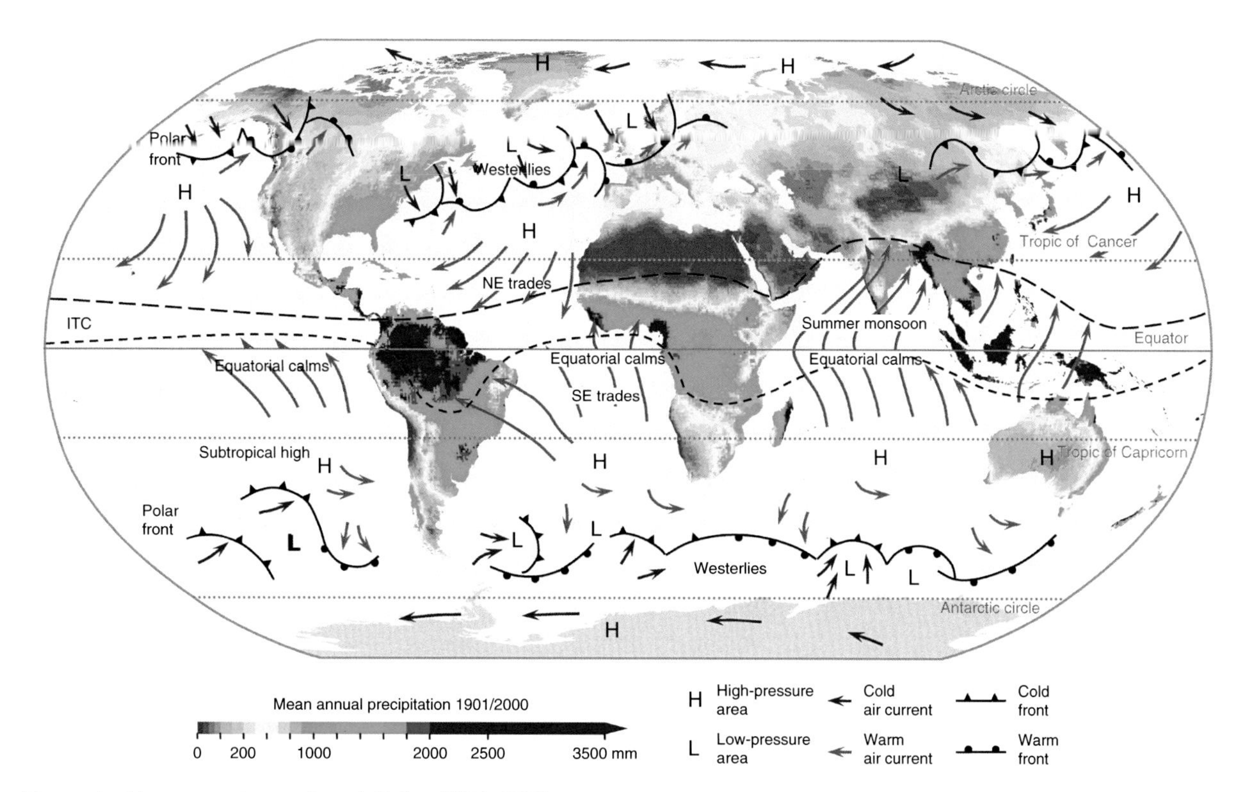

Figure 1 Mean annual sum of precipitation (1901–2000).

shifting of the rainy period over the year in tropical regions in conjunction with the Sun's peak position (see location of areas with greatest precipitation in the tropics, **Figures 2** and **3**).

The Availability of Water for the Evaporation

In general, more precipitation falls at or near the coasts than in the interior of continents; this can be seen clearly in the decrease of precipitation from west to east in Europe as, for example, shown in **Figure 4**.

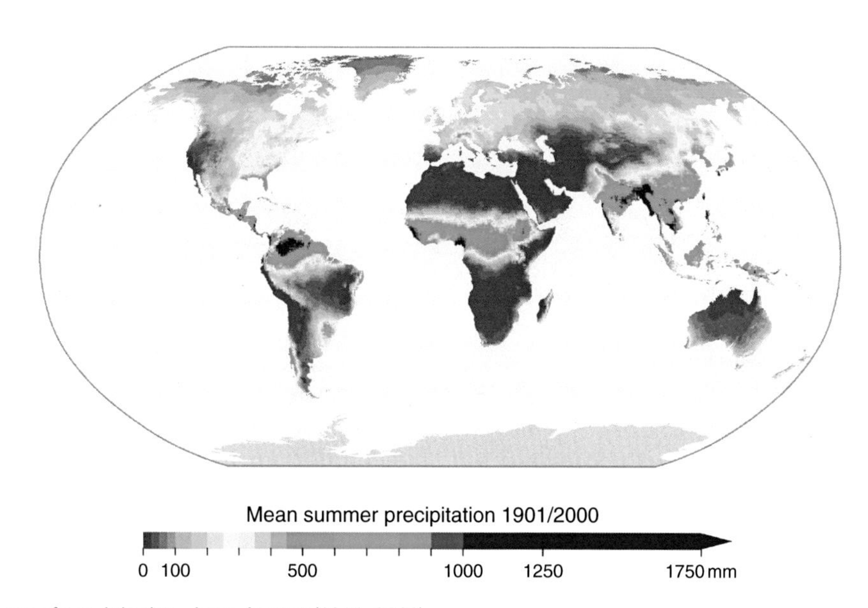

Figure 2 Mean sum of precipitation, June–August (1901–2000).

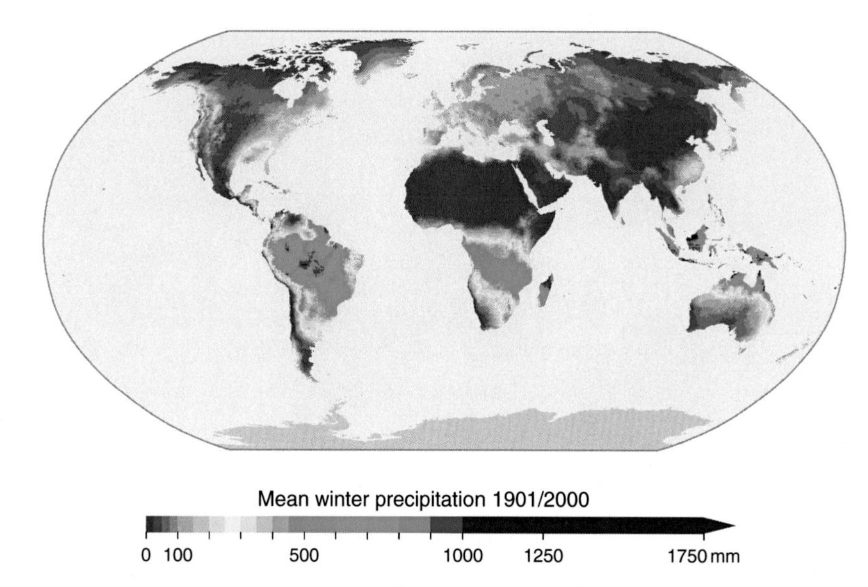

Figure 3 Mean sum of precipitation, December–February (1901–2000).

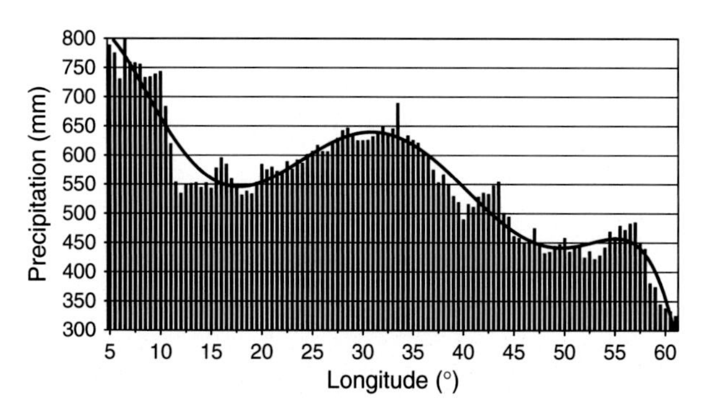

Figure 4 Mean annual sum of precipitation on 53° N from the Netherlands up to the Ural mountains.

Stability or Instability of Atmospheric Layering

In areas of low pressure, where there is instable layering, there is an upward current with corresponding formation of clouds and precipitation. High-pressure areas on the other hand are characterized by sinking air, which causes clouds to disperse. Often high-pressure areas also have an inversion (the normal decrease of temperature with altitude changes to an increase of temperature) at the upper limit of the ground layer (roughly 1.5 km), which again inhibits the formation of clouds. This is a typical phenomenon in the trade-wind zone (**Figure 1**). These two contrasting types of air pressure conditions result in contrasting precipitation sums, relatively high precipitation in the middle latitudes where areas of low pressure are carried by westerly air streams and little or no precipitation in the subtropical high-pressure belt.

Convection and thus cloud formation can be hindered; layering is extremely stable (inversion) due to the ground surface being colder than the air above it. This phenomenon is found particularly at coasts adjacent to cold ocean currents (e.g., Southwest Africa – Benguela current → Namibian desert; Chile – Humboldt current → Atacama desert; West Australian current → dry areas of the Australian interior, which reach to the coast).

Mountain Ranges

On the windward side of mountain ranges air is forced upwards, contributing powerfully to the formation of clouds and precipitation. On the leeward side the air sinks and the clouds dissolve. This can be observed well as a large-scale phenomenon along parts of the west coast of America: west of the Cordilleras there is higher precipitation, to the east lower.

Transport of Masses of More Humid Air from Sea to Land

A typical example of this are the bands of westerly winds in the Northern and Southern Hemispheres (see **Figure 1**) which are responsible for the relatively even distribution of precipitation over the year in the middle latitudes. In the winter it also covers parts of the subtropics (e.g., the Mediterranean), so that most precipitation in these areas falls in winter.

The seasonal variation in transportation of air masses is linked with the position of the Intertropical Convergence Zone (ITCZ) or with the varying warming of land and sea. This process is known as the monsoon. The Indian monsoon is the most distinctive. At the beginning of June the first warm air masses from the southwest cross to the Indian subcontinent. This pattern of currents continues until the

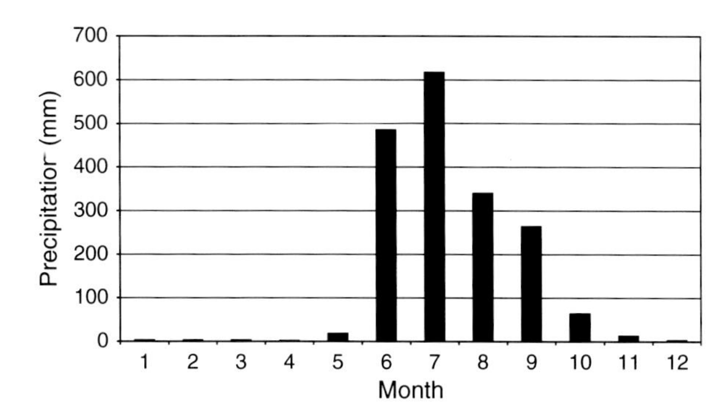

Figure 5 Monthly mean of precipitation, Mumbai, 18°58′ N, 72°50′ E (annual sum = 1815 mm).

beginning of October and is accompanied by intensive rainfall (see also **Figure 2**). In winter the northeast trade winds bring dry air from central Asia and precipitation tends to zero (see also **Figure 3**), as can be seen from the annual course of precipitation for Mumbai (see **Figure 5**). A similar pattern can be observed on the west coast of Africa (Sierra Leone, Liberia) and in Southeast Asia.

Self-Intensification Effects Caused by the Ground Surface

If the ground surface over large areas is waterless, no evaporation can take place and so no formation of precipitation can occur. The clearest example of this is the Sahara. The converse is true for tropical evergreen rainforests, where evaporation is very great and formation of clouds is thus fostered.

Seasonal Variance of Solar Radiation

As the Sun's position changes during the course of the year, the regional heat balance of the atmosphere also changes. This influences the position of pressure centers and large-scale wind patterns, as well as evaporation and the content of water vapor in the atmosphere. In some regions this is the cause of seasonally contrasting precipitation conditions (e.g., monsoon, semihumid tropics, and subtropics).

Where several factors influencing precipitation occur in conjunction in a region, special – sometimes extreme – precipitation conditions can arise (see also **Table 1**).

Equatorial Areas

Radiation and humidity conditions make the atmospheric layering extremely unstable. Huge cumulus clouds which can reach a height of 14–20 km form in the rising air masses. Evaporation is high due to the very wet underlying surface. Rainfall, which occurs almost daily, is spread relatively evenly over the year (but is stronger when the ITCZ, which has lower pressure than the surrounding area, lies over the region). On the large scale, the humid tropics are the areas with the highest precipitation on Earth. The Amazon Basin, parts of equatorial West Africa and some parts of Southeast Asia, are typical representatives of this precipitation regime. An example is given in **Figure 6**, which shows the monthly mean of precipitation for Singapore, which has an annual sum of 2423 mm.

Sea–Land Currents and Mountain Ranges

When very humid air masses coming from the sea meet a mountain range and are forced to rise, very intensive precipitation can occur, often reaching record levels. The best-known case is the precipitation at Cherrapunji in the Khasi mountains in the approaches to the Himalayas. Massive amounts of precipitation fall due to the interplay of the summer monsoon and the effect induced by the mountains. The other precipitation extremes shown in **Table 1** are also a consequence of humid ocean air flowing against mountains. In addition to these examples, one can say that the windward side of mountain ranges near the sea are in general characterized by high levels of precipitation (examples are the Cordilleras (Canada, South Chile), the Scandinavian mountains, the Alps, and the Tsaratanana massif on

Table 1 Extremes of precipitation

Kind of extreme	*Location*	*Value (mm)*
Highest annual sum	Debundscha, Cameroun (Africa)	10 287
	Lloro, Colombia (America)	13 299
	Cherrapunji, India (Asian)	26 461
	Bellender Ker, Queensland (Australia)	8 636
	Crkvice, Croatia (Europe)	4 648
Highest monthly sum	Cherrapunji, India	9 300
Highest daily sum	Cilaos, La Reunion	1 870
Highest annual sum of snow fall	Paradise, Mt. Rainier, Washington, USA	31 100
Highest snow cover	Sierra Nevada, California, USA	11 455

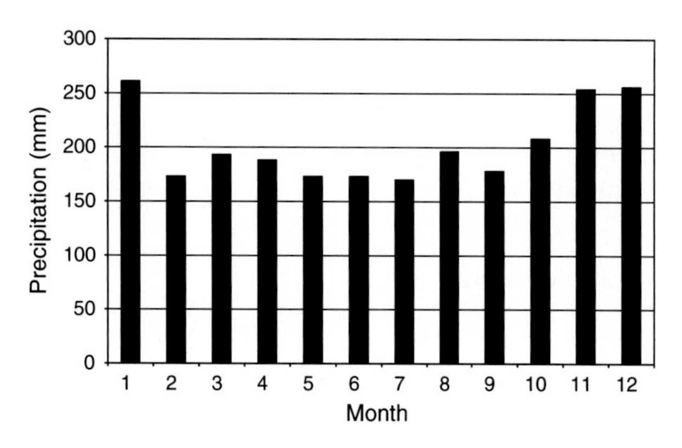

Figure 6 Monthly mean of precipitation, Singapore, 1°17′ N, 103°50′ E (annual sum = 2423 mm).

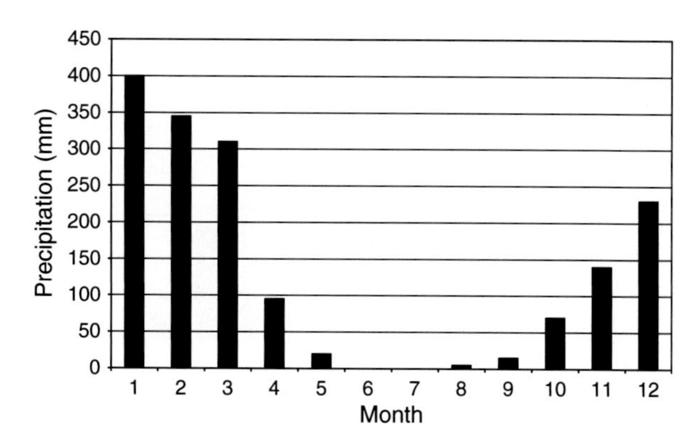

Figure 7 Monthly mean of precipitation, Darwin, 12°28′ S, 130°50′ E (annual sum = 1630 mm).

Madagascar). These regions are mostly not large in extent, but their annual sums are significant (Sitka, Canada: 2454 mm; Bergen, Norway: 1958 mm). The example of Hawaii shows that when humid air masses are forced to rise through meeting a mountain barrier, high values for precipitation can also be produced even in the trade-wind zone which is otherwise comparatively dry due to inversion. The 1569 m high Mt. Waialeale on Hawaii shows a mean annual sum of 11 684 mm.

High Pressure

The interaction of a belt of high pressure on a large land mass whose surface has very little moisture leads to the formation of a desert, such as the Sahara, where there are places which have had no rainfall throughout the whole of the twentieth century.

In addition to the regions with extreme precipitation conditions described here (extreme means that they show the whole extent of precipitation conditions on Earth), other regions with a particular precipitation regime can be identified.

Regions with High Annual Sums of Precipitation

In addition to the humid tropics, these include the semihumid tropics. In these regions there is a dry season and a rainy season according to the Sun's position. The maximum precipitation is directly connected with the point in time of the Sun's highest position (zenith). A typical example for this precipitation pattern is given by the Darwin Station in North Australia (see **Figure 7**).

The humid subtropics also have annual precipitation sums of more than 1300 mm. The southern United States (New Orleans 1373 mm), parts of the Caribbean and Central America, Southeast China (Taipei 2100 mm), and southern Japan (Tokyo 1520 mm) fall under this precipitation regime. The largest area in the Southern Hemisphere is in South America and covers southern Brazil (Sao Paulo 1317 mm) and parts of Paraguay, Argentina, and Uruguay.

Areas with Little or no Precipitation

Very dry areas with an annual sum of precipitation of less than 300 mm occur in different regions of the world. In addition to the typical areas of desert in the subtropics such as the Sahara or Central Australia, such very dry areas are found in the interiors of the continents of Asia and North America. Other such regions are coastal areas adjacent to cold ocean currents, or in the lee of large mountain ranges (see above). Large parts of the polar regions are also among these very dry areas, since due to the low temperatures there is only a small amount of moisture in the atmosphere.

Areas with Moderate Precipitation

Zones of moderate precipitation are spread over all the continents and naturally enough form a boundary between extremely dry and extremely wet regions. However, the precipitation falls in greatly varying amounts over the course of the year. The following basic types can be observed:

- precipitation falls during the whole of the year, but more in summer than in winter; this type is widespread in the moderate climate zone (e.g., Berlin, **Figure 8**);
- precipitation falls during the whole of the year, but more in winter than in summer (e.g., Tromsö, northern Norway, **Figure 9**); and
- hardly any precipitation occurs during summer, countries north of the Mediterranean Sea, North California, central Chile, the southern tip of Africa and Southwestern Australia (e.g., Rome, **Figure 10**).

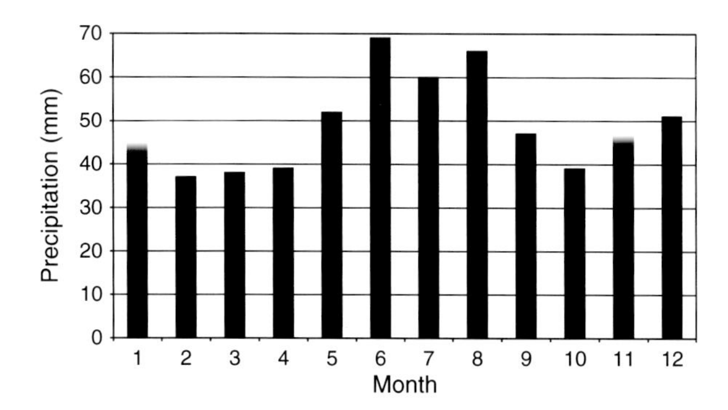

Figure 8 Monthly mean of precipitation, Berlin, 52°28′ N, 13°18′ E (annual sum = 586 mm).

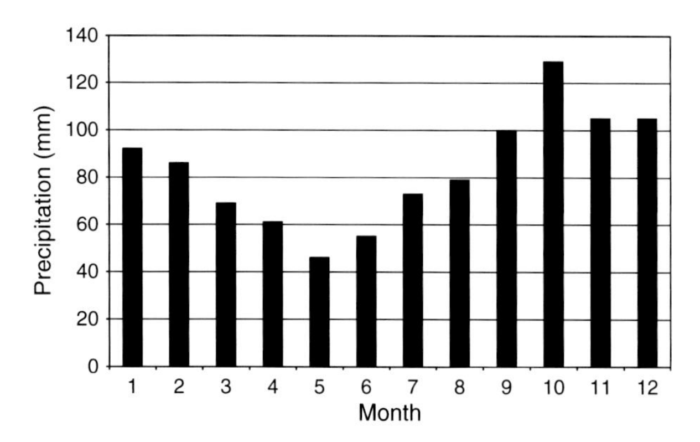

Figure 9 Monthly mean of precipitation, Tromsoe, 69°40′ N, 18°56′ E (annual sum = 1000 mm).

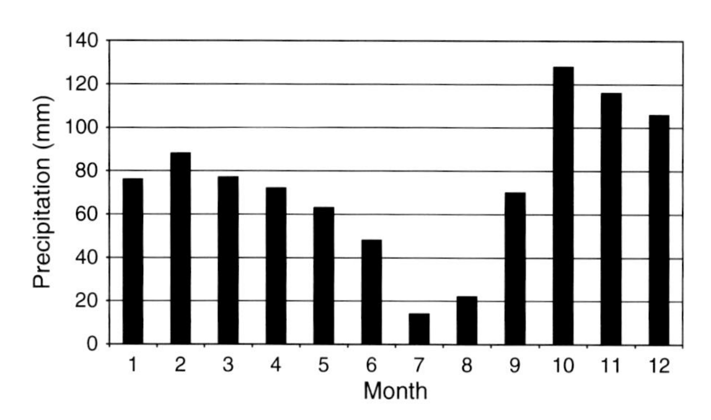

Figure 10 Monthly mean of precipitation, Rome, 41°32′ N, 12°17′ E (annual sum = 880 mm).

Variation of Global Precipitation Patterns

As has been mentioned several times, the position of the precipitation zones is dependent on the season or, in other words, on the Sun's position. Thus, the heavy-rainfall areas in the tropics shift significantly to the north in July–August (**Figure 2**). At the same time the dry areas in the subtropics also extend toward the north, so that they cover, for example, the Mediterranean region.

In the period from December to February (**Figure 3**), considerably less precipitation occurs in the Northern Hemisphere than in the period July–August. This is due to the shift of the heavy-rainfall areas to the Southern Hemisphere, caused again by the Sun's position. Exceptions to this general rule are, for example, the western coast of the United States and Canada and the Mediterranean region, which have more precipitation due to the intensified westerly wind circulation.

Trends of Global Precipitation

During the twentieth century, the atmosphere has warmed up by around 0.6 K in global mean. This has increased the water vapor content of the atmosphere and altered general circulation patterns and the distribution of pressure systems. This is confirmed in **Figure 11**. The blue areas represent regions with increased precipitation, whereas the red areas are those with decreased precipitation. In the gray-colored areas no secure precipitation trend has been proved. The overall area with increased precipitation is, as might be expected, greater than that with decreased precipitation. Regions with a mainly positive precipitation trend are, for example, large parts of Canada, Western Europe, Scandinavia, the northeastern part of Europe, eastern Siberia, a large part of South America, eastern Central Africa, and Northwest Australia. A marked negative trend is present in the African savannas north of the equator. This development with its varying regional effects is first and foremost the result of altered pressure conditions and currents. Moreover, there are also areas where both positive and negative trends take place within a relatively small area, such as in the Mid-West region of the USA.

Figure 12 shows an example of precipitation increase in Scandinavia. One can see that in Goetheburg, precipitation has increased on average by around 250 mm in recent decades. This represents an increase of about 135% on the previous mean figure.

The Antarctic

The Antarctic plays a particular role in the global water budget because it is the biggest reservoir of freshwater on Earth. Because of the climatic situation there are only very few measurement stations on the continent, and observations from these, moreover, can be made only very infrequently. These difficult conditions make the measurements relatively imprecise. For this reason, **Figure 13** shows monthly mean values for two points on the continent which were generated through modeling. At

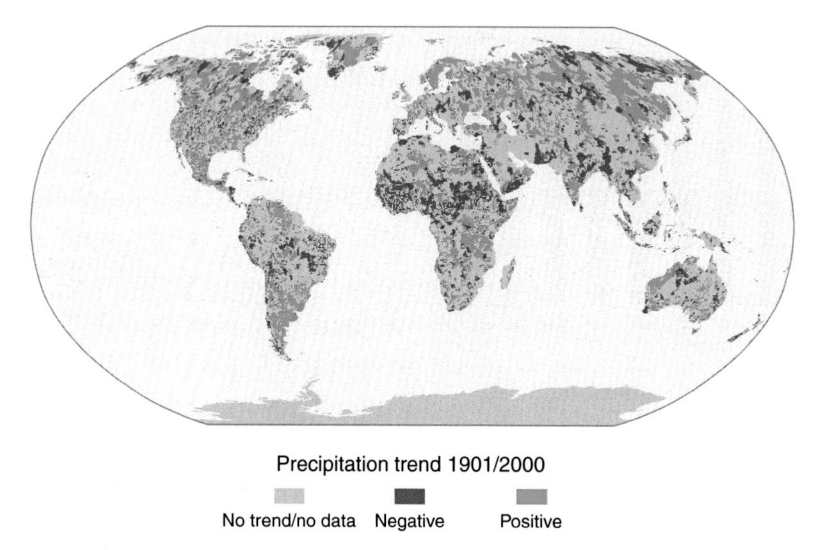

Figure 11 Trend of the annual sum of precipitation, 1901–2000.

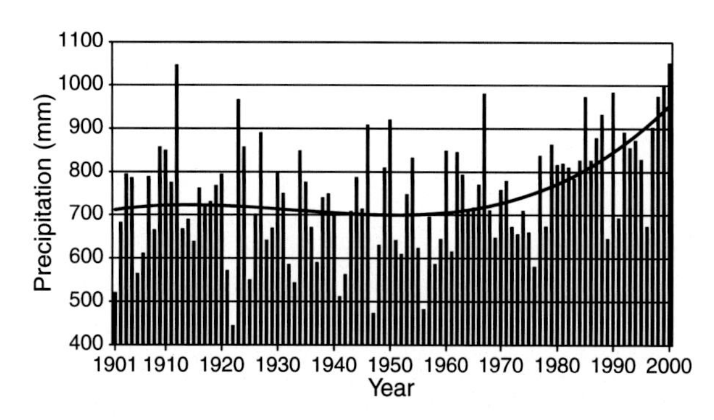

Figure 12 Annual sums of precipitation, Goeteborg, Sweden.

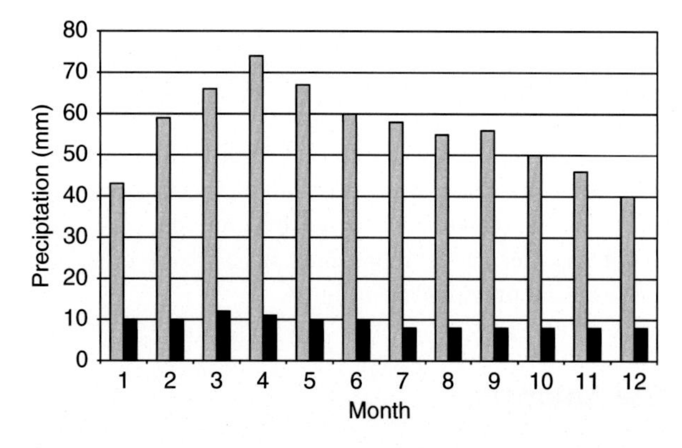

Figure 13 Monthly mean of precipitation, Antarctic, west coast (annual sum = 674 mm), South Pole (annual sum = 111 mm).

the South Pole itself the annual precipitation is very low, at 111 mm, and hardly varies from month to month. On the west coast of the continent the annual sum increases to 674 mm and shows a distinct seasonality with a maximum in April and a minimum in December.

Particular Precipitation Phenomena

Besides the precipitation patterns described above, there are events which are caused by particular phenomena.

El Niño

El Niño is a climate anomaly which occurs in the Pacific between the west coast of South America and Southeast Asia (Indonesia, Australia) at intervals of 2–7 years. During an El Niño event, the prevailing trade winds weaken and the equatorial countercurrents strengthen. This causes warm surface waters from the Indonesian region to flow eastward, displacing the cold water of the Humboldt Current, which in turn leads to disruption of the temperature and moisture regime. Large amounts of precipitation (up to 4000 mm in a month) occur in the coastal deserts of Peru and northern Chile. More rainfall than normal also falls on the islands east of Indonesia, while Indonesia itself and Northern Australia suffer drought. In some cases, further-ranging effects of El Niño have been observed: increased precipitation in southern Argentina, the southern states of the USA and northern Mexico, decreased precipitation in southeast Africa, Korea, southern Japan, and northeast Brazil.

Tropical Cyclones

These storms (e.g., hurricane, typhoon, and cyclone) get their energy from the process of evaporation over the warm seas north or south of 5° of latitude and the accompanying condensation in the atmosphere. Typical is the formation of huge clouds, from which massive amounts of rain fall (up to several 100 mm in a few hours), frequently causing flooding.

Data Basis

The description of the precipitation regime is based on data from chosen meteorological stations. In addition, a grid database with a resolution of 0.5° × 0.5° for the period 1901–2000 was used, which was made available by the Potsdam Institute for Climate Impact Research. The values for the Antarctic were taken from a model calculation of IPCC SRESA1B for the period 1980–99.

See also: Temperature Patterns; Water Cycle.

Further Reading

Hastenrath S (1991) *Climate Dynamics of the Tropics*. Dordrecht: Kluwer Academic Publishers.
Linacre E and Geerts B (1997) *Climates & Weather Explained*. London: Routledge.
Rudloff W (1981) *World Climates*. Stuttgart: Wiss. Verlagsges.
Wang B (2006) *The Asian Monsoon*. Berlin: Springer.

Temperature Patterns

I I Mokhov and A V Eliseev, AM Obukhov Institute of Atmospheric Physics RAS, Moscow, Russia

Introduction

Temperature is a basic climate variable. Being subjected to wide spectrum of variability, temperature exhibits changes at scales ranging from short-term subsynoptic (diurnal or shorter-term) through seasonal to interannual, interdecadal, and longer-term variability. Some of this variability is forced by external sources, for example, by diurnal and annual cycles of insolation, changes in Earth orbital parameters (Milankovitch cycles), volcanism, and anthropogenic influence. Other temperature variations (e.g., related to El Niño/Southern Oscillation (ENSO), North Atlantic/Arctic Oscillation) are manifestations of the internal variability of the climate system.

Meteorological Temperature Measurements

Meteorological measurements cover only the last few centuries. The longest instrumental temperature records exist for central England (monthly means are available since 1659), Berlin, Germany (available since 1701), and de Bilt, Netherlands (available since 1706). Instrumental meteorological data became more or less common to the mid-nineteenth century. Since that time, the number of meteorological stations has increased, especially in Europe and North America. However, the coverage of the meteorological stations was still sparse until the International Geophysical Year (1957–58). In this year, numerous meteorological stations were established. It was followed by the First Global GARP (Global Atmosphere Research Program) Experiment in 1978–79. In addition, rawinsondes and rocket measurements were organized to probe the upper layers of the troposphere, stratosphere, and mesosphere. Over the oceans, these measurements are supported by the sea-surface temperature (SST) measurements from commercial ships. The latter is the reason why it is common to represent global fields of temperature in a blended format employing surface-air temperature (SAT, measured at a 2 m height over the surface) over land, and SST over the oceans. Currently, an estimated uncertainty for global and annual averages of the near-surface temperature is equal to 0.05 °C.

A very important period in the climatic measurements started in the 1970s when meteorological satellites with globally covered data became common. However, despite the progress in understanding of the physical laws governing the Earth's climate, the satellite-derived data still suffer from the uncertainties related to the retrieval of the typical climate characteristics (e.g., SAT).

In the last decade, an additional source of temperature data (so-called 'reanalysis data') has emerged. It is produced by the weather forecast models forced by the instrumental meteorological measurements and satellite retrievals employing respective data assimilation. The reanalysis data sets are free from nonhomogeneity induced by the changes in assimilation scheme and dynamically consistent with other meteorological fields. They are still not free from nonhomogeneity induced by the changes in the observing system (e.g., establishment or closure of particular meteorological stations, shifts of their locations, and inclusion of satellite observations in the last few decades). In addition, one notes that, in reanalysis, there is a strong difference between SAT and temperature at different heights in the atmosphere. If the latter is strongly influenced by the assimilated meteorological data, the former is a pure model output with a corresponding influence by the model's physical packages, dynamic core, etc. Currently, two basic reanalysis data sets are used: the reanalysis provided by the National Center for Environmental Prediction and the reanalysis prepared by the European Centre for Medium-Range Weather Forecasts.

According to the definition by the World Meteorological Organization, climatological (i.e., representative of the typical climate conditions) temperature fields are determined for the period 1961–90.

Global Patterns of the SAT: Climatology

Annual Mean Temperature

The pattern of the annually averaged blended SAT/SST is shown in **Figure 1**. Its zonal near-symmetry is an indication of the differential heating of the Earth by the Sun; the planet receives more solar energy at lower than at higher latitudes. Temperature changes from high values (above 25 °C) in the tropical area to the negative values −15 to −20 °C in the northern polar latitudes, and to the lower values −40 to −50 °C in the southern polar area. This large difference between the Arctic and the Antarctic is due to the huge ice sheet covering the Antarctic continent with a thickness about 3 km. The blended SAT/SST value for 1961–90, averaged over the globe, is estimated to be 14.0 °C (14.6 and 13.4 °C in the Northern and Southern Hemispheres, respectively).

Regionally, temperature is determined by orography, atmospheric centers of action, urbanization, etc. The lowest annual mean temperatures (*c.* −60 °C) are reached in the Antarctic interior. The highest annual mean temperature is observed over the Sahara in the northern Africa (*c.* 30 °C), and over the Indian–Pacific warm pool (*c.* 26–28 °C) located in the tropical Indian and western Pacific oceans.

Annual Cycle of the Global Temperature

The dominant mode of the climatic variability is an annual cycle.

Averaged over the whole Northern Hemisphere, the amplitude of the annual harmonics $T_{s,1}$ (**Figure 2a**) amounts to about 7 °C (~11–13 °C over land and 4 °C over ocean). In the Southern Hemisphere, its mean value is 3–4 °C (~7 °C over land and 2–3 °C over ocean). Regionally, it is quite small over the oceans and over the low-latitude land (with typical values of a few centigrades). Over the extratropical land, it is larger (commonly ~10–20 °C). The maxima of the amplitude of this harmonics are observed over Siberia (~35 °C) and North America (22 °C).

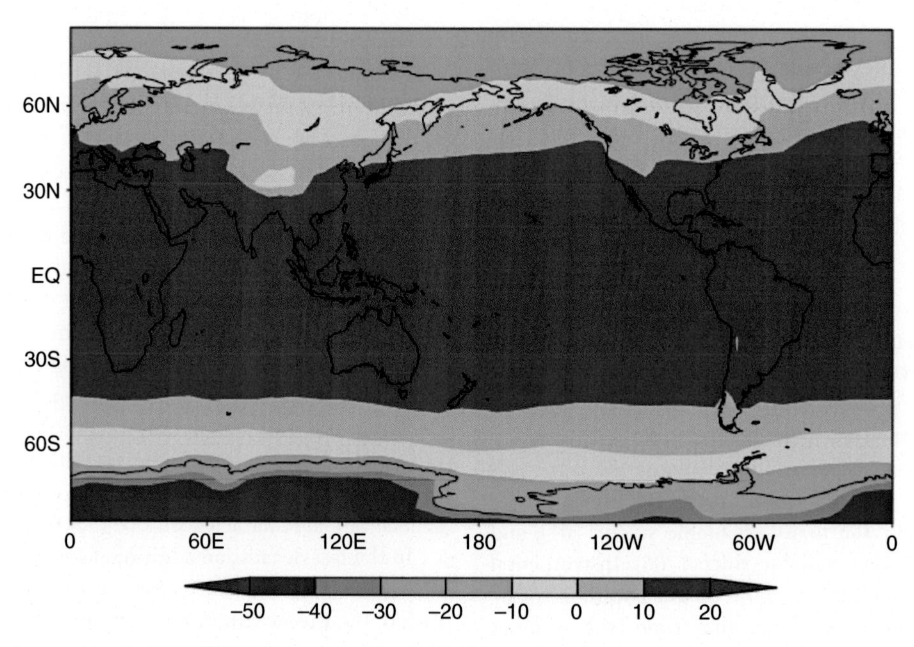

Figure 1 Annual mean blended SAT/SST (°C) averaged for 1961–90 based on the 5° × 5° data prepared by Climate Research Unit of the University of East Anglia (UEA CRU, Norwich, UK).

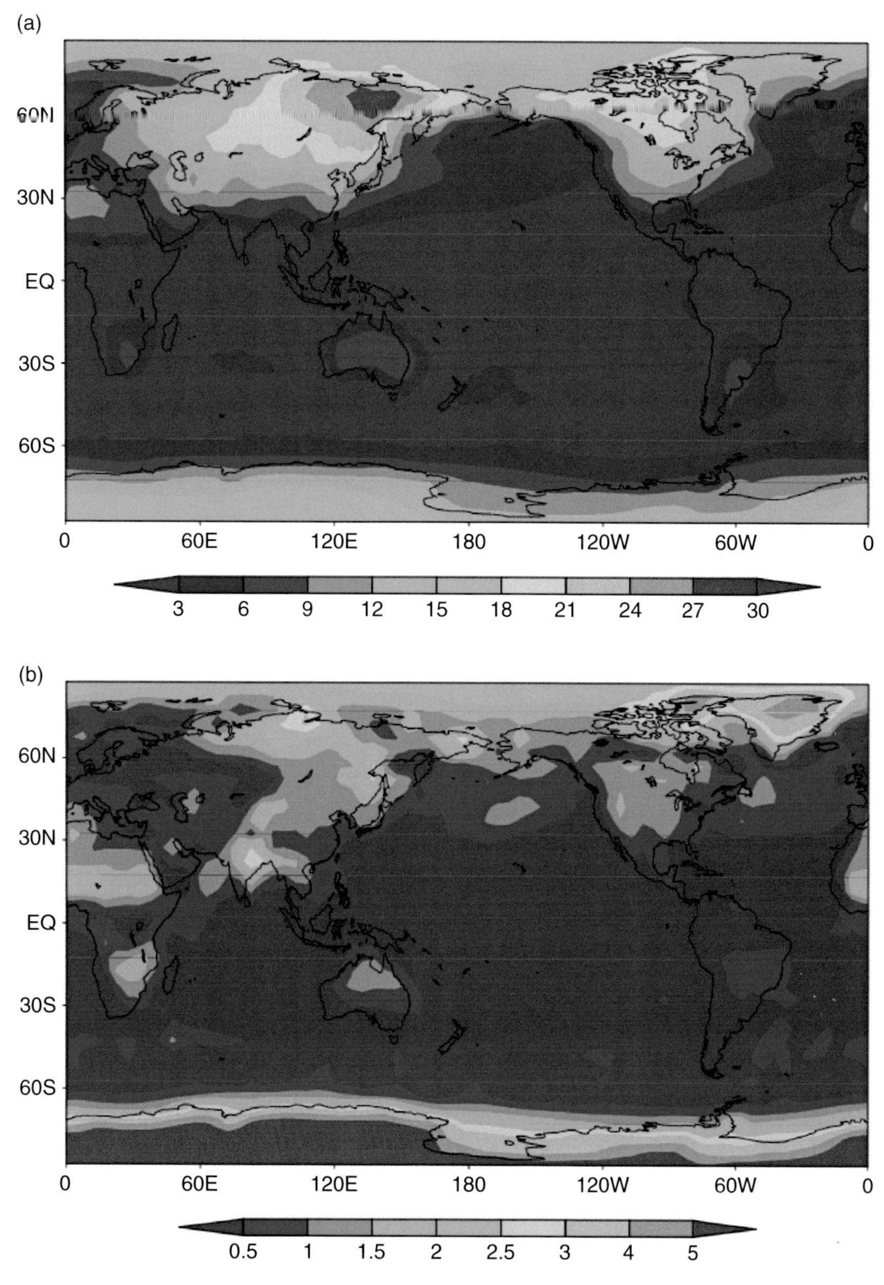

Figure 2 Amplitudes of the (a) annual and (b) semiannual harmonics (°C) of the blended SAT/SST temperatures computed based on the UEA CRU gridded data set for 1961–90.

Amplitude of the semiannual harmonics $T_{s,2}$ (**Figure 2b**) is generally smaller than its annual counterpart. Extremes of the semiannual harmonics amplitude occur in spring and in autumn when the annual harmonics is near nodal points. The maxima of $T_{s,2}$ are observed near the poles with the largest values above 4 °C over the Greenland and the Antarctica. Another maximum of the amplitude of semiannual harmonics is in the tropics (up to 2 °C). Both polar and tropical maxima of $T_{s,2}$ are related to the respective latitudinal extremes of the semiannual harmonics of insolation. An additional maximum of $T_{s,2}$ is over the continents in mid-latitudes. This maximum results an amplification of the respective insolation harmonics due to snow albedo–temperature feedback.

Generally, annual cycle of temperature over the oceans lags in comparison to that over the land by about a month. This reflects higher thermal inertia of oceanic water than inertia of the seasonally active layer of the land.

Global Patterns of the SAT: Variability and Trends

Subseasonal Variability and Temperature Extremes

The coldest temperature ever recorded instrumentally at the Earth's surface, −89.4 °C, was registered at the station Vostok (Antarctica) in 1983. In the Northern Hemisphere, the temperature minimum at −67.8 °C was recorded in 1892 in Verkhoyansk (Siberia). The highest surface air temperature, +57.7 °C, was measured in Al'Aziziyah (Libya) in 1922.

In summer, extreme warmth lasting at least two consecutive days is called as 'heat wave'. One of the strongest heat waves that has occurred in 2003 in Europe has caused numerous human deaths and vast ecological disasters. Cold spells may occur both in summer and in winter. A marked cold spell occurred in the Eastern Europe in winter 2006.

In autumn, a period of unusually warm and sunny weather lasting about 1–2 weeks (or longer) is poetically named either the 'Indian summer', or 'Old Wives' summer', and 'Tiger season'. At the higher latitudes, Indian summer starts in early autumn, and at the lower latitudes, the climatological dates of its onset are in late autumn or even in early winter. Interannual variability may result in the shift of the dates of the Indian summer onset, duration, and withdrawal. Multiple Indian summers are also observed.

In spring, unseasonal cold weather is termed 'cold snap' (or 'blackberry snap', and 'dog winter'). If sufficiently strong, this cold snap may damage vegetation in its early photosynthetic phase. Typical duration of spring cold snap is about a week.

Interannual Variability

Standard deviation σ of the detrended (linear trend is removed) anomalies of the blended SAT/SST anomalies from the respective long-term mean annual cycle, on a global basis, is about 0.1–0.2 °C. Regionally, it is relatively small (<0.5 °C) over the tropical and subtropical oceans. However, there is an important exception from this rule. In the eastern equatorial Pacific, there is a tongue with higher values of σ (slightly above 1 °C). This tongue is a manifestation of the ENSO process. This process is an irregular oscillation with periods from 2 to 7 years exhibiting alternating warm (El Niños) and cold (La Niñas) events in this basin. ENSO is the strongest mode of the global SAT interannual variability. This process has large impacts not only in the tropics (during El Niño events, fishery near the coast of Peru collapses, and agriculture in the tropical America suffers from diminished productivity), but also in the remote regions of the world. In particular, in winters, during these warm events, surface temperature is above normal in the Canadian-Greenland region, and below normal over the central north Pacific, central north Atlantics, and over the United States. These warm and cold events are phase-locked to the seasonal cycle with largest anomalies tied to the northern winter. The most intensive (in terms of anomalies from a fixed climatological state) El Niños occurred in 1982–83 and 1997–98.

Over the continents, interannual temperature variability is stronger than over the oceans. Over the subtropical land, the corresponding standard deviation is above 1 °C. In the mid-latitudes, it amounts to 3 °C or larger. Interannual variability of SAT is at its maximum over the northern subpolar land (with maxima about 4 °C) near land snow boundary.

Decadal-Scale Variability

A prominent mode of decadal-scale variability is the North Atlantic Oscillation which is manifested through coherent variability in main quasi-stationary Atlantic centers of action: Icelandic Low and Azores High. In winter, corresponding changes in zonal winds lead to either enhanced or suppressed advection of warm moist oceanic air to Europe with respective regional temperature changes. The advection is stronger (weaker) than normal, the European winters are milder (more severe) than usual. At their peaks, associated monthly mean SAT anomalies over Europe reach 2 °C. The North Atlantic Oscillation corresponds to the zonal symmetric mode of the climate variability while the Arctic Oscillation does to the coherent variations between the subtropics and the higher latitudes.

In the Pacific, in addition to ENSO, the Pacific Decadal Oscillation (PDO) is observed. In a warm PDO phase, positive temperature anomalies about 0.4 °C develop in the central equatorial Pacific, and negative anomalies about −0.6 °C develop in the northern Pacific. Smaller anomalies (about −0.2 °C) occur in the southern Pacific. In this phase, fishery conditions are favored in the Pacific high latitudes, and suppressed in the subtropical Pacific. In the cold phase, the associated anomalies are reversed in sign with a reversed impact on fishery. The mean period of PDO is about 20–30 years. Its cold phase has persisted since the middle of the twentieth century to the late 1970s, and its warm phase persists since that latter time (while the short-termed excursion into negative phase has been observed in mid-1990s).

Trends

The twentieth century was unprecedented in terms of change in the surface's air temperature. On a global basis, linear trend of SAT was about 0.6 °C per 100 years. In the last decades, the change in temperatures has become even more prominent.

Figure 3a presents annual temperature changes explained by linear trends in 1951–2000 (based on the

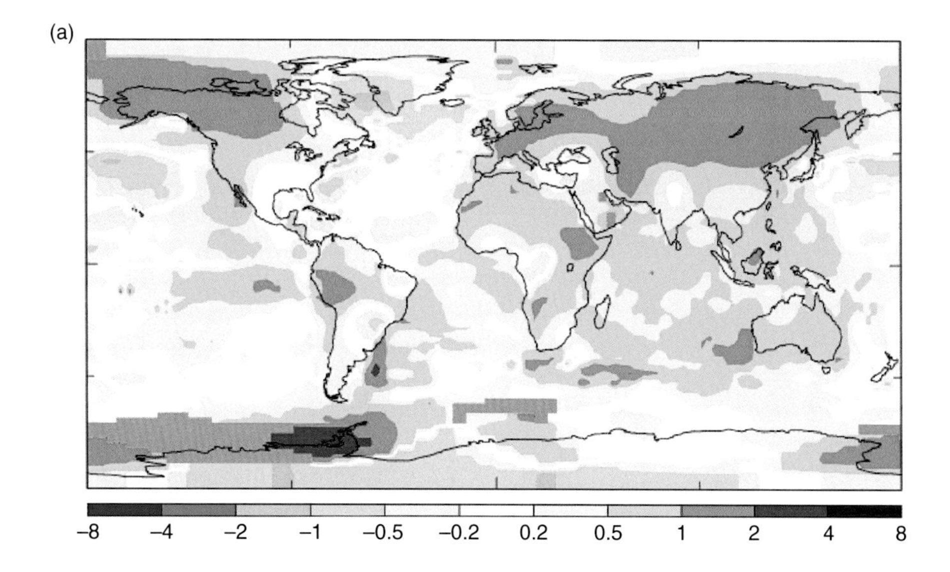

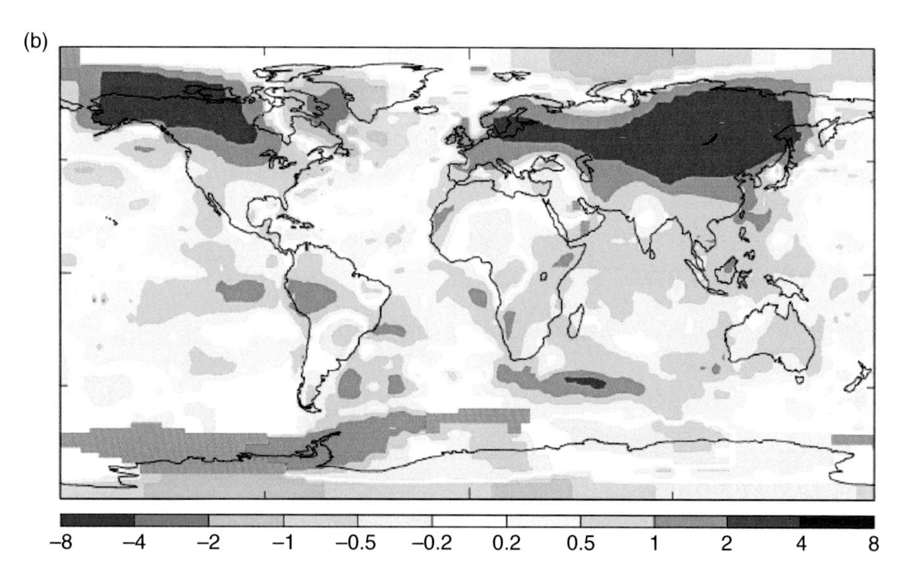

Figure 3 (a) Changes of annual SAT explained by linear trends (°C) in 1951–2000 computed on the base of gridded analysis of the Goddard Space Flight Center (GSFC, National Aeronautics and Space Administration, USA). (b) The same as (a) but for the Northern Hemisphere winter (December–February). In both plots, gray shades correspond to the areas of missing data. The pictures were downloaded from http://data.giss.nasa.gov/gistemp/.

Goddard Space Flight Center gridded data set). Positive trends of SAT are observed over the most land and oceanic areas. The largest rates of linear trends are over the subpolar and mid-latitudinal land (above 1 °C per 50 years). However, negative trends are exhibited over the vast oceanic areas (north Pacific, northernmost Atlantic, and near the Antarctic). During the last decades, the warming has accelerated.

Generally, temperature trends are more prominent during the cold part of the year. This is shown in **Figure 3b** for the change in temperatures of the Northern Hemisphere winters (December–February) during 1951–2000. In this season, the areas of the most pronounced warming over the northern extratropical continents exhibit changes, which are larger by a factor of 1.5 than their annual mean counterparts. This corresponds to the decrease of the temperature annual cycle amplitude in recent decades.

Summary

The annual mean field of the SAT is basically zonal indicating differential solar heating of the Earth's climate. By being averaged globally, the SAT for 1961–90 is estimated to be 14.0 °C (14.6 and 13.4 °C in the Northern and Southern Hemispheres, respectively).

A dominant mode of temperature variations is an annual cycle. It is more prominent over the high-latitude land. In the Northern Hemisphere, the range of annual cycle is about 14 °C (approximately 25 °C over the land and 8 °C over the oceans). In the Southern Hemisphere, its mean value is about 8 °C (approximately 15 °C over the land and 5 °C over the oceans).

Interannual and interdecadal variability is manifested via dominant modes, such as ENSO, North Atlantic/Arctic Oscillation, or PDO. Standard deviation of the detrended anomalies of the globally averaged temperature is about 0.1–0.2 °C

On a global basis, linear trend explains about 0.6 °C of the rise in SAT in the twentieth century. The corresponding temperature changes are more prominent in the cold part of the year.

See also: Climate Change 3: History and Current State.

Further Reading

Anderson DLT and Willebrand J (eds.) (1996) *Decadal Climate Variability: Dynamics and Predictability (NATO ASI Series/Global Environmental Change)*. Berlin: Springer.

Budyko MI and Izrael YA (eds.) (1991) *Anthropogenic Climate Change*. Tucson, AZ: Arizona University Press.

Eliseev AV, Mokhov II, Rubinshtein KG, and Guseva MS (2004) Atmospheric and coupled model intercomparison in terms of amplitude–phase characteristics of surface air temperature annual cycle. *Advances in Atmospheric Sciences* 21(6): 837–847.

Hansen J, Ruedy R, Sato M, *et al.* (2001) A closer look at United States and global surface temperature change. *Journal of Geophysical Research* 106(D20): 23947–23963.

Houghton JT, Ding Y, Griggs DJ, *et al.* (eds.) (2001) *Climate Change 2001: The Scientific Basis Contribution of Working Group I to the Third Assessment Report of the Intergovernmental Panel on Climate Change (IPCC)* Cambridge: Cambridge University Press.

Jones PD, New M, Parker DE, Martin S, and Rigor IG (1999) Surface air temperature and its changes over the past 150 years. *Reviews of Geophysics* 37(2): 173–199.

Mokhov, II (1993) *Diagnostics of Climate System Structure (in Russian)*. St.Petersburg: Gidrometeoizdat.

Peixoto JC and Oort AH (1992) *Physics of Climate*. New York: American Institute of Physics.

Philander SG (1990) *El Niño, La Nina and the Southern Oscillation*. San Diego: Academic Press.

Simmons AJ, Jones PD, da Costa Bechtold V, *et al.* (2004) Comparison of trends and low-frequency variability in CRU, ERA-40, and NCEP/NCAR analyses of surface air temperature. *Journal of Geophysical Research* 109(D24): D24115 (doi: 10.1029/2004JD005306).

Urbanization as a Global Process

A Svirejeva-Hopkins, Potsdam Institute for Climate Impact Research, Potsdam, Germany

Introduction and Definitions

The most concise definition of urbanization is given by *Encyclopaedia Britannica*: "this is the process by which large numbers of people become permanently concentrated in relatively small areas, forming cities and their suburbs." Population, P, and territory, S, of these formations are changing in time, so that urbanization is a dynamic process controlled by these main variables. The degree of people concentration is determined by the 'population density', $D=P/S$, where P is population size and S is the area it occupies.

Different national statistics operate with different definitions and meanings of the terms 'urban', 'urban area', or 'urbanized territory'. For instance, the United Nations defines all places with more than 20 000 inhabitants living close together as urban, while the US Census Bureau uses 'urban area' as a densely populated area (built-up area) with $D>1000$ inhabitants per square mile and $P>50\,000$. Thus, the minimal urban area is equal to 50 square miles. At the same time, settlements with more then 2500 inhabitants are considered to be urban areas in the USA; while 2000 inhabitants living in contiguous housing form an urban area in France, and in the Netherlands it is municipalities with 2000 or more inhabitants. Settlement densities can be orders of magnitude higher than agricultural rates, although residential densities in some urban areas are only marginally higher than the farmland densities in the most intensively cultivated agricultural areas (compare 2500 people per km^2 in Los Angeles suburbs with 2000 peasants per km^2 of arable land in Sichuan, China). However, maximum residential densities, c. 90 000 people per km^2 (the center of Hong Kong), translate into an anthropomass of 36 MJ m^{-2}. This is roughly 200 times higher than the density of large herbivorous ungulates in Africa's richest ecosystem.

The definition of urban area in some other regions differs significantly from the above-mentioned one, with the concept of urban area somewhat based on the ancient structure of a city. All this shows a significant uncertainty in the term 'urban(ized) area' and which is necessary to take into account in any quantitative estimations.

Past, Present, and Future of Urbanization

Two thousand years ago, there were about a quarter of a billion people living on our planet. The global population doubled to about half a billion by the sixteenth to seventeenth centuries. The next doubling required two centuries (from the middle of the seventeenth century to 1850, when the size of two European cities, Paris and London, exceeded 1 million inhabitants); the following doubling occurred just over the next 100 years, while the last one took only 39 years. The year 1650 is named as the start of 'the urban explosion'. Generally speaking, beginning from this date, the enormous population growth started.

Nevertheless, it is a common practice to perceive the beginning of urbanization to coincide with the start of the agricultural revolution (7000–5000 BCE). It was at this time that nomadic hunters settled down and began to grow their food. A food surplus was created, and the division of labor made it possible to evolve gradually into the complex, interrelated social structures we know now as cities. The first cities were located along the Tigris and Euphrates Rivers (4000–3000 BCE) in contemporary Iraq, with urbanization then occurring also in Egypt, North Africa, India, China, Japan, and Europe – the Americas being the regions of most recent urbanization. Environmental factors were the major driving forces in the development of earlier cities. Fertile soils and easy access to water bodies, as well as adequate water supply, were essential. The first environmental disaster was triggered by the deforestation of the Middle East that led to soil degradation in the area, followed by a collapse of irrigations systems, and, as a consequence, to famine. Ancient cities were extremely dependent on the surrounding ecosystems, in particular, on agricultural lands. In Europe, since the eleventh century, there has been a historical continuing flow of people from the countryside to the cities, although the 'Black Death' in the fourteenth century impaired the process of urbanization. Europe recovered from the effect of this pandemic only by the middle of the sevententh century, when the 'urban explosion' occurred. Urbanization had also been occurring worldwide for at least two centuries. During the eighteenth century we have seen modern urbanization due to technological development, while earlier the process was driven by the migration of people from rural areas, since they were not needed in farming anymore.

However, in the last decades of the last century, we observed the unprecedented global population growth and the accompanying process of urbanization (**Figure 1**).

Generally speaking, this enormous population growth was accompanied by other significant changes:

1. the rise of each person's ability to affect the natural environment through energy sources' exploitation;
2. the rise of the unevenness of the spatial distribution of people through development of 'cities';
3. migration and travels' increase, while contacts between cultures also rise.

Although only 12% of the world's population lived in urban centers in 1940, this percentage had risen to 33% by 1980. After World War II, a 2% urbanization rate was

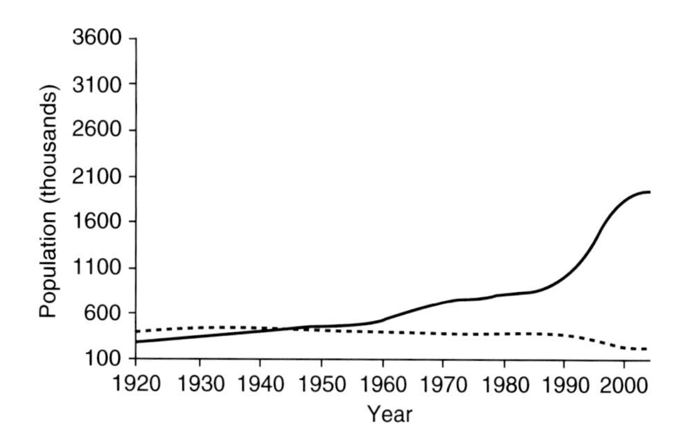

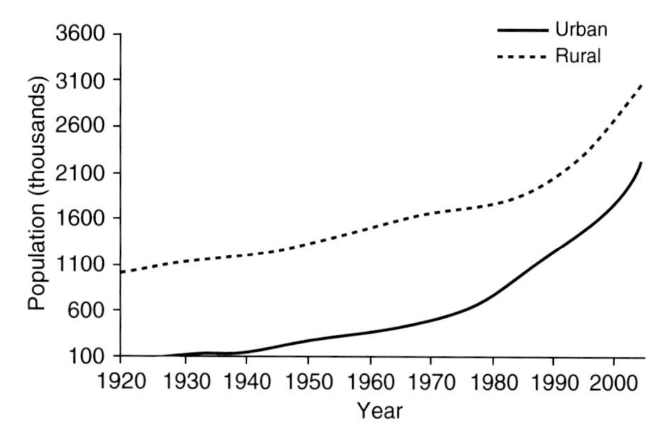

Figure 1 Comparison of urban and rural population growth in developed and in developing countries (urban is defined as settlements of 20 000 people and above). Source: UN, World Urbanization Prospects, 1990, 1991.

observed in the developed world, while it was almost 4% in the developing countries.

However, urban growth rates doubled that of the total population. And while the total population growth rate in the developed world has been decreasing, the urban population's proportion has increased from 55% to 70% of the total population. The major reason is the decline in rural population, as well as the arrival of new immigrants to the cities of some countries.

Since the middle of the last century, the following trend is observed in the percentages of urban population: 1950, 29%; 1960, 34%; 1970, 37%; 1980, 40%; 1990, 43.5%, and in 1995 45%. However, the definition of what is urban varies greatly, which is why the estimations differ as well. For example, another source estimates 20% in 1950.

Urbanization growth rate has significantly left total population growth behind. From the year 1800 to 1990, the absolute number of city dwellers increased from 18 million to 2.3 billion, a 128-fold increase, while the total population has increased by only 6 times (from 0.9 billion to 5.3 billion). Furthermore, more then 1.4 billion city dwellers live in the less-developed world. If we look at the rates of urban area growth and compare them with urban population growth, we find that the first grows faster than the population, which in turn grows at a faster rate than the total country population, and this is a common phenomenon.

In 1990–95, the world's urban population grew by 2–4% per year, while rural population only grew by 0.7% per year. Urban population increased by 2.1% during the 1995–2000 period, while rural population only by 0.7%. Today, 75% of the world's population lives in the less-developed countries, and 58% in Asia. In 1999, 19 urban settlements had 10 million or more inhabitants, and 47% of all people lived in cities. The number of 'megacities', that is, giant urban agglomerations with a densely settled urban core of the original city, is increasing. Around this core, the satellite cities have grown, either planned or unplanned, linked to the central core by transport, communication, economic interdependence, and political-administrative structures. This tendency is confirmed by the following statistics: from 1950 until 1975, many cities with population of 5 million people have doubled in total urban population, while at the same time, cities with less than 100 000 people declined in their relative importance. In 1992, there were 23 megacities with populations greater than 8 million: 6 in the developed world (Tokyo, New York, Los Angeles, Osaka, Paris, and Moscow), and 17 in the developing world (from which 11 were in Asia). For most Asian cities, the shortage of water will be the most critical issue and is the limiting factor for the further growth of Beijing, Manila, Bangkok, Jakarta, and other cities. Also, while during the nineteenth century water and air pollution were associated with only a few larger cities, they are now becoming a global problem due to the rapid industrialization and the simultaneous concentration of people in cities.

While the past and current demographic situations are estimated more or less accurately, future dynamics are forecast with a very high level of uncertainty. The UN vividly illustrates that if current exponential and hyperbolic growth continues in each major region and at the current rates, then the population will increase by more than 130-fold in 160 years, from 5.3 billion in 1990 to 694 billion in 2150. However, eventually, the problem will be how to feed these people, since food and water limitations will certainly arise. The UN also shows that future global population size is very sensitive to the future level of average fertility.

Projections of global population dynamics are also uncertain, because external factors such as climate may change unexpectedly. Furthermore, even if external factors change as expected, the relationship between those factors and demographic rates may change.

We have the following hypothetical picture for the next half of the century. The global population will grow by 2–4 billion people, mostly in poor, but not rich, countries. It will also increase less rapidly than before and will become more urban than now. Hence, "from here on it is an urban world." Most of all, the additional people will be living in cities in poor countries, which can become an epidemiological danger. Population of the more developed countries will decline slightly, but increase substantially in less-developed countries.

In this century, global urban population will increase 1.8 times by 2020 (relative to 1990), while the total population will grow by only 1.4 times, and almost all population growth will be associated with cities in the developing countries. By the year 2030, the world urban population will reach 4.9 billion (1 billion in developed countries and 3.9 billion in developing countries). The global rural population will remain constant at 3.2 or 3.3 billion, although in the developed countries the rural population will decline. The trend in the developing countries is that rural population will slowly rise through the next couple of decades, reaching 3.1 billion, and then will slowly decline.

Present and Future Dynamics of Urban Areas

In 1985, 43% of the world population lived in cities while urban settlements covered just over 1% of the Earth's surface. In 1990, 50% of the global population of *Homo sapiens* inhabited less than 3% of the Earth's ice-free land area. However in the near future we may expect a further growth of the urban territories' area. World dynamics of this process is shown in **Figure 2**. All these prognoses are based on two

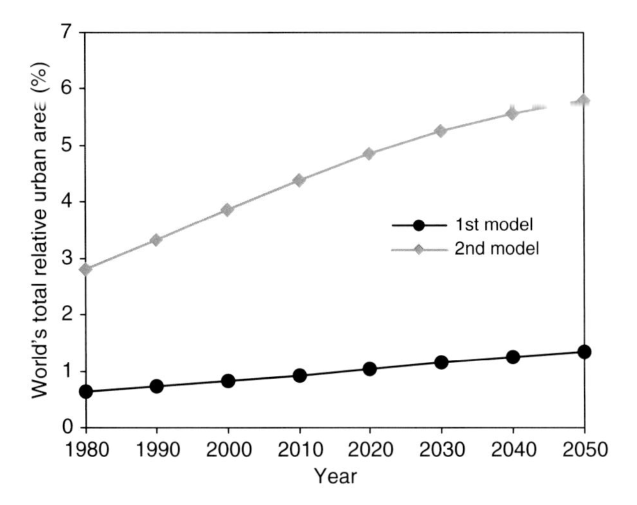

Figure 2 Dynamics of the relative world urban area (in percentage of the total world area) between 1980 and 2050.

models: the first is a regression model, connecting urban population and urban area, and giving a minimal estimation, and the second, uses the spatial distribution of population density and gives a maximal estimation.

Note that the dynamics of urban areas are significantly different in the major world regions (**Figure 3**). For instance, the fast growth in African region differs from almost constant dynamics in the highly industrialized countries.

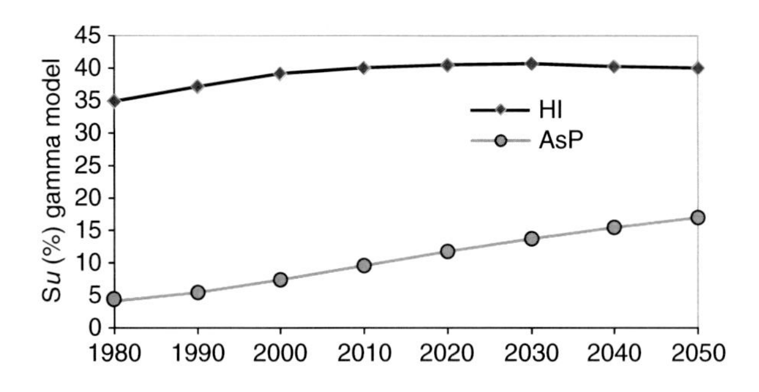

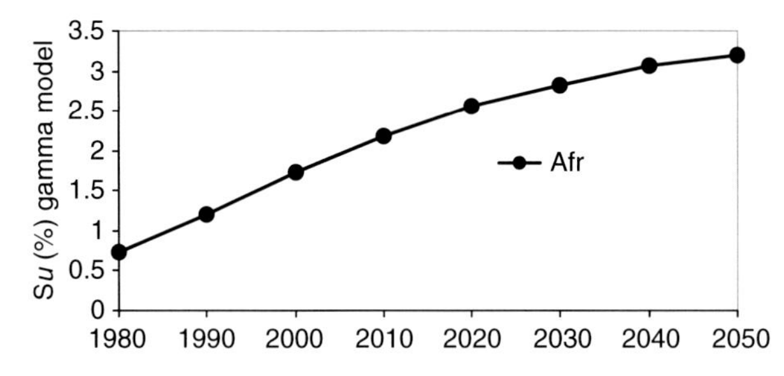

Figure 3 Dynamics of the relative urban areas (% of the total regional area) for the three world regions: HI, highly industrialized European; AsP, Asia and Pacific; and Afr, African.

The City as a Specific Heterotrophic Ecosystem

From an ecological point of view, any city is a heterotrophic system maintained by external inflows of energy, food, water, and other substances. Thermodynamically any city (and generally, any urbanized territory) is an open system that is far from thermodynamic equilibrium. All matter and energy needed for a city's functioning are collected from external territories that are significantly larger than the area of the city itself and very often are located quite far away.

The heterotrophic ecosystem 'city' differs very much from a natural heterotrophic ecosystem. In fact, a city has a more intensive metabolism per area unit, requiring a significant inflow of artificial energy. Its consumption per urban area unit may be 3–4 orders of magnitude higher than the same for rural area. For instance, the annual subsidies in fuel, fertilizers, labor, etc., required to maintain a lawn in the Madison metropolitan area (Wisconsin, USA) is equal to 22 GJ ha^{-1}, which is approximately equal to the artificial energy input for a maize field. During the process of its own metabolism, a city consumes large amounts of various materials: food, water, wood, metals, etc., all that we call 'gray energy'. Products of city's metabolism have larger volumes of, and more toxic, substances than the same of natural ecosystems.

If we compare cities and natural forests in Wisconsin, USA, we can see that the number of species in a city forest (75 tree and 74 bush species) is more than in natural one (10 tree and 20 bush species). The annual net production and the amount of living biomass (in carbon units) in city's forests are equal to $500\,t\,C\,km^{-2}\,yr^{-1}$ and $7000\,t\,C\,km^{-2}$; while in a natural forest the corresponding values are $400\,t\,C\,km^{-2}\,yr^{-1}$ and $13\,000\,t\,C\,km^{-2}$. The greater values of species diversity and production are provided in city's forests at the expense of additional inflow of 'gray energy'. For instance, the annual import of fertilizers is about $140\,t\,km^{-2}$.

While natural forest is almost a closed system, a city forest is a typical through-flow system with 'gray energy' input and output in the form of dead organic matter: about half of the annual accretion is exported from the city to waste treatment plants. This is one possible explanation why the amount of biomass in the city forest is lesser than in the natural ecosystem. The carbon storage in urban forests with their relatively low tree cover ($25.1\,t\,C\,ha^{-1}$ in average for US) is less than in natural forest stands ($53.5\,t\,C\,ha^{-1}$). The gross sequestration rate, that is, the fraction of the gross annual production accumulated in wood, in urban forests is equal to $0.8\,t\,C\,ha^{-1}\,yr^{-1}$, which is also less than in natural ones (for instance, $1.0\,t\,C\,ha\,yr^{-1}$ for a 25-year-old natural regeneration spruce–fir forest with $0.1\,kg\,C\,m^{-2}$ cover), although the difference is insignificant. However, on a per-unit tree cover basis, carbon storage by urban tree and gross sequestration may be greater than in natural forests, $92.5\,t\,C\,ha^{-1}$ and $3.0\,t\,C\,ha^{-1}\,yr^{-1}$, due to a larger proportion of large trees and the more open structure (that leads to the weakness of competition) in urban forests.

Environmental Effect of Urbanization and Ecological Footprint

Retrospectively, most humans have lived in small settlements dispersed within larger biomes (**Figure 4**). We see there is a certain correlation between the type of biomes and the degree of urbanization, some biomes being more preferable for cities.

The growth of cities in these 'patches', absorbing and transforming nearby natural ecosystems and agricultural lands, modifying energy and matter flows, typical for these ecosystems, negatively influences the local and regional biodiversity, increasing fragmentation of large rural areas and natural zones. This process (and contamination of the atmosphere) leads to the changing of the nature of land surfaces and near-surface atmospheric layer, and therefore its reflection and absorption of solar radiation and aerodynamic properties. This in turn leads to raising urban temperatures and the changing of the local climate, creating so-called 'urban heat islands', which is warmer by 1–2 °C than surrounding territories. The 'urban heat island' effect occurs mainly at night, when the buildings, etc., release heat absorbed during day.

In addition, metropolitan agglomerations influence the local and global environment through their consumption of non-native resources and their concentrated production of waste and consumables.

The footprint is the quantitative conversion of the material and energy flows required to support human population in cities into the land area required to produce these flows. Although cities occupy a relatively small area on the planet, they are the dominant human ecosystem and the ecological space taken up by humans as a species is much higher. Every city depends (for its existence and growth) on a globally diffuse productive hinterland up to 200 times the size of a city itself. To illustrate this, let us examine the following case studies.

For instance, one person of the USA urban population consumes daily (1) food produced by 0.75 ha of agriculture land, (2) paper and wood by 0.4 ha of forest, (3) water by about $7.5\,m^3$. So, a 1-million-populated city with the population density of 4000 persons per km^2 (city area is $250\,km^2$ correspondingly) needs a significantly larger area for its support: $7500\,km^2$ of agriculture land and $4000\,km^2$ of

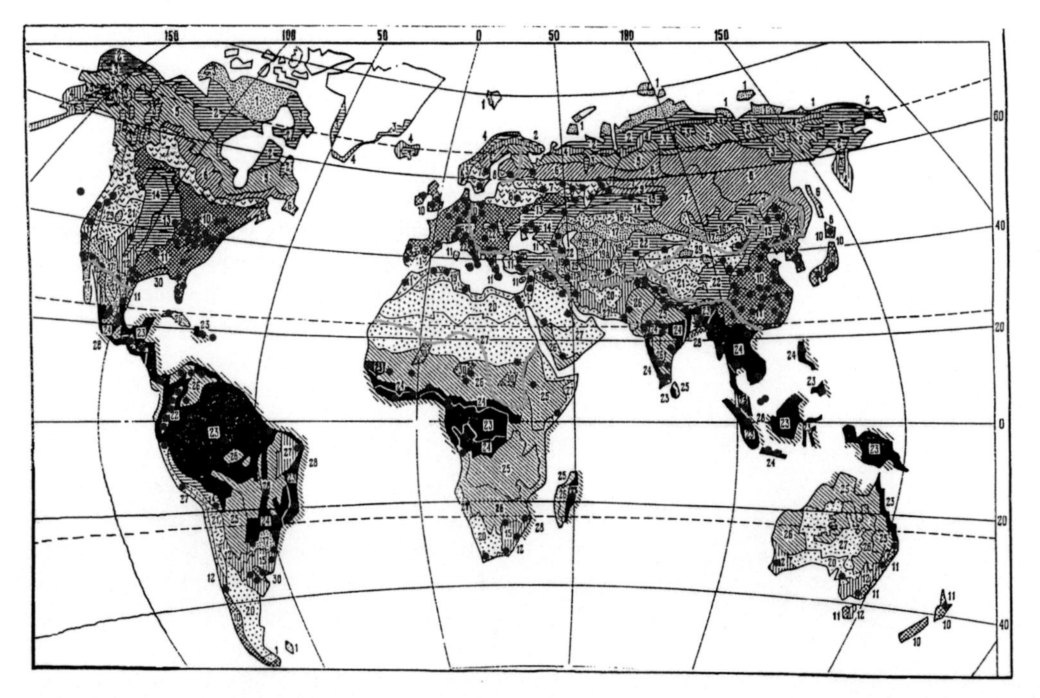

Figure 4 Different types of global vegetation (biomes). Red dots represent the concentration of cities; the green line is the border of the regions. 1, Polar desert, polar tundra; 2, tundra; 3, mountainous tundra; 4, forest tundra; 5, north taiga; 6, middle taiga; 7, south taiga; 8, temperate mixed forest; 9, aspen–birch lower taiga; 10, deciduous forest; 11, subtropical deciduous and coniferous forest; 12, xerophyte woods and shrubs; 13, forest steppe; 14, temperate dry steppe (including mountainous); 15, savanna; 16, dry steppe; 17, sub-boreal desert; 18, sub-boreal saline 'desert'; 19, subtropical semidesert; 20, subtropical desert; 21, mountainous desert; 22, alpine and subalpine meadows; 23, evergreen tropical rainforest; 24, deciduous tropical forest; 25, tropical xerophyte woodland; 26, tropical savanna; 27, tropical desert; 28, mangrove forest; 29, saline land; 30, subtropical and tropical woodland and Tugay shrubs.

forest. A rather large river watershed (presuming abundant precipitation) can provide inflow of 7.5 million m^3 of water daily. If we take into account that the area of such watershed is about 15 000 km^2 then the total footprint area will be about 26 500 km^2, that is, 106 times the city area.

The city of Vancouver (Canada) had in the year 1991 a population of 47 2000 living in an area of 11 400 ha. If we assume that the per capita land consumption rate is 4.3 ha, then the people in this city would require 2.03 Mha of land. Hence, the inhabitants would require a land area 180 times larger then their habitat. Furthermore, adding a marine footprint of 0.7 ha per person, the total area needed to support the city becomes 2.36 Mha, or 200 times larger than the geographic area of the city. For London, the equivalent footprint is 120 times the area of the city itself. The New York metropolitan area annually consumes the equivalent of 800 000 ha of wheat, or approximately the total amount of wheat grown yearly in the state of Nebraska.

So, if we presume that the world urban area constitutes 1% of the total land area, and the footprint is 100 times this area (note that these are minimal estimations), even in this case the urban footprint exceeds all the world land area.

Carbon Balance in Urbanized Territories

Therefore, we can say that although the total area of urbanized territories is relatively small (~1–2% in 1990s), they play an ever-increasing role in global change in general and in the global carbon cycle, the main biogeochemical cycle of the biosphere, in particular.

Urban areas emit (in accordance with different estimations) 78–97% of the total anthropogenic carbon emission. Up to 60% of this emission comes from the transportation and building sectors, while the rest are from industry. Of course, all of these emissions are 'spread' and mixed in the entire atmosphere over 3–4 months period, but they are generated in particular by urban point sources.

Cities transform the natural territories they occupy, partially obliterating vegetation and soil, partially modifying them. Similarly, urbanization changes the structure and function of the local carbon flows within these territories. Note that the process often involves considerably larger territories than the exact city areas.

Cities consume a lot of organic carbon in the form of food and other agricultural products, as well as wood, etc., produced, as a rule, far from the urban territories, transforming them into other forms of carbon (feces, exhaled CO_2, residues of food processing, dead organics of 'green zones', etc.) in the process of urban and purely physiological human metabolism. In other words, cities destroy the spatial entity of the processes of production and decomposition of living matter that is typical for natural ecosystems. Note that this entity provides the closure of any local carbon cycle.

Urban territories have more carbon stored per unit area than natural ecosystems. Organic carbon is stored in soils and vegetation of urban territories, and also in people and pets; nonorganic carbon includes carbon transported into the cities and stored in buildings, etc., but most of this carbon is transformed into waste. Processes similar to those in peatlands accompany carbon fluxes from mineralization, incineration (rapid oxidation of carbon), and landfilling of solid waste. For instance, the global input of carbon into solid waste (sludge and industrial waste) is estimated to be 0.16 Gt yr^{-1} (1 Gt = 10^9 t).

Long-term organic carbon in urban territories is accumulated in

1. *Biomass in humans and animals.* For the world population of 6 billion, the total amount of carbon is equal to 45 million t of carbon that constitutes about 10% of the total biomass of land animals. A biological metabolism of 6 billion people is accompanied by exhalation of 0.34 Gt C yr^{-1} and secretion of 0.18 Gt yr^{-1} with feces and other discharges that, respectively, give a total 0.52 Gt C yr^{-1}. The value is entirely comparable with components of the global carbon cycle. For instance, soil erosion gives 0.98 Gt C yr^{-1}.
2. *Biomass in trees and other plants.* The mean global value of living plant biomass in cities is 3500 t C km^{-2}, and the mean net primary production is 500 t C km^{-2} yr^{-1}. By taking the global urban area in 1980 as 2×10^6 km^2 and assuming that 50% of it is covered by city vegetation, we find that urban territories contain 3.5 Gt C in living vegetation biomass, while a global figure for net plant assimilation of carbon in urban territories is approximately 0.5 Gt C yr^{-1}.
3. *Carbon in construction material, furniture, books.* Extensive amounts of carbon are accumulated for long time period in building constructions, furniture, books, and other articles made of organic materials. For instance, *c.* 3 Gt C is fixed in houses in the whole of Europe, North America, Japan, and Australia, and about 0.4 Gt C in other regions.
4. *Carbon in solid waste.* Most products of forestry and agriculture are turned eventually into waste. Solid waste is either deposited in sanitary landfills or incinerated. Carbon stored in landfills experiences slow decomposition rates and is gradually released due to microbiological activity. The annual world solid-waste production is equal to 170–180 million t of carbon. Approximately 60–70 million t is released into the atmosphere by burning, while about 110–120 million t is deposited in landfills followed by slow release into the atmosphere.

Landfills are often regarded as long-term accumulators of carbon and in this respect can be compared with natural peatland ecosystems (even after 30 years one-third of the organic carbon remains nonmineralized). This carbon is bound in long-lived humus and is not mineralized for a very long time.

Conclusion

Urbanized territories dominate the surrounding environment in a number of ways – the growth of cities, and absorbing and transforming nearby natural ecosystems and agricultural lands. This process leads to the changing of the nature of land surfaces. Therefore, we can say that although the total area of urbanized territories is relatively small (~1–2% in 1990s), they play an ever-increasing role in global change in general and in the global carbon cycle in particular.

We can summarize this influence as the following:

1. Cities transform the natural territories they occupy, partially obliterating vegetation and soil, partially modifying them. By the same token, urbanization changes the structure and function of the local carbon flows within these territories. Note that the process often involves considerably larger territories than the exact city areas.
2. Cities consume a lot of organic carbon in the form of food and other agricultural products, as well as wood, etc., produced, as a rule, far from the urban territories, transforming it into other forms of carbon (feces, exalted CO_2, residues of food processing, dead organics of 'green zones', etc.) in the process of urban and purely human metabolism. In other words, cities destroy the spatial entity of the processes of production and decomposition of living matter that is typical for natural ecosystems. Note that this entity provides the closure of any local carbon cycle.

See also: Biosphere. Vernadsky's Concept; Carbon Cycle; Climate Change 1: Short-Term Dynamics.

Further Reading

Brundtland's World Commission on Environment and Development (1987) *Our Common Future*. Oxford: Oxford University Press.
Encyclopædia Britannica, Inc. (2005) *Encyclopædia Britannica 2005*. CD-ROM/Ultimate Reference Suite 2005 DVD.
Hauser JA (1992) Population, ecology and the new economics: Guidelines for a steady-state economy. *Futures* 24(4): 364–387.
Heinke G W (1997) The challenge of urban growth and sustainable development for Asian cities in the 21st century. *Environmental Monitoring and Assessment* 44: 155–171.
Miller GT (1988) *Living in the Environment*, 6th edn. Belmont, CA: Wadsworth.
Small C and Cohen JE (1999) Continental physiography, climate and the global distribution of human population. In: Svirezher Yu (ed.) *Proceedings of the International Symposium on Digital Earth*, pp. 965–971.Beijing: Chinese Academy of Science.
Stempell D (1985) *Weltbevölkerung 2000*. Leipzig: Urania-Verlag.
Svirejeva-Hopkins A and Schellnhuber H-J (2006) Modelling carbon dynamics from urban land conversion: Fundamental model of city in relation to a local carbon cycle. *Carbon Balance and Management* 1: 8.
Svirejeva-Hopkins A, Schellnhuber H-J, and Pomaz VL (2004) Urbanised territories as a specific component of the global carbon cycle. *Ecological Modelling* 173: 295–312.
United Nations (UN) (1999) *Prospects for Urbanization – 1999 Revision*. New York: United Nations (ST/ESA/SER.A/166), Sales No. E.97.XIII.3.
United Nations (UN) (2000) *The State of the World Cities 2001*, 121pp. Nairobi: United Nations Centre for Human Settlements (UNCHS).

Weathering

S Franck, C Bounama, and W von Bloh, Potsdam Institute for Climate Impact Research, Potsdam, Germany

Introduction

The idea of cyclic processes in geology was very important for the development of scientific thinking. The eighteenth century was a time of rationalism and discovery. The Neptunists developed the idea that in the early stages of its evolution the Earth was covered by a universal ocean and the present continents have emerged by secular lowering of the sea level. A famous Neptunist was A. G. Werner according to whom the Earth's crust has been laid down as a series of worldwide formations by primeval ocean. Thus, the face of the Earth had been built and shaped mainly by the agency of water. In the opinion of the Plutonists, rocks were mainly the results of cycling of material derived from erosion of older rocks. According to J. Hutton the products of erosion accumulated on the seafloor where they became hardened by the Earth's internal heat. In this way, the Neptunist–Plutonist controversy of the eighteenth century already covered two general features of scientific thinking, the repetitious cycling of planetary matter on the one side and the directional evolution of the Earth on the other side. V. Vernadsky was the first who created the concept of biogeochemistry as the intersection of biological, geological, and chemical processes with the cycling of elements through the biosphere as the central process.

In this article, we discuss the importance of weathering in the framework of global geochemical cycling. After reviewing the global volatile cycle and the weathering process, we present a global carbon cycle model that includes both silicate and carbonate weathering. Furthermore, we describe in detail the biotic enhancement of weathering and the importance of weathering after heavy impacts.

Parametrized Convection Model with Volatile Exchange

The global volatile cycle is based on the interaction of the Earth's mantle and the surface reservoirs of volatiles. The transport of volatiles to and from the mantle is via mantle convection that can be described by parametrized convection models. Parametrized convection models have been developed for more than 30 years. They are applied to study the temporal variations of quantities such as average mantle temperature and heat flow by parametrizing the heat flow in terms of the Rayleigh number Ra:

$$N \propto Ra^{\beta} \qquad [1]$$

where N is the ratio of the total heat flow and the conductive transported heat (Nusselt number) and β is an empirical constant, usually equal to 0.3.

The effect of volatile-dependent rheology on the thermal evolution of the Earth was first analyzed by showing that the existence of volatiles has a significant effect on the thermal history of the mantle.

In order to calculate the exchange of volatiles and to investigate the feedbacks existing between heat transport and volatile-dependent viscosity, a self-consistent model was first presented by McGovern and Schubert. The main idea is a coupling of the thermal and degassing history of the Earth with the help of simple relations from boundary layer theory:

$$q_m = \frac{\sqrt{S_A}\, 2k\,(T_m - T_s)}{\sqrt{\pi \kappa\, A_o(t)}} \qquad [2]$$

where q_m is the mean heat flow from the mantle, S_A is the areal seafloor spreading rate, k is the thermal conductivity, κ is the thermal diffusivity, T_m is the average mantle temperature, T_s is the surface temperature, and $A_0(t)$ is the area of ocean basins at the time t. $A_0(t)$ can be used to introduce different continental growth models. In the evolution model, water from the mantle degases at mid-ocean ridges from certain volume (degassing volume) that depends on S_A and the melt generation depth d_m:

$$(\dot{M}_{mv})_d = \rho_{mv}\, d_m\, S_A \qquad [3]$$

where $(\dot{M}_{mv})_d$ is the degassing rate of water at mid-ocean spreading centers and ρ_{mv} is the density of water in the mantle (mass of mantle volatiles M_{mv} per mantle volume).

The rate of regassing of water at subduction zones F_{reg} is proportional to the water content on the basalt layer f_{bas}, the mass fraction of water in the oceanic crust ρ_{bas}, the thickness of oceanic crust d_{bas}, and S_A:

$$F_{reg} = f_{bas}\, \rho_{bas}\, d_{bas}\, S_A\, R_{H_2O} \qquad [4]$$

where R_{H_2O} is the regassing ratio, representing the fraction of water that actually enters the deep mantle instead returning to the surface through back-arc or andesitic volcanism. R_{H_2O} is very important for the participation of volatiles between the mantle and the surface reservoirs.

The degassing and regassing of water representing the main volatile in the evolution model are presented in **Figure 1**.

The balance equation for the mass of mantle water is given as

$$\dot{M}_{mv} = F_{reg} - (\dot{M}_{mv})_d \qquad [5]$$

First we can solve eqns [3]–[5] for the steady-state, that is, the global volatile cycle proceeds in the same way, without any long-term evolutionary change. Balancing the degassing flux [3] with the regassing flux [4] gives

$$\rho_{mv}\, d_m = f_{bas}\, \rho_{bas}\, d_{bas}\, R_{H_2O} \qquad [6]$$

which can be used to estimate the steady-state value of R_{H_2O}. Using the numerical values of $d_m = 56$ km, $\rho_{mv} = 4.898$ kg m^{-3} (present value for 3 ocean masses of water in the mantle), $f_{bas} = 0.03$, $\rho_{bas} = 2950$ kg m^{-3}, and $d_{bas} = 5$ km, we can find that the steady-state value for R_{H_2O} is 0.62.

The initial value of M_{mv} is the number n of ocean masses M_{ocean} originally present in the mantle:

$$M_{mv}(t = 0) = n\, M_{ocean} \qquad [7]$$

The ratio of the water amount still remaining within the Earth to those in the atmosphere, hydrosphere, and sediments can be estimated between 3 and 20. In the sense of a conservative approximation, we assume this ratio to be 3 for the present Earth, that is, there are altogether 4 ocean

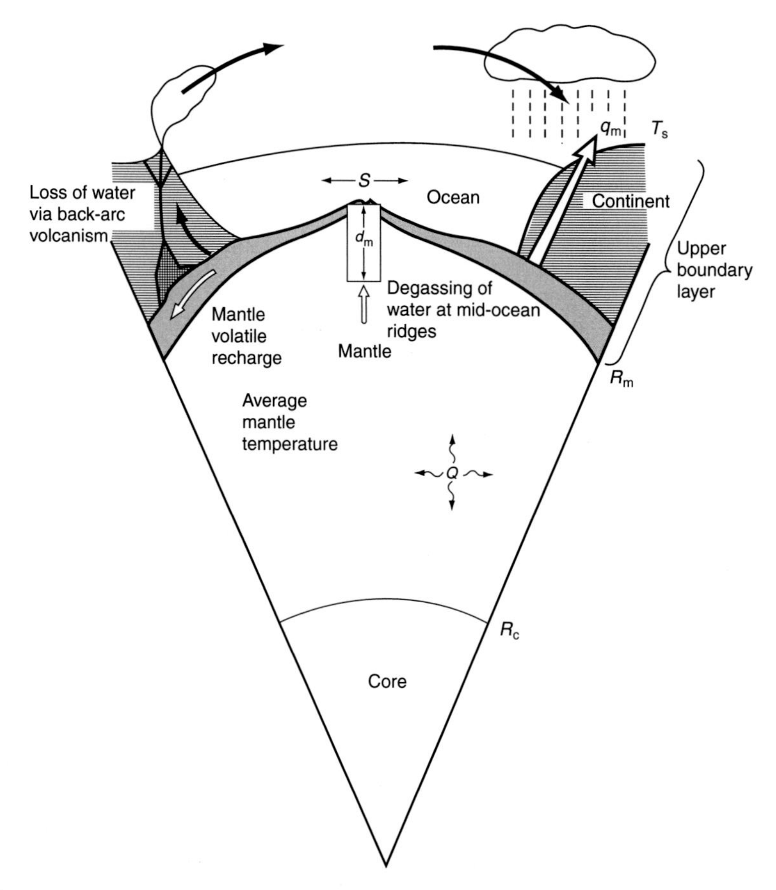

Figure 1 Schematic view of the parametrized convection model with volatile exchange between mantle and surface reservoirs (not in scale). The heat flow q_m at the base of the lithosphere (upper boundary layer) results from cooling of the mantle with an average mantle temperature t_m and from a radiogenic energy production rate Q. The core radius, the mantle radius, and the surface temperature are R_c, R_m, and T_c, respectively. The mantle degases at mid-ocean ridges from a partial melting depth d_m and an area defined by the areal spreading rate S. At subduction zones the volatiles are partly regassed into the mantle ant partly recycled to the atmosphere via volcanism. Reproduced from Franck S (1998) Evolution of the global mean heatflow over 4.6 Gyr. *Technophysics* 291:9–18, with permission from Elsevier.

masses of water in the Earth system. No evidence exists so far about its distribution at the beginning of planetary evolution. Assuming that the total amount of water in the system is constant during Earth's history, we attempt to investigate different distributions of water at the beginning of planetary evolution ($t=0$).

The Weathering Process

The role of weathering for the Earth's climate was first described by Walker *et al.* In particular, the potential of weathering to stabilize the Earth's surface temperature by a negative feedback mechanism that is strongly modulated by the biosphere has gained recent interest. Compared to subareal weathering, silicate-rock weathering on land primarily controls long-term atmospheric CO_2 content. The question of to what extent the biota are actually able to play an active role in stimulating the strength of the main carbon sink through weathering is crucial for an understanding of the dynamic properties of the overall Earth system.

The total process of weathering embraces first the reaction of silicate minerals with carbon dioxide, second the transport of weathering products, and third the deposition of carbonate minerals in sediments. The availability of cations plays the main role in these processes and is the limiting factor in the carbonate-sediments-forming reaction (third process) between cations (Ca^{2+} and Mg^{2+}) and carbonate anions (CO_3^{2-}). Therefore, for the mathematical formulation we have only to take into consideration the amount of released cations and their runoff (first and second process), respectively. The weathering rate F_{wr}, as a global average value, is the product of cations concentration in water (in mass per unit volume) and runoff (in volume per unit area per unit time). Therefore, the weathering rate is the mass of cations formed per unit area and unit time. Combining the direct temperature effect on the weathering reaction, the weak temperature effect on river runoff, and the dependence of weathering on soil CO_2 concentration, the global mean silicate-rock weathering rate can be formulated via the following implicit equation:

$$\frac{F_{wr}}{F_{wr,0}} = \left(\frac{a_{H^+}}{a_{H^+,0}}\right)^{0.5} \exp\left(\frac{T_s - T_{s,0}}{13.7\,K}\right) \qquad [8]$$

Here the prefactor outlines the role of the CO_2 concentration in the soil, P_{soil}, a_{H^+} is the activity of H^+ in fresh soil-water and depends on P_{soil} and the global mean surface temperature T_s. The quantities $F_{wr,0}$, $a_{H^+,0}$, and $T_{s,0}$ are the present-day values for the weathering rate, the H^+ activity, and the surface temperature, respectively. The activity a_{H^+} is itself a function of the temperature and the CO_2 concentration in the soil. The equilibrium constants for the chemical activities of the carbon and sulfur systems involved have been taken from Stumm and Morgan. Note that the sulfur content in the soil also contributes to the global weathering rate, but its influence does not depend on temperature. It can be regarded as an overall weathering bias, which has to be taken into account for the estimation of the present-day value.

Equation [8] is the key relation for the models. For any given weathering rate the surface temperature and the CO_2 concentration in the soil can be calculated self-consistently. P_{soil} can be assumed to be linearly related to the terrestrial biological productivity Π and the atmospheric CO_2 concentration P_{atm}. Thus,

$$\frac{P_{soil}}{P_{soil,0}} = \frac{\Pi}{\Pi_0}\left(1 - \frac{P_{atm,0}}{P_{soil,0}}\right) + \frac{P_{atm}}{P_{soil,0}} \qquad [9]$$

where $P_{soil,0}$, Π_0, and $P_{atm,0}$ are again present-day values. Biologically enhanced Hadean and Archaean weathering processes would have been very different from the modern ones, although the purely inorganic processes are the same. Nevertheless, one can assume that at least as far back to the Proterozoic, the biosphere generates the same effects as today, namely the enhancement of CO_2 concentration in soil compared to the atmospheric value.

Weathering and the Global Carbon Cycle

The global carbon cycle model describes the evolution of the mass of carbon in the mantle, C_m, in the combined reservoir consisting of ocean and atmosphere, C_{o+a}, in the continental crust, C_c, in the ocean crust and floor, C_f, in the kerogen, C_{ker}, and in the biosphere, C_{bio}. The equations for the efficiency of carbon transport between reservoirs take into account mantle de- and regassing, carbonate precipitation, carbonate accretion, evolution of continental biomass, the storage of dead organic matter, and weathering processes:

$$\frac{dC_m}{dt} = \tau_f^{-1}(1-A)RC_f - S_A f_c d_m C_m/V_m \qquad [10]$$

$$\begin{aligned}\frac{dC_{o+a}}{dt} = {} & \tau_f^{-1}(1-A)(1-R)C_f + S_A f_c d_m C_m/V_m + \\ & + F_{weath}(C_{o+a}, C_c) + (1-\gamma)\tau_{bio}^{-1} C_{bio} + \tau_{ker}^{-1} C_{ker} \\ & - \Pi_{bio}(C_{o+a}) - F_{prec}(C_{o+a}, C_c) - F_{hyd}\end{aligned} \qquad [11]$$

$$\frac{dC_c}{dt} = \tau_f^{-1} AC_f - F_{weath}(C_{o+a}, C_c) \qquad [12]$$

$$\frac{dC_f}{dt} = F_{prec}(C_{o+a}, C_c) + F_{hyd} - \tau_f^{-1} C_f \qquad [13]$$

$$\frac{dC_{bio}}{dt} = \Pi_{bio}(C_{o+a}) - \tau_{bio}^{-1} C_{bio} \qquad [14]$$

$$\frac{dC_{ker}}{dt} = \gamma\tau_{bio}^{-1} C_{bio} - \tau_{ker}^{-1} C_{ker} \qquad [15]$$

where t is the time, τ_f the residence time of carbon in the seafloor, A the accretion ratio of carbon ($A > 0$ for $A_o < A_e$), R the regassing ratio, f_c the degassing fraction of carbon, V_m the mantle volume, F_{weath} the weathering rate, F_{prec} the rate of carbonate precipitation, F_{hyd} the hydrothermal flux, γ the fraction of dead biomass transferred to the kerogen, τ_{bio} the residence time of carbon in the biosphere, Π_{bio} the productivity of the continental biosphere, and τ_{ker} is the residence time of carbon in the kerogen. The accretion ratio, A, is defined as the fraction of seafloor carbonates accreted to the continents to the total seafloor carbonates. The regassing ratio, R, is defined as the fraction of seafloor carbonates regassed into the mantle to the total subducting carbonates. S_A and d_m are calculated from a parametrized thermal evolution model of whole mantle convection including the water exchange between mantle and surface reservoirs. The values of the constants are summarized in **Table 1**.

τ_f is the residence time of carbon in the seafloor and is parametrized with the help of the heat production rate Q:

$$\tau_f(t)\frac{Q^*}{Q(t)}\sqrt{\frac{A_o}{A_o^*}} \cdot \tau_f^* = e^{\lambda(t-4.6Ga)}\sqrt{\frac{A_o}{A_o^*}} \cdot \tau_f^* \qquad [16]$$

where Q^* is the present heat production rate of the mantle, A_o^* the present-day area of the ocean basins, τ_f^* the present seafloor residence time, and λ the decay constant (see **Table 1**).

The total amount of carbon in the system, C_{tot}, is conserved, that is,

$$\frac{dC_m}{dt} + \frac{dC_{o+a}}{dt} + \frac{dC_c}{dt} + \frac{dC_f}{dt} + \frac{dC_{bio}}{dt} + \frac{dC_{ker}}{dt} = 0 \qquad [17]$$

Therefore it is possible to reduce the set of equations for the global carbon cycle from six to five independent equations. The box model including the pertinent fluxes is sketched in **Figure 2**.

Table 1 Constants used in the carbon cycle model

Constant	*Value*	*Remarks*
A	0.7	Accretion ratio
A_e	$5.1 \times 10^{14}\,m^2$	Surface area of the Earth
A_o^*	$3.63 \times 10^{14}\,m^2$	Present area of ocean basins
C_{tot}	$7.4 \times 10^{20}\,kg$	Total mass of carbon in the system
f_c	0.194	Degassing fraction of carbon
g	$9.81\,m\,s^{-2}$	Acceleration of gravity
$p_{1/2}$	210.8 ppm	Michaelis–Menten parameter
pH^*	8.2	Present day ocean pH
p_{min}	10 ppm	Minimum CO_2 partial pressure allowing photosynthesis
R	0.7	Regassing ratio
S_A^*	$2.7\,km^2\,a^{-1}$	Present day areal spreading rate
V_m	$8.6 \times 10^{20}\,m^3$	Mantle volume
V_o	$1.1687 \times 10^{18}\,m^3$	Volume of the ocean
λ	$0.34\,Ga^{-1}$	Decay constant
Π_{max}	$180 \times 10^{12}\,kg\,a^{-1}$	Maximum biosphere productivity
τ_{f^*}	77 Ma	Present day seafloor residence time
τ_{bio}	12.8 a	Residence time of carbon in the biosphere
τ_{ker}	1.58 Ga	Residence time of carbon in the kerogen

Data from Franck S, Kossacki KJ, von Bloh W, and Bounama C (2002) Long-term evolution of the global carbon cycle: Historic minimum of global surface temperature at present. *Tellus* 54B: 325–343.

There are two main types of weathering processes: silicate weathering and carbonate weathering. Carbonate weathering is the main process accounting for the flux of carbon from the continents to the ocean as described in eqns [11] and [12]. Silicate weathering is only important for the flux of CO_2 out of the combined 'ocean + atmosphere' pool and does not influence the continental carbon reservoir (see Precipitation Pattern). The description of the weathering process is simplified in the sense of a minimal model. Detailed models of weathering reactions can be found in Berner and Kothavala.

The rate of carbon transport from continents to the ocean driven by carbonate weathering is

$$F_{weath} = F_{weath}^{cc} + F_{weath}^{mc} \qquad [18]$$

where F_{weath}^{cc} and F_{weath}^{mc} are the weathering rates of $CaCO_3$ and $MgCO_3$, respectively. The weathering of calcium and magnesium carbonates is determined using the following formulas:

$$F_{weath}^{cc} = k_w^{cc} \left(\frac{a_{H^+s}}{a_{H^+s}^*} \right)^{0.5} \exp\left(\frac{T_s - 288\,K}{13.7\,K} \right) C_c \qquad [19]$$

$$F_{weath}^{mc} = k_w^{mc} \left(\frac{a_{H^+s}}{a_{H^+s}^*} \right)^{0.5} \exp\left(\frac{T_s - 288\,K}{13.7\,K} \right) C_c \qquad [20]$$

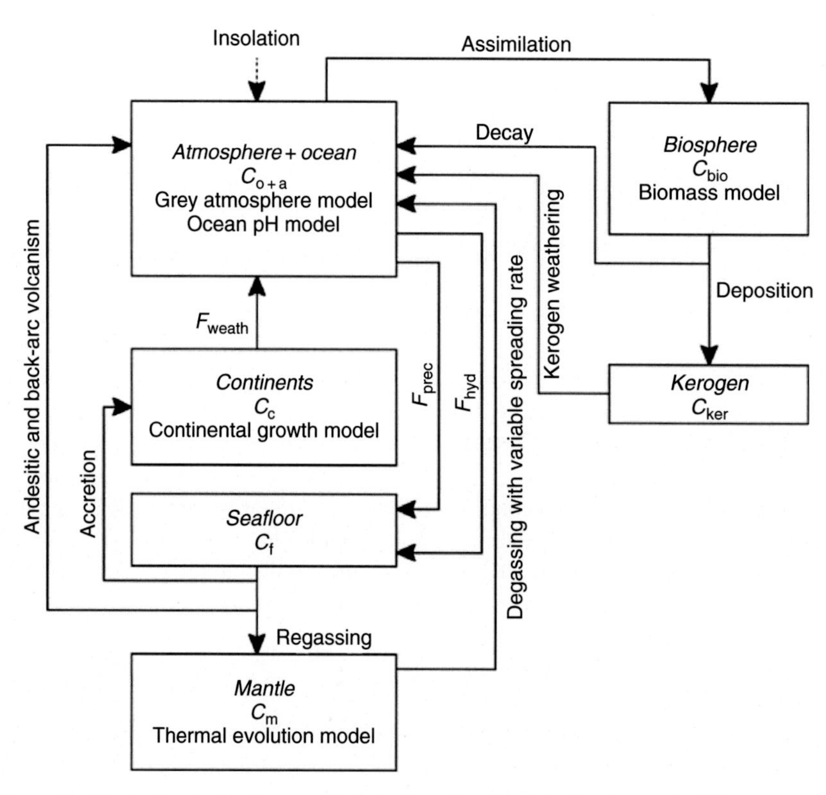

Figure 2 Diagram illustrating the basic mechanisms and interactions for the global carbon cycle. The fluxes from and to the different pools are indicated by arrows.

The prefactor outlines the role of CO_2 partial pressure in the soil, p_{soil}, a_{H^+s} is the activity of H^+ in fresh soil-water. The quantity $a^*_{H^+s}$ is the present value for the H^+ activity in the soil. The activity a_{H^+s} itself is a function of the surface temperature and the CO_2 partial pressure in the soil, where the equilibrium constants for the chemical activities of the carbon and sulfur systems have been taken from Stumm and Morgan. p_{soil} depends on the terrestrial biological productivity per area, Π_A, the atmospheric CO_2 partial pressure, p_{CO_2}, and their corresponding (long-term mean pre-industrial) present values ($p^*_{soil}, \Pi^*_A, p^*_{CO_2}$):

$$\frac{p_{soil}}{p^*_{soil}} = \frac{\Pi_A}{\Pi^*_A}\left(1 - \frac{p^*_{CO_2}}{p^*_{soil}}\right) + \frac{p_{CO_2}}{p^*_{soil}} \quad [21]$$

It is assumed that $p^*_{soil} = 10p^*_{CO_2}$. The constants in the weathering functions, k^{cc}_w and k^{mc}_w, are shown in **Table 2**.

Weathering products are transported to the ocean and, depending on the solubility product, precipitated to the ocean floor. The rate of precipitation is

$$F_{prec} = \left(F^{cc}_{weath} + F^{mc}_{weath}\right) + \left(F^{sc}_{weath} + F^{sc}_{weath}\right)\mu_c - \left(\frac{dm^{ca}_{eq}}{dt} + \frac{dm^{mg}_{eq}}{dt}\right)V_{shallow}\mu_c \quad [22]$$

with $V_{shallow}$ being volume of shallow water where precipitation to the seafloor can take place. About 8% of the Earth's area is covered with ocean less shallow than 10^3 m, that is, $V_{shallow} = 0.08\,A_e\,10^3$ m. This value is a first approximation, because the shallow ocean area might have changed during Earth's history. F^{sc}_{wealth} and F^{sm}_{wealth} are the weathering rates of $CaSiO_3$ and $MgSiO_3$. The biotic enhanced silicate weathering rate is defined similar to the carbonate weathering (see eqn [20]):

$$F^{sc}_{weath} = k^{cs}_w\left(\frac{a_{H^+s}}{a^*_{H^+s}}\right)^{0.5}\exp\left(\frac{T_s - 288\,K}{13.7\,K}\right)\left(\frac{A_e - A_o}{A_e - A^*_o}\right) \quad [23]$$

$$F^{sm}_{weath} = k^{ms}_w\left(\frac{a_{H^+s}}{a^*_{H^+s}}\right)^{0.5}\exp\left(\frac{T_s - 288\,K}{13.7\,K}\right)\left(\frac{A_e - A_o}{A_e - A^*_o}\right) \quad [24]$$

where k^{cs}_w and k^{ms}_w are given in **Table 2**.

Table 2 Constants in the weathering functions, equilibrium constants, and activity coefficients

k^{cc}_w	$175 \times 10^{-5}\,Ma^{-1}$	α_{H_2O}	0.967
k^{mc}_w	$142 \times 10^{-5}\,Ma^{-1}$	$\gamma_{Ca^{2+}}$	0.203
k^{cs}_w	$28 \times 10^{17}\,mol\,Ma^{-1}$	$\gamma_{Mg^{2+}}$	0.260
k^{ms}_w	$31 \times 10^{17}\,mol\,Ma^{-1}$		
K_0	$3.48 \times 10^{-2}\,bar^{-1}$		
K_1	$4.45 \times 10^{-7}\,mol\,l^{-1}$		
K_2	$4.69 \times 10^{-11}\,mol\,l^{-1}$		
k^{cc}_{sp}	$3.60 \times 10^{-9}\,mol^2\,l^{-2}$		
k^{mc}_{sp}	$1.00 \times 10^{-5}\,mol^2\,l^{-2}$		

m^{ca}_{eq} and m^{mg}_{eq} are the equilibrium concentrations of Ca and Mg cations in ocean water, which are always in a saturated state. The change of equilibrium concentrations results in a change of solubility of carbonates in ocean water and therefore influences the precipitation flux. The equilibrium concentrations of Ca and Mg cations are given by

$$m^{ca}_{eq} = \frac{k^{cc}_{sp}}{\gamma_{Ca^{2+}}K_0K_1K_2\alpha_{H_2O}} \times \frac{a^2_{H^+}}{p_{CO_2}} = c_1 \times \frac{a^2_{H^+}}{p_{CO_2}} \quad [25]$$

$$m^{mg}_{eq} = \frac{k^{mc}_{sp}}{\gamma_{Mg^{2+}}K_0K_1K_2\alpha_{H_2O}} \times \frac{a^2_{H^+}}{p_{CO_2}} = c_2 \times \frac{a^2_{H^+}}{p_{CO_2}} \quad [26]$$

where k^{cc}_{sp} and k^{mc}_{sp} are equilibrium constants and $\gamma_{Ca^{2+}}$ and $\gamma_{Mg^{2+}}$ are activity coefficients (**Table 2**).

Biogenic Enhancement of Weathering

One role of the biosphere is to increase the soil CO_2 partial pressure C_{soil} in relation to the atmospheric one. This effect is supposed to be mainly proportional to Π. This is rather a weak functional dependence of the weathering rate on biological productivity. A tenfold increase in soil p_{CO_2} relative to atmosphere gives only a 1.56-fold amplification of weathering rate due to the present biota. This is a significant underestimate, indicating that much of the observed biotic amplification of weathering is due to processes other than increased soil p_{CO_2}. The total amplification due to land life is at least a factor of 10 and may exceed 100. Therefore, one can introduce in eqn [8] an additional functional dependence of weathering on biologic productivity by the factor F_{bio}:

$$F_{bio} \equiv F_{bio}(\Pi) = \left(1 - \frac{1}{\alpha}\right)\frac{\Pi}{\Pi_0} + \frac{1}{\alpha} \quad [27]$$

For the parameter $\alpha = 1$, there is no additional amplification of weathering ($F_{bio}(\Pi) \equiv 1$). Equation [6] is a mechanistic attempt to describe the microbiological-driven processes in the sense of a first-order approximation. The enhancement factor is assumed to be linearly related to the biological productivity on a global spatial and a geological timescale.

The biological productivity Π is itself a function of various parameters such as water supply, photosynthetic active radiation (PHAR), nutrients (e.g., N and C), C_{atm}, and T_s. In the framework of this model, Π is considered to be a function of the temperature and CO_2 partial pressure in the atmosphere exclusively. According to Liebig's principle, Π can be set to a multiplicative form

$$\Pi \equiv \Pi(T_s, C_{atm}) = \Pi_{max}\cdot\Pi_T(T_s)\cdot\Pi_C(C_{atm}), \quad 0 \le \Pi_T(T_s), \Pi_C(C_{atm}) \le 1 \quad [28]$$

where Π_T describes the temperature dependence and Π_C the CO_2 dependence. The maximum productivity Π_{max} is estimated to be twice the present value, that is, $\Pi = 2\Pi_0$. Michaelis–Menten hyperbolas are suitable for describing the functional behaviour of Π_C:

$$\Pi_C(C_{atm}) = \begin{cases} \dfrac{C_{atm} - C_{min}}{C_{1/2} + (C_{atm} - C_{min})} & \text{for } C_{atm} > C_{min} \\ 0 & \text{else} \end{cases} \quad [29]$$

where $C_{1/2} + C_{min}$ is the value at which $\Pi_C = 1/2$, $C_{min} = 10$ ppm. Equation [8] tends to 1 for $C_{atm} \rightarrow \infty$.

The temperature dependence Π_T is described by a parabolic function having a maximum at $T_s = 25\,°C$:

$$\Pi_T(T_s) = \begin{cases} 1 - \left(\dfrac{T_s - 25\,°C}{25\,°C}\right)^2 & \text{for } 0°C < T_s < 50\,°C \\ 0 & \text{else} \end{cases} \quad [30]$$

The resulting function Π (T_s, C_{atm}) is a good description of the so-called net primary productivity (NPP) for the present biosphere.

Weathering depends strongly on the continental area A_c. The formation of continents is not well understood because of the diverse and heterogeneous nature of the continental crust. In modern studies, there is a growing appreciation of the observation that the continental crust grows episodically. One can use a continental growth model that is based on geological investigations of the best-studied regions, North America and Europe, which formed a single landmass for most of the Proterozoic. The model is linearly extrapolated for the future evolution of the continental area of the Earth.

One can find that biogenic enhancement of weathering has a strong influence on the overall evolution of the Earth system. Such microbiologically driven processes are able to extend the life span of the biosphere significantly up to 1.2 Gyr (see **Figure 3**). The Sun is becoming more luminous with time and will eventually overheat the biosphere. However, life cools the Earth by amplifying the rate of silicate rock weathering and maintaining a low level of atmospheric CO_2. For large enough amplification factors the global carbon cycle exhibits bistability: biotic and abiotic states of the Earth system are both stable, perturbations of the cycle are able to provoke extinctions of the biosphere if the stability domain of the biotic state is left. However, the global carbon cycle is quite resilient to random perturbations for almost the full life span of the biosphere.

See also: Climate Change 2: Long-Term Dynamics; Precipitation Pattern.

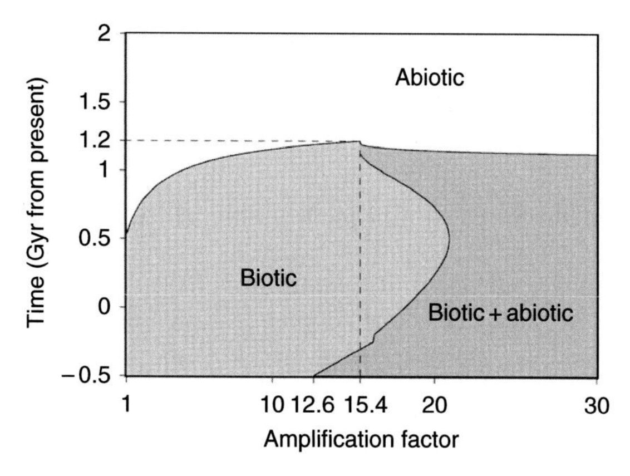

Figure 3 Earth system state as a function of biotic amplification factor α and time. In the 'biotic' region, only the solution with life is stable. In the 'abiotic' region only the solution without life is stable. 'Biotic + abiotic' indicates the bistable regime in which solutions with and without life are both stable. Reproduced from Lenton TM and von Bloh W (2001) Biotic feedback extends the life span of the biosphere. *Geophysical Research Letters* 28: 1715–1718, with permission.

Further Reading

Berner RA and Kothavala Z (2001) GEOCARB III: A revised model of atmospheric CO_2 over Phanerozoic time. *American Journal of Science* 301: 182–204.
Davis GF (1999) *Dynamic Earth, Plates, Plumes and Mantle Convection*, 458pp. Cambridge: Cambridge University Press.
Ernst WG (ed.) (1999) *Earth Systems, Processes and Issues*, 566pp. Cambridge: Cambridge University Press.
Franck S (1998) Evolution of the global mean heatflow over 4.6 Gyr. *Technophysics* 291: 9–18.
Franck S, Kossacki KJ, von Bloh W, and Bounama C (2002) Long-term evolution of the global carbon cycle: Historic minimum of global surface temperature at present. *Tellus* 54B: 325–343.
Gregor CB, Garrels RM, Mackenzie FT, and Maynard JB (1988) *Chemical Cycles in the Evolution of Earth*, 376pp. New York: Wiley.
Kump LR, Kasting JF, and Crane RG (1999) *The Earth System*, 351pp. Upper Saddle River: Prentice-Hall.
Lenton TM and von Bloh W (2001) Biotic feedback extends the life span of the biosphere. *Geophysical Research Letters* 28: 1715–1718.
Lerman A and Meybeck M (1988) *Physical and Chemical Weathering in Geochemical Cycles*, 375pp. Dordrecht: Kluwer Academic Publishers.
McGovern PJ and Schubert G (1989) Thermal evolution of the Earth: Effects of volatile exchange between atmosphere and interior. *Earth and Planetary Science Letters* 96: 27–37.
Schopf JW (1999) *Cradle of Life*, 367pp. Princeton: Princeton University Press.
Schwartzman D (1999) *Life, Temperature and the Earth*, 241pp. New York: Columbia University Press.
Stumm W and Morgan JJ (1981) *Aquatic Chemistry*, 780pp. New York: Wiley.
Walker JCG, Hays PB, and Kasting JF (1981) A negative feedback mechanism for the long-term stabilization of the Earth's surface temperature. *Journal of Geophysical Research* 86: 9776–9782.

PART D

Climate Change

Climate Change 1: Short-Term Dynamics

G Alexandrov, National Institute for Environmental Studies, Tsukuba, Japan

Introduction

Short-term dynamics of the global carbon cycle is closely related to the concept of climate system: the totality of the atmosphere, hydrosphere, biosphere, and their interactions. Human activities have been substantially increasing the concentrations of carbon dioxide and other greenhouse gases in the atmosphere and thus inducing potentially adverse changes in the climate system. This tendency has become of public concern that led to the United Nations Framework Convention on Climate Change (UNFCCC). This convention suggests protection of carbon pools, enhancement of carbon sinks, and reduction of emissions from carbon sources.

Carbon Pools

Carbon pool (or reservoir, or storage) is a system that has the capacity to accumulate or release carbon. The absolute quantity of carbon held within at a specified time is called carbon stock. Transfer of carbon from one carbon pool to another is called carbon flux. Transfer from the atmosphere to any other carbon pool is said to be carbon sequestration. The addition of carbon to a pool is referred to as uptake.

Carbon Sink

Carbon sink is a process or mechanism that removes carbon dioxide from the atmosphere. A given carbon pool can be a sink, during a given time interval, if carbon inflow exceeds carbon outflow.

Carbon Source

Carbon source is a process or mechanism that releases carbon dioxide to the atmosphere. A given carbon pool can be a source, during a given time interval, if carbon outflow exceeds carbon inflow.

Carbon Budget

The estimates of carbon stocks and carbon fluxes form the carbon budget, which is normally used as a kind of diagnostic tool in the studies of the short-term dynamics of the global carbon cycle.

Carbon Budget Components

The components of the global carbon budget may be subdivided into fossil and dynamic categories (**Figure 1**).

Fossil Components

The fossil components are naturally inert. The stock of fossil organic carbon and mineral carbonates (estimated at 65.5×10^6 PgC) is relatively constant and would not dramatically change within a century. The lithospheric part of the carbon cycle is very slow; all the fluxes are less than 1 PgC yr^{-1}. For example, volcanic emissions are estimated at 0.15–0.25 PgC yr^{-1}. Therefore, the turnover time of the storage amounts to millions or hundred millions of years.

Dynamic Components

The dynamic components are not inert. The carbon stocks in the atmosphere, ocean, soil, and vegetation remain constant as long as they are at dynamic equilibrium.

Atmosphere

The turnover time of atmospheric carbon is very short. It is less than 5 years. Therefore, the balance of atmospheric carbon strongly depends on the gas exchange between the atmosphere and ocean as well as on the gas exchange between the atmosphere and terrestrial ecosystems.

Atmospheric CO_2 content. The atmospheric content of carbon dioxide gradually increased from 680 PgC in 1960–69 to 760 PgC in 1990–99. The rate of increase was 3.3 ± 0.1 PgC yr^{-1} during 1980–89 and 3.2 ± 0.1 PgC yr^{-1}

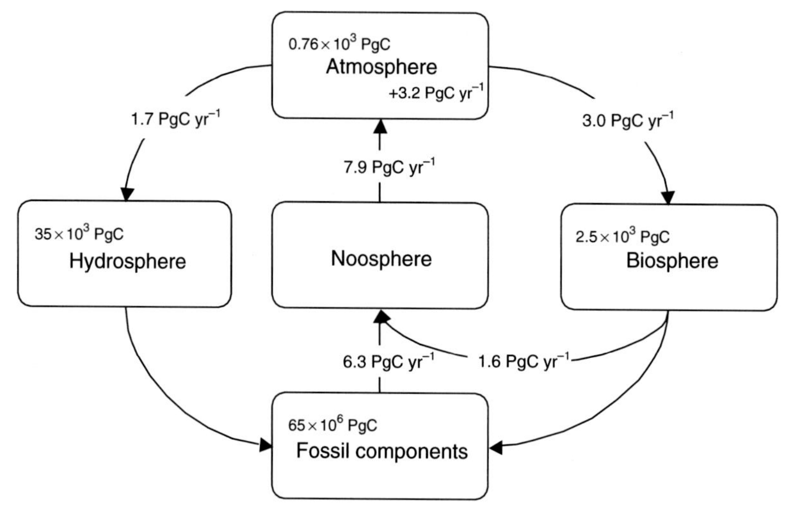

Figure 1 Dynamic components of the global carbon cycle.

during 1990–99. Interannual variations were significantly wider. The lowest rate of 1.9 PgC yr^{-1} was observed in 1992, and the highest rate of 6.0 PgC yr^{-1} was observed in 1998. Since fossil fuel emission does not show short-term variability of this magnitude, the interannual variations are normally attributed to the climate-induced variations in the land–atmosphere flux, or the ocean–atmosphere flux, or both.

Air/sea exchange. The gross carbon exchange between the atmosphere and ocean is estimated at 90 PgC yr^{-1}. Since atmospheric CO_2 concentration is increasing, there is net uptake of carbon by the ocean, driven by the atmosphere–ocean difference in partial pressure of CO_2. The magnitude of the uptake slightly varies on decadal scale: 1.9 ± 0.6 PgC yr^{-1} for 1980–89, 1.7 ± 0.5 PgC yr^{-1} for 1990–99. This variability is attributed to El Niño events. The natural efflux of CO_2 from equatorial Pacific is reduced during El Niño mainly due to the reduced upwelling of CO_2-rich waters. The decrease in the efflux varies between 0.2 and 1.0 PgC yr^{-1}.

Air/land exchange. The gross carbon exchange between the atmosphere and land is estimated at 60 PgC yr^{-1}. Despite the net release of CO_2 associated with land use, the balance between emissions from fossil fuel combustion, net ocean uptake, and accumulation in the atmosphere is positive suggesting net uptake of carbon by the land. The net terrestrial uptake widely varies on decadal scale: 0.2 ± 0.7 PgC yr^{-1} for 1980–89 and 1.4 ± 0.7 PgC yr^{-1} for 1990–99.

Ocean

The total amount of inorganic carbon in the sea is about 35 000 PgC. This includes dissolved carbon dioxide, HCO_3^-, and CO_3^{2-}. The dissolved carbon dioxide forms only one per mille of the total. Most of carbon is stored in the form of HCO_3^-. The share of CO_3^{2-} is less than 15% (about 13%).

The ocean storage is divided into surface water and deep sea. The surface water is in turn divided into cold surface water and warm surface water, both extending to 75 m depth – that is, the depth of seasonal thermocline. Warm surface water is a part of ocean between 40° N and 40° S where a well-defined permanent thermocline exists. Cold surface water is the rest of the ocean, which exchanges with deeper layers of the ocean by convection.

The surface water contains about 10% more carbon than the atmosphere that constitutes about 2% of the total carbon content in the ocean.

The concentration of dissolved carbon dioxide is proportional to its partial pressure above ocean surface (P_{CO_2}). The coefficient of proportionality is called solubility coefficient. The solubility of CO_2 varies with temperature suggesting a transfer of CO_2 in the atmosphere from warm regions to polar regions where CO_2 solubility is higher. This atmospheric transfer is balanced by the backward net oceanic transfer, the so-called 'conveyer belt'.

The dissolved carbon dioxide forms about 0.5% of the total dissolved inorganic carbon (DIC). The concentration of DIC also depends on P_{CO_2}, but in a more complicated way:

$$\frac{\Delta \mathrm{DIC}}{\mathrm{DIC}} = \frac{1}{\xi}\frac{\Delta P_{CO_2}}{P_{CO_2}}$$

where ξ is the buffer factor.

The surface water contains a significant amount of organic carbon: about 50 PgC of dissolved organic carbon (DOC), 30 PgC of particulate organic carbon (POC), and 3 PgC of plankton. The rate of photosynthesis varies from 0.06 gC m^{-2} d^{-1} in the desert regions which are characterized by downwelling and lack of nutrients to 0.6 gC m^{-2} d^{-1} in areas of intense upwelling. The total primary production amounts to 40 PgC yr^{-1}. About 10% of the primary production reaches the deep water that contains about 900 PgC of DOC.

Land

Terrestrial ecosystems store about 2500 PgC: 500 PgC in vegetation and 2000 PgC in soil. The amount of carbon stored in vegetation varies significantly depending on vegetation type. Forests generally store ten times more carbon than grasslands. However, forest soils do not necessarily contain more carbon than grassland soils, for carbon stock in soil depends on the factors that control the rate of organic matter decomposition.

The global net production of organic matter by plants, net primary production (NPP), is about 60 PgC yr^{-1}. This input to the pool of living organic matter is compensated by litter fall. The residence time of the pool is 1 year in case of annual grasslands, but it may be 100 years in case of pristine forests.

The litter fall in its turn is the input to the pool of nonliving organic matter, which is compensated by heterotrophic respiration (i.e., by CO_2 released with the respiration of soil biota decomposing organic debris). The net accumulation of carbon by ecosystem, including both soil and vegetation, is called net ecosystem production (NEP).

NEP of an undisturbed ecosystem should be close to 0. However, most of ecosystems are disturbed in some way (harvesting, fire, etc.). Therefore, global NEP is estimated at 10 PgC yr^{-1}. This value characterizes nonrespiratory losses such as release of carbon due to forest fires, or relocation of carbon to wood products and other components of urban metabolism.

The net accumulation of carbon that includes both respiratory and nonrespiratory losses is called net biome production (NBP). NBP for a relatively short period of time may differ from 0, reflecting continued effect of the losses occurred in the past. Thus, global NBP (i.e., net terrestrial uptake) for the decade 1990–99 has been estimated to be positive (1.4 ± 0.7 PgC yr^{-1}) due to decrease in the nonrespiratory losses.

Human Intervention

Fossil Fuel Combustion and Cement Production

The main anthropogenic source of CO_2 emissions is the combustion of carbon-based fuels. Since 1860, industrialization has progressed with an increase in the use of fuels – especially fossil fuels – and a corresponding increase in CO_2 emissions. The growth has been steady and exponential, interrupted only by the two World Wars and the Great Crash of 1929. In 1997 the emissions reached a maximum of 6.6 PgC yr^{-1} (0.2 PgC yr^{-1} of this was from cement production). The average value of emissions increased from 5.44 ± 0.3 PgC yr^{-1} in 1980s to 6.3 ± 0.4 PgC yr^{-1} in 1990s.

Land-Use Change

About 10–30% of the current total anthropogenic emissions of CO_2 are estimated to be caused by land-use conversion. In the historical perspective, the share of land use is larger. From 1850 to 1998, 270 ± 30 PgC has been emitted as a result of fossil fuel burning, and 136 ± 55 PgC as a result of land-use change.

The current net land-use flux comprises the balance of positive terms due to deforestation and negative terms due to regrowth on abandoned agricultural land. During 1980s the net land-use flux of 1.7 ± 0.8 PgC yr^{-1} was almost entirely due to deforestation of tropical regions. Temperate forests show approximate balance between carbon uptake in regrowing forests and carbon lost in oxidation of wood products. Since the rates of deforestation are declining, the net land-use flux was slightly smaller in 1990s: 1.6 ± 0.8 PgC yr^{-1}.

Biosphere Response

Growth Rate of Atmospheric CO_2

From 1850 to 1998, atmospheric concentration of CO_2 increased by 28%, from 285 to 366 ppm, that corresponds to 176 ± 10 PgC increase in the atmospheric content of CO_2 and to 43% of the total anthropogenic emission over this time. More than half (57%) of the anthropogenic emission has been taken up in the ocean and the terrestrial ecosystems.

Net Oceanic Uptake

The cumulative ocean uptake during the period from 1850 to 1998 is estimated to be 120 ± 50 PgC – that is, about 30% of anthropogenic emission has been taken up in the ocean.

The fraction of anthropogenic CO_2 that is taken up in the ocean declines with increasing CO_2 concentration in the atmosphere. Increasing atmospheric CO_2 concentration maintains the atmosphere–ocean difference in partial pressure of CO_2 that causes net uptake of carbon by the ocean. However, increasing DIC reduces the buffer capacity of the carbonate system, and thus weakens the capacity of the oceanic uptake.

The capacity of the ocean uptake is also sensitive to the rate of increase of atmospheric CO_2. The uptake is limited with the rate of mixing between deep water and surface water, and hence the lower the growth rate of atmospheric CO_2, the higher the rate of CO_2 sequestration.

Residual Terrestrial Uptake

Balancing the carbon budget for the period from 1850 to 1998 yields a global net terrestrial source of 26 ± 60 PgC. Hence, a residual terrestrial sink of 110 ± 80 PgC is required to reconcile the difference between the net terrestrial source and the larger terrestrial source resulted from land-use change (136 ± 55 PgC). This residual terrestrial uptake is often referred to as the 'missing carbon sink', although it can be attributed to well-known biophysical mechanisms.

Carbon dioxide fertilization

Stimulation of photosynthesis at higher CO_2 is one of the well-known mechanisms to which the 'missing carbon sink' is normally attributed. Carbon dioxide fertilization effect on plant productivity is not linear:

$$\frac{\Delta \mathrm{NPP}}{\mathrm{NPP}} = \gamma \ln\left(1 + \frac{\Delta P_{\mathrm{CO_2}}}{P_{\mathrm{CO_2}}}\right)$$

It is weakening at high atmospheric concentration of CO_2. If $\gamma = 0.35$, then NPP increases by 24%, when CO_2 increases two times. Thus, the growth of atmospheric concentration of CO_2 enhanced plant productivity by less than 10% in comparison to 1850.

Carbon dioxide fertilization produces only an excess amount of organic matter. The mechanism of carbon sequestration associated with CO_2 growth is more complicated. Carbon sequestration stems from disbalance between production and decomposition of organic matter – that is, from NEP. NEP approaches to naught when ecosystem approaches to equilibrium. Continuous growth of CO_2 maintains ecosystem disequilibrium. Therefore, the rate of carbon sequestration is determined by the rate of atmospheric CO_2 growth and the rate of carbon turnover in terrestrial ecosystems.

Nitrogen deposition

Production of organic matter is generally limited with nitrogen and some other nutrients. Therefore, NPP is expected to increase with a rapid growth in reactive nitrogen deposition over the last 150 years. Reactive nitrogen is released into the atmosphere in the form of nitrogen oxides (NO_x) during fossil fuel and biomass combustion. The annual deposition even in rural areas may amount to $40\,\mathrm{kg\ ha^{-1}}$. Another source (mainly of ammonia) is animal husbandry and fertilizer use. The nitrogen deposition mainly affects ecosystems close to cities and industrial centers as well as in the vicinity of intensive agricultural enterprises.

Land-use management

Agricultural and forest management practices significantly affect carbon stocks in managed ecosystems. These practices dramatically changed since 1850; they began to be oriented to the sustainable use of natural resources. A forest that is managed in sustainable manner operates as a machine that removes carbon from atmosphere and exports it as forest products. Similarly, alteration of tillage practices leads to protection and increase of soil carbon content.

Concluding Remarks

This article presents an overview of basic biogeochemical concepts related to short-term dynamics of global carbon cycle. It is intended to make short-term dynamics of global carbon cycle understandable to ecologists who are involved in carbon management and related ecological applications and targeted to a reader who has basic background in ecology or is familiar with the basic ideas of sustainable development, biosphere equilibrium, and environmental protection. The 'Further reading' list and cross-reference links provide the information on a broad context in which the basic concepts overviewed in this article (or their modifications) normally appear.

See also: Carbon Cycle; Global Change Impacts on the Biosphere; Ocean Currents and Their Role in the Biosphere.

Further Reading

Bolin B, Degens ET, Kempe S, and Ketner P (eds.) (1979) *The Global Carbon Cycle, SCOPE 13*. Chichester: Wiley.
Bolin B, Döös BR, Warrick RA, and Jäger J (1986) *The Greenhouse Effect, Climatic Change, and Ecosystems, SCOPE 29*. Chichester: Wiley.
Field C and Raupach MR (eds.) (2004) *The Global Carbon Cycle: Integrating Humans, Climate and the Natural World, SCOPE 62*. Washington: Island Press.
Global Carbon Project (2003) Science framework and implementation. Earth System Science Partnership Report no. 1. Canberra: Earth System Science Partnership (IGBP, IHDP, WCRP, DIVERSITAS).
Houghton JT, Ding Y, Griggs DJ, *et al.* (eds.) (2001) *Climate Change 2001: The Scientific Basis*. New York: Cambridge University Press.
Jørgensen SE and Svirezhev YM (2004) *Towards a Thermodynamic Theory for Ecological Systems*. Amsterdam: Elsevier.
Melilo JM, Field CB, and Moldan B (eds.) (2003) *Interactions of the Major Biogeochemical Cycles: Global Change and Human Impacts, SCOPE 61*. Washington: Island Press.
Schellnhuber HJ, Crutzen PJ, Clark WC, *et al.* (eds.) (2004) *Earth System Analysis for Sustainability*. Cambridge: MIT Press.
Watson RT, Noble IR, Bolin B, *et al.* (eds.) (2000) *Land Use, Land-Use Change, and Forestry*. New York: Cambridge University Press.

Climate Change 2: Long-Term Dynamics

W von Bloh, Potsdam Institute for Climate Impact Research, Potsdam, Germany

Introduction

The main component of the early Earth's atmosphere after the formation of the oceans was CO_2 as the second most abundant volatile in the accreting material. On planetary timescales the variation of the CO_2 content in the atmosphere is most important in investigating the relation between the evolution of the Sun as a main-sequence star and the stabilization of the Earth's surface temperature. During the history of the Earth the luminosity of the Sun has increased to the present level starting with a 30% lower value. There must be a mechanism that provides a feedback, generating a high concentration of atmospheric CO_2 in the past to prevent the Earth from freezing while solar luminosity was low. On the other hand a progressive lowering of CO_2 concentration is necessary for an increasing solar luminosity. Such a negative feedback mechanism is provided by the global carbon cycle among the surface reservoirs of carbon (atmosphere, ocean, crust) and the mantle reservoir. The overall chemical reactions for the weathering processes are

$$CO_2 + CaSiO_3 \leftrightarrow CaCO_3 + SiO_2$$

$$CO_2 + MgSiO_3 \leftrightarrow MgCO_3 + SiO_2$$

The main idea of this abiotic feedback is the interplay between weathering rate, surface temperature, and atmospheric CO_2 pressure. An increase of the luminosity leads to a higher mean global temperature causing an increase in weathering. Then more CO_2 is extracted from the atmosphere weakening the greenhouse effect. Overall the temperature is lowered and homeostasis, that is, self-stabilization of the global surface temperature is achieved. Plate tectonics is a necessary condition for closing the carbon cycle. Without spreading and subduction carbonates would be buried on the seafloor and not brought back to the atmosphere via regassing at mid-ocean ridges or andesitic/back-arc volcanism.

During the evolution of the Earth the global carbon cycle has been mediated by the biosphere. The occurrence of microfossils and stromatolites in rocks has shown that life had originated at least 3.5 Gyr ago. Isotopic signatures of $^{13}C/^{12}C$ suggest the presence of organic carbon in 3.8 Gyr rocks from Greenland. Photosynthetic fixation leads to buildup of organic material consuming CO_2 from the atmosphere. On the other hand, organic material is decomposed and CO_2 is remobilized:

$$CO_2 + H_2O \leftrightarrow CH_2O + O_2$$

Carbon isotopes imply that the enzyme Rubisco preferring the lighter isotope ^{12}C and therefore oxygenic photosynthesis controlled the global distribution of carbon in the atmosphere–ocean system for at least 3.5 Gyr and gave an imprint on the isotopic signature of the carbon reservoirs. Direct evidence of oxygenic photosynthesis is given by the steep rise of oxygen in the atmosphere 2.2 Gyr ago. Life is unable to influence Earth's carbon cycle in the absence of photosynthesis.

The biosphere can in principle be divided into three life forms that appeared on Earth in consecutive order (see **Figure 1**). First Archaean and prokaryotic bacteria appeared, second eukaryotic life, and third complex multicellular life. In particular complex multicellular organisms have a strong influence on weathering by amplifying the weathering rates.

Global Carbon Cycle and the Biosphere

The global carbon cycle can be described by a box model of the surface reservoirs of carbon and the mantle reservoir. **Figure 2** denotes the most pertinent fluxes between the storage of carbon in the mantle, the combined reservoir consisting of ocean and atmosphere, the continental crust, the ocean crust and floor, the kerogen, and the three different biospheres. The efficiency of carbon transport between the reservoirs takes into account mantle de- and regassing, carbonate precipitation, carbonate accretion, hydrothermal reactions at mid-ocean ridges, decay of dead organic matter, and weathering processes. The Earth's crust contains the carbon storage of the continents, kerogen, and the ocean crust and seafloor. Biomass is accumulated from the atmosphere by photosynthesis. A fraction of the dead organic matter is buried in the kerogen pool. Additionally to these direct effects of the biosphere on the surface reservoirs of

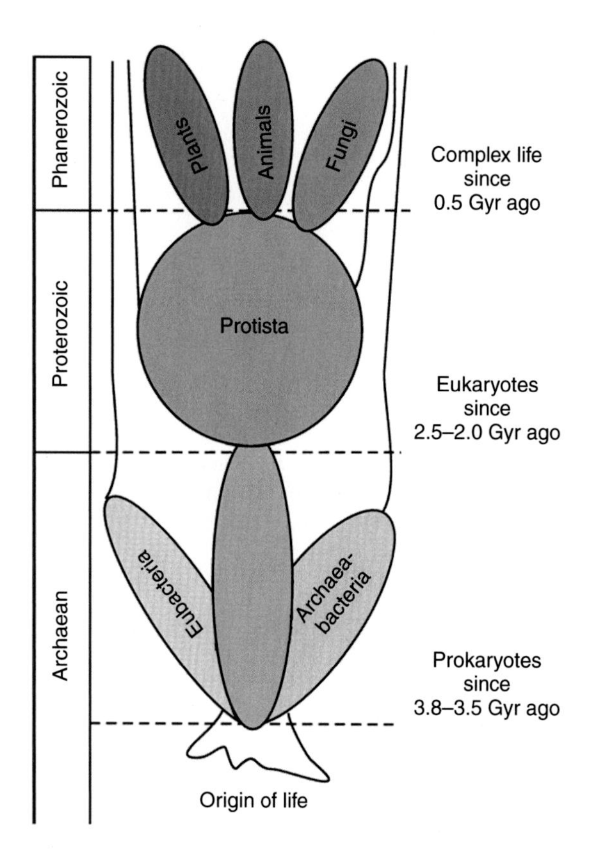

Figure 1 Evolutionary path of life on Earth.

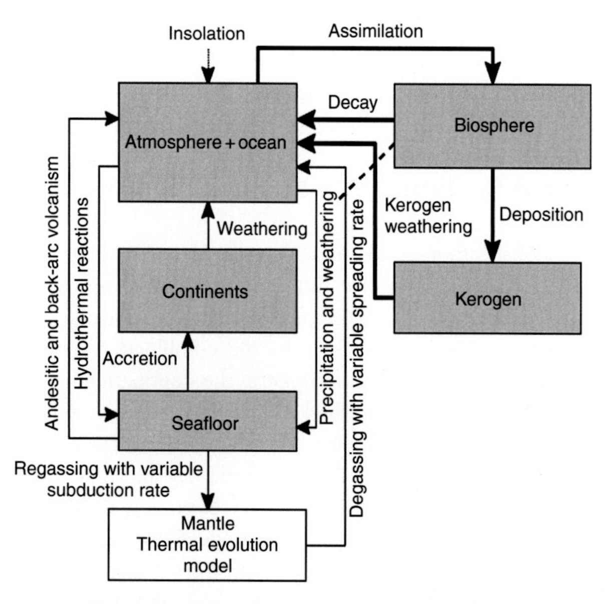

Figure 2 Box diagram illustrating the basic mechanisms and interactions of the global carbon cycle. Gray boxes are the surface reservoirs of carbon. The fluxes from and to the different pools are indicated by arrows. Bold arrows denote fluxes affected by the biosphere.

carbon there is an indirect effect due to the increase in weathering rates denoted by the dashed arrow in **Figure 2**.

Continental Growth

The continental crust is especially diverse and heterogeneous, and its formation is less understood than that of the geologically relatively simple oceanic crust. In contrast to the seafloor which is primarily made of basalts it is granitic in composition. Two different hypotheses have been suggested to explain the evolution of the continental crust. The first proposes that the present continental crust formed very early in the Earth's history and has been recycled through the mantle in steadily decreasing fashion such that new additions are balanced by losses, resulting in a steady-state system. The return of the continental material to the mantle and its replacement by new younger additions reduces its mean age, both of which keep the mass of the continents constant. The second hypothesis proposes crustal growth throughout geological time without recycling into the mantle.

Figure 3 summarizes the different classes of continental growth scenarios. The continental area can be assumed to be fixed at its present value, can be grown linearly with a constant growth rate, or linearly with a delay. Geological investigations of the best-studied regions, North America and Europe, which formed a single land mass for most of the Proterozoic suggests that the continental crust grows episodically, and it is concluded that at least 60% of the crust has been replaced by the late Archaean. This data-based description is clearly more realistic than the theoretical models. The continental area is directly related to the

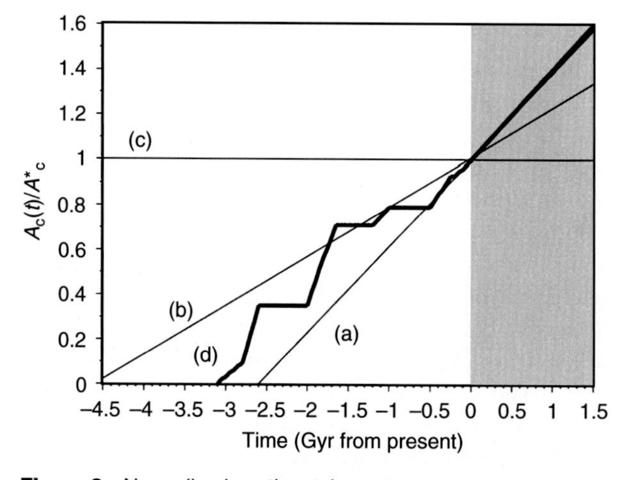

Figure 3 Normalized continental area as a consequence of the following continental-growth scenarios: (a) delayed growth, (b) linear growth, (c) constant area, (d) growth derived from geological investigations. From Condie KC (1990) Growth and accretion of continental crust: Inferences based on Laurentia. *Chemical Geology* 83: 183–194.

weathering process. A larger continental area leads to proportionally higher weathering rates.

Atmospheric Carbon Dioxide and Climate

The climate of the Earth is governed by the energy-balance equation between incoming and outgoing radiation and depends on the concentration of greenhouse gases and the solar luminosity. CO_2 and H_2O are the most abundant greenhouse gases in the atmosphere. The partial pressure of H_2O, p_{H_2O}, can be expressed as a function of temperature and relative humidity using the Clausius–Clapeyron equation. Therefore, the global mean temperature of the Earth depends primarily on the CO_2 concentration in the atmosphere and the solar luminosity. During the Earth's history the luminosity of the Sun has increased at a rate of about 10% per Gyr and will increase in the future (up to the next 5 Gyr) at approximately the same rate.

The efficiency of weathering processes and the biosphere productivity strongly depend on the partial pressure of atmospheric CO_2, p_{CO_2}. On geological time-scales atmospheric carbon content is always in chemical equilibrium with the carbon content in the ocean. In the case of a constant density profile of carbon dioxide in the atmosphere the distribution of carbon can be calculated from the condition of equal partial pressures of CO_2 at the interface between atmosphere and ocean.

Biotic Enhancement of Weathering

The rate of weathering is greatly amplified by a range of biological processes that respond to photosynthetic productivity. First, there is an increase of soil CO_2 partial pressure due to respiration of soil organisms and due to the respiration from the roots of vascular plants. Furthermore litter is decomposed by microorganisms, through providing organic matter for the formation of humic acids, and through mycorrhizal and fungal digestion. This can be expressed by a direct dependence of weathering on biological productivity by a factor β mediating the weathering rate, F_{weath}:

$$F_{weath} \propto \beta \cdot \left(\frac{a_{H^+,s}}{a^*_{H^+,s}}\right)^{0.5} \exp\left(\frac{T_s - 288\,K}{13\,K}\right)$$

where $a_{H^+,s}$ denotes the activity of fresh soil water $a^*_{H^+}$ is the corresponding present-day values. The activity $a_{H^+,s}$ itself is a function of the surface temperature and the CO_2 partial pressure in the soil, p_{soil}, where the equilibrium constants for the chemical activities of the carbon and sulfur systems have been taken into account. p_{soil} depends on the terrestrial biological productivity Π, the atmospheric CO_2 partial pressure, p_{CO_2}, and their corresponding (long-term mean pre-industrial) present values p^*_{soil}, Π^*, $p^*_{CO_2}$:

$$\frac{p_{soil}}{p^*_{soil}} = \frac{\Pi}{\Pi^*}\left(1 - \frac{p^*_{CO_2}}{p_{CO_2}}\right) + \frac{p_{CO_2}}{p^*_{soil}}$$

It is assumed that $p^*_{soil} = 10 p^*_{CO_2}$. The parametrization of weathering that considers only the variation of soil carbon dioxide levels gives a biotic amplification of weathering of 1.56 for the present state, which is a significant underestimate. Furthermore vascular land plants increasing the partial pressure of CO_2 in the soil appeared on Earth only 0.35 Gyr ago. The total weathering amplification due to land life is at least a factor of 10. This indicates that much of the observed biotic amplification of weathering is due to processes other than increased soil p_{CO_2}. This is considered by the prefactor β that reflects the biotic enhancement of weathering by the biosphere types, i:

$$\beta = \prod_{1=1}^{3}\left(\frac{1}{\beta_i} + \left(1 - \frac{1}{\beta_i}\right)\frac{\Pi_i}{\Pi^*_i}\right)$$

The factor β_i denotes the specific biotic amplification of weathering, Π_i the specific biological productivity, and Π^*_i the respective present-day value of biosphere type i. Biotic enhancement of weathering is only affected by complex multicellular life ($\beta_1 = \beta_2 = 1$, $\beta_3 > 1$). Complex multicellular life contributes about 10–100 times more to the biotic enhancement of weathering than primitive life.

Biological Productivity

The biological productivity Π is the amount of biomass that is produced by photosynthesis per unit time. In reality, Π is a function of various parameters as water supply, photosynthetically active radiation (PAR), nutrients (N, P, etc.), atmospheric CO_2 content and surface temperature:

$$\frac{\Pi_i}{\Pi_{max,i}} = f_{T_{s,i}}(T_s) \cdot f_{CO_2,i}(p_{CO_2}) \cdot f(N, P, H_2O, PAR, \ldots)$$

where $\Pi_{max,i}$ is the maximum productivity of biosphere type i. For simplification biological productivity should depend only on the mean global surface temperature, T_s, and on the CO_2 partial pressure of the atmosphere, p_{CO_2}. Both variables are affected by the global carbon cycle. The qualitative dependence on CO_2 partial pressure and temperature is shown in **Figure 4**. The function for the temperature dependence, $f_{Ts,i}$, can described by a parabola and the function for the p_{CO_2} dependence is an increasing function with a saturation level. A minimum CO_2 atmospheric partial pressure, $p_{min,l}$, allowing photosynthesis is necessary for all biosphere types. A biosphere based on C3 photosynthesis has a minimum value of 150×10^{-6} bar, while C4 photosynthesis results in a value of 10^{-5} bar. The interval $[T_{min,i} \ldots T_{max,i}]$ is the temperature tolerance window for the biosphere. If the global surface temperature is inside this window a global

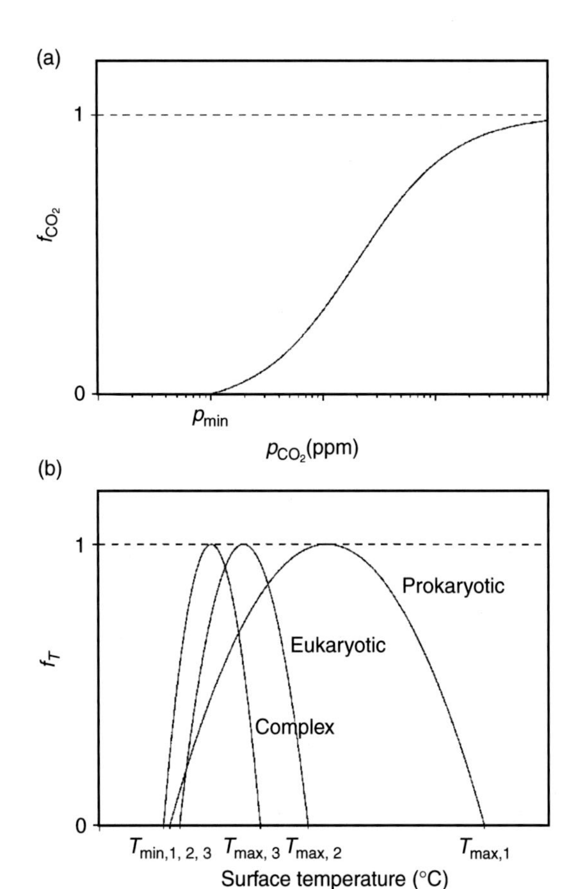

Figure 4 (a) The dependence of biosphere productivity on CO_2 partial pressure in the atmosphere. (b) The dependence of biosphere productivity on global surface temperature for prokaryotes, eukaryotes, and complex multicellular life.

abundance of biosphere type *i* is possible. It must be emphasized that this window is related to the mean global surface temperature. Latitudinal differences in temperature decrease as global mean temperature increases and might vanish for $T > 30\,°C$. **Table 1** contains estimated parameter ranges for the prokaryotic, eukaryotic, and complex multicellular biosphere, respectively.

Carbonate Precipitation

Weathering products are transported to the ocean and, depending on the solubility product, precipitated to the ocean floor. Because there exists a calcium carbonate compensation depth level in the present ocean, carbonates can precipitate only in the shallower regions such as around the mid-ocean ridges and the continental shelves. A total of 8% of the Earth's area is covered with ocean less shallow than 10^3 m. The change of equilibrium concentrations of Ca and Mg in water results in a change of solubility of carbonates in ocean water. Furthermore, oceanic photosynthesis provides an additional way to sequester carbon on the seafloor.

Hydrothermal Reactions

Due to hydrothermal reactions CO_2 dissolved in the oceans reacts with fresh mid-ocean basalts and precipitates in the form of carbonates to the ocean floor. Therefore it is an additional sink in the atmosphere-ocean reservoir. The hydrothermal flux is proportional to the production of fresh basalt at mid-ocean ridges, which in turn is proportional to the areal spreading rate. The area around the spreading centers is likely to be one of the most habitable environments for a subsurface biosphere. It is porous and characterized by extensive hydrothermal circulation. Such hydrothermal systems provided a site for the rapid emergence of life through a sequence of abiotic synthesis.

Kerogen

Kerogen comprehends the dispersed, insoluble, organic carbon in rock including coal and mineral oil deposits. It is probably the least important reservoir from the point of view of carbon cycling because it is relatively inert. However, there are processes of kerogen weathering and kerogen formation. Kerogen is formed from ~0.1% of the dead biomass that is not returned to the atmosphere through litter decomposition. The present size of the kerogen reservoir of 10–20% of the surface reservoirs is obviously the net result of these processes. The main constraint for the reservoir size results from isotopic geochemistry. Since kerogen is isotopically light due to its biological origin it sequesters preferentially ^{12}C, while the continental carbon reservoir must get enriched in the heavier isotope ^{13}C. The

Table 1 Parameter estimates for the three different life forms (prokaryotes, eukaryotes, complex multicellular life)

Life form	*Prokaryotes*	*Eukaryotes*	*Complex multicellular*
T_{min} (°C)	2	5	0
T_{max} (°C)	100–130	45–60	30–45
P_{min} (10^{-6} bar)	10^a–150^b	10^a–150^b	10^a–150^b
β	1	1	4–20

[a]For C4 plants.
[b]For C3 plants.

isotopic signature is measured as a difference to a standard sample:

$$\delta^{13}C = \left[\frac{(^{13}C/^{12}C)\text{sample}}{(^{13}C/^{12}C)\text{standard}} - 1\right] \times 1000‰$$

The kerogen has a δ^{13}C value of $\sim -20‰$. The isotopic composition of the two carbon reservoirs kerogen and continental crust might have been constant over the last 3.5 Gyr. The ratio of kerogen carbon to continental carbon would also have been constant at a value of 1:4 taking into account the isotopic signature of the mantle carbon of $\delta^{13}C \sim -5‰$.

Atmospheric Oxygen

The evolution of the atmospheric partial pressure of oxygen, p_{O_2}, can be derived from the evolution of the kerogen pool C_{ker}, that is, the long-term deposition of reduced organic carbon. Between about 2.2 and 2.0 Gyr ago there was a global oxidation event in which atmospheric p_{O_2} rose from <0.0008 to >0.002 bar. Under the assumption that before 2.2 Gyr all oxygen had been chemically bound it is possible to make the following simple estimate:

$$p_{O_2}(t) = p^*_{O_2} \cdot \frac{C_{ker(t)} - C_{ker}(t = -2.2\,\text{Gyr})}{C^*_{ker} - C_{ker}(t - 2.2\,\text{Gyr})}$$

where $p^*_{O_2}$ is the present atmospheric O_2 level and C^*_{ker} is the size of the present kerogen pool.

Coevolution of the Biosphere–Geosphere System

The feedback between the biosphere and the surface reservoirs of carbon leads to several bifurcation points in Earth's history. In particular the evolution of the climate is affected by the change in CO_2 concentration in the atmosphere. Atmospheric CO_2 concentration is regulated by biologically mediated weathering processes and is driven by an increase in solar luminosity, continental growth, and lowering mantle temperatures. The decline of mantle temperature is causing a decrease in the spreading rate with lower outgassing at mid-ocean ridges.

Evolution of the Climate

Figure 5a shows the results for the evolution of the mean global surface temperature (solid line). The figure has been derived from a coupled model of the global carbon cycle including the biosphere. The modeled surface temperature curve is in good agreement with the ^{18}O chert thermometer. According to these data, the ocean surface water has cooled from 70 °C (±15 °C) in the Archaean

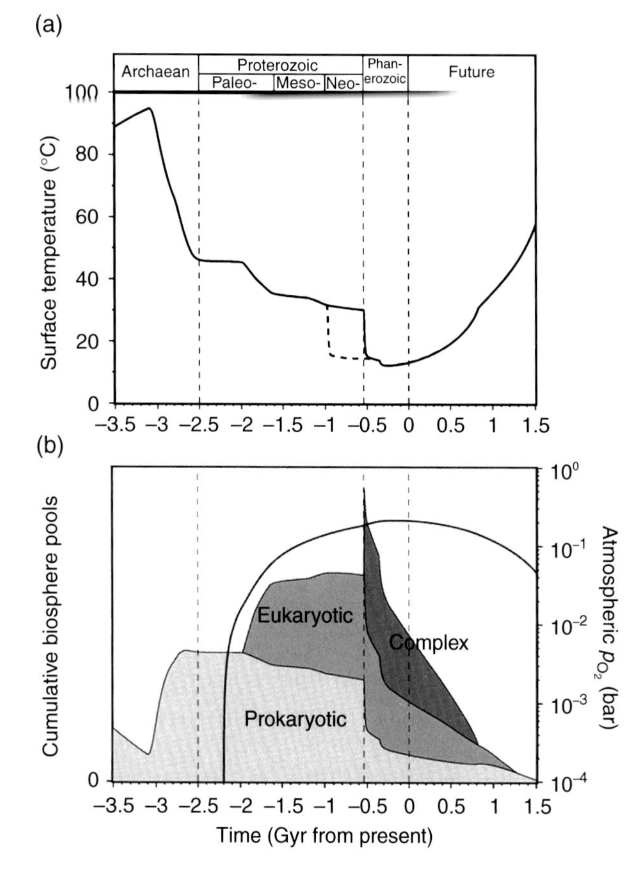

Figure 5 (a) Evolution of global surface temperature (solid line). The dashed line denotes a second possible evolution path triggered by a temperature perturbation in the Neoproterozoic era. (b) Evolution of the cumulative biosphere pools for prokaryotes, eukaryotes, and complex multicellular life. From Von Bloh W, Bounama C, and Franck S (2003) Cambrian explosion triggered by geosphere–biosphere feedbacks. *Geophysical Research Letters* 30(18): 1963–1967.

to the present value. This is caused by the growth of the continental area increasing the weathering processes and decreasing spreading rates lowering CO_2 outgassing at mid-ocean ridges. There was a drop in temperature 0.54 Gyr ago due to an increase in weathering rates caused by the first occurrence of complex life. After that event temperatures have roughly stabilized around the optimum growth temperatures for complex life. In the future the global surface temperature will rise because the increase in solar luminosity cannot be balanced by intensified weathering rates.

Evolution of the Biosphere

Figure 5b shows the cumulative biomasses for the three life forms. From the Archaean to the future there always exists a prokaryotic biosphere. At 2 Gyr ago eukaryotic life first appears because the global surface temperature reaches the tolerance window for eukaryotes. This moment correlates with the onset of a temperature fall caused by an increasing continental area. The resulting enlargement in the weathering flux takes out CO_2 from the atmosphere. In contrast to the eukaryotes the first appearance of complex multicellular life starts with an explosive increase in biomass connected with a strong decrease in Cambrian global surface temperature at about 0.54 Gyr ago. The biological colonization of land surface by metaphyta and the consequent increase in silicate weathering rates caused a reduction in atmospheric CO_2 and planetary cooling. Protein sequence analysis has shown that a first appearance of land plants at this time was already possible. Metazoan fecal pellets supplied a new and important transport mechanism of organic carbon to the deep ocean. This provides an additional sink for CO_2 in the atmosphere–ocean system.

Cambrian Explosion

The Cambrian explosion is known as the Big Bang in biology. It began 542 million years ago and ended about 40 million years later. This period is characterized by the first appearance of abundant skeletonized metazoans, a sudden increase in biodiversity, and the emergence of most modern lines. In the Vendian (0.56–0.54 Gyr ago) first animals with soft bodies appeared announcing the Cambrian explosion. Before the Vendian period life was microscopic, vegetative, and mainly prokaryotic and eukaryotic.

There is still a lot of speculation about what caused the Cambrian explosion and why it happened when it did after 3 billion years of potential evolutionary time. The approaches that have been put forward to solve the puzzle of what triggered the explosion can be split into extrinsic (environmental) factors, intrinsic (biological) factors, or a mixture of both. Extrinsic factors are physical changes in the Precambrian environment. Among these changes are the breakup of the supercontinent Rodinia and the Neoproterozoic glaciations known as snowball Earth events. The snowball Earth events and the continental breakup are associated with genetic isolation but also with a reorganization of oceanic flow patterns causing upwelling, with increasing primary production, and with a consequently higher atmospheric oxygen level. Another cause is given by the rise of atmospheric oxygen as a trigger of the Cambrian explosion. This higher oxygen level can be caused by an intensified phosphorus flux into the ocean. The phosphorus is released by weathering rates biotically enhanced by the first colonization of continents. Intrinsic causes involve some mechanisms within the Precambrian biosphere itself, which enabled evolution and diversification to start. An example is the finding in developmental genetics that the mutation of an ancestral metazoan could potentially initiate a large morphological change in the animal.

The dashed line in **Figure 5a** shows a possible second evolutionary path. A cooling event can cause a premature rising of complex life. Up to 1.75 Gyr ago there is only a unique evolutionary path. After that time more than one stable state of the Earth system exists (bistability). It depends on the environmental conditions which state (with or without complex life forms) is realized.

Phanerozoic Time

After the Cambrian explosion, there was a continuous decrease of biomass in all pools. At 0.35 Gyr ago there was a slight drop in the global temperature connected with the rise of vascular plants. At this time weathering rates were increased due the elevated partial pressure of CO_2 in the soil by root respiration. The continuous decrease in biomass of primitive life forms (prokaryotes and eukaryotes) since the Cambrian explosion is related to the fact that Phanerozoic surface temperatures are below the optimum for these life forms. The decrease in biomass of complex life forms is due to the fact that there is a continuous decrease in Phanerozoic atmospheric carbon content. At present the biomass is almost equally distributed between the three pools and the mean global surface temperature of about 15 °C is near the optimum value for complex multicellular life.

Future Evolution

In the future we can observe a further continuous decrease of biomass with the strongest decrease in complex multicellular life. The life spans of complex multicellular life and of eukaryotes end at about 0.8 Gyr and 1.3 Gyr from present, respectively. In both cases the extinction will be caused by reaching the upper limit of the temperature tolerance window. In contrast to the first appearance of complex multicellular life via the Cambrian explosion, its extinction proceeds more or less continuously. In the future there will be no bistability, that is, the extinction of complex multicellular life will not proceed as an implosion. Comparing these results with the life span for an Earth without biotic enhancement of weathering ($\sim$0.5 Gyr) the life spans are extended.

Summary

The global temperature on Earth is regulated by the global carbon cycle. The main negative feedback is provided by the weathering processes mediated by the biosphere. In the past the Earth system was characterized by lowering temperatures caused by continental growth and declining outgassing at mid-ocean ridges. In the future, however, temperatures will rise due to the increase in solar luminosity.

The Cambrian explosion can be explained by extrinsic environmental causes, that is, a gradual cooling of the Earth. The Cambrian explosion was so rapid because of a positive feedback between the spread of biosphere, increased silicate weathering, and a consequent cooling of the climate. The environment itself has been actively changed by the biosphere maintaining the temperature conditions for its existence. Therefore, this explanation of the Cambrian explosion is in line with the Gaia theory of the Earth as a self-regulating system.

Prokaryotes, eukaryotes, and complex multicellular life forms will become extinct in reverse sequence of their appearance. Nonlinear interactions in the biosphere–geosphere system caused bistability during the Neo- and Mesoproterozoic era. There is no bistability in the future solutions for complex life. Therefore, complex organisms will not become extinct by an implosion (in comparison to the Cambrian explosion). Eukaryotes and complex life become extinct because of too high surface temperatures in the future. The time of extinction is mainly determined by the upper temperature tolerance limit of these life forms. The ultimate life span of the biosphere is defined by the extinction of prokaryotes in about 1.6 Gyr because of CO_2 starvation. Only in a small fraction (1.3–1.7 Gyr) of its habitability time (6.2 Gyr) can our home planet harbor advanced life forms.

See also: Weathering.

Further Reading

Berner RA and Kothavala Z (2003) GEOCARB III: A revised model of atmospheric CO_2 over phanerozoic time. *American Journal of Science* 301: 182–204.

Caldeira K and Kasting JF (1992) The life span of the biosphere revisited. *Nature* 360: 721–723.

Franck S, Kossacki KJ, von Bloh W, and Bounama C (2002) Long-term evolution of the global carbon cycle: Historic minimum of global surface temperature at present. *Tellus* 54B: 325–343.

Knauth LP and Lowe DR (2003) High Archean climatic temperature inferred from oxygen isotope chemistry of cherts in the 3.5 Ga Swaziland supergroup, South Africa. *GSA Bulletin* 115(5): 566–580.

Condie KC (1990) Growth and accretion of continental crust: Inferences based on Laurentia. *Chemical Geology* 83: 183–194.

Lenton TM and Watson AJ (2004) Biotic enhancement of weathering, atmospheric oxygen and carbon dioxide in the Neoproterozoic. *Geophysical Research Letters* 31(5) L05202, dio: 10.1029/2003GLO18802.

Lenton TM and von Bloh W (2001) Biotic feedback extends the life span of the biosphere. *Geophysical Research Letters* 28(9): 1715–1718.

Lovelock JE and Watson A (1982) The regulation of carbon dioxide and climate. *Planetary and Space Science* 30: 795–802.

Nisbet EG, Cann JR, and Dover CL (1995) Origins of photosynthesis. *Nature* 373: 479–480.

Schidlowski M (2001) Carbon isotopes as biogeochemical recorders of life over 3.8 Ga of Earth history: Evolution of a concept. *Precambrian Research* 106: 117–134.

Schwartzman D (1999) *Life, Temperature and the Earth*. New York: Columbia University Press.

Schwartzman D and Volk T (1991) Biotic enhancement of weathering and the habitability of Earth. *Nature* 340: 457–460.

Smil V (2002) *The Earth's Biosphere: Evolution, Dynamics, and Change*. Cambridge, MA: MIT Press.

Tajika E and Matsui T (1990) The evolution of the terrestrial environment. In: Newsom HE and Jones JH (eds.) *Origin of the Earth*, pp. 347–370. Oxford: Oxford University Press.

Volk T (1987) Feedbacks between weathering and atmospheric CO_2 over the last 100 million years. *American Journal of Science* 287: 763–779.

Von Bloh W, Bounama C, and Franck S (2003) Cambrian explosion triggered by geosphere–biosphere feedbacks. *Geophysical Research Letters* 30(18): 1963–1967.

Westbroek P (1991) *Life as a Geological Force: Dynamics of the Earth*. New York: W. W. Norton & Company.

Climate Change 3: History and Current State

I I Mokhov and A V Eliseev, AM Obukhov Institute of Atmospheric Physics RAS, Moscow, Russia

Introduction

The term 'climate' has originated from the Greek word *klima*, meaning inclination. In ancient Greece, difference in weather conditions was associated with a different inclination of solar rays to the surface of the Earth. At present, the term 'climate' commonly means typical weather conditions for a given area. More formally, climate is determined as a statistical ensemble of weather for sufficiently long, usually a few (frequently, three) decades of years, time intervals.

The basic compartments of the Earth's climate system are 'atmosphere', oceans, land, 'cryosphere', and 'biosphere' ('biota'). For the atmosphere, only typical timescales are up to a few months. For the ocean, the timescale is of order 10^0–10^3 years, depending on the layer involved. For the land compartment, appropriate timescales are between a few days up to several decades of years. For the terrestrial biosphere, a typical timescale is about 10^0–10^1 years. Very large timescale of order 10^3–10^4 years is associated with the terrestrial ice sheets.

External forcing also operates at different timescales. For instance, orbital forcing exhibits strongest changes at periods about 20 ky (precessional forcing), 40 ky (obliquity forcing), and 100 ky (eccentricity forcing); these are the so-called Milankovitch periods. According to the currently accepted astrophysical theory, solar irradiance has increased by about 2% over the last 200 million years. Earth volcanism, very intensive in the past, has gradually diminished over the last about 6 ky, while explosive eruptions have still occurred during the last few centuries.

As meteorological instrumental measurements cover only the last few centuries, to quantify past climates indirect (proxy) data are used. Among those data, terrestrial and marine sediments cover very old paleoepochs up to 1 My BP. Epochs that are more recent may be reconstructed based on the flora pollen and oceanic corals. The latter are based mostly on the fossil 'foraminifera' samples. Ice boreholes serve as a very important source of our knowledge about the last several hundreds of thousands of years, especially the deep drilling holes at the Antarctic sites Vostok and Dome C providing information of the last 420 ky and 720 ky, respectively. For the last few thousand years, important climatic information comes from tree annual rings.

The longest instrumental temperature record exists for central England (monthly means are available since 1659). Instrumental meteorological measurements have become routine since the middle of the nineteenth century.

Pre-Quaternary Climates

The Earth formed as a planet about 4.6 billion years BP.

The Precambrian climates (older than 570 My BP; here and thereafter see **Table 1**) are poorly covered by the proxy data and are the most uncertain. While solar irradiance was lesser by about one-third in this period than at present, there is an evidence of the existence of liquid water on the Earth's surface. Such a low solar input into the climate system could be compensated by high concentrations of the greenhouse gases, basically CO_2 and CH_4, in the atmosphere. It is likely that part of the Precambrian period was extremely cold, as glacial deposits were found even at the sites located in the tropical latitudes.

For the Paleozoic era (570–236 My BP), more reliable data on the oceanic temperatures exist only for the late period (the Permian). However, there are evidences of large climate variations during the Paleozoic. In particular, the supercontinent Gondwana (located in the high southern latitudes) is supposed to be covered by ice in the late Paleozoic era (the so-called 'Permian glaciation').

Table 1 The geochronological chart

Era	*Period*	*Epoch*
Cenozoic (65 My BP–present)	Quaternary (1800 ky BP–present)	Holocene (11 ky BP–present)
		Pleistocene (1800–11 ky BP)
	Tertiary (67–1.8 My BP)	Pliocene (9–1.8 My BP)
		Miocene (25–9 My BP)
		Oligocene (37–25 My BP)
		Eocene (56–37 My BP)
		Paleocene (67–56 My BP)
Mesozoic (236–67 My BP)	Cretaceous (133–67 My BP)	
	Jurassic (186–133 My BP)	
	Triassic (236–186 My BP)	
Paleozoic (570–236 My BP)	Permian (282–236 My BP)	
	Carboniferous (346–282 My BP)	
	Devonian (402–346 My BP)	
	Silurian (435–402 My BP)	
	Ordovician (490–435 My BP)	
	Cambrian (570–490 My BP)	
Precambrian (beginning of the Earth–570 My BP)		

The Mesozoic era (236–67 My BP) was extremely warm, ~10–15 °C warmer than at present, with only small temperature difference (about 15 °C) between equatorial and polar belts (currently, the value for this difference is about 46 °C). There is no evidence of ice during the Mesozoic. According to the vegetation proxy data, even in winter the land interiors were under the positive temperatures. Currently, it is unknown what induced this warming.

In the Late Mesozoic era, geographical distribution of continents has become similar to that at present. However, large channels between North and South Americas, and between Africa and Eurasia, allowed the forming of an intensive circumequatorial current.

Atmospheric circulation was less intense than at present due to smaller meridional temperature gradients. It is likely that passat belts were extended further to the poles, and midlatitudinal westerlies were shifted poleward.

Since the Mesozoic, the climate cooled. This cooling started in the Eocene (56–37 My BP) and has continued to the present. In the Oligocene (37–25 My BP), the Antarctic ice sheets were formed, but the Northern Hemisphere was free from ice sheets. The glaciation in the Northern Hemisphere was started only in the Miocene (25–9 My BP) in Greenland. The Late Pre-Quaternary period, the Pliocene (9–1.8 My BP) was characterized by freezing in the Arctic.

Quaternary Climates

The Pleistocene

The Early Quaternary, the Pleistocene (1800–11 ky BP), is characterized by successive glaciations and interglaciations. The timescales of their recurrence correspond to the Milankovitch periods. In the Early Pleistocene (older than 700 ky), the dominant period of recurrence was about 40 ky, and it is shifted to about 100 ky lately.

Among the Late Pleistocene glacials/interglacials, a warm period around 120 ky BP (the 'Eemian interglacial'), is very notable. Changes in orbital parameters in comparison to those at present (greater obliquity and eccentricity, and perihelion) have led to greater seasonal temperature variations in the Northern Hemisphere. Sea level is supposed to be about 4–6 m higher than today, with much of this additional liquid water coming from Greenland.

At the end of the Eemian interglacial, temperature dropped rapidly during about 10 ky and then cooled more gradually, leading to the development of the last major glaciation. The peak of this glaciation was at 18–22 ky BP, the Last Glacial Maximum (LGM). At this time, huge ice sheets covered Northern Europe, Canada, and the northern half of the West Siberian Plain. In the South America, the Patagonian Ice sheet developed. The flows in the Ob and Yenisei Rivers were stopped due to ice sheets, creating large lakes. Permafrost covered most part of Europe. In warmer regions, LGM climates were dry. Sea level dropped by about 120 m due to the huge amount of water stored in ice sheets with thickness up to 3 km. Global mean temperature is estimated to be about 5° lower than today. The concentration of the carbon dioxide in the atmosphere was about 200 ppmv, and the concentration of methane was about 350 ppbv.

In the period 18–11 ky BP, the climate warmed gradually. However, this gradual warming was interrupted, for example, during the Younger Dryas (about 12 ky BP). This cold event is attributed to the large influx of freshwater to the North Atlantics from the melting Laurentide Ice Sheet. Freshening of the North Atlantic surface water may have resulted in the weakening of the thermohaline circulation which transports warm water from the low latitudes to the northern subpolar belt. Lacking this heat transport, the North Atlantics had to cool.

The Holocene

After termination of the latter major glaciation, about 11.5 ky BP, the climate became milder. The period after this termination and continuing to the present is called the Holocene. The onset of this interglaciation was not synchronous in different regions with a time difference probably up to 4 ky. In particular, the Laurentide Ice Sheet in the North America existed during 11–9 ky BP, chilling the continent. In the Holocene, the climate optimum occurred at 6–5 ky BP. At this time, temperature increased up to 4 °C in the northern high latitudes. Averaged globally, the plausible warming was 0.5–2 °C in comparison to the mid-twentieth century. In the Southern Hemisphere (e.g., the Antarctic), the warmest period during the Holocene occurred at ~10.5–8 ky BP. There is evidence that in the Early Holocene (older than 6 ky), Africa was more humid than today, with Sahara, presently the great desert, covered by numerous lakes.

The Holocene is an interglacial, which is unprecedently long in the Quaternary. This is much longer than other interglacials in the Quaternary, which have lasted a few thousand years.

The Climate of the Last Millennium

Climate of the Preindustrial Era

The last millennium is presented by relatively numerous proxy data, derived from trees, ice cores, and corals better than previous ones.

There is a lot of regional evidence, that, in the ninth to eleventh centuries, climate was warmer than today, the 'Medieval climatic optimum'. Amount of ice in the North Atlantics diminished. Glaciers in the Alps retreated. European winters were milder than the present-day ones. However, due to limitations of the proxy data, quantitative reconstructions are uncertain, especially if they are interpreted in terms of the globally averaged annual mean temperature. While some studies conclude that global temperature in this period was slightly higher than today, others argue that the Medieval climate optimum was limited for the Northern Hemisphere extratropics.

The period following the Medieval warm optimum was colder than this optimum. In the mid-latitudinal and subpolar Eurasia, winters became more severe. In the Alps, the glaciers transgressed again. Sea ice was more abundant, appearing earlier in autumn and withdrawing later in spring in the North Atlantics. According to the presently available data, global temperature in the seventeenth century was lower than that in the mid-twentieth century by about 0.5 °C. This extreme cooling is related to the diminished solar irradiance (the Maunder Minimum). In Europe, the seventeenth and early eighteenth centuries were very dry. There is also evidence that, in the fifteenth to nineteenth centuries, precipitation was diminished in Asia as well. This latter dry anomaly is usually interpreted as a result of changes in the Asian monsoon. Since the seventeenth century, climate gradually warmed, while the mean temperature was below that for the mid-twentieth century. This warming was followed by another cool period at the turn of the nineteenth century, again attributed to the diminished solar irradiance (the Dalton Minimum). This latter cold event was weaker and shorter than that occurred in the seventeenth century. The seventeenth to nineteenth centuries are known as the Little Ice Age (while sometimes this term is extended for the whole interval of thirteenth to nineteenth centuries). However, proxy data for the second part of the last millennium, in the Southern Hemisphere (while limited to the tropics), do not show any substation coldness.

Climate of the Last 150 Years

The period since the mid-nineteenth century is a period of instrumental meteorological observations. Since this time, these instrumental observations become routine and the number of stations steadily increases until the late twentieth century. In addition, measurements from commercial ships and from the island stations cab are used for the oceanic regions.

In the last 150 years, humankind has exerted an unprecedented influence on the environment. During this period, emissions of the main anthropogenic greenhouse gas, CO_2, amount to about 420 GtC (in carbon units). These emissions have led to the growth of the concentration of carbon dioxide in the atmosphere from about 275–285 ppmv (this value was quite stable since the beginning of the Holocene) to about 370 ppmv in year 2000 (and about 380 ppmv in 2005). Concentrations of other greenhouse gases have increased as well, for example, for CH_4 from about 830 to 1760 ppbv in year 2000. Atmospheric burdens of the sulfate aerosol also have increased several folds (to 0.52 MtS in year 2000). In addition, natural forcing has contributed significantly to the total external forcing of the climate system. In particular, linear increase of the total solar irradiance in 1900–2000 has amounted 1.6 W m^{-2}, and volcanic forcing was temporarily inhomogeneous during the century.

According to the analysis of the blended land air/marine surface records, on the whole, temperature has increased during the course of the twentieth century by about 0.6 °C. Generally, warming is more manifest in winter than in summer, over land than over ocean, and in the higher latitudes than in the tropics. In addition, decreases of the surface air temperature diurnal range have been noted to be typical over land. For the period 1961–90, the annual mean in globally averaged surface air temperature is estimated to be 14.0 °C (14.6 and 13.4 °C in the Northern and Southern Hemispheres, respectively).

The twentieth-century temperature change was not monotonic. Three main periods may be distinguished. During the 1900s to the early 1940s, global blended temperature has increased by 0.4–0.5 °C. Since this period, climate was cooling until the second half of the 1970s, with a total temperature decrease about 0.1–0.2 °C. Afterward, global temperature rose almost monotonically until the end of the twentieth century by about 0.3–0.4 °C. The last decade of the twentieth century was the warmest during the period of the instrumental meteorological measurements, and this warming continues in the early twenty-first century. Linear trend of warming increases steadily during the last decades.

Possible causes of coldness in the middle of the twentieth century are related to the enhanced volcanism experienced by the Earth during this period. In some studies, the mid-twentieth century cooling is related to the dynamical mechanisms associated with interdecadal climate variability.

The most prominent mode of internal interannual variability, the El Nino/Southern Oscillation (ENSO), is manifested by the recurrent warm and cold events experienced by the eastern equatorial Pacific with strong impacts on the climate in remote regions. This mode has intensified during the second part of the century, with the strongest warm events occurring in 1982–83 and 1997–98. Short-term external forcing has also contributed to the corresponding interannual variability. In particular, after the strong eruption of the Mt. Pinatubo in Indonesia in 1991, the estimated annual mean cooling in the lower troposphere was about 0.7–0.8 °C.

Along with the trends in the mean characteristics of the climate, changes in other climate statistics are noted for the twentieth century and for the beginning of the twenty-first century, for example, in those related to the climate extremes. The latter are defined as unusual climate or weather events (strong cold spells in winter and heat waves in summer, heavy rains, strong droughts, floods, etc.). While the data for these extreme statistics are inherently limited, it can be inferred that heavy rains become more frequent in different parts of the world. The number of extremely cold nights has diminished, and the number of extremely warm days has increased. One of the strongest heat waves occurred in the summer of 2003 in Europe.

Summary and Discussion

Being subjected to different natural and anthropogenic forcing, the Earth's climate changes at a variety of timescales. In the course of the history of the Earth, a general cooling trend can be observed. The turn from the generally warm Paleozoic and, especially, Mesozoic epochs to the Quaternary is a marked example of this general cooling. This latter turn has been followed by the periodic variations of the Earth's climate in the Quaternary with its alternate glacials and interglacials. The current interglacial, the Holocene, is lasted unprecedently long, about 11 ky. In the last millennium, substantial variations related to natural and anthropogenic forcing were observed. The last century and a half are unprecedented in terms of the rate of the climate change.

To apply the knowledge about past climates for future climate changes, a method of paleontology's analogs has been suggested. The backbone of this method is an assumption that for future warmer periods it is possible to find a past warm epoch, which may approximate the climate state for this future epoch. This method has been proved to be useful as a diagnostic tool in inferring about possible climate changes. However, quantitative results from this method are to be treated with caution. The causes involved in an expected greenhouse warming and in the formation of the past warm epochs may be quite different. In particular, the Eemian interglacial and the Holocene optimum are cited frequently as analogs of future greenhouse warming. However, for both these paleoepochs, only summer warming has been demonstrated while the values for annual mean temperature change remain uncertain (see above). In addition, reconstructions for these warm paleoepochs represent mean conditions for rather long intervals (centuries or millennia) while the current warming progresses at much shorter timescales (decades) as well.

As the past epochs represent climate states quite different from the present-day one, they may be useful for validating climate models. For instance, if we compare model simulations and available reconstructions of past epochs with high atmospheric concentration of carbon dioxide, we get that the most probable range of the climate model sensitivity to the doubling of the carbon dioxide in the atmosphere is $3.0 \pm 1.5\,^{\circ}\mathrm{C}$.

See also: Temperature Patterns.

Further Reading

Alexander LV, Zhang X, Peterson TC, *et al.* (2006) Global observed changes in daily climate extremes of temperature and precipitation. *Journal of Geophysical Research* 111(D5): D05109.

Berger A (1988) Milankovitch theory and climate. *Reviews of Geophysics* 26: 624–657.

Brohan P, Kennedy JJ, Harris I, Tett SFB, and Jones PD (2006) Uncertainty estimates in regional and global observed temperature changes: A new data set from 1850. *Journal of Geophysical Research* 111(D12): D12106.

Budyko MI and Izrael YA (eds.) (1991) *Anthropogenic Climate Change*. Tucson, AZ: Arizona University Press.

Budyko MI, Ronov AB, and Yanshin AL (1987) *History of the Earth's Atmosphere*. Heidelberg: Springer.

COHMAP members (1988) Climatic changes of the last 18 000 years: Observations and model simulations. *Science* 241: 1043–1052.

Crowley TJ and North GA (1991) *Paleoclimatology*. New York: Oxford University Press.

EPICA community members (2004) Eight glacial cycles from an Antarctic ice core. *Nature* 429: 623–628.

Hart MH (1978) The evolution of the atmosphere of the Earth. *Icarus* 33: 23–39.

Houghton JT, Ding Y, Griggs DJ, *et al.* (eds.) (2001) *Climate Change 2001: The Scientific Basis Contribution of Working Group I to the Third Assessment Report of the Intergovernmental Panel on Climate Change (IPCC)*. Cambridge: Cambridge University Press.

Jones PD and Mann ME (2004) Climate over last millennia. *Reviews of Geophysics* 42(2): RG2002 (doi:10.1029/2003RG000143).

Kiktev D, Sexton DMH, Alexander L, and Folland CK (2003) Comparison of modeled and observed trends in indices of daily climate extremes. *Journal of Climate* 16(22): 3560–3571.

Lambeck K and Chapell J (2001) Sea level change through the last glacial cycle. *Science* 292: 679–686.

Monin AS and Shishkov YA (1979) *History of Climate*. (in Russian). Leningrad: Gidrometeoizdat.

Peixoto JC and Oort AH (1992) *Physics of Climate*. New York: American Institute of Physics.

Peltier WR and Solheim LP (2001) Ice in the climate system: Paleoclimatological perspectives. In: Matsuno T (ed.) *Present and Future of Modeling Global Environmental Change: Toward Integrated Modeling*, pp. 221–241. Tokyo: Terrapub Inc.

Petit JR, Jouzel J, Raynaud D, *et al.* (1999) Climate and atmospheric history of the past 420 000 years from the Vostok ice core, Antarctica. *Nature* 399: 429–436.

Valdes P (2000) Paleoclimate modeling. In: Mote P and O'Neil A (eds.) *Numerical Modeling of the Global Atmosphere in the Climate System*. Dordrecht, The Netherlands: Kluwer Academic Publishers.

Coevolution of the Biosphere and Climate

D W Schwartzman, Howard University, Washington, DC, USA

Introduction

This article summarizes the current state of knowledge of global biogeochemical cycles that are especially relevant to climate and its coevolution with life. Following basic definitions is a short history of the essential concepts, including Gaia theory, followed by a survey of the carbon biogeochemical cycle operating on different timescales, concluding with a discussion of those aspects of the cycles of oxygen, phosphorus, and sulfur relevant to climate, and overall implications to biospheric evolution.

What Is a Biogeochemical Cycle, the Biosphere, Climate?

A biogeochemical cycle of a chemical element is the natural and human-modified circuit including all aspects of its chemical/physical interactions through the biosphere. The Earth's biosphere is the interactive whole consisting of the living organisms themselves plus the atmosphere, hydrosphere (oceans, rivers, lakes, glaciers, icecaps, and subsurface water), soil and subsurface of the crust down to life's upper temperature limit. The climate is defined here as the physical characteristics of the atmosphere, particularly the surface temperature.

History of the Concept of Coevolution of Biosphere and Climate

Vladimir Vernadsky (1863–1945) should be regarded as the father of biogeochemistry, having coined the word in 1926 in his book on the biosphere. For Vernadsky, the central concept of biogeochemistry, the intersection of the biological, geological, and chemical, is the cycling of elements through the biosphere. Vernadsky had a profound influence on the subsequent development of biogeochemistry in the West since his son George was a professor at Yale, and a friend of G. E. Hutchinson, an influential force in the development of ecology, biogeochemistry, and limnology.

Ecologists have long recognized the coevolution of life and its environment, following Darwin's theory, but the 'environment' with rare exceptions refers to the immediate local influences on the organism. This insight has now been developed in depth with the theory of niche construction, which follows up Richard Lewontin's concept of organism/environment coevolution.

A much more radical conception of coevolution was put forward by James Lovelock, an atmospheric chemist, soon joined by Lynn Margulis (the biologist best known for her theory of endosymbiogenesis), namely the Gaia hypothesis. Lovelock argued that the classical view of coevolution of climate and life neither captures the richness of interactive processes and feedbacks, nor recognizes that planetary biota actively determines its planetary environment. But Lovelock goes even further, asserting that in some sense the Earth (surface) is living, with its own physiology, a 'geophysiology' of a superorganism,with even homeostasis, a general characteristic of metabolism. Biotic regulation of its global external environment leading to, for example, stable climate is for Lovelock homeostasis on a planetary scale. The Earth as a 'superorganism' resonates with the conception of Hutton, the eighteenth-century Scottish doctor and farmer, generally regarded as the founder of modern geology.

Lovelock argued that the Earth's habitability for the last 3 billion years (now accepted for at least 3.5 billion years based on fossil evidence) was a result of continuous biotic interaction with the other components of the biosphere: the atmosphere, ocean, and soil/upper crust. The requirements of habitability include favorable temperatures, ocean salinity, and, at least for the last 2 billion years, atmospheric oxygen at aerobic levels. In Lovelock and Margulis' early papers, we find a formulation of Gaia as a homeostatic system.

> From the fossil record it can be deduced that stable optimal conditions for the biosphere have prevailed for thousands of millions of years. We believe that these properties of the terrestrial atmosphere are best interpreted as evidence of homeostasis on a planetary scale maintained by life on the surface. (Lovelock and Margulis, 1974)

However, the concept of optimality is highly problematic: optimal for the persistence of planetary biota, but which components? Is optimality to be measured in the number of species? Did the anaerobic prokaryotes of the Archean have the foresight to optimize atmospheric conditions for their successors, the aerobes?

As a result of sustained challenges to 'homeostatis' Gaia, Lovelock reformulated his original concept as 'geophysiological' Gaia, "a theory that views the evolution of the biota and of their material environment as a single, tightly coupled process, with the self-regulation of climate and chemistry as an emergent property" (Lovelock, 1989). Thus, the biosphere is now seen as an evolving system with negative feedback such as climatic stabilization.

However, biological regulation may well be limited to restricted 'phase space' – the matrix of physicochemical variables – in biospheric evolution, affecting some conditions but not others (e.g., ocean pH but not salinity), without constituting a global homeostatic system. Another possibility is that for global or regional ecosystems, homeostasis alternates with periods of drift for a given regulated parameter in phase space. This mode has been called 'intermittent Gaia'. Alternatively, homeostatic regulation in some habitats may have persisted since the origin of life. Could this be the case for the 'deep hot biosphere' of the subsurface?

In 'progressive' Gaia, the biota mediates, but both the biota and biosphere coevolve, with no real optimization for the biota at any time. The Gaian research program has proceeded vigorously in its search for self-regulation of other global effects of biotic/biospheric evolution besides the variation of atmospheric oxygen levels, for example, surface temperature, atmospheric composition, as well as for self-regulatory mechanisms operating at smaller scales in the biosphere's subsystems.

The Carbon Biogeochemical Cycle

The cycling of carbon through the biosphere is of critical concern today because of accelerated global warming and its impacts on global society. The enhanced greenhouse is a result of the additional warming from infrared radiation to the Earth's surface by the anthropogenic greenhouse gases, carbon dioxide being the largest trace gas contributor (water vapor actually accounts for most of the greenhouse effect, but its level is dependent on the independent variation of, first of all, atmospheric carbon dioxide). Of critical concern is where the carbon dioxide emitted to the atmosphere ends up, and how this pattern might change as global surface temperature increases. Thus, knowledge of the multifold fluxes in and out of the systems and subsystems of the biosphere and their temporal and spatial variation is essential. Anthropogenic methane also contributes to global warming as a trace greenhouse gas, much more potent than carbon dioxide although at a level *c.* 1% of the latter's concentration in the atmosphere.

A summary of the global carbon cycle is shown in **Figure 1**. First, the natural fluxes are discussed. The total photosynthetic flux is about 170 $Pg\,C\,yr^{-1}$ (the prefix 'P' stands for peta-, 10^{15}), 50 for marine biota, 120 for terrestrial. This flux is almost exactly balanced by a respiration and decay flux of C back into the atmosphere/ocean pool, mainly as carbon dioxide. Only a small flux of organic C and carbonate of about 0.2 $Pg\,C\,yr^{-1}$ is buried, constituting a 'sink' with respect to the atmosphere/ocean pool. This latter flux balances the net source of C to the atmosphere, namely the volcanic (about 0.1 $Pg\,yr^{-1}$) and a roughly equal flux of C from the oxidation of organic C present in exposed terrestrial rocks.

The anthropogenic fluxes to the atmosphere consist of the carbon dioxide from fossil fuel burning (about 5 $Pg\,C\,yr^{-1}$) and deforestation from the burning and decay of organic C (1–2 $Pg\,C\,yr^{-1}$). Note that this sum (6–7 $Pg\,C\,yr^{-1}$) is roughly 60 times the natural flux from volcanism, and of course accounts for the well-known rise of carbon dioxide in the atmosphere in the last 100 years, and most of the enhanced greenhouse effect. One critical flux to the long-term carbon cycle consists of 'riverborne material', the flux of bicarbonate and calcium/magnesium ions derived from the weathering of CaMg silicates on land, mainly following minerals: plagioclase (an NaCa feldspar, which are aluminosilicates), biotite (sheet silicate containing Mg), pyroxenes (single-chain silicates), and amphiboles (double-chained silicates).

Carbon in the crust occurs mainly in the form of limestone and its metamorphic equivalent marble and

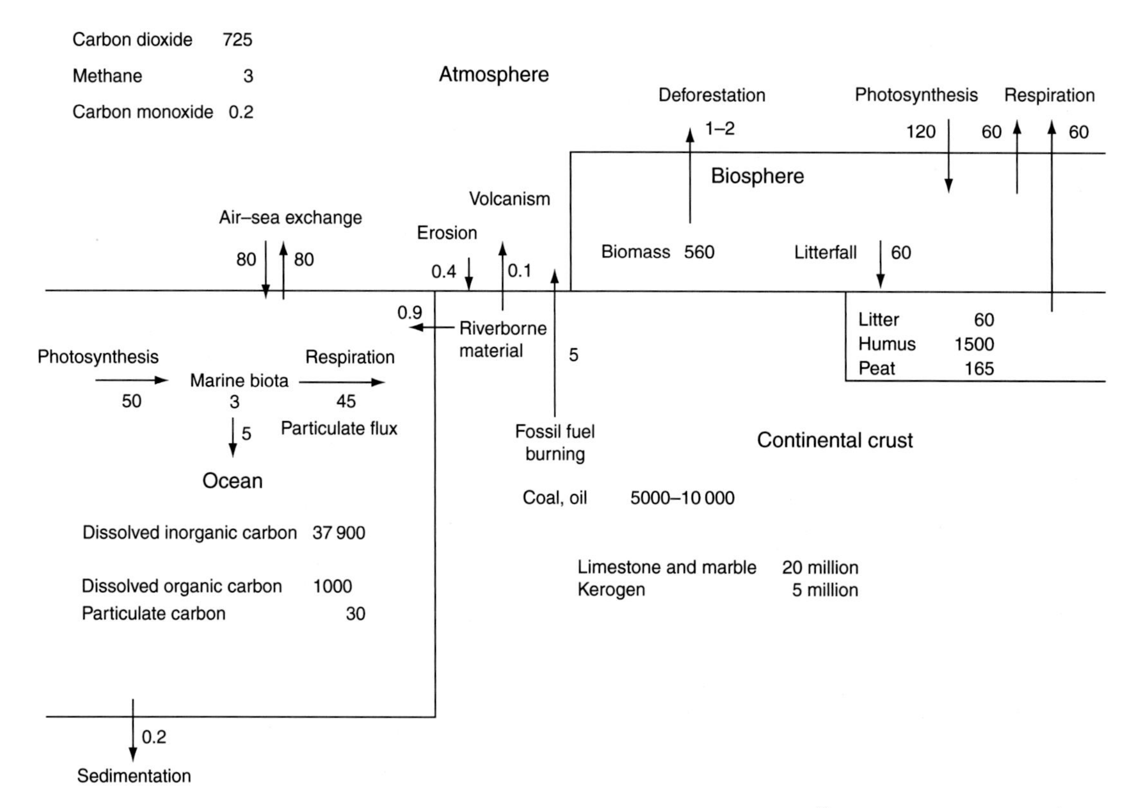

Figure 1 The global biogeochemical carbon cycle. Reservoir contents are given in Pg C (10^{15} g) and fluxes in Pg C yr^{-1}. Modified from Holmen K (1992) The global carbon cycle. In: Butcher SS, Charlson RJ, Orians GH, and Wolfe GV (eds.) *Global Biogeochemical Cycles*, pp. 239–262. London: Academic Press.

is some 500 times the mass of the total C in the atmosphere, biosphere, and ocean combined. Oceanic C, mainly as bicarbonate ions, is some 50 times the mass in the atmosphere, while soil C is some 2 times the atmospheric C mass. While the terrestrial biomass is over 1000 times that of the oceanic, its rate of assimilation (photosynthesis) of carbon dioxide from the atmosphere is little more than 2 times the oceanic rate; most of the terrestrial biomass is in the form of dead wood in trees.

The photosynthetic flux of oxygen is almost exactly balanced by the respiration and decay flux (the small burial of organic carbon is thus an oxygen source balancing natural sinks such as oxidation of ferrous iron in rocks). The flux of fossil fuel burning is about 50 times that of the natural flux of volcanic/metamorphic release of carbon dioxide to the atmosphere. Hence, the anthropogenic rise of atmospheric carbon dioxide and global warming occur on a timescale of decades to hundreds of years.

The short-term carbon cycle is illustrated in **Figure 2**. In the soil pool, we see a large potential positive feedback of global warming, the release of soil carbon into the atmosphere (**Figure 1**). Note that the ratio of soil organic carbon to atmosphere carbon is about 2:1. The global

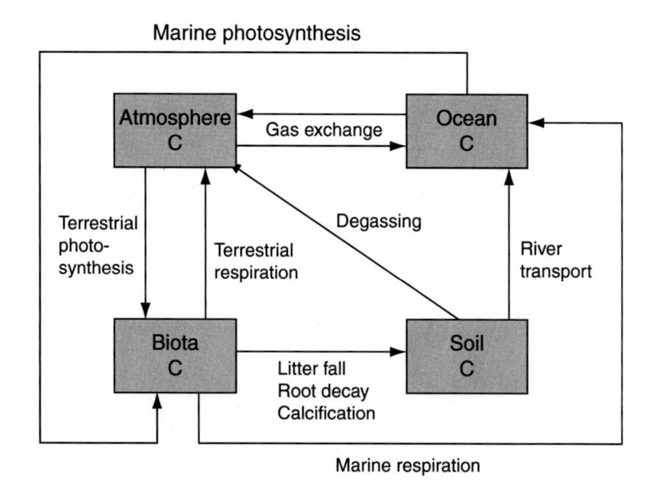

Figure 2 Short-term carbon cycle. Box model diagram. Boxes represent reservoirs and arrows represent mass fluxes between reservoirs. Human effects are not shown; deforestation would be an acceleration of terrestrial respiration, but fossil fuel burning is an acceleration of sedimentary organic matter weathering, a flux from the long-term carbon cycle. From Berner RA (1999) A new look at the long-term carbon cycle. *GSA Today* 9: 1–6.

estimate of soil organic matter divided by the carbon deposited as litter gives a mean residence time of about 25 years. However, the residence time is apparently significantly reduced as temperature increases with huge releases of soil carbon to the atmosphere expected from global warming, although the kinetics and feedbacks of this process are still under active investigation.

What Regulates the Long-Term Climate?

On short timescales of $<10^4$ years, the cycling between the atmosphere/ocean and surface pools such as organic carbon can have significant impact on atmospheric carbon dioxide levels (e.g., the glacial/interglacial cycles of the last 2 million years, anthropogenic impacts), the long-term cycle ($>10^5$ years) is controlled by the silicate–carbonate geochemical cycle. This cycle entails transfers of carbon to and from the solid Earth, that is, the crust and mantle. In the modern era, this cycle was first described by Harold Urey over 50 years ago:

$$CO_2 + CaSiO_3 = CaCO_3 + SiO_2 \quad [1]$$

The reaction to the right corresponds to chemical weathering of Ca silicates on land ($CaSiO_3$ is a simplified proxy for the diversity of rock-forming CaMg silicates such as plagioclase and pyroxene which have more complicated formulas, for example, Ca plagioclase: $CaAl_2Si_2O_8$), while the reaction to the left corresponds to metamorphism ('decarbonation') and degassing returning carbon dioxide to the atmosphere.

This cycle is really biogeochemical. While decarbonation and outgassing is surely abiotic, taking place at volcanoes associated with subduction zones and oceanic ridges, chemical weathering involves active biological participation. Chemical weathering requires a flow of water and carbon dioxide through a layer of soil, with a high reactive surface area of CaMg silicates if consumption of atmospheric carbon dioxide is to occur at a rate similar to that on today's Earth. Thus, most chemical weathering occurs on vegetated continental surface in temperate and tropical climates because of moderate to high rainfall and temperatures. Naturally, higher temperatures, with other conditions constant, mean higher rates of reaction, because of the normal temperature effect on chemical reactions. In general, rocks that formed at high temperatures from cooling magma, such as basalt, weather fastest with respect to atmospheric carbon dioxide consumption. Taking in carbon dioxide and water in the weathering equation for a very common rock-forming mineral in basalt, Ca-rich plagioclase (simplifying by leaving out the Na component), gives the following:

$$CaAl_2Si_2O_8 + 3H_2O + 2CO_2 \rightarrow Ca^{2+} + 2HCO_3^- + Al_2Si_2O_5(OH)_4 \quad [2]$$

$Al_2Si_2O_5(OH)_4$ is kaolinite, a clay mineral. The products include dissolved Ca cations and bicarbonate anions, which ultimately wind up in the ocean from river input, and kaolinite, left in the soil.

Only the reaction of calcium and magnesium silicates with carbonic acid results in a carbon sink via the formation of bicarbonate, its transfer to the ocean, and the precipitation of $CaCO_3$; the weathering of limestone produces no net change in carbon dioxide in the atmosphere/ocean system, nor as a first approximation at least does the weathering of NaK silicates, since no Ca or Mg is supplied to the ocean.

On land:

$$CaSiO_3 + H_2O + 2CO_2 \rightarrow 2HCO_3^- + Ca^{2+} + SiO_2 \quad [3]$$

In ocean:

$$2HCO_3^- + Ca^{2+} \rightarrow CaCO_3 \downarrow + CO_2 + H_2O \quad [4]$$

(reverse reaction is weathering of limestone on land).

Note that for each mole of $CaSiO_3$ reacting, there is a net consumption of 1 mole of CO_2, which is buried on the ocean floor as limestone.

Surface temperature is stabilized because of the dependence of the weathering rate on surface temperature, itself controlled by carbon dioxide in the atmosphere. This rate increases with increasing temperature because of two main effects: the speedup of chemical reactions and the increase in rainfall and therefore river runoff with increasing temperature. These effects are empirically supported by global studies of weathering rates and solute levels in rivers, as well as theoretically by models. Negative feedback resulting in temperature stabilization is obtained because the carbon sink increases as temperature (carbon dioxide) increases.

Steady-state levels of carbon dioxide in the atmosphere are achieved on timescales of the order of 10^5–10^6 years, since as a first approximation, the time needed to reach steady-state level is the residence time of carbon in the atmosphere/ocean pool with respect to the volcanic source, or $(40\,000/0.1) = 4 \times 10^5$ years. Note that carbon rapidly equilibrates within the atmosphere/ocean pool ($\leq 10^3$ years), with about 50 times as much carbon in the present ocean as in the atmosphere. Short times are also entailed for the achievement of equilibrium between the atmosphere/ocean and the carbon in the biota and soil, so that for the long-term C cycle, the atmosphere/ocean/biota/soil C can be considered as one pool at equilibrium.

Departures from the carbon dioxide steady state in the atmosphere/ocean pool can occur as a result of Earth's orbital variations, fluctuations in organic carbon burial, pulses in volcanic outgassing, etc.

Sources and Sinks: Equal at Steady State

A steady-state carbon dioxide level in the atmosphere/ocean system is achieved by the equality of the sink fluxes (removal processes) and the sources (supply processes) to this 'pool'. At a given time, the partitioning of carbon between atmosphere and ocean is determined by global temperature (or the corresponding atmospheric pCO_2) and either the pH or degree of carbonate saturation of the ocean (i.e., the carbonate and bicarbonate level). Now most of the carbon in the atmosphere/ocean pool is in the ocean (**Figure 1**) with a ratio of atmospheric to oceanic carbon of 1:50, while earlier in Earth's history, especially in the Early Precambrian, the ratio was likely closer to 1:1, since the speciation of carbon in the ocean shifts from bicarbonate to dissolved carbon dioxide with rising pCO_2 and the solubility of carbon dioxide in water is relatively small.

For the atmosphere/ocean 'pool' to remain at steady state on a timescale of 10^5–10^6 years or more, the sums of the input fluxes must equal the output fluxes (see **Figure 3**). F_1 corresponds to the flux of carbonate deposited in the ocean, derived from the reaction of atmospheric carbon dioxide with CaMg silicates and carbonates in weathering reactions on land. F_2 is the flux of carbon from the chemical weathering of land carbonates alone. Thus $(F_1 - F_2)$ corresponds to the CaMg silicate weathering sink alone since a flux equal to F_2 is deposited as carbonate in the ocean. F_3 is the flux of organic carbon into the sedimentary reservoir from the net deposition in the ocean from both terrestrial and oceanic sources. F_4 is the carbon flux back into the pool derived from the weathering of organic carbon in exposed sedimentary rock on land (e.g., oxidation of coal). V is the volcanic/metamorphic C source flux. Then $V = (F_1 - F_2) + (F_3 - F_4)$. The evolution of carbon sinks with respect to the atmosphere/ocean reservoir over geologic time remains uncertain despite inferences made from the sedimentary carbon isotopic record. Low-temperature alteration of seafloor basalt by seawater and the weathering of NaK silicates have also been proposed as long-term carbon sinks with respect to the atmosphere/ocean pool but their importance is still unclear. The weathering of CaMg silicates and marine organic carbon deposition likely dominate this sink.

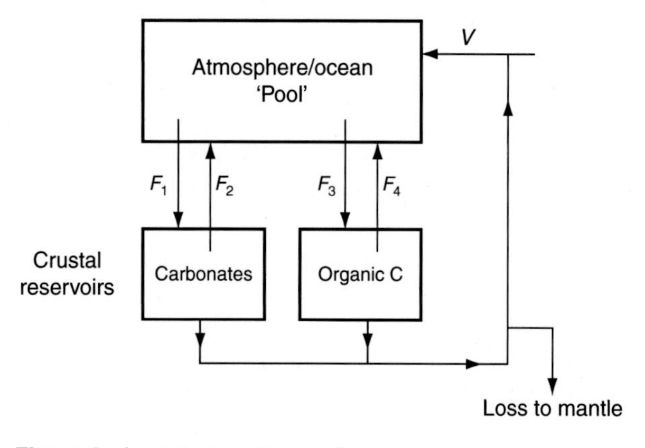

Figure 3 Long-term carbon cycle.

The Biotic Enhancement of Weathering and the Long-Term Carbon Cycle

The biotic enhancement of weathering (BEW) is defined as how much faster the silicate weathering carbon sink is under biotic conditions than under abiotic conditions at the same atmospheric pCO_2 level and surface temperature. If BEW is significantly greater than 1, on an abiotic Earth the steady state would occur with higher atmospheric carbon dioxide levels and surface temperatures doing the 'work' generating an equal flux as the outgassing flux, than the biotic cover on land does at lower atmospheric carbon dioxide levels and temperatures by the multifold processes included in BEW.

The likely contributors to the present BEW with forest and grassland ecosystems/soils as the main sites include:

1. soil stabilization (likely the most important, and likely more than a tenfold contribution to reactive mineral surface);
2. pCO_2 elevation in soil from root/microbial respiration and decay (Lovelock and Watson's original BEW);
3. organic acids/chelators in soil, the multifold interactions in the rhizosphere including biogeophysical weathering, biogeophysical/biochemical weathering by soil fungi leading to the breakup and digestion of CaMg silicate mineral particles;
4. evapotranspiration contribution to maintaining soil water and runoff;
5. diffusion of oxygen into soil contributing to oxidation of CaMgFe silicates and production of sulfuric and nitric acids;
6. water retention by soil organics;
7. turnover of soil by ants, earthworms, mixing organic and mineral particles (e.g., Darwin's pioneering research); and
8. biotic sink effect for mineral nutrients.

And if the cumulative global BEW is on the order of 10–100, then frost wedging contributing to greater surface area of reactive minerals is a component as well, since the progressive increase of BEW over geologic time cooled the climate so much that by about 1.5 Ga ice wedging emerged as a factor.

There are also biotic effects that likely slow down weathering:

1. accumulation of thick, depleted soils acting as a barrier to water flow to fresh bedrock;

2. macropores in soils that allow water flow bypassing saturated pore space.

Lichen coverage of bedrock in most cases promotes chemical denudation, so it is not an example of biotic retardation of weathering (BRW).

But even with these retarding effects, BEW dominates globally. An abiotic Earth surface has no soil, little regolith, so as a first approximation the land surface potential reacting with water/carbon dioxide is two-dimensional. The likely cumulative BEW at present is on the order of 100, with vascular plant ecosystems adding roughly an order of magnitude on the previous BEW of microbial/lichen/bryophyte Earth (there is a good case for lichen weathering being a BEW of at least 10 times).

Since the emergence of higher plant ecosystems in the Paleozoic, positive feedbacks may have temporarily dominated the negative even on geologic timescales on the order of a million years. For example, this may have been the case when very warm global climates occurred, such as during the Triassic–Jurassic boundary, leading to potential lethal leaf overheating (e.g., several pathways such as a–b–c–d–u–q–r–g in **Figure 4**). On medium timescales (>100 to 100 000 years), both negative and positive feedbacks are also possible.

Another cycle, the carbonate (see eqn [4]), with biotic mediation involved in the precipitation of marine calcium carbonate and the weathering of limestone and marble on land, also has a role in regulating atmospheric carbon dioxide levels and surface temperature on timescales less than *c.* 100 000 years. Any process which increases the rate of marine calcium carbonate deposition thereby leads to an increase in atmospheric carbon dioxide (eqn [4]), but on the longer timescale the carbonate–silicate cycle dominates with its negative feedback creating a steady-state atmospheric carbon dioxide level. The carbonate cycle is likely important in the glacial/interglacial cycle (e.g., during the Pleistocene), along with changes in oceanic circulation and terrestrial organic C cycling.

The influence of life on marine calcium carbonate precipitation also plays a role in the long-term C cycle by its influence on the site of this precipitation, the continental shelves, or deep ocean floor. Calcium carbonate deposited in the deep ocean is more likely to be subducted as the oceanic crust with its sediment cover plunges down the oceanic trenches. There, decarbonation occurs, thus releasing carbon dioxide to the atmosphere and increasing the source flux in the carbonate–silicate cycle. With the spread of the deep-water protists, coccolithophores, and foraminifera in the Mesozoic, this deep-

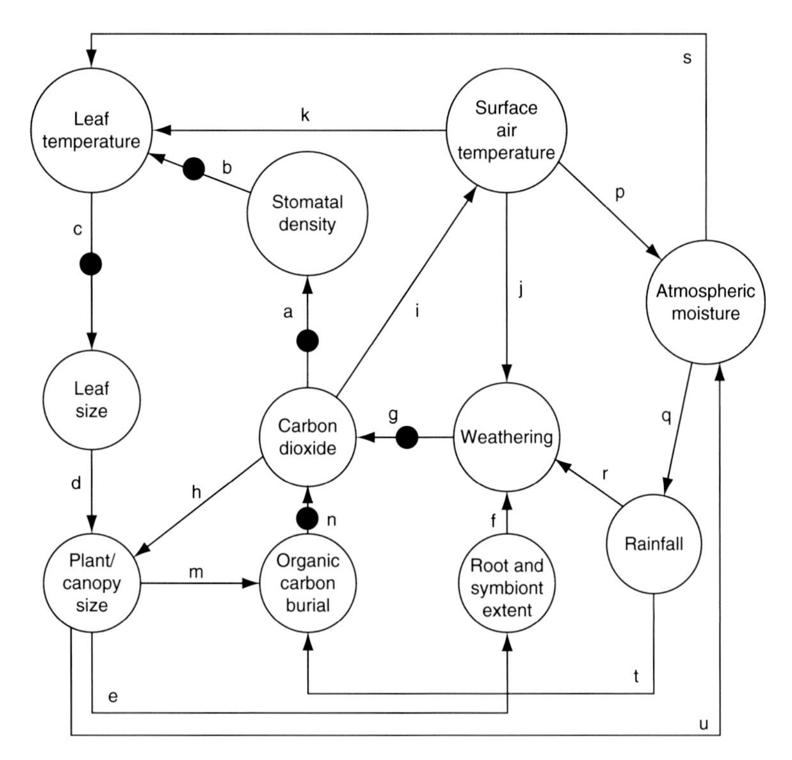

Figure 4 Geophysiological feedbacks between plants and carbon dioxide. It applies only when times of potential overheating of leaves due to high CO_2 levels may have occurred. Arrows originate at causes and end at effects. Arrows with bull's-eyes represent inverse responses. Arrows without bull's-eyes represent direct responses. Letters adjacent to arrows designate paths followed by feedback loops. The timescales over which the paths operate are as follows: a, b, i, k, p, q, s, and u, 1–10 years; c and d, 10–10^3 years; f–h, j, m, n, r, and t, 10^3–10^6 years; e, $\geq 10^6$ years. From Beerling DJ and Berner RA (2005) Feedbacks and coevolution of plants and atmospheric CO_2. *Proceedings of the National Academy of Sciences of the United States of America* 102: 1302–1305.

sea deposition increased, thereby setting the stage for an increased source flux of carbon dioxide to the atmosphere, raising the steady-state atmospheric level of this greenhouse gas.

Biogeochemical Cycles of Oxygen, Phosphorus, and Sulfur and Their Influences on the Coevolution of Life and Climate

The biogeochemical carbon cycle is linked to the oxygen cycle and of course to the cycles of all the necessary nutrient elements for life. The rise of atmospheric oxygen by 1.9 Ga to approximately 20% present atmospheric level (PAL) predated the emergence of Metazoa, which may have required less than ~2% PAL (e.g., mud-dwelling nematodes). This history is consistent with a temperature constraint on the timing of the emergence of Metazoa. On the other hand, the atmospheric oxygen level plausibly constrained the emergence of megascopic eukaryotes, particularly Metazoa, as originally argued by the paleobiologist Preston Cloud 30 years ago, with the explanation being the diffusion barrier of larger organisms. The rise of atmospheric oxygen in the Carboniferous, driven by a burst of organic carbon burial (hence organic carbon not oxidized) leading to the great coal deposits of this age, was by no coincidence the epoch of giant insects. Likewise, the apparent doubling of atmospheric oxygen levels in the last 200 million years has been linked to the evolution and increase in size of placental mammals.

Changes in the level of atmospheric carbon dioxide may also be linked to events in the history of the Earth's biota. An atmospheric pCO_2 constraint on evolution has been suggested for the emergence of leaves (megaphylls) in Devonian triggered by sharp decline in atmospheric pCO_2 as well as for macroevolutionary events in marine fauna in the Phanerozoic and emergence of terrestrial C_4 photosynthesis in the Paleogene. The rise of a methane-dominated greenhouse and concomitant drop in atmospheric carbon dioxide level at about 2.8 Ga has been proposed as the trigger for the emergence of cyanobacteria.

Phosphorus is an example of a minor nutrient element essential for several biomolecules, in particular, nucleic acids and ATP/ADP. Phosphorus plays an interesting role in the long-term C cycle (see **Figure 5**) because of its essentiality and its relatively insoluble state in the continental crust in phosphate minerals. Both negative and positive feedback loops are potentially important, such as A–H–Q–C (negative) and D–F–M (positive). The weathering flux of phosphorus is very small compared to the recycled flux in the biotic loop both on the land and in the sea.

The sulfur biogeochemical cycle is very complex, entailing several possible oxidation states (e.g., hydrogen sulfide, sulfates), hence linked to the oxygen cycle. In the ocean, dimethyl sulfide (DMS) produced by algae oxidizes in the atmosphere to form sulfate aerosols, which are cloud condensation nuclei (CCNs) over the ocean, raising cloud albedo and leading to surface cooling. The possibility of a negative feedback loop was originally proposed, where DMS production varies directly with temperature. However, the overall global feedback may be positive except during the coldest glacial episodes. The key feedback is the apparent positive link between cold water and marine productivity and DMS production. Apparently only eukaryotic algae produce DMS. Thus, in their absence, during most of the Archean, without comparable alternative sources of CCNs, the Earth's albedo might have been significantly lower, approaching a cloud-free value of around 0.1. This effect alone would have resulted in higher surface temperatures. However, there were other possible contributors to cloud albedo, such as intermittent organic haze produced by the reaction of methane and carbon dioxide.

The Habitability of Earth

The astronomical and cosmogeochemical context for Earth formation created the necessary conditions for the origin of life (biogenesis), and the emergence of biogeochemical cycles. This context included the mass of the Sun (and hence its luminosity history as a main sequence star), the Earth's distance from the Sun, the Earth's composition, and early impact history. The presence of liquid water was likely a critical necessary condition for biogenesis, which occurred plausibly during the Hadean, from 4.4 to 3.8 billion years ago, since there is suggestive evidence for liquid water from the oxygen-isotopic record of mineral inclusions in Hadean zircons, dated by the U–Pb method.

Would the Earth's surface have been habitable if it were lifeless (i.e., abiotic)? Surely, it was at the time of biogenesis, but would it have remained habitable over geologic time if life did not emerge? The likely magnitude of the present BEW, about 100, the culmination of the progressive increase in BEW over geologic time, can be used to estimate the likely abiotic temperature history of the Earth's surface. The result is yes, the Earth would have remained habitable for thermophilic life (growing above 50 °C) but not for mesophiles, that is, low-temperature life, including plants and animals. In the future, with rising solar luminosity, the biosphere will terminate in some 1–2 billion years, long before the Sun becomes a red giant star.

The Faint Young Sun Paradox and Its Solution

Sagan and Mullen proposed that the Sun's luminosity (energy flux) was probably low enough to freeze the oceans some 2 billion years ago if the atmospheric composition

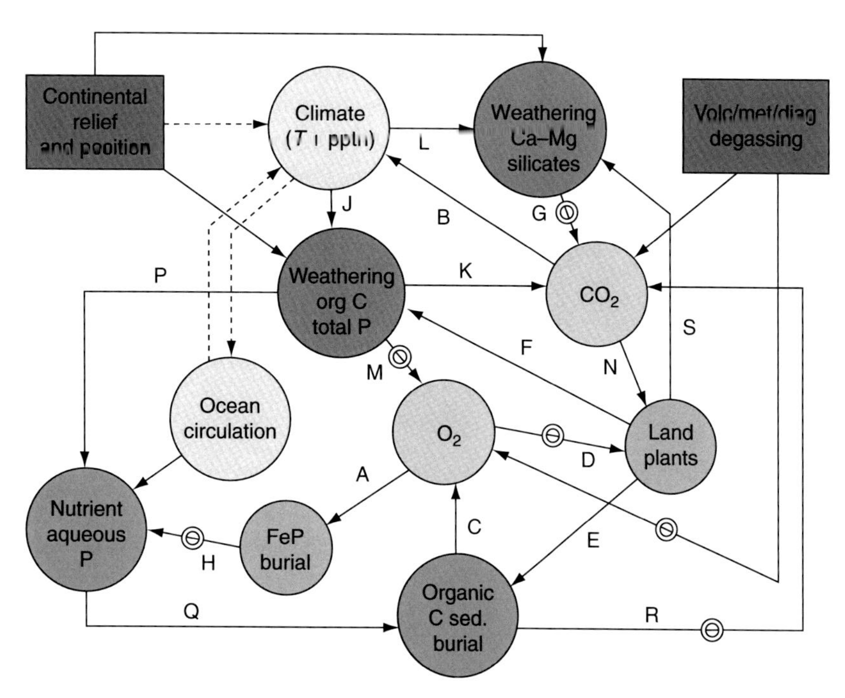

Figure 5 Cause–effect feedback diagram for the long-term carbon cycle. Arrows originate at causes and end at effects. The arrows do not simply represent fluxes from one reservoir to another. Arrows with small concentric circles represent inverse responses; for example, as Ca–Mg silicate weathering increases, CO_2 decreases. Arrows without concentric circles represent direct responses; for example, as organic C burial goes up, O_2 goes up. Letters adjacent to arrows designate paths followed by feedback loops. *T* is temperature; pptn is precipitation. Tectonic processes include volcanic, metamorphic, diagenetic degassing; continental relief and position. Dashed lines between climate and tectonics or ocean circulation refer to complex combinations of physical processes. Nutrient aqueous P is phosphorus dissolved in natural waters that is available for uptake via photosynthesis, both continental and marine; FeP represents phosphate adsorbed on hydrous ferric oxides. Organic C and P burial includes that on the continents and in marine sediments. For diagrammatic clarity, arrows from organic carbon burial to organic weathering or degassing (i.e., recycling of carbon) are not shown. There is no arrow going directly from O_2 to the weathering of organic carbon because of evidence that changes in atmospheric O_2 probably do not affect organic carbon weathering rate. From Berner RA (1999) A new look at the long-term carbon cycle. *GSA Today* 9: 1–6.

were the same as now (the 'faint young Sun paradox'), yet for the last 3.8, possibly even the last 4.4, billion years the Earth has had liquid oceans, a fact established from the fossil and marine sedimentary record and the isotopic record of Hadean zircons. The solution to this paradox is that atmospheric compositions have not been constant over geologic time; rather, higher levels of the greenhouse gases carbon dioxide and likely methane were present in the past, especially the Early Precambrian, to generate warming compensating the fainter Sun, even overcompensating, generating higher temperatures than at present.

The Temperature Constraint on Biologic Evolution

A temperature constraint on biologic evolution was first apparently proposed over 30 years ago by the astronomer Fred Hoyle, who suggested that a warm early Earth held back the emergence of low-temperature life. If a robust climatic record can be inferred from the paleotemperatures derived from the oxygen-isotopic record of marine cherts, then important implications to our understanding of the evolutionary history of the biosphere are implied, first of all that climatic (surface) temperature itself was plausibly the determining constraint with respect to the timing of major events in microbial evolution. These events included the emergence of photosynthesis, eukaryotes, and Metazoa/fungi/plants (see **Figure 6**).

If surface temperature was a critical constraint on microbial evolution, then the approximate upper temperature limit for viable growth of a microbial group should equal the actual surface temperature at the time of emergence, assuming that an ancient and necessary biochemical character determines the presently determined upper temperature limit of each group, T_{max} (see **Table 1**). The latter assumption is supported by an extensive database of living thermophilic organisms. No phototroph has been found to grow above about 70 °C in spite of a likely age of ≥3.5 billion year for this metabolism, similarly for eukaryotes with an upper limit of just over 60 °C, with at least 2 billion years for the possibility of adaptation to life at higher temperatures.

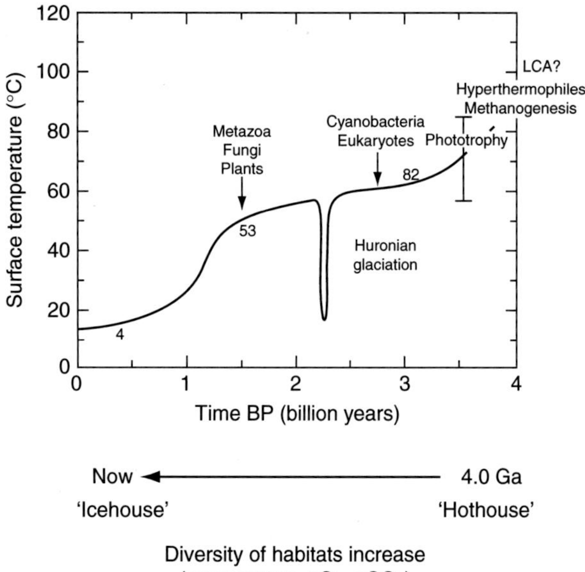

Now ← 4.0 Ga
'Icehouse' 'Hothouse'

Diversity of habitats increase
(temperature, pO_2, pCO_2)

Increase in biotic productivity on land

Progressive increase in biotic
enhancement of weathering

Figure 6 Temperature of the surface of the Earth as a function of time. Emergence times of major organismal groups and long-term trends in biospheric evolution are shown. Numbers near curve are the computed model ratios of the present BEW to the value at that time. From Schwartzman D (2002) *Life, Temperature, and the Earth: The Self Organizing Biosphere* (first publ. 1999). New York: Columbia University Press. The temperature for 3.5 Ga is from Knauth LP and Lowe DR (2003) High Archean climatic temperature inferred from oxygen isotope geochemistry of cherts in the 3.5 Ga Swaziland supergroup, South Africa. *Geological Society of America Bulletin* 115(5): 566–580.

Table 1 Upper temperature limits for growth of living organisms and approximate times of their emergence

Group	*Approximate upper temperature limit (°C)*	*Time of emergence (Ga)*
Plants	50	0.5[a]–1.5[b]
Metazoa	50	0.6[a]–1.5[c]
Fungi	60	0.8[a]–1.5[b]
Eukaryotes	60	2.1[c]–2.8[e]
Prokaryotic microbes		
Phototrophy	70	≥3.5[a]
Hyperthermophiles	≥80	≥3.7[d]

[a]Fossil evidence.
[b]Molecular phylogeny.
[c]Problematic fossil evidence, molecular phylogeny.
[d]Possible biogenic carbon-isotopic signature.
[e]Problematic organic geochemistry (biomarkers) in Archean sediments.

The upper temperature limit for viable growth is apparently determined by the thermolability of biomolecules (e.g., nucleic acids), organellar membranes, and enzyme systems. For example, the mitochondrial membrane is particularly thermolabile, apparently resulting in an upper temperature limit of about 60 °C for aerobic eukaryotes (note that the presumed ancestor of the mitochondrion, the proteobacteria, has an upper temperature limit of 60 °C). Eukaryotes most likely emerged as the endosymbiogenic product of archaeal and bacterial cells associating because of complementary metabolisms, with the acquisition of the mitochondrion close in time with the initial emergence of the eukaryotic cell.

The upper temperature limit of 50 °C for Metazoa may be linked to the thermolability of proteins essential to blastula formation, or to the synthesis of collagen, an essential structural protein. Clearly, fundamental research is needed for a better understanding of the biochemical and biophysical basis for the upper temperature limits of normal metabolism of the organismal groups.

Now we return to the contention that the actual surface temperatures were thermophilic (greater than 50 °C) for the first two-thirds of the history of the biosphere, that is, from 4 Ga to *c.* 1.5 Ga. This high-temperature scenario for the Earth's climatic history remains controversial. The strongest evidence for this temperature scenario comes from the inferred paleotemperatures from the oxygen-isotopic record of ancient cherts and carbonates. Paleotemperatures have been derived from the oxygen-isotope record of pristine cherts. The fractionation of the stable isotopes of oxygen is a function of the phases involved and temperature. A necessary assumption for these paleotemperature calculations supporting the inference of a much warmer Archean and Early Proterozoic climate, that the Precambrian oxygen-isotopic composition of seawater was close to the present, has received strong support from studies of ancient seawater-altered oceanic crust in the geologic record going back to the Archean. Robust new evidence for a very warm Archean comes from a detailed study of the Onverwacht cherts, South Africa, centering on their oxygen-isotopic record in the context of their geologic characteristics, and recently from the silicon isotopes in the same cherts with a preserved paleotemperature record. The inferred paleotemperatures of ancient seawater are consistent with the inferred Precambrian temperature history illustrated in **Figure 6**.

From 'Hothouse' to 'Icehouse': The Increase in Diversity of Habitats for Life

The progressive increase in the BEW as a product of biotic and biospheric evolution intensified the carbon sink with respect to the atmosphere/ocean system, leading to the transition of climate from a 'hothouse' in the Early Precambrian to an 'icehouse' in the Phanerozoic. By the Late Proterozoic, the rise of atmospheric oxygen may have resulted in a substantial increase in terrestrial biotic productivity, which along with the onset of frost wedging substantially increased the BEW. Climate and life coevolved as a tightly coupled system, constrained by abiotic factors (varying solar luminosity, and the crust's tectonic and impact history). Self-regulation of this coupled system is a property of geophysiology.

With the long-term drop of surface temperature over the last 3.8 billion years, a progressive expansion of the diversity of habitats occurred (i.e., first hyperthermophiles, then thermophiles, mesophiles, finally psychophiles, the organisms living near 0 °C). Since the first appearance of ice on high mountains, likely by the Proterozoic, the diversification of habitats opened up new ecological niches, with a cumulative retention of older habitats (e.g., there are still hot springs with thermophiles and the 'deep hot biosphere' persisted since the Hadean). As global temperatures dropped, particularly in the Mid- to Late Proterozoic, latitudinal differences in temperature and therefore a zonal differentiation of ecosystems followed roughly parallel to the equator. Of course, the position and configuration of the continents also changed over time, affecting then, as now, the distribution of ecosystems.

As the atmosphere changed its composition in the Mid-Proterozoic, becoming aerobic with the rise of oxygen, the area of surface aerobic environments expanded from the microenvironments (e.g., cyanobacterial mats) previously present. Anaerobic environments of course persisted in soils, deep in the ocean and in stratified bodies of freshwater, and finally in the gut of nearly emerged animals. If biotic productivity increased with the rise of macroeukaryotic algae in the oceans and on land, then the deposition of organic carbon in stratified water bodies could have maintained anaerobic conditions.

Further, new pCO_2 (high, then low) and total pressure (first possibly $\geq$10 bars, then low) environments progressively emerged since the origin of the biosphere at no later than 3.8 Ga. Again, the diversity increased as new environments were added, while old conditions persisted (e.g., high pCO_2 and total pressure in the deep hot biosphere). In the case of pH, both very acid and basic and intermediate habitats were likely present even before the origin of life at hydrothermal vents. In Michael Russell's brilliant scenario for biogenesis, hot alkaline springs, generated in the early ultramafic crust by interaction with circulating seawater, react with acidic seawater at the seafloor to produce iron sulfide and clay membranes, the postulated sites for catalyzed reactions producing protocells. A primordial pH diversity is consistent with the present wide range of optimal pH for growth of thermophilic prokaryotes. The coupling of the full spectrum of pH habitats with lower temperatures and higher oxygen levels occurred with long-term cooling and the rise of atmospheric oxygen.

With the progressive increase in diversity of habitats, the evolution of life naturally quickly filled these habitats with novel varieties. These new habitats of course included those within newly emerging organisms and ecosystems (e.g., within the digestive tracts of Metazoa, the forest canopy, and soil). The evolution of the biosphere does not optimize conditions for existing biota, unlike the original Gaia hypothesis. At least two catastrophes for existing life occurred; the well-known oxygen catastrophe and an earlier temperature catastrophe for thermophilic bacteria, once the likely colonizers of the ocean and land, now restricted to living in hot springs, hydrothermal vents on the ocean floor, porosity in the first few kilometers of the crust.

Surface temperature is one parameter which is regulated by the biosphere, since progressive cooling arises from the circularity of the feedbacks where BEW intensifies from both the direct effects of biotic evolution and regional/global effects (e.g., frost wedging in mountains), the latter themselves being a product of the evolving biosphere as a whole. The main abiotic constraints on the feedback of biospheric evolution to surface temperature are solar history, and the Earth's tectonic and impact histories. The main factors of tectonics relevant to the first-order surface temperature history are the variation over time of volcanic/metamorphic outgassing of carbon dioxide and land area. The feedbacks from biospheric to biotic evolution include this long-term decrease in surface temperature, constraining microbial evolution, and the process of endosymbiogenesis, including the history of emergence of cell types (e.g., eukaryotes). Endosymbiogenesis begins with parasitic relations between prokaryotes, progresses to symbiotic, and finally culminates in the intimacy of new cells and organisms.

Besides changes in productivity and oxygen-producing capacity, these changes in biotic evolution feed back into biospheric evolution, which likely includes the progressive increase in biomass on the planet's surface, particularly with the rise of higher plant forests.

A continuing debate centers around a challenge to a long-held view of evolutionary biology, namely, that if the 'tape' of life's history were to be played again, the results would be radically different owing to the stochastic nature of evolutionary emergence. The alternative is that the evolution of the biosphere is roughly deterministic, that is, its history and the general pattern of biotic evolution and the tightly coupled evolution of biota and climate are very probable, given the same initial conditions. Major events in biotic evolution were likely forced by environmental physics and chemistry, including photosynthesis as well as the merging of complementary metabolisms that resulted in new types of cells (such as eukaryotes) and multicellularity. Determinism likely breaks down at finer levels. Critical constraints in this deterministic aspect of biotic evolution have likely included surface temperature, along with oxygen and carbon dioxide levels. In addition to events that were probably inevitable, it is likely there is a role for randomness in both abiotic and biotic evolution even on the coarsest scale, for example, including the influence of large impacts on the history of life, the possible multiple attractor states in mantle convection and therefore plate tectonic history, even multiple attractors for steady-state climatic regimes.

Summary

Over geologic time, climate and life coevolved as a tightly coupled system, constrained by abiotic factors (varying solar luminosity, and the crust's tectonic and impact history). Self-regulation of this coupled system is a property of geophysiology, a central aspect of the evolution of biogeochemical cycles including the coevolution of the biosphere and climate over geologic time.

See also: Climate Change 1: Short-Term Dynamics; Climate Change 2: Long-Term Dynamics; Climate Change 3: History and Current State; Phosphorus Cycle.

Further Reading

Beerling DJ and Berner RA (2005) Feedbacks and coevolution of plants and atmospheric CO_2. *Proceedings of the National Academy of Sciences of the United States of America* 102: 1302–1305.
Berner RA (1999) A new look at the long-term carbon cycle. *GSA Today* 9: 1–6.
De La Rocha CL (2006) In hot water. *Nature* 443: 920–921.
Falkowski PG, Katz ME, Milligan AJ, *et al.* (2005) The rise of oxygen over the past 205 million years and the evolution of large placental mammals. *Science* 309: 2202–2204.
Holmen K (1992) The global carbon cycle. In: Butcher SS, Charlson RJ, Orians GH, and Wolfe GV (eds.) *Global Biogeochemical Cycles*, pp. 239–262. London: Academic Press.
Kasting JF and Catling D (2003) Evolution of a habitable planet. *Annual Review of Astronomy and Astrophysics* 41: 429–463.
Knauth LP and Lowe DR (2003) High Archean climatic temperature inferred from oxygen isotope geochemistry of cherts in the 3.5 Ga Swaziland supergroup, South Africa. *Geological Society of America Bulletin* 115(5): 566–580.
Kump LR, Brantley SL, and Arthur MA (2000) Chemical weathering, atmospheric CO_2, and climate. *Annual Review of Earth and Planetary Science* 28: 611–667.
Lewontin RC (1983) Gene, organism and environment. In: Bendall DS (ed.) *Evolution from Molecules to Men*, pp. 273–285. Cambridge: Cambridge University Press.
Lovelock JE (1989) Geophysiology, the science of Gaia. *Reviews of Geophysics* 27: 215–222.
Lovelock JE and Margulis L (1974) Homeostatic tendencies of the Earth's atmosphere. *Origins of Life* 5: 93–103.
Odling-Smee FJ, Laland KN, and Feldman MW (2003) *Monographs in Population Biology 37: Niche Construction. The Neglected Process in Evolution*. Princeton, NJ: Princeton University Press.
Robert F and Chaussidon M (2006) A palaeotemperature curve for the Precambrian oceans based on silicon isotopes in cherts. *Nature* 443: 969–972.
Russell MJ, Hall AJ, Cairns-Smith AG, and Braterman PS (1988) Submarine hot springs and the origin of life. *Nature* 336: 117.
Schneider SH, Miller JR, Crist E, and Boston PJ (eds.) (2004) *Scientists Debate Gaia: 2000*. Cambridge, MA: MIT Press.
Schwartzman D (2002) *Life, Temperature, and the Earth: The Self Organizing Biosphere* (first publ. 1999). New York: Columbia University Press.
Schwartzman D and Lineweaver CH (2005) Temperature, biogenesis and biospheric self-organization. In: Kleidon A and Lorenz RD (eds.) *Non-Equilibrium Thermodynamics and the Production of Entropy: Life, Earth, and Beyond*, pp. 207–221. New York: Springer.
Smil V (2003) *The Earth's Biosphere: Evolution, Dynamics, and Change*. Cambridge, MA: MIT Press.
Vernadsky VI (1998) *The Biosphere*. New York: Copernicus (Springer).
Volk T (1998) *Gaia's Body: Toward a Physiology of Earth*. New York: Copernicus (Springer).

Global Change Impacts on the Biosphere

W Cramer, Potsdam Institute for Climate Impact Research, Potsdam, Germany

Introduction

Due to the direct dependence of all living organisms on their physicochemical environment, the biosphere has always responded to changes in climate, at all scales from local to global. On geological timescales, a quasi-equilibrium between ecosystem (biome) distribution and climate exists, although the slow growth of forests and the slow buildup of organic matter in soils and peat bogs create substantial nonlinear responses at the timescale of centuries. The current interest in global change 'impacts' on the biosphere is due to the novel, potentially devastating nature of 'anthropogenic' climate change in many ecosystems. Human society associates high values with the biosphere, due to its intrinsic value and also as provider of 'ecosystem services' such as food, fiber, clean water, as well as spiritual and other values. The concept of 'vulnerability' has been employed to describe this human view on biospheric change. It allows the consistent assessment of biospheric change involving a common metric. It also includes the formulation of future 'risk', expressed by the scenario-based analysis of losses that may arise if trends of global change are extrapolated into the future.

This presentation of impacts focuses not only on changes brought about by anthropogenic perturbations due to global processes such as the emission of greenhouse gases (GHGs) but also the increasing pressure from land-use change including land degradation and deforestation. Global change impacts are, on the one hand, documented from observations and careful attribution to driving forces, and, on the other hand, recognized by consideration of 'expected impacts' due to future trends in drivers, expressed as scenarios. These scenarios are not predictions, because the precise nature of future drivers cannot be known as they depend on human demography, economic development, application of technology, climate protection policy, and other factors. Scenarios are not speculative, however, because they rely on established causal relationships between driving forces and ecosystem response. For example, since the link between GHG emissions, global temperatures, and sea-level rise is well established, and since quick stabilization of GHG concentrations is very unlikely, it can therefore be inferred with high confidence that coastal ecosystems are at risk worldwide, even in places where currently no damage is observed yet. Numerical models for the assessment of future impacts (developed during 25 years by many groups worldwide) are now a mature technique for the consideration of nonlinear impacts and the combination of multiple forcings – as is amply illustrated by the recent assessments of the Millennium Ecosystem Assessment and the Intergovernmental Panel on Climate Change.

In the following, a summary will first be given about the nature of currently observed changes in the biosphere that can be attributed to global change. Then a scenario-based presentation of expected future impacts will be given.

Observed Impacts on Ecosystems

Numerous recent changes in the biotic and abiotic environment can be attributed to recent global change processes. Among the most important drivers of change are climate (temperature, rainfall, extreme events), land-use change (deforestation, urbanization, land degradation), the invasion of alien species, and pollution. Many different data sources are used for statistical inference of impacts, which in many cases yields robust results. A bias might be suspected because there are many more studies of changing systems than there are of stable systems, but the majority of correlations are well supported by explanations of the underlying mechanisms.

Organisms react to changes in the environment in multiple ways: Their phenology (the timing of critical stages in their life cycle) is affected by the seasonal development of environmental conditions such as ambient temperature and moisture availability. Their spatial distribution is affected by the geographical range of suitable conditions. Their morphology and physiology are responsive to temperature, moisture, and the availability of nutrients (including CO_2), while their risk of extinction is a result of both average and extreme event conditions in climate (e.g., frost or drought). Because these drivers show different changes in different regions, the actually observed attributable changes differ between major ecosystems and are presented by major biomes in the following subsections.

Marine Ecosystems

Worldwide, the marine environment is characterized by increasing temperatures and decreasing pH. In some (subtropical) areas, increased intensity and frequency of tropical storms also play a major role. These changes cause loss or decline of ecosystems, or large changes in their geographic distribution.

The majority of tropical coral reefs are affected widely by mortality due to 'bleaching' (expulsion of their symbiotic algae) which has been linked to rapid warming of sea water. There is further damage to some coral reefs due to increased hurricane force in some, mostly subtropical regions. Ocean acidification is expected to affect reefs as well, but the evidence for this is not yet conclusive. In Antarctic waters, warming has led to a decline in krill populations and, as a further consequence, to declining seabird and seal populations. Virtually all fish populations of commercial interest are heavily affected by overexploitation and in many areas at risk of total failure, irrespective of climate change. In some cases, however, part of their decline is attributed to climate-caused impacts on fish larvae, or on organisms at lower levels of the food web, for example, through changes in marine plankton which is a food source of cod larvae in the N. Atlantic.

Major changes in geographic distribution can be observed for several marine organisms, such as intertidal communities (e.g., kelp forest fish) in several regions. Subtropical fish are observed more frequently in northern waters. There appears to be a general northward move of planktonic communities in the N. Atlantic, by about 10° latitude over the past 40 years, which is much more pronounced than anything documented from terrestrial ecosystems. Different species and functional groups change their seasonal cycle in different ways as response to warming, leading to a disruption of the interactions between different trophic levels.

Land Ecosystems

Widespread changes in phenology (particularly leaf canopy development) have been observed both from ground-based networks and from satellites for over two decades, indicating earlier onset of spring, later autumn, and a lengthening of the vegetative season in most northern areas – trends that are well correlated with observed warming. Many animal species (e.g., birds, butterflies) have adapted their life cycle (migration, breeding, etc.) in similar ways, either as a consequence of earlier vegetation development or triggered by temperature directly. Several meta-analyses of these observations demonstrate that, in the Northern Hemisphere, spring has arrived between 2.3 and 5.2 days earlier for every decade during the last 30 years. For example, swallows have arrived 2–3 days earlier for an increase in March temperature of 1 °C in England. Plants and birds adapted to early arrival generally show stronger reactions to warming than others – probably because they are more strictly limited by temperature than by other factors. Autumn arrives somewhat later, although the trend is less clear. Phenological indicators have a particular significance due to the wide range of organisms they cover and also because they are largely unbiased, that is, numerous species are observed irrespective of their expected rate and direction of change.

There are some counterexamples to these trends (lack of response, or response in a direction from climate change) in species of birds and small mammals, demonstrating that the linkage between environment and phenology is complex and involves many confounding factors. One such complication arises when the potential climatic range of a species shifts due to an atmospheric warming trend, but where no suitable habitat exists for whatever reason. In such cases, the low-latitude boundary may shift as expected, but similar area is not gained at higher latitude.

In some mountainous areas, shifts toward higher altitudes have been observed for some species and communities. These have mostly been detrimental for high-alpine species for which no further habitat exists. At lower altitudes, the effect is less clear, mostly because trends in tourism, agriculture, and forestry affect ecosystems as well.

The attribution of individual species' extinctions to climate change has been made for some amphibians and butterflies – in a statistical sense, it is very likely, however, that climate change has increased the extinction risk (and the extinction rate) for many species worldwide. A proportion of current extinctions is due to invasive species. While most invasions occur due to intended or unintended introduction by people, some cases of invasion have likely been enhanced by climate change.

Attribution of changes in agricultural and forestry systems to climate change during recent decades is difficult due to the intensive transformations of most systems due to improved technology and management during recent decades. Studies in Europe have nevertheless documented advancements of the agricultural calendar for certain crops (notably grapes) that are mostly due to climate change (**Figure 1**). There are also indications of pests responding positively to warming requiring greater effort for pest control. In India, a reduction in expected rice production increases has been attributed to increasing night-time temperatures.

In forests, part of the recent increase in productivity in Northern Europe and North America can be attributed to warming and enhanced atmospheric CO_2 concentrations. Semiarid regions in Southern Europe have seen declines in forest productivity, which can be linked to higher temperatures and moisture losses, a particular example being the extreme summer of 2003. Mediterranean woodlands experience strong increases in wildfire frequency and intensity, which is partly due to land abandonment but also to higher temperatures and drier soils. Also for trees, pests increase in some areas due to warming.

Lakes and Rivers

Freshwater ecosystems are directly affected by higher temperatures and the impacts of changing thermal structure and lake chemistry. At high latitudes and/or altitudes, ice cover is reduced and productivity increases, leading to increased algal abundance and in some cases fish production. In tropical lakes, in contrast, reduced productivity has been observed due to more stable stratification and reduced nutrient upwelling. As a consequence, freshwater ecosystems have changed in composition, which sometimes has led to detrimental consequences for human health caused by increased cyanobacteria populations. In river systems, numerous diverse changes have been observed in response to warming, including displacement of thermophilic species to the north (Rhone) and shifts in the timing of fish migration (North American rivers).

Vulnerability of Ecosystems to Future Change

State-of-the-art assessments of likely future changes in ecosystems due to global change are typically not extrapolated from the observations reviewed above. This is because most ecosystem processes are driven by multiple factors, making the net result a highly nonlinear function of GHG increase or warming. A key issue for many ecosystems is whether there is a threshold that might be reached at which the intrinsic capacity of the ecosystem to adapt to the changing environment is exceeded. Another problem occurs where time lags of response exist, after which particularly strong impacts must be expected (i.e., due to failed

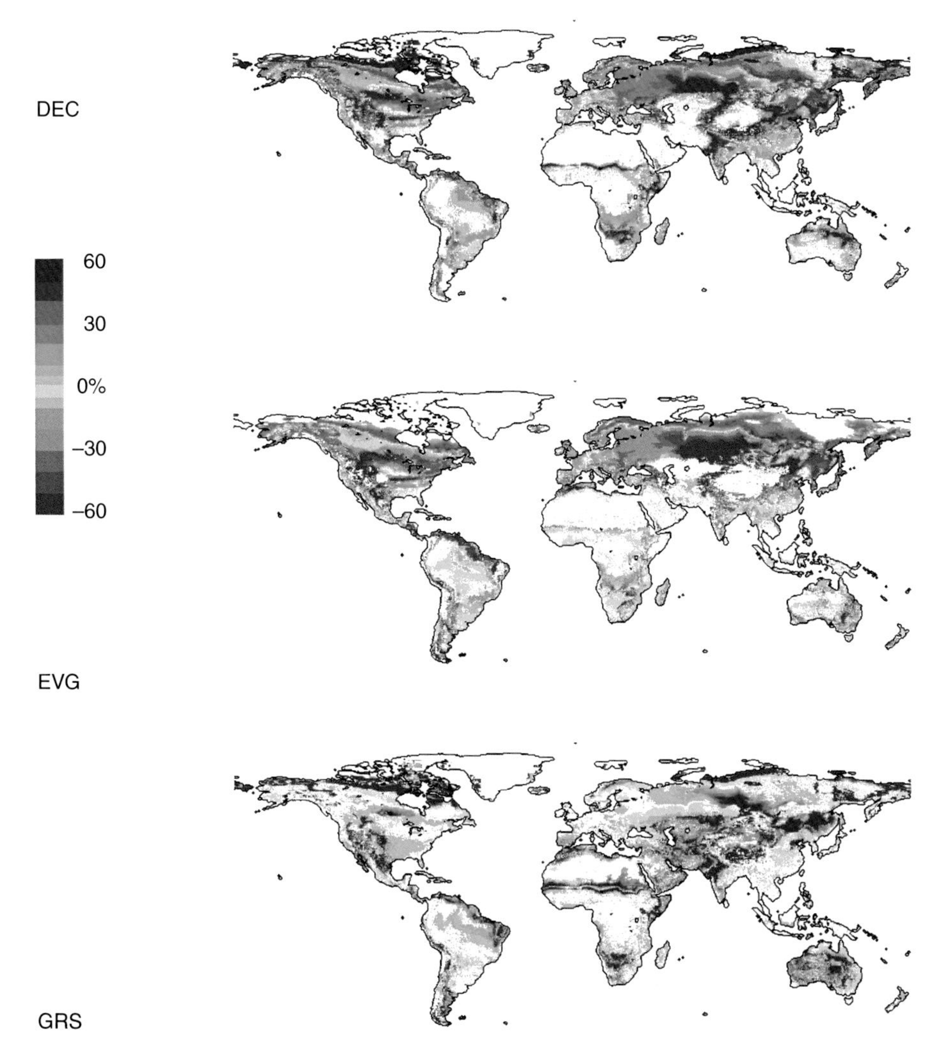

Figure 1 Changes in global vegetation distribution 2100–2000. Simulated changes (2100–2000) of the fractional cover of deciduous woody (top), evergreen woody (middle), and nonwoody (bottom) vegetation functional types for a strong climate change scenario (SRES-A2, HadCM3). Simulations with the LPJ-DGVM. Reproduced from Lucht W, Schaphoff S, Erbrecht T, Heyder U, and Cramer W (2006) Terrestrial vegetation redistribution and carbon balance under climate change. *Carbon Balance and Management* 1: 6.

reproduction of some important species). Many ecosystem processes show strong hysteresis, for example, the rate of response to reduced resources may be quick, while the recovery with increased resources could be much slower. Finally, changes in variability (e.g., extreme weather event frequencies) may be more important for ecosystems than gradual changes toward warmer or drier/wetter conditions. For all these and other reasons, the different drivers of change are typically investigated using numerical models, mostly based on experimental or observational evidence of key mechanisms controlling ecosystem functioning. The accumulated observations of recent and historical changes serve as important data sources for the validation of these models. Only where process understanding is mostly lacking, correlative approaches are still being used, for example, for the estimation of changing geographic distribution of many species.

Deserts and Dryland Ecosystems

A first-order impact of warming on deserts and drylands is that evapotranspiration increases and thereby reduces plant productivity. Experimental and modeling studies indicate that many plants might respond with increased water-use efficiency due to higher amount of atmospheric CO_2, and this could make deserts and dryland ecosystems benefit from GHG increase. Whether this effect might provide greater ecosystem service provision is difficult to assess, since it could be counterbalanced by higher water loss due to warming in some areas, and also because many dry rangeland systems are currently overexploited. Moreover, while warming must be expected in most dry regions of the planet, even the sign of changing precipitation is often uncertain and little is known about expected extreme event frequencies. A key conclusion however is that unchanged conditions are highly unlikely for most regions. Deserts are expected to increase in area due to extended drought periods in the Americas and elsewhere.

Grasslands and Rangelands

Grass-dominated ecosystems generally share the feature that moisture is limiting vegetation production, that gains in water-use efficiency due to increased atmospheric CO_2 may occur, and that – in most regions – great uncertainty exists with regard to future quantities and variability of rainfall. Vulnerability to climate change is therefore high, and many people depend directly on rangeland productivity for their livelihood, but global impacts cannot be generalized due to the uncertainty of future rainfall patterns.

Mediterranean Ecosystems

Shrub-dominated ecosystems in mid-latitudes with typical winter rainfall pattern are considered the ecosystems with the strongest expected impacts from climate change. This is because they occupy regions for which the global climate model now consistently projects substantial reductions in rainfall, with direct detrimental consequences for plant productivity. In addition, fire disturbance increases due to a combination of greater drought and land abandonment in many regions, causing widespread damage and reduction of ecosystem services for forestry and tourism.

Tropical Forests

Deforestation is the dominating threat to many tropical forest ecosystems, on all continents and across the gradient from semiarid to perhumid climate. Nevertheless, there is concern about climate change impacts in several areas, particularly in Amazonia. Some climate models project dramatic losses in rainfall during the twenty-first century, associated with increased interannual variability. This translates to a risk for widespread dieback of these forests, although the picture is currently inconclusive due to differences between climate models. Independent of this, increased fragmentation of tropical forests has been shown to lead to regional changes in rainfall which in turn reduces forest stability.

Temperate and Boreal Forests

Forests outside the tropical regions are affected by a wide range of different climatic conditions and differences in projected change. There are regions where water availability will likely decrease (e.g., SE Europe) leading to reduced forest growth and biomass. In other regions, growth is expected to increase due to increasing temperatures, higher CO_2, and also the deposition of nitrogen-rich pollutants (N. Europe). Typical boreal forests can be expected to expand to regions further north, but it is not always clear what the response is at their southern boundary, where either a transformation to different forest types could occur or a loss of forest cover due to drought. In most forests, dynamics of disturbances (fire, wind storms, insects) are highly sensitive to changes in climate.

Tundra Ecosystems

Tundras exist in subarctic and Arctic climates at high southern and northern latitudes, as well as in high mountain areas. Particularly in the north, these are the regions where observed and expected warming reach the highest values. Physical impacts occur predominantly through the destabilization of permanently frozen soils and subsequent erosion. The next step of ecosystem change is due to the encroachment of taiga plants into the tundra, changing the entire species pool. The time lags for these responses may differ widely in different regions.

Freshwater Ecosystems

Rivers and lake ecosystems are affected by changing hydrologic conditions (e.g., due to melting mountain glaciers) as well as by increasing temperatures. For many lakes, this may imply reductions in oxygen concentrations and subsequent changes in species composition and water quality. In northern lakes and rivers, productivity may increase. Wetlands are affected by local to regional changes in water levels, in the case of warming implying the risk of drier conditions and loss of peat. Many Arctic lakes are at risk to dry out, with negative consequences for many species (fish, birds, etc.).

Marine Ecosystems

Observed trends in marine ecosystems (coral bleaching, disruptions of food web due to differential responses of species to warming) are expected to continue and accelerate during coming decades. A particular concern is the rapid loss of sea ice, affecting krill, fish, seals, and polar bears negatively to an extent that makes polar bears a species threatened with extinction. The melting of ice sheets and changes in ocean circulation both reduce salinity in high-latitude oceans, which affects plankton communities and threatens polar species such as the narwhal and migratory whales. Sea-level rise is expected to put a third of sea turtle nesting beaches at risk.

Increased atmospheric CO_2 concentrations affect ocean pH negatively, a trend which is expected to continue by 0.3–0.4 units until 2100. This trend is expected to impact almost all life in the ocean, but particularly benthic calcifying organisms. Particularly, organisms using aragonite for their shells will be at risk of extinction (these are widespread and important for the food web in Arctic and Southern Oceans).

Coastal Ecosystems

Many coastal areas are directly at risk from sea-level rise, either through submergence or due to the risk of frequent storm damage and saltwater intrusion. This affects mangroves and other coastal wetlands around the world.

Summary of Future Risks for the Biosphere

Impacts of global change on the biosphere are generally the sum of numerous local to regional changes, and confounded by other drivers of global change. From reviewing the entire current literature about this topic, the following picture emerges:

- Even if some ecosystems may temporarily benefit from increasing atmospheric CO_2 and/or the associated climatic warming, few of these advantages will remain during the latter half of the twenty-first century.
- The known time lags in the response of ecosystems generally imply greater risk of devastating impacts later during the twenty-first century.
- Many impacts on the biosphere are strictly irreversible on a timescale of decades to millennia, such as the loss of species and soil organic matter pools.
- Ecosystems are impacted by climate change that are already currently under substantial pressure from land use, pollution, and other nonclimatic drivers.
- Several options for the mitigation of climate change have direct additional consequences for the biosphere, particularly the development of biofuels.
- The biosphere continues to provide essential services for a growing global population – these services can only be sustained if the risks due to climate change are reduced to a much lower level.

See also: Temperature Patterns.

Further Reading

Carpenter SR, Pingali PL, Bennett EM, and Zurek MB (eds.) (2005) *Ecosystems and Human Well-Being, Vol. 2: Scenarios*. London: Findings of the Scenarios Working Group of the Millennium Ecosystem Assessment, Island Press.

Casassa G, Rosenzweig C, Imeson A, *et al.* (2007) Assessment of observed changes and responses in natural and managed systems. In: Parry ML, Canziani OF, Palutikof JP, Hanson CE, and van der Linden PJ (eds.) *Climate Change 2007: Impacts, Adaptation and Vulnerability*. Contribution of Working Group II to the Fourth Assessment Report of the Intergovernmental Panel on Climate Change. Cambridge: Cambridge University Press.

Fischlin A, Midgley GF, Price JT, *et al.* (2007) Ecosystems, their properties, goods, and services. In: Parry ML, Canziani OF, Palutikof JP, Hanson CE, and van der Linden PJ (eds.) *Climate Change 2007: Impacts, Adaptation and Vulnerability*. Contribution of Working Group II to the Fourth Assessment Report of the Intergovernmental Panel on Climate Change. Cambridge: Cambridge University Press.

Hassan R, Scholes R, and Ash N (eds.) (2005) *Ecosystems and Human Well-Being, Vol. 1: Current State and Trends*. London: Findings of the Condition and Trends Working Group of the Millennium Ecosystem Assessment, Island Press.

Lucht W, Schaphoff S, Erbrecht T, Heyder U, and Cramer W (2006) Terrestrial vegetation redistribution and carbon balance under climate change. *Carbon Balance and Management* 1: 6.

PART E

Ecological Stoichiometry

Ecological Stoichiometry: Overview

R W Sterner, University of Minnesota, St. Paul, MN, USA

J J Elser, Arizona State University, Tempe, AZ, USA

Introduction

Most ecological analyses have been constructed from single-currency descriptions, ones based on single dimensions such as biomass, carbon, nitrogen, or energy. For example, the carbon budget of a forest or a lake would include the inputs and outputs and relevant interchanges of C within the ecosystem. This means of analysis has enabled a great deal of progress. However, a multivariate approach that looks simultaneously at multiple dimensions can provide additional insight and predictive power. For example, both carbon and nitrogen can be studied at once and such a description would have to include knowledge of the C:N ratio of different ecosystem components. These multivariate approaches considering multiple currencies are useful because each individual currency interacts with others. Sometimes these interactions are simple, sometimes they are complex. Some pairs of measures are tightly linked in certain locations. For example, the Ca:P ratio in different fish species is almost exactly equal to 2.3, which is the same ratio as Ca:P in bone. Linkages such as these result from the fact that energy and multiple substances do not flow independently through ecosystems. In other cases, element ratios can exhibit a great deal of flexibility with and across species. Understanding these linkages is the goal of ecological stoichiometry.

'Ecological stoichiometry' is a relatively new term, but stoichiometric approaches were used in some of ecology's classic studies. Considerations of limiting factors, as in Liebig's law of the minimum, are inherently stoichiometric. Resource ratio approaches to competition explicitly include ratios of nutrient supply and nutrient content of competitors, making these models stoichiometric. The classical view of the oceans as having N and P in balance with biotic demand, as first articulated by A. C. Redfield in the middle of the twentieth century, is a stoichiometric view. Ecological stoichiometric principles are involved in studies of food quality, nutrition, nutrient recycling, and others, which have long histories of investigation.

To be more explicit, ecological stoichiometry is defined as the study of the balance of energy and multiple chemical elements in ecological interactions. Most will be familiar with the concept of stoichiometry from introductory chemistry classes where it is used to analyze chemical reactions to identify, for example, the reactant that might limit the formation of some specific chemical product in a chemical reaction. In any chemical reaction when reactants combine to form products, mass must balance during these atomic rearrangements. Ecological stoichiometry seeks to apply this same line of reasoning to understand some of the factors regulating ecological processes such as trophic interactions (herbivory, predation, detritivory), competition, energy flow, and biogeochemical cycling. In considering the stoichiometry of a single chemical reaction, knowing the elemental composition of the reactants and products is essential. Similarly, in ecological stoichiometry, one must know the elemental composition of the organisms involved, and their abiotic world.

Consider the familiar example of the enzyme-catalyzed and light-driven reaction of photosynthesis:

$$6CO_2 + 12H_2O \rightarrow C_6H_{12}O_6 + 6H_2O + 6O_2$$

This reaction involving carbon dioxide and water as reactants, and glucose, water, and oxygen as products in actuality is the net outcome of dozens of individual reactions. It summarizes the overall requirements for CO_2 and H_2O needed for the formation of a glucose molecule by this complex, multistep biochemical process. Our concern here is that the fixed chemical structure of glucose firmly establishes the chemical bounds of the system's behavior. If only one carbon dioxide molecule is added to the above, CO_2 will be in excess, and no more glucose molecules can be produced due to limitation by the other reactant, water. Considering the stoichiometry of this reaction calls attention to the powerful ways in which chemistry imposes constraints on biology. All chemical elements on the left side of the reaction must be accounted for on the right, thus explaining a very important outcome of photosynthesis: the production of O_2. Also, the specific chemical formula of the desired product determines what mixture of reactants can be optimally used.

Ecological stoichiometry takes the same approach. However, instead of considering one to perhaps dozens of reaction steps, it summarizes the chemical balance of ecological transactions that are the net outcome not of dozens but perhaps of tens of thousands of reactions that comprise an organism's entire metabolism (its 'metabolome'). Despite this leap in reaction number, the law of mass balance for all constituent elements must still be obeyed. Functional organisms cannot be constructed with arbitrary proportions of chemical elements. Thus, stoichiometric constraints impose order on ecological interactions same as they do on individual chemical reactions. Most ecological interactions involve some form of transfer of matter. These fluxes strongly control productivity and community structure both in terms of absolute magnitude (the fertility or richness of the habitat) and in terms of relative abundance (the ratios of limiting resources). Elements such as nitrogen and phosphorus act like limiting reagents in the highly complex set of chemical reactions that take place during organism assimilation, growth, and decay.

Two classical papers in ecological stoichiometry in particular provided groundwork for subsequent studies. The oceanographer A. C. Redfield noted the quantitative similarity in terms of N:P ratios between the mean chemical composition of deep oceanic waters and the chemical composition of active plankton in surface waters. He hypothesized that the similarity was not accidental. This 'Redfield ratio' of 16 N:1 P atoms was considered by Redfield to be evidence that the biota exerted a large-scale influence on the chemical composition of the sea, a new perspective relative to prevailing views that did not assign the biota with a major causative role in global chemical cycling. This stoichiometric process, operating over extremely large spatial and temporal scales, imposes a biotic fingerprint on the chemistry of the abiotic world. The influence of the Redfield ratio on oceanic biogeochemistry is difficult to overestimate but more recent developments have come to place this finding in a broader context.

The terrestrial biogeochemist W. A. Reiners proposed that the study of the chemical signatures of living things (such as their C:N:P ratios) and of their ecological coupling in nature provides a 'complementary' perspective on ecosystem dynamics, supplementing understanding derived from the then-dominant single currency bioenergetics perspective. He argued that, at the core of living things, what he called 'protoplasmic life', chemical composition was relatively constrained. However, around their protoplasmic core, living things deployed drastically different materials for structural support. These structural materials may have dramatically different elemental composition, for example, C-rich cellulose in plants, Ca-rich shells in mollusks, Si-rich frustules in diatoms, or P-rich bones in vertebrates. The evolution of these major structural adaptations, and the subsequent proliferation of biota bearing them, in turn had major impacts on ecological dynamics and ultimately on large-scale biogeochemical cycling. Reiners's argument highlights the importance of protoplasmic versus structural allocations in determining organismal elemental composition. While indeed this is important, it is also true that even 'protoplasmic' life can vary in elemental composition in important ways due to differences in biochemical allocations connected to growth status and life-history strategy (as described below).

Homeostasis and the Constraints of Mass Balance and Chemical Proportions

A key aspect to ecological stoichiometry is the degree of variation of elemental composition of an organism or species. Is elemental composition a fixed trait characteristic of that species, or is it a flexible parameter that largely reflects local resource and environmental conditions? This contrast is captured in a quantitative way in the concept of 'stoichiometric homeostasis', the degree to which the elemental composition of an organism maintains a strict chemical composition (like a glucose molecule) despite variation in the chemical composition of resources. Stoichiometric homeostasis can be parametrized in the form of a homeostasis coefficient, H (eta), as:

$$H = m^{-1}$$

where m is the slope of a plot of the elemental composition of the consumer under consideration against the elemental composition of the resource being consumed. Both the x- and y-variables should be log-transformed in this measure. In this approach, if the consumer shows no variation in its elemental composition despite wide variation in its resource supply, then the slope of such a line (m) approaches zero and H approaches infinity (strict homeostasis). On the other hand, if the consumer's elemental composition exactly tracks that of its resource supply, then m takes the value of 1 and H has a value of 1 (no homeostasis, complete plasticity).

Data for real organisms range between these extremes and can depend on the experimental conditions imposed. Strict homeostasis does not appear to be reached in many situations, although strong homeostasis appears to be the rule for many metazoans. For example, in experiments in which the P-content of the algae on which it is raised varies from 1% to 0.05%, the P-content of the crustacean herbivore *Daphnia* varies little (only from 1.5% to 0.9%, corresponding to H of ~7, strong but not strict homeostasis). In contrast, the green alga *Scenedesmus*, when grown on inorganic nutrient supplies with N:P ratios ranging from 5 to 80, shows almost a 1:1 correspondence between its biomass N:P ratio and that of its medium. In this case of weak homeostasis, H approaches a value of 1. These examples illustrate a general pattern in which photoautotrophic organisms (cyanobacteria, algae, vascular plants) are generally thought to exhibit great plasticity in elemental composition. In contrast, heterotrophic organisms (including bacteria but especially metazoans) regulate their elemental composition more strictly around particular values. The contrast of these physiological strategies has profound consequences for food web dynamics and energy flow and nutrient cycling in food webs.

The Biology of Elements

Each of the 100+ elements in the periodic table has characteristic chemical and physical features governing properties like the elements it will bond with, how easily it will ionize, etc. These properties arise mostly from the atomic size of each element and the arrangement of electrons within shells surrounding the nucleus. Metals such as Fe or Mo for instance can exist in different oxidation–reduction states, and thus these elements are involved in electron transport in membranes and in other biological locations. Other elements such as C or N are very useful for making highly complex three-dimensional (3-D) shapes. Thirty or more elements are thought to be essential to the growth of at least some organisms. This number and range of reactivities give biological systems a great range of chemical behaviors to choose from in order to organize their activities. The physical and chemical properties of elements have a great bearing on ecological stoichiometry.

As discussed above, differences in organism elemental content between species are very relevant in ecological stoichiometry and ecology in general. Nevertheless, some elements are consistently high in abundance in biological systems while others are consistently rare. For example, the chemical composition for a living human can be written:

$$H_{375000000}O_{132000000}C_{85700000}N_{6430000}Ca_{1500000}P_{1020000}S_{206000}$$
$$Na_{183000}K_{177000}Cl_{127000}Mg_{40000}Si_{38600}Fe_{2680}Zn_{2110}Cu_{76}I_{14}$$
$$Mn_{13}F_{13}Cr_{7}Se_{4}Mo_{3}Co_{1}$$

This formula combines all the countless different compounds in a human being into a single abstract 'molecule'. Ecological stoichiometry asks how far this analogy to a single complex molecule will take us.

Ecological stoichiometric principles should apply to any element, and possibly to some of the less-reactive biochemicals as well. However, stoichiometric analysis has focused most on several elements that make up moderate to large proportions of living biomass and that also may become limiting to organism growth. The four elements considered below relate strongly to ecological dynamics and evolutionary fitness.

Carbon

Carbon (C) is the third or fourth most abundant element in the universe. Though it is not especially abundant on the whole of the Earth, it is a major component element of life. C is a highly mobile element in ecosystems, with common forms in gaseous, liquid, as well as solid phases. C has a valence state of 4 and can form four strong bonds with four other atoms; this allows it to produce virtually limitless arrangements of chains, rings, and other highly complex 3-D shapes. Its importance in biological systems is indicated by the fact that the soft tissue of living organisms is generally 40–55% C by dry mass. This large amount and generally tightly bounded range reflects the constant, ubiquitous, structural role of C throughout living systems.

Most biochemicals fall within a narrow range of C contents. Carbohydrates are similar to many other types of molecules in being almost 50% C. Of all common biological molecules, fats have the highest carbon content, with about 75% of their mass contributed by C. Organic structural matter that may make up large amounts of organism biomass includes the cellulose and lignin of plants and the chitin of arthropod exoskeletons. Even most inorganic structural matter used in living things contains considerable carbon, for example, the calcium carbonate of mollusk shells and the apatite of vertebrate bones both have large amounts of C. An exception is the silicon-containing shells of diatoms; these are nearly carbon free.

Nitrogen

Nitrogen (N) neighbors carbon in the periodic table and has many similarities in terms of biological chemistry. It too is highly abundant in the universe and is similarly abundant as carbon on Earth. It too is a major element of life and is found in all three phases in ecosystems. Its redox reactions include N-fixation, denitrification, and nitrification. With a valence state of 3, it is not quite as capable as C in forming complex 3-D shapes. The soft tissue of living organisms is generally 2–10% N by dry mass. This relatively large value is due to the fact that molecules needed in large quantities in the cell, such as proteins, nucleic acids, and pigments, contain N.

Chief among the important N-contained biomolecules are proteins, which are about 15% N by mass. Proteins themselves can serve a structural role but, as enzymes, they also drive nearly all the biochemical activities of a cell. Nucleic acids also are about 15% N by mass. In the form of DNA, nucleic acids make up the genetic code, though DNA is not a large fraction of cell mass. Even more important to ecological stoichiometry is the more abundant RNA (especially rRNA), which is critically important to cell growth and thus evolutionary fitness. All organisms require abundant proteins and nucleic acids to live and grow. The lowest N-content in living things is generally found in very metabolically inactive creatures.

Phosphorus

Phosphorus (P) has a larger atomic mass than C or N, and it is less abundant in the universe than those two elements. However, it is more abundant on the whole of the Earth than is either C or N. P lacks a significant gaseous phase, imparting very different biogeochemical behaviors to this element compared to C or N. Its abundance in the aqueous phase is strongly controlled by the presence of other elements, most notably oxygen (O) and iron (Fe). Most of the P at Earth's surface is found in the oxidized form of PO_4^{3-} (phosphate). Phosphates can polymerize, and the reaction

$$ATP + H_2O \rightarrow ADP + H_2PO_4^-$$

has a high free energy change but occurs slowly without a catalyst; these characteristics make P a highly suitable element to be involved in transfer of chemical energy around a cell. The soft tissue of living organisms is generally 0.3–2% P by dry mass.

Phosphorus is widespread in biochemistry, being found in relatively high abundance in membranes and being involved in many biochemical reactions throughout a cell. However, P is thought to have two focal roles in whole-organism stoichiometry. First, P alternates with sugars in forming the backbone of nucleic acids. As rRNA, nucleic acid serves a key role in allowing a cell to make proteins. Eukaryotic ribosomes are ~50% rRNA while prokaryotic ribosomes are ~65% rRNA. In this way, we can say that phosphorus is the elemental engine for protein manufacture. Second, P in the form of the mineral apatite is a major component of bone and thus has a major structural role in vertebrates. In small, unicellular heterotrophs such as bacteria and in rapidly growing larvae of insects, generally around 50% and sometimes as much as 90% of total cellular phosphorus may be contained in nucleic acids, especially rRNA. In larger heterotrophs, structural support tissues such as bones take on greater importance in body P-pools and in accounting for differences between species. These interspecific differences can be ancient in evolutionary origin and represent major differences in body plans comparable to taxonomic levels such as family or above.

Iron

Iron is a considerably larger atom than even P. It is as much abundant in the universe as C and N, but it is more abundant than C and N on Earth. It is the first or second most abundant element in the whole of the earth. A majority of the Fe in the Earth's crust is in the form Fe^{2+} (ferrous) but this is quickly oxidized at the oxygen-rich surface to Fe^{3+} (ferric). Though making up 30% of the mass of the whole Earth (a very large fraction), concentrations of Fe in oxic waters can be extremely low due to the low solubility of ferric iron. Low iron concentrations have been shown to limit primary production rates, biomass accumulation, and ecosystem structure in a variety of open-ocean environments, including the equatorial Pacific, the subarctic Pacific, and the Southern Ocean, and even in some coastal areas. Binding of Fe with other molecules is very important to governing its overall solubility. Organically bound Fe may be the dominant chemical form in aqueous solutions.

Iron is chemically versatile. It can coordinate with O, N, or sulfur (S), and it can bind to small molecules. Iron-containing proteins are key features of many energy-transducing biological reactions in processes such as photosynthesis, respiration, and others. Iron serves no major structural role in biological systems. Instead, it is a renewable reaction center. Thus, the soft tissue of living organisms is generally <1% Fe by dry mass.

Molecules containing Fe include hemoglobin, cytochrome *c*, and ferridoxin.

Contrasting Homeostasis in Plants and Animals

Autotrophs rely on either light or chemical energy to turn CO_2 into organic carbon molecules. Photoautotrophs are photosynthesizing organisms such as algae and higher plants that use light for this process. Heterotrophs, in contrast, obtain their chemical energy from preexisting organic molecules. Examples of heterotrophs include bacteria, which absorb organic substances from their surroundings, and many different animals, which consume and digest other organisms. These two major contrasting nutritional strategies of autotrophy and heterotrophy also contrast in their stoichiometric flexibility. Autotrophs obtain carbon, energy, and nutrients from different, somewhat independent sources, whereas many heterotrophs obtain all of these at once from the same food parcels. This contrasting flexibility in turn has a great bearing on the specifics of how stoichiometry enters into ecology.

Photosynthesis relies on light energy to fix CO_2 into organic molecules such as sugars. From these building blocks many other biochemicals can be made. Carbon:nutrient stoichiometry (C:N or C:P ratios) in individual autotroph species can be quite variable. Biochemicals such as carbohydrates and many lipids, which contain only C, H, and O, are made without incorporation of nutrients such as N or P. An autotroph in the light and with adequate access to CO_2 can make a plentiful supply of these compounds (starches, oils, organic acids, etc.) without investment of other critical resources. It is often observed that autotrophs growing in high-light, low-nutrient environments will possess a great abundance of these molecules, so much so in fact that the C-content of the autotroph will be elevated under those types of conditions. Carbon:nutrient ratios within such plants can be exceedingly high (>1500 C:P, for example). When a slow-growing, nutrient-limited autotroph suddenly is exposed to high nutrient availability, it will take up those nutrients much faster than its growth rate. That is, nutrients are taken up in excess compared to growth requirements and in some extreme cases stored in specialized structures such as vacuoles or in specialized molecules such as polyphosphate. High carbon:nutrient ratios are also characteristic of large autotrophs such as trees, which require substantial investment in wood and ancillary tissues having high C:nutrient ratio. Ecological implications of these stoichiometric responses to light:-nutrient ratios are discussed below.

Autotroph nutrient content is related to growth rate (μ, g g^{-1} d^{-1}). A quota (Q) is the mass or molar quantity of nutrients per cell (this discussion assumes a constant cell size). In unicellular autotrophs, the 'cell quota' concept relates these two variables. The quota of the element that regulates growth rate will be very tightly related to growth rate by a relationship referred to as the Droop formula:

$$\mu = \mu'(1 - k/Q)$$

where μ' is a theoretical maximum growth, never attained, associated with infinite quota, and k is the minimum quota occurring at zero growth.

Under strongly nutrient limiting conditions where growth rate is low, quota of the limiting nutrient will be low, meaning a low nutrient:C or high C:nutrient ratio (see cellular C:P, **Figure 1**, top panel). The minimum cell quota (k) is set by the level of nutrient-containing biochemicals necessary for basic metabolism, and nutrient requirements for growth are added to this basal level. A true upper level for nutrient content (less than μ') will be set by some combination of the composition of protoplasm at high growth rate or the ability of an autotroph to store excess quantities of any nutrient not currently needed for growth. In autotrophs, growth involves at least two specific major stoichiometric components, and probably more. The first is N for proteins involved in photosynthesis, especially the enzyme RUBISCO, which can be a major portion of cellular biomass. Metabolism in vascular plants relates more strongly and consistently to N than biomass or C. The second is P for ribosomes, which are needed to manufacture additional proteins.

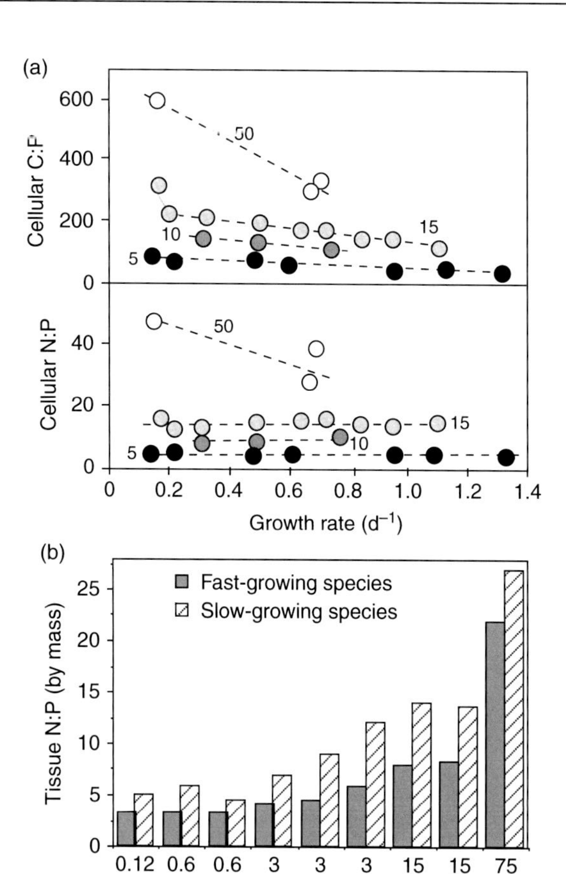

Figure 1 Autotroph nutrient content as a function of both growth rate and nutrients in the external environment. (a) Experiments with the unicellular alga *Dunaliella tertiolecta*. Symbols refer to different N:P in the growth medium (5–50). (b) Experiments with two species of grasses, one (*Dactylis glomerata*) fast-growing and the other (*Brachypodium pinnatum*) slow-growing. In the upper part of (a), note that cellular C:P declines with increasing growth rate, and is highest at low growth rate and where the environmental N:P is greatest. Similarly, both panels of (a) show that environmental N:P has a positive effect on algal N:P at all growth rates. In panel (b), note again that environmental N:P has a positive influence on tissue N:P. Panel (b) also shows that for any given environmental N:P, the fast-growing species has lower N:P than the slow-growing species.

In addition to these patterns relating content of the limiting nutrient to growth rate, the ratio of nutrient elements in an autotroph varies positively with the ratio of those nutrients in the environment. Soils or water of high N:P ratio will generally support plants or algae with high N:P ratio. This positive relationship derives in part from shifts in species across gradients such as these, with competition favoring species that have similar nutrient ratios as the supply ratio in the environment. It also derives from intraspecific, physiological shifts associated with differing storage and utilization of the two nutrients similar to those described for quota above. **Figure 1** summarizes these different influences on autotroph nutrient content.

Samplings of whole assemblages of autotroph biomass have been examined in terrestrial, freshwater, and marine ecosystems, and have included microscopic as well as macroscopic species. Terrestrial ecosystems, with their larger, cellulose-rich, and woody plant species have higher and more variable C:P and C:N ratios than aquatic ecosystems. In the aquatic realm, offshore marine environments characteristically have low and less variable C:P and C:N ratios in their suspended matter, which contains a strong signal of autotroph biomass. We saw this relative constancy in the offshore marine realm when we discussed the Redfield ratio above. Redfield described the marine plankton to have a C:N:P ratio of 106:16:1. Today there is continuing interest in the Redfield ratio in the ocean, and it is known that it is not a true constant but rather varies with several factors, including climate. Freshwater ecosystems can be thought of as being intermediate in their stoichiometric patterns of C:N:P between terrestrial ecosystems and offshore marine ecosystems.

Animals and other heterotroph species also vary in their chemical content. Large shifts in C:N or C:P ratios in heterotrophs can follow from storage of large amounts of chemical energy in the form of lipids. Some invertebrates in seasonal environments, for instance, may assimilate and store lipids to the point where they are approximately half of organism mass. When those lipids are subsequently catabolized, dramatic shifts in C:N or C:P result. However, in contrast to the great stoichiometric flexibility often observed in autotrophs, unicellular and multicellular heterotrophs come closer to approaching an idealized, strictly homeostatic, abstract 'molecule' of defined chemical composition. Reasons for this contrast between plants and animals are not well understood but might involve lack of specialized storage vacuoles in animal cells and the fact that animals obtain carbon, energy, and nutrients from living or recently living material, which is less chemically variable than the abiotic sources of carbon, energy, and nutrients used by plants.

Metazoan animal species exhibit a wide range of N:P ratios. Small, poorly skeletonized organisms such as tadpole stages of amphibians have N:P of $\sim$20 whereas some fish species that are heavily endowed with calcium phosphate apatite mineral both in their internal skeleton and in their scales have N:P of $\sim$5. Fish in fact are a highly stoichiometrically variable group. From that minimum N:P of about 5, different species of lower structural P content range upward to N:P of 15. Within fish, the Ca:P ratios are highly constrained, indicating that most of the stoichiometric differences in this group result from evolutionary pressures on structure and hardness of the integument.

These inter- and intraspecific patterns of elemental content combine in food webs of many species. Stoichiometric imbalance, where resource and consumer differ radically in their nutrient content, generates interesting ecological dynamics that we will consider next.

Stoichiometry of Limiting Elements

The earliest application of stoichiometric reasoning in ecology probably was that in the work by Justus von Liebig (1803–73), who was concerned with factors influencing crop growth yield. His studies led to what we now call 'Liebig's law of the minimum'. One way to phrase Liebig's law is that growth is controlled not by the total of all resources available, but by the one scarcest resource. This is a direct analogy to the concept of a limiting reagent in a chemical reaction producing a stoichiometrically fixed product. 'Scarcest' is a relative term and refers to the availability of a resource in the environment relative to biological demand. For example, chemical resources needed in large quantities such as C or N might be 'scarce' in this context even if they are at higher concentration than elements such as Fe or Mo needed in much smaller quantities. Liebig's law does not apply to all possible resources. Ones that fit it most precisely are referred to as 'nonsubstitutable' resources, ones which organisms cannot trade one for another. Individual elements are largely nonsubstituable with each other whereas many (though not all) organic molecules have substitutable alternatives.

If one knows the stoichiometry of resources and consumers, one can make predictions about which resource will run out and therefore become the 'limiting reagent', regulating growth and production. We see this reasoning in such recently studied phenomena as the Southern Ocean, high-nutrient/low-chlorophyll (HNLC) region. This portion of the ocean is notable because of not only a simultaneous occurrence of low algal biomass but also high levels of the nutrient elements N and P. In most marine and freshwaters, one or both of N and P become exhausted and become limiting. However, in the HNLC regions, it is usually Fe that becomes exhausted first. Iron limitation in HNLC regions has been confirmed in small-scale bottle experiments, and in some spectacular, large-scale open-ocean fertilization experiments where Fe was introduced from a ship and the subsequent development and decay of a high algal patch was followed. Fe-limitation has also been described in near-shore oceans and even in lakes.

Liebig's law is most often thought of in terms of plant population dynamics, where individual elements form the set of possible resources. There are, however, also circumstances where Liebig's law has been applied to animal growth. Liebigian dynamics are strongly suggested, for instance, in the section below on stoichiometry and animal growth.

The stoichiometry of limiting factors varies considerably in space and time. Over long timescales associated with soil development during primary succession on a mineral soil, the N:P balance shifts because P is present in the original parent material, but N derives largely from biological processes including N-fixation. Thus, a new mineral soil will contain P but almost no N, and N will be limiting to plants growing on that site. N-fixers are favored. As N in the soil builds up over successive production/decay cycles, N comes into approximate balance with P. Over long timescales, highly weathered soils can have most of their P removed or bound into inaccessible forms, and P can become limiting. Soil development fitting this general pattern has been observed on the Hawaiian islands. Ancient tropical soils are often highly deficient in P, and there is a general pattern of an increased N:P in the foliage of coniferous trees, grasses, herbs, shrubs, and trees as one goes from the poles to the tropics.

When one seeks to apply Liebig's law in any individual case, complications often arise. For example, different species of an assemblage may be limited by different nutrients simultaneously, meaning multiple resources might limit a community. This form of multiple resource limitation is very well studied in theoretical models and has been the foundation of some recent elegant experimental studies as well; theory and experiment together suggest that the number of elements that are simultaneously limiting to different species has a large influence on the biodiversity of a community. Where multiple resources become limiting, biodiversity is higher. In this form of multiple-element limitation, one might potentially see the community as a whole respond to single additions of more than one resource. The largest response is expected where multiple resources are added, because this will stimulate the greatest number of species. In other cases, due to biochemical reasons, availability of one element might aid in the acquisition of another element. For example, the enzyme that allows cells to make use of organically bound P contains Zn; thus, potentially adding or subtracting Zn might be functionally similar to adding or subtracting P. Ecologists today do not have a comprehensive, widely accepted terminology for dealing with cases of multiple-nutrient limitation.

Liebig's law appears today in work referred to as resource competition theory (RCT). RCT provides a predictive, mechanistic framework for studying how community structure relates to resource availability. Good competitors for a resource are defined as those that require relatively small amounts of a resource in order to maintain themselves in a community in the face of mortality losses. Stochiometrically, good competitors often are those that themselves have low nutrient contents or high C:P or C:N ratios. However, both differential resistance to mortality and differing ability to acquire nutrients can also play strong roles in determining competitive ability.

Various tests of this theory have been performed and support for RCT predictions have come from the laboratory and field in experiments involving microorganisms (bacteria, cyanobacteria, algae) and vascular plants and even metazoan animals. For example, resource conditions with high Si:P ratios tend to favor diatoms (which have relatively low P-requirements but require silica), while other algal taxa (which have no Si-requirement) dominate over diatoms when Si:P ratios are low. Coexistence is predicted, and observed, when Si:P ratios are intermediate; in such a situation, the diatoms are limited by Si while its competitor would be limited by P. Similarly, the relative abundance of cyanobacteria (and especially N-fixing cyanobacteria) is often higher when environmental nutrient supplies occur at low N:P ratios. Recent studies have also shown that the nonlinear resource utilization functions of species allow for multispecies coexistence due to the complex, and perhaps chaotic, dynamics that ensue when multiple resources (light, nutrient elements) can be limiting to multiple species ('coexistence by chaos').

Nutrient limitation by N or P interacts with the carbon cycle in interesting ways. Ecosystems that are strongly nutrient limited are often observed to have primary producers with elevated C:N or C:P ratios. Those nutrient-limited ecosystems can be said to have high nutrient-use efficiency. They make much biomass with each unit of nutrient acquired. As discussed above, in some systems it is the overall availability of light relative to nutrients that controls C:N or C:P ratios. No matter which terminology or thought process is used to study these relationships, the functional outcome is that growth rate and stoichiometry are related in nonhomeostatic organisms such as autotrophs.

Life Histories

Chemical composition is part of an organism's phenotype. It is therefore molded by natural selection and other evolutionary forces like any other aspect of an organism. Thus, the role of the organism in nutrient fluxes in ecosystems is shaped in evolutionary time. As we saw in our discussion of resource competition theory, ecological success or failure, and therefore evolutionary fitness, is tied directly to chemical content. Organism stoichiometry affects fitness. There are also more complex connections between stoichiometry and evolutionary pressures.

Because of the distinct stoichiometry of structural matter, evolutionary pressures on body size or major body plans involving structure will have stoichiometric implications. In fact, anywhere organism function has some kind of distinct stoichiometric signature, evolution will be steered by and will in turn alter stoichiometry. Ecological stoichiometry can help us better understand some of the many selective forces on organism function. A good example is in life histories. A life history for a given species is its schedule of vital rates including growth, reproduction, and mortality. Life-history theory typically considers these schedules to be phenotypic traits exposed to selection and shaped by ecological forces. However, these vital schedules, though measurable, do not exist independent of the physical, inorganic world. Because certain vital rates have distinct chemical signature, ecological stoichiometry can illuminate key aspects of life-history evolution.

Growth Rate

The growth rate hypothesis (GRH) links elemental content to biochemistry and in turn links both of those to life history. It concerns the specific growth rate of body mass (μ, g g^{-1} d^{-1}). The GRH comes about because of the almost uniquely low N:P ratio of nucleic acids compared to all other major biochemicals. According to the central dogma of molecular biology, information flows from DNA to proteins by way of several forms of RNA. Protein is a large fraction of cell mass and to achieve high specific growth rates of cell mass, a great deal of ribosomal RNA is needed to manufacture large amounts of protein at high rates. In rapidly growing *Escherichia coli*, RNA makes up approximately 35% of total cell mass, and as much as 90% cellular P is in RNA. Such a large proportion of cell mass in a single biochemical of unusually low N:P generates a distinct stoichiometric signature for the whole organism. Due to this stoichiometric imprint of RNA, the GRH posits positive relationships among three variables: specific biomass growth rate, P-content, and RNA content. Support for the GRH comes from multiple groups of organisms ranging from unicells to small invertebrates such as *Drosophila* and *Daphnia* (**Figure 2**).

Applicability of the GRH to larger organisms is still unclear. When organisms are examined over a range of many orders of magnitude in body size, RNA content is observed to decline greatly with body size, as does growth rate. Whole-organism P content, however, does not decline as strongly with body size in such a range. Hence, in considering a range of organisms from microbes to large multicellular animals, there must be a shift in P-containing pools along body-size gradients from RNA to other substances. As we will see in more detail below, in vertebrates bone replaces RNA as a major P-reservoir.

There is evidence too that the GRH applies to autotrophs as well as heterotrophs. In a study with the unicellular alga *Selenastrum*, high growth rate was associated with a shift in the boundaries between N- and P-limitation; the N:P of the boundary was lower at high growth rate. This is consistent with a need for larger quantities of low N:P RNA at high growth rate. Similarly, a model of unicellular algal growth was analyzed where cellular investment in two broad classes of biochemical machinery was examined: (1) assembly machinery, for example, ribosomes with low N:P ratios, and (2) resource-acquisition machinery, which is made up mainly of protein containing N but little or no P. Shifts in optimal N:P ratios in the model were associated with growth rate. Low N:P was favored at high growth, whereas high N:P was favored under low growth and strong resource scarcity. These studies of microscopic autotrophs are consistent with the GRH. In one study looking at foliar N and P in more than 100 vascular plants, N:P was lower in species that have higher growth rate. **Figure 1b** showed one such contrast.

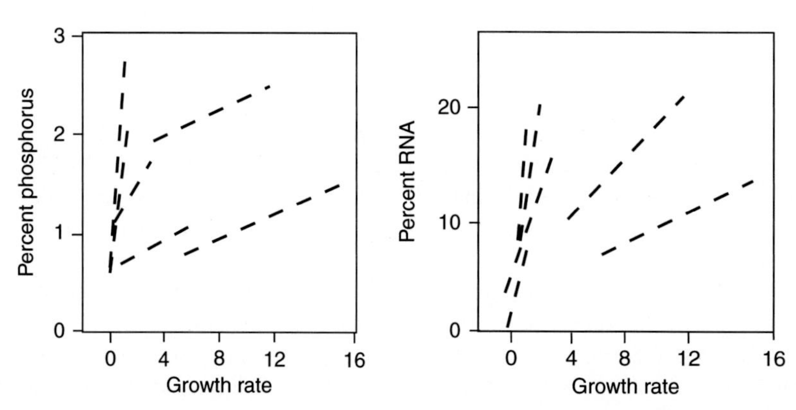

Figure 2 Positive relationships among specific biomass growth rate (μ, d^{-1}) and both P-content and RNA content. Each dotted line represents a different species. Though species differ from one another, these variables are often linked in a way consistent with the GRH.

The GRH thus proposes a material basis to a key life-history parameter: growth rate. All else being equal, specific biomass growth rate will be closely coupled to such life-history parameters as age and size at first reproduction, and therefore is directly tied to fitness itself. Because it is especially P among all the elements that is disproportionately required for high growth, we can further hypothesize that high-growth phenotypes might be particularly sensitive to the presence or absence of phosphorus from the environment. A fascinating biomedical example of the GRH in action is in cancerous tumors, which in many cases are high in RNA. P-content of tumors has not often been studied, but evidence for high P-content in tumors has been seen.

Growth–Competition Tradeoffs

The GRH provides an explanation for why an organism might be selected to have high P content. Why then don't all organisms have high P content? As elsewhere in evolutionary biology, the answer lies in understanding key tradeoffs. What disadvantage might there be to having a high-growth, high-P lifestyle? Ecologists have long recognized a general syndrome of life-history characteristics, those that promote rapid-growth, colonist-type, life histories from those that promote success in the face of extreme competition due to presence of many competitors. The former are referred to as *r*-selected species and the latter are referred to as *K*-selected species. Stoichiometry has its own version of *r*–*K* selection theory.

We have already seen that high-growth (*r*-selected) life histories require high P. However, it is equally the case that species of low nutrient content will outcompete species of high nutrient content during exploitative competition for resources. Stoichiometry predicts that this tradeoff will be particularly evidenced under conditions of P-limitation, because of the close tie between P and high growth rate. High-growth-rate species are anticipated to be poor competitors. According to stoichiometric theory, this tradeoff results from the material basis of a high-growth-rate lifestyle.

Nutrient Cycling, Flux, and Dynamics

The rate of cycling of potentially limiting elements is a critical feature of ecosystems, determining important aspects of their structure and function. This cycling includes movement of elements between abiotic and biotic pools as well as among living species. Autotrophs take up resources from their environment, and they may also leak some nutrients back into the surrounding soil or water media; they also produce litter which is broken down, and decays and releases nutrients into abiotic pools. Heterotrophic consumers participate in nutrient cycling when they ingest food and release wastes back to the abiotic pool. If we understand the controls on the rate that nutrients are made available or reavailable for living organisms to take up, we will have a powerful tool to understand and predict many features of ecosystems. Stoichiometric approaches to nutrient cycling are based on the conservation of matter and consider patterns among multiple elements.

Under strict homeostasis, we consider individual species to have fixed stoichiometric coefficients in their chemical makeup. Thus, the chemical flexibility needed to balance a reaction where resources are reactants and organism biomass is a product comes from the composition of waste products, another product in the reaction:

$$\text{Resources} \rightarrow \text{Organism biomass} + \text{Wastes}$$

If we know both the chemical content of organisms and the chemical content of their resources, we can use the conservation of matter to calculate the chemical composition of wastes. Note that variation in the chemical content of either resources or organism biomass can influence the chemical content of wastes.

One aspect of stoichiometry is merely the use of a set of tools for balancing chemical reactions involved in nutrient cycling and maintaining appropriate relationships among all the elements. However, stoichiometric homeostasis also has a characteristic and somewhat peculiar effect on nutrient cycling. A homeostatic consumer alters the fraction of ingested nutrients that are retained for growth in response to the chemical content of the food. A homeostatic consumer must generally retain most stringently the scarcest element. For example, when ingesting food of relatively low P-content, a homeostatic consumer must retain P with elevated efficiency. In other words, it must retain that substance which already is in scarce supply in the ecosystem while it resupplies the nonlimiting elements back to abiotic pools (**Figure 3**). Nutrients that are found in relative excess tend to be recycled back to the environment while nutrients that are scarce in the food relative to a consumer's requirements are retained for growth. Thus, when put into the context of whole ecosystems, interesting dynamics result from homeostatic consumers.

These processes have been measured in many different situations. **Figure 4** shows the effect of chemical variation of both resources and consumers in one set of aquatic organisms. Because natural ecosystems are composed of numerous species all with their own characteristic stoichiometry, natural food webs consist of many different resource–consumer pairs. Some of these will be stoichiometrically similar while others may be stoichiometrically dissimilar.

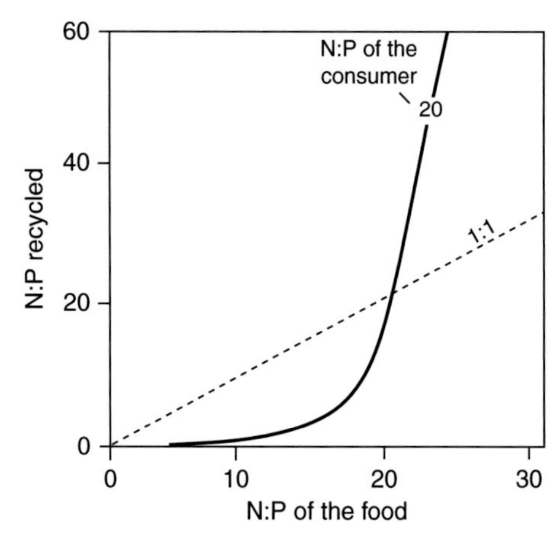

Figure 3 The ratio of nutrient elements recycling by a feeding homeostatic consumer as a function of the ratios of elements in its food. At low N:P in the food, the homeostatic consumer retains N with high efficiency, recycling excess P, and hence the N:P recycled is very low. In contrast, at high N:P in the food, the homeostatic consumer retains P with high efficiency, releases relatively more N, and thus the N:P recycled is very high. Homeostasis in the consumer causes this relationship to curve and the greater the ability of the consumer to retain the element limiting its own growth, the tighter the bend in the function.

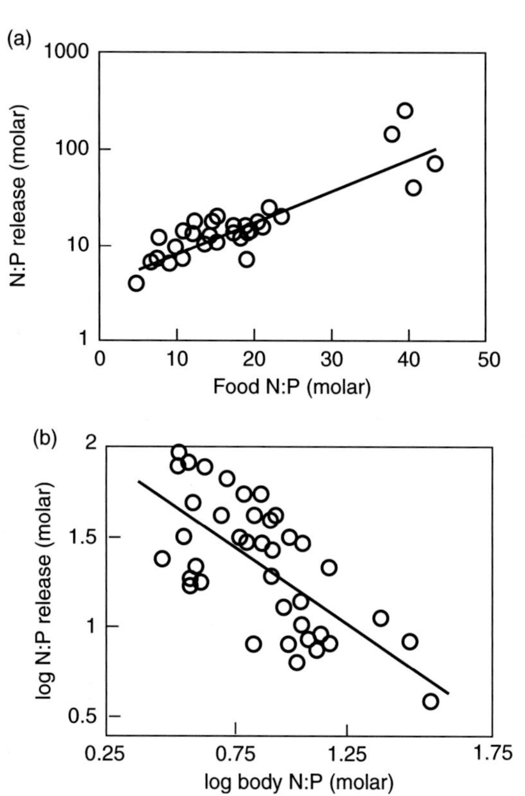

Figure 4 Two aspects of stoichiometric determination of nutrient cycling, the nutrient content of the resources (a) and the nutrient content of the consumers (b). Both examples consider homeostatic animal consumers feeding upon chemically variable resources. In (a), the N:P ratio of nutrients released by freshwater zooplankton is positively related to the N:P ratio of the food they eat. This example shows that when consuming foods relatively high in either N or P, homeostatic consumers dispose of the excess nutrient and retain the nutrient most limiting to their own needs. Note the log scale of the *y*-axis, meaning a curvilinear function on linear axes. In (b), the stoichiometric variability of different aquatic vertebrate consumers generates a negative relationship between the N:P ratios of nutrients released and the N:P ratio of the consumers themselves. This example shows that consumers with bodies containing either relatively high N or P must retain that element high in their biomass and further that they will dispose of the element relatively scarce in their own biomass.

We also see a strong stoichiometric control on nutrient recycling in terrestrial systems. Leaves exhibit a wide range in C:N:P ratios, which is largely a function of the species or functional group of the plant, but as we have already seen it also depends on growth conditions and other factors. When leaves die, there is a corresponding wide range in nutrient content of the detritus they form. Due to nutrient resorption prior to leaf abscission, detritus often has a very low N- and P-content, a factor which emphasizes the stoichiometric dissimilarity between resources (detritus) and living consumers (microbial decomposers and detritivorous animals). Ecological stoichiometry is most easily revealed in systems where there is great chemical variability, like in this consideration of detritus and detritivore. **Figure 5** shows how strongly the nutrient content of detritus corresponds to the rate that detritus breaks down, at least at this large, cross-ecosystem scale.

This range in mineralization rate results from the stoichiometric match between the chemical composition of the litter and the needs of the organisms feeding on that litter (microorganisms, detritivores), so it combines stoichiometric food-quality effects, such as will be described below, with stoichiometric nutrient cycling rates. The highly biodiverse soil food web can grow and metabolize more rapidly when supported by high-nutrient litter instead of low-nutrient litter.

Another example of stoichiometric recycling effects involving plants, soils, and decomposers involves Ca-content. In a long-term (30-year) study where 14 different tree species were raised in monoculture, it was found that soil properties came to strongly depend on the Ca-content of the leaves of the tree species growing on that plot. Soil under tree species with high-Ca leaves, roots, and litter (e.g., maples, basswoods) was higher in pH, higher in C, lower in C:N, and higher in exchangeable Ca than was soil under tree species low in Ca (e.g., oaks, pines). Earthworm biomass also was higher under high-Ca tree species. The chemical parameter Ca was better related statistically to these and other similar relationships than was identity of the tree species themselves or the identity of functional groups like angiosperms versus gymnosperms, suggesting that it was indeed the stoichiometry of Ca in the litter that was the key factor involved.

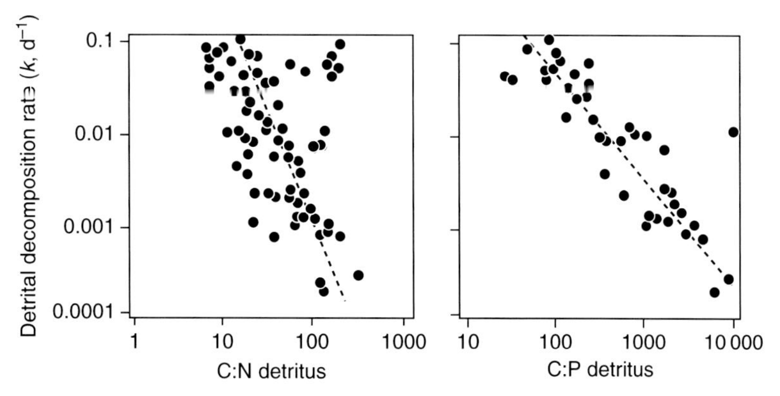

Figure 5 The rate of breakdown of terrestrial detritus depends strongly on stoichiometry. The higher the C:N or C:P ratio, the slower the leaf litter breaks down (lower rate coefficient, *k*). As a consequence, the rate that nutrients are recycling back into the soil also depends on stoichiometry, with higher mineralization rates for detritus with low C:N or C:P.

Alteration of Community Composition and Ecosystem Dynamics through the Stoichiometry of Recycling

When we place these specific stoichiometric recycling effects into the context of complete, complex ecosystems, a variety of interesting dynamics result. For example, as we just saw, terrestrial plant species that produce low-nutrient litter will have a depressive effect on mineralization rates in their vicinity. A low mineralization rate will reduce primary productivity, lower plant biomass, and raise light:nutrient ratios. These conditions in turn will favor particular plant species that are good nutrient competitors. Because one factor that can improve a species' competitive ability for a given nutrient is a low content of that nutrient in its leaves, a cycle of positive feedback is favored: with good nutrient competitors altering nutrient cycling in their vicinity in such a way as to improve their chances of success against other species.

Shifts among consumer species differing in stoichiometry can also have ecosystem-level consequences. In freshwater plankton food chains, large-body-size, high-growth-rate *Daphnia* are favored under particular regimes of fish predation, for example, when there are many piscivores that deplete the planktivorous fish populations, releasing *Daphnia* from fish predation. Because *Daphnia* have a characteristic high-P stoichiometric signature, changes in fish predation regime can realign pelagic nutrient pools and fluxes. It has been observed that phytoplankton in the presence of abundant *Daphnia* grazers will be relatively more P- than N-limited and N-fixation will be reduced. These examples illustrate the close connections between community structure and ecosystem nutrient fluxes that are easily understood by consideration of stoichiometric principles.

Animal Growth: Stoichiometry and Food Quality

Ecological theory often incorporates the effects of food supply on consumer growth and reproduction by a hyperbolic functional response between animal growth and reproductive rates and the density of food. This is referred to as a 'numerical response'. In addition, in the growth of an individual, a minimal quantity of food, alternatively known as the 'threshold food concentration' or the 'subsistence food level', is defined as that needed to offset respiratory losses. At the population level, a higher threshold is needed for growth to be sufficient to offset various sources of mortality. These effects of food quantity have been studied for a long time; however, incorporating the effects of the quality of food on consumer performance and dynamics has been much more difficult yet is done where ecological stoichiometry becomes very relevant.

While there are many known aspects of food quality (digestibility, palatability, toxicity, content of nutritive biochemicals such as essential amino acids and fatty acids), ecological stoichiometry measures food quality as the content of potentially important nutrient elements, especially N and P. This measure is in keeping with the extensive literature on the importance of N in affecting consumers in terrestrial and marine environments. It is also consistent with the increasing knowledge of the role of dietary P in freshwaters and perhaps in terrestrial ecosystems as well. In general, the quality of a food item is inversely related to its C:nutrient ratio, where the nutrient can be N or P. This effect can be understood to be an outcome of the fact that the nutrient element is increasingly diluted in the dominant C-biomass, making it difficult for the animal to extract sufficient limiting element from the food.

Stoichiometric theory gauges the onset of such limiting effects via calculation of the threshold elemental ratio

(TER). The TER is defined as the food C:nutrient ratio above which the animal's growth becomes limited by nutrient X rather than C (carbon), where X might be N, P, or another element. In the simplest models, the TER is based on only three traits of the consumer: its own C:X ratio and its maximal gross growth efficiencies for C (E_C) and the nutrient (E_X). Gross growth efficiency is defined as the rate at which mass accumulates in production (or new growth) divided by the rate at which mass is ingested by the consumer. The maximum gross growth efficiency for an element is assumed to occur when that element is limiting growth. Assuming strict stoichiometric homeostasis of the consumer, the TER is calculated as

$$\mathrm{TER}_{\mathrm{C:X}} = (E_\mathrm{C}/E_\mathrm{X})\mathrm{C:X}_{\mathrm{consumer}}$$

Noting again that the TER is the C:X ratio in the food above which the consumer's growth should become limited by element X, a consumer with low TER is more likely to be limited by element X than a consumer with a high TER. Thus, an animal becomes less likely to be limited by nutrient X (its TER increases) as the consumer's C:X ratio increases (its nutrient content declines), as its ability to sequester element X increases (E_X increases), or as its carbon-assimilation efficiency decreases (E_C decreases).

The dependence of the TER on net C-growth efficiency implicitly means that the TER increases as food concentration decreases. This is because under such strong food-limited conditions, the respiratory demands of maintenance metabolism dominate the energy or C-balance, and C-growth efficiency is reduced. Thus, TER theory predicts that stoichiometric food quality should be of greatest importance under conditions of high food abundance. **Figure 6** illustrates this in a general

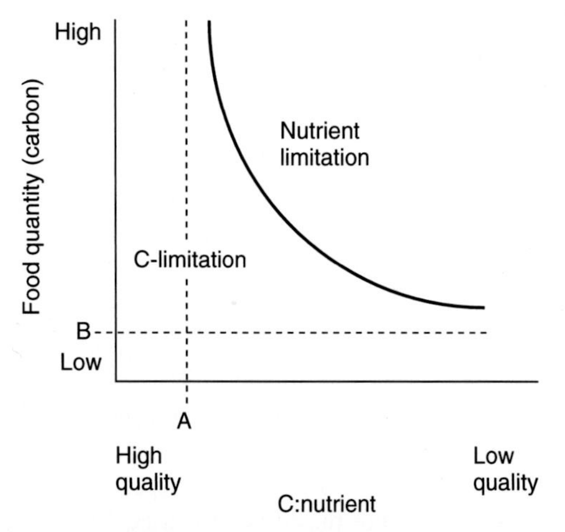

Figure 6 Boundaries for limitation by food quantity (energy) vs. food quality (stoichiometry) for a homeostatic consumer as a function of food quantity. At high food abundance and high food C:nutrient ratios (low food quality), the consumer should be limited by the nutrients in its food. At low food abundance or low food C:nutrient, the consumer should be limited by total food quantity, or total energy content.

case, in which the line labeled 'A' is the minimum TER for the consumer in question, asymptotically achieved at maximum E_C at infinitely high food quantity. The line labeled 'B' is the minimum amount of food that the consumer needs in order to maintain a constant body mass.

The effects of stoichiometric food quality have been extensively tested in the laboratory and the field, especially for freshwater zooplankton under conditions of potential P-limitation. For example, *Daphnia* growth commonly declines when food C:P ratio is above ~250. P-assimilation efficiency increases to high values as the TER is approached. Lab experiments involving short-term manipulation of food P-content or direct P-supply to the *Daphnia* have confirmed that this growth decline is at least partially due to a direct P-limitation of the animal. Furthermore, field studies have shown that *Daphnia* abundance is low under lake conditions in which seston C:P ratios are above 250. Further, short-term amendment of seston P-content increases *Daphnia* growth when seston C:P is above the TER, showing that the predictions of stoichiometric theory are confirmed not only in the lab but also under natural conditions. Also consistent with TER theory, animals with low body P-content, such as the crustacean *Bosmina*, appear to be relatively insensitive to food P-content, both in the lab and in lakes. Recent studies outside of the plankton have also provided evidence of dietary P-limitation in benthic (stream insects, snails) and terrestrial animals (caterpillars, weevils), adding to previous evidence of dietary N-limitation from diverse habitats.

In sum, these studies show that stoichiometry is an important axis of food quality affecting basal consumers in diverse food webs. Relatively simple physiological processes and species traits can be used to predict their operation and impact.

Complex System Dynamics Driven by Stoichiometry

Theoretical ecology uses mathematical tools to place the nature of interactions into a formal, analytical framework. A goal is to understand features governing population dynamics and species diversity. Special emphasis has been placed on ecological competition and predator–prey dynamics. These studies often focus on the presence or absence of equilibrium points in model solutions, which indicate the possibility of species coexisting with each other over the long term. Moreover, a fundamental difference between stable equilibrium points and unstable equilibrium points is recognized. In the former, dynamics tend to restore the system to its equilibrium point after a small perturbation, whereas in the latter, like a ball rolling off the peak of a hill, small displacements from equilibrium can result in large changes and even be catastrophic to coexistence.

Herbivore Extinction Due to Poor Food Quality

Stoichiometrically implicit predator–prey and especially plant–herbivore interactions have been built and analyzed, providing some clues as to the role of stoichiometry in system dynamics and species diversity. In these models, consumers respond to food quality determined by element ratios, and nutrients are recycled according to stoichiometry and strict homeostasis. Recycling provides a feedback between a herbivore and its algal or plant prey. The herbivore's growth and nutrient release rates decline as the prey's C:nutrient ratio increases above its TER. These analyses involve development of versions of the classic Lotka–Volterra predator–prey models in which the homeostatic regulation of nutrient content varies between predator and prey. When the prey is modeled as a photoautotroph, its nutrient content will be variable and the predatory herbivore will have fixed nutrient.

The effects of introducing stoichiometric constraints to such models of trophic interaction are dramatic (**Figure 7**). In classical Lotka–Volterra-type models, the predator can always grow whenever prey abundance is above some finite level. This is seen as a vertical, linear nullcline for the predator; the predator's nullcline is the set of points where the predator's net growth rate is zero. There is thus only one possible equilibrium point of plant and herbivore density for the system. However, in a stoichiometric version of these equations, both prey and predator nullclines are hump shaped. The prey's nullcline intersects the x- (prey-) axis due to nutrient limitation. The intersection point is determined by the maximal C:nutrient ratio the prey achieves under nutrient limitation. Stoichiometric models are unusual in that the predator's nullcline also is hump shaped, in this case due to the negative effects of poor food quality when the plant achieves high biomass on a fixed and limited amount of nutrient in the system. At high plant biomass, plants have high C:nutrient ratio, and further additions of biomass (C) can have a depressive effect on herbivore dynamics, a paradoxical result. The herbivore population is inhibited by the addition of bulk food. In fact, the herbivore can shift between positive growth and negative growth by the addition of plant biomass to the system! This counterintuitive aspect of the model has been termed the 'paradox of energy enrichment'.

Stoichiometry alters nullclines so that both are nonlinear, meaning they can intersect each other at more than one point. Depending on the parameters used, there is potential for multiple equilibria, limit cycles, and potentially chaotic dynamics when one includes stoichiometry in such a system. A particularly interesting equilibrium is illustrated in **Figure 7b**, in which conditions are such that the prey nullcline intersects the prey axis at a high value, beyond the upper intersection of the predator's nullcline. In this situation, a stable equilibrium exists at the prey's intersection on the x-axis, a point where the grazer is extinct and cannot invade from low population levels. This is a grazer extinction point occurring in a 'world' of very high food (plant) biomass! In ecophysiological terms, this situation is more likely to occur for grazers with high body nutrient content and for environmental

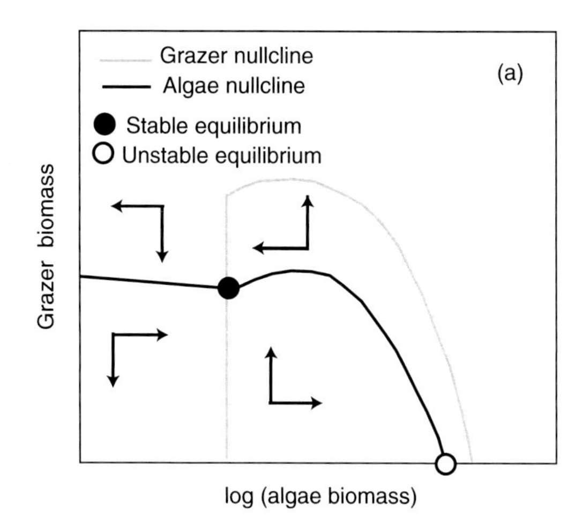

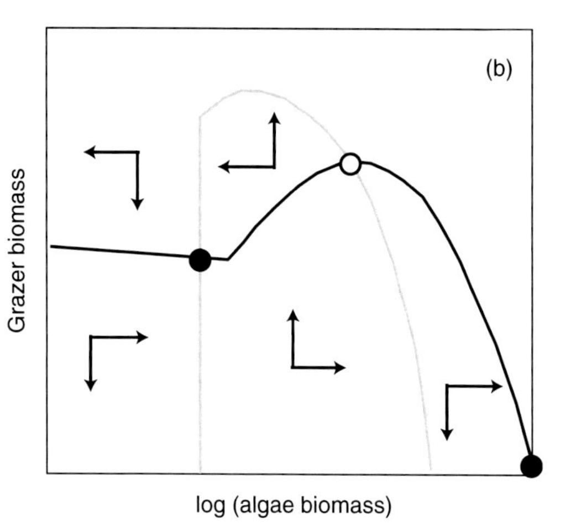

Figure 7 Predator (grazer) and prey (algae) nullclines in a theoretical model. Panels (a) and (b) use different parameters for the same model. Stoichiometry makes the grazer nullcline hump shaped, which alters predicted dynamics and makes it possible even for a stable equilibrium point at zero grazer biomass.

conditions in which autotrophs develop very high biomass C:nutrient ratios. As described earlier, the latter occurs when nutrients are severely limiting and when light intensities and perhaps pCO_2 levels are high. Indeed, the theory predicts that local deterministic extinction of a herbivore can occur with increased light intensity or with any ecological change that induces high C:nutrient ratio in plant biomass.

A similar model has been built and analyzed where the prey is made to be stoichiometrically variable, but the predators are held to be homeostatic. An analogous situation to this would be the case of bacteria (homeostatic) being fed upon by Protozoa (stoichiometrically variable). When we change the point of strict homeostasis from the predator to the prey, the complex dynamics discussed for **Figure 7** disappear. It is both the presence of strict homeostasis and its location within a food web that influences system dynamics.

Increased Primary Production Can Reduce Food Chain Production

The idea that increased food abundance might have negative effects on consumers is a surprising stoichiometric prediction. Numerous lab and field studies have tested predictions of stoichiometric theory for trophic interactions with a special emphasis on examining how light and nutrients can jointly regulate herbivore production. For example, a laboratory study involving *Daphnia* and a green alga showed that growth of *Daphnia* was maximal at intermediate light. At low light, algal biomass was limiting to *Daphnia* growth. At very high light intensities, algal P-content was limiting (**Figure 8**). Consistent with stoichiometric theory, the light intensity supporting maximal growth moved to higher light intensities as nutrient content in the system increased. In a longer-term study also involving *Daphnia* and a green alga (**Figure 9**),

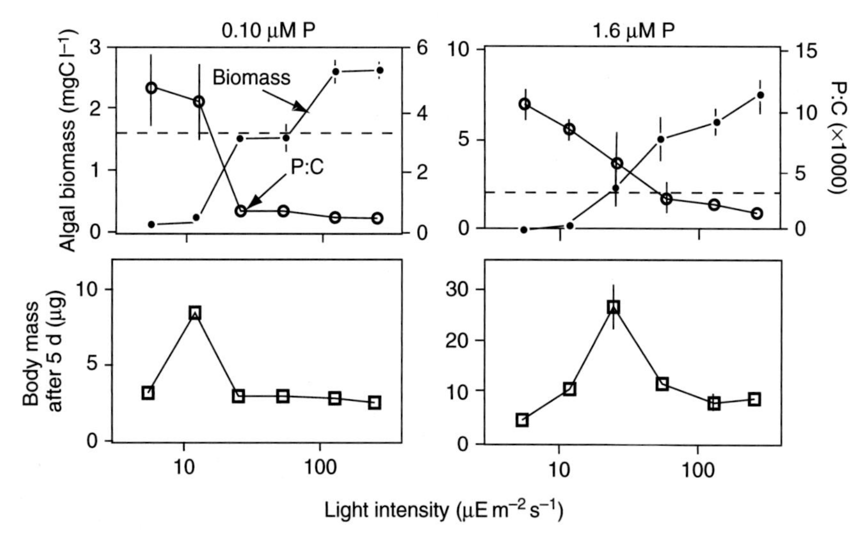

Figure 8 Algae and grazers in light gradients. As light increases, algal biomass increases and algal P:C decreases, worsening food quality for grazers. Maximal grazer growth (highest mass) occurs at intermediate light levels.

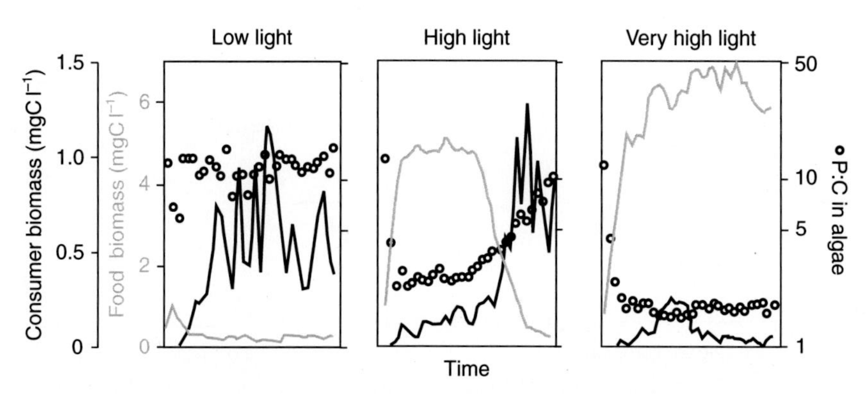

Figure 9 Algal (food) and grazer (consumer) dynamics over time (>90 days) at three light levels. At low light, algal biomass remained low, P:C in algae was high, and grazers were variable but often abundant. At high light, algal abundance was initially high and P:C was low, but eventually the grazer became abundant and the system resembled the low-light condition. Finally, at very high light, algal abundance remained high, P:C remained low, and grazer abundance remained low.

increased light intensity was shown to inhibit *Daphnia* population growth and trophic efficiency but eventually nutrient recycling by the *Daphnia* was able to increase algal nutrient content so that high grazer densities eventually were achieved. In a result that supports the idea that increased light intensity can result in grazer extinction, in this study one replicate vessel received especially high light intensity, which resulted in unusually high algal C:P ratio and a *Daphnia* population that was never able to increase in abundance and was undergoing significant decline at the end of the experimental period.

Several studies involving field mesocosms have also shown that manipulation of light-nutrient balance has a strong effect on secondary production in nutrient-limited ecosystems. For example, a field experiment in a P-limited Canadian lake reduced light intensity to mescosms by more than tenfold. After 30 days, seston C:P ratio had decreased significantly and zooplankton production was increased nearly fivefold. Thus, there is strong evidence that secondary production in food webs is strongly influenced not only by overall rates of ecosystem productivity ('food quantity') but also by the quality of that production as indexed by its nutrient content.

Large-Scale Stoichiometry

Ecological stoichiometry also offers insights into factors regulating the multiple pathways by which energy and multiple chemical elements move or are stored at the scale of ecosystems and above. For example, whole-ecosystem manipulations of food web structure have shown how ecosystem nutrient levels and stoichiometric constraints influence the operation of cascading trophic interactions in which alterations in top trophic levels impinge on ecosystem productivity and nutrient cycling. In an experimentally fertilized, P-rich lake, introduction of a fourth trophic level (piscivorous northern pike, *Esox lucius*) led to strong reductions in planktivorous minnows and thus a major increase in zooplankton biomass and especially *Daphnia*. Consistent with stoichiometric nutrient recycling theory, increased *Daphnia* abundance was associated with increased N:P ratios in nutrient pools and a major reduction in the previously dominant N-fixing cyanobacteria. In a parallel experiment in which pike were introduced to a similar minnow-dominated but unfertilized lake, reduced minnow abundances did not result in increased *Daphnia* abundance. Instead, *Daphnia* abundances declined in parallel with changes in an unmanipulated control lake. Both *Daphnia* declines were associated with major increases in seston C:P ratios, suggesting that the *Daphnia* success in these lakes was strongly controlled by 'bottom-up' food-quality constraints and relatively unaffected by 'top-down' effects of food web interactions. These observations are consistent with an emerging view that strong trophic cascades may be confined to ecosystems having nutrient-rich autotrophic production at their base, as ecosystems with poor quality, nutrient-limited autotroph biomass may be unable to support high biomasses of herbivores capable of exerting significant grazing pressure.

Comparative studies across multiple ecosystems have suggested that stoichiometric constraints play an important role in regulating the fate of organic matter and the cycling of nutrients at the ecosystem scale. As mentioned earlier (**Figure 5**), when considered across multiple studies in diverse terrestrial ecosystems, the rate constant of detrital breakdown correlates negatively with detritus C:nutrient ratio, although considerable variation can exist within particular studies. That is, nutrient-rich detritus (low C:nutrient ratio) breaks down rapidly, returning nutrients to available pools for reuptake by plants, while low-nutrient detritus (high C:nutrient ratio) breaks down slowly and may enhance immobilization of soil nutrients by microbiota, slowing the reuse by plants. Other large-scale patterns in the fate of C in ecosystems appear to be tied to autotroph nutrient content. When data for numerous ecosystems across diverse habitats (oceanic, limnetic, terrestrial) were compiled, strong positive correlations are seen between the percentage of primary production consumed by herbivores and plant nutrient content and turnover rate: ecosystems with fast-growing, nutrient–rich plant production support significant populations for herbivores which consume important quantities of that production. In contrast, in ecosystems with slow-growing, low-nutrient plant biomass, significant quantities of primary production escape consumption in the grazing food chain and enter detrital pathways and long-term C storage. Likewise, the release or retention of nitrogen or organic C appears to be a function of stoichiometric balance in watersheds. Several studies have shown that stream NO_3 concentrations or export rates are negatively correlated with average watershed soil C:N ratios, consistent with stoichiometric theory. Conversely, other studies have shown that concentrations or export rates of dissolved organic C from watersheds are positively correlated with watershed soil C:N ratios. Thus, ecophysiological limitations on the processing of organic C and nutrients by soil microorganisms ramify to affect the fluxes of materials at watershed and regional scales.

Stoichiometric constraints also play a role in the regulation of biosphere-scale processes governing oceanic and terrestrial C-cycling, atmospheric CO_2 concentrations, and thus global climate. In the open ocean, biogeochemical processes appear to be closely linked to multiple limitations associated with light intensity (as mediated by water column mixing processes), macronutrients (N, P), and micronutrients (especially iron). In the enormous central Pacific gyre, long-term studies have shown climatic variations associated with El Niño that reduce the intensity of vertical mixing processes and cross-thermocline nutrient transfers that enhance the success of light-limited, N-fixing cyanobacteria. Dominance of these algae, in turn, increases the C:P ratio of primary production and thus increases net carbon sequestration in deep waters. In other parts of the ocean, low iron supplies rather than light may limit primary production and N-fixation and thus stoichiometric coupling of C- and macronutrient cycles can ultimately depend on iron supply, itself regulated by long-term processes associated with delivery of continental dust. Since the production of continental dust is itself closely regulated by climatic conditions (rainfall patterns), a set of complex feedbacks that plays out over tens of thousands of years is established.

These feedbacks have been incorporated into global biogeochemical models that investigate the autoregulatory capability of the biosphere ('Gaia'). In these studies, it has been shown that a self-regulating system at the scale of the biosphere emerges if the organisms driving the Earth's biogeochemical systems operate under functional constraints on their processing of energy and matter. In other words, no higher-level selection at the scale of the biosphere ('Gaia' as organism) is required for planetary self-regulation to operate. Instead, 'Gaia' is a complex interactive system with dynamics determined as an emergent property of selection operating primarily on the ecophysiological traits and constraints of individual organisms.

See also: Ecosystem Patterns and Processes; Evolution of Oceans; Evolutionary and Biochemical Aspects; Organisal Ecophysiology; Population and Community Interactions; Trace Elements.

Further Reading

Elser JJ, Acharya K, Kyle M, *et al.* (2003) Growth rate – stoichiometry couplings in diverse biota. *Ecology Letters* 6: 936–943.
Elser JJ and Urabe J (1999) The stoichiometry of consumer-driven nutrient recycling: Theory, observations, and consequences. *Ecology* 80: 735–751.
Redfield AC (1958) The biological control of chemical factors in the environment. *American Scientist* 46: 205–221.
Reiners WA (1986) Complementary models for ecosystems. *American Naturalist* 127: 59–73.
Sterner RW and Elser JJ (2002) *Ecological Stoichiometry: The Biology of Elements from Molecules to the Biosphere*. Princeton, NJ: Princeton University Press.
Sterner RW and Hessen DO (1994) Algal nutrient limitation and the nutrition of aquatic herbivores. *Annual Review of Ecology and Systematics* 25: 1–29.
Williams RJP and Fraústo Da Silva JJR (1996) *The Natural Selection of the Chemical Elements: The Environment and Life's Chemistry*. Oxford: Clarendon.

Ecosystem Patterns and Processes

S A Thomas, University of Nebraska, Lincoln, NE, USA
J Cebrian, Dauphin Island Sea Laboratory, Dauphin Island, AL, USA

Introduction

Ecosystems are complex entities composed of diverse organisms expressing a myriad of life histories, body sizes, and metabolic pathways embedded in a dynamic physical and chemical environment. A suite of internal processes create an intricate web of relationships that control the flow of energy and materials within and between ecosystems. Contributing to the complex nature of ecosystems is the breath of spatial and temporal scales over which ecosystems operate. Ecosystems occupy spatial scales from meters to hundreds of kilometers and temporal scales from days to centuries owing to contrasting organismal life spans. For instance, consider boreal forests. There, organism life histories range from a few days (i.e., bacteria in the soil) to centuries (i.e., trees) generating a plethora of timescales over which ecosystem function can be addressed. However, the most characteristic aspect of ecosystem research is that it explicitly incorporates the chemical and physical environment, often requiring multidisciplinary approaches to characterize patterns and identify controlling mechanisms.

Ecosystem studies are largely organized around two complementary themes, energy flow and elemental cycling. Because ecological stoichiometry examines the nature, control, and implications of elemental balances in ecological processes, it also provides an appropriate framework for linking elemental cycles in nature. Elements required for life, such as carbon (C), nitrogen (N), phosphorus (P), calcium (Ca), and trace metals, are conserved as they move between organisms and their environment. That is, elements cannot be destroyed and are instead used, released back into the environment, reused again by organisms, and so forth, cycle after cycle. Continuous elemental flow between the environment and organisms with specific elemental compositions creates important ecological constraints, the implications of which represent one aspect of ecological stoichiometry.

Ecological stoichiometry offers insight into ecosystem dynamics because it readily applies to the composite nature of ecosystems. Stoichiometric constraints affecting primary producers and higher trophic levels have consequences for the relative cycling rates of elements. This condition has an important implication; if we understand the stoichiometric interactions between organisms, we should be able to predict and understand stoichiometric outcomes at the ecosystem level. However, elemental cycling is not simply the sum of assimilative growth, trophic transfer and mineralization (conversion of organic compounds to inorganic forms). Rather, a suite of other biogeochemical processes also influence the availability and cycling of elements. In this chapter, we discuss the consequences of relative nutrient availability for ecosystem processes and review research that has used a stoichiometric approach to investigate the coupling of nutrient cycles in nature.

Nutrient Limitation of Ecosystem Production

Nutrient limitation is a fundamental concept in ecological research and, at its core, is a question of ecological stoichiometry. Primary production is usually stimulated by nutrient addition. Increased production following fertilization has been identified by some as one of the most repeatable and predictable features of nature and one humans rely upon for their food supply. Modern research into nutrient limitation can be traced to Justus Liebig, who developed the "law of the minimum" to explain why crop yield was most often stimulated by a single element, though that element may change from one system to the next. Simply put, the law of the minimum states that the nutrient present in the lowest amount relative to organism demand will limit growth and production. Liebig's law rests on an implicit assumption that an organism's elemental composition is relatively constrained and that its production will reduce elemental pools until one element becomes rare enough to limit production.

As a modern research topic, nutrient limitation spans levels of organization from individuals to ecosystems. At the ecosystem scale, investigations of nutrient limitation tend to focus on primary rather than secondary production because it represents the interface between the physical and chemical environment and the resident biological community. Rates of material flow across this boundary and the factors that influence those rates are fundamental topics in ecosystem science. From a stoichiometric perspective, limitation status is a product of relative nutrient supply and demand for those elements by biota capable of 'using' inorganic nutrient forms (e.g., plants, bacteria, and fungi). At the scale of an organism, relative nutrient requirements are controlled by growth rate, specific physiological processes, and plasticity therein. At the ecosystem scale, nutrient demand is also influenced by trophic structure (e.g., relative abundance of primary producers), nonassimilative biological processes (e.g., denitrification) and various physical and chemical processes (e.g., precipitation, weathering, leaching). Nevertheless, we frequently speak of biological production in forests, grasslands, lakes, and streams as nitrogen or phosphorus limited because nutrient imbalances between inorganic pools and organism stoichiometry are pronounced or experimental manipulations have indicated that the system responded in some way to nutrient addition.

In terrestrial ecosystems, N limitation is most common though P-limitation has been observed. Global scale patterns in soil nutrient availability suggest that N limitation is most pronounced in recently glaciated temperate areas with P-poor soils dominating tropical regions. More current research indicates that foliar chemistry tracks this pattern with lower C:N and higher N:P ratios typical of leaf tissue from tropical versus temperate ecosystems. A frequent explanation for N limitation is the rarity of N in mineral substances relative to P and the difficulty of transforming dinitrogen gas to forms available to plants and microorganisms (biologically mediated N-fixation). A secondary explanation of N limitation in terrestrial ecosystems relates to the relative mobility of dissolved forms of N and P. Phosphorus mobility in soils is poor due to sorption kinetics of dissolved P-forms. Therefore, as P cycles from organic compartments to inorganic forms (via mineralization), there is a tendency for it to remain in the local environment. Nitrogen is mineralized from organic material as ammonium (NH_4^+), which like dissolved P has poor mobility in soils. However, microbial activity rapidly converts ammonium to a much more mobile form, nitrate (NO_3^-), via nitrification, which can be rapidly leached from soils. In addition, loss of N in dissolved organic forms (e.g., humic and fulvic acids) has been shown to be an important avenue of N loss. The development of P-poor soils and the potential for P-limitation become pronounced as landscapes and associated ecosystems age. Research on geological chronosequences in Hawaii and New Zealand indicates that terrestrial ecosystems move toward P-limitation with age due to the establishment of N-fixing organisms, accumulation of organic N in soils, and long-term losses of P via weathering and subsequent leaching. Tropical ecosystems have also been shown to tend toward P-limitation for similar reasons (i.e., highly weathered soils and abundant N-fixation). In the modern era, a human-induced change in the abundance of biologically available N is also likely to push terrestrial systems away from N limitation (potentially toward P-limitation) even in relatively undisturbed ecosystems (via atmospheric deposition).

In aquatic habitats, the prevailing contention has historically been that phosphorus tends to limit gross primary production in freshwater ecosystems, whereas nitrogen tends to limit production in marine environments. However, recent reviews indicate that this conclusion is an oversimplification and the nature of nitrogen and phosphorus limitation of gross primary production in aquatic ecosystems appears more complicated than previously thought. Unlike the terrestrial case presented above, anthropogenic nutrient enrichment of lakes indicates greater relative additions of P versus N. In contrast, human impacts on stream and river ecosystems tend to increase nitrogen availability relative to phosphorus. Recent reviews also suggest gross primary production in marine offshore ecosystems is often limited by phosphorus or micronutrients, whereas gross primary production in coastal marine ecosystems tends to be N limited. An unfortunate stoichiometric outcome of human activity has arisen from elevated N in rivers alleviating primary producer limitation in coastal waters. Excess production in many estuaries has led to extensive 'dead zones' created by anaerobic conditions that develop as excess production is mineralized.

Efficiency in Elemental Use

In the previous section, we discussed the stoichiometric patterns in nutrient limitation. However, there is considerable variation in published relationships between biomass accrual and nutrient supply such that the relationship between nutrient availability and production is often not linear. We now turn our attention to examining how primary producers differentially accumulate biomass under nutrient-limiting conditions. The nutrient use efficiency (NUE) concept describes the amount of producer biomass generated per unit of nutrient taken up by organisms. Where conclusions about limitation are categorical, NUE is a continuous variable that reflects the relative influence of nutrient availability on net production.

NUE is normally expressed as the ratio of producer growth to the amount of nutrients assimilated by the organism. More explicitly, NUE is the ecosystem-scale expression of the ratio between individual photosynthetic nutrient use efficiencies (PNUE; flux of nutrient relative to carbon fixation) and individual growth rates (μ) such that $\mathrm{NUE} = \mathrm{PNUE}/\mu$. Relationships between net photosynthesis and organism N content demonstrate that PNUE can vary by an order of magnitude (25–200). Combined with variation in organism growth rates, it is not surprising that NUE varies considerably among ecosystems.

In stoichiometric terminology, variation in NUE indicates that ecosystems are nonhomeostatic, or plastic, with respect to nutrient use. Why should biomass production become less efficient with increased nutrient availability? A frequent explanation is that strong selective pressures for greater efficiency at low-nutrient concentrations drive the observed pattern. However, there is no obvious explanation for why efficiency would be reduced under high nutrient conditions from either a competition or natural selection perspective. The likely explanation relates to the multiplicity of factors that can limit primary production. For example, under conditions of rapid biomass accumulation, light limitation of producer growth through self-shading increases, leading to smaller producer growth per unit of nutrient absorbed (i.e., reduced producer NUE) if resident organisms are plastic in their elemental composition. Another explanation presented in the literature relates to trophic structure and the strength of top-down control of primary producer biomass. When top-down control of primary producers is strong, as in systems with an even number of trophic levels, rapid consumption maintains primary producers in a state of rapid growth with consequent increased nutrient content (as described elsewhere in this encyclopedia).

Nutrient use efficiency has largely been a primary producer concept though there is no conceptual barrier to its application to heterotrophic organisms. In part, this may stem from the historical perspective that higher trophic level organisms are strongly homeostatic in their elemental composition (and thus have constant NUE). However, by now it should be apparent that heterotrophic stoichiometry is not fixed. This observation has recently led to a carbon-based corollary of NUE. Carbon use efficiency (CUE) has been defined as secondary production divided by net primary production. Many people will recognize this as the stoichiometric repackaging of trophic transfer efficiency (based on C rather than energy). In lakes, CUE has been demonstrated to vary by more than two orders of magnitude (0.002–0.4%). We are unaware of similar data from terrestrial systems. Stoichiometric theory predicts that consumers will use nutrients more efficiently when C:nutrient ratios are high.

Consumers tend to use carbon inefficiently under these conditions because they will need to release excess carbon as they accrue biomass. All else being equal, it follows that CUE and NUE should be inversely related, a prediction that has yet to be assessed as far as we are aware.

Up to this point, we have focused our discussion on ecosystems where nutrients enter biotic pools through autotrophic production. However, heterotrophic microbial production is also an important avenue through which nutrients must flow in most, if not all, ecosystems. For example, in headwater streams heterotrophic production often exceeds photoautotrophy. As a result, nutrient assimilation through microbial heterotrophs has received considerable attention. Lotic ecosystem ecologists have a rich history of measuring ecosystem-scale metabolic rates (respiration and gross primary productivity) and nutrient uptake. Recently, stoichiometric relationships have been combined with metabolic measurements to predict nutrient demand. In the Lotic Intersite Nitrogen experiment (LINX), algal and bacterial C:N ratios and carbon use efficiencies (though not expressed as such) were combined with estimates of heterotrophic respiration and gross primary production to predict nitrogen uptake in several stream ecosystems. Using whole-stream ^{15}N–NH_4 experiments and an oxygen mass balance approach for measuring ecosystem metabolism, these researchers found that predicted and measured rates of N uptake were similar and straddled the 1:1 line, though considerable variation existed (**Figure 1**). Given the number of assumptions in this analysis, the observed variance was not surprising and they concluded that continued application and improvement of this approach holds promise for illuminating links between carbon and nutrients as they move from inorganic to biotic pools.

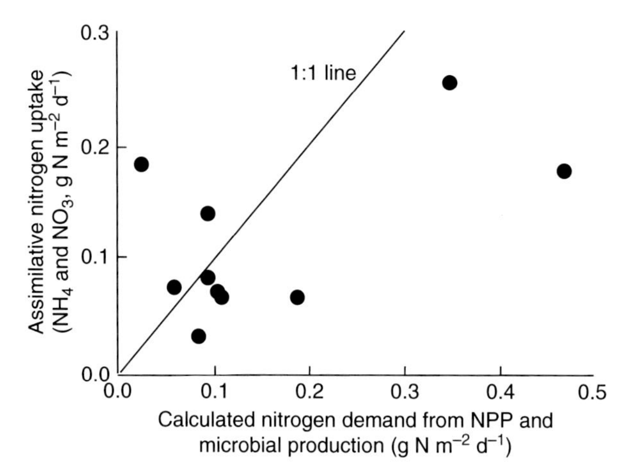

Figure 1 The relationship between measured assimilative nitrogen uptake and calculated nitrogen demand for biomass production. The line indicates where equal values would fall. Adapted from Webster JR, Mulholland PJ, Tank JL, *et al.* (2003) Factors affecting ammonium uptake in streams: An interbiome perspective. *Freshwater Biology* 48: 1329–1352.

Primary Producers and Consumers

Material flow in ecosystems incorporates more than the fluxes between inorganic nutrient pools and organisms capable of accessing these pools. In addition, material fluxes between organisms, like those discussed in previous entries, coalesce with important ecosystem-scale consequences. One stoichiometric characteristic that is consistently important across scales of biological organization is the contrast between first-order consumers (i.e., organisms that obtain all or part of their energy requirements from consuming primary producer biomass or detritus) and the primary producers they feed upon. As discussed in prior sections, primary producers tend to have higher carbon content and lower N and P content than their consumers. The extent of elemental differences between primary producers and first-order consumers depends on the ecosystem considered. In freshwater and marine pelagic systems where primary production is dominated by algae and cyanobacteria, differences between producers and first-order consumers in body nitrogen and phosphorus concentrations are smaller than the differences observed in benthic and terrestrial ecosystems dominated by vascular plants.

Despite this heterogeneity, the trend for higher body nitrogen and phosphorus concentrations in first-order consumers relative to their diet is consistent across ecosystems. In this article, we are less interested in the causes for these differences (e.g., differences in nutrient homeostasis) than in their implications. The first implication is that the potential exists for first-order consumers to be nutrient rather than carbon (C) limited if food C:nutrient ratios provide carbon in excess of the consumer's energy requirements. In this situation, we expect an increase in the N or P concentrations in the producer tissue to cause higher growth rates in the consumer. This hypothesis has been validated with many experimental manipulations including microbial, invertebrate, and vertebrate first-order consumers. The second implication is that nutrient limitation of first-order consumer growth should be greater in terrestrial than in aquatic ecosystems, since C:nutrient imbalances are greatest where vascular plants dominate primary production.

Given these two implications, what kind of patterns of matter flow in ecosystems can we expect? One expectation is that as we move from ecosystems composed of primary producers with poorer nutrient quality (lower body nitrogen and phosphorus concentrations) to those with greater N and P content, we should observe higher rates of herbivory (i.e., intake of producer biomass by herbivores) and first-order detritivory (i.e., intake of producer detritus by detritivores) and resulting decomposition (**Figure 2**). That should be the case whether we compare aquatic ecosystems only, terrestrial ecosystems only, or we include both types of ecosystems in the

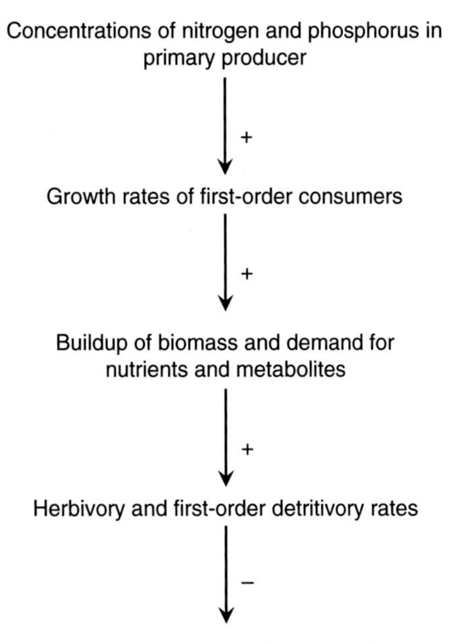

Figure 2 Dependence of growth rates of first-order consumers on nitrogen and phosphorus concentrations in the diet and expected patterns of herbivory, first-order detritivory, and accumulation of producer biomass and detritus across ecosystems.

comparison. In addition, because aquatic first-order consumers should have faster growth rates than their terrestrial counterparts, we should also observe higher rates of herbivory and first-order detritivory in aquatic than in terrestrial ecosystems. The hypothesized higher herbivory and first-order detritivory rates in ecosystems composed of richer primary producers suggests another important trend; because losses of producer biomass and detritus would be larger for richer producers than for poorer producers due to promoted herbivory and decomposition, we could also expect smaller pools of producer biomass and detritus in ecosystems composed of richer producers.

Herbivory

Humans have known for sometime that herbivores prefer more nutritional plants and that soil fertilization, by increasing plant nutritional value, usually leads to increased foraging by herbivores. However, only recently has it been demonstrated that a strong empirical association between herbivory rates, expressed as the percentage of gross primary production consumed by herbivores, and food N and P concentrations spans a broad range of ecosystem types (**Figure 3**). This association holds whether only aquatic ecosystems are compared, only terrestrial ecosystems are compared, or both types of ecosystems are included in the comparison. In general, aquatic ecosystems have a larger percentage of gross primary production consumed by herbivores than do terrestrial ecosystems (**Table 1**), supporting the expected stoichiometrically derived pattern of higher herbivory rates in ecosystems composed of richer producers (algae vs. vascular plants). The strength of the relationship is surprising in view of the many other factors that can also affect herbivory, such as herbivore type and size, feeding specificity, migratory behavior, and intensity of predation on herbivores.

The relationship described above suggests important consequences for trophic functioning and elemental cycling. By removing a larger percentage of gross primary production, aquatic herbivores should exert greater control of producer biomass than their terrestrial counterparts. On this basis, it can also be expected that ecosystems with highly nutritional producers will tend to have relatively small pools of producer biomass compared to ecosystems with similar values of gross primary production but of poorer nutritional quality. Finally, herbivores in nutritionally rich ecosystems have a greater impact on elemental recycling in ecosystems with richer producers since a larger fraction of producer biomass passes through herbivores before being recycled back into inorganic nutrient pools.

In contrast, when herbivory is expressed as an absolute flux (i.e., quantity of producer biomass consumed per unit area per unit time), consumption rates do not increase with primary producer N and P content when examined over a broad range of ecosystems (**Figure 3**). In part, this relationship results from the poor relationship between gross primary production and N and P concentrations when analyzed over a broad range of ecosystems. This may seem counterintuitive until one considers the suite of environmental factors known to influence gross primary production (e.g., temperature, humidity, and soil redox conditions among others). Differences in the relationship between relative (percentage) and absolute herbivory rates and ambient food quality are also an artifact of broad variation in primary production rates. Primary production rates can vary by several orders of magnitude. Therefore, even modest relative consumption rates (percentage of production consumed) like those expected under low-nutrient content can result in large absolute values of herbivory. That is, the relationship between primary production and absolute herbivory rates trumps the relationship between food N and P content and relative herbivory rates. Interestingly, relative herbivore production (i.e., ratio of herbivore biomass produced per unit of producer biomass ingested) does not appear to vary systematically among ecosystems.

First-Order Detritivory

First-order detritivory is the consumption of primary producer detritus. First-order detritivores include microbial

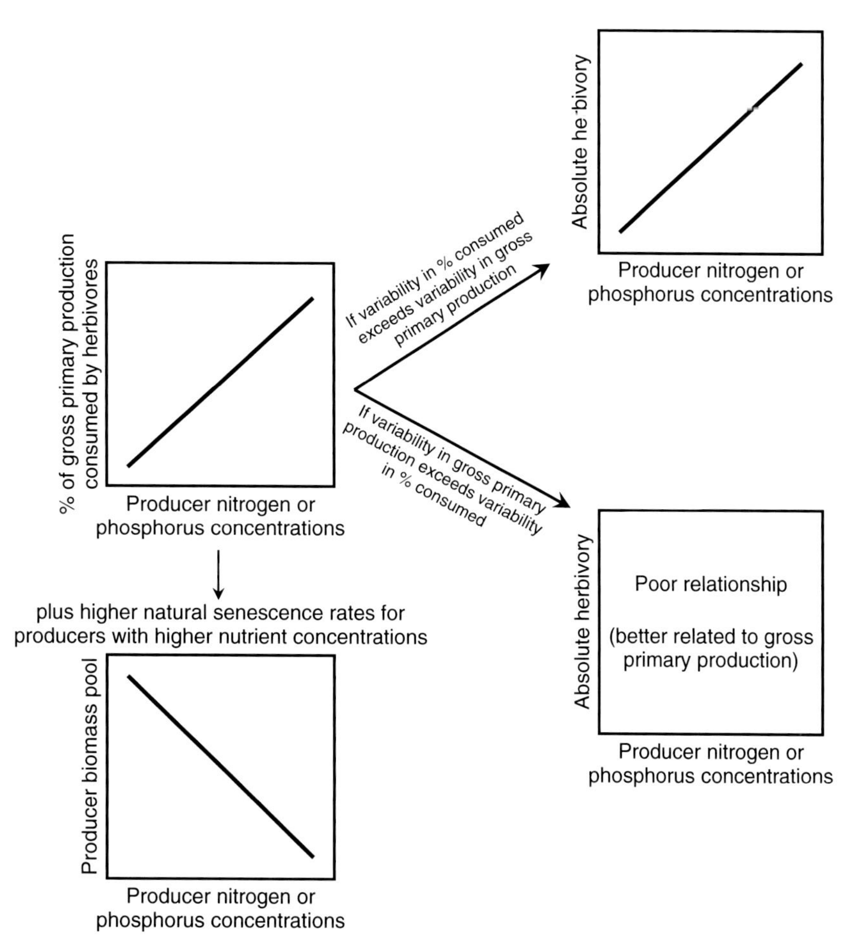

Figure 3 Patterns of herbivory, and implications on the size of the producer biomass pool, observed across a broad range of ecosystems.

Table 1 Overall differences in producer nutrient concentrations and matter flow and storage between aquatic and terrestrial ecosystems. Differences are expressed in relative terms

	Aquatic ecosystems	*Terrestrial ecosystems*
Nitrogen and phosphorus concentrations in producer biomass/detritus	High	Low
Herbivory rates (% of gross primary production consumed)	High	Low
Pool of producer biomass	Low	High
First-order detritivory rates (% of producer detritus mass lost per day)	High	Low
Pool of producer detritus	Low	High

decomposers (i.e., bacteria and fungi) and detritivorous invertebrate (and very rarely vertebrate) organisms that feed upon primary producer detritus and attached microbial decomposers. The ingestion of producer detritus by microbial decomposers and invertebrates is difficult to measure and thus, direct values of first-order detritivory are rare in the literature. Alternatively, first-order detritivory is often estimated as the temporal rate of loss of producer detritus mass incubated in containers or mesh bags (i.e., 'detritus incubation' method), which can be measured much more easily. This method has received criticisms, mainly because some of the producer detritus mass lost in mesh bags is not consumed by the first-order detritivores enclosed in the bags but instead is flushed out of the bags. However, rates derived in this manner are suspected to hold for comparisons across ecosystems because natural variation in first-order detritivory is expected to exceed the methodological error.

For many decades it has been clear to researchers that richer producer detritus (i.e., producer detritus with higher concentrations of nitrogen and phosphorus) decays faster than poorer detritus, which indicates that richer producer detritus is subject to higher rates of consumption by detritivores. Nevertheless, comparisons between producer detritus decay rates and nitrogen and phosphorus concentrations in the detritus encompassing a broad range of ecosystems have not been done until recently. The comparisons have shown that faster decay rates, expressed as the percentage of detritus mass lost per day, are associated with higher nitrogen and phosphorus concentrations in the detritus, albeit not strongly, across ecosystems (**Figure 4**). The unexplained variance in this relationship is not surprising given the many environmental factors that can affect detritus decay rates, such as temperature, humidity, redox conditions, type and size of the detritivore population, and predation intensity on the detritivore population. These broad comparisons demonstrate that detritus in aquatic ecosystems tends to have a higher percentage of mass lost per day than do detrital pools in terrestrial ecosystems (**Table 1**).

As with herbivory, when rates of producer detritus decay are expressed in absolute terms (i.e., quantity of producer detritus mass lost per square meter per year), decay rates and nitrogen and phosphorus concentrations in the detritus are often unrelated across a broad range of ecosystems (**Figure 4**). The reasons are twofold: first, absolute producer detritus production (i.e., quantity of producer detritus generated per square meter per year) is unrelated to the nitrogen and phosphorus concentrations in the detritus across a broad range of ecosystems; second, absolute producer detritus production varies more broadly than does the percentage consumed by detritivores. Thus, absolute rates of producer detritus decay, which result from the product between absolute producer detritus production and the percentage consumed by detritivores, will often be more closely associated with absolute producer detritus production than with the percentage consumed when a broad range of ecosystems are compared and, by extension, will often be unrelated to the nitrogen and phosphorus concentrations in the producer detritus. This logic is identical to that presented above for herbivory since it is the lack of a relationship between nutrient content and primary production driving the differences between relative and absolute consumption rates (as herbivory or detritivory).

Pools of Producer Biomass and Detritus

The tendency toward larger percentages of gross primary production consumed by herbivores with higher producer nitrogen and phosphorus concentrations suggests that, all else being equal, ecosystems composed of nutrient-rich primary producers will have lower producer biomass and accumulate primary detritus (derived from primary producer tissue) more slowly than systems dominated by nutrient-poor producers. This expectation will hold if

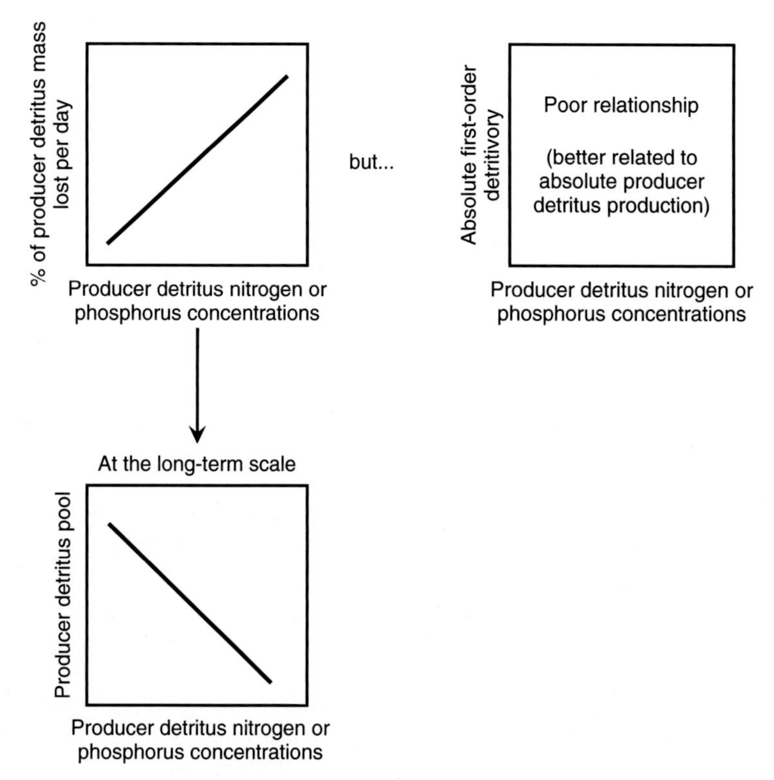

Figure 4 Patterns of first-order detritivory and implications on the size of the producer detrital pool, observed across a broad range of ecosystems.

gross primary production varies to a lesser extent than does the percentage consumed by herbivores across the ecosystems compared. If, in contrast, gross primary production varies to a greater extent than does the percentage consumed by herbivores across the ecosystems compared, then the quantity of matter available for storage as producer biomass will be more closely associated with rates of gross primary production than with the percentage consumed by herbivores. Unfortunately, too few comparisons across ecosystems have been done to conclude with rigor whether these expectations hold. However, one recent comparison including a broad range of aquatic and terrestrial ecosystems demonstrated that ecosystems with nutrient-rich producers tended to have smaller pools of producer biomass (**Figure 3**). Interestingly, in the ecosystems compared in this study gross primary production varied more broadly than did the percentage of production consumed by herbivores. This observation is a direct contradiction to the arguments presented above which predict that biomass pools should track primary production rates rather than patterns in producer nutrient content.

Evolutionary constraints may provide an answer to this apparent contradiction. Primary producers with higher nitrogen and phosphorus content tend to grow faster, have shorter life spans, and higher natural mortality rates (i.e., senescence) than their low-nutrient counterparts. Thus, when a broad range of ecosystems are compared, nutrient-rich producers have both larger proportions of their biomass consumed by herbivores and higher natural mortality rates (**Figure 3**). Both mechanisms lead to smaller pools of producer biomass under high-nutrient conditions across a broad range of aquatic and terrestrial ecosystems despite large differences in gross primary production. Indeed, recent modeling advances have confirmed this prediction.

The moderate tendency toward higher percentages of producer detritus mass lost per day with higher nitrogen and phosphorus concentrations in the detritus observed when a broad range of ecosystems are compared suggests that, if the differences in absolute producer detritus production do not exceed the differences in the percentage consumed by detritivores, ecosystems with richer producer detritus should also tend, at least moderately, to have smaller pools of producer detritus. The rationale is analogous to the case of producer biomass pools discussed above. The observation that aquatic ecosystems, which tend to have richer primary detritus and higher percentages of producer detritus production consumed by detritivores, generally have smaller pools of producer detritus when compared to terrestrial ecosystems is consistent with this hypothesis (**Table 1**).

Large- Scale Implications of Ecological Stoichiometry

In this and other entries, ecological stoichiometry has been presented as a theoretical and empirical approach for linking organism biochemical properties to larger-scale ecological patterns and processes. A logical framework for ecological stoichiometry (though the term was not used) was initially formalized by W. A. Reiners in the mid-1980s. Up to that time, ecosystem science had been largely organized around energetics. Reiners offered his framework as a complementary view of ecosystems in which the flow of matter, rather than energy, was the organizing feature. Using a limited number of logical steps, Reiners combined biochemical and ecological axioms (true statements) to derive a series of theorems that describe how biological processes control biogeochemical cycling from local to global scales (**Figure 5**). One of Reiners' principal axioms (axiom 2) was that protoplasmic life is consistent across organisms. We now know that this axiom is not strictly true and that cellular stoichiometry can vary, with important ecological implications. Interestingly, this shortcoming has not diminished the utility of this logical sequence because it still remains useful for understanding how stoichiometric variation at basal levels of biological organization can cascade to have ecosystem and even global scale implications.

Prior to Reiners' theoretical work, A. C. Redfield found a surprising congruence in C:N:P ratios in plankton from widespread regions of the world's oceans. In this seminal work, Redfield determined that ocean seston had a consistent C:N:P stochiometry equal to 106:16:1. The 'Redfield ratio', as it has become known, is perhaps the most famous result in ecological stoichiometry and many have come to revere it as a rare ecological constant, analogous to better-known constants in chemistry and physics. Perhaps more surprising, and more interesting in the context of this entry, seston C:N:P mirrors the ratio of these elements in dissolved pools. Redfield interpreted the equivalence between N:P ratios in seston and dissolved pools as evidence that N and P had balanced flow into and out of biotic pools. Furthermore, regressions of ocean N and P concentrations pass through the origin, indicating that these elements are depleted from ocean water simultaneously. Given the myriad of geological and meteorological factors that could influence nutrient availability in the world's ocean, there is no *a priori* expectation for this equivalence. Redfield's explanation for this finding was that the biota controlled the relative availability of these elements in the ocean. He went on to suggest that P, rather than N availability limited ocean production because biological processes

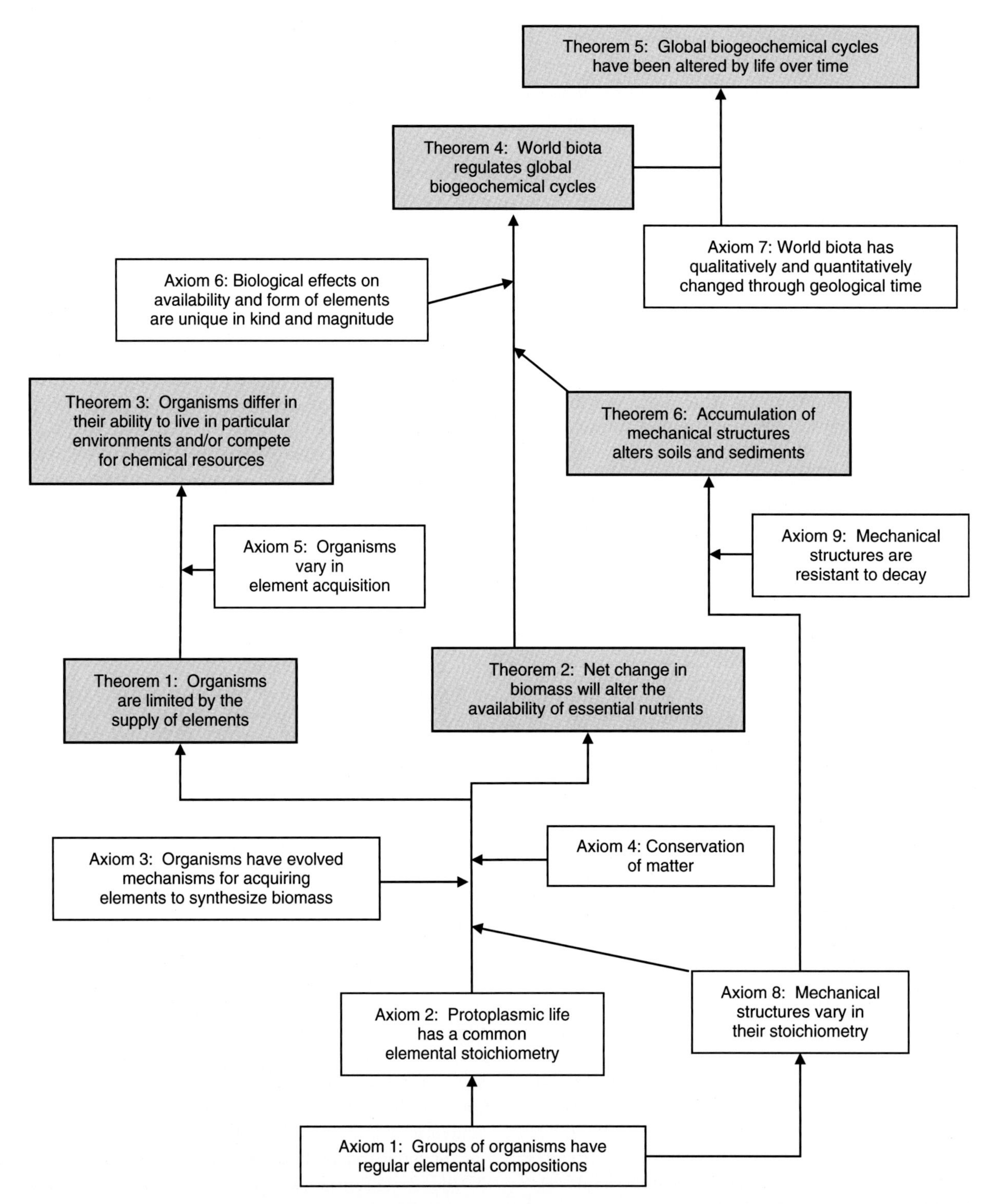

Figure 5 A logic flowchart of stoichiometric theory illustrating how a small number of axioms from biochemistry (clear boxes) lead to theorems describing global biogeochemical cycles (shaded boxes). Adapted from Reiners WA (1986) Complementary models for ecosystems. *American Naturalist* 127: 59–73.

(e.g., N-fixation) are capable of adjusting N availability to match P constraints. Contemporary support for Redfield's contention is found in recently published models of N-fixation in the open ocean. In addition, extensive datasets on N-fixation in lakes clearly indicates that overall N:P ratio constrains when and where N-fixation is active (restricted to conditions where molar $N{:}P < 20$). Interestingly, similar mechanisms have been invoked to explain the development of P-limitation in terrestrial soils as they age (as discussed earlier in this article). Redfield's findings have proved important beyond the realm of oceanography because they suggest that biotic processes control element cycles at global scales, a world view that Reiners and others would further develop.

Global scale coupling of C, N, and P cycles involve a highly complex set of interactions between Earth's subsystems (biosphere, atmosphere, and geosphere). Feedbacks among these systems operate over a suite of spatial and temporal scales and include interactions between terrestrial, freshwater, and ocean ecosystems that remain poorly understood. However, this complexity has not hindered researchers from exploring the role of biological processes in structuring biogeochemical cycles and their evolution through geological time. Perhaps the best example of such work relates to maintenance of global atmospheric oxygen concentration. Oxygen has remained between 15% and 35% of the atmosphere for the last several hundred million years. Under current biogeochemical conditions, atmospheric oxygen turns over every 4000 years. Combined, these observations suggest that a dynamic system of feedbacks may exist to stabilize oxygen content. Biotic activity is a principle driver of modern oxygen cycling. The production and breakdown of organic matter produces and consumes equal amounts of oxygen and as a result does not perturb the existing oxygen levels. However, biomass standing stock has not been constant through time. In addition, slow cycling of oxygen through nonbiotic processes such as carbonate precipitation, oxidation of uplifted iron, photolysis of water, and oxidation of ammonium also influence long-term patterns in oxygen concentration. Combined, it is not obvious why these processes should combine to create stability in the atmospheric pool. One mechanism for the maintenance of this stability is variation in the efficiency of P burial in the world's oceans. A thorough discussion of this mechanism is beyond the scope of this text. Suffice it to say that the crucial feature of this mechanism is the positive relationship between oxygen content and the efficiency of P burial via precipitation with iron hydroxides. Efficient P burial ultimately leads to reduced delivery to the photic zone, reduced autotrophic activity, and consequent oxygen production. Other examples of biologically driven feedback mechanism have been identified by several authors. Feedbacks, and their role in autoregulation of Earth system properties require constraints and ecological stoichiometry suggests that the elemental composition of core biomolecules and the processes that create them provide these constraints.

Given the pervasiveness of human activity and its impact on element abundances, efforts to understand the complexity associated with biological and physical controls on biogeochemical cycling are of paramount importance if society is to forecast ecological conditions and design strategies for responsible stewardship. Though a vigorous debate continues over the relative importance of biological and geological processes in controlling global scale elemental cycles, it is nevertheless critical to continue to explore the stoichiometric underpinnings of biological processes and their consequences for ecosystem and biosphere structure and function.

See also: Ecological Stoichiometry: Overview; Evolution of Oceans; Evolutionary and Biochemical Aspects; Organisal Ecophysiology; Population and Community Interactions; Trace Elements.

Further Reading

Cebrian J (2004) Role of first-order consumers in ecosystem carbon flow. *Ecology Letters* 7: 232–240.
Cebrian J and Lartigue J (2004) Patterns of herbivory and decomposition in aquatic and terrestrial ecosystems. *Ecological Monographs* 74: 237–259.
Downing JA (1997) Marine nitrogen:phosphorus stoichiometry and the global N:P cycle. *Biogeochemistry* 37: 237–252.
Elser JJ, Sterner RW, Gorokhova E, *et al.* (2000) Biological stoichiometry from genes to ecosystems. *Ecology Letters* 3: 540–550.
Enriquez S, Duarte CM, and Sand-Jensen K (1993) Patterns in decomposition rates among photosynthetic organisms: The importance of detritus C:N:P content. *Oecologia* 94: 457–471.
Reiners WA (1986) Complementary models for ecosystems. *American Naturalist* 127: 59–73.
Schade J, Espeleta J, Klausmeier CA, *et al.* (2005) A conceptual framework for ecosystem stoichiometry: Balancing resource supply and demand. *Oikos* 109: 40–51.
Sterner RW and Elser JJ (2002) *Ecological Stoichiometry: The Biology of Elements from Molecules to Biosphere*. Princeton, NJ: Princeton University Press.
Vitousek PM (1982) Nutrient cycling and nutrient use efficiency. *The American Naturalist* 119: 553–572.
Webster JR, Mulholland PJ, Tank JL, *et al.* (2003) Factors affecting ammonium uptake in streams: An interbiome perspective. *Freshwater Biology* 48: 1329–1352.

Evolutionary and Biochemical Aspects

A D Kay, University of St. Thomas, St. Paul, MN, USA

T Vrede, Uppsala University, Uppsala, Sweden

Introduction

Darwin's theory of evolution by natural selection provides a general explanation for the diversity of life. If organisms differ in traits that are heritable, then differences in reproductive success among individuals will result in changes in the relative prevalence of traits in subsequent generations; over time, this process of natural selection leads to adaptation, an improved fit between the features of organisms and the demands of the environment.

Adaptations may often be manifested in the biochemical makeup of organisms. This is because an organism's biochemistry has functional and economic consequences that influence reproductive success. From a functional perspective, an organism's makeup affects its ability to meet environmental challenges because compositionally distinct biomolecules, cellular structures, and tissues have different chemical properties that determine their biological roles. From an economic perspective, an organism's makeup reflects its demand for structural resources, and environmental scarcity of particular substances may constrain the evolution of reliance on those substances. The composition favored by natural selection should reflect the optimal compromise between these functional and economical considerations.

Ecological stoichiometry, the study of the balance of energy and chemical elements in living systems, provides a general structure for examining how the composition of life has evolved. Clearly, organisms are not bags of independently functioning atoms, and a focus on elements will often miss the consequences of how elements are arranged in biochemicals. However, this focus has significant advantages because it provides a currency that facilitates comparisons across diverse taxonomic groups and levels of biological organization.

Consider the following examples that illustrate some of the ways in which elemental composition can have adaptive significance.

The water flea, *Daphnia*, is a common herbaceous crustacean found in lakes, ponds, and quiet streams. There is now ample evidence that the rate at which a *Daphnia* can grow is functionally related to the concentration of phosphorus (P) in its tissue when P is scarce in the environment. The connection between body P levels and growth rate exists because growth rate depends on the concentration of RNA in *Daphnia* tissue, and RNA is an abundant cellular component that contains more P than other major biomolecules. This linkage among whole-body P concentration, RNA levels, and growth rate is the central prediction of the growth rate hypothesis (GRH), which is described more thoroughly in Organismal Ecophysiology.

Aquatic mollusks depend on calcium (Ca)-rich shells for protection against fish and crustacean predators. The protective benefit of a shell for mollusks depends on its thickness, form, and Ca content, which affects hardness and other physical properties that help shells maintain structural integrity in the face of attack. Because Ca influences shell function, it is not surprising that Ca availability is thought to have played an important role in the evolution of molluskan shell morphology. Ca availability can also induce short-term changes in shell investment, which has been demonstrated in the freshwater snail *Lymnaea stagnalis*. The relationship between Ca availability and shell investment suggests that the evolutionary significance of predation pressure will be mediated by the types of resources available in the environment.

Aphids are insect herbivores that tap into a plant's phloem and feed on sap. Solutes in sap tend to be rich in carbon (C) but contain few nitrogen (N)-rich amino acids, P-containing molecules, or other important minerals and vitamins. Sap thus creates a significant dietary challenge for an insect because its elemental composition (high C:N and C:P ratios) differs substantially from the balance of these elements in insect tissue. Adaptations to this imbalance include phenotypes that promote mutualistic interactions with bacteria and ants. For example, aphids such as the greenbug aphid, *Schizaphus graminum*, house intracellular bacteria which supply essential amino acids to their hosts after receiving C and nonlimiting amino acids. Aphids have also evolved the ability to produce a carbon-rich exudate called 'honeydew', which some species present to ants in exchange for protection from aphid predators and parasites. Thus, the compositional disparity between aphids and their food may have been a key factor promoting the evolution of mutualistic interactions.

These examples illustrate how elemental composition can play a role in a wide range of phenomena that ultimately affect an organism's reproductive success. It follows that conditions favoring particularly strategies for maximizing reproductive success will also have resulted in evolutionary changes in elemental composition. If so, organisms should differ in elemental composition, and these differences should be associated with biochemical differences that affect functional capabilities. Let us review some of the available evidence for evaluating these predictions.

Diversity of Organismal Stoichiometry

All organisms contain the same major elements (e.g., C, N, P, Ca, hydrogen (H), oxygen (O), S (sulfur)), but existing information indicates that the concentrations of these elements in organisms can differ substantially among and within taxonomic groups. Patterns of elemental abundance are best described for C, N, and P (**Table 1**), which have received most of the attention in ecological stoichiometry because of their importance in biological structures and because N and P commonly limit production in nature.

First, let us examine the relative amounts of C, N, and P in plants and other autotrophic organisms. The C:N:P composition of individual autotrophs is generally quite variable because molecules containing these elements can be stored in large quantities in vacuoles or in the cytoplasm. Despite this variability, there are still clear differences in C:N:P composition among species. One way to categorize this variation is by making comparisons among major habitats, such as oceanic, freshwater, and terrestrial systems. In oceans, particulate matter (which provides a measure of phytoplankton biomass) tends to be rich in N and P relative to autotrophic biomass in freshwater and terrestrial systems. Seminal work in the mid-1900s by Alfred Redfield showed that the relative amount of C, N, and P in marine particulate matter was 106:16:1 (molar ratio). This description is referred to as the Redfield ratio. More recent findings have shown that although the average C:N:P composition of marine particulate matter tends to be relatively homogeneous across sampling locations, there is more than threefold variation in C:P and N:P ratios across different phytoplankton phyla and superfamilies. In lakes, particulate matter tends to contain much less P than it does in oceans. On average, C:P ratios in freshwater particulates are about 300, and N:P ratios are about 30; these values are substantially higher than the Redfield values of 106 and 16. Phosphorus content in lake particulates is also highly variable: N:P ratios in one survey ranged from 6.5 to 125 across sites. Mean C:N ratio in lake particulates is about 10, similar to values for ocean particulate matter. Terrestrial autotrophs tend to have much higher C:N and C:P ratios than oceanic and freshwater autotrophs: the mean C:N ratio in leaves is around 36, and the mean C:P ratio is near 1000. C:nutrient ratios in whole plants are likely even higher than ratios in leaves because wood, bark, and other secondary growth generally contain low concentrations of N and P. Concentrations of N and P in leaves also differ substantially among plant species (**Table 1**), showing 25- and 50-fold ranges of variation, respectively.

Next, consider the elemental composition of animals. Because animals have limited capacity for nutrient

Table 1 Approximate ranges for C, N, and P concentrations across species in select taxa

Taxon	*% C*	*% N*	*% P*
Terrestrial plants (leaves)	36–64[a]	0.25–6.4[b]	0.02–1.0[b]
Benthic invertebrates	35–57[c]	6–12.0[c]	<0.2–1.8[c]
Zooplankton		7–12.5[a]	0.5–2.5[a]
Terrestrial insects	36–61[a]	6–12.0[d]	0.35–1.5[e]
Freshwater fish	~40–50[f]	8–12.0[f]	1.5–4.5[f]
Birds and mammals			0.7–3.7[g]

[a]Elser JJ, Fagan WF, Denno R F, *et al.* (2000) Nutritional constraints in terrestrial and freshwater food webs. *Nature* 408: 578–580.
[b]Güsewell S (2004) N:P ratios in terrestrial plants: Variation and functional significance. *New Phytologist* 164: 243–266.
[c]Cross WF, Benstead JP, Rosemond AD, and Wallace JB (2003) Consumer-resource stoichiometry in detritus-based streams. *Ecology Letters* 6: 721–732.
[d]Fagan WF, Siemann E, Mittler C, *et al.* (2002) Nitrogen in insects: Implications for trophic complexity and species diversification. *American Naturalist* 160: 784–802.
[e]Woods HA, Fagan WF, Elser JJ, Harrison JF (2004) Allometric and phylogenetic variation in insect phosphorus content. *Functional Ecology* 18: 103–109.
[f]Sterner RW and George NB (2000) Carbon, nitrogen and phosphorus stoichiometry of cyprinid fishes. *Ecology* 81: 127–140.
[g]Gillooly JF, Allen AP, Brown JH, *et al.* (2005) The metabolic basis of whole-organism RNA and phosphorus content. *Proceedings of the National Academy of Sciences of the United State of America* 102: 11923–11927.
Element concentrations are reported as % dry mass. Data are for whole organisms unless otherwise noted.

storage and possess mechanisms for selectively acquiring and retaining particular substances, individual animals tend to maintain their elemental composition within limited bounds. However, there are often significant differences in whole-body stoichiometry among animal species.

Patterns of elemental abundances in invertebrates have been most thoroughly cataloged for two groups: zooplankton and insects. Zooplankton species, which consist mostly of rotifers, cladocerans, copepods, and other small crustaceans, vary considerably in P content, but are less variable in terms of N and C (**Table 1**). For example, P concentration (% dry mass) in zooplankton varies fivefold across species, whereas zooplankton N content varies less than twofold. Terrestrial insect species also vary considerably in P concentration, which can range from ~0.35% to ~1.5% dry mass in adults (**Table 1**). As in zooplankton, N concentration in insect species is more tightly constrained, varying only about twofold (~6–12% dry mass) among adults across species. Zooplankton and insects contain similar concentrations of C, N, and P, but both groups contain considerably more N and P than autotrophs (**Table 1**). The compositional disparity between invertebrate animals and autotrophs is particularly striking in terrestrial systems.

Less information is available about whole-body C:N:P stoichiometry for vertebrates, although descriptions are available for some fish, birds, and mammals (**Table 1**). In fish, P content can range from 1.5% to 4.5%. The N content in fish is again less variable (8–12%) than P content, and mean values are similar to levels found in invertebrates. P content appears also to be high and variable in terrestrial vertebrates. For example, reported levels for P content range from 0.67% in pigs to 3.67% in humans. P concentrations in vertebrates are considerably higher than in invertebrates due in large part to the high P content of bone.

These comparisons illustrate that there are clear compositional differences among major taxonomic groups: the low N and P concentrations in terrestrial plants and the high P content in vertebrates are particularly distinctive. Perhaps more importantly, there is also considerable variation in C:N:P stoichiometry within groups such as zooplankton, insects, and fish, whose members have many morphological and ecological similarities. It is thus likely that the adaptations distinguishing species are reflected in the elemental composition of organisms. If so, the chemical composition of organisms may be a key factor guiding the evolutionary dynamics of populations. To get a better understanding of the evolutionary relevance of stoichiometric variation, we must first explore how elemental composition is connected to functional capabilities that ultimately affect an organism's reproductive success. Useful information for making this connection comes from a multilevel analysis of organismal biochemistry.

Elemental Composition and Biological Function

The elements themselves have distinct properties that influence their roles in biological systems. For example, the ability of a C atom to form strong, stable covalent bonds with C itself and with other major elements (e.g., H, N, O, S, and P) explains why C forms the backbone of a variety of kinetically stable polymers that serve as membranes and other structural biomolecules. In addition, C–C bonds have the highest bond energy of all solid nontransitional elements, which makes C atoms well-suited for their prominent role in energy storage molecules. These biochemical features help to explain why C is often the most abundant element in the dry mass of organisms. Nitrogen is typically the fourth most abundant component of organism dry mass (after C, H, and O). Biological N tends to occur in a reduced state bound to C and H; the resulting amine group ($R{-}NH_2$) is basic. The combination of this basic amine group with an acid carboxyl group is the peptide bond that links amino acids in proteins. At neutral pH, the amine group is protonated, providing one of the few sources of positive charge in biomolecules. In addition, N-H bonds are polar and readily form H bonds to other polar groups; these bonds are critical for determining the structure of proteins and nucleic acids. Phosphorus is typically the fifth or sixth most abundant element in organisms. It occurs almost exclusively in the oxidized phosphate form (HPO_4^{2-}) in biological systems. As bound phosphate, it can function in several important ways, including (1) as a link between subunits in large molecular assemblies (e.g., DNA, RNA, and phospholipids), (2) as a carrier of chemical energy (e.g., adenosine 5′-triphosphate, ATP) and substrates (e.g., glucose phosphate), (3) as a signaling mechanism (e.g., cyclic-adenosine monophosphate, c-AMP), and (4) as a component of biominerals (e.g., calcium phosphate). Phosphate serves as an effective link in nucleic acids and a center of mobility in lipids because it readily reacts with attacking reagents. This flexibility is essential to RNA, which must have the capacity to turn over quickly in order to be effective in carrying rapidly reproducible information.

The distinctive chemical properties of C, N, and P suggest major differences in elemental composition among organisms which will be associated with differences in capacities for meeting environmental challenges. However, the connection between elemental composition and biological function becomes much more tangible with a general description of how elements are distributed among biomolecules, cellular components, and tissues.

Elemental Composition of Biomolecules

Variation in elemental composition among organisms can be driven by stoichiometric differences at many levels of internal organization. As outlined below, major classes of biomolecules such as lipids, carbohydrates, protein, and nucleic acids contain different concentrations of C, N, P, and other major elements such as H, O, and S (**Table 2**). As a result, differences in biomolecular mixtures result in stoichiometric variation at higher levels of organization, such as among cellular structures or tissues. Variation in elemental composition among organisms can potentially be explained by stoichiometric differences at any level of organization, provided components at that level (1) have distinctive elemental signatures, (2) vary significantly in concentration among organisms, and (3) contain a sizable fraction of the whole-organism pool of some element. Determining which components explain the variation in whole-organism stoichiometry is a key step in relating elemental composition to biological functions that affect fitness.

What then is known about variation in elemental composition among major biomolecules, and how might this variation be related to biological function?

Lipids

Lipids are a diverse group of chemicals with a variety of biological functions. Triacylglycerols, waxes, and other fats and oils are the principal mechanism for storing energy in most organisms; sterols and phospholipids serve as structural elements in membranes; and other less-abundant lipids act as electron carriers, pigments, and enzyme cofactors. Triacylglycerols, waxes, and phospholipids are the primary components of lipid biomass in organisms, and are thus more likely than other lipids to contribute to a distinctive whole-body elemental composition.

Triacylglycerols and waxes consist of only C, H, and O; they contain neither N nor P. Carbon content in these molecules is generally high. For example, a triacylglycerol with three 16-carbon unsaturated fatty acids will contain 76% C; myricyl palmitate, the primary component of beeswax, contains about 82% C. The C content of triacylglycerols is substantially higher than levels in other major biomolecules (**Table 2**) and in animal bodies as a whole (see **Table 1**).

Phospholipids contain a glycerol molecule attached to two fatty acids and a functional group. Phospholipids consist almost entirely of C, H, O, and P, which is contained in a phosphodiester linkage between the functional group and the glycerol molecule. The functional group in some phospholipids does contain N, although the N content in phospholipids is generally low. For example, phosphatidylcholine, a common phospholipid, contains 67% C, 1.9% N, and 4.2% P.

Storage lipids are likely the only form of lipids that will contribute significantly to whole-organism stoichiometry. Storage lipids can make up a substantial fraction of organism biomass. For example, mean lipid concentration

Table 2 Approximate C:N:P:H:O:S stoichiometry (% of mass) of selected macromolecules and other organic compounds

	%C	*%N*	*%P*	*%H*	*%O*	*%S*
Lipids						
Glycerol (triacylglyceride)[a]	75	0	0	14	12	0
Cholesterol	84	0	0	12	4	0
Phosphatidylcholine (phosphoglyceride)[a]	67	2	4	11	16	0
Carbohydrates						
Starch and cellulose	46	0	0	3	51	0
Lignin	63	0	0	6	31	0
Chitin	44	7	0	7	42	0
Peptidoglycan[b]	49	12	0	7	30	1.5
Proteins						
Protein[c]	54	17	0	7	20	2.7
Rubisco (large subunit)	54	17	0	7	21	1.2
Myosin (heavy chain)	53	17	0	7	22	0.8
Collagen	54	18	0	7	21	0.5
Nucleic acids						
DNA[d]	38	17	10	4	31	0
RNA[d]	36	16	10	3	35	0
Other organic compounds						
ATP	24	14	18	2	41	0
Chlorophyll	74	6	0	8	9	0

[a]Fatty acid chains consisting of stearic acid (saturated fatty acid with 18 C atoms).
[b]Oligopeptide side chain with five amino acids, 5% each of all 20 amino acids.
[c]Hypothetical nonphosphorylated protein consisting of 5% each of all 20 amino acids.
[d]Assuming 50% G–C base pairs.

in insects has been measured to be about 25%; similar levels have been measured in marine calanoid copepods and albacore tuna. Similarly, triacylglycerols in adipose tissue make up about 21% of the mass of a nonobese 70 kg man. Storage lipid levels can also vary substantially among taxa and among individuals within taxa. In whole tuna, lipid content can vary from 1% to 43% dry mass. Lipid stores are known to vary with fluctuations in activity demands and resource availability. Because storage lipids are rich in C but contain no N or P, increases in storage lipid levels will increase C:N and C:P ratios in whole organisms, but will have no effect on body N:P ratios. Directional selection on energy storage could thus result in evolutionary increases in whole-body C:N and C:P ratios. Unlike triacylglycerols, phospholipids contribute little to the total biomass of most organisms. For example, in crustacean zooplankton phospholipids make up only about 6% of total body mass; in the mussel *Mytilus edulis*, phospholipid content ranged from only 0.36% to 0.64% in soft tissue.

Carbohydrates

Carbohydrates generally serve as fuel, energy stores, and building materials, although specific carbohydrate-containing molecules also act in cell–cell recognition and information transmission. Sugars and starches are carbohydrates involved in energy transport and storage; they have a general chemical formula of $(CH_2O)_n$. Thus, like storage lipids, carbohydrates with this basic formula contain neither N nor P. However, they contain less C than lipids. For example, glucose contains 40% C and glycogen contains 52% C, while triacylglycerols typically contain more than 70% C. Carbohydrate fuels generally make up only a small fraction of organism biomass: in humans, glycogen is only about 225 g of the mass of a nonobese 70 kg man. They are thus unlikely to contribute substantially to variation in whole-body C:N:P stoichiometry.

Structural carbohydrates are more significant contributors to organismal stoichiometry, especially in plants. Insoluble carbohydrate polymers serve structural and protective roles in the cell walls of plants and bacteria and in animal connective tissue. Examples of common structural polysaccharides include cellulose and lignin, which are tough, fibrous polymers in plant cell walls; chitin, which strengthens insect exoskeleton; and hyaluronate, which provides viscosity and lubrication in vertebrate joints. Cellulose $(C_6H_{10}O_5)_n$ contains 46% C and no N or P. It makes up most of the mass of wood, and will thus largely determine the elemental composition of large terrestrial plants with extensive support structure. Selection favoring investment in cellulose will increase C:N and C:P ratios in whole plants but will not affect plant N:P ratios. Some other structural polysaccharides do contain N. For example, chitin has an acetylated amino group (–NH) in place of the hydroxyl group (–OH) in cellulose. As a result, it contains 6.9% N, which is within the range (6–12% N) found in the bodies of insects. Chitin generally makes up <10% of insect biomass, but chitin investment can differ among species. Because chitin itself contains no P, variation in chitin investment could explain some of the variation in insect C:P and N:P ratios.

Protein

Proteins are versatile nitrogenous biomolecules that serve structural, signaling, and catalytic roles in cells. They occur in thousands of varieties ranging in size from small peptides to large polymers with molecular weights in the millions. The 20 amino acids that are the monomers or building blocks of proteins all have a carboxyl group and an amino group bonded to a carbon atom, but they have different side chains which determine their distinct chemical properties. Differences in side-chain composition are also what determine differences in the elemental composition among amino acids. On average, the 20 common amino acids contain 53% C, 17% N, and no P; greater investment in protein will thus increase levels of N and reduce P content in organisms. Proteins constitute the largest fraction of the biomass in most cells. For example, proteins make up 50% of the dry mass in the bacteria *Escherichia coli*, 30–50% of the dry mass in crustacean zooplankton, 30% of plant leaf biomass, and about 40% of the dry mass in a nonobese human male. Protein investment thus is a major determinant of whole-body elemental composition.

There is also considerable variability in elemental composition among amino acids. C content ranges from less than 30% in cysteine to 65.5% in phenylalanine, and N content ranges from 7.7% in tyrosine to 32.1% in arginine. This variation has some relationship to amino acid function. For example, amino acids with aromatic side chains (phenylalanine, tyrosine, tryptophan) are relatively nonpolar and participate in hydrophobic interactions; they also have high C (mean = 63.2%) and low N (mean = 10%). In contrast, amino acids classified by their positively charged side chains (lysine, arginine, histidine) contain low C (mean = 45.7%) and high N (mean = 26.2%). These amino acids are important for enzyme catalysis and are often involved in weak electrostatic interactions with negatively charged biomolecules like nucleic acids. Selection on traits biasing the overall amino acid composition of protein could thus be manifested in the elemental composition of whole organisms.

Sulfur is not present in most macromolecules in organisms, but does occur in protein (**Table 2**). It is present in two amino acids: methionine and cysteine. The average S content of the 20 commonly occurring amino acids is 2.7%, but protein S content is generally lower (on average 1.3%). This can at least partly be explained by the relatively low S content in some of the most commonly occurring proteins, that is, rubisco (ribulose bisphosphate carboxylase, the enzyme catalyzing C fixation in plants, and thus probably the most common protein in the world), myosin (a major protein in muscle tissue), and collagen (a major protein in the extracellular matrix of metazoan animals).

Proteins not only make up a substantial fraction of intracellular mass, they are also key components of important extracellular excretions such as hair, nails, skin, feathers, horns, spider webs, poisons, and venoms. Most of the dry mass of hair, claws, hooves, tortoise shells, and horns consists of α-keratin, a strong fibrous protein. Fibroin, the protein of silk, is a polypeptide rich in N-rich alanine and glycine. These excretions clearly have important functional consequences, and thus provide a rich set of opportunities for determining how elemental composition relates to the evolution of traits.

Nucleotides

Nucleotides are the constituents of nucleic acids (DNA, RNA), which store and transmit genetic information. They also act as carriers of chemical energy in cells, as enzyme cofactors, and as secondary messengers. Nucleotides consist of a nitrogenous base, a five-carbon sugar, and a phosphate group. In DNA and RNA, nucleotides are covalently linked by the phosphate group; the negative charge of the phosphate group at neutral pH is essential for stabilizing nucleotides against hydrolysis and for retaining them within a lipid membrane. Differences in the nitrogenous base determine the elemental and functional variation among nucleotides. Nucleotides, as they appear in RNA, contain on average 36.2% C, 16% N, and 9.6% P; the elemental composition of nucleotides in DNA is very similar. Thus, nucleotides in nucleic acids have N levels that are similar to those found in protein (16% vs. 17%), which are in turn much higher than levels in most organisms. Most notably however, nucleotides contain very high levels of P and low C:P and N:P ratios relative to other major biomolecules. The P content of an average nucleotide is also an order of magnitude higher than the P content of most insects, marine invertebrates, and plants.

RNA can comprise a large, but variable fraction of organism biomass. For example, RNA content can range from 12% to 30% of cell dry mass in *E. coli*, from 0.1% to 14% dry mass in invertebrates, and from 0.02% to 9% dry mass in birds, mammals, and fish. This substantial variation coupled with the high-P, low-N content of RNA make this molecule a major source of variation in organismal C:N:P ratios.

DNA levels are known to be much lower than RNA levels in most organisms. Genome size varies by up to five orders of magnitude among taxa, but this variation is accompanied by corresponding changes in cell size. As a result, DNA levels as a fraction of cellular biomass appear to be quite consistent across organisms and thus likely explain very little of the variation in C:N:P stoichiometry among taxa.

ATP is a nucleoside that is widely used for transporting energy in cells. An ATP molecule contains 24% C, 14% N, and 18% P; it is thus even more P-rich than the nucleotides in nucleic acid. However, ATP generally makes up only a small fraction of the total biomass of most organisms. For example, ATP levels range from 0.3% to 1.8% dry mass in marine copepods, and from only 0.02% to 2% dry mass in insects. As a result, variation in ATP content is unlikely to explain much of the variation among organisms in C:N:P stoichiometry.

Biominerals

Biominerals are inorganic solids produced by a wide variety of organisms to harden and stiffen tissues. Hard tissues containing biominerals are used for support, protection, and resource acquisition. The physical properties of these tissues depend on the identity of biominerals and the degree of biomineralization; thus, there is often a clear link between tissue elemental composition and function. Biominerals also make up a large fraction of the biomass in some organisms.

The three principal classes of skeletal biominerals are calcium carbonates, silica, and calcium phosphates. Calcium carbonate is the most abundant and widespread biogenic mineral. It makes up a large component of mollusk shells. Bird egg shells also consist mainly of calcium carbonate; hen egg shells contain about 95% calcium carbonate. Investment in thick, calcium carbonate-rich egg shells is thought to be an adaptation to hard nesting substrates or other conditions where egg damage is likely. Silica-based scales and skeletons are common in several ameba groups, but are found in only a few animals (sponges, a few copepods, some brachiopod larvae). Diatoms contain uniquely high concentrations of silica (Si:N ratios in diatoms are typically about 1), which serves as a main constituent of diatom cell walls. Silica is also an important component of some grasses and sedges. For example, plants in the genus *Equisetum* (horsetails) use silicic acid to maintain stem erectness; levels of silica in *Equisetum palustre* have been reported to be as high as 7.4% of dry mass. Calcium phosphates serve as skeletal material in vertebrates and a few brachiopods. Bone consists mainly of a calcium phosphate known as hydroxyapatite [$Ca_{10}(PO_4)_6$], which is also prominent in teeth and antlers. Although calcium-, silica-, and phosphorus-containing molecules are the most common biominerals, other elements such as iron and zinc form the basis of structural molecules in some organisms. For example, the marine bloodworm *Glycera dibranchiate* contains a copper-based biomineral

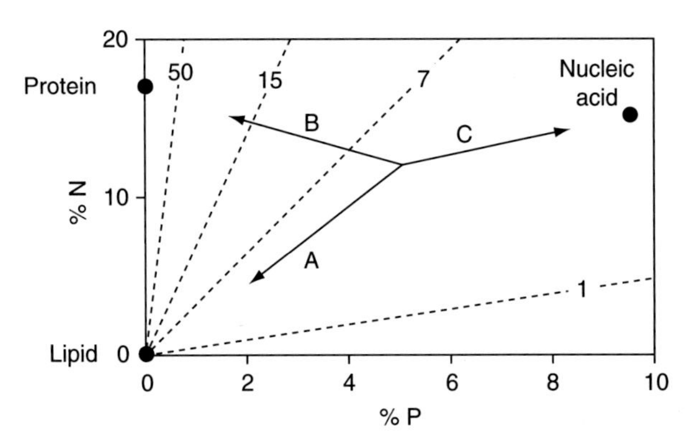

Figure 1 Stoichiometric diagram illustrating how altering storage lipid allocation proportionally affects %N and %P, leaving N:P constant but changing C:P and C:P (arrow A); how altering protein allocation increases %N while lowering %P, increasing N:P (arrow B); and how altering nucleic acid allocation disproportionately affects % P, lowering N:P (arrow C). Dashed lines indicate particular N:P ratios. Sterner RW and Elser JJ; *Ecological Stoichiometry*. © 2002 Princeton University Press. Reprinted by permission of Princeton University Press.

(atacamite, $Cu_2(OH)_3Cl$) in their jaws, which are extremely resistant to abrasion.

In summary, storage lipids, structural carbohydrates, protein, RNA, and biominerals are the biomolecular constituents of organisms that will likely account for most of the variation in elemental composition among organisms. Greater investment in lipids, for instance, will increase body C but lower N and P contents, higher protein levels will increase body N but lower body P content, and greater investment in RNA will substantially increase body P levels (**Figure 1**). These differences show how selection for functions met by particular biomolecular mixtures can influence the evolution of body composition. Conversely, an organism's elemental composition should reflect its ability to respond to specific adaptive challenges.

Elemental Composition at Higher Levels of Organization

Stoichiometric variation also occurs at higher levels of internal organization because organisms have different allocations to cellular structures and tissues having distinct biomolecular mixtures. A description of this variation is valuable because it helps to link elemental composition more directly to phenotypes.

A focal point in ecological stoichiometry has been the compositional difference between ribosomes and other cellular components. Ribosomes are the centers of protein synthesis; they catalyze peptide bond formation between amino acids in sequences determined by genetic information received from messenger RNA. The rate of protein synthesis often depends more on the number of ribosomes in cells than on the efficiency of individual ribosome molecules. As a result, ribosome concentration may largely determine protein synthesis rate, a trait with important evolutionary consequences (see below).

Ribosomes have the highest P concentration of any organelle. Ribosomes consist almost entirely of RNA and protein, with RNA making up a sizable fraction of ribosome mass (the RNA:protein ratio in eukaryotic ribosomes is about 1.2; in prokaryotes, it is 1.8). Because RNA is P rich (~10%), the biomolecular makeup of ribosomes results in a particularly P-rich structure (eukaryotes: 41.8% C, 16.3% N, 5% P; prokaryotes: 40% C, 16.1% N, 5.6% P).

Other cellular components likely also influence whole-organism stoichiometry. For example, the mammalian nucleus contains about 12.8% N and 2.3% P, which reflects an abundance of high-N nuclear proteins and the high P content of DNA. Mitochondria and chloroplast are also high-N organelles, as each contain about 11% N. They also contain very little P (0.31% and 0.32% P, respectively). Animal cell membranes have a fairly high amount of phospholipids relative to protein, and as a result have moderately high P content (59.5% C, 9.5% N, 1.5% P). Finally, plant cell walls consist mostly of N- and P-free cellulose and lignin, although they do contain small amounts of proteins and lipids. The estimated composition of plant cell walls is 35–38% C, <0.5% N, and ~0% P. These calculations suggest that adaptive allocation to particular cellular components will create a link between elemental composition and biological function.

Clear compositional differences also exist among different tissues. For instance, the N content in the leaves of apple trees (~1.2% dry mass) is considerably higher than N levels in stems and roots, which contain higher concentrations of cellulose and lignin (**Table 3a**). N content is particularly low in a tree's older woody tissue. In the crayfish, *Astacus astacus*, there are substantial differences in C:P and N:P ratios among the hepatopancreas (a digestive organ) and other major

Table 3a N concentrations in different components of apple trees[a]

Tissue	*Age (year)*	*% N*
Leaves		1.23
Spurs		1.04
Wood	1	0.93
	11–18	0.16
Roots	1–6	1.24
	14–18	0.32

[a]Data from Murneek AE (1942) Quantitative distribution of nitrogen and carbohydrates in apple trees. *Missouri Agricultural Experimental Station Research Bulletin* 348, as reported in Sterner RW and Elser JJ (2002) *Ecological Stoichiometry*. Princeton, NJ: Princeton University Press.

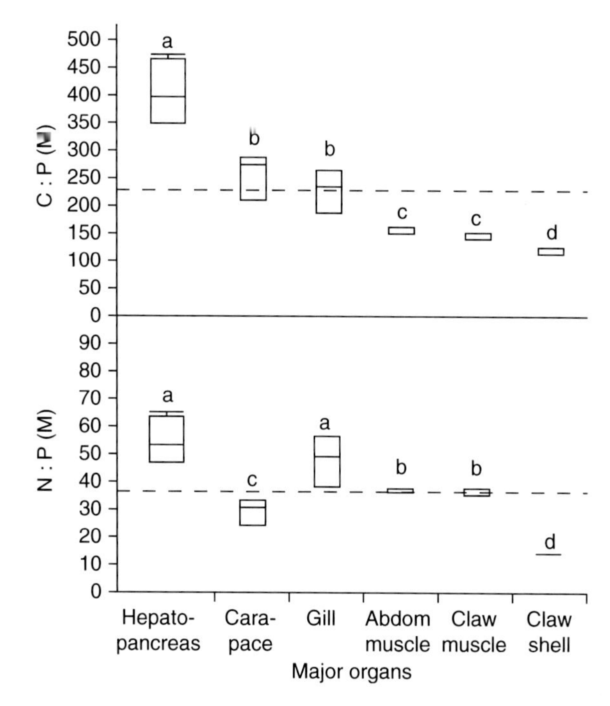

Figure 2 C:P and N:P ratios for major tissue of adult *Astacus astacus* crayfish. Extensions of boxes represent the 25th and 75th quartiles. Different characters on top of each box indicate statistically significant difference. From Faerovig PJ and Hessen DO (2003) Allocation strategies in crustacean stoichiometry: The potential role of phosphorus in the limitation of reproduction. *Freshwater Biology* 48: 1782–1792.

tissues (**Figure 2**). Finally, N and P levels differ among human tissues, and similar differences are likely in most other mammals (**Table 3b**). P content is particularly high in bone and other bony tissues such as teeth. Bone makes up roughly 10% of the biomass in mammals, and the skeleton may contain as much as 85% of a mammal's P content.

Table 3b N and P concentrations (% dry mass) in human tissues

Tissue	*%N*	*%P*
Kidney	7.2[a]	0.70[a]
Liver	7.2[a]	0.94[a]
Muscle	7.2[a]	0.3–0.85[a]
Bone	4.3[a]	~12.00[a]
Skin	16.0[a]	~0.10[a]
Hair	15.7[b]	0.01[b]
Nails	14.0[c]	0.01[c]

[a]Bowen HJM (1979) *Environmental Chemistry of the Elements*. London: Academic Press.
[b]Johnston FA, Debrock L, and Diao EK (1958) The loss of calcium, phosphorus, iron, and nitrogen in hair from the scalp of women. *American Journal of Clinical Nutrition* 6: 136–141.
[c]Iyengar GV (1978) *The Elemental Composition of Human Tissues and Body Fluids*. New York: Verlag Chemie.

Stoichiometry and Phenotypic Evolution

The previous section illustrated some of the compositional differences among structures that serve different biological roles. Ecological stoichiometry has also explored the evolutionary implications of these differences from both functional and economic perspectives.

A Functional Perspective: The GRH

Establishing the relationship between elemental composition, phenotypes, and functional performance is a key step toward understanding the role of stoichiometry in adaptive evolution. The clearest example of such a relationship is the positive association among P content, RNA content, and growth rate described by the GRH (see Organismal Ecophysiology). The GRH has obvious evolutionary implications because growth rate influences an organism's age and size at maturity, traits that often affect survivorship and fecundity.

A test of the GRH in an evolutionary context comes from a comparison between *Daphnia* species from different geographic regions. This comparison found that *Daphnia* from the arctic have higher P content and faster growth rates than *Daphnia* species from temperate lakes. These results suggest that the short arctic growing season selects for rapid growth, which is reflected in whole-body stoichiometry. Indirect support for the GRH comes from studies on zooplankton that show positive associations between body P concentration and growth rate across ontogenetic stages.

An Economic Perspective: Stoichiometrically Explicit Tradeoffs

Growth rate and other life-history attributes have been a focal area of theoretical and empirical research in evolutionary biology for more than 30 years. This work has shown that the evolution of life histories is guided by tradeoffs, in that costs associated with the expression of one trait limit the expression of another. Examples of life-history tradeoffs include those between current reproduction and future reproduction, reproduction and growth, or growth and defense.

Evolutionary tradeoffs will often be mediated by the availability of resources that must be allocated between traits. Life-history theory has often been based on the assumption that a single substance underlies such resource-based tradeoffs. For example, a model might be based on the assumption that a consumer has access to a pool of food energy that can be used equally well to meet its demands for growth or defense. Under this assumption, an increase in one trait (e.g., growth) leads to a proportional decrease in the expression of the other (defense). However, as outlined above, there are key elemental differences among biomolecules that should lead to differences in resource requirements between traits. In fact, resource requirements of traits such as growth, reproduction, and defense may be similar to the elemental composition of those traits because most of the substrate is used for structure rather than the metabolic requirements of trait assembly.

The existence of stoichiometrically distinct demands leads to an important prediction for the evolution of life histories: an optimal investment strategy will depend on the stoichiometry of each trait and the relative supply of different elements in the environment. Investment in a trait that requires an element that is difficult to acquire will be more costly than a trait requiring the same amount of another element that is relatively available. As a result, selection may favor traits (e.g., rapid growth) that rely heavily on particular substances (e.g., P) only when those substances are readily available in the environment.

Several empirical patterns suggest that economic considerations have influenced the evolution of organismal stoichiometry. For example, leaf nutrient concentrations are generally lower in plant species that dominate nutrient-poor sites than they are in species from fertile areas. Similarly, insect herbivores tend to contain less N than insect predators; one explanation for this pattern is that N scarcity in plant tissue has selected for lower N dependence in herbivores. Economic constraints may also have influenced the evolution of stoichiometry at finer scales. For example, enzymes used by the bacterium *E. coli* and the eukaryote *Sacharomyces cerevisiae* to assimilate C contain significantly less C than do typical proteins in these organisms. This result is intriguing because C assimilatory enzymes are probably most valuable to the organisms when C is scarce in the environment. The low C content in these enzymes may thus reflect selection for higher enzyme production in the face of a particular scarcity. Similar patterns have also been documented for the S content of S-assimilating enzymes of *E. coli* and *S. cerevisiae.*

Economic constraints can also be integrated with functional considerations to predict the evolution of organismal composition. Consider the evolutionary tradeoff between growth rate and competitive ability in marine phytoplankton. Such a tradeoff is likely in many organisms because the high resource demands associated with rapid growth will often put an organism at a competitive disadvantage when resources are scarce. Ecological stoichiometry provides a basis for determining the mechanisms underlying this tradeoff. In phytoplankton, response to this tradeoff can be characterized by investments in assembly machinery for biosynthesis and in resource-acquisition machinery. Assembly machinery consists of P- and N-rich ribosomes, while resource-acquisition machinery consists of chloroplasts and nutrient-uptake proteins that are rich in N but contain little P. When resources are abundant, selection favors greater investment in ribosomes for supporting rapid growth, leading to lower optimal N:P ratios. In resource-poor environments, phytoplankton with low N:P ratios are less successful because of their limited investment in resource acquisition and their high P demands for assembly. The N:P ratio in phytoplankton that is predicted to evolve by natural selection will thus depend on the availability of resources in the environment because of the different functions of P- and N-rich structures.

See also: Ecological Stoichiometry: Overview; Ecosystem Patterns and Processes; Organisal Ecophysiology; Population and Community Interactions; Trace Elements.

Further Reading

Acharya K, Kyle M, and Elser JJ (2004) Biological stoichiometry of *Daphnia* growth: An ecophysiological test of the growth rate hypothesis. *Limnology and Oceanography* 49: 656–665.
Baudouin-Cornu P, Surdin-Kerjan Y, Marliere P, and Thomas D (2001) Molecular evolution of protein atomic composition. *Science* 293: 297–300.
Bowen HJM (1979) *Environmental Chemistry of the Elements*. London: Academic Press.
Cross WF, Benstead JP, Rosemond AD, and Wallace JB (2003) Consumer-resource stoichiometry in detritus-based streams. *Ecology Letters* 6: 721–732.
Elser JJ, Acharya K, Kyle M, *et al.* (2003) Growth rate–stoichiometry couplings in diverse biota. *Ecology Letters* 6: 936–943.
Elser JJ, Dowling T, Dobberfuhl DA, and O'Brien J (2000) The evolution of ecosystem processes: Ecological stoichiometry of a key herbivore in temperature and arctic habitats. *Journal of Evolutionary Biology* 13: 845–853.
Elser JJ, Fagan WF, Denno RF, *et al.* (2000) Nutritional constraints in terrestrial and freshwater food webs. *Nature* 408: 578–580.
Faerovig PJ and Hessen DO (2003) Allocation strategies in crustacean stoichiometry: The potential role of phosphorus in the limitation of reproduction. *Freshwater Biology* 48: 1782–1792.
Fagan WF, Siemann E, Mittler C, *et al.* (2002) Nitrogen in insects: Implications for trophic complexity and species diversification. *American Naturalist* 160: 784–802.
Güsewell S (2004) N:P ratios in terrestrial plants: Variation and functional significance. *NewPhytologist* 164: 243–266.
Gillooly JF, Allen AP, Brown JH, *et al.* (2005) The metabolic basis of whole-organism RNA and phosphorus content. *Proceedings of the National Academy of Sciences of the United State of America* 102: 11923–11927.
Iyengar GV (1978) *The Elemental Composition of Human Tissues and Body Fluids*. New York: Verlag Chemie.
Johnston FA, Debrock L, and Diao EK (1958) The loss of calcium, phosphorus, iron, and nitrogen in hair from the scalp of women. *American Journal of Clinical Nutrition* 6: 136–141.
Kay AD, Ashton IW, Gorokhova E, *et al.* (2005) Toward a stoichiometric framework for evolutionary biology. *Oikos* 109: 6–17.
Klausmeier CA, Litchman E, Daufresne T, and Levin SA (2004) Optimal nitrogen-to-phosphorus stoichiometry of phytoplankton. *Nature* 429: 171–174.
Murneek AE (1942) Quantitative distribution of nitrogen and carbohydrates in apple trees. *Missouri Agricultural Experimental Station Research Bulletin* 348, as reported in Sterner RW and Elser JJ (2002) *Ecological Stoichiometry.* Princeton, NJ: Princeton University Press.

Reiners WA (1986) Complementary models for ecosystems. *American Naturalist* 127: 59–73.
Rundle SD, Spicer JI, Coleman RA, Vosper J, and Soane J (2004) Environmental calcium modifies induced defences in snails. *Proceedings of the Royal Society of London B* 271: 567–570.
Sterner RW and Elser JJ (2002) *Ecological Stoichiometry*. Princeton, NJ: Princeton University Press.
Sterner RW and George NB (2000) Carbon, nitrogen and phosphorus stoichiometry of cyprinid fishes. *Ecology* 81: 127–140.
Williams RJP and Fraústo da Silva JJR (1996) *The Natural Selection of the Chemical Elements: The Environment and Life's Chemistry*. Oxford: Clarendon.
Woods HA, Fagan WF, Elser JJ, and Harrison JF (2004) Allometric and phylogenetic variation in insect phosphorus content. *Functional Ecology* 18: 103–109.

Organisal Ecophysiology

T Vrede, Umeä University, Umeä, Sweden

A D Kay, University of St. Thomas, St. Paul, MN, USA

Introduction

Physiological processes are responsible for the multitude of functions that are essential for living organisms. These processes include the acquisition, incorporation, turnover, and release of energy and nutrients (**Figure 1**), and they ultimately affect the elemental composition (stoichiometry) of organisms. Whereas some of the processes are regulated so as to maintain a homeostatic stoichiometry (e.g., selective nutrient uptake or excretion, and synthesis of basic macromolecular structures), other processes will invariably result in changes in stoichiometry (e.g., excess nutrient uptake and storage). Rates of the physiological processes are affected by environmental factors (both abiotic and biotic) and are of great importance for the ecological role and fitness of organisms. In addition to the impact of these physiological processes on organismal stoichiometry, they are also subject to evolution by natural selection and ultimately affect the turnover of nutrients in entire ecosystems. Many physiological processes are ubiquitous in organisms because they share basic metabolic pathways and cell structures. Each of these metabolic pathways and cell structures involves and/or produces a specific setup of macromolecules or other compounds, and because these different types of compounds are characterized by specific elemental stoichiometries, the allocation to different compounds will affect organismal stoichiometry. Thus, there are inevitably strong links between ecological performance, physiological function, allocation patterns, and elemental stoichiometry of organisms.

In this article, basic principles regarding regulation of the elemental stoichiometry of organisms and their molecular constituents (both organic and inorganic) are presented in the first section. Then the acquisition, incorporation, turnover, and release of elements are discussed with special emphasis on physiological processes that link molecular processes and structures with elemental stoichiometry and the maintenance of homeostasis.

Organismal and Molecular Stoichiometry

Stoichiometric Homeostasis in Organisms

All organisms are made of organic matter containing the same basic set of elements, but there is a fair amount of variation in elemental stoichiometry. Some of this variation is due to interspecific differences among organisms with a homeostatically regulated stoichiometry. However, there are also organisms that have a chemical composition that is not homeostatically regulated, and in which the elemental content closely reflects the availability of elements. An example of this lack of homeostatic control is the N:P ratio of the green alga *Scenedesmus*, which shows a very tight correlation between its cellular N:P ratio and the N:P ratio of the medium (**Figure 2a**). Although there certainly are limits to this lack of homeostatic regulation, the investigated range in N:P ratios is relevant from an ecological point of view, and thus indicates that the N:P stoichiometry of this autotroph varies substantially within this range of nutrient supply ratios. Other photoautotrophic organisms, for example, vascular plants, microalgae, and cyanobacteria, show a similar lack of homeostatic regulation

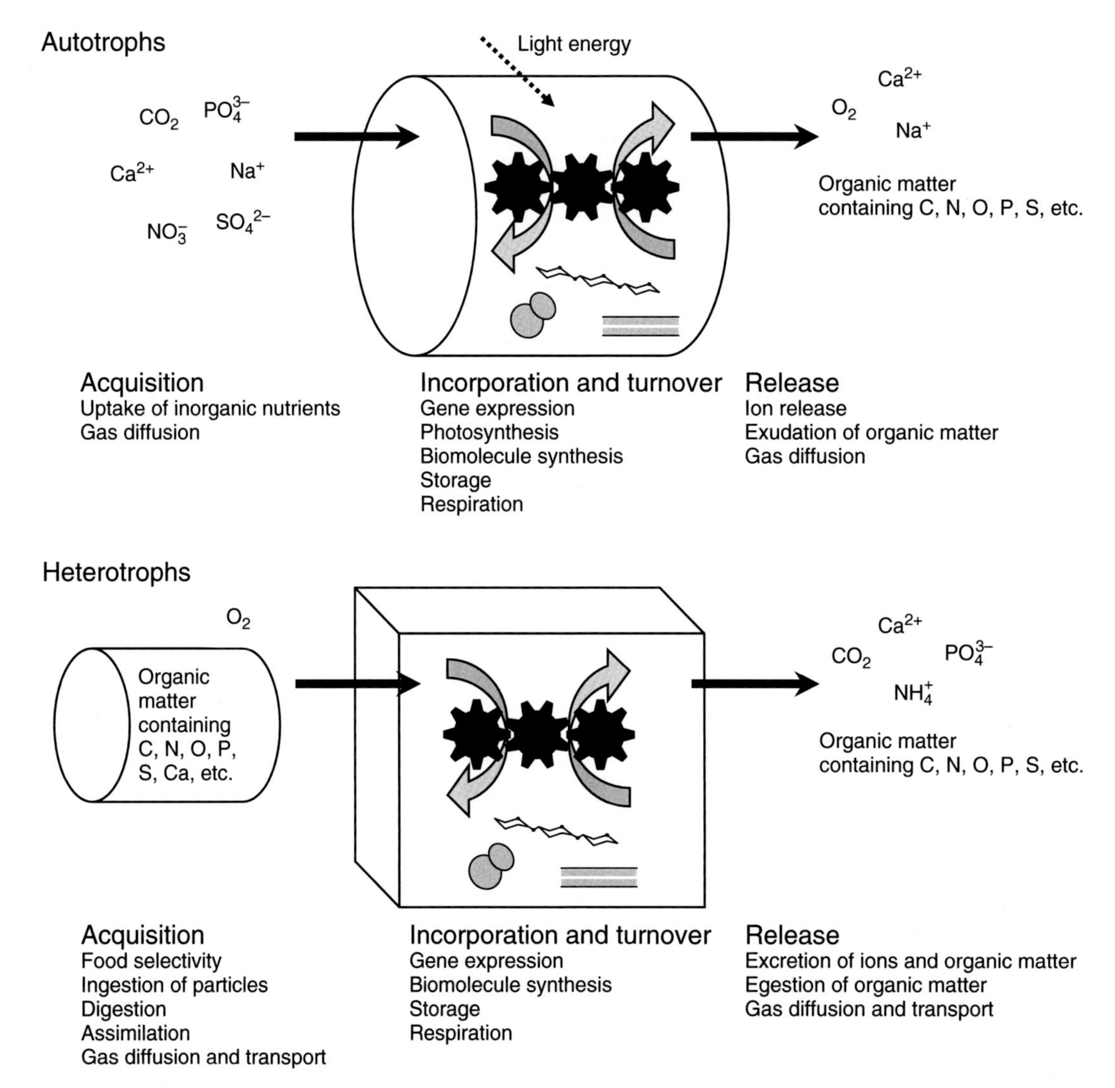

Figure 1 Conceptual outline of the acquisition, incorporation, turnover, and release of nutrients by autotrophs (plants) and heterotrophs (metazoan animals). The physiological processes of the organisms involve key metabolic pathways and molecular structures that are ubiquitous (e.g., synthesis of nucleic acids and proteins, and respiration), restricted to autotrophs (e.g., uptake of inorganic nutrients, and photosynthesis), or restricted to heterotrophs (e.g., feeding on particles). The incorporation and turnover comprise a biosynthetic machinery required for anabolic and catabolic processes, as well as organic and inorganic compounds that are synthesized and/or transformed. The physiological processes are similar in that they are performed in response to abiotic and biotic environmental stressors, and they affect the ecological performance of the organism, as well as their elemental stoichiometry. Modified from Frost PC, Evans-White MA, Finkel ZV, Jensen TC, and Matzek V (2005) Are you what you eat? Physiological constraints on organismal stoichiometry in an elementally imbalanced world. *Oikos* 109: 18–28.

of their N:P, and C:nutrient ratios, with variation within species that can be of a similar magnitude as the total variation across autotroph taxa.

In contrast to the large intraspecific variation in autotroph C:N:P stoichiometry, the elemental composition of metazoan animals varies much less. Even when the food is stoichiometrically imbalanced compared to the nutritional demands, the stoichiometry of the animals varies only within a relatively narrow range. Thus, metazoan elemental composition appears to be homeostatically controlled (albeit not strictly) and there are large interspecific differences. Likewise, the stoichiometry of bacterial cells varies much less than that of their substrates, suggesting a homeostatic control of bacterial stoichiometry. This is exemplified by the bacterium *Pseudomonas fluorescens* which does not change its biomass C:N ratio even when its substrate C:N stoichiometry varies more than one order of magnitude (**Figure 2b**).

Although the P and N content of resources may be high under conditions of high nutrient availability, the C:N and C:P ratios of food resources are frequently higher than those of herbivores, detritivores, and bacteria, both in terrestrial and aquatic systems. This elemental imbalance, as well as the limited intraspecific variation

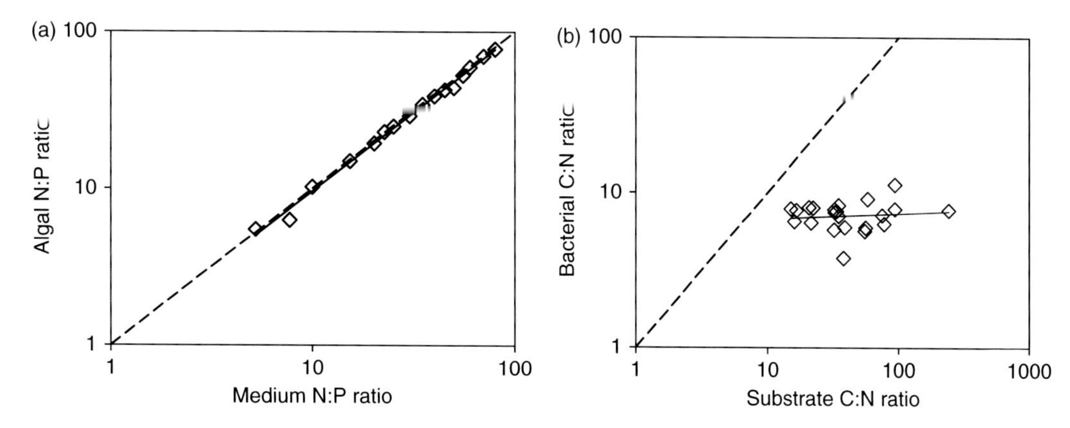

Figure 2 Relationships between nutrient supply ratios and cellular nutrient ratios (molar ratios) showing either lack of homeostatic control or strong homeostatic control. Broken lines indicate a 1:1 ratio between resource and cell nutrient ratios. Solid lines show linear regression lines of log-transformed data. If the slope of the solid line is equal to 1, there is a complete lack of homeostatic regulation, and if the slope is 0, the nutrient ratio is constant. (a) The N:P ratio of the green alga *Scenedesmus* closely resembles that of its nutrient medium, indicating a lack of homeostatic regulation. (b) The C:N ratio of the bacterium *Pseudomonas fluorescens* does not depend on the substrate C:N ratio, indicating a homeostatic regulation of its C:N ratio. (a) Data from Rhee G-Y (1978) Effects of N:P atomic ratios and nitrate limitation on algal growth, cell composition and nitrate uptake. *Limnology and Oceanography* 23: 10–25. (b) Data from Chrzanowski TH and Kyle M (1996) Ratios of carbon, nitrogen and phosphorus in *Pseudomonas fluorescens* as a model for bacterial element ratios and nutrient regeneration. *Aquatic Microbial Ecology* 10: 115–122.

in animal and bacterial elemental composition, suggests that there is a potential for resource quality limitation of consumer growth. Consequently, there is a need for physiological adaptations helping consumers to cope with their elementally imbalanced diet. To achieve an understanding of the mechanisms behind the stoichiometric patterns among organisms, we need to consider the physiological processes that organisms use for acquiring, incorporating, and releasing nutrients, and how these are connected to biochemical allocation patterns.

Allocation Patterns Affect Organism Stoichiometry

Organisms consist mainly of macromolecules, but there are also monomeric precursors to the macromolecules, a multitude of other small organic metabolites, and inorganic compounds. The organic matter contains the elements hydrogen (H), carbon (C), nitrogen (N), oxygen (O), phosphorus (P), and sulfur (S) in various proportions depending on the type of molecule. The macromolecules have functions as structural components of the cells (e.g., cellulose, phospholipids, and some proteins), metabolically active substances (e.g., enzymes and ribosomal RNA), carriers of genetic information (e.g., DNA and messenger RNA), or storage products (e.g., starch and some lipids). In addition to organic compounds, organisms also contain inorganic compounds such as free ions involved in osmoregulation, signal transduction, and other electrochemical reactions, or larger inorganic molecules used for nutrient storage, as structural components of cell walls, or vertebrate and invertebrate supportive tissues. Each of these molecules has a specific elemental composition (reviewed in more detail in Evolutionary and Biochemical Aspects). The most conspicuous differences are that proteins are rich in N but contain little P, nucleic acids are rich in both N and P, and carbohydrates and lipids contain only minor fractions of N and P. A consequence of the distinct patterns in elemental stoichiometry of these organic and inorganic molecules is that the allocation of these will have a strong influence on the elemental stoichiometry of organisms. For example, storage of lipids or starches as energy reserves, and allocation to structural tissues containing cellulose, lignin, and chitin, will result in high C:N, C:P, and C:S ratios of the organism. Likewise, a large allocation to protein results in high N content and intermediate C content, and thus low C:N ratio, and high C:P and N:P ratios. Allocation to large quantities of nucleic acids (which contain the major part of the organism P) will result in high P and N contents, and low C:P, C:N, and N:P ratios. Therefore, any stressors that change the allocation to different macromolecules and inorganic compounds will also affect the elemental stoichiometry of organisms.

Acquisition of Nutrients

While all biota contain the same elements, the mode of acquisition differs in a fundamental way between plants and animals. The dominating uptake mechanism in plants is uptake of inorganic ions, whereas metazoan animals typically consume particles that can be considered as parceled nutrients (**Figure 1**). Bacteria, archaea, fungi, and many other eukaryotic microorganisms are capable of uptake of inorganic compounds, but they can also process small organic molecules that can be used either as an energy source or as a source of nutrients for biosynthesis. Even though only small organic molecules are transported across the cell membranes of these organisms, microorganisms have a potential to utilize large molecules or even particles because these can be digested by extracellular enzymes. Some unicellular eukaryotes have the capacity to ingest particles such as bacteria or small algae by means of phagocytosis, a mode of acquisition that resembles animal feeding in the sense that whole packages of nutrients are ingested. This facultative phagotrophy can be an important mechanism for incorporating iron(Fe) and/or P in environments with low dissolved nutrient concentrations, as well as an important contribution of organic matter.

Nutrient Acquisition in Plants

The uptake of inorganic nutrients from the environment across the plant cell membrane is either by a transport through selective ion channels or by active transport by carrier proteins. While the ion channels allow rapid uptake of, for example, Na^+, K^+, Ca^{2+}, and Cl^-, and require no energy because the transport is downhill across electrochemical gradients, the carrier proteins may transport compounds against concentration gradients, using energy from electrochemical gradients, ATP, or light to drive the reactions. The activity of ion channels and carrier proteins can change in response to energy availability, ion gradients, and specific activators and inhibitors. Uptake capacity also changes as an effect of changes in gene expression, and as a consequence, changes both the type and number of ion channels and carrier proteins. For example, plant nitrate uptake is performed by several different transport systems with different affinities for nitrate. Some of these systems are facultative whereas others are induced as a result of N stress. This large flexibility in physiological capacity, as well as the large variations in environmental concentrations of the ions, results in large differences in uptake rates both within and among species. Because the uptake of different ion species is not directly coupled, uncoupled nutrient uptake rates for different elements enhance the variation in elemental stoichiometry of plants.

Under nutrient-limited conditions, plants face problems in acquiring enough of the limiting nutrient to sustain growth, and both intra- and interspecific competition may be intense. Consequently, high-affinity nutrient uptake systems have evolved as an adaptive strategy to cope with low nutrient availability. Species with efficient high-affinity uptake systems may eventually suppress the concentration of the limiting nutrient to the extent that its competitors cannot subsist. The uptake of nutrients generally follows Michaelis–Menten kinetics and is dependent on both the nutrient concentration and the physiological capacity of the organism. Under steady-state conditions in chemostat cultures, the growth rate of microalgae is proportional to the uptake rate of the limiting nutrient. In that case, the specific growth rate (biomass increase rate per unit biomass) can be described as a function of the limiting nutrient concentration in the medium according to the Monod model:

$$\mu = \mu_{\max}\frac{S}{S + k_s}$$

where μ is the achieved specific growth rate at the concentration S of the limiting nutrient, $\mu_{\max}$ is the maximum specific growth rate of the organism, and k_s is the half-saturation constant, that is, the nutrient concentration that sustains a growth rate equivalent to $\mu_{\max}/2$. The growth rate approaches the maximum physiological growth capacity of the organism asymptotically at high nutrient concentrations, whereas the growth rate declines at lower nutrient concentrations, finally reaching zero growth rate when the resource is absent (**Figure 3**). Apart from a qualitative similarity among species in the shape of these growth curves, there are species-specific differences both in the maximum specific growth rate as well as in the affinity for nutrients. For example, the green alga *Chlorella* has a higher $\mu_{\max}$ than the diatom *Synedra*, and thus grows more rapidly at high P concentrations (**Figure 3**). At low P concentrations, the low k_s (i.e., a high affinity and thus an efficient nutrient uptake) of *Synedra* makes it grow at rates close to its $\mu_{\max}$ already at low P concentrations, making it a superior competitor for P compared with *Chlorella*. Such physiologically based differences among phytoplankton species in maximum growth rates and nutrient uptake capability are an important part of the explanation of species

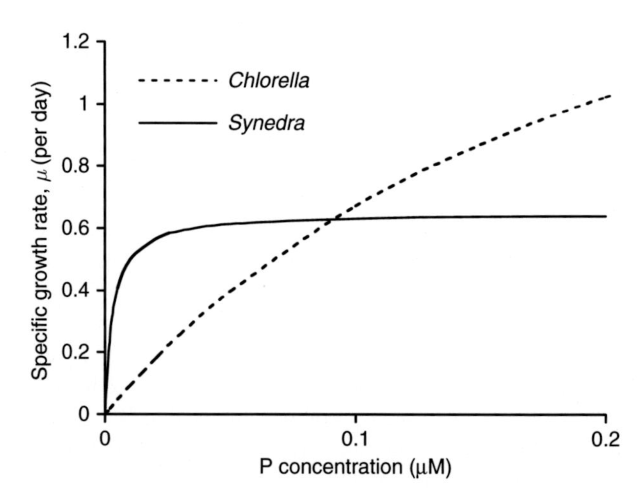

Figure 3 Specific growth rates of the green alga *Chlorella* and the diatom *Synedra* across a supply gradient of phosphorus. The lines show Monod curves fitted to experimental data obtained from chemostat experiments. *Chlorella* has a $\mu_{\max}$ of 2.15 per day and a k_s of 0.22 μM P, and *Synedra* has a $\mu_{\max}$ of 0.65 per day, and a k_s of 0.003 μM P. Data from Tilman D, Kilham SS, and Kilham P (1982) Phytoplankton community ecology: The role of limiting nutrients. *Annual Review of Ecology and Systematics* 13: 349–372.

succession patterns in phytoplankton communities that typically occur in temperate lakes.

In addition to the uptake of inorganic ions, some prokaryotes have evolved the physiological capacity to assimilate N_2 which is ubiquitous in the atmosphere. Because N fixation is energetically costly compared to ammonium and nitrate uptake, it is not an efficient strategy in environments with high availability of nitrate or ammonium and/or low energy supply. On the other hand, N-fixing cyanobacteria are superior competitors in aquatic environments with low concentrations of inorganic N and high concentrations of other nutrients. N fixation is also important in terrestrial habitats, where N-fixing *Rhizobium* bacteria live in symbiosis with some species of vascular plants (e.g., legumes). *Rhizobium* provides the plant with N, and gets energy in the form of organic carbon from the plant.

Nutrient Acquisition in Metazoan Animals

Animals feed on particles which are parcels containing a mix of nutrients. This means that they can fulfill their nutritional requirements by ingesting nutritionally balanced particles. However, consumers that do not feed on a nutritionally balanced diet will encounter difficulties in obtaining nutrients that are deficient in the food. In this context, it is also important to note that in addition to elements, animals also need to ingest some essential biochemical compounds that they cannot synthesize. These biochemicals include some amino acids, polyunsaturated fatty acids, vitamins, and sterols.

Carnivores often feed on prey with a similar nutrient and biochemical composition as their own, suggesting that the diet often is close to balanced. In contrast, herbivores and detritivores frequently encounter a nutritionally imbalanced diet both in terms of elemental and biochemical composition. They, therefore, need to maximize the uptake of the limiting nutrient. This can be achieved either by selective feeding or by selective digestion and absorption of the limiting substance. Many animals are capable of selecting food particles of high quality. A prerequisite for this ability is that they must have a sensory system that enables them to assess the quality of the food particles as well as an ability to capture suitable particles. Such a system may be energetically costly to produce, maintain, and use. This appears to be a drawback, but this energetic cost can be counteracted by a higher growth efficiency when feeding on a high-quality diet compared to feeding on a low-quality diet. Food particles may also have imbalanced but complementary nutritional values to consumers, and in such a scenario the consumer has to be able to mix the diet in order to balance its nutrient intakes and to obtain the optimum intake ratio, given the available choices.

Once ingested, the food particles are digested, that is, the organic matter is broken down to monomers. This is achieved by enzymatic hydrolysis, which can be both intracellular (in food vacuoles) or extracellular (in a gastrovascular cavity or a digestive tract). Ions and organic monomers can then be absorbed by the gastrodermic cells by similar uptake mechanisms as involved in the uptake of nutrients in autotrophs. Both in the digestive and absorption processes, dietary imbalances may be compensated for by regulation of the expression and/or the activity of the hydrolytic and uptake enzymes. Although such regulatory mechanisms have been documented, it is, at present, unclear to what extent this mechanism actually is used to balance the intake of nutrients.

Incorporation and Turnover of Elements

After the nutrients have entered into the cells, they are processed or used in different anabolic and catabolic pathways depending on the physiological status and biosynthesis and energy requirements of the cell. These processes include biological reactions such as photosynthesis, respiration, biosynthesis, and accumulation of storage compounds. With the exception of photosynthesis, which only takes place in photoautotrophs, there is a backbone of major biochemical pathways common to all biota. The regulation and coordination of these biochemical processes is under genetic control by numerous feedback mechanisms that includes inhibition or activation of enzymes as well as changes in gene expression. This basic similarity in biochemistry among organisms puts constraints on variation in elemental stoichiometry, but changes in the rates of these processes ultimately affect allocation patterns of major biochemical compounds and thus, overall elemental composition. The focus of the following sections will be on growth of heterotrophs and autotrophs and on nutrient storage in autotrophs. The reasons for this are that these processes have a profound impact on organismal stoichiometry, and these topics have also been the subject for some detailed study. However, conceptually similar connections between elemental stoichiometry and other physiological processes involved in nutrient turnover and biosynthesis can be hypothesized to exist.

Stoichiometry of Heterotroph Growth

Growth (i.e., the biosynthesis of new organic matter that is allocated either to biomass increase or to reproductive output) is a fundamental process in all biota, and it is important for the retention and turnover of nutrients in ecosystems as well as for population growth. Protein synthesis makes up a significant proportion of this biosynthesis simply because proteins are so abundant, both as structural components and enzymes, constituting as much as 30–75% of the dry mass of organisms. Consequently, the synthesis of proteins is a core biosynthetic pathway in all organisms. Protein synthesis takes place in ribosomes, which consists of ribosomal RNA (rRNA) and proteins, approximately 50% by weight of each. The protein synthesis rate of an organism is proportional to the number of ribosomes. Growth rate should, therefore, be closely related to protein synthesis rate as well as to its number of ribosomes. In each cell, there are thousands of ribosomes, which makes rRNA the most abundant nucleic acid. Because RNA is very rich in P (~10% by weight), a large number of ribosomes should also be reflected in a high P content. These mechanistic links between organism P content, RNA content, and growth rate have been proposed as the growth rate hypothesis (GRH).

Empirical data from metazoan animals, bacteria, and eukaryotic microorganisms support the GRH, although there are some exceptions. A positive relationship between growth rate and RNA content is observed among species representing as phylogenetically diverse organisms as bacteria, eukaryotic microorganisms, crustaceans, insects, and mollusks (**Figure 4**). Similar positive relationships between RNA content and growth rate have also been observed within species, for example, in the bacterium *Escherichia coli*, the insect *Blatella*, and the crustacean *Daphnia*. However, the relationship between RNA and growth rate may break down under some circumstances. This can be the case, for example, if the N supply is limited and therefore the supply of amino acids constrains protein synthesis rate.

There is also a positive relationship between organism RNA and P content (**Figure 5**). It should be noted that the slope of the relationship between organism total P content

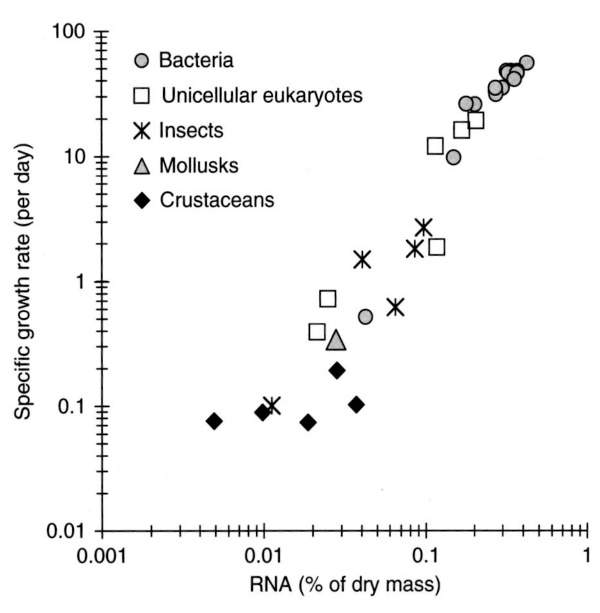

Figure 4 Organismal RNA content and specific growth rate in a phylogenetically diverse set of taxa. Adapted from Vrede T, Dobberfuhl DR, Kooijman SALM, and Elser JJ (2004) Fundamental connections among organism C:N:P stoichiometry, macromolecular composition, and growth. *Ecology* 85: 1217–1229.

and RNA-P content is close to unity both within and among taxa. This means that differences in P content among organisms can be interpreted as an effect of differences in allocation to RNA. The distance (parallel to the *y*-axis) from a data point to the 1:1 line represents the organism content of non-RNA-P (e.g., DNA, ATP, and phospholipids), and it is evident from the graph that RNA constitutes a major fraction of P in many organisms, in particular, in those with high growth rate. Along with this increase in cellular P content with increasing growth rate, the N content also increases because RNA is also rich in N. However, this increase is smaller relative to overall N content because a significant fraction of the cellular N is allocated to protein. An effect of this less-pronounced increase in N is that the N:P ratio decreases with increasing growth rate.

Despite the wide diversity in phylogenetic origin of the organisms studied, there are apparently fundamental similarities in basic biosynthetic machinery and processes, which are associated with different stoichiometric patterns as well as organism life history.

Stoichiometry of Autotroph Growth

The patterns described above are valid for organisms that regulate their stoichiometry homeostatically, but is the same also true for nonhomeostatic organisms, that is, plants? Although less is known about RNA content in relation to growth of vascular plants, observed N:P ratios, and growth rates of autotrophs covering the whole range from microalgae to trees are consistent with the GRH: the growth rate is positively correlated with plant P and N content, and the N:P ratio decreases with increasing growth rate (**Figure 6**).

Similar relationships between plant growth and elemental stoichiometry have also been modeled for microalgae. In the Droop model, the relationship between the cellular content of the limiting nutrient and the growth rate is expressed as

$$\mu = \mu'_{\mathrm{m}}\left(1 - \frac{Q_{\mathrm{min}}}{Q}\right)$$

where the specific growth rate, μ, is a function of the cell quota, Q (the intracellular amount of the limiting nutrient, expressed as nutrient per cell or per unit biomass), the minimum cell quota, Q_{min} (below which growth and cell division is not possible), and the theoretical asymptotic maximum growth rate, μ'_{m}, that is reached at infinite Q. From a physiologial point of view, the minimum cell quota can be interpreted as the minimum amount of P that is needed for the genome (DNA), phospholipids in cell membranes, and other P compounds (including rRNA and ATP) required for retaining a basal metabolism that permits survival. Because Q cannot be infinite (the organism cannot consist of 100% of any element, and even then Q is not greater than 1 if it is expressed as a fraction of total

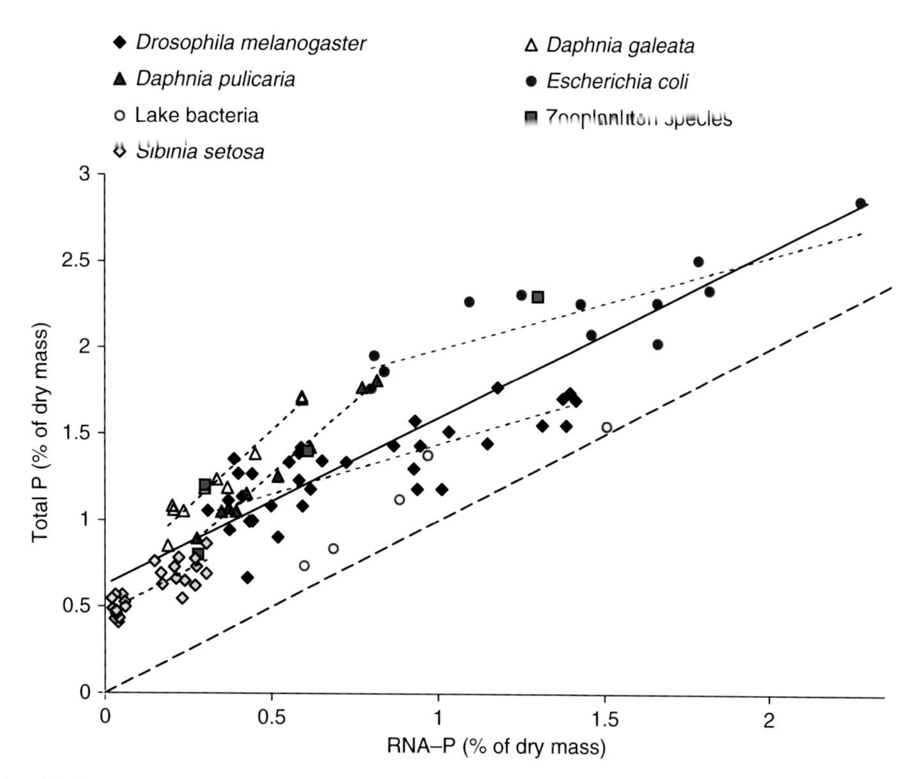

Figure 5 Relationship between organismal RNA-P content and total P content. The broken line indicates a 1:1 relationship between RNA-P and total organismal P, that is, that all cellular P is allocated to RNA. The dashed lines are linear regression curves fitted to each species, and the solid line is a linear regression curve fitted to all data. Because the P content of RNA is ~10% by weight, the RNA content of the organism can be calculated as 10 × RNA-P. Adapted from Elser JJ, Acharya K, Kyle M, *et al.* (2003) Growth rate–stoichiometry couplings in diverse biota. *Ecology Letters* 6: 936–943.

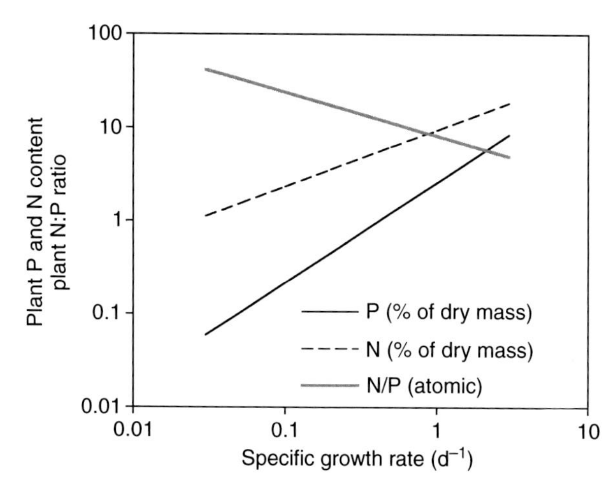

Figure 6 Correlations between plant N and P content and N:P ratio with specific growth rate. Based on data from Nielsen SL, Enriquez S, Duarte CM, and Sand-Jensen K (1996) Scaling maximum growth rates across photosynthetic organisms. *Functional Ecology* 10: 167–175.

biomass), μ'_m cannot be reached. However, for elements that can be stored in large quantities in relation to the cellular demands (e.g., micronutrients), μ_{max} approaches μ'_m. Another way of expressing the Droop model is to calculate the relative growth rate (RGR) as a function of the inverse of Q. RGR is the ratio between μ and μ'_m, $1/Q$ is equal to $M{:}X$, the biomass:nutrient ratio of nutrient X in the cell (or the C:nutrient ratio), and the inverse of Q_{min} is equal to $M{:}X_{max}$:

$$\mathrm{RGR} = \frac{\mu}{\mu'_m} = 1 - \frac{Q_{min}}{Q} = 1 - \frac{M:X}{M:X_{max}}$$

In this form, the model predicts a linear decrease in RGR with increasing $M{:}X$, and the slope is equal to $-1/M{:}X_{max}$. The Droop model has been developed and tested in chemostat systems with steady state, and there are many empirical examples where it produces a good fit to data. It has also been shown that the Droop model successfully predicts growth rates of natural phytoplankton that presumably are not at steady state. However, there are examples when algal growth deviates from the predictions of the Droop model, and the RGR is nonlinearly related to $M{:}X$. Such deviations point to the fact that autotroph stoichiometry is related not only to growth, but that other processes such as nutrient storage also may have a large influence on their stoichiometry. Although not mechanistically connecting growth with cellular nutrient content, the predictions of the Droop model when growth is

P-limited are qualitatively similar to the relationship between cellular P content and growth rate of the GRH.

Growth rate is also affected by temperature because enzymatic process rates generally are temperature-dependent. Growth rate therefore co-varies with temperature even if the nutrient supply rate and the macromolecular allocation pattern are constant. The availability of other resources such as light and inorganic carbon also affects the elemental stoichiometry of autotrophs. At a given supply rate of the limiting nutrient, the C:nutrient ratio increases with increasing light intensity (but only at light intensities below those where photoinhibition occurs). There are physiological adaptations to high light intensities, such as downregulation of chlorophyll synthesis and consequently lower chlorophyll:biomass ratios. However, this decrease in photosynthetically active systems only partially compensates for the increased C fixation, and therefore the C yield per nutrient atom increases with increasing light intensity. At low light intensities, the C:nutrient ratio decreases. This can be understood as an effect of lower photosynthetic rates because of lower C assimilation rates, but it can also be an effect of increased allocation to photosynthetic machinery (i.e., chloroplasts, which are rich in N-rich chlorophyll and proteins like rubisco and cytochromes). Because chloroplasts contain a high fraction of N but little P, this allocation pattern will result in low C:N and high N:P ratios, and is consistent with observed C:N and N:P ratios of phytoplankton cells. The high N:P ratios of N-fixing cyanobacteria, which have been explained as an effect of their ability to take up N in an otherwise N-deficient environment, may also be explained as an effect of allocation patterns. Because N fixation is very energy demanding, N-fixing cyanobacteria have to allocate considerable resources to photosynthetic machinery, and therefore their N:P ratio can be expected to be high.

The C:nutrient ratio of terrestrial plants increases also with increasing partial pressure of carbon dioxide (pCO_2). Similar patterns may prevail in phytoplankton too, especially under nutrient-limited conditions, but consistent data are at present lacking. From a physiological point of view this increase in plant C content at elevated pCO_2 can be understood as an effect of increased availability of substrate for rubisco, and hence increased C assimilation rates.

Nutrient Storage

Animals generally have a limited ability to store chemical elements for later use. The main exception to this is the storage of lipids (with high C:nutrient ratios) in lipid droplets or adipose tissue for subsequent use as energy reserves. In contrast to the limited nutrient storage capacity of animals, this ability is very well developed in autotrophs, thus increasing the range of variation in elemental stoichiometry of plants. Nutrient storage does not normally take place under rapid growth, when biosynthetic requirements balance nutrient uptake, but when growth is resource-limited there is a luxury uptake of nutrients that are available in excess. Even though nutrient uptake systems of nonlimiting nutrients are generally downregulated, nutrient uptake continues in facultative uptake systems. These excess nutrients, which cannot be used for immediate growth and biosynthesis, can be stored in many different forms in plant cells: as ions or organic compounds in the cytoplasm, as large intracellular deposits of macromolecules, or in the vacuole. C that is produced in excess can be stored as nonpolar lipids, starch, glycogen, or other low-N and low-P organic molecules. A large store of nutrient-poor organic matter can be of great importance for the plants because this makes C:nutrient ratios so high that herbivore growth cannot be sustained on such low-quality food. P can be stored as increased concentrations of vacuolar phosphate and small P-rich organic molecules or as polyphosphate granules. N is stored as, for example, nitrate or amino acids. The presence of vacuole in autotrophs is of special importance in the context of nutrient storage. This cell compartment can store large quantities of nutrients that can be transported back to the cytoplasm when the nutrients are needed. This provides autotrophs with an alternate strategy to the high-affinity uptake system (see the Monod model discussed above) to compete for nutrients. When nutrient availability is high, luxury uptake provides an internal nutrient store that can be used for growth when the resource is limiting. This strategy can be observed in some phytoplankton that are capable of vertical migration between nutrient-rich but dark deep water layers and well-illuminated but nutrient-depleted surface waters. In higher plants, the metabolism and storage of nutrients and organic compounds is even more complex and multifaceted, since it both involves reallocation of storage products among different tissues, and differences in nutritional requirements among tissues. Thus, nutrient storage products and strategies specific for plants have evolved. For example, storage proteins and phytate deposited in seeds provide germinating seeds with N and P.

Nutrient Release

In autotrophs, excess nutrients are normally stored, but there are also disposal mechanisms present that enable them to dispose of excess organic carbon. First, organic matter can be released by diffusion or active transport across the cell membrane. This release of dissolved organic matter may be significant; phytoplankton release on average 13% of the assimilated C as dissolved organic carbon, but release rates as high as 80% of total primary production have been reported. In lacustrine plankton, the percentage of total C assimilation that is released increases with decreasing nutrient levels, suggesting that the release may serve as a way of disposing of excess C. Plants with symbiotic N-fixing *Rhizobium* bacteria release carbohydrates to the bacteria and receive N in exchange. The release of excess C thus serves as a way to promote the uptake of a limiting nutrient. Second, there are metabolic pathways, futile cycles, that provide alternatives to normal catabolic pathways. In these cycles, excess C is respired without producing new biomass or performing biochemical work. For example, in the alternative oxidase (AOX) pathway, the enzyme AOX is an electron acceptor that is not coupled to the generation of a proton motive force, which is generated by the normal oxidative phosphorylation pathway. The seemingly wasteful AOX pathway allows the mitochondrion to modulate its ATP production rate and to reduce the rate of production of reactive oxygen species. The AOX activity is induced by N and P deficiency, thus increasing the respiration rate and decreasing the C-use efficiency.

Animals homeostatically regulate their nutrient balance not only by selective uptake, digestion or absorption, but also by selectively releasing nutrients in excess. The main excretory products containing P and N are phosphate (P), ammonium (N), and urea (N). In this way, the C:N:P ratios of the animal are regulated at relatively fixed levels. When the C:nutrient ratio of the assimilated food is higher than the demands of the consumer, the excess C has to be expelled either by increased respiration or by selective excretion of dissolved organic matter. Increased respiation using the oxidative phosphorylation pathway produces an energy surplus that has to be used in some way. Although data are rather anecdotal, increased physiological activity such as intensified filtration and swimming may account for some extra energy consumption. However, similar to the AOX pathway in plants, there are also futile cycles in animal metabolism that provide a decoupling of energy production and respiration. The other way of disposing excess C is to excrete it. Both these mechanisms lead to a depression of the energetic growth efficiency, but there is a tradeoff because the stoichiometry can be maintained at a level balancing the nutritional demands. Although data are scarce both on respiration rates and excretion rates as a function of food quality, there is considerable evidence that growth efficiency varies inversely with food C:nutrient ratio, and that the disposal of C at high C:nutrient ratios is due to both increased respiration and excretion. Similar results have also been obtained from a modeling study of C, N, and P turnover in animals, based on analysis of major physiological processes including assimilation, maintenance metabolism, growth, respiration, and excretion. Thus, physiological processes and associated allocation patterns are able to explain observed patterns in elemental stoichiometry of nutrient release and C metabolism in animals.

See also: Ecological Stoichiometry: Overview; Evolutionary and Biochemical Aspects; Population and Community Interactions.

Further Reading

Andersen T (1997) *Pelagic Nutrient Cycles: Grazers as Sources and Sinks of Nutrients*. Berlin: Springer.
Anderson TR, Hessen DO, Elser JJ, and Urabe J (2005) Metabolic stoichiometry and the fate of excess carbon and nutrients in consumers. *American Naturalist* 165: 1–15.
Chrzanowski TH and Kyle M (1996) Ratios of carbon, nitrogen and phosphorus in *Pseudomonas fluorescens* as a model for bacterial element ratios and nutrient regeneration. *Aquatic Microbial Ecology* 10: 115–122.
Droop MR (1974) The nutrient status of algal cells in continuous culture. *Journal of the Marine Biological Association of the UK* 54: 825–855.
Elser JJ, Acharya K, Kyle M, *et al.* (2003) Growth rate–stoichiometry couplings in diverse biota. *Ecology Letters* 6: 936–943.
Elser JJ, Dobberfuhl D, MacKay NA, and Schampel JH (1996) Organism size, life history, and stoichiometry: Towards a unified view of cellular and ecosystem processes. *BioScience* 46: 674–684.
Frost PC, Evans-White MA, Finkel ZV, Jensen TC, and Matzek V (2005) Are you what you eat? Physiological constraints on organismal stoichiometry in an elementally imbalanced world. *Oikos* 109: 18–28.
Grossman A and Takahashi H (2001) Macronutrient utilization by photosynthetic eukaryotes and the fabric of interactions. *Annual Review of Plant Physiology and Plant Molecular Biology* 52: 163–210.
Nielsen SL, Enriquez S, Duarte CM, and Sand-Jensen K (1996) Scaling maximum growth rates across photosynthetic organisms. *Functional Ecology* 10: 167–175.
Rhee G-Y (1978) Effects of N:P atomic ratios and nitrate limitation on algal growth, cell composition and nitrate uptake. *Limnology and Oceanography* 23: 10–25.
Sterner RW and Elser JJ (2002) *Ecological Stoichiometry*. Princeton: Princeton University Press.
Tilman D, Kilham SS, and Kilham P (1982) Phytoplankton community ecology: The role of limiting nutrients. *Annual Review of Ecology and Systematics* 13: 349–372.
Urabe J and Watanabe Y (1992) Possibility of N or P limitation for planktonic cladocerans: An experimental test. *Limnology and Oceanography* 37: 244–251.
Vrede T, Dobberfuhl DR, Kooijman SALM, and Elser JJ (2004) Fundamental connections among organism C:N:P stoichiometry, macromolecular composition, and growth. *Ecology* 85: 1217–1229.

Population and Community Interactions

J P Grover, University of Texas at Arlington, Arlington, TX, USA

Introduction

Ecological stoichiometry is relatively new as a recognized area within ecology, but the broad implications for population and community interactions were recognized by one of the discipline's modern founders, Alfred Lotka. Stoichiometry deals with quantitative relations among the masses of elements and compounds, so the law of mass conservation is a primary tool. It dictates that masses of substances appearing during a process (products of a chemical reaction, production of new organisms or tissues, excretory wastes, etc.) must be balanced by masses of substances disappearing (reactants, dissolved nutrients, food or prey eaten, etc.). Inevitably, there are mass-conservation constraints or mass-balance constraints on the dynamics of such transformations.

Lotka made a concise formal statement of such constraints for ecological communities where n populations with densities N_i interact according to the governing equations

$$\frac{dN_i}{dt} = f_i(N_1, N_2, \ldots, N_n) \quad \text{for } i = 1, \ldots, n \qquad [1]$$

Each organism in one of these interacting populations will contain a certain mass of element j, called the quota, Q_{ij}. Supposing that biomass is composed of m elements, which are assumed for the moment not to appear in nonliving forms, conservation of mass leads to the following constraints on population densities:

$$\sum_{i=1}^{n} N_i Q_{ij} = \text{const.} \quad \text{for } j = 1, \ldots, m \qquad [2]$$

These constraints reduce the dynamical degrees of freedom among the interacting species from n (the number of species) to $n - m$. Although this general principle has long been known, it was many decades after Lotka's first statement before ecologists explored the detailed consequences of such mass-conservation constraints for population and community interactions.

The contemporary approach to studying these consequences appends to Lotka's equations another set of equations for the dynamics of free elements, because most nutrient elements are present in the environment in at least one dissolved or gaseous form. The dynamics of free elements typically depend on densities of at least some of the populations, and in turn population dynamics often depend on concentrations of nutrient elements. Thus the set of equations [1] is expanded to two sets:

$$\frac{dN_i}{dt} = f_i(N_1, N_2, \ldots, N_n; R_1, R_2, \ldots, R_m) \quad \text{for } i = 1, \ldots, n \qquad [3a]$$

$$\frac{dR_j}{dt} = g_j(N_1, N_2, \ldots, N_n; R_1, R_2, \ldots, R_m) \quad \text{for } j = 1, \ldots, m \qquad [3b]$$

where R_j is the concentration of nutrient element or resource j, free in the environment. The mass-conservation constraints are correspondingly modified to read

$$R_j + \sum_{i=1}^{n} N_i Q_{ij} = \text{const.} \quad \text{for } j = 1, \ldots, m \qquad [4]$$

At first sight, such constraints might appear unimportant, since we have now restored the dynamical degrees of freedom to n, the number of interacting populations in the community. Nevertheless, mass-conservation constraints have proven helpful in analyzing model ecological communities represented by equation system [3].

A potential drawback of Lotka's perspective is that mass-conservation constraints of the form of equation [4] most clearly apply to idealized, closed systems. In contrast, natural systems are open, with materials entering and exiting at their boundaries. However, constraints of the form of eqn [4] apply to long-term dynamics for an important class of open systems, those that resemble a laboratory chemostat, a well-mixed culture device into which a medium containing nutrient elements is pumped, and from which all organisms and dissolved nutrients are removed indiscriminately by outflow. Such devices are often used for experiments on nutrients and microbial population dynamics, and caricaturize some of the open systems found in nature. More generally, mass-conservation constraints often emerge in the long-term dynamics of open systems that achieve a steady state.

Classification of Population Interactions

Ecologists have developed a traditional classification of population interactions, closely tied to a particular form of equation system [1], in which

$$\frac{\mathrm{d}N_i}{\mathrm{d}t} = f_i(N_1, N_2, \ldots, N_n) = \beta_i N_i + \sum_{i=1}^{n} \alpha_{ij} N_i N_j \qquad [5]$$

Here, the parameter β_i is the capacity of population i to increase in the absence of interactions, and the interaction coefficients α_{ij} quantify the impact of interactions among populations on the dynamics of population i. In most cases the intraspecific coefficient, α_{ii}, is negative due to competition among individuals of the same species. But the interaction coefficients among different species take various signs, leading to a canonical classification of pairwise interactions (**Table 1**). Though well entrenched in ecological thinking, this classification has at least two potential drawbacks. First, the nature of an interaction can depend on densities of the interacting species or other species, or upon other environmental factors. For example, an interaction might sometimes be competitive, and at other times mutualistic. Second, eqn [5] and its α_{ij} coefficients describe the effects of interactions on population dynamics, but not the mechanisms of those interactions.

Table 1 Classification of pairwise population interactions

Signs of interaction coefficients	*Type of interaction*
$\alpha_{ij} < 0$ and $\alpha_{ji} < 0$	Competition (through resource exploitation, allelopathy, aggression, or interference)
$\alpha_{ij} > 0$ and $\alpha_{ji} < 0$ or $\alpha_{ij} < 0$ and $\alpha_{ji} > 0$	Predation (including herbivory, parasitism, and disease)
$\alpha_{ij} > 0$ and $\alpha_{ji} = 0$ or $\alpha_{ij} = 0$ and $\alpha_{ji} > 0$	Commensalism
$\alpha_{ij} < 0$ and $\alpha_{ji} = 0$ or $\alpha_{ij} = 0$ and $\alpha_{ji} < 0$	Amensalism
$\alpha_{ij} > 0$ and $\alpha_{ji} > 0$	Mutualism
$\alpha_{ij} = 0$ and $\alpha_{ji} = 0$	None

Theoretical models incorporating nutrient resource dynamics have been an important source of mechanistic knowledge enriching ecologists' understanding of population interactions. Much progress has been made by coupling equation system [3] with stoichiometric constraints (eqns [4]), for special cases with small numbers of populations and nutrient resources. Although the traditional classification of population interactions has limitations, it provides a convenient way to catalog such efforts.

Competition

Competition among populations often arises from exploitation of common resources. In the simplest situations where a single nutrient resource is involved, a well-known result is that only one population persists at steady state while all others are excluded from the habitat. The winner is the population that reduces the nutrient to the lowest concentration at steady state when growing in the absence of other populations. The steady-state nutrient concentration when population i grows alone is denoted R_i^*, and the prediction that it wins is called the R^* rule. This prediction, in principle, is easily tested: grow all populations on their own and record R_i^*, then grow them all together and see whether the population with the lowest value excludes others. Such tests have been conducted many times, mostly with bacteria or microalgae growing in laboratory cultures. The R^* rule has been verified in virtually all cases where other complications can safely be ruled out.

Although stoichiometry can play a role in determining the R^* value expressed by a population, stoichiometric constraints are especially evident in competition for multiple nutrient resources. Consider adapting equation system [3] to represent two populations that compete for two nutrient resources. With no other intraspecific or interspecific interactions, the reproductive rates of the two populations depend only on the concentrations of the two nutrient resources. The total concentration of nutrient j in the habitat consists of its dissolved concentration plus the amounts bound in individuals of both populations, leading to the mass-conservation constraints

$$S_j = R_j + \sum_i N_i Q_{ij}, \quad \text{for } j = 1, 2 \qquad [6]$$

where S_j denotes the total concentration.

This mass-conservation constraint reduces the competition system to two dynamical degrees of freedom, permitting graphical analysis on the resource plane of nutrient concentration R_1 versus concentration R_2. On this plane, the 'zero net growth isocline' (ZNGI) for population i is a graph showing the combinations of nutrient concentrations supporting reproduction that exactly balances mortality. Steady states can only occur at these nutrient concentrations. The shape of a ZNGI graph depends on the physiological roles of the nutrient resources. Out of many possibilities, two suffice for illustration. For essential resources, each nutrient plays a unique physiological role that cannot be substituted by the other. Different nutrient elements used by autotrophs, such as nitrogen and phosphorus, are a good example. The absolute requirement for each nutrient to support reproduction leads to a ZNGI graph that is rectilinear, with a limb parallel to each nutrient concentration axis (**Figure 1a**). For substitutable resources, each nutrient can substitute for the other's physiological role. Different chemical forms of the same element, such as organic carbon substrates used by bacteria, are a good example. The availability of one nutrient at a given concentration reduces proportionally the amount of the other required to support reproduction, leading to a linear ZNGI graph with negative slope (**Figure 1b**).

If the ZNGI graphs for the two competing populations intersect, coexistence is possible. Biologically, such an intersection implies that one population is a superior competitor for the first resource and an inferior competitor for the second, while the other population is superior for the first resource and inferior for the second. **Figure 2a** shows an example of competition for essential resources, where species 1 is a superior competitor for resource 1 and an inferior competitor for resource 2, while species 2 is a superior competitor for resource 2 and an inferior competitor for resource 1.

When coexistence of competitors at steady state is possible, it will not be achieved if inadequate nutrient supplies render coexistence infeasible, or if it is dynamically unstable. These questions are strongly affected by the competitors' stoichiometry for the nutrient resources in question. The mass-conservation constraints (eqns [6]) imply that on the plane of resource concentrations (R_1 vs. R_2), feasible steady states lie between two lines of positive slope passing through the supply point (S_1, S_2). If the intersection of the competitors' ZNGI graphs lies within these boundaries, then steady-state coexistence is feasible (**Figure 2b**). Biologically, the supplies of both nutrients are high enough to support both populations at steady state. The slopes of these mass-conservation constraints are the whole-organism stoichiometric ratios of the two nutrients for the two competitors. For example, if two

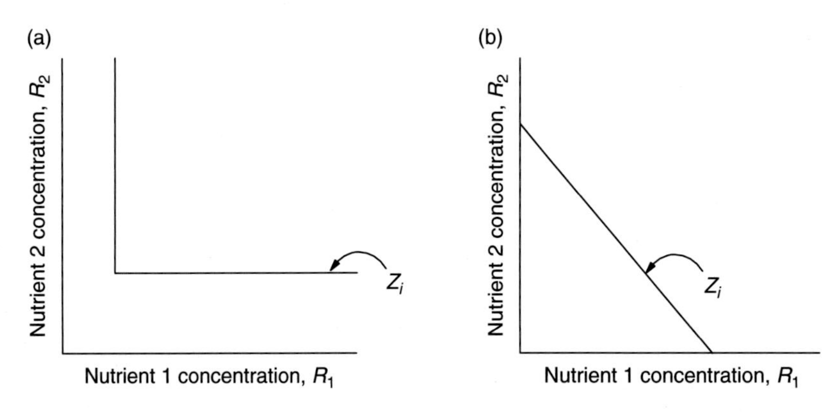

Figure 1 'Zero net growth isocline' (ZNGI) of a population i for two nutrient resources (indicated by Z_i). (a) Essential resources; (b) substitutable resources.

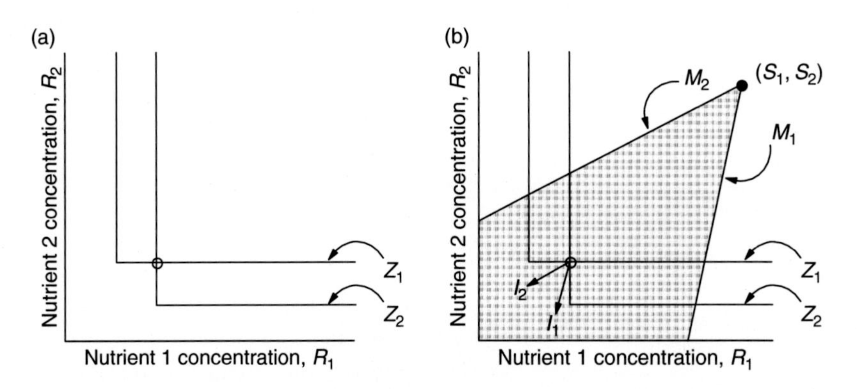

Figure 2 Competition between two populations for two essential nutrient resources. (a) ZNGI graphs (indicated by Z_i) intersect at the circled point, making steady-state coexistence possible. (b) The shaded feasible region for steady states is superimposed, bounded by mass-conservation constraints for populations 1 and 2 (indicated by M_i), as are impact vectors implying stable coexistence (indicated by I_i).

algal populations compete for dissolved nitrogen and phosphorus, the slopes are their cellular N:P ratios.

The dynamical stability of coexistence depends on the same stoichiometric ratios, and can be analyzed graphically by plotting impact or consumption vectors for each population. These portray the relative impacts of the competitors on the two resources through consumption, at steady state. Only the slopes of these vectors affect stability of the steady state, and these slopes are the same as those of the mass-conservation constraints. Biologically, these vectors represent consumption of the two resources in the proportion required to maintain organismal stoichiometry. The arrangement of the impact vectors in **Figure 2b** shows an example of stable coexistence. Stability arises because each of the competitors consumes proportionally more of the resource for which it is the inferior competitor, enhancing the negative intraspecific effect of a population on its own growth rate, and ameliorating the negative impact on its competitor's growth rate. This result illustrates a general feature of competition theory that is easily derived from analyzing eqn [5] for two competing populations (the famous Lotka–Volterra equations of competition). Stable coexistence of competitors at steady state requires that intraspecific competitive effects be stronger than interspecific effects. Explicitly including the stoichiometry of resource consumption brings some biology into this rather abstract statement.

Explicitly including the stoichiometry of resource consumption also leads to an important testable prediction. When stable coexistence is possible (e.g., **Figure 2b**), manipulating the resource supply stoichiometry by moving the position of the supply point (S_1,S_2) around the resource plane alters the outcome of competition. When the supply point is moved down and to the right, increasing the supply ratio S_1:S_2, the shaded feasible region is moved beyond the intersections of the competitors' ZNGI graphs. Coexistence is then no longer feasible, and the relatively low supply of resource 2 leads species 2, the superior competitor for this resource, to exclude species 1. Moving the supply point in the opposite direction, up and to the left, has the opposite effect, producing exclusion of species 2 by species 1.

The predicted dependence of competitive outcomes on resource supply stoichiometry is called the resource ratio hypothesis, and it has been tested many times, primarily with diatoms or other microalgae competing for dissolved nutrient elements in laboratory cultures (e.g., **Figure 3**). Nearly all of these tests have verified the prediction, and some of the patterns of competitive dominance observed in the laboratory appear in the natural distributions of algae. For example, some cyanobacteria are better competitors for nitrogen than are certain eukaryotic algae, which are better competitors for phosphorus. Distributional data show an association of cyanobacterial dominance with a low N:P concentration ratio in lakes.

What happens when coexistence of two populations is feasible but unstable? For competitive interactions, instability usually implies that one competitor will persist while the other is excluded. The identity of the winner will depend on initial conditions, usually going to the advantage of the population with greater initial abundance, leading the term priority effect to be applied to such situations. However, instability and priority effects are unlikely in competitive interactions, if competitive

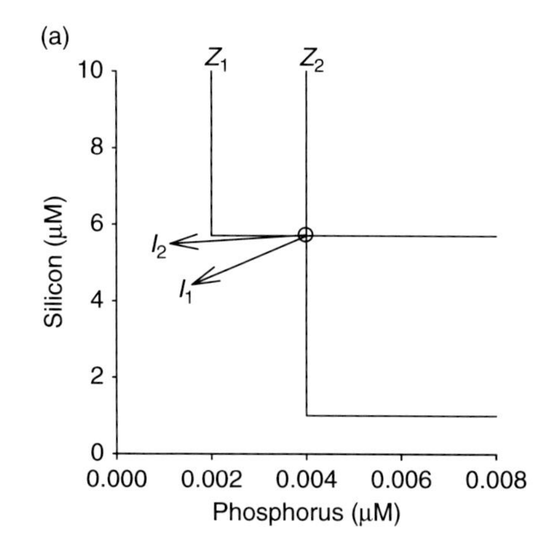

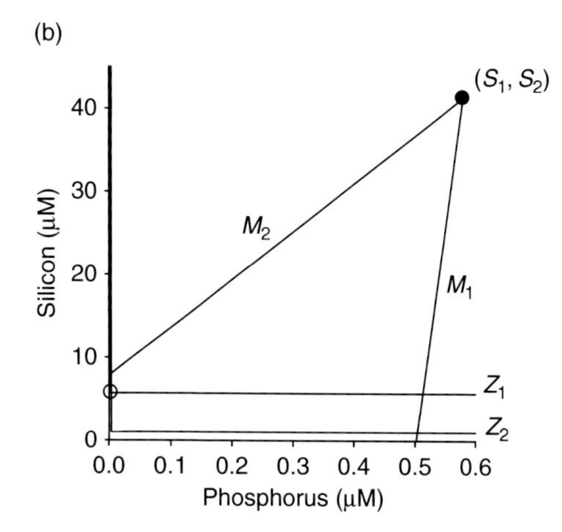

Figure 3 Competition between two diatoms, *Synedra filiformis* (species 1) and *Asterionella formosa* (species 2), for the essential nutrients phosphorus (resource 1) and silicon (resource 2) in laboratory cultures. (a) The ZNGI graphs (Z_i) and impact vectors (I_i) predict that stable coexistence is possible. (b) The supply point (S_1,S_2) and mass-conservation constraints (M_i) for one of the nutrient supply conditions was tested, for which stable coexistence was predicted and observed. Other supply conditions were tested for which exclusion of an inferior competitor was predicted, and trends toward this outcome were observed. Four other pairs of competing diatom species were tested in similar experiments, and observed results matched predictions based on parametrized, stoichiometric models. Data from Tilman D (1981) Tests of resource competition theory using four species of Lake Michigan algae. *Ecology* 62: 802–815.

abilities depend on organismal stoichiometry as explained below (see the secion titled 'Physiological variability in stoichiometry').

Commensalism

Before its connections with other stoichiometric approaches were fully recognized, the theory of competition sketched above was well verified with experiments on microalgae and bacteria. Experimental studies also turned up some exceptions. Many studies of interacting microbial populations reveal an important role for excreted substances that act as either resources or toxins. Such interactions mediated by excreted substances are also amenable to stoichiometric approaches.

For example, one such interaction involves two populations that compete for one resource, while one of the populations excretes a substance used as a second resource by the other population. This complex interaction is common in aquatic ecosystems: autotrophic algae require inorganic nutrients, and are often limited by the supply of either nitrogen or phosphorus. Heterotrophic aquatic bacteria require the same inorganic nutrients, but also need organic carbon, which is excreted by nutrient-limited algae as a by-product of excess photosynthesis. So it is common to find algae and bacteria competing for an inorganic nutrient, while the bacteria rely on dissolved organic carbon produced by the algae, an interaction combining competition and commensalism.

To represent this interaction graphically on the resource plane, the ZNGI of the algae is a line perpendicular to the axis for the first resource, the inorganic nutrient, expressing the fact that they do not require the second resource, organic carbon (**Figure 4**). The ZNGI of

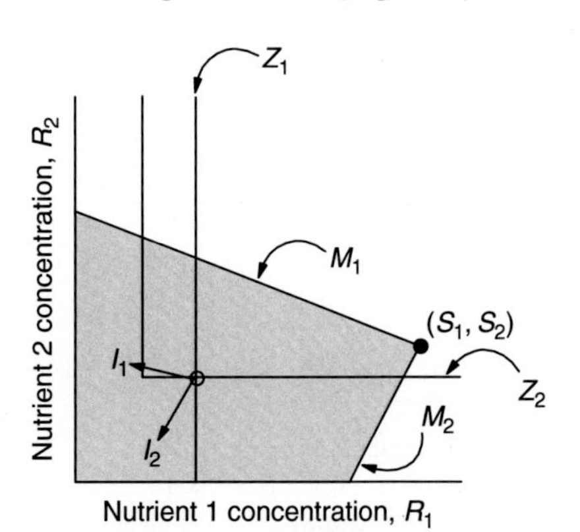

Figure 4 Competition and commensalism involving two populations. ZNGI graphs (indicated by Z_i) intersect at the circled point, making steady-state coexistence possible. The shaded feasible region for steady states is superimposed, bounded by mass-conservation constraints for populations 1 and 2 (indicated by M_i), as are impact vectors implying stable coexistence (indicated by I_i).

the bacteria is rectilinear, because inorganic nutrient and organic carbon are essential resources for them. The heterotrophic bacteria in aquatic ecosystems are usually better competitors for inorganic nutrients than are the algae, implying that the vertical limb of their ZNGI falls to the left of the algal ZNGI, an arrangement that makes coexistence possible.

The bacteria consume both of these resources, so their mass-conservation constraint and impact vector have slopes representing the organismal stoichiometry of the bacteria, as did these graphical constructs for the competition example presented above. But these constructs differ for the algae. Their mass-conservation constraint again passes through the supply point (S_1,S_2), but has a negative slope expressing the stoichiometry of organic carbon excretion in relation to inorganic nutrient consumption. The impact vector shares this slope, representing consumption of one resource, and production of another. The arrangement shown in **Figure 4** is stabilizing, because the underlying consumption and production dynamics amplify the negative effects that each population has on its own growth rate.

The supply point in **Figure 4** is drawn with the S_2 component positive to illustrate an abiotic supply of organic carbon in addition to algal excretion, because many inland and coastal aquatic ecosystems receive organic carbon washed off the nearby landscape. Offshore in the deep ocean, this supply point moves close to the axis for resource 1, the inorganic nutrient, because here production by resident algae is virtually the only source of organic carbon. Experiments to test this view of algal–bacterial interactions go the other direction by enriching aquatic microbial communities with organic carbon. This moves the supply point and the feasible region upon the resource plane, until the feasible region moves beyond the intersection of the ZNGI graphs and the bacteria are predicted to competitively exclude the algae, an outcome which has been observed.

Although many experiments with algae and bacteria appear consistent with the scenario sketched above, none permits a complete parametrization of the stoichiometric model of competition and commensalism. However, sufficient information is available for a similar interaction between yeast (*Saccharomyces cerevisiae*) and a bacterium (*Lactobacillus casei*) (**Figure 5**). The yeast consumed glucose while excreting the vitamin riboflavin, and the bacteria required both glucose and riboflavin as essential nutrients. Stable coexistence was observed experimentally, as predicted. Commensalism based on excreted substances is common in microorganisms, and theoretically tends to produce stable coexistence under broad conditions, suggesting that this interaction could be an important source of microbial diversity.

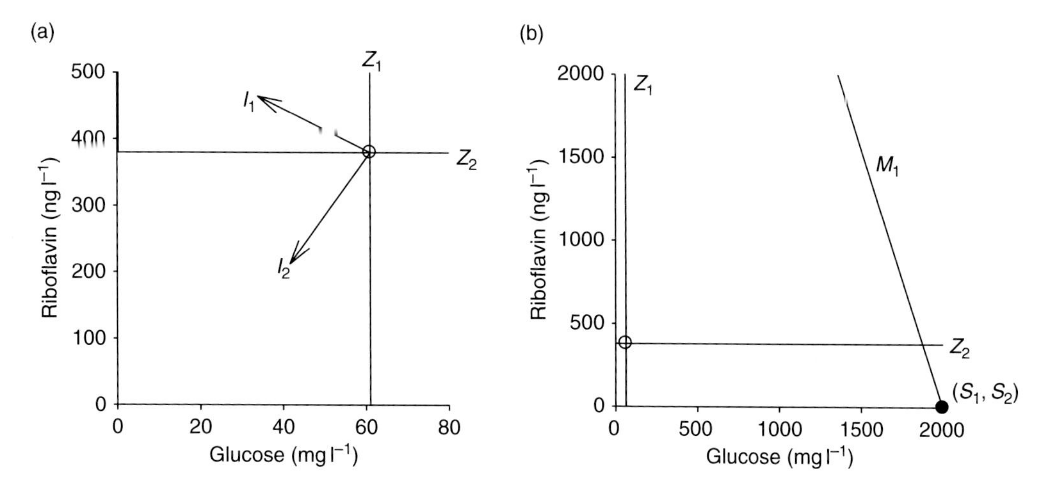

Figure 5 Competition and commensalism between a yeast, *Saccharomyces cerevisiae* (species 1) and a bacteriaum, *Lactobacillus casei* (species 2). Both species compete for glucose (resource 1), and the yeast produce riboflavin, which bacteria require as an essential nutrient (resource 2). (a) The ZNGI graphs (Z_i) and impact vectors (I_i) predict that stable coexistence is possible. (b) The supply point (S_1,S_2) and mass-conservation constraints (M_i) for the nutrient supply tested, for which stable coexistence was predicted and observed. Drawn from the data of Megee RD, Drake JF, Frederickson AG, and Tsuchiya HM (1971) Studies in intermicrobial symbiosis: *Saccharomyces cerevisiae* and *Lactobacillus casei*. *Canadian Journal of Microbiology* 18: 1733–1742.

Allelopathy

Allelopathy involves negative interactions mediated by substances released by the competing populations. There are many possible configurations of such interactions. A population can excrete a substance inhibitory only to a single competitor species, or one that is inhibitory to many species, including perhaps itself. In addition, the populations involved might compete for one or more nutrient substances. To illustrate the stoichiometric approach to allelopathic interactions, consider two populations competing for a nutrient that both require, while each population also excretes a substance that inhibits both populations. For example, some of the bacterial strains inhabiting mammalian guts might compete for one of the dietary nutrients present, while excreting a common and inhibitory metabolic by-product, such as hydrogen sulfide or a volatile fatty acid.

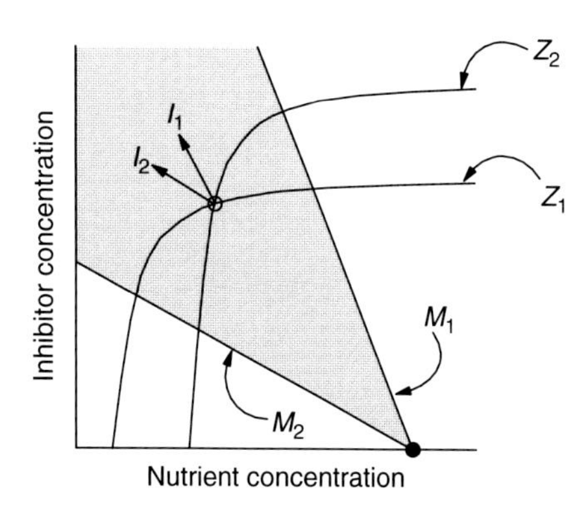

Figure 6 Allelopathy involving two populations that compete for one nutrient. ZNGI graphs (indicated by Z_i) intersect at the circled point, making steady-state coexistence possible. The shaded feasible region for steady states is superimposed, bounded by mass-conservation constraints for populations 1 and 2 (indicated by M_i), as are impact vectors implying stable coexistence (indicated by I_i).

On the plane of inhibitor concentration versus resource concentration, the ZNGI graphs of each population are now positively sloping curves (**Figure 6**). Increasing concentration of the inhibitor reduces the growth potential of a population, so to maintain a steady state the population must consume the nutrient resource more rapidly to restore its growth rate. When the ZNGI graphs of the populations intersect (as in **Figure 6**), steady-state coexistence is possible. Biologically, such an intersection arises from a tradeoff between the ability to compete for the nutrient in the absence of the inhibitor, and susceptibility to the inhibitory effects on growth rate.

As drawn in **Figure 6**, species 1 is a superior nutrient competitor but also more sensitive to the inhibitor. Feasibility of its coexistence with a more resistant, inferior competitor for the nutrient depends on mass-conservation constraints that express the stoichiometry of inhibitor production in relation to nutrient consumption for each population. These constraints are negatively sloping lines that pass through a supply point on the *R*-axis when the populations themselves are the only source of the inhibitor (**Figure 6**). As in other examples, feasibility of steady-state coexistence requires that these bounds enclose the intersection of the ZNGI graphs. Finally, the populations again have impact vectors displaying stability properties, and also expressing the stoichiometry of inhibitor production relative to nutrient consumption. When the more sensitive species produces relatively more of the inhibitor, its coexistence with the resistant species is stable because each species then has a stronger negative impact on its own growth rate than that of the other species.

Mutualism

Mutualism has been somewhat neglected by ecologists, compared to competition, but stoichiometric approaches again apply, at least for mutualisms based on production of substances beneficial to growth of another population. Consider two populations in a symmetrical mutualism where each population consumes a resource produced by the other population: species 1 consumes resource 1 and produces resource 2, while species 2 consumes resource 2 and produces resource 1. Assume also that the two resources are also supplied by other processes.

The graphical display of this interaction follows a now familiar path: plot ZNGI graphs, mass-conservation constraints, and impact vectors. For each population, its ZNGI graph is a line parallel to the axis for the resource it produces but does not require (**Figure 7**), an arrangement that is guaranteed to make coexistence possible. For each population, its mass-conservation constraint expresses the stoichiometry of the resource it produces relative to the resource it consumes, and steady-state coexistence is feasible when these bounds enclose the intersection of the ZNGI graphs. For each population, its impact vector expresses the same stoichiometry of production and consumption, and stability in this example is guaranteed because each population has a negative impact on its own growth through consumption of its own required resource.

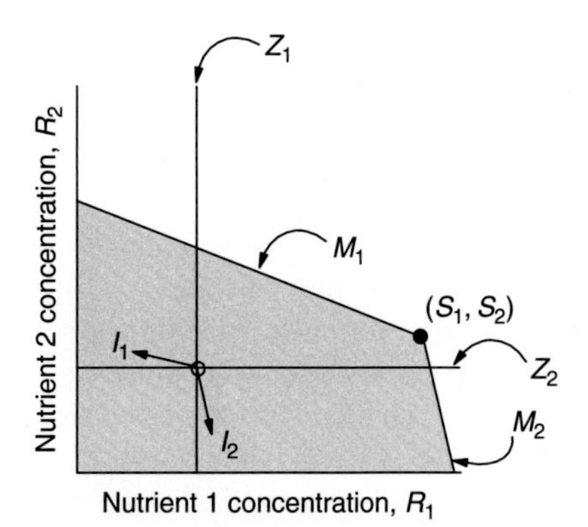

Figure 7 A symmetric mutualism involving two populations. ZNGI graphs (indicated by Z_i) intersect at the circled point, making steady-state coexistence possible. The shaded feasible region for steady states is superimposed, bounded by mass-conservation constraints for populations 1 and 2 (indicated by M_i), as are impact vectors implying stable coexistence (indicated by I_i).

Predator–Prey Interactions

Predator–prey interactions, especially between planktonic herbivores and algae, have been studied extensively from the perspective of ecological stoichiometry. Therefore, this presentation will emphasize only those principles strongly related to those elaborated above for other interactions. There is a close analogy between a population that produces a substance inhibitory to itself and other populations, and one that supports a predator. In either case, the population produces something that decreases its net potential to grow, and as a result the population must consume more of whatever resource(s) limit its growth. The stoichiometry relating production of the growth-reducing factor to consumption of the resource then has profound effects on how a focal population interacts with others.

Consider a prey population (N) whose growth is limited by a nutrient resource (R) and which supports a predator population (P). In a closed environment where recycling of nutrients is the only supply process, a simple mass-conservation constraint applies to this interaction. This constraint partitions total nutrient (S) into free nutrient, nutrient bound in prey, and bound in predators:

$$S = R + NQ + Pq \qquad [7]$$

where Q is the nutrient content of prey and q the nutrient content of predators. It is often convenient to measure prey and predator populations as biomass densities of carbon, so that Q and q are nutrient:carbon ratios.

The dynamics of this predator–prey interaction can be conveniently represented on the RP-plane (**Figure 8**). The ZNGI graph for the prey population is a forward sloping curve: as predator density and hence mortality increase, maintaining a steady-state population of prey, requires more nutrient consumption to support higher reproduction. Given certain simplifying assumptions about predation, the steady-state prey density, N^*, is fixed. Then, at steady state, the mass-conservation constraint (eqn [7]) plots as a straight

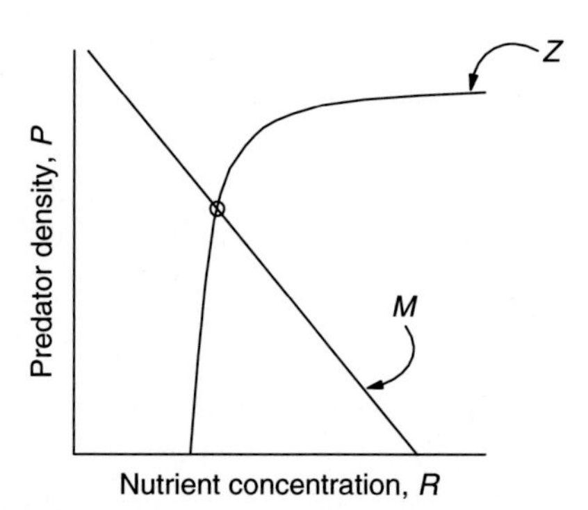

Figure 8 A predator–prey interaction. The ZNGI graph of the prey population (Z) and the mass-conservation constraint (M) intersects at the steady state.

line with elevation proportional to the total nutrient supply, and negative slope equal to the predator's organismal carbon:nutrient ratio. The intersection of the ZNGI graph and this line shows the steady-state nutrient concentration and predator density. These quantities are thus predicted to increase with increasing nutrient supply while prey density remains constant.

Adding another prey species to this system, which depends on the same nutrient resource and is preyed upon by the same predator, generates a potentially complex interaction. The two prey populations compete by consuming the nutrient, but also engage in apparent competition by supporting the predator which attacks their competitor. Intersection of the ZNGI graphs for the two prey populations implies that coexistence of all three species is possible, and requires a tradeoff in which one prey species is a superior nutrient competitor in the absence of the predator, but is also more susceptible to predation.

In **Figure 9**, prey species 1 plays this role, and additional conditions for feasibility and stability of steady-state coexistence are shown. The feasible region for steady-state coexistence lies between bounds derived from the mass-conservation constraints associated with each prey species. Each prey species also has impact vectors related to dynamical stability, but in contrast to interactions examined previously, these vectors are not parallel to the mass-conservation constraints. Slopes of the latter express only the predator's organismal stoichiometry, but slopes of the impact vectors are also determined by the predator's conversion efficiencies when consuming each prey species. Stability arises when, as illustrated, the more susceptible prey population (species 1) supports greater predator productivity in proportion to prey nutrient consumption, a relationship depending on both stoichiometry and trophic efficiency.

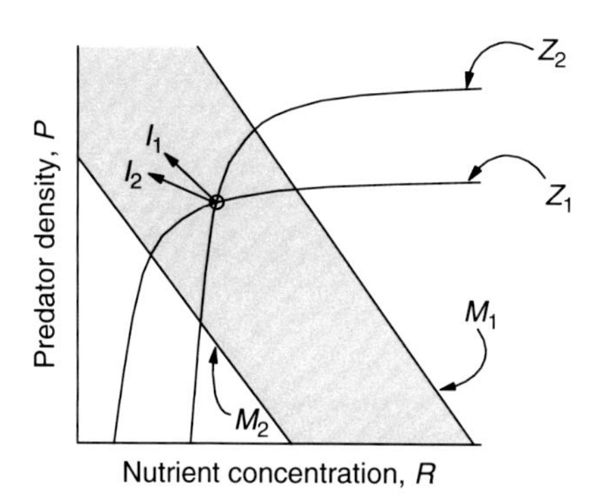

Figure 9 Two prey populations that share a predator and compete for a nutrient. ZNGI graphs (indicated by Z_i) intersect at the circled point, making steady-state coexistence possible. The shaded feasible region for steady states is superimposed, bounded by mass-conservation constraints for populations 1 and 2 (indicated by M_i), as are impact vectors implying stable coexistence (indicated by I_i).

Stable coexistence of two competing prey as illustrated here cannot occur if we take away the predator, because then the prey compete for one resource, and the competitor with the lower R^* drives the other to exclusion. A predator that mediates coexistence of its competing prey species, increasing the diversity of the prey community, is called a keystone predator. Although many observations suggest that keystone predators are common in ecological communities, few studies have sufficient information to construct and test fully parametrized models. An exception involves the phage viruses that prey on bacteria (**Figure 10**).

Community Assembly

Graphical models of small numbers of interacting species at steady state might seem to offer limited insights for ecological communities composed of many more species. However, the contributions of the theory presented above become clearer in the contexts of community assembly, succession, and metacommunities. These related concepts recognize that many communities begin their history in sparsely populated habitats created by a major

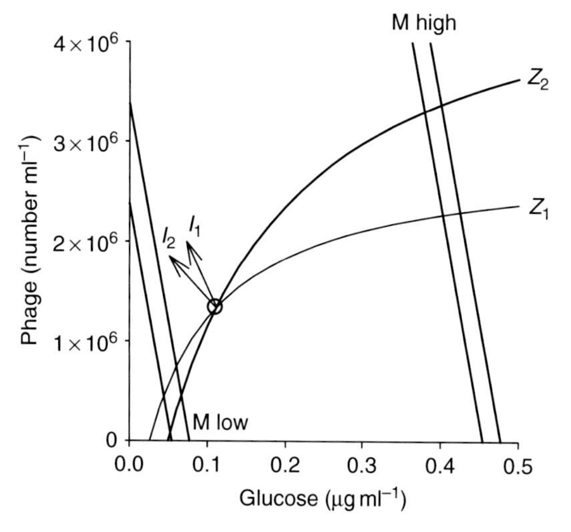

Figure 10 Competition for glucose between two strains of *Escherichia coli*, one more susceptible to predation by phage (strain 1), and one less susceptible (strain 2). The ZNGI graphs (Z_i) and impact vectors (I_i) predict that stable coexistence is possible. The mass conservation constraints labeled 'M low' represent one of the nutrient supply conditions tested, for which exclusion of the less susceptible strain was predicted, while the mass conservation constraints labeled 'M high' represent the other nutrient supply condition tested, for which exclusion of the more susceptible strain was predicted. Trends toward these predictions were observed, but were often slower than predicted, probably due to adaptive evolution of the inferior strain during the experiments. Adapted from Bohannan BJM and Lenski RE (2000) The relative importance of competition and predation varies with productivity in a model community. *American Naturalist* 156: 329–340, with permission from University of Chicago Press.© 2000 by the University of Chicago Press.

disturbance, and are colonized over time by species arriving from established communities in similar habitats. As this succession proceeds, species are sorted into those that persist locally, and those that do not.

Ecologists have long recognized that inhibitory and facilitative interactions such as competition, commensalism, and mutualism play a part in this sorting. Stoichiometric theory and related approaches can identify properties of the species that ultimately persist. Basically, species that coexist stably must have tradeoffs in their abilities to use different resources, between resources that they require and the substances they produce as resources or toxins for other species, or between abilities to use resources and withstand predation or allelopathy. Stable coexistence based on the right tradeoffs and the right stoichiometry imparts a measure of determinism to local community assembly and species composition. As a result, nonrandom patterns are predicted in the species properties summarized by their ZNGI graphs, for example, and in their distributions in relation to environmental gradients of resources.

Stable coexistence is not the only possible consequence of the interactions presented above. Especially for predation and allelopathy, scenarios consistent with priority effects are plausible. The stoichiometric relationships that impart stability are not biologically necessary, permitting destabilization leading to priority effects. When priority effects characterize interspecific interactions, community assembly becomes stochastic. There are multiple endpoints, that is, alternative stable states, and long-term species composition depends on the order in which colonists of different species arrive. Even so, on the larger scale of the metacommunity – the collection of similar communities linked by dispersal and colonization – it is possible that patterns with stoichiometric signatures emerge. When examining a productivity gradient with increasing supplies of nutrient resources, priority effects in the interactions discussed above would make species composition in more productive local communities more stochastic, and reduce compositional similarity among localities. Higher species diversity would also be associated with higher productivity at the metacommunity scale, while local communities would display a peak in diversity at intermediate productivity.

Physiological Variability in Stoichiometry

The theory sketched above treats each population's stoichiometry as fixed. This is often a reasonable approximation for animals that exhibit homeostatic composition, but autotrophs typically have highly variable composition. When a population's stoichiometry varies physiologically, the stoichiometric quantities Q_{ij} are variables rather than constants. Their dynamics can be described by equations relating their time derivatives to the consumption, release, and use in net production of the nutrient in question. In a steady-state situation, these process balance, and the stoichiometric quantities Q_{ij} are constant. As a result, algebraic constraints representing mass conservation apply, and eqns [3] and [4] apply as steady-state approximations. The rigor of graphical analyses is somewhat diminished, but the key insights remain. In particular, graphical theory has been applied very successfully to algae competing for dissolved nutrients, despite demonstrations of potentially large physiological variations in the competitors' stoichiometry.

Physiological variation in stoichiometry can be an important determinant of a population's ability to compete for a nutrient, however. The growth rate of a population generally increases with organismal nutrient content, that is, with Q for that nutrient. At some low value Q_{min}, population growth ceases, and as Q becomes very large, growth is maximal. Consider two populations that differ in Q_{min}, but are otherwise equivalent (**Figure 11**). When population growth balances mortality, the population with lower Q_{min} will have a lower steady-state quota. This in turn reduces the rate of nutrient consumption required for growth that balances mortality. The rate of nutrient consumption generally increases with nutrient concentration, so reducing this rate ultimately lowers steady-state nutrient concentration required by the species with the lower Q_{min}; that is, the quantity R^* is reduced, making this species a better competitor.

Now, if one species evolves a lower Q_{min} to become better for one nutrient, but owing to physiological constraints it then requires greater amounts of another nutrient, Q_{min} for that nutrient rises and this species becomes a worse competitor for it. In the graphical theory of competition for two resources (**Figure 2**), this makes intersections of ZNGI graphs likely for different species, and also imparts stabilizing slopes to their impact vectors. Thus, selection on the ability to compete for one resource

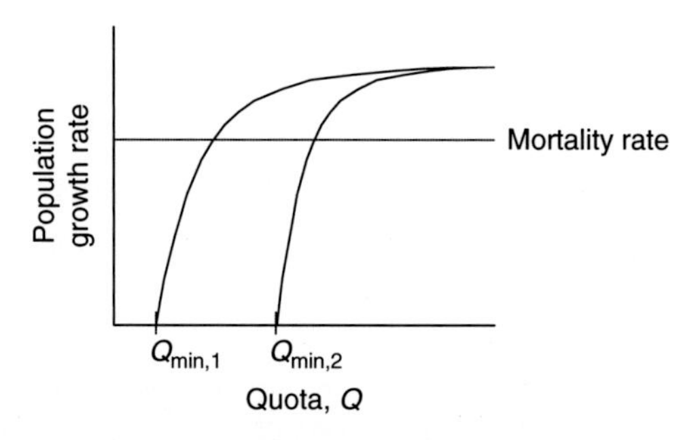

Figure 11 Two populations with physiologically variable stoichiometry that differ only in minimal nutrient quota (Q_{min}). Each population's growth rate is an increasing function of nutrient quota, and at steady state each must grow at a rate balancing mortality.

likely produces the tradeoffs required for stable coexistence of populations competing for two resources. Experimentally, it has proven relatively easy to construct stable coexistence of algae competing for two resources, suggesting that the underlying tradeoffs are common for this group of autotrophs.

It is likely that physiological variations in stoichiometry contribute to a population's ability to compete for a nutrient in variable environments, as well as those at steady state. Often, there is a physiological limit on the quota for a nutrient, Q_{max}, and the ratio of Q_{max}:Q_{min} then measures the capability to store the nutrient. Theoretical studies suggest that populations with high storage capability are favored when competition occurs in variable environments characterized by large, infrequent pulses of nutrient concentration. Such storage specialists can dominate even when they suffer disadvantages in their rates of nutrient consumption and population growth, compared to other species.

The physiological variation in stoichiometry characteristic of autotrophs plays many roles in their interactions with herbivores. Interestingly, it can theoretically permit stable coexistence of two herbivore species that compete by exploiting the same autotroph population. This result arises in mathematical models of two herbivore species consuming one algal species, whose nutrient:carbon ratio varies physiologically in response to the supply of the nutrient relative to the light intensity that supports photosynthetic carbon fixation. As is commonly observed for aquatic herbivores, their nutrient composition is homeostatic, and due to the varying composition of their food, the growth of herbivore populations can be either carbon- or nutrient-limited.

The steady states of such an interaction can be analyzed graphically through an adaptation of **Figure 2**. The resource axes represent the biomass densities of carbon and nutrient embodied in the algal population. Feasibility bounds and impact vectors arise in part from mass-conservation constraints, but also from other feedbacks in the interaction. When herbivores feed on algae whose composition does not meet their stoichiometric requirement, recycling of nutrient occurs, altering the supply experienced by algae and thus their nutrient:carbon ratio. Remarkably, theory predicts that these complex processes can stabilize with the algal nutrient:carbon ratio poised at a value permitting coexistence of both herbivores. In such a steady state, growth of one herbivore population is carbon-limited and that of the other is nutrient-limited. Representing this coexistence graphically requires some simplifying assumptions, but relaxing these does not change the essential prediction that two herbivores can coexist while exploiting one autotroph population. The surprise in this result is that such coexistence occurs due to stoichiometry, without any of the factors classically proposed to explain herbivore coexistence, such as spatial processes or differentiation of the autotroph individuals into various physical structures (e.g., leaves, stems, etc.).

Extensions

Most theoretical and experimental work on stoichiometry and population interactions deals with situations where growth of the focal populations depends on two nutrient resources, or one nutrient resource and another factor such as a predator. Some progress has been made in understanding situations where two populations compete for two nutrients, and a predator attacks both of them while recycling a proportion of one or both nutrients. Such a scenario is common for planktonic algae, where zooplankton herbivores recycle dissolved nitrogen and phosphorus required by the algae. Without the predator, stable coexistence of two competitors on two resources is likely, in a range of nutrient supplies that make this feasible (e.g., **Figure 2**). In some theoretical scenarios, the predator merely shifts the range of nutrient supplies where such stable coexistence is feasible. In other cases, the predator destabilizes coexistence, so that priority effects occur. Another theoretical possibility is destabilization resulting in persistent periodic or chaotic dynamics.

When the number of nutrient resources increases to three, dynamical complexity is a possibility even in the absence of predators. If the stoichiometry of consumption for the three nutrients follows the right relationships among competing population, periodic dynamics or chaotic dynamics can arise with more than three populations persisting.

Summary

Stoichiometry integrates readily into mechanistic approaches to species interactions that have become popular among ecologists in recent decades. Mechanisms of interaction often involve consumption and transformation of materials, processes that must follow mass-conservation laws. As anticipated by Lotka, mass conservation often constrains dynamics and simplifies analysis in theoretical ecology. Beyond convenience, stoichiometry provides insights into how the physiology and metabolism of consumption and transformation scale up to population and community levels. Ultimately, stoichiometry helps ecologists to understand better the range of mechanisms contributing to population persistence and community diversity.

See also: Ecological Stoichiometry: Overview; Ecosystem Patterns and Processes; Evolutionary and Biochemical Aspects; Organisal Ecophysiology; Trace Elements.

Further Reading

Bohannan BJM and Lenski RE (2000) The relative importance of competition and predation varies with productivity in a model community. *American Naturalist* 156: 329–340.

Bratbak G and Thingstad TF (1985) Phytoplankton–bacteria interactions: An apparent paradox? Analysis of a model system with both competition and commensalism. *Marine Ecology – Progress Series* 25: 23–30.

Butler GJ and Wolkowicz GSK (1987) Exploitative competition in a chemostat for two complementary, and possibly inhibitory, resources. *Mathematical Biosciences* 83: 1–48.

Chase JM and Leibold MA (2003) *Ecological Niches*. Chicago, Illinois: Chicago University Press.

DeFreitas MJ and Frederickson AG (1978) Inhibition as a factor in the maintenance of the diversity of microbial ecosystems. *Journal of General Microbiology* 106: 307–320.

Grover JP (1997) *Resource Competition*. London: Chapman and Hall.

Grover JP (2002) Stoichiometry, herbivory and competition for nutrients: Simple models based on planktonic ecosystems. *Journal of Theoretical Biology* 214: 599–618.

Grover JP (2004) Predation, competition and nutrient recycling: A stoichiometric approach with multiple nutrients. *Journal of Theoretical Biology* 229: 31–43.

Hall SR (2004) Stoichiometrically explicit competition between grazers: Species replacement, coexistence, and priority effects along resource supply gradients. *American Naturalist* 164: 157–172.

Holt RD, Grover P, and Tilman D (1994) Simple rules for interspecific dominance in systems with exploitative and apparent competition. *American Naturalist* 144: 741–771.

Huisman J and Weissing FJ (1999) Biodiversity of plankton by species oscillations and chaos. *Nature* 402: 407–410.

Loladze I, Kuang Y, Elser JJ, and Fagan WF (2004) Competition and stoichiometry: Coexistence of two predators on one prey. *Theoretical Population Biology* 65: 1–15.

Lotka AJ (1925) *Elements of Physical Biology*. Baltimore, MD: Williams & Wilkins.

Megee RD, Drake JF, Frederickson AG, and Tsuchiya HM (1971) Studies in intermicrobial symbiosis: *Saccharomyces cerevisiae* and *Lactobacillus casei*. *Canadian Journal of Microbiology* 18: 1733–1742.

Miller TE, Burns JH, Munguia P, *et al.* (2005). A critical review of twenty years' use of the resource-ratio theory. *American Naturalist* 165: 439–448.

Sterner RW and Elser JJ (2002) *Ecological Stoichiometry*. Princeton, NJ: Princeton University Press.

Tilman D (1981) Tests of resource competition theory using four species of Lake Michigan algae. *Ecology* 62: 802–815.

Tilman D (1982) *Resource Competition and Community Structure*. Princeton, NJ: Princeton University Press.

Trace Elements

A Quigg, Texas A&M University at Galveston, Galveston, TX, USA

Introduction

Of the 90 or so natural elements in the periodic table, life has chosen about 20 to perform essential structural and metabolic functions. Carbon (C), hydrogen (H), oxygen (O), nitrogen (N), phosphorus (P), magnesium (Mg), sodium (Na), copper (Cu), iron (Fe), manganese (Mn), cobalt (Co), zinc (Zn), and molybdenum (Mo) are required by all aquatic organisms, while sulphur (S), potassium (K), and calcium (Ca) are required but can be partially replaced (e.g., Ca by strontium (Sr)). Because of selective preferences by some but not all aquatic organisms, the following elements also commonly occur in cells: vanadium (V), selenium (Se), silica (Si), boron (B), chromium (Cr), iodine (I), bromide (Br), fluoride (F), aluminum (Al), cadium (Cd), lead (Pb), and rubidium (Rb). However, the mere presence of an element in an organism is not evidence that is required. Absolute requirement for an element necessitates that

1. a lack of the element makes it impossible for the organism to grow or reproduce;
2. it cannot be replaced by another element; and
3. the element's effect is direct, and not due to the interaction with another element, with associated epiflora, or the like.

Trace elements are those required in minute quantities (typically $<0.01\%$ of the organism or its environment) to maintain proper physiological function. They are used in redox reactions (e.g., Fe, Mn, Cu), acid–base catalysis (e.g., Zn, nickel (Ni)), transmission and storage of information and energy (K, Ca), and in structural cross-links (e.g., S, Si). The characteristic nature of their evolved uses reflects a refined selection of elements within biology.

While this selection may have been biased from life's beginnings by abundance, later energy costs and functional advantages would have become more important. Physiological role(s) for many trace elements remain undefined (e.g., Cr in microalgae and Br and I in macroalgae). Others, such as arsenic (As), mercury (Hg), and Pb, are nonselectively accumulated and have no known functions. A variety of biotic and abiotic factors affect trace element accumulation in organisms, such as organism size, sexual maturity, season, feeding habits, trophic position, water quality, and environmental contamination.

Essential Trace Elements

The most important trace elements in all organisms are Fe (major electron transfer agent), Mn (O_2 release), Mo (N uptake), and Zn (major acid catalyst). Major roles for these and other trace elements are summarized in **Table 1**, and briefly below. They are found in association with enzymes and proteins, nucleic acids, and a plethora of other biologically critical molecules.

Iron

Cell metabolism and growth are dependent on iron because it is at the center of cytochromes and ferredoxin in the respiratory and photosynthetic metabolic reactions, respectively. Iron is also the most common metallic component of membranes, usually present as Fe/S proteins. It is required for nitrate, nitrite, and sulfate reduction and for nitrogen fixation. Fe acts as an acid catalyst in hydrolytic enzymes, oxidation–reduction catalysis, as well as acid–base reactions and much more (see **Table 1**). Its unique chemical properties – the ability to change valance between Fe^{2+} and Fe^{3+} – allows the reduced ion to form stable complexes facilitating electron transfer, thereby making it well suited for many biological reactions (**Table 1**).

While Fe is not part of chlorophyll, iron deficiency leads to a chlorotic effect in phototrophs because it is required as a cofactor in pigment synthesis. Nitrogen metabolism is also extremely sensitive to Fe stress (limitation) because many of the major enzymes of this pathway contain Fe. Cyanobacteria and chlorophytes can avoid Fe limitation by replacing Fe-containing ferredoxin and cyctochrome c_6 with iron-free flavodoxin and plastocyanin, respectively. Mollusks and crustaceans use the Cu-based hemocyanin while other invertebrates possess the iron-based version of this respiratory protein. There are likely to be other replacement strategies for this element.

Manganese

Mn plays a vital role in the oxygen-evolving complex of photosynthesis; four atoms are required to split water. Mn(II) forms strong nitrogen (N) and S ligands and is a cofactor in several Krebs-cycle enzymes. Mn(III) is used in superoxide dismutases, acid phosphatases, and ribonucleotide reductases and in transfers across membranes (**Table 1**). Evolutionarily, Mn was thought to be selected for some functions because of its greater overall availability (reduced and oxidized) relative to Fe, rather than because of its unique chemistry.

Zinc

Zn is an essential element with a long list of important metabolic functions (over 150 enzymes are known), including protein and nucleic acid synthesis enzymes, the activator of several dehydrogenases, and a cofactor in superoxide dismutase. **Table 1** lists the most common ones. Without Zn, chlorophyll and phycobilin production may be hindered in phototrophs. Zn can bind to the same centers as Fe but it does not exhibit any redox chemistry. This makes Zn a useful catalyst center and so it is typically employed in signaling systems. Zn plays a critical role in the enzyme carbonic anhydrase, one of the most catalytically active enzymes known across plant and animal kingdoms. All organisms rely on Zn as a critical structural element in the 'zinc finger' thiolates that control DNA expression.

Copper

Cu has three accessible oxidation states (I, II, III) in biological systems, giving it a redox potential range from +0.2 to 0.8 V. Cu is present in the photosynthetic electron-transport chain in the metalloprotein plastocyanin (**Table 1**). Cytochrome oxidases in the respiratory chain require two Cu ions. Unlike Fe, Mn, and Zn, copper is toxic if overaccumulated. Its internal concentration is tightly regulated (homeostasis) by many ingenious cellular processes. Cells will internalize Cu to vesicles or externalize it on cell membranes where it plays an important role as a redox element, replacing Fe as a component of cell surface oxidases. High Cu efflux rates have been measured within seconds of elevated Cu exposures in phytoplankton and some animals.

Cobalt, Nickel, Selenium, and Molybdenum

Little is known about the precise catalytic activity of nickel and cobalt in biology. The redox functions of Ni are replaceable by Fe, Cu, or Mn while those of Co may be replaceable by Zn (and some evidence suggests Cd). Nickel acts as the cofactor for urease (**Table 1**), which catalyzes the hydrolysis of urea and ammonium. Cobalt is the metallocenter for vitamin B12. The specialized roles for these elements do not explain their cellular concentrations or stoichiometric relationships with other trace elements or nutrients.

Table 1 Major trace elements, functions and examples of uses in cells

Element	*Probable functions*	*Examples of uses*
Cadmium[a]	May substitute into enzymes that typically use Zn	[a]Carbonic anydrase; [b]alkaline phosphatase (?)
Cobalt	Core component of vitamin B12 C and H transfer reactions with glycols and ribose Rearrangements and reduction reactions	Vitamin B12
Chromium[c] (trivalent form)	[a]Maintain normal blood-sugar metabolism, acting as partner to insulin; enzyme activation, food metabolism, and cholesterol regulation; helps cells (e.g., in the heart) absorb energy	[c]Insulin pathways (fat storage and carbohydrate metabolism)
Copper	Mitochondrial electron transport Laccase, oxidases including amine oxidase [b]Electron transport in photosynthesis and respiration Disproportionation of O_2 radicals to O_2 and H_2O_2 [c]Production of red blood cells, involved in maintenance of cardiovascular and skeletal systems	Cyctochrome oxidase [b]Plastocyanin and cyctochrome *c* oxidase, Superoxide dismutase (alternatively with Fe or Mn) [c]Works with vitamin C in the production of collagen and elastin
Iron	Active groups in porphyrin molecules and enzymes; component of cytochromes and certain non-heme iron proteins and a cofactor for some enzymatic reactions Reactions with oxygen and nitrogen – e.g., N_2 fixation, N uptake, photosystems, blood [b]Cyctochrome oxidase [c]Cytochrome P450	Nitrate reductase, nitrite reductase, catalase, cytochromes (e.g., *b* and *c* for electron transport in photosynthesis and respiration; *f* for photosynthesis electron transport) Ferredoxin, part of hemoglobin [b]Reduction of O_2 to water [c]O-insertion from O_2, detoxification
Manganese	[b]Electron transport in PSII, maintenance of chloroplast membrane structure Breakdown reactions and those involving halogens [c]Supports the immune system, regulates blood sugar levels, is involved in the production of energy and cell reproduction; important for bone growth	[b]O_2 evolving complex Catalases, peroxidases (alternatively with Fe, Se, or V) [c]Works with vitamin K to support blood clotting; with the B-complex vitamins to control the effects of stress
Molybdenum	Nitrogen reduction (nitrate and nitrite reduction to ammonium), ion absorption	Nitrate and nitrite reductase [d]Nitrogenase enzyme
Nickel	[b]Hydrolysis of urea [c]Cofactor in enzymes	[b]Urease
Selenium	[b]Cofactor in enzymes [c]Protects blood cells from certain damaging chemicals; together with vitamin E, helps immune system produce antibodies; makes tissues elastic	[b]Glutathione peroxidase which acts to detoxify lipid peroxides and maintain membrane integrity
Vanadium	[c]Controls blood sugar levels; assists in the development of bones and teeth; increases liver glycogen (stored glucose) and improves utilization of glucose by muscle tissues; stabilize the body's insulin production	[c]Glucose
Zinc	Enzymes – *c.* 150 known Nucleic acid replication and polymerization Hydration and dehydration of CO_2 Hydrolysis of phosphate esters [c]The creation, release and use of hormones >100 reactions in the body; reactions that construct and maintain DNA; needed for growth and repair of tissues specifically connective tissue; teeth, bones, nails, skin; healthy immune system; developing fetus and brain's function	 DNA and RNA polymerases, structures carbonic anhydrase Alkaline phosphatase [c]Peptidases

[a]While definitely believed not to be essential for plant and animal life processes, there is some evidence of a role in phytoplankton.
[b]Plant cells.
[c]Animal cells.
[d]Bacterial cells.

Experiments in which Se was omitted from growth media have invariably shown that it is required by various micro- and macroalgae. Glutathione peroxidase is a Se-containing enzyme that occurs in the mitochondria (and chloroplasts of phototrophs) where it acts to detoxify lipid peroxides, which are potentially dangerous to cellular function, and to maintain membrane integrity. Selenium, like S, has advantages over trace metals in that it transfers redox equivalents openly in the cytoplasm without generating free radicals. Evolutionary selection of this element may primarily reflect this property of avoiding the risks associated with free radicals.

Mo is most important in microbial N fixation, as it is a component of enzymes in the nitrogen reduction and fixation processes (**Table 1**). In general, marine plants (seaweed and phytoplankton) have lower amounts of Mo than their freshwater relatives even though the concentration of dissolved Mo is about 20 times greater in seawater than in freshwater. It has been shown that sulfate inhibits Mo assimilation by microalgae, making Mo less available in seawater than freshwater because of differences in sulfate distributions in these water types. Therefore, N assimilation may require greater energy expenditure in seawater than freshwater. Mo also catalyzes many multielectron oxidation–reduction electron-transfer reactions such as SO_4^{2-} to SO_3^{2-} and –CHO to –COOH. Nitrate reductase used ubiquitously by all organisms in N metabolism; it has a Mo metallocenter.

Stoichiometric Relationships between Essential Trace Elements

Trace element concentrations ($\mu g\,g^{-1}$ dry weight) in microalgae (marine and freshwater), seagrasses and macroalgae (the most extensively examined of all organisms with respect to metal content), invertebtrates (e.g., plankton, mussels, oysters, shrimp), and vertebrates (e.g., fish, seabirds, seals, turtles, dolphins, whales) growing in healthy environments have been measured, allowing a broad-scale consideration of their stoichiometric ratios (**Table 2**).

Stoichiometrically, the most abundant trace element on Earth is iron (**Figure 1**). Collectively, microalgae, seagrasses, and invertebrates have the lowest Fe quotas (median 185 $\mu g\,g^{-1}$ dry weight), while vertebrates have 2.5 times this amount (**Table 2**). Macroalgae have 10 times more Fe than all the other aquatic organisms examined (**Figure 1**), reflecting their ability to accumulate trace metals to concentrations thousands of times higher than in seawater (**Table 2**). This, however, does not necessarily reflect higher elemental requirements, as the cell wall of macroalgae consists of a variety of polysaccharides and proteins, some of them containing anionic carboxyl, sulfate, or phosphate groups that are excellent binding sites for metal retention. Binding of metals by macroalgae has been shown to be strong, with only a minimal exchange between bound metals and ambient water.

Manganese and zinc are required (accumulated) in high concentrations second only to iron across all organisms examined (**Figure 1**). While phototrophs (micro- and macroalgae, seagrasses) have higher Mn contents than animals (about 20 times), Zn was present ubiquitously in relatively high concentrations in all organisms (median 40 $\mu g\,g^{-1}$ dry weight). Differences between organisms in terms of cellular concentrations of Cu, Co, Ni, Se, and Mo (**Figure 1**; **Table 2**) are not readily explained given the current state of knowledge of the roles and behavior of these elements. While there is a developing understanding of antagonistic and/or synergistic relationships among trace elements in aquatic environments, how different organisms respond to and interact with bioavailable trace elements in various combinations is less well understood.

In the early 1940s, Arthur Redfield published a stoichiometric relationship of macroelements in microalgae (C:N:P = 106:16:1) which he found to have a similar composition to seawater (C:N:P = 105:15:1). The Redfield ratio is a capstone concept in biological oceanography, used to interconnect the nutritional status of microalgae and their environment. It should be possible to define an extended Redfield ratio that includes trace elements. Currently data exist to do so for a bacterium, some phototrophs, but not for animals. Few studies measure C, N, or P along with trace elements. One of the greatest hurdles in defining the ecological stoichiometry of trace elements across all organisms has been the development of instruments with sufficient sensitivity and precision. Simply by definition, trace elements have been difficult to measure at biologically and environmentally appropriate concentrations. The technology to measure ultra-trace amounts of elements in single cells has only recently become available. The development of 'clean' techniques several decades ago for sample collection, preparation, and processing was a critical but relatively recent step forward by trace elemental chemists. In any case, based on available data, an extended Redfield ratio might look like the equation below where X represents the average concentration for a particular organism:

$$\begin{aligned}&\mathrm{C}\,(500-100):\ \mathrm{N}\,(30-15):\mathrm{P}\,(1):\\&\quad \mathrm{Fe, Mn, Zn}\ (X/10^{3}):\mathrm{Cu, Co, Ni, Se, Mo}\ (X/10^{6-9}):\\&\quad \mathrm{others}\ (X/10^{12-21})\end{aligned}$$

This is consistent with the mean (and median) trace elemental concentrations provided in **Table 2** and earlier multielemental studies on organisms (e.g., bacteria and microalgae). Fe, Mn, and Zn are required (accumulated) by all organisms, at concentrations at least 1–2 orders of magnitude higher than those for the other essential elements (Cu, Co, Ni, Se, Mo), and several orders of

Table 2 Summary of trace metal concentrations (μg g^{-1}, dry weight) in different subgroups of aquatic organisms. The average of *n* samples is included, along with the median and the range of values reported for organisms growing in pristine conditions

					Invertebrates			*Vertebrates*		
Element	*Concentration* (*μg g^{-1}*, **dry wt.**)	*Microalgae*	*Seagrasses*	*Macroalgae*	*Mollusks*	*Crustacea*	*Polychaete worms*	*Fish*	*Seabirds*	*Mammals*
As	Mean (*n*)	0.76 (*3*)	1.4 (*3*)	47 (*34*)	3.87 (*5*)	8 (*6*)	5.8 (*1*)	5.2 (*8*)	2.56 (*8*)	0.3 (*1*)
	Median	0.33	1.2	15.3	2.65	5.2		4.7	2.45	
	Range	0.2–1.74	1.04–1.97	0.97–230	1.2–8.4	1.1–27		1.2–11	0.48–5.4	
Cd	Mean (*n*)	0.43 (*3*)	1.58 (*11*)	0.66 (*50*)	2.30 (*15*)	1.84 (*10*)	3.2 (*1*)	0.41 (*10*)	0.44 (*8*)	0.58 (*12*)
	Median	0.29	0.57	0.36	1.12	1.5		0.05	0.41	0.2
	Range	0.21–0.78	0.16–10	0.05–10.4	0.03–14	0.83–7.15		0.001–29	0.08–0.54	0.1–1.1
Co	Mean (*n*)			28 (*65*)	0.64 (*3*)	2.78 (*4*)		0.1 (*1*)		
	Meadin			1.55	0.70	0.82				
	Range			0.24–389	0.13–1.08	0.13–9.4				
Cr	Mean (*n*)		1.5 (*1*)	26 (*81*)	1.49 (*9*)	1.25 (*4*)	16 (*3*)	1.06 (*5*)		
	Median			4.26	0.85	0.79	12.9	0.67		
	Range			0.44–775	0.3–7.45	0.2–3.2	11–24	0.23–2.2		
Cu	Mean (*n*)	0.84 (*1*)	7.96 (*13*)	4.8 (*50*)	17 (*16*)	49 (*10*)	11.2 (*4*)	7.1 (*11*)	6.37 (*8*)	4.9 (*13*)
	Median		6.7	4.1	8.14	27	11.2	1.96	6.32	5
	Range		3.9–17.3	0.2–15.2	0.83–61	1.6–77	8.15–14	0.27–48	5.17–7.43	1.1–7.8
Fe	Mean (*n*)		228 (*2*)	2950 (*67*)	100 (*1*)	594 (*2*)		135 (*2*)		692 (*2*)
	Median		228	2109		594		135		692
	Range		169–287	23.6–22725		217–970		21–248		485–900
Hg[a]	Mean (*n*)		0.003 (*11*)		0.011 (*1*)	0.017 (*3*)		0.36 (*2*)		12 (*3*)
	Median					0.02		0.36		5.6
	Range					0.006–0.025		0.04–0.69		0.7–30
Mn	Mean(n)	16.4(*3*)	176 (*2*)	104 (*66*)	12.4 (*3*)	4.24 (*4*)	80 (*3*)	17.3 (*4*)	0.56 (*8*)	0.19 (*1*)
	Median (*n*)	13	176	67.9	4.08	0.45	61	7.3	0.53	
	Range	0.52–35.6	95–256	0.33–514	1.14–32	0.29–15.8	57–121	1.2–53	0.49–0.68	

Mo	Mean (*n*)	0.02 (*1*)		0.66 (*11*)	0.08 (*1*)	0.04 (*3*)		1.27 (*2*)	0.05 (*8*)	0.01 (*1*)
	Median			0.6		0.03		1.27	0.04	
	Range			0.6–1.02		0.03–0.07		0.24–2.3	0.03–0.12	
Ni	Mean (*n*)	1 (*3*)		4.7 (*14*)	1.87 (*1*)	7.56 (*7*)		1.23 (*4*)	0.07 (*8*)	0.05 (*1*)
	Median	1.38		3.7	1.9	0.8		1.17	0.04	
	Range	0.11–1.5		1.8–8.8	1.2–2.3	0.04–38.7		0.65–1.9	0.02–0.29	
Pb	Mean (*n*)	2.44 (*3*)	2.33 (*13*)	2.6 (*50*)	1.12 (*15*)	3.4 (*9*)	12 (*1*)	0.87 (*9*)	0.05 (*7*)	0.01 (*1*)
	Median	2.46	2.1	1.59	0.95	0.4		0.54	0.04	
	Range	0.06–4.8	0.8–5.4	0.09–7.6	0.2–3.1	0.13–23.9		0.1–2.2	0.01–0.12	
Se	Mean (*n*)	0.15 (*1*)	0.38 (*1*)	0.72 (*35*)	3.74 (*6*)	3.08 (*5*)	1.3 (*4*)	4.7 (*9*)	2.58 (*8*)	[illegible].15 (*3*)
	Median			0.63	4.05	3.6	0.7	4.1	1.9	[illegible].5
	Range			0.05–2.28	1.22–4.71	1.79–1447	0.1–4.8	0.9–9.3	1.5–5.8	[illegible].44–7.5
Sr	Mean (*n*)	3.46 (*1*)		611 (*61*)	74 (*1*)	52 (*5*)		97 (*2*)	0.11 (*8*)	[illegible]06(*1*)
	Median			472		7.05		97	0.09	
	Range			6.05–2103		5.8–141		22–173	0.02–0.23	
V	Mean (*n*)	0.06 (*1*)		7.25 (*28*)	0.042 (*1*)	0.43 (*5*)		0.96 (*2*)		
	Median			5.75		0.29		0.96		
	Range			0.1–38.4		0.05–0.97		0.62–1.3		
Zn	Mean (*n*)	37 (*3*)	40 (*13*)	33 (*85*)	84 (*15*)	58 (*10*)	47 (*4*)	28 (*12*)	17.8 (*8*)	5[illegible] (*13*)
	Median	28	24	22.5	94	29	42	23.5	16	4[illegible]
	Range	2.9–79	15–133	0.1–307	5–150	14.3–152	18–86	19–58	11.2–27	2[illegible]–126

[a]Only 'total' Hg concentrations were considered.

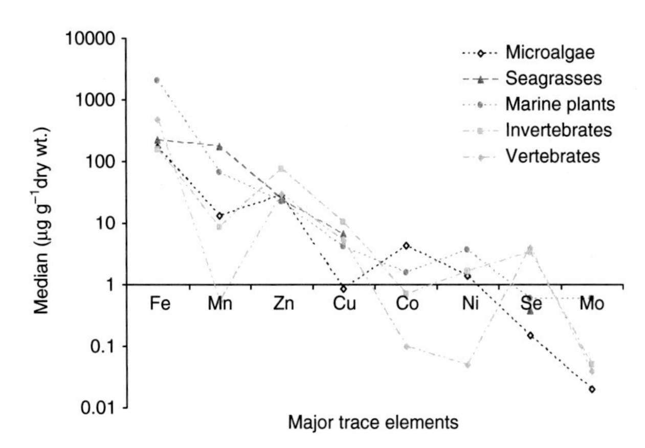

Figure 1 Median concentrations ($\mu g\, g^{-1}$ dry weight) of the essential trace elements Fe, Mn, Zn, Cu, Co, Ni, Se, and Mo in various marine organisms: microalgae, seagrasses, macroalgae, invertebrates (e.g., oysters, mussels) and the muscle tissue of vertebrates (e.g., seabirds, fish, dolphins). Only values relating to organisms growing in pristine conditions were considered.

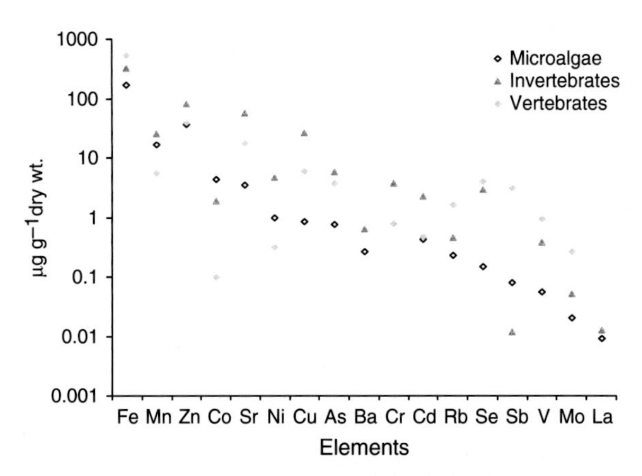

Figure 2 Bulk elemental composition (median $\mu g\, g^{-1}$ dry weight) of microalgae, invertebrates, and vertebrates collected from healthy environments. Seventeen elements were considered including those classified essential for growth and reproduction (Fe, Mn, Zn, Cu, Co, Ni, Se, and Mo), those with few known but very specific roles (Sr and V), and those whose roles remain undefined (As, Ba, Cr, Cd, Rb, Sb, and La). Of those trace elements in the latter group, some have been shown to replace other metals as metallocenters in enzymes (e.g., Cd) while others have no known roles, but their complete absence reduces the viability of certain organisms (e.g., Cr in microalgae). Only values relating to organisms growing in pristine conditions were considered.

magnitude higher than elements whose roles remain poorly defined (e.g., I, Cr) (see **Table 2** and **Figure 2**).

Multielemental signatures of microalgae, invertebrates, and vertebrates are shown in **Figure 2**. The extended Redfield equation holds with a few exceptions in which some trace elements are found in higher concentrations than one might predict based on their cellular functions. Macroalgae have metal specificities thought to be related to their unique outer-wall materials and metal uses (**Table 1**). Phaeophyta (brown macroalgae) have a clear tendency to concentrate I, As, and Cr relative to Rhodophyta (red) and Chlorophyta (green) while the highest concentration of Br is typically found in red macroalgae (not shown). For those elements that have no known role, or are likely to be toxic rather than beneficial to organisms (e.g., As, Sb, lanthanide (La)), the mechanisms for uptake, storage, and efflux should be elucidated in future studies. Such elements may replace essential elements in metallocenters, as has been reported for Cd in carbonic anhydrase (Cd replaces Zn and/or Co). Others may act as antagonists and prevent nutrient (essential) elements from carrying out their normal roles. Many of these nonessential elements are often detected in aquatic organisms such as oysters and mussels. This selectivity may encourage their use as biomonitors for pollution. The appearance of elevated concentrations of nonessential trace elements in tissues may be taken as evidence of deteriorated ecological conditions.

The majority of animal (e.g., oysters, fish, seabirds, seals) studies to date have focused on only one (e.g., Hg) or a small selection of four to six elements (e.g., Cd, Se, Zn). In the future, it would be beneficial to expand such studies to a greater number of elements (at least 20). Difficulties associated with defining an extended Redfield ratio or trace elemental stoichiometry for animals arise principally because most investigators do not (or cannot) measure whole animals and examine only muscle, liver, and/or kidney tissues. Tissue flesh (muscle) is the final target tissue for trace elements and the main edible component of animals. Thus, its study serves for the protection of public health.

Uptake, Storage, and Efflux of Trace Elements

The acquisition and retention of trace elements in organisms is strongly affected by internal (physiological) and external (physicochemical) factors. If available, organisms will concentrate trace elements to concentrations up to 10^{3-5} higher than in their environment. This process, 'bioaccumulation', concerns essential elements such as Cu, Fe, Se, or Zn, as well as toxic elements such as Cd or Hg. Major pathways for the uptake of trace metals by aquatic organisms include direct uptake from solution through permeable surfaces such as the cell walls of phytoplankton and the gills of crustacea and fish. Physicochemical parameters, such as pH, salinity, temperature, light, and particulate and organic matter concentrations, influence the accumulation ('bioconcentration') of metals in organisms.

Uptake kinetics for trace elements have been measured but absolute comparisons are difficult because of differences among studies in

1. the physiological status of the organism;
2. the environmental conditions (laboratory vs. field) and the nature (oxidized or reduced; with or without chelators) of the element used;
3. the experimental protocol; and
4. the diversity of transporting surfaces in organisms.

The last aspect, for example, has been used to explain why barnacles, which filter-feed via cirri with large surface areas, have a higher uptake rate constants for Zn and Cd from solution than decapods, which have mostly impermeable surfaces and restricted gill areas. Actual uptake mechanisms for trace elements in seagrasses are difficult to assess because exposure involves both above- and belowground parts and because the bioavailable fraction of elements in ambient versus interstitial water in the sediment differs. Despite decades of research on uptake kinetics of specific trace metals such as Fe and Cu, the exact mechanics of the processes have not been fully resolved. We do however know that rates of uptake of trace metals from solution are directly proportional to the bioavailable concentration of the metals, which in turn is highly dependent on the redox potential of the environment. Iron, for example, is the fourth most abundant element in the Earth's crust, yet also one of the least soluble metals in oxygenated marine waters, and so least 'bioavailable' to organisms in the ocean. 'Free metal ion activity models' define the bioavailability of a trace metal in solution.

Chelators act as buffers, releasing some ions into the medium while internalizing others. The strength with which chelators bind metal ions is important in determining trace elemental availability outside cells and/or to organisms. For example, in response to recently upwelled waters, marine microorganisms produce specific chelators to reduce trace metal concentrations (e.g., Cu ions) in their immediate surroundings, thereby diminishing their immediate potential toxicity. Siderophores (low molecular weight compounds) are produced by bacteria to aid in the assimilation of otherwise difficult to obtain elements such as Fe. Production of siderophores is induced along with specific membrane uptake systems. Sequestration mechanisms appear to be as unique as the trace elements being sequestered and the organisms sequestering.

Active regulation, via negative feedback, of the uptake of essential elements has been demonstrated. There is a strong internal economy for the use and reuse such that cells have a 'homeostatic system' to maintain appropriate cellular concentrations of trace elements. As the inherent biochemistry of organisms optimizes the acquisition and utilization of multiple elements for growth and reproduction, stoichiometric analysis can be used to describe mass balances and transfers in ecological systems. Many organisms can store surplus trace elements (for future requirements or, via sequestration, for detoxification) or reduce their quotas for trace metals when their availability is limited. Phytochelatins (in algae) and metallothioens (other organisms) are part of the regulatory system. This approach may fail if organisms are overwhelmed by the influx or ingestion of some (usually toxic) elements. Trace element accumulation affects the ecology of organisms as a consequence of the energy and nutrient costs associated with excreting and/or detoxifying incoming metals. If the costs are significant, then this may reduce growth, reproduction, and/or competitive ability.

Food Webs and Movement of Elements across Trophic Levels

At higher trophic levels, consumers have stoichiometric requirements for multiple elements that are not always satisfied by their food. Poor food quality can limit growth and reproduction of an organism and consequently also affect demography and population dynamics, species interactions, and community structure. Studies of stoichiometric mismatches at ecological interfaces (plant vs. inorganic resources, herbivore vs. carnivore) have focused on the ecophysiology and life-history traits at the level of the individual organism, and on the role of C, N, and P but rarely on the consequences of mismatches between demand and supply of trace elements.

All elements move through trophic levels in defined modes following the laws of mass balance and thermodynamics. 'Biomagnification' occurs when an increase in trace metal concentration occurs through at least two trophic levels in a food chain. Biomagnification through the food chain starts with positive 'biotransference', which is the transfer of trace metals from a food source to consumer. Unlike the nutrient elements C, N, and P, if biomagnifcation of trace metals occurs, their elevated concentrations in higher trophic organisms could pose a threat to the organisms themselves or to subsequent consumers, including humans. 'Biodilution' is the decrease in concentration of an element with increasing trophic level. The trace elemental accumulation patterns in aquatic organisms (**Figures 3** and **4**) reflect active mechanisms for bioaccumulation, biomagnification, and biodilution at each trophic level. Only values relating to organisms growing in pristine conditions were considered.

Hg and Se and are examples of elements whose concentrations increase up food chains (**Figures 3a** and **3b**, respectively). This positive biotransference of Hg (4000-fold; **Figure 3a**) is globally well documented; Hg is associated with neurological damage in humans and nonhumans. Though excretory processes are known, this level of biomagnification suggests that homeostatic systems can be overwhelmed in highly contaminated environments. Se concentrations increase 30-fold (**Figure 3b**) along trophic

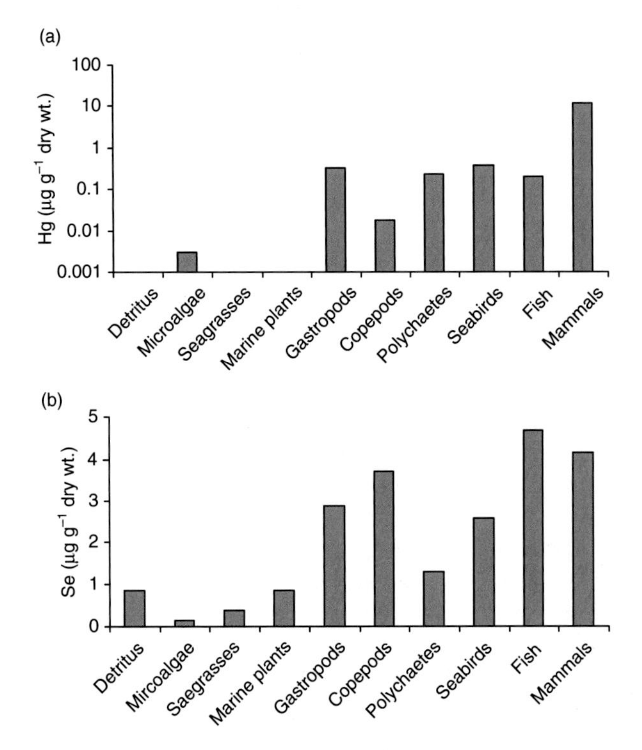

Figure 3 Median Hg (log scale; $\mu g\, g^{-1}$ dry weight) and Se concentrations ($\mu g\, g^{-1}$ dry weight) across trophic levels from detritus to phototrophs to invertebrates and then vertebrates. While Hg and Se are bioaccumulated by aquatic organisms, their concentrations are also biomagnified through the food web as a result of positive biotransference, primarily through dietary consumption. Only values relating to organisms growing in pristine conditions were considered.

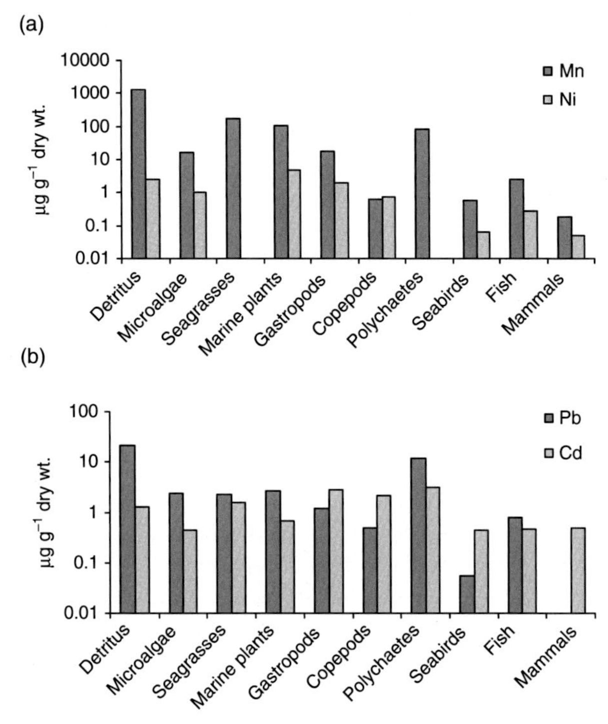

Figure 4 Positive biotransference can occur for essential (Mn and Ni) and potentially toxic elements (Pb and Cd) in the food web but without concurrent biomagnification. Median elemental concentrations ($\mu g\, g^{-1}$ dry weight) were compared across trophic levels from detritus to vertebrates. Only values relating to organisms growing in pristine conditions were considered.

levels; biomagnification factors for other trace elements typically fall between these two extremes (**Table 2**).

Marine mammals, the top predators in many ocean food webs, bioaccumulate metals from their diet (e.g., fish; **Figure 3**). Cd accumulation (not shown) is linked to diet (cephalopods) while other factors such as age or body condition can also explain elevated quotas in marine mammals (**Table 2**). Unlike other long-lived marine vertebrates, sea turtles do not accumulate Hg. The absence of long-term Hg accumulation suggests that this animal, which is an opportunistic predator, feeds generally on prey at low (bivalves, gastropods) trophic levels.

Benthic organisms (phototrophs and invertebrates) can have the highest Cu (not shown), Mn, and Ni concentrations relative to vertebrates such as birds, fish, and mammals (**Figure 4a**). Organisms belonging to higher trophic levels are clearly better able to regulate internal concentrations (homeostasis) of these metals. Similarly to these nutrient-type trace elements, the potentially toxic elements Cd, Pb, silver (Ag), uranium (U), and La can have biotransference factors that are greater between lower (e.g., autotrophs to herbivores) than between higher trophic levels (e.g., herbivores to carnivores) (**Figure 4b**). These elements are not easily transferred to upper trophic biota muscle tissue, even when that biota is exposed to elevated levels of these elements in their diet. Dilution of the metal burden by body growth and cell turnover ensures reduction via trophic transfer. Uptake from food is thus unimportant for these elements; kinetic modeling indicates uptake from surrounding environments (sediment, suspended matter, or water) is more important.

Ecological stoichiometry seeks to track the mass flow of the elements through ecosystems. While trace elements generally occur at very low concentrations in the environment, most organisms bioaccumulate them intracellularly to perform a wide range of functions. Using a range of mechanisms, homeostatic concentrations of trace elements can be maintained in organisms, allowing a multielemental stoichiometric signature of organisms to be defined. Variations of trace elemental concentrations in the environment by local processes (e.g., volcanism or upwelling) or human activities (e.g., mining, urbanization) often lead to increases in the metal burdens of biota. Processes such as biodilution diminish the potential negative impacts of biomagnification as a result of positive transference across multiple trophic levels. While there is a considerable knowledge base for trace elements such as Fe, Mn, Zn, Cd, and Hg, very few studies have investigated Co, Cr, Ni, or V in detail. Trace elements that are selectively accumulated by specific organisms can be used as tracers of contamination (biomonitors). Studies of ecological stoichiometry as they pertain to trace elements are vital in the future if we are to deepen our understanding of homeostatic regulation of trace metals in biota and of trace metal cycling in ecosystems and food webs.

See also: Material and Metal Ecology.

Further Reading

Al-Masri MS, Mamish S, and Budier Y (2003) Radionuclides and trace metals in eastern Mediterranean Sea algae. *Journal of Environmental Radioactivity* 67: 157–168.

Barwick M and Maher W (2003) Biotransference and biomagnification of selenium, copper, cadmium, zinc, arsenic, and lead in a temperate seagrass ecosystem from Lake Macquarie Estuary, NSW, Australia. *Marine Environmental Research* 56: 471–502.

Blackmore G and Morton B (2001) The interpretation of body trace metal concentrations in neogastropods from Hong Kong. *Marine Pollution Bulletin* 42: 1161–1168.

Bohn A (1979) Trace metals in fucoid algae and purple sea urchins near a high Arctic lead/zinc ore deposit. *Marine Pollution Bulletin* 10: 325–327.

Bustamante P, Garrigue C, Breau L, *et al.* (2003) Trace elements in two odontocete species (*Kogia breviceps* and *Globicephala macrorhynchus*) stranded in New Caledonia (South Pacific). *Environmental Pollution* 124: 263–271.

Campanella L, Conti ME, Cubadda F, *et al.* (2001) Trace metals in seagrass, algae and molluscs from an uncontaminated area in the Mediterranean. *Environmental Pollution* 111: 117–126.

Campbel LM, Norstrom RJ, Hobson KA, *et al.* (2005) Mercury and other trace elements in a pelagic Arctic marine food web (Northwater Polynya, Baffin Bay). *The Science of the Total Environment* 351–352: 247–263.

Catsiki V-A and Florou H (2006) Study on the behavior of the heavy metals Cu, Cr, Ni, Zn, Fe, Mn and ^{137}Cs in an estuarine ecosystem using *Mytilus galloprovincialis* as a bioindicator species: The case of Thermaikos gulf, Greece. *Journal of Environmental Radioactivity* 86: 31–44.

Catsiki V-A and Strogyloudi E (1999) Survey of metal levels in common fish species from Greek waters. *The Science of the Total Environment* 237–238: 387–400.

Conti ME and Cecchetti G (2003) A biomonitoring study: Trace metals in algae and molluscs from Tyrrhenian coastal areas. *Environmental Research* 93: 99–112.

Das K, Beans C, Holsbeek L, *et al.* (2003) Marine mammals from northeast Atlantic: Relationship between their trophic status as determined by $d^{13}C$ and $d^{15}N$ measurements and their trace metal concentrations. *Marine Environmental Research* 56: 349–365.

Farýas S, Arisnabarreta SP, Vodopivez C, *et al.* (2002) Levels of essential and potentially toxic trace metals in Antarctic macro algae. *Spectrochimica Acta Part B* 57: 2133–2140.

Hou X and Yan X (1998) Study on the concentration and seasonal variation of inorganic elements in 35 species of marine algae. *The Science of the Total Environment* 222: 141–156.

Ip CCM, Li XD, Zhang G, *et al.* (2005) Heavy metal and Pb isotopic compositions of aquatic organisms in the Pearl River Estuary, South China. *Environmental Pollution* 138: 494–504.

Karez CS, Magalhaes VF, Pfeiffer WC, *et al.* (1994) Trace metal accumulation by algae in Sepetiba Bay, Brazil. *Environmental Pollution* 83: 351–356.

Marsden ID and Rainbow PS (2004) Does the accumulation of trace metals in crustaceans affect their ecology – the amphipod example? *Journal of Experimental Marine Biology and Ecology* 300: 373–408.

Ndiokwere CL (1984) An investigation of the heavy metal content of sediments and algae from the river Niger and Nigerian Atlantic Coastal Waters. *Environmental Pollution (Series B)* 7: 247–254.

Nguyen HL, Leermakers M, Osan J, *et al.* (2005) Heavy metals in Lake Balaton: Water column, suspended matter, sediment and biota. *Science of the Total Environment* 340: 213–230.

Nienhuis PH (1986) Background levels of heavy metals in nine tropical seagrass species in Indonesia. *Marine Pollution Bulletin* 17: 508–511.

Quigg A, Finkel ZV, Irwin AJ, *et al.* (2003) The evolutionary inheritance of elemental stoichiometry in phytoplankton. *Nature* 425: 291–294.

Raven JA, Evans MCW, and Korb RE (1999) The role of trace metals in photosynthetic electron transport in O_2-evolving organisms. *Photosynthesis Research* 60: 111–149.

Reinfelder JR, Fisher NS, Luoma SN, Nichols JW, and Wang W-X (1998) Trace element trophic transfer in aquatic organisms: A critique of the kinetic model approach. *The Science of the Total Environment* 219: 117–135.

Saha M, Sarkar SK, and Bhattacharya B (2006) Interspecific variation in heavy metal body concentrations in biota of Sunderban mangrove wetland, northeast India. *Environment International* 32: 203–207.

Szefer P, Geldon J, Ali AA, *et al.* (1998) Distribution and association of trace metals in soft tissue and byssus of *Mytella strigata* and other benthal organisms from Mazatlan Harbour, mangrove lagoon of the northeast coast of Mexico. *Environment International* 24: 359–374.

Whelan T, Espinoza J, Villarreal X, *et al.* (2005) Trace metal partitioning in *Thalassia testudinum* and sediments in the Lower Laguna Madre, Texas. *Environment International* 31: 15–24.

Whitfield M (2001) Interactions between phytoplankton and trace metals in the ocean. *Advances in Marine Biology* 41: 3–128.

Williams RJP and Fraústo Da Silva JJR (1996) *The Natural Selection of the Chemical Elements: The Environment and Life's Chemistry*. Oxford: Clarendon Press.

INDEX

NOTES:

Cross-reference terms in italics are general cross-references, or refer to subentry terms within the main entry (the main entry is not repeated to save space). Readers are also advised to refer to the end of each article for additional cross-references - not all of these cross-references have been included in the index cross-references.

The index is arranged in set-out style with a maximum of three levels of heading. Major discussion of a subject is indicated by bold page numbers. Page numbers suffixed by *t* and *f* refer to Tables and Figures respectively. *vs.* indicates a comparison.

This index is in letter-by -letter order, whereby hyphens and spaces within index headings are ignored in the alphabetization. Prefixes and terms in parentheses are excluded from the initial alphabetization.

A

C

D

E

F

G

J

K

L

M

N

P

Q

R

T

X

Y

Z